Lecture Notes in Electrical Engineering

Volume 639

AF806700

The book series *Lecture Notes in Electrical Engineering* (LNEE) publishes the latest developments in Electrical Engineering—quickly, informally and in high quality. While original research reported in proceedings and monographs has traditionally formed the core of LNEE, we also encourage authors to submit books devoted to supporting student education and professional training in the various fields and applications areas of electrical engineering. The series cover classical and emerging topics concerning:

- Communication Engineering, Information Theory and Networks
- Electronics Engineering and Microelectronics
- Signal, Image and Speech Processing
- Wireless and Mobile Communication
- Circuits and Systems
- Energy Systems, Power Electronics and Electrical Machines
- Electro-optical Engineering
- Instrumentation Engineering
- Avionics Engineering
- Control Systems
- Internet-of-Things and Cybersecurity
- Biomedical Devices, MEMS and NEMS

For general information about this book series, comments or suggestions, please contact leontina.dicecco@springer.com.

To submit a proposal or request further information, please contact the Publishing Editor in your country:

China

Jasmine Dou, Associate Editor (jasmine.dou@springer.com)

India, Japan, Rest of Asia

Swati Meherishi, Executive Editor (Swati.Meherishi@springer.com)

Southeast Asia, Australia, New Zealand

Ramesh Nath Premnath, Editor (ramesh.premnath@springernature.com)

USA, Canada:

Michael Luby, Senior Editor (michael.luby@springer.com)

All other Countries:

Leontina Di Cecco, Senior Editor (leontina.dicecco@springer.com)

**** Indexing: The books of this series are submitted to ISI Proceedings, EI-Compendex, SCOPUS, MetaPress, Web of Science and Springerlink ****

More information about this series at http://www.springer.com/series/7818

Yong Qin · Limin Jia · Baoming Liu ·
Zhigang Liu · Lijun Diao · Min An
Editors

Proceedings of the 4th International Conference on Electrical and Information Technologies for Rail Transportation (EITRT) 2019

Rail Transportation System Safety and Maintenance Technologies

Springer

Editors
Yong Qin
State Key Laboratory of Rail Traffic Control and Safety
Beijing Jiaotong University
Beijing, China

Limin Jia
State Key Laboratory of Rail Traffic Control and Safety
Beijing Jiaotong University
Beijing, China

Baoming Liu
National Innovation Center of High Speed Train
Qingdao, China

Zhigang Liu
Beijing Jiaotong University
Beijing, China

Lijun Diao
Beijing Jiaotong University
Beijing, China

Min An
School of Science, Engineering and Environment
University of Salford
Salford, UK

ISSN 1876-1100 ISSN 1876-1119 (electronic)
Lecture Notes in Electrical Engineering
ISBN 978-981-15-2868-2 ISBN 978-981-15-2866-8 (eBook)
https://doi.org/10.1007/978-981-15-2866-8

This Springer imprint is published by the registered company Springer Nature Singapore Pte Ltd.
The registered company address is: 152 Beach Road, #21-01/04 Gateway East, Singapore 189721, Singapore

Contents

Research on Monitoring Technology of Pantograph Sliding Plate Abrasion Based on Sub-pixel Edge Extraction

Wang Dai, Zhaoyi Su, Lichao Fang, Shibo Li, Yong Zhang and Zongyi Xing

Abstract Aiming at the fault of the urban rail train caused by the excessive abrasion of the pantograph slide plate, a pantograph sliding plate abrasion monitoring algorithm based on sub-pixel edge extraction is designed in this paper. This paper mainly analyzes the process of the pantograph slide's wear monitoring algorithm. The experimental results show that the method proposed in this paper can effectively detect the abrasion curve of the pantograph slider and get the minimum residual abrasion value.

Keywords Pantograph · Adaptive Canny edge detection · Sub-pixel edge extraction · Camera calibration

1 Introduction

The urban rail train obtains the electrical energy required for the operation of its electrical system from the contact network via the pantograph slide plate. Therefore, the state of the abrasion of the pantograph slide plate is closely related to the power supply of the urban rail train. Studying the pantograph slide monitoring technology can effectively avoid the accident caused by the excessive abrasion of the pantograph slide plate of the urban rail train, which has great practical significance for ensuring the safe operation of the urban rail train [1].

This paper proposes a pantograph skater wear detection algorithm based on sub-pixel edge extraction. The experiment proves that the algorithm has good feasibility and reliability.

W. Dai · L. Fang · S. Li · Y. Zhang · Z. Xing (✉)
School of Automation, Nanjing University of Science & Technology, Nanjing, China
e-mail: xingzongyi@163.com

Z. Su
Guangzhou Metro Group Co., Ltd., Haizhu District, China

Y. Qin et al. (eds.), *Proceedings of the 4th International Conference on Electrical and Information Technologies for Rail Transportation (EITRT) 2019*, Lecture Notes in Electrical Engineering 639, https://doi.org/10.1007/978-981-15-2866-8_1

2 Research on Monitoring Technology of Pantograph Sliding Plate Abrasion Based on Sub-pixel Edge Extraction

The method of this paper is to process the left and right half-bow images of the pantograph detected by the acquisition system to obtain the residual wear curve of the pantograph slide. First, image filtering is performed on the original image to reduce noise. The pixel edge is detected by the adaptive Canny edge detection algorithm. Then, sub-pixel edge is extracted by Gaussian fitting sub-pixel edge extraction based on sliding edge points. Finally, the complete residual wear curve of the pantograph slide is obtained by camera calibration and curve fusion and obtains the minimum residual wear value in the worn area. The process of the pantograph slide's wear monitoring algorithm is shown in Fig. 1. This paper uses Gaussian algorithm as the filtering algorithm in the pantograph slide monitoring algorithm [2].

2.1 *Edge Detection Algorithm*

This paper proposes an edge detection algorithm combining adaptive Canny edge detection algorithm and Gaussian fitting sub-pixel edge extraction algorithm based on sliding edge points to get a more accurate edge [3].

2.1.1 Adaptive Canny Edge Detection Algorithm

The adaptive Canny edge detection algorithm proposed in this paper is an improved edge detection method based on gradient histogram adaptive setting high and low thresholds. The steps are as follows:

First, gradient calculation:

Calculate the difference between the x-direction, the y-direction, the 135° direction, and the 45° direction by using the eight neighborhood templates for the filtered image to obtain the gradient amplitude.

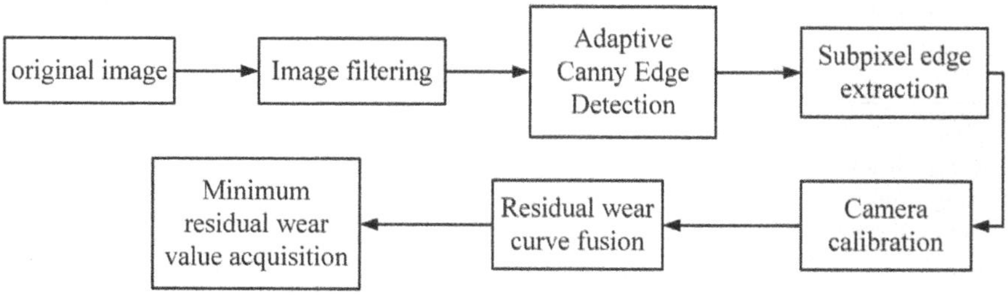

Fig. 1 Flowchart of pantograph skateboard wear monitoring algorithm

$$M(i,j) = \sqrt{(f_x(i,j))^2 + (f_y(i,j))^2 + (f_{135^\circ}(i,j))^2 + (f_{45^\circ}(i,j))^2} \quad (1)$$

The gradient direction is

$$\theta(i,j) = \arctan\left(f_y(i,j)/f_x(i,j)\right) \quad (2)$$

where $f_x(i,j)$, $f_y(i,j)$, $f_{45^\circ}(i,j)$, $f_{135^\circ}(i,j)$ the difference between the x direction, the y direction, the 45° direction, and the 135° direction by using the 8 neighborhood template for the filtered image respectively.

Second, non-maximum suppression:

(1) Divide the gradient direction into four sectors by size and number four sectors, corresponding to the four possible combinations of fields as shown in Fig. 2.
(2) Determine two adjacent pixel of the gradient direction according to the gradient direction and sector division of the pixel (i, j), and compare the gradient magnitudes with the two adjacent pixels. If the gradient of the pixel (i, j) is greater than the gradient of the adjacent pixel, then the pixel (i, j) may be an edge point, and the gradient magnitude of the pixel (i, j) is retained. If the gradient of the pixel (i, j) is not greater than the gradient of the adjacent pixel, then the pixel (i, j) is not an edge point and is marked as a non-edge point.

Third, high and low threshold selection based on gradient histogram:

(1) Calculate the result of the gradient magnitude calculation to obtain the gradient histogram.
(2) The gradient amplitude H_{max} with the largest number of pixels is obtained by comparison, and the mean square error σ_{max} corresponding to the gradient amplitude is further calculated.

$$\sigma_{max} = \sqrt{\sum_{i=0}^{N} (H_i - H_{max})^2 / N} \quad (3)$$

where N is the total number of image pixels.

Fig. 2 Gradient direction sector division

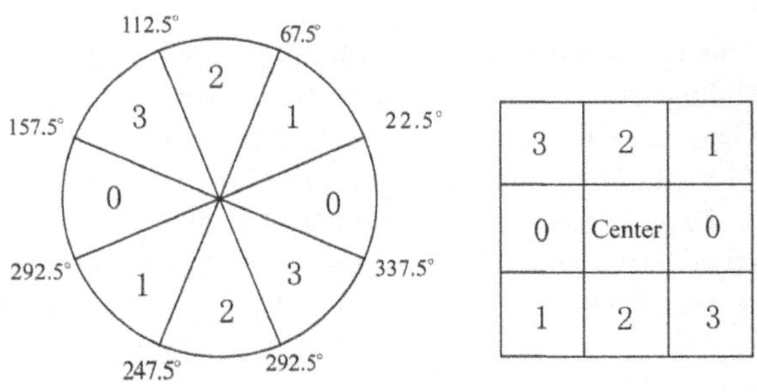

(a) Sector division map (b) Neighborhood division map

(3) Calculate the high threshold τ_{High} based on the gradient magnitude $H_{\max}$ and the mean square error $\sigma_{\max}$.

$$\tau_{\text{High}} = H_{\max} + \sigma_{\max} \tag{4}$$

(4) Clear the gradient magnitude greater than τ_{High} in the gradient histogram obtained in Step (1) and recount to obtain a new gradient histogram.
(5) Obtain the gradient amplitude $H_{\max}$ with the largest number of pixels from the gradient histogram in Step (4), and further, calculate the mean square error $\sigma_{\max}$ corresponding to the gradient amplitude.

$$\sigma'_{\max} = \sqrt{\sum_{i=0}^{N'} \left(H_i - H'_{\max}\right)^2 / N'} \tag{5}$$

where N' is the total number of pixels corresponding to the new gradient histogram.
(6) The low threshold τ_{Low} can be calculated according to the gradient amplitude $H'_{\max}$ with the largest number of pixels and its corresponding gradient amplitude mean square error $\sigma'_{\max}$.

$$\tau_{\text{Low}} = H'_{\max} + \sigma'_{\max} \tag{6}$$

Fourth, edge connection:

The high and low thresholds divide the single pixel width edge in the entire gradient magnitude image $M_1(i, j)$ into three parts. $T_h(i, j)$ is an area larger than the high threshold, $T_l(i, j)$ is an area smaller than the high threshold but larger than the low threshold, and a part is the part smaller than the low threshold. The pixels in $T_h(i, j)$ are edge points. Part of the pixel in $T_l(i, j)$ is the edge point. Therefore, it is necessary to select edge points from $T_l(i, j)$ by means of recursive tracking to achieve edge connection [4].

2.1.2 Sub-pixel Edge Extraction Algorithm

This paper proposes a Gaussian fitting sub-pixel edge extraction method based on sliding edge points. This method consists of fitting point determination and Gaussian curve fitting of sliding edge points.

First, determination of the fit point:

Determination of the fit point of each edge point is obtained by adaptive Canny edge detection according to the gradient direction. The specific fitting point process is described by taking one point P_0 as an example. Take the point P as the center and take the left and right six adjacent points along the gradient direction of this point, as shown in Fig. 3 [5]. The distance between every two adjacent neighbors is l. According to the angle θ of the gradient direction of the point P_0 and the horizontal direction, the l can be divided into two cases. When $|\theta| \leq 45^{\circ}$, $l = 1/\cos(\theta)$ and the difference between the abscissas between adjacent points is 1; When $|\theta| \geq 45^{\circ}$, $l = 1/\cos(\theta)$, and the difference between the ordinates between adjacent points is 1.

Determine the gray value of the determined fitting point. When $|\theta| \leq 45^{\circ}$, take point A as an example for explanation. The coordinates of point P_0 are (x_0, y_0), The abscissa of point A: $x_1 = x_0 + 1$, and the ordinate of point A: $y_1 = y_0 + \tan(\theta)$. The linear interpolation process is performed by the gray values of the upper and lower whole pixels of point A, that is, the gray value at the star point in Fig. 4 is processed, and the gray value of point A is further obtained.

$$f(x_0, y_0) = f([x_0], [y_0]) \tag{7}$$

$$f(x_1, y_1) = \lambda \times f([x_1], [y_0] + 1) + (1 - \lambda) \times f([x_1], [y_0]) \tag{8}$$

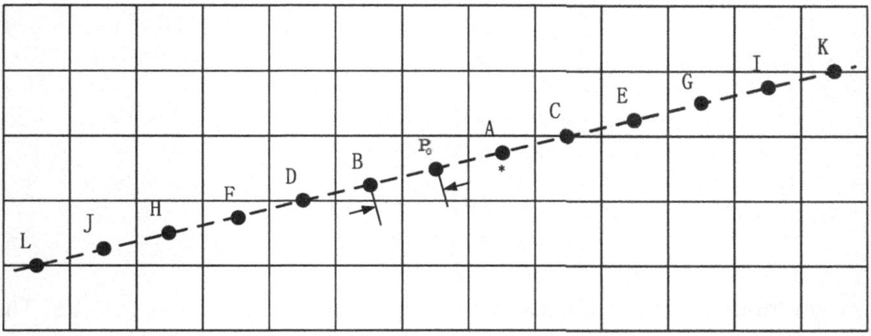

Fig. 3 Determination of the fit point

Fig. 4 Original image

where $f(x_0, y_0)$ and $f(x_1, y_1)$ are the gray values of point P and point A, respectively. λ is the difference of y_1 and y_0. The gradation value of B to L is obtained by the above method, and the gradation difference of each point is determined by the mean of the forward difference and the backward difference.

Second, Gaussian curve fitting of sliding edge points:

Due to the grayscale variation of the edge points of the image along the gradient direction, it approximates the Gaussian distribution and the quadratic curve can approximate the Gaussian curve with high order. So we can use the quadratic curve $y(x) = Ax^2 + Bx + C$ instead of the Gaussian curve. x is the distance from the corresponding fitting point to point P, and $y(x)$ is the corresponding gray point difference [6].

$$\begin{aligned}
f_{-5} &= 25Al^2 - 5Bl + C \\
f_{-4} &= 16Al^2 - 4Bl + C \\
f_{-3} &= 9Al^2 - 3Bl + C \\
f_{-2} &= 4Al^2 - 2Bl + C \\
f_{-1} &= Al^2 - Bl + C \\
f_0 &= C \\
f_1 &= Al^2 + Bl + C \\
f_2 &= 4Al^2 + 2Bl + C \\
f_3 &= 9Al^2 + 3Bl + C \\
f_4 &= 16Al^2 + 4Bl + C \\
f_5 &= 25Al^2 + 5Bl + C
\end{aligned} \tag{9}$$

Converting the above formula into a matrix can be expressed as $U = VE$. The vector U is formed by the gradation difference of each fitting point, The coefficient matrix V is composed of the coefficients of the formulas in the Formula (9), and the vector E is composed of the parameters A, B, and C of the quadratic curve.

Each time participating in the quadratic curve fitting is five adjacent fitting points; the following formula is obtained.

$$\begin{cases} U_s = [U(4+n), U(5+n), U(6+n), U(7+n), U(8+n)]^T \\ V_s = [V(4+n,:); V(5+n,:); V(6+n,:); V(7+n,:); V(8+n,:)] \\ U_s = V_s E \end{cases} \tag{10}$$

where $|n|$ represents the number of slides, and the positive and negative of n represent the sliding direction. $U(4+n)$ represents the $(4+n)$th row element in the vector U, and $V(4+n,:)$ represents the $(4+n)$th row element in the coefficient matrix V. By formula (10) and least squares calculation, the following formula can be obtained.

$$E = \left(V_s^T V_s\right)^{-1} V_s^T U_s \tag{11}$$

Through the above formula, the three parameters of the quadratic curve: $y(x) = Ax^2 + Bx + C$ can be obtained, and the extreme points of the curve can be obtained by derivation, $\delta = -B/(2A)$, $|\delta|$ is the distance between the highest point P of the curve and the point P_0. The coordinates of the highest point P of the fitted curve are $(x_0 + \delta \times \cos(\theta), y_0 + \delta \times \sin(\theta))$. The condition for determining the success of the fitting is that the distance $|\delta'|$ between the highest point of the fitting curve and the current edge point is less than *l*, and five fitting points satisfy the condition: $U_s(1) < U_s(2)$, $U_s(1) < U_s(3)$, $U_s(3) < U_s(5)$, and $U_s(4) > U_s(5)$.

2.2 Complete Residual Wear Curve Fusion Algorithm

Camera calibration using Zhang Zhengyou calibration algorithm [7]. Therefore, the remaining wear curve of the skateboard can be obtained by the above algorithm.

The image used in this article is the left and right half-bow images of the pantograph skateboard. To obtain the complete residual wear curve, the curve of the remaining wear curve of the half-bow skateboard processed by the above algorithm needs to be curve-fused. The curve of the remaining wear curve of the left half-bow slide and the remaining wear curve of the right half-bow slide are analyzed by the following formula [8].

$$\begin{cases} x_m = x_{ml}, y_m = y_{ml} \\ x_m = x_{mr} + \Delta x, y_m = y_m \end{cases} \tag{12}$$

where (x_{ml}, y_{ml}) represents the coordinates of the left half-bow skate curve and (x_{mr}, y_{mr}) represents the coordinates of the right half-bow skate curve. Since the camera is designed to have overlapping parts on the left and right, the upper and lower sides do not overlap, so there is only an offset Δx.

The steps to determine the offset Δx are as follows:

(1) According to the length of the skateboard and the length of the remaining curve of the left and right half-bow skateboards detected, the offset is initially determined.
(2) Move the remaining wear curve of the right half-bow slide to the corresponding position according to the initially determined offset Δx_1. The remaining wear curve of the right half-bow slide is moved left and right at this position, the corresponding moving distance $d\Delta x$ is recorded, and the square sum e_{sum} of the difference between the two curved portions of the corresponding overlapping portion is calculated. Take the moving distance $d\Delta x$ of the square and e_{sum}, and finally determine the value of the offset Δx.

$$\Delta x = \Delta x_1 + d\Delta x \tag{13}$$

According to Formula (12), the complete slippery wear curve of the skateboard can be obtained. By analyzing the curve, the minimum residual wear value in the wear region can be obtained.

3 Testing and Analysis

This section analyzes the above algorithms and verifies the feasibility of this algorithm by performing a skateboard wear monitoring on a set of images. Multiple sets of images are processed, and the minimum residual wear value obtained by processing is compared with the manually detected value to verify the reliability.

The original image is shown in Fig. 4.

The filtering result is shown in Fig. 5.

The edge image of the pantograph is obtained by the above adaptive Canny edge detection algorithm as shown in Fig. 6.

The half-bow skateboard edge image is processed by Gaussian fitting sub-pixel edge extraction algorithm based on sliding edge points to obtain a residual wear curve of the half-bow slide, as shown in Figs. 7 and 8.

The residual wear curve of the left and right skateboards obtained above is curve-fused by the complete residual wear curve fusion algorithm, and the complete residual wear curve is shown in Fig. 9.

The above-mentioned pantograph skateboard wear monitoring algorithm was used to process the images collected by the urban rail trains with the numbers 7888, 8788, and 8990 every three days, and the corresponding 12 received. The ten minimum residual wear values of the electric bow skate are as shown in Fig. 10. The corresponding manual detection value is shown in Fig. 11.

The distribution of the minimum residual wear deviation of the skateboard obtained by the above is shown in Fig. 12. Analyze this graph, and get the following conclusions: All deviations are within ±1 mm, of which 116 groups are within ±0.8 mm, accounting for 96.7% of the total; 88 groups are within ±0.5 mm, accounting for 73.3% of the total.

Fig. 5 Filtered image

Fig. 6 Half-bow skateboard edge image

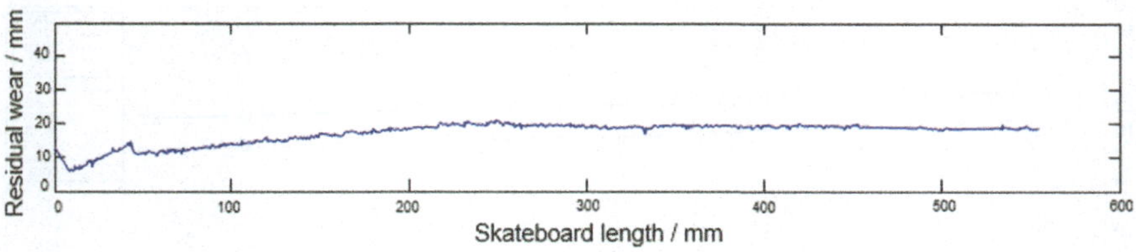

Fig. 7 Left half-bow skateboard residual wear curve

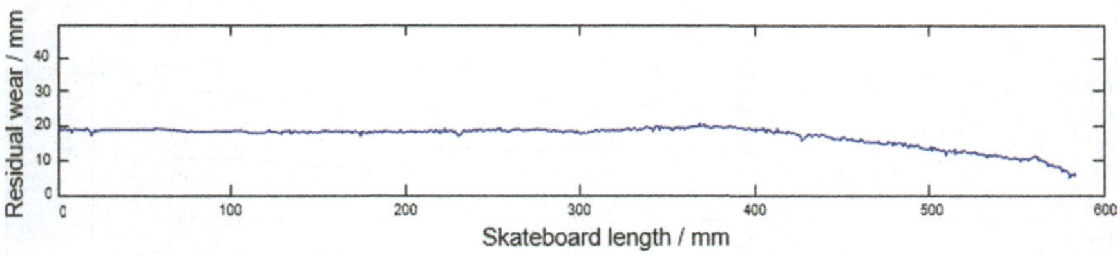

Fig. 8 Right half-bow skateboard residual wear curve

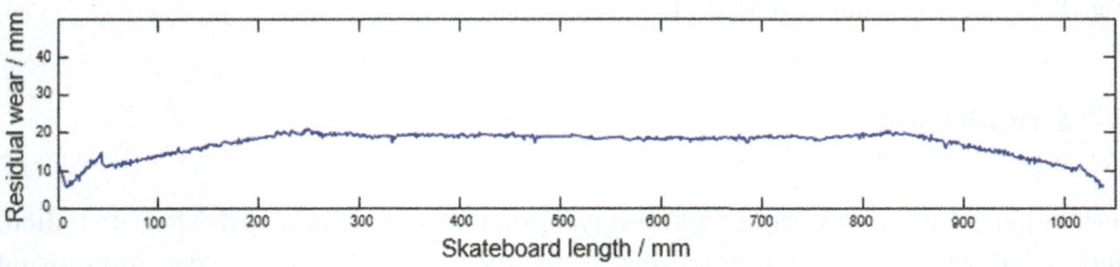

Fig. 9 Complete skateboard residual wear curve

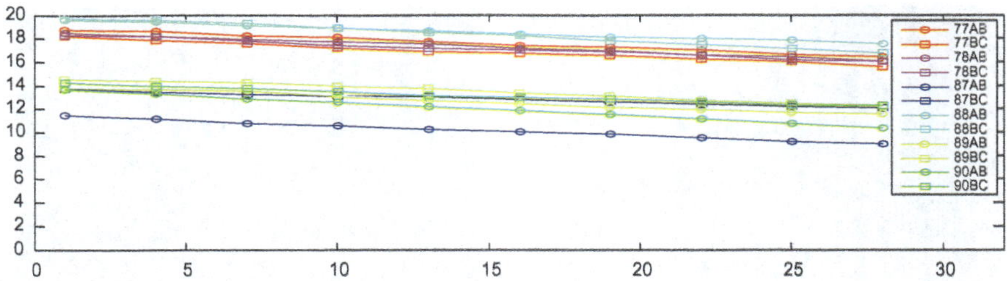

Fig. 10 Skateboard minimum residual wear algorithm detection value graph

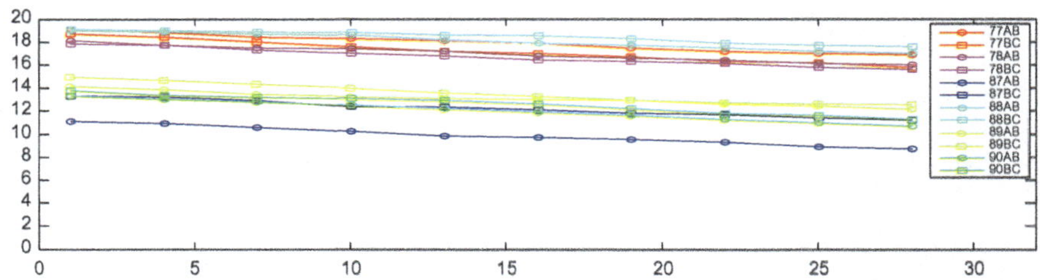

Fig. 11 Skateboard minimum residual wear manual detection value graph

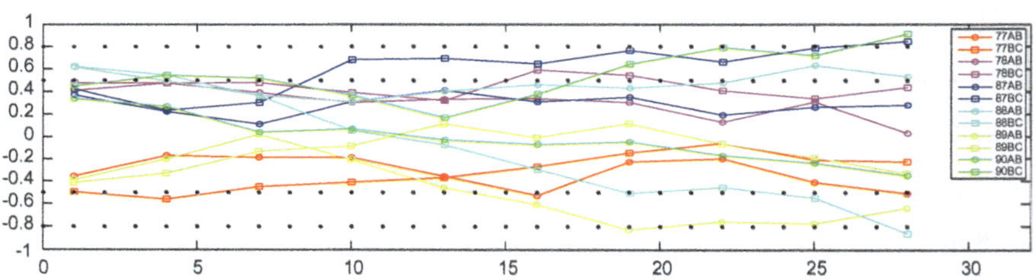

Fig. 12 Skateboard minimum residual wear deviation distribution graph

4 Conclusion

In this paper, the monitoring technology of pantograph skateboard wear is studied, and a sub-pixel edge extraction-based pantograph skateboard wear monitoring algorithm is designed. In this paper, the monitoring technology of pantograph skateboard wear is studied, and a sub-pixel edge extraction-based pantograph skateboard wear monitoring algorithm is designed. An edge detection algorithm based on adaptive Canny edge detection algorithm and Gaussian fitting sub-pixel edge extraction algorithm based on sliding edge points is used to detect the upper and lower edge point sets of the pantograph, which can detect the edge points of the pantograph more accurately point sets. The test verifies the feasibility and reliability of the algorithm. The experimental results show that the proposed algorithm can detect the residual wear curve of the pantograph slide more accurately and effectively, and obtain the minimum residual wear of the wear zone, which is of great significance.

Acknowledgements This work is supported by National Key R&D Program of China (2017YFB1201201).

References

1. Zheng J (2013) Discussion on the synchronous planning of urban rail transit and urban coordination. J Railway Eng Soc 12(9):116–120 (in Chinese)
2. Yang C (2016) Research on several problems of image denoising and its effect evaluation. Jilin University (in Chinese)
3. Liu R, Sun Q, Wang C (2015) Research on edge detection algorithm in digital image, vol 23, issue 12. Science Press, pp 17–18 (in Chinese)
4. Wang Z (2004) Research on multiscale adaptive edge detection method. Wuhan University (in Chinese)
5. Zhang L (2014) Research on precision assembly method of quartz flexible accelerometer components. Tianjin University (in Chinese)
6. Zhao A (2006) Image sub-pixel edge localization algorithm based on quadratic curve fitting. J Harbin Univ Sci Technol 11(3):68–70 (in Chinese)
7. Xu D, Tan M, Li Y (2016) Robot vision measurement and control. National Defense Industry Press (in Chinese)
8. Yin Y, Xu D, Zhang Z (2013) Plane measurement based on monocular vision. J Electr Measur Instrum 27(4):347–352 (in Chinese)

Partial Discharge Detection Technique Using Polar Coordinate Phase-Resolved Pattern in Rail Transportation Gas-Insulated Substation

Rui Yao, Meng Hui, Lin Bai, Zhen Ye and Qisheng Wu

Abstract It is known that there are several kinds of PD patterns to be applied to describe the insulation defects in rail transportation GIS. In order to solve the problems of discontinuous discharge region when using the traditional PRPD pattern, the polar coordinate phase-resolved PD pattern approach to characterize insulation defect states is presented. PRPS data sets are accessed through the experiments on the insulation defects physical models in the GIS test sample. The data collecting uses the UHF measuring setup. The steps to plot the proposed polar coordinate pattern based on PRPS data sets are described in details. It is proved through the experiments that, a much clearer characterization of the defect types can be made by this new proposed pattern. And the distinguishing characters can be extracted from the new pattern.

Keywords Gas-insulated metal-enclosed switchgear (GIS) · Ultra-high frequency (UHF) · The phase-resolved pulse sequence (PRPS)

1 Introduction

Gas-insulated metal-enclosed switchgear (GIS) compared with the traditional substation has advantages of occupying little space, assembling in short time, long maintenance period, and high operational reliability [1, 2]. It has been widely used in rail transportation. There may be insulation faults in the GIS. If the fault cannot be detected and effectively processed, it will lead the equipment to insulation failure and even been developed into a more serious power grid accident. Partial discharge detection is used to evaluate the insulation state and can provide preventive solutions because the partial discharge is an early indication of the potentially insulation defect [3–6].

R. Yao (✉) · M. Hui · L. Bai · Z. Ye · Q. Wu
School of Electronic and Control Engineering, Chang'an University, Xi'an 710064, China
e-mail: emilyyao@chd.edu.cn

Y. Qin et al. (eds.), *Proceedings of the 4th International Conference on Electrical and Information Technologies for Rail Transportation (EITRT) 2019*, Lecture Notes in Electrical Engineering 639, https://doi.org/10.1007/978-981-15-2866-8_2

The phase-resolved pulse sequence (PRPS) data collecting is accessed by the ultra-high frequency (UHF) setup, when the discharge is stable at different test voltages [7–10].

It is known that there are several kinds of PD patterns to be applied to describe the insulation defects in rail transportation GIS. In particular, the estimation by distribution (statistical) graphs is commonly used [11]. The partitioning method according to the positive and negative for the distribution(statistical) graphs makes the discharge area discontinuous, and the extracted characteristic parameters are not consistent with the physical meaning and statistical significance [12, 13]. In using the traditional phase-resolved partial discharge (PRPD) pattern to characterize the insulation defect, there is a problem that the discontinuous discharge region is for the entire AC cycle. To remedy this problem, the polar coordinate phase-resolved PD pattern approach is presented accordingly in this paper. And the steps in detail to plot the proposed polar coordinate pattern based on PRPS data sets are shown in the paper.

2 Test Sample

Three kinds of physical models are designed. There are corona discharges, surface discharges, and internal discharges. The metal protrusion on the enclosure is used to simulate corona discharge. It is a 1.7 cm long single metal fixed on the enclosure as shown in Fig. 1a. Surface contamination on the spacer is used to simulate the surface discharge. As shown in Fig. 1b, the 5.3 cm long metal wire is placed close to the high conductor. The metal wire is along the radial direction of the basin-shaped insulator inner wall. Through artificial making of a cavity in the insulator, the internal discharges can be achieved. The cavity making process is described as follows. Firstly, a hole is drilled on the insulator. Then, the epoxy resin, which is the same material with the insulator, is used to fill the hole. Finally, the cavity is formed after the coagulation. The X-ray flaw detection is shown in Fig. 1c.

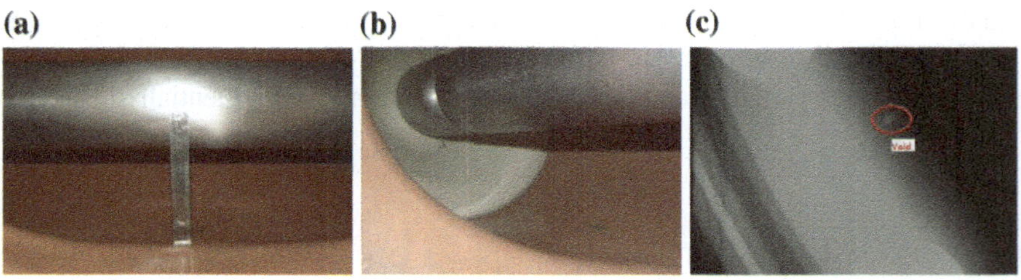

Fig. 1 Physical models: **a** a fixed protrusion on the enclosure; **b** surface contamination on the spacer; and **c** the X-ray flaw detection of the cavity

3 The Proposed PD Pattern

3.1 The Propose Reasons

The partitioning method according to the positive and negative of the traditional distribution (statistical) graphs makes the discharge area discontinuous, and the extracted characteristic parameters are not consistent with the physical meaning and statistical significance. The discontinuous discharge region forms the bank areas. As a result, the mathematical meaning of the skewness, kurtosis, and asymmetry extracted from the traditional PRPD pattern is not precise.

The new proposed polar coordinate phase-resolved partial discharge pattern is used to solve this problem. The polar axis at the horizontal placement makes the observation more intuitive. Additionally, there are two discharge clusters of the typical insulation defects in the proposed pattern. According to this feature, the based-on discharge clustering partitioning method is proposed.

3.2 The Plotting Method

1. Firstly, the data sets preparation. The data sets are a sequence of PD pulses $u_s(t_s, U(t_s))$, of which t_s is the time of PD pulses occurrence, $U(t_s)$ is the applied voltage, and u_s is the discharge amplitude of PD pulses.
2. The PD amplitude is linear normalized by the following equation:

$$u_{\mathrm{norm}} = (u_s - u_{\min})/(u_{\max} - u_{\min}) \tag{1}$$

   where $u_{\min}$ and $u_{\max}$ are the minimum and maximum amplitudes of the discharge separately.
3. The angle to radius conversion.
4. Drawing the polar radius and the polar angle using u_{norm} and φ_{norm}.
5. In the polar coordinate, classifying the partial discharge points into two clusters through K-means clustering algorithm. The centroid of each cluster is calculated by the following:

$$c_i = 1\Big/N_i \sum_{x \in C_i} x \tag{2}$$

   where C_i is the ith cluster, x is a point in C_i, c_i is the mean of the ith cluster, and N_i is the PD numbers of each PD cluster.

4 Experimental Results

The new proposed pattern of three kinds of insulation defects at different voltage levels is shown in Figs. 2, 3, and 4.

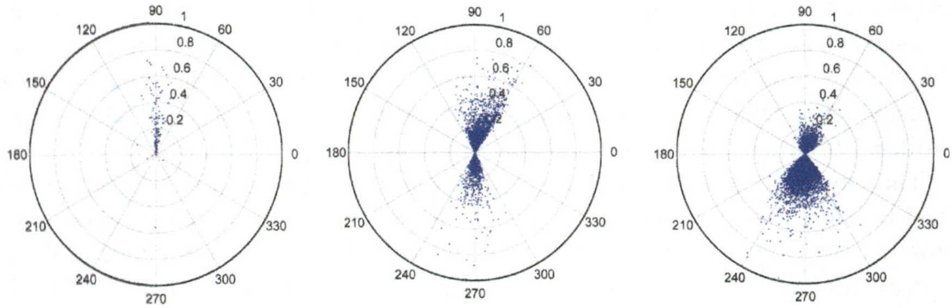

(a) 210kV (b) 271kV (c) 383kV

Fig. 2 Polar coordinate pattern of the fixed protrusion on the enclosure defect at three voltage levels

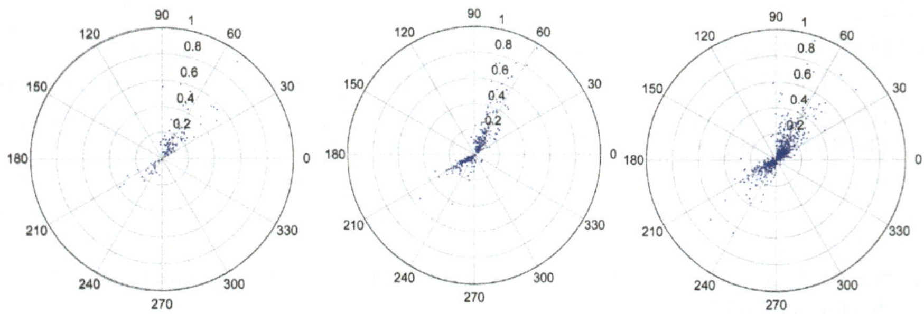

(a) 46kV (b) 58kV (c) 69kV

Fig. 3 Polar coordinate pattern of surface contamination on the spacer defect at three voltage levels

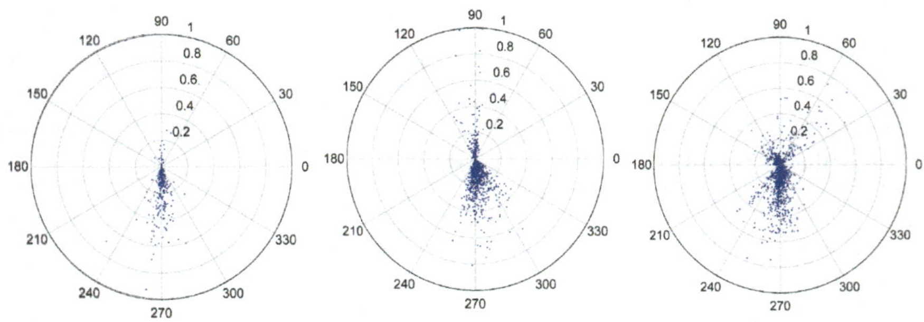

(a) 247kV (b) 340kV (c) 385kV

Fig. 4 Polar coordinate pattern of a void in the insulator defect at three voltage levels

It can be seen from the figures that the discharges of these defects have two discharge clusters in the polar coordinate spectrum. According to this characteristic, this pattern analyzes the discharge characteristics according to clustering results.

The discharges of the fixed protrusion first occur at around 90° and 270°, respectively. A void in the insulator defect concentrates near 90° and 270°. The two discharge clusters of a void in the insulator defect are asymmetrical. The discharge near 270° is much denser.

Surface contamination on the spacer defect is distributed at −30° to 90° and 180°–270°. Under the lower voltage, the defect is much denser in the range of 0–0.3.

It can be seen that the polar coordinate pattern can effectively characterize the discharge characteristics of different types of insulation defects.

5 Conclusions

The aim is to remedy the problems of discontinuous discharge region and extracted parameters not representing the original physical meaning and statistical significance according to the positive and negative for the traditional distribution (statistical) graphs.

As a result, the polar coordinate phase-resolved PD pattern approach to characterize corona discharge, surface discharge, and internal discharge is presented. There are two innovations of this paper. Firstly, the polar axis is joined from the beginning to end. Secondly, the discharge region is classified into two clusters. In this way, the discontinuous discharge region does not exist and the discharge characteristics can be distinguished from the two clusters.

Acknowledgements This research was funded by the National Natural Science Foundation of China through grants 51407012, the Nature Science Foundation of Shaanxi Province through grants 2016JQ5047, and the Fundamental Research Funds for the Central Universities through grants 300102329104 (Chang'an University).

References

1. Tian Y, Lewin PL, Davies AE (2002) Comparison of on-line partial discharge detection methods for XLPE cable joints. IEEE Trans Dielectr Electr Insul 9:604–615
2. Wu M, Cao H, Cao J, Nguyen HL (2015) An overview of state-of-the-art partial discharge analysis techniques for condition monitoring. IEEE Electr Insul Mag 31:22–35
3. Kreuger FH (1989) Partial discharge detection in high voltage equipment. Butterworth & Co. (Publishers) Ltd
4. Tang J, Jin M, Zeng F (2017) Feature selection for partial discharge severity assessment in gas-insulated switchgear based on minimum redundancy and maximum relevance. Energies 10(10):1516

5. Hsieh J, Tai C, Su M, Lin Y (2014) Identification of partial discharge location using probabilistic neural networks and the fuzzy c-means clustering approach. Electr Power Compon Syst 42:60–69
6. Sharkawy RM, Mangoubi RS, Abdel-Galil TK, Salama MMA, Bartnikas R (2007) SVM classification of contaminating particles in liquid dielectrics using higher order statistics of electrical and acoustic PD measurements. IEEE Trans Dielectr Electr Insul 14:669–678
7. Putro WA, Nishigouci K, Khayam U, Suwarno, Kozako M, Hikita M, Urano K, Min C (2012) PD pattern of various defects measured by UHF external sensor on 66 kV GIS model. In: International conference on condition monitoring and diagnosis (CMD), 23–27 Sept 2012, pp 954–957
8. Wang YN, Liu GZ, Ma XG (2018) Application of online partial discharge detection of Switchgear based on Ultra-High Frequency (UHF) method. In: IEEE 2nd international conference on dielectrics (ICD), 1–5 July Budapest, Hungary
9. Pearson JS, Hampton BF, Sellars AG (1991) A continuous UHF monitor for gas-insulated substations. IEEE Trans Electr Insul 26:469–478
10. Bo L, Li G, Min C (2019) Simulation on propagation characteristics of Ultra-High Frequency (UHF) electromagnetic wave inside typical Gas Insulated Switchgear (GIS) structures. J Nanoelectr Optoelectr 14(4):551–562
11. Zhu Y, Jia Y, Wang L (2016) Partial discharge pattern recognition method based on variable predictive model-based class discriminate and partial least squares regression. IET Sci Meas Technol 10(7):737–744
12. Mazroua AA, Salama MMA, Bartnikas R (1994) PD pattern recognition with neural networks using the multilayer perceptron technique. IEEE Trans Electr Insul 28:1082–1089
13. Su MS, Chen JF, Lin YH (2013) Phase determination of partial discharge source in three-phase transmission lines using discrete wavelet transform and probabilistic neural networks. Int J Electr Power 51:27–34

Method of Wheel Out-of-Roundness Detection Based on POVMD and Multinuclear LS-SVM

Lichao Fang, Shibo Li, Wang Dai, Yong Zhang, Zongyi Xing and Yulin Han

Abstract With the continuous increase of the running mileage and speed of the train, more and more wheels become out-of-roundness due to the collision and friction between the wheels and track. It has great significance to detect wheel polygon in order to ensure the safe operation of trains. The wheel out-of-roundness detection method based on POVMD and multinuclear LS-SVM is investigated by using POVMD algorithm to decompose the vibration signal, and then PSO to get optimal parameters which takes VMD algorithm into consideration. Such a method extracts some features from IMF components. Finally, Gaussian kernel function and directed acyclic graph classification method are chosen to build multinuclear classifier to detect wheel out-of-roundness. The experiment results show that the proposed method is effective to analyze wheel out-of-roundness.

Keywords Wheel out-of-roundness · Vibration signal · POVMD · Multinuclear LS-SVM

1 Introduction

The wheel is an important factor that affects the safety operation of trains. Due to the constant impact and collision between wheels and tracks, the wheels will appear non-circular phenomenon, causing the strong vibration of the train which may cause trains shake violently and even lead to derailment [1, 2]. Since wheel detection has such a great significance, in this paper we propose a wheel non-circular detection method based on variational mode decomposition based on parameter optimization (POVMD) and multi-core least squares support vector machine (LS-SVM). The feasibility of the method is verified by experiments.

L. Fang · S. Li · W. Dai · Y. Zhang (✉) · Z. Xing · Y. Han
School of Automation, Nanjing University of Science & Technology, Nanjing, China

Y. Qin et al. (eds.), *Proceedings of the 4th International Conference on Electrical and Information Technologies for Rail Transportation (EITRT) 2019*, Lecture Notes in Electrical Engineering 639, https://doi.org/10.1007/978-981-15-2866-8_3

2 Algorithm

In the running process of urban rail vehicles, if wheels are not round, the vibration signals generated between wheels and tracks will change, so we can get the wheels' specific conditions by analyzing their signal characteristics.

2.1 POVMD Algorithm

However, traditional variational mode decomposition (VMD) algorithm [3] needs to set the parameters K artificially. In order to set K and α more reasonably, we combine VMD with particle swarm optimization (PSO) algorithm [4], and we call this algorithm POVMD.

Assume that the search space is D-dimension, a particle swarm has n particles. In the tth iteration, the ith particle's position is $x_i(t) = (x_{i1}, x_{i2}, \ldots, x_{iD})$, search speed is $v_i(t) = (v_{i1}, v_{i2}, \ldots, v_{iD})$, and local optimal position is $p_i(t) = (p_{i1}, p_{i2}, \ldots, x_{iD})$. The speed and position of each particle are continuously updated according to Eq. 1 where w is inertia weight, c_1 and c_2 are leaning factors, and r_1 and r_2 are independent sequence.

$$\begin{aligned} v_i(t+1) &= wv_i(t) + c_1 r_1 (p_i(t) - x_i(t)) + c_2 r_2 (p_g(t) - x_i(t)) \\ x_i(t+1) &= x_i(t) + v_i(t+1), \quad c_1, c_2 \in [0, 2], r_1, r_2 \in [0, 1] \end{aligned} \tag{1}$$

Here we introduce envelope entropy E_p proposed by Tang [5]. Given a time signal $x(i)$, its E_p value can be described as Eq. 2 where $a(i)$ is the envelope signal obtained by Hilbert demodulation. $x(i)$ is inversely proportional to E_p, so we can analyze E_p instead of $x(i)$.

$$E_p = -\sum_{i=1}^{n} \left(\left(a(i) \Big/ \sum_{i=1}^{n} a(i) \right) \lg \left(a(i) \Big/ \sum_{i=1}^{n} a(i) \right) \right) \tag{2}$$

When ith particle's position is x_i, computing E_p of all IMF components obtained by VMD decomposition, and the smallest E_p $(\min E_p)$ is called local minimum entropy value. The IMF component corresponding to $\min E_p$ has the most feature information in this group, which is called the local optimal component. In order to obtain the global optimal component, $\min E_p$ is used as the fitness value.

2.2 The Basic Principle of Multi-core LS-SVM

Wheel non-circular can be divided into first-to-fourth-order non-circular. The first-order non-circular refers to the eccentricity during production; the second-order non-circular means the wheels become elliptical; the third-order non-circular means wheels appear triangle; the fourth-order non-circular means wheels appear rectangular.

We select three IMF components with the smallest envelope entropy for analysis. Then, we calculate the five features: kurtosis, root mean square (RMS), energy, energy moment (EM), and Shannon entropy for each selected IMF component. Their definitions are shown in Eq. 3, where $\bar{x}$ is the average of the signal amplitude, Δt is sampling period, and $P(x_i)$ is probability of signal x_i.

$$\begin{aligned} \text{Kurtosis} &= E(x^4) - 3\left[E(x^2)\right]^2 \\ \text{RMS} &= \sqrt{\frac{1}{N}\sum_{i=1}^{N}(x_i - \bar{x})} \\ \text{Energy} &= \sum_{i=1}^{N}|x_i|^2 \\ \text{EM} &= \sum_{i=1}^{N}(i \cdot \Delta t)|x_i|^2 \\ \text{Entropy} &= -\sum_{i=1}^{N}P(x_i)\log P(x_i) \end{aligned} \tag{3}$$

LS-SVM [6] uses the equality constraint to transform the quadratic optimization problem into linear equations. The multi-core algorithm [7, 8] combines different types of kernel functions to form a new kernel function.

Suppose that the multi-core LS-SVM has m kernel functions $K_1, K_2, \ldots, K_m$, and the training dataset is D described as Eq. 4 where $x_i \in R_n$.

$$D = \{(x_1, y_1), (x_2, y_2), \ldots, (x_l, y_l)\} \tag{4}$$

According to the principle of structural risk minimization and the principle of multi-core algorithm, the optimization problem can be transformed into:

$$\begin{cases} \min \Phi(w, \mu, b, \xi) = \frac{1}{2}\sum\limits_{t=1}^{m}\mu_t w_t^T w_t + \frac{C}{2}\sum\limits_{i=1}^{l} e_i^2 \\ \text{s.t. } y_i\left[\sum\limits_{t=1}^{m}\mu_t w_t^T \phi_t(x_i) + b\right] = 1 - e_t, \sum\limits_{t=1}^{m}\mu_t = 1, \quad \mu_t \geq 0 \end{cases} \tag{5}$$

In the formula, μ_t is the weight coefficient of the tth kernel function, w_t is the weight coefficient, b is the threshold, e_i is the slack variable, C is the penalty coefficient, and $\phi_t(x_i)$ is the mapping relationship.

Using Lagrange multiplier method, we get Eq. 6, where α_i is Lagrange operator.

$$L(w,\mu,b,e,a) = \Phi(w,\mu,b,\xi) - \sum_{i=1}^{l}\alpha_i\left\{y_i\left[\sum_{t=1}^{m}\mu_t w_t^T\phi_t(x_i) + b\right] + e_t - 1\right\} \quad (6)$$

According to the *KKT* theorem, we can get Eq. 7. The matrix form of b and α are shown as Eq. 8, where $\vec{\alpha} = [\alpha_1, \alpha_2, \ldots, \alpha_l]^T$.

$$\begin{cases} \frac{\partial L}{\partial w_i} = 0 \Rightarrow w_i = \sum_{i=1}^{l}\alpha_i y_i \phi_i(x_i) \\ \frac{\partial L}{\partial w_i} = 0 \Rightarrow \sum_{i=1}^{l}\alpha_i y_i = 0 \\ \frac{\partial L}{\partial e_i} = 0 \Rightarrow \alpha_i = Ce_i \\ \frac{\partial L}{\partial \alpha_i} = 0 \Rightarrow y_i\left[\sum_{t=1}^{m}\mu_i w_t^T\phi_t(x_i) + b\right] + e_i - 1 \end{cases} \quad (7)$$

$$\begin{bmatrix} 0 & y^{\mathrm{T}} \\ y & ZZ^{\mathrm{T}} + C^{-1}I \end{bmatrix}\begin{bmatrix} b \\ \vec{\alpha} \end{bmatrix} = \begin{bmatrix} 0 \\ \vec{1} \end{bmatrix} \quad (8)$$

Using a kernel function $K(x_i, x_j)$ that satisfies the Mercer theorem and substituting it into $\Omega = ZZ^{\mathrm{T}}$, we obtain:

$$\Omega_{ij} = y_i y_j \sum_{t=1}^{m}\mu_i w_t^T\phi_t^T(x_i)\phi_t(x_j) = y_i y_j \sum_{t=1}^{m}\mu_t K_t(x_i, x_j) \quad (9)$$

The optimal classification decision function for multi-core LS-SVM is:

$$f(x) = \mathrm{sgn}\left[\sum_{i=1}^{l}\alpha_i y_i \sum_{t=1}^{m}\mu_t K_t(x, x_i) + b\right] \quad (10)$$

In order to calculate the weight coefficients (μ_t) of each kernel function, the nuclear polarization [9] was introduced; it is defined as Eq. 11. When samples of the same class approach each other, the value of P will increase. Therefore, P can be used to determine the value of μ_t (Eq. 12).

$$P = \sum_{i=1}^{l}\sum_{j=1}^{l}y_i y_j K(x_i, x_j) = \sum_{y_i = y_j}K(x_i, x_j) - \sum_{y_i \neq y_j}K(x_i, x_j) \quad (11)$$

In the formula, K is the kernel function, $y = [y_1, y_2, \ldots, y_l]^{\mathrm{T}}$, yy^{T} is the ideal kernel matrix.

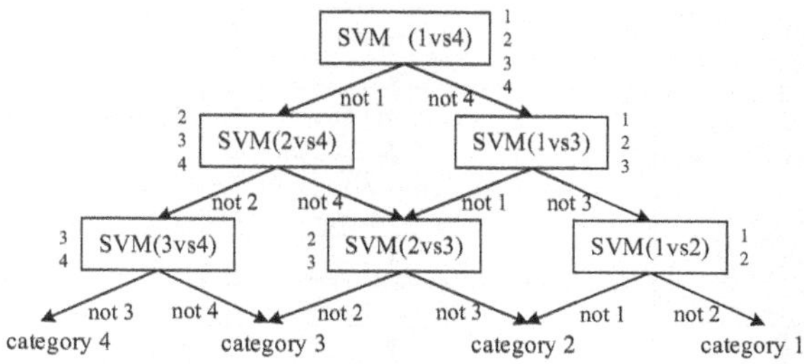

Fig. 1 Schematic diagram of directed acyclic graph classification

$$\mu_t = P_t \Big/ \sum_{t=1}^{m} P_t \tag{12}$$

2.3 *Selection of Classifier and Kernel Function*

In this paper, we select Gaussian function [10] as kernel function and a directed acyclic graph to construct classifier. Gaussian function is defined as Eq. 13.

$$K(x_i, x_j) = \exp(-\|x_i - x_j\| / (2\gamma^2)), \ \gamma > 0 \tag{13}$$

The directed acyclic graph classification constructs a binary classifier; it starts from the root node in the form of a directed acyclic graph. Fig. 1 shows a four-category classification example. Each binary classifier acts as a node. It starts from the root node and classifies the samples one layer at a time until the last layer, and the result of the last layer is its corresponding category.

Combined with the previous content, Fig. 2 shows the implementation steps of the method proposed in this paper in detail.

3 Experiments and Results

3.1 *Constructing Multi-core LS-SVM Classifier Model*

The first-order non-circular is caused during production. Here we focus on analyzing second-to-fourth-order non-circular. We select 720 sets of rail vibration signals which include first-to-fourth-order non-circular, and each class has 180 sets. We select 440 sets as training samples, and the other 280 sets are used as test samples.

For the LS-SVM model, we set the penalty parameter $C = 38$ and select the Gaussian kernel function and directed acyclic graph to construct a multi-class classifier.

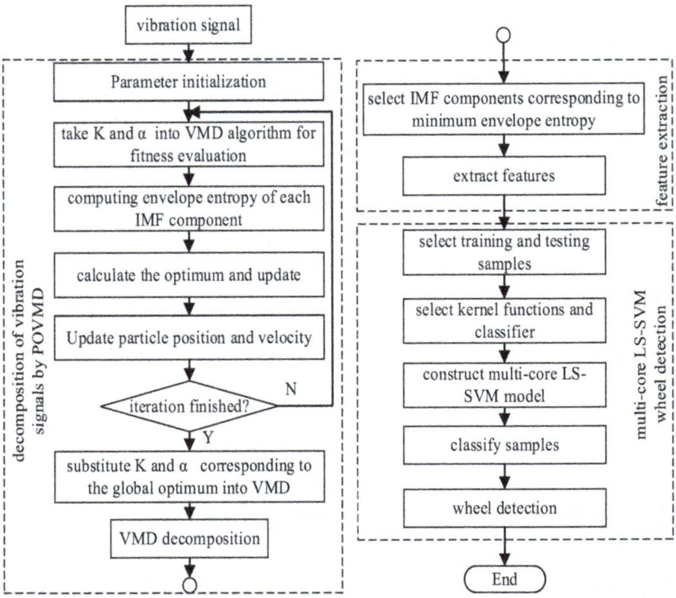

Fig. 2 Wheel non-circular detection based on POVMD and multi-core LS-SVM

Fig. 3 Wheel classification results

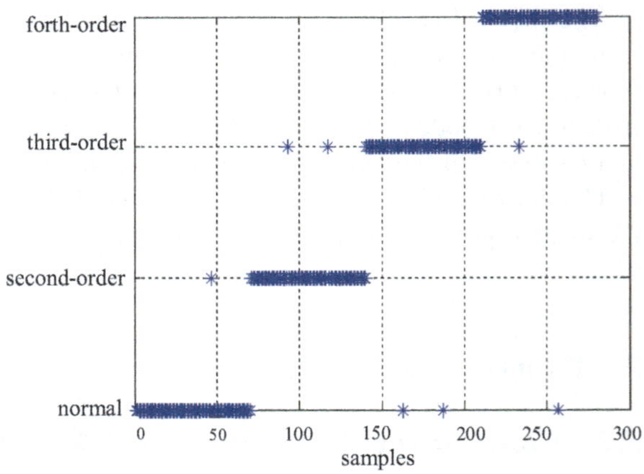

The specific classification results are shown in Fig. 3. The abscissa is the number of test samples, numbers 1–70 are normal wheels, 71–140 are second-order wheels, 141–210 are third-order wheels, and 211–280 are fourth-order wheels. We can see that the result of normal wheels is the best and the correct rate is 0.9857; the results of second-to-fourth-order wheels are all 0.9714. Using multi-core LS-SVM model to detect wheel non-circle, we can get a high accurate rate, which meets the actual engineering requirements.

3.2 Field Data Experiment

After the model has been built, we take the single-side wheel data collected from carriage 3A85 of 8586 train in Guangzhou Metro as an example. The vibration accelerometer was installed beside tracks, and vibration signal is collected at 10 kHz frequency. The wheel numbers are 2, 4, 6, and 8. After filtering, we can get the time domain of the vibration signal which is shown in Fig. 4.

The filtered four wheels signals are decomposed by POVMD algorithm. We use PSO algorithm to get the optimal VMD combination parameters K and α of the second, fourth, sixth, and eighth wheels. They, respectively, are ($K = 7$, $\alpha = 1471$), ($K = 7$, $\alpha = 2729$), ($K = 7$, $\alpha = 1953$), and ($K = 7$, $\alpha = 1275$). We substitute these parameters into the VMD to obtain the IMF components. The time domain diagram is shown in Fig. 5. It can be seen that the waveforms of different wheels after POVMD decomposition show different distribution, which indicates that different wheels have different state. Also, after decomposition, the noise components in the original signal are effectively suppressed.

We select three IMF components with the smallest envelope entropy for feature calculation. Table 1 shows the values of five main extracted features (kurtosis, root mean square value, energy, energy moment, and information entropy). Then, the multi-core LS-SVM model trained in 3.1 segments is used to classify the four wheels according to their features in Table 1. The results are: Nos. 2, 6, and 8 wheels are normal wheels, and the No. 4 wheel has a second-order non-circular fault, which is consistent with the on-site manual test results. Thus, the feasibility of our method has been verified.

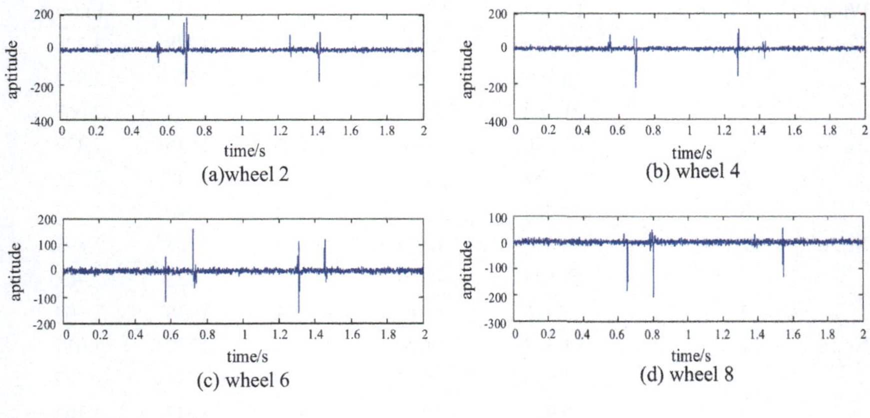

Fig. 4 Wheel vibration signal

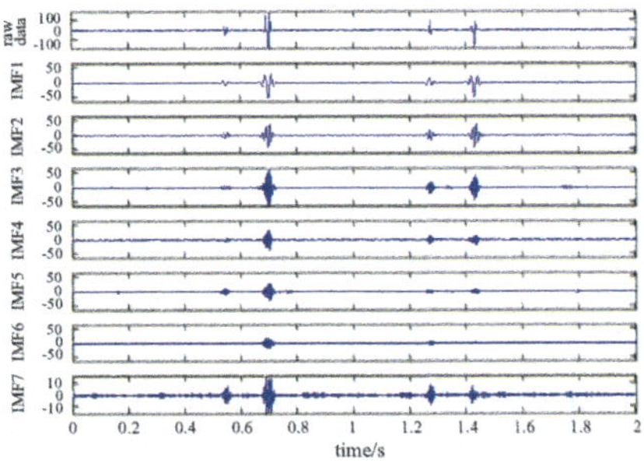

(a) wheel 2

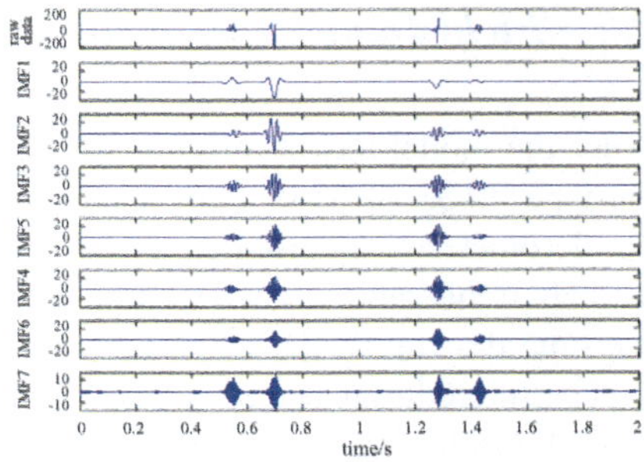

(b) wheel 4

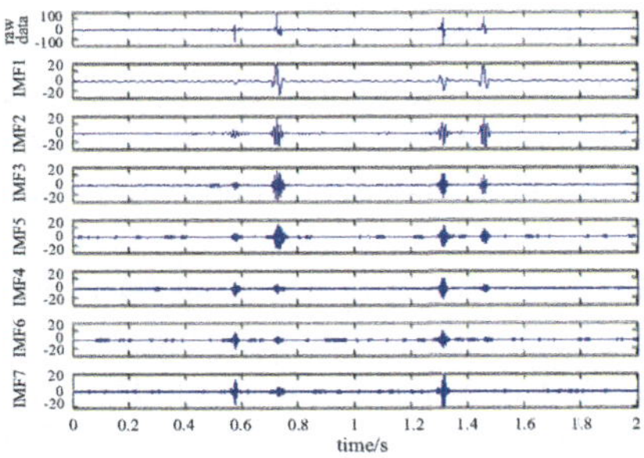

(c) wheel 6

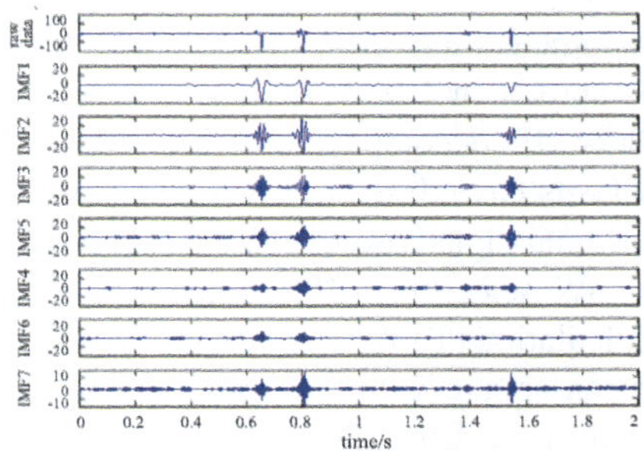

(d) wheel 8

Fig. 5 Time-domain waveform of POVMD decomposition of wheel vibration signal

Table 1 Wheel vibration signal characteristic quantity

Wheel	Component	Kurtosis	RMS	Energy	EM	Entropy
No. 2	1	48.579	3.547	31,426	1548	3.376
	2	46.793	3.586	31,538	1576	2.873
	3	51.762	4.139	34,739	1727	2.667
No. 4	1	61.326	3.846	28,626	1536	4.437
	2	51.257	3.631	24,689	1248	2.824
	3	32.923	3.397	21,693	1090	2.733
No. 6	1	48.893	3.829	27,539	1365	2.897
	2	41.825	3.836	27,672	1423	2.768
	3	48.376	3.716	19,131	1039	2.604
No. 8	1	53.368	3.624	24,635	1237	4.617
	2	42.575	3.983	31,739	1551	2.892
	3	46.991	3.328	21,465	1042	3.264

4 Conclusion

The method of wheel out-of-roundness based on POVMD and multi-core LS-SVM has been described in detail in this paper. By combining PSO algorithm with traditional VMD algorithm, we overcome the shortcoming that VMD algorithm needs to set the decomposition parameter *K* manually according to experience, and make the value of parameter *K* more reasonable. We use the field data for verification. The experimental results show that this method can effectively decompose the vibration signal and analyze the wheel condition according to its characteristics. In the future, we will further improve the algorithm and apply the detection equipment to the actual scene.

Acknowledgements This work is supported by National Key R&D Program of China (2017YFB1201102).

References

1. Hong L, Shen G (2009) A portable measuring device for circular profile and wheel diameter. Urban Rail Transit Res Inst 12(2):53–55 (in Chinese)
2. Lee KY, Lee KK, Ho SL (2004) Exploration of using FBG sensor for derailment detector
3. Yang H, Jiang C et al (2016) Fault information extraction method for rolling bearing based on VMD parameter estimation. Bearing 10:49–52 (in Chinese)
4. Kennedy J, Eberhart R (1995) Particle swarm optimization. In: Proceedings of IEEE international conference on neural networks, vol IV, pp 1942–1948
5. Tang G, Wang X (2015) Parameter optimized variational mode decomposition method with application to incipient fault diagnosis of rolling bearing. J Xi'an Jiaotong Univ 49(5):73–81 (in Chinese)
6. Li H, Li X et al (2016) Fault diagnosis of vibration for hydropower units based on empirical mode decomposition and support vector machine. J Hydroelectr Eng 35(12):105–111 (in Chinese)
7. Wang H, Sun F et al (2010) On multiple kernel learning methods. Acta Automation Sin 36(8) (in Chinese)
8. Gönen M, Alpaydın E (2011) Multiple kernel learning algorithms. J Mach Learn Res 12:2211–2268
9. Baram Y (2005) Learning by kernel polarization. Neural Comput 17(6):1264–1275
10. Vapnik VN (2005) Universal learning technology: support vector machines. NEC J Adv Technol 2(2):137–144

Analysis of Impact of Guiding Signs on Evacuation Decision-Making Process

Yiwen Chen, Jie Xu, Hui Zhang and Mengdi Liang

Abstract The efficiency of emergency evacuation signs is influenced by its visibility catchment area (VCA) and pedestrian behavior and route selection in the evacuation. In order to describe the pedestrian route selection process in evacuation in specific subway station signs' VCA, we identify the signs' properties which influence the efficiency of emergency evacuation signs and check the parameters of facility layout structure which affects VCA through investigation and analysis of test results, then constructing decision-making behavior model of pedestrian evacuation route selection base on the analysis of different route selection behavior during evacuation. This paper simulates the process of route selection in evacuation through AnyLogic and then analyzes the influence of the emergency signs' location, the initial evacuation volume and the success rate of pedestrian interaction on the evacuation time by simulating the process of route selection in evacuation.

Keywords Urban Rail Transit station · Signs · Emergency evacuation · Visibility catchment area · Decision-making behavior model

1 Introduction

More and more attention has been paid to the attributes and layout of signs in railway station, and some achievements have been made [1–4], such as the basic design of guiding sign [3, 5], the setting location and number of guiding sign [4, 6], the VCA of guiding signs [2], the effective distance calculation model [7],

Y. Chen · J. Xu (✉) · H. Zhang · M. Liang
State Key Laboratory of Rail Control and Safety, Beijing Jiaotong University, Beijing, China
e-mail: jxu1@bjtu.edu.cn

School of Transportation and Traffic, Beijing Jiaotong University, Beijing, China

Beijing Research Centre of Urban Traffic Information Sensing and Service Technologies, Beijing, China

Research and Development Center of Transport Industry of Technologies and Equipment of Urban Rail Operation Safety Management, MOT, PRC, Beijing, China

Y. Qin et al. (eds.), *Proceedings of the 4th International Conference on Electrical and Information Technologies for Rail Transportation (EITRT) 2019*, Lecture Notes in Electrical Engineering 639, https://doi.org/10.1007/978-981-15-2866-8_4

the utility coordination model of guiding signs [8], and the influence of poor visibility and path's geometric attributes on the efficiency of evacuation signs [5, 9]. In addition, some researchers study the evacuation efficiency of guiding signs in specific scenarios by simulating the process of evacuation through dynamic model such as CA model or social force model [10]. To sum up, existing research which considers the optimization of guiding sign's attributes and evacuation efficiency has been carried out, and however, there is still lack of related research which confiders the impact of signs on passenger's decision-making behavior in evacuation. This paper analyzes the VCA of guiding signs based on the identification and calibration of factors which affect the recognition degree of guiding signs. Then we construct the decision-making behavior model for different decision-making tendencies, and simulating by AnyLogic. Finally, this paper discusses the influence of guiding signs' setting location, initial evacuation passenger flows, and the impact rate of crowd on the time which is required for decision making.

2 Analysis of Sign's VCA

2.1 The VCA of Common Guiding Signs

This paper focuses on 4 kinds of guiding signs in railway station. The material of sign 2 (Fig. 2) and 3 (Fig. 3) is phosphorescence. Usually, sign 2 sets on the ground, and sign 3 sets on the wall of escaping walkway, stairs, and pillars. The VCA of signs is ellipse [11], and by data analysis, we can get the relationship between the longer axis a_1 (shorter axis b_1), high x_1, and brightness x_2, which is (Eq. 1)

$$\begin{cases} a_1 = 15.8118x_1 + 0.0407x_2 \\ b_1 = 5.6176x_1 + 0.0545x_2 \end{cases} \quad (1)$$

Taking the sign as the origin of coordinates, the VCA of phosphorescent sign is

$$\left(\frac{x}{15.8118x_1 + 0.0407x_2}\right)^2 + \left(\frac{y}{5.6176x_1 + 0.0545x_2}\right)^2 \leq 1 \quad (2)$$

Fig. 1 Signage 1: safety exit

Fig. 2 Signage 2: short arrow

Fig. 3 Signage 3: long arrow

Sign 1 (Fig. 1) and 4 (Fig. 4) are electric. Usually, sign 1 sets between facilities (e.g., stair and pillar), and sign 4 sets in the station hall. And the relationship between electrical sign's longer axis a_2 (shorter axis b_2), high x_1, and brightness x_2 is (Eq. 3)

$$\begin{cases} a_2 = 5.1765x_1 + 0.3936x_2 \\ b_2 = 1.2118x_1 + 0.1725x_2 \end{cases} \tag{3}$$

Taking the sign as the origin of coordinates, the VCA of electric sign is

$$\left(\frac{x}{5.1765x_1 + 0.3936x_2}\right)^2 + \left(\frac{y}{1.2118x_1 + 0.1725x_2}\right)^2 \leq 1 \tag{4}$$

2.2 The Influence of Sign's Setting on VCA

Walkway

In walkway, there are mainly 2 kinds of guiding sign which are sign 2 and 3, and they are set on the wall and ground (Fig. 5). The VCA is influenced by the layout. Assuming that the width of the walkway is W_1 and the length is L_1, and there are n signs in the walkway, and the distance between each sign is *k*, and *a*(*b*) is the long (short) axis of single sign's VCA. The VCA in the walkway is related to the *k* (Figs. 6 and 7).

Fig. 4 Signage 4: exit sign

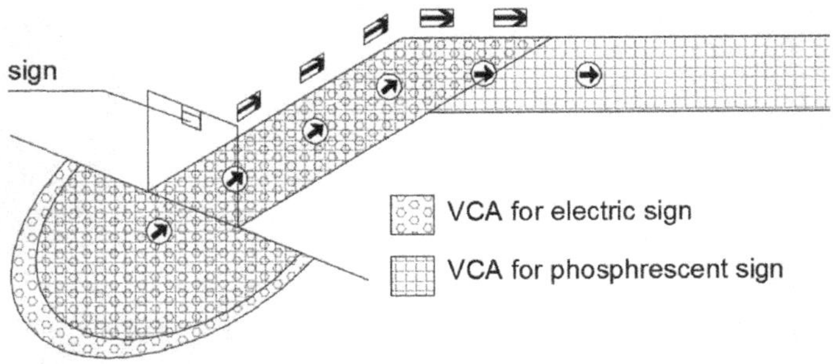

Fig. 5 VCA within walkway

Fig. 6 Diagram of walkway's VCA coverage ($k \leq 2a$)

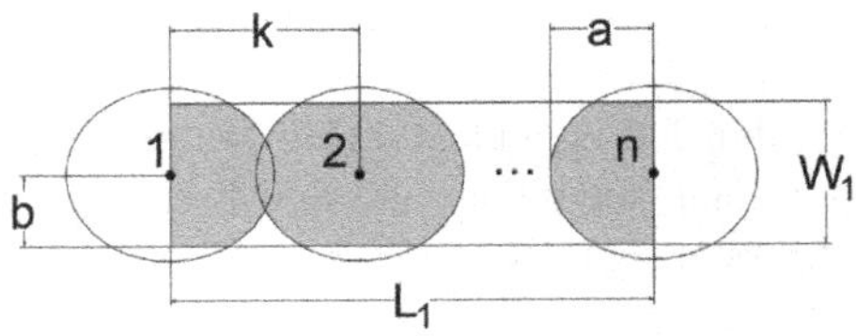

Fig. 7 Diagram of walkway's VCA coverage ($k > 2a$)

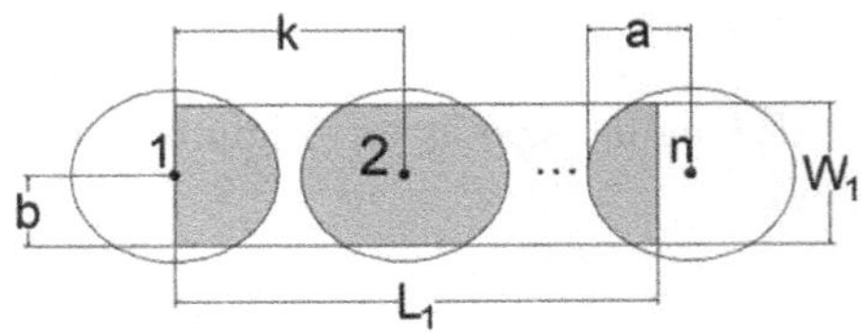

$$S = \min\left\{L_1 \times W_1, L_1 \times W_1 - 2(n-1)k\left(\frac{W_1}{2} - b\sqrt{1 - \frac{k^2}{4a^2}}\right)\right\}, k \leq 2a \quad (5)$$

$$S = W_1 \times [L_1 - (n-1)k], k > 2a \quad (6)$$

Stairs

Near the stairs, there are mainly 3 kinds of guiding sign which are sign 1, 2, and 3. Sign 1 is set above the stair, sign 2 is set on the ground, and sign 3 is set on the pillars. Affected by the layout, the shape of VCA is irregular (Fig. 8). Assuming that the width of stair is W_3, the length is L_3, and the length of obstacle beyond stair is L_2 (Fig. 9). Then the VCA is,

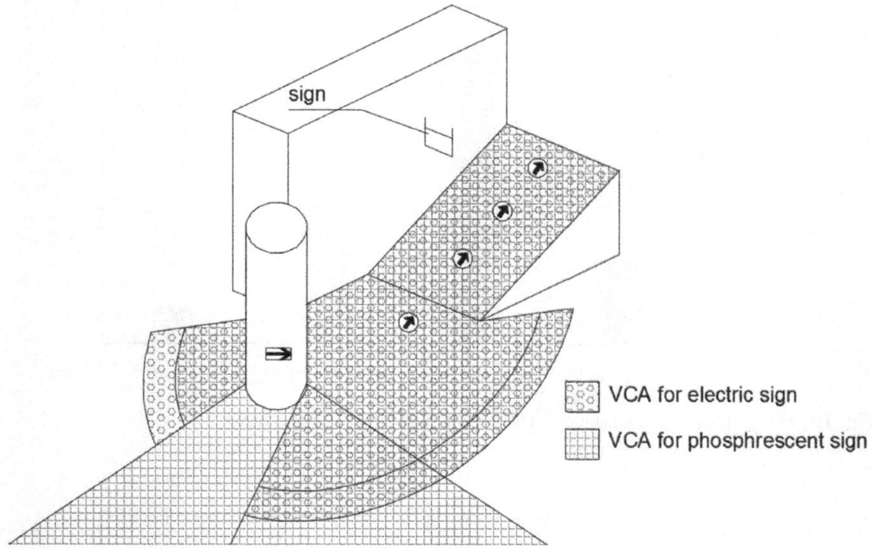

Fig. 8 VCA within stair

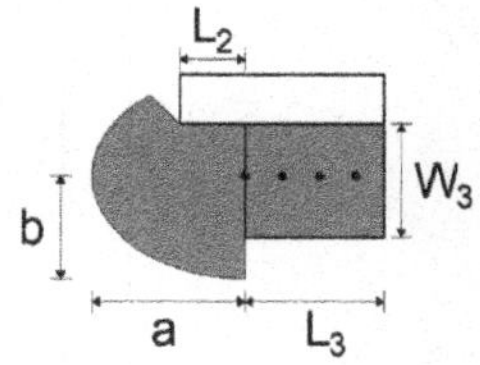

Fig. 9 Diagram of stair's VCA coverage

$$S = W_3L_3 + \frac{1}{2}\pi ab - \frac{L_2b\sqrt{1 - \frac{L_2^2}{a^2}}}{2} + \frac{W_3L_2}{2} \quad (7)$$

3 Passenger Flow Evacuation Path Selection Tendency

In order to obtain the decision-making behavior of crowd, we investigate about 600 passengers randomly by questionnaire. The decision-making behavior can be divided into three categories [12]: (1) selecting a path immediately based on his/her experience or memory; (2) following the guidance of signs because they can comprehend and trust the sign, but they will follow other passengers if they cannot comprehend or trust the sign; (3) following other passengers because they are not familiar with the layout of station.

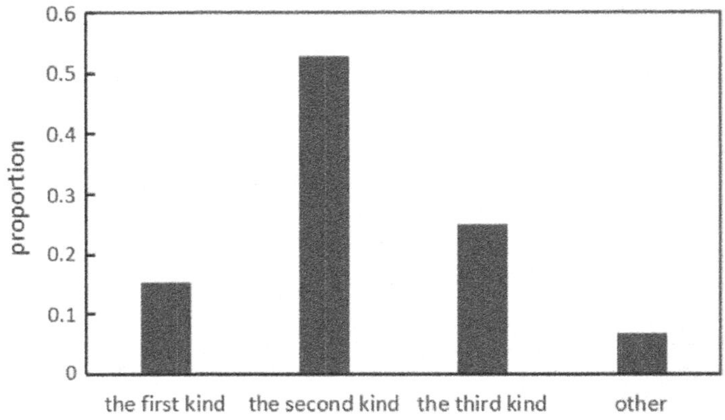

Fig. 10 Decision-making tendency of different type of passenger flow

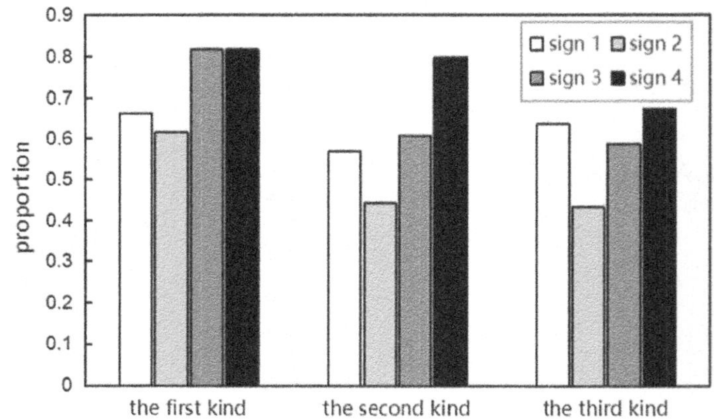

Fig. 11 Comprehension rate of different guiding sign

The proportion of 3 kinds of passenger flow is shown in Fig. 10. In addition, the 3 kinds of passenger flow's comprehension rate of guiding signs are the difference (Fig. 11). Assuming at time t, the number of the first type of passenger flow which has already made decision (passengers who make decisions based on experience or memory) is N_t^1, and the initial size of the first type of passenger flow is m_1, then

$$N_t^1 = N_{t-1}^1 = \cdots = N_1^1 = m_1 \tag{8}$$

In some cases (the layout of station is changed due to some accidents), the first type of passenger flow may turn into the second type of passenger flow (the decision-making passenger flow based on guiding information) or the third type of passenger flow (choose to follow the crowd), in which proportion is q_{12} and q_{13}.

The number of the second type of passenger flow which has already made decision at time t is N_t^2, and it includes 3 parts: (1) the N_{t-1}^2; (2) making decision by following the guidance of signs at time t; and (3) making decision by following crowd at time t. Assuming p_2 is the probability of passenger who may within the scope of VCA (it also represents the coverage rate of VCA). Assuming p_1 is the probability of passenger who can comprehend and trust the sign, ∂_1 is the invalid probability of sign, the initial size of passenger flow is m, and the initial size of the second type of passenger flow is m_2, then

$$\partial_1 = 1 - p_1 \times p_2 \tag{9}$$

Passenger may be influenced by the crowd, and the influence rate is p_3, and the initial size of passenger flow is m, and the initial size of the second type of passenger flow is m_2, and the number of passengers within the visual field is n, then N_t^2 is

$$N_t^2 = N_{t-1}^2 + \left[\left(m_2 - N_{t-1}^2\right) \times p_1 \times p_2\right] + \left[\left(m_2 - N_{t-1}^2\right) \times \partial_1 \times n \times p_3 \times \frac{N_{t-1}}{m}\right] \tag{10}$$

In some cases (large passenger flow), the second type of passenger flow has no time to obtain guiding information, and turn into the third type of passenger flow, and the proportion is q_{23}. The third type of passenger flow which has already made decision at time t is N_t^3, and it includes 2 parts: (1) the N_{t-1}^3 and (2) passengers which make decision at time t. Assuming the initial size of the third type of passenger flow is m_3, then the N_t^3 is

$$N_t^3 = N_{t-1}^3 + \left[\left(m_3 - N_{t-1}^3\right) \times \partial_1 \times n \times p_3 \times \frac{N_{t-1}}{m}\right] \tag{11}$$

In some cases (small passenger flow), the third type of passenger flow cannot obtain the decision-making tendency of crowd, and they will turn into the second type of passenger flow, and the proportion is q_{23}.

In summary, when the passenger flow is small, the third type of passenger flow will turn into the second type of passenger flow. When the layout changes, the first type of passenger flow will turn into the second type of passenger flow. When the passenger flow is large, the first and second type of passenger flow will turn into the third type of passenger flow.

4 Simulation of Passenger Flow's Decision-Making Behavior

4.1 Construction of Decision-Making Process Under Evacuation Based on AnyLogic

The process of passenger's decision making is shown in Fig. 12. Assuming the proportion of the first, second and third type of passenger flow is ρ_1, ρ_2, and ρ_3. Other parameters include p_1 (the comprehension rate of guiding signs), p_2 (the coverage rate of VCA), p_3 (the influence rate), and so on.

4.2 Simulation of Decision-Making Process Based on AnyLogic

The appropriate coverage of VCA (p_2*)

If p_2 is higher, the passenger flow is easier to obtain guidance information, but the decision-making time will increase too. The purpose of this simulation is to discuss the appropriate value of p_2*. The simulation environment is $m = 3000$, $n = 10$, $p_3 = 0.1$, and the value of p_2 is shown in Table 1. The decision-making total time

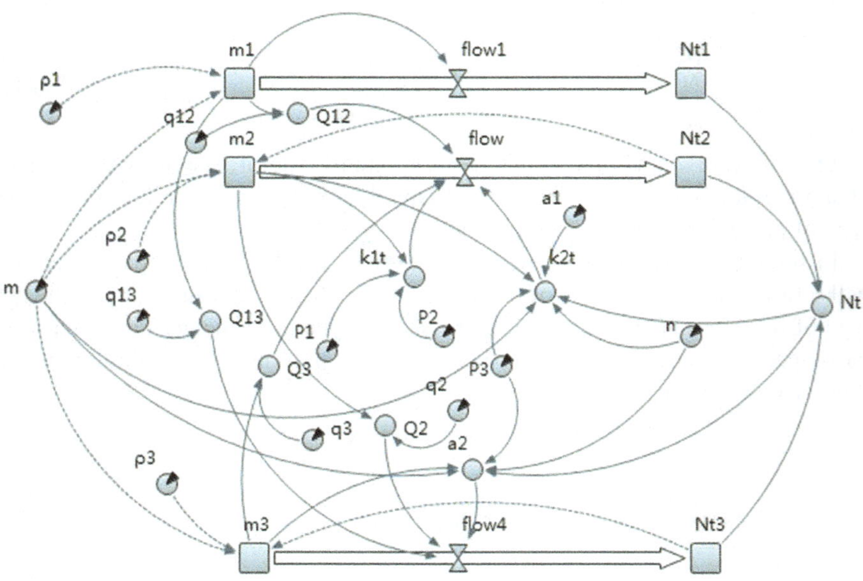

Fig. 12 Passenger's decision-making process model

Table 1 Value of VCA coverage (p_2)

No.	p_2	No.	p_2
1	0.2	5	0.6
2	0.3	6	0.7
3	0.4	7	0.8
4	0.5	8	0.9

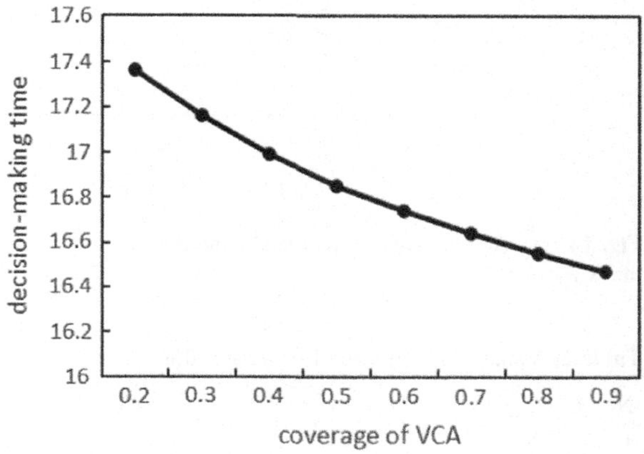

Fig. 13 Decision-making time under different coverage of VCA

(Fig. 13) and the decision-making time of the second and third type of passenger flow (Fig. 14) under different p_2 are obtained by simulation.

The initial size of passenger flow (m)

The purpose of this simulation is to discuss the relationship between m and the decision-making time. In this simulation, the simulation environment is $p_2 = p_2$*, n = 10, p_3 = 0.1, and the value of m is shown in Table 2. Assuming t_i is the decision-making time of experiment No. i and Δt is the difference between t_i and t_{i-1}, then

$$\Delta t = t_{i-1} - t_i \tag{12}$$

and the result is shown in Fig. 15.

The influence rate of crowd (p_3)

The influence rate of crowd represents the probability which passenger will be affected by others during decision-making process. In this simulation, the simulation environment is m = 3000, $p_2 = p_2$* and n = 10, and the value of p_3 is shown in Table 3. The decision-making time under different p_3 is shown in Fig. 16.

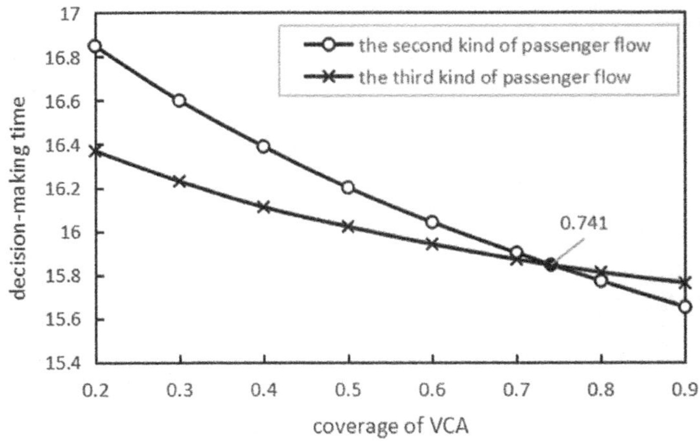

Fig. 14 Decision making of the second and third type of passenger flow under different coverage of VCA

Table 2 Value of initial size of passenger flow (m)

No.	m	No.	m
1	3000	5	7000
2	4000	6	8000
3	5000	7	9000
4	6000	8	10,000

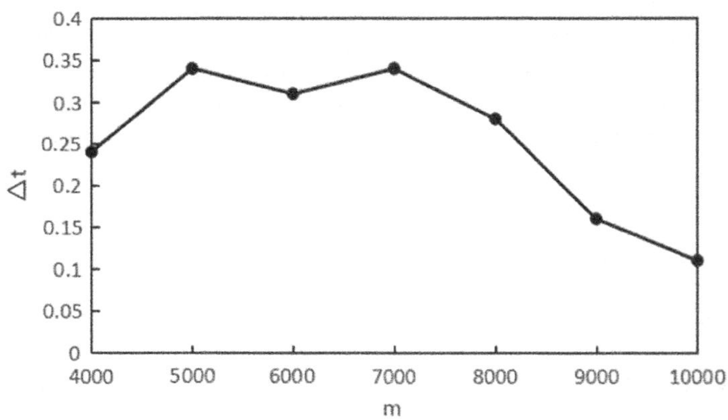

Fig. 15 Δt under different initial size of passenger flow (m)

Table 3 Value of the influence rate (p_3)

No.	p_3	No.	p_3
1	0.1	5	0.5
2	0.2	6	0.6
3	0.3	7	0.7
4	0.4	8	0.8

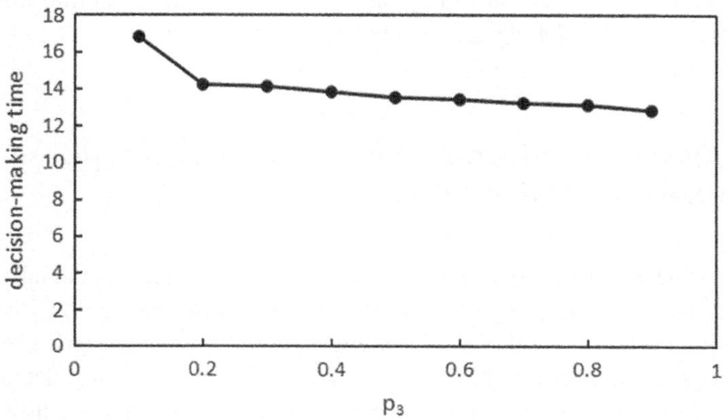

Fig. 16 Decision-making time under different influence rate ($p3$)

5 Discussion

5.1 The Appropriate Coverage of VCA (p_2*)

Figure 13 shows that the total decision-making time reduces gradually and the absolute value of slope decreases too. Further analysis shows that only the second type and the third type of passenger flow are influenced by VCA. When p_2 is greater than 0.741, the decision-making time of the second type of passenger flow is less than the third type of passenger flow (Fig. 14), so the total decision-making time depends on the third type of passenger flow, and the total decision-making time will not be obviously reduced even if p_2 increases, so the appropriate coverage of VCA (p_2*) is 0.741.

5.2 The Influence of Initial Size of Passenger Flow (m) on Path Selection Time

Figure 15 shows that the total decision-making time increases at first and then decreases gradually. The initial size of passenger flow (m) has an influence on the passenger flow density and the decision-making behavior. When m is small, the number of passengers can be seen within the visual field decreases, so part of the third passenger flow will need more time to obtain the decision-making tendency of the crowd. Therefore, the total decision-making time increased at first. When m increases to a certain extent, the crowd density increases, and the first type and second type of passenger flow are influenced by the density and cannot maintain their original decision, namely some of the first type of passenger flow questions their decision and chose to follow the crowd or the second type of passenger flow is washed away by crowd because they have no time to obtain guiding information. So, the total decision-making time decreases in the later period.

5.3 Influence of Interaction Between Passengers (p_3) on Decision-Making Time

Figure 16 shows that the total decision-making time decreases gradually with the increase of the p_3. Comparing with other two types of passenger flow, the p_3 has a greater influence on the third type of passenger flow. With the increase of the p_3, the decision-making time of the third type of passenger flow gradually decreases. The third type of passenger flow is the smallest part of the total passenger flow, so it has less influence on the crowd, and with the increase of the p_3, the decision-making time of the third type of passenger flow gradually decreases and the decreasing trend is not obvious.

6 Conclusion

Guiding signs directly determine the evacuation efficiency in emergency situation. Based on investigation and theoretical foundation, this paper identifies and calibrates the influencing factors, analyzes decision-making behavior of the three types of passenger flow, and builds the decision-making behavior model. The simulation results show that the appropriate coverage of VCA is 0.741. With the increase of initial size of passenger flow, the total decision-making time increases at first and then decreases, and the influence rate of the crowd has a positive impact on the total decision-making time. However, due to the lack of time, data, and other factors, this paper is still quite limited, and it is suggested that some factors (such as age and gender) should be taken into account.

Acknowledgements The authors gratefully acknowledge the support provided by China National "13th Five-Year" key research project "Safety assurance technology of urban rail system" (Grant No. 2016YFB1200402).

References

1. Chen HT, Yang P, Yu RC (2014) The influence of evacuation signs on evacuation for bad visibility. Appl Mech Mater 472:574–578
2. Filippidis L, Galea ER, Gwynne S, Lawrence PJ (2006) Representing the influence of signage on evacuation behavior within an evacuation model. J Fire Prot Eng 16(1):37–73
3. Filippidis L, Galea ER, Blackshields D, Lawrence P (2009) Experimental study of the effectiveness of emergency signage. In: Fourth international symposium on human behaviour in fire
4. Zhang Z, Jia L, Qin Y (2017) Optimal number and location planning of evacuation signage in public space. Saf Sci 91:132–147 (Elsevier Ltd)
5. Collins BL, Dahir MS, Madrzykowski DM (1992) Visibility of exit signs in clear and smoky conditions (75 lx)
6. Liu M, Zheng X, Cheng Y (2011) Determining the effective distance of emergency evacuation signs. Fire Saf J 46(6):364–369 (Elsevier)
7. Yue H, Shao C, Guan H, Gui D (2013) Location method of emergency evacuation signs in large pedestrian facilities. J Beijing Univ Technol 39(6):914–917
8. Zhang Z, Jia L, Qin Y (2016) Optimization of signage system configuration on metro platform based on cooperative guidance. J Transp Syst Eng Inf Technol 16(4):146–152 (in Chinese)
9. Heliövaara S, Kuusinen J-M, Rinne T, Korhonen T, Ehtamo H (2012) Pedestrian behavior and exit selection in evacuation of a corridor—an experimental study. Saf Sci 50(2):221–227
10. Li D, Han B (2015) Behavioral effect on pedestrian evacuation simulation using cellular automata. Saf Sci 80:41–55 (Elsevier Ltd)
11. Song B, Wu Z, Xie Y, Dai B, Liu C (2012) Experimental study on the visibility catchment area of direction sign. J Saf Sci Technol 8(2):16–22 (in Chinese)
12. Guo J (2015) Study on the mode of digital library promotion based on BASS model. New Century Libr 11:39–43 (in Chinese)

Research on the Construction of Big Data Management Platform of Shuohuang Railway Locomotive Operation and Maintenance

Zhaohui Zhang and Fei Li

Abstract In recent years, along with the emergence, development, and application of big data technology, data has become strategic resources of country, also commanding point of improving competitiveness and seizing market share for companies. This paper analyzes the application status of the existing information systems of the Shuohuang Railway and sorts out the needs of data sharing, business coordination, and mining and analysis. It proposes an overall plan and technical plan for the big data management platform of Shuohuang Railway locomotive operation and maintenance, and determines the data source of the platform, and designs seven functions, such as locomotive fault prediction and health management, personnel health management, transportation environment monitoring, and intelligent locomotive maintenance. This paper has certain guiding significance for the construction of big data management platform for the operation and maintenance of the Shuohuang Railway, and it also provides a template for the promotion and application of big data technology in railway.

Keywords Big data · Locomotive · Operation · Maintenance · Shuohuang Railway

1 Introduction

With the development of information system construction in recent years, Shuohuang Railway informatization has basically covered the transportation and production, operation and management, maintenance and reconditioning, education and training, etc., which strongly supports the development of various fields of business, and provides an important guarantee for strengthening the core compet-

Z. Zhang
Shuohuang Railway Development Co., Ltd, Suning 062350, Hebei, China

F. Li (✉)
Zhuzhou CRRC Times Electric Co., Ltd, Zhuzhou 412001, Hunan, China
e-mail: hunterlee222@163.com

Y. Qin et al. (eds.), *Proceedings of the 4th International Conference on Electrical and Information Technologies for Rail Transportation (EITRT) 2019*, Lecture Notes in Electrical Engineering 639, https://doi.org/10.1007/978-981-15-2866-8_5

itiveness of Shuohuang Railway and improving economic benefits. However, due to historical reasons and former technology conditions, earlier railway business systems are in a state of scattered way for a long time, and all kinds of existing information systems focus more on solving problem in a single business area, with various barriers for information isolated islands. There is big difficulty in information sharing, low level in data utilization, without crossing overall view. And it is difficult to carry out business coordination between departments, as well as many other problems [1, 2]. In particular, after years of accumulation, a large amount of historical data has been accumulated in various business fields. Due to the lack of effective technical means and methods, along with data fragmentation, data inconsistency, and defects in master data among systems, a large amount of valuable information is hard to be mined, and multi-dimensional, multi-service aggregation and comprehensive analysis cannot be carried out [3].

In this background, research on the construction of big data management platform for the locomotive operation and maintenance of Shuohuang railway is carried out. With "data concentration, information sharing, comprehensive analysis, security assurance" as the construction goal, "human, loco, environment, management" as four elements for the research, the big data management platform for the locomotive operation and maintenance of Shuohuang Railway is to be set up. To realize the full coverage of the four business areas of "operation, overhaul, maintenance and comprehensive guarantee" in Shuohuang Railway Locomotive Branch, collect, integrate, analyze, and apply the massive traffic data, production data, operation and maintenance data in the railway transportation process, so as to give full play to the maximum value of the data, to provide decision-making basis and intelligent analysis as strong guarantee for safe transportation and loco maintenance.

2 Research Contents

In this paper, research on the construction of big data management platform of Shuohuang Railway locomotive operation and maintenance will be carried out to realize data collection and integration of various application objects and information systems, and to support data mining, correlation analysis, and trend analysis. Develop health management application oriented to the four elements of "human, loco, environment and management," to construct the Maintenance, Repair, and Operation (MRO) application framework of Shuohuang Railway locomotive operation and maintenance big data, to cover the four systems of "operation, overhaul, maintenance and comprehensive guarantee," to realize the closed-loop management of locomotive and personnel, with the ability to predict the development direction in business, assist in warning of potential safety hazards, and provide decision-making basis for safe transportation and loco maintenance. The logical architecture is shown in Fig. 1.

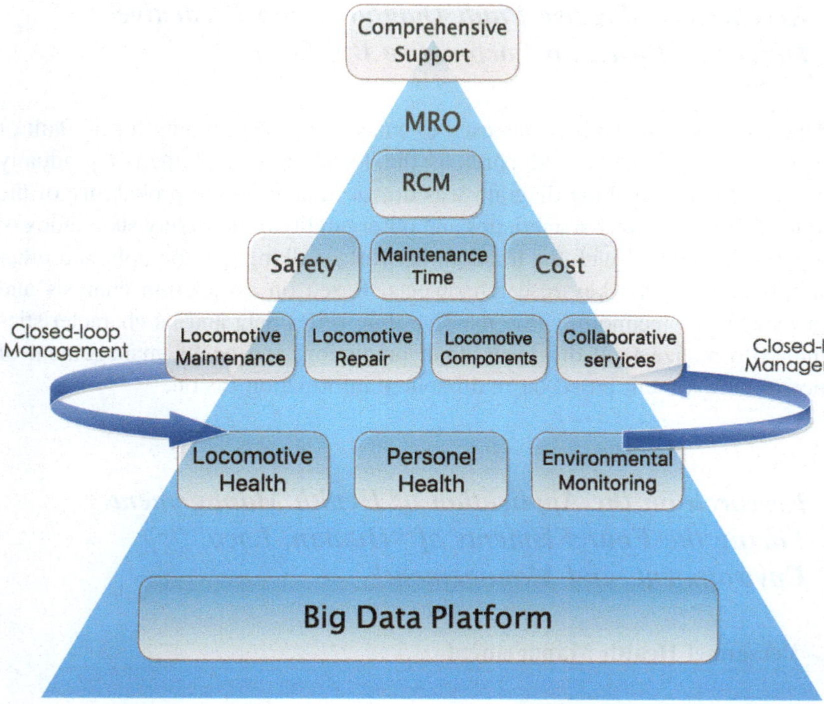

Fig. 1 Overall framework of big data management platform of locomotive operation and maintenance

2.1 Research on Big Data Platform Supporting Data Fusion of Multi-service System

Connect existing business systems to realize multi-dimensional heterogeneous data fusion, to cover seven categories of data, including common basic data, archives and technical data, detection and test data, plan and command data, business operation and process monitoring data, statistical analysis and report data and security data, and to open up the information island. Provide data analysis and mining platform to support data correlation analysis, mining and visual display of cross-service systems within the four systems of "operation, repair, maintenance, and comprehensive guarantee."

2.2 Research on Precise Fault Diagnosis and Predictive Diagnosis Based on Locomotive Big Data

Accident Tree Analysis (ATA) was used to analyze the "descending form" fault of core systems, components, and parts of the locomotive, and then a gradually expanded tree-like branching diagram was drawn to analyze the probability of the top event. The evolution characteristics and observability of the safety state index of each system are studied with the traction, network, walking, pantograph, and other core components as the key research objects. Based on association analysis and pattern recognition technology, the massive time series data and its characteristics are studied to realize fault diagnosis, fault prediction, and health management of locomotive core system based on multi-source information fusion.

2.3 Research on the Application of Health Management Facing the Four Elements of "Human, Loco, Environment, and Management"

2.3.1 Personnel Health Management

Based on the safety monitoring data of locomotive operation, identify the illegal operation of drivers, present early warning of dangerous items, and eliminate potential safety hazards. Based on the management data of loco drivers, multi-dimensional records such as personnel assessment and safety files, a personnel assessment and comprehensive evaluation system is formed. Based on the attendant status monitoring data, the real-time monitoring data of driver mental state (based on iris recognition) was obtained to alert the fatigue state. Based on the data of smart health mattress, the sleep quality report of drivers was obtained, and the sleep quality was evaluated. Based on the data of drivers' on and after work, the drivers' working time and overwork trend can be accurately calculated to realize the early warning of strain.

2.3.2 Locomotive Health Management

The influence mechanism and propagation model of unit performance degradation and failure on the safety of locomotive electric drive system, running gear and pantograph under multi-failure modes, and dynamic working conditions in complex operating environment were studied, and the health state index system of locomotive was formed to meet the needs of system safety state assessment and early warning. Complete the establishment of electronic resume management system for locomotives and large parts, and realize the management of large parts exchange, maintenance, inventory management, and quality evaluation. The expert knowledge base of fault diagnosis was constructed to form the fault knowledge map combining the experience of the construction and repair enterprises and field-workers. Through continuous improvement and modification, the systematic and online support for common fault diagnosis and the experience support for difficult fault diagnosis were realized.

2.3.3 Locomotive Running Environment Monitoring

Locomotive running environment, line status and the data access, state of the natural environment perception, comprehensive evaluation and early warning were studied. Based on wide-area environment monitoring under the locomotive state intellisense technology, space and time dimension on multiple state information fusion and identification technology, loco/line/power/signal/weather and the mapping relation of train running safety and risk chain construction and system decoupling, risk identification and early warning for hidden safety trouble were **discovered**.

2.4 Application Research of Reliability-Centered MRO Maintenance and Repair System

Maintenance and repair management process is optimized to realize closed-loop management and intelligent operation in all operation processes of small & medium-sized repair, technological upgrading and temporary repair, from the planning, operation, acceptance to delivery and other steps. To improve precision of maintenance management, to ensure the quality of locomotive overhaul, and to realize application for strategy of improving maintenance efficiency and resources allocation optimization. The intelligent maintenance knowledge base is constructed to realize multi-level locomotive fault data acquisition and input, support closed-loop tracking and automatic report generation based on single-fault-related factor correlation analysis and processing flow, support batch fault analogy analysis and hidden trouble troubleshooting, and support fault statistical analysis and report presentation within a variable time period.

To take on research of whole life cycle of locomotive and large replaceable parts, research on the reliability of the equipment, research of equipment failure consequence as the main basis for maintenance strategies, and research of comprehensive information about fault consequences and failure mode. To improve the security and economic benefit as a starting point, we take effort to optimize the repair time and repair process, to strenthen integration of the locomotive servicing, maintenance management, collaborative services, so as to realize the reliability-based maintenance. Improve the management of locomotive parts, to support the multi-level parts library and the overall visualization of parts resources, and to provide an integrated platform for parts demand, procurement, acceptance, storage, use, management and other activities.

2.5 Research on the Construction of Comprehensive Support System of Big Data for Loco Operation and Maintenance

Realize the correlation and reconstruction of vehicle-mounted fault record data, LKJ (train operation monitoring and recording device) record data, trackage data and environmental data, conduct intelligent screening of fault data, and generate fault analysis reports. Take data mining, association analysis, and visualization display in "operation, repair, maintenance, and integrated safeguard" four business types. Carry on the multiple correlation analysis to find the dependence of the data dimension and coupling relationship in single event. Multi-source spatiotemporal data and spatiotemporal process are analyzed by group collaboration for batch events, to mine hidden associations and development trends. The comprehensive guarantee system of big data for locomotive operation and maintenance is constructed to realize the process design, closed-loop management, and comprehensive information release that support multi-department collaboration [4].

3 The Overall Plan

Big data management platform of Shuohuang Railway locomotive operation and maintenance adopts the mode of "1 + 7 + 7" [5], that is, one public big data foundation platform (including big data analysis engine), seven business function modules and access to seven categories of data sources. Among them, the seven business function modules include locomotive fault prediction and health management, personnel health management, locomotive running environment monitoring, intelligent locomotive maintenance, locomotive electronic resume, comprehensive analysis and release, and comprehensive display for big data. The standard and specification system in big data platform, including data access technical specifications, data interface specifications, and platform data specifications, shall be constructed synchronously to ensure the standardization of platform construction and define data standards for subsequent platform extension. Information security assurance system, including operating system security, network security, big data component security, and data security, is built to ensure the security and reliability of the platform and data [6]. See in Fig. 2.

The seven categories of data sources to be accessed include locomotive operation data, locomotive monitoring data, locomotive detection data, equipment maintenance data, personnel operation data, basic general data, and other third-party data sources. Data of locomotive operation include daily running, tonnage of transport, daily running distance, and daily output. Locomotive monitoring data include data from LKJ, main and auxiliary transformer, network, OCE, running gear, braking, rear-end, video, voice and other real-time monitoring data of locomotive state and vehicle-mounted record files. Locomotive detection data

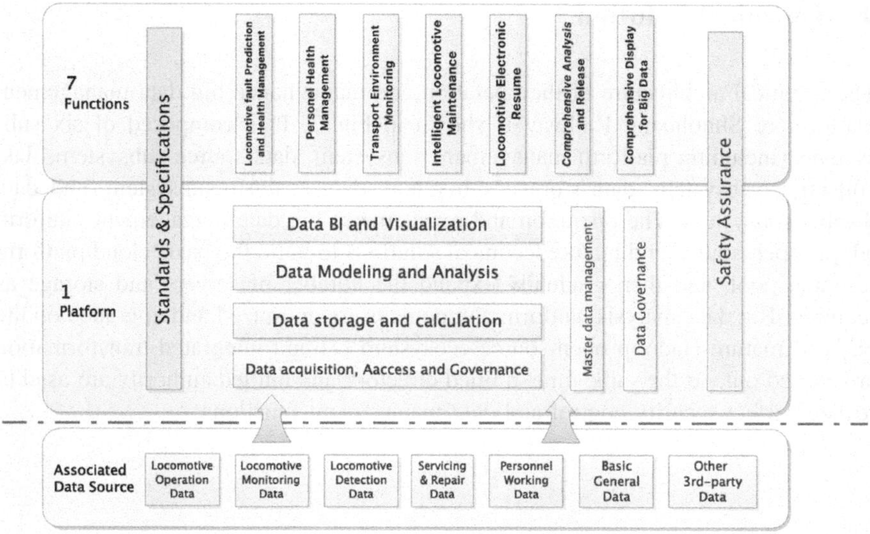

Fig. 2 Function block of big data management platform of locomotive operation and maintenance

include data from pantograph/roof detection, wheelset detection, running gear detection, and other indicators. The reconditioning maintenance data includes the production data of reconditioning process, repair ticket activity, temporary repair, technical reform, small- and medium-sized minor repair, function experiment, and accessories circulation. Data used by personnel include attendance, leave, illegal operation, safety assessment, duty state, and sleep quality. Basic general data includes locomotive files, big equipment files, steward files, and lines.

The main users of the platform include leaders and decision-makers of Shuohuang Railway company, heads of each center/department, chief engineers of each center/department, and grassroots managers/ordinary employees. Leaders and decision-makers of Shuohuang Railway Company have the highest authority to view and browse all information, especially all kinds of statistics, statements, and other decision-making auxiliary information. The person in charge of each center/department can view and review all information resources within the business jurisdiction. The chief engineer of each center/department can view, review, and edit all information resources within the jurisdiction of the post. Basic managers/ordinary employees can input and maintain the relevant business data and information of the post and view the designated information.

4 Technical Solution

The technical architecture of the operation and maintenance big data management platform of Shuohuang Railway is shown in Fig. 3. It is composed of six subsystems, including platform management subsystem, data source subsystem, data collection subsystem, data storage subsystem, data analysis subsystem, and data display subsystem. The operation and maintenance big data management platform adopts open-source and mature technical solutions to build the basic cloud platform resource pool, and then gradually expand the number of servers and storage as required. Big data analysis platform components are optimized and upgraded on the basis of mature Hadoop open-source ecosystem [7] and integrated transformation are carried out. At the same time, unified directory and unified authority are used to realize perfect security control and data management functions.

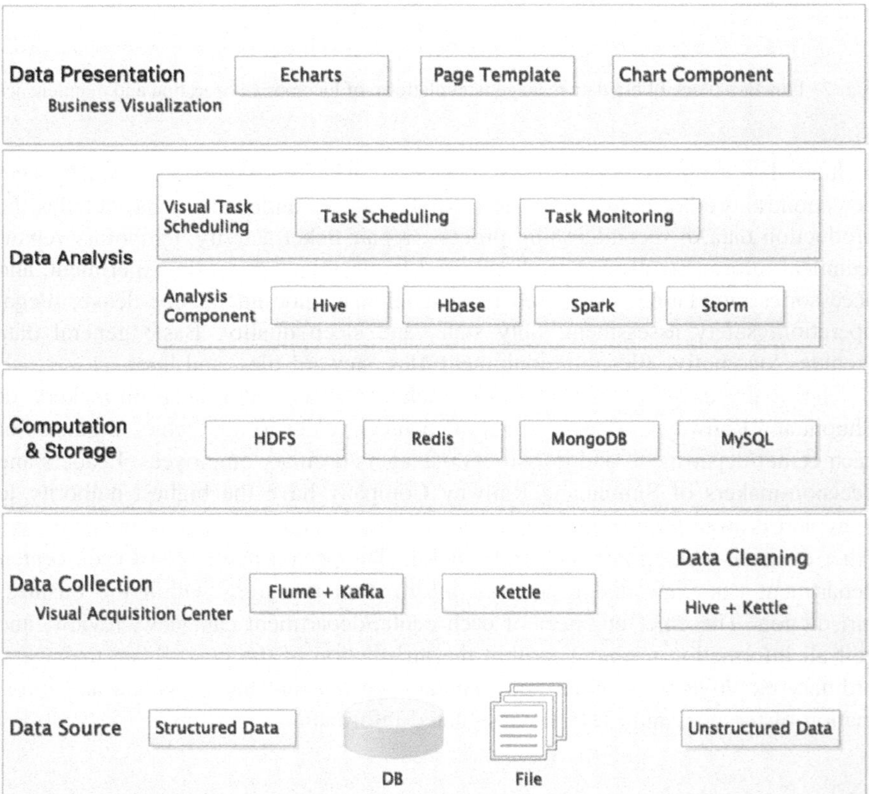

Fig. 3 Big data infrastructure platform architecture

4.1 Platform Management

This module mainly provides operation and management functions for big data platform operators and users, including cluster deployment, cluster management, visual integration, cluster monitoring, multi-tenant management, audit log, and platform security.

Cluster deployment automatically allocates computing resources required by the big data platform, checks the platform operating environment, automatically installs components required by the big data platform, and adjusts the parameters of components already installed. Cluster management manages asset allocation and resources of the big data platform and multiple clusters and supervises the overall state. Visual integration provides visual management of the platform: component integration and functional module integration. Cluster monitoring monitors the big data platform: resource monitoring, business system monitoring, alarm and automatic processing when problems are found. Multi-tenancy management assigns individual customer accounts to the business unit and manages users under each unit account, including role management and authority management. Audit log audits the log generated by the big data platform, including operation log, business log, component log, and system log. Platform security provides comprehensive security management of big data platform, including operating system security, network security, big data component security, multi-tenant security, and data security.

4.2 Data Source

The data sources of the big data management platform for computer operation and maintenance support all kinds of structured data (database, etc.) and unstructured data (binary files, document files, audio and video files, etc.), and the specific sources include existing comprehensive reports and statistical data, existing information system business data, and existing audio and video data.

4.2.1 Structured Data

Structured data mainly includes all kinds of database data of existing information systems, including operation and security system, intrinsical security system, locomotive health management system, real-time prevention and control system of maintenance safety, one-stop ground handling management system of locomotive reconditioning, and ground handling analysis system of smooth operation.

The main business data of the operation and safety system include personnel planning, attendance, leave, operation disclosure, and overwork. The security system includes operating environment and safety hazards. Locomotive health management system includes real-time status data of locomotive, real-time fault data, LKJ/TCMS/OCE/train tail, and other vehicle-mounted record files [8]. The real-time prevention and control system of aircraft maintenance safety includes LKJ alarm/early warning data, alarm/early warning item point statistical data, LKJ/ voice/column tail/JK430, and other vehicle-mounted record files. Locomotive equipment one-stop ground handling management system includes locomotive basic information, out and in storage data, fragmentary repair data, and fault statistics data. Smooth operation ground analysis system includes driver operation score and driver's hand account data.

4.2.2 Unstructured Data

Unstructured data mainly includes document files of various business ports and records files of various devices.

The documents are mainly related to the statistical analysis reports of various business outlets, including locomotive operation, locomotive maintenance, personnel operation, and safety assurance. The main report data of locomotive operation includes Statistical Report of Locomotive Operation (SRLO) 1/SRLO 27 (quarterly report), locomotive daily mileage, monthly platinum report, and monthly communication report. Personnel use includes piecework, safe 100 trips, standby sleep, out and out of attendance statistics, leave statistics, overwork statistics, training and examination, and add report form. Locomotive maintenance includes SRLO 6/SRLO 28 (quarterly report), locomotive broken repair, out and in storage statistics, and production plan. Safety assurance code includes safety equipment quality analysis daily, monitoring comprehensive analysis daily, column end analysis daily, abnormal parking monthly statistics, and five safety monthly statistics.

Equipment record files mainly include record files from vehicle-mounted equipment, ground core electrical equipment, security equipment, such as security video of Shenchi south, Suning north, Huanghua port station, AC, DC, diesel locomotive audio and video, vehicle-mounted network, traction, train tail, running gear and other records, and log files of ground big data platform.

4.3 Data Collection

According to the differences in the data interface of the operation and maintenance platform and the differences in the top-level business types, the secondary data acquisition of the big data platform can adopt real-time or offline mode, as is shown in Fig. 4. Real-time stream processing mode: through the k-v query interface, in stream processing mode, after itemized transmission, ETL (Extract-Transform-Load), integration and finally stored in the business platform. Offline batch processing mode: Through the Secure File Transfer Protocol (SFTP) transmission interface, the data is

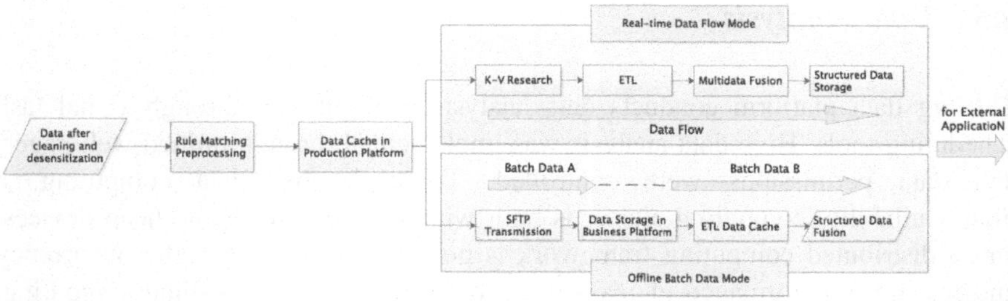

Fig. 4 Data collection flowchart

collected offline in batch to the business platform cache, and then batch ETL is conducted, merged, and stored in the business platform.

The collected data shall be cleaned before entering the platform for storage, with the purpose of detecting errors and inconsistencies in the data, eliminating or correcting them, filtering or modifying data that does not meet the requirements (including incomplete data, wrong data, repeated data, etc.), so as to improve the quality of data. The principle is to use statistical methods, data mining methods, pattern rules, and other methods to transform dirty data into data to meet the data quality requirements. The tools used are mainly ETL components (Extract, Transform, and Load).

4.4 Data Storage

The big data basic platform adopts distributed storage + distributed database to form a complete storage scheme.

Distributed storage uses a large number of ordinary PC servers through Internet interconnection, external as a whole to provide storage services, to meet the needs of large-scale data storage at a low cost. It includes distributed block storage, which transforms distributed large number of server hard disks into a unified by distributed block storage software, distributed file storage, which divides a large file into several small file blocks as distributed storage on server nodes, controlled by the metadata server, suitable for the storage and processing of big data files, distributed object storage, which is characterized by flatness, conducive to the rapid search of the corresponding file (object), suitable for various sizes of massive files based on the Internet online storage, access, and backup.

Distributed NoSQL database, NoSQL = Not Only SQL, generally refers to the non-relational database, reduces the constraints on the traditional two-dimensional relational database, and can provide a kind of flexible and extensible storage and query mode for large-scale, multi-type data. It can process extremely large amounts of data. The data relation is flexible and has less limitation on the size and type of data, and the execution is fast.

4.5 Data Analysis

The big data platform conducts data analysis configuration through visual task scheduling tools. Hive data analysis was implemented through unified web pages, and data permissions were controlled. To apply big data computing, a high-availability computing cluster is built with a large number of cheap devices, and a distributed computing framework is provided to achieve high concurrency and real-time performance of mass data processing, including offline computing, flow computing, real-time computing, and graph computing.

4.5.1 Offline Calculation: MapReduce

Offline batch processing of massive data with low real-time requirement, such as DPI analysis and ETL, data analysis and report form, can process massive data at low cost and has high scalability. The disadvantage is that the disk IO consumption is large, it is difficult to carry on Streaming data calculation and iteration calculation, and the real-time is low.

4.5.2 Streaming Data Calculation: Storm

Streaming data processing with high real-time requirements, which can process millions of messages per second, is often used in real-time analysis, online machine learning, continuous computing, distributed remote call, ETL, and other fields, such as real-time diagnosis of locomotive faults, real-time health assessment, with low delay (second level) and high real-time performance.

4.5.3 Real-Time Computing: Spark

Spark is suitable for iterative algorithms that require repeated access to the same data in data mining and machine learning, improving the performance of iterative data processing and providing a unified data processing platform that supports batch processing, interaction, streaming, machine learning, and graph computing.

4.6 Data Presentation

The relationship between data is expressed in the form of graphics and images, and the meaning of graphics and indicators is associated with providing users with deep and intuitive data and its structural relationship representation, so that users can have insight into the rules and essential characteristics.

Data presentation integrates data analysis, image processing, and other technologies to support the analysis and presentation of data in distributed storage, large database, or data warehouse. It can be widely used in mind mapping, trend analysis, data display, connection display, feature display, and so on.

5 Conclusion

Big data technology plays a very important role in improving the level of railway informatization in China. Aiming at the Shuohuang Railway demand for information resource sharing and data analysis, we study the core demand and key technology of big data management platform of Shuohuang Railway operation and maintenance and put forward the "1 + 7 + 7" overall construction scheme and technical requirement in consistent standards and gradual extension. We analyze the main data sources, data collection, data storage, data analysis, and data demonstrating as four key technologies, discuss seven function modules of the platform in detail, and clear main functional requirements of each module as well as relevant railway information systems with specific data processing, analysis, query, and presentation. This scheme has important value in reference for the application and implementation of the big data management platform of Shuohuang Railway locomotive operation and maintenance.

References

1. Hu W (2016) The thinking in promoting railway innovation development with big data. Theor Study Explor (1)
2. UN Global Pulse. Big data for development: challenges and opportunities [EB/OL]. (2012-05) [2016-12-30]. http://www.unglobalpulse.org/sites/default/files/BigDataforDevelopment-UNGlobalPulseJune2012.pdf
3. Wang T (2017) Top-level design research and practice in China railway big data application. China Railway 1:8–16
4. Gu X, Dai F, Yang Q et al (2015) The content and method of the top-level design of big data in manufacturing industry. Group Technol Prod Modernization (4):12–17
5. Duan J, Zhang N, Zhao B et al (2015) Power big data basic architecture and application research. Power Inf Commun Technol 13(2):92–95
6. Liu H, Li J, Zhang R (2016) Research on three-dimensional and dynamic big data security protection system architecture. Inf Netw Secur (9):18–25
7. Hadoop. Hadoop distributed file system: architecture and design [EB/OL]. http://hadoop.apache.org/common/docs/r0.18.2/cn/hdfs_design.html, 2013-9-10
8. Tang G, Li G, Wang X (2007) LAIS train operation status information system. Electr Drive Locomotives 4:52–56

A Global Non-roundness Detection Algorithm for Urban Rail Vehicle Wheels

Zhihui Zhou, Xinhai Liu, Yuanyuan Zhou, Gaopeng Zhao and Zongyi Xing

Abstract Based on the frequent occurrence of the global non-roundness phenomenon of urban rail vehicle wheels, a global non-roundness detection algorithm for urban rail vehicle wheels based on vibration signals is proposed. Firstly, extract valid data segments from the vibration signals and denoise the vibration signals. Secondly, use the feature extraction algorithms to obtain the non-roundness characteristics of the actual wheels. Finally, the faults and the global non-roundness order are identified according to the characteristics. The feasibility and effectiveness of the algorithm are proved by simulation and field data analysis.

Keywords Global non-roundness of wheel · Vibration signal · Wavelet threshold denoising · Multi-scale permutation · Extreme learning machine

1 Introduction

The wheel transmits the driving force and braking force between the rail and the vehicle while carrying the weight of the entire vehicle [1]. At present, static detection and dynamic detection are the two most important methods in the detection of wheel non-roundness.

In this paper, a global non-roundness detection algorithm for urban rail vehicle wheels based on vibration signals is proposed. The vibration acceleration sensor is installed on the outer side of the track to detect the vibration between the vehicle and the track. Analyze the vibration signal, extract the characteristics and identify the faults to achieve the detection.

Z. Zhou (✉) · X. Liu · Y. Zhou · G. Zhao · Z. Xing
School of Automation, Nanjing University of Science and Technology,
No. 200 Xiao Ling Wei, Xuanwu District, Nanjing 210094, China

Y. Qin et al. (eds.), *Proceedings of the 4th International Conference on Electrical and Information Technologies for Rail Transportation (EITRT) 2019*, Lecture Notes in Electrical Engineering 639, https://doi.org/10.1007/978-981-15-2866-8_6

2 Detection Algorithm

The algorithm circuit of the detection system is as shown in Fig. 1.

2.1 Extract Valid Data Segments

Non-roundness is the difference between the largest and smallest diameters on the cross-section of the wheel tread [2]. The global non-roundness ranges from 14 cm to the circumference of a wheel, and the amplitude does not exceed 1 mm [3].

Currently, urban rail vehicles have wheel diameters ranging from 780 to 840 mm. The effective acquisition area of a single vibration acceleration sensor $l = 700$ mm, and then, the number of required vibration acceleration sensors is at least:

$$N_{\min} \geq S_{\max}/l = 2 \cdot \pi \cdot 840/700 = 3.7 \tag{1}$$

where $N_{\min}$ is the minimum number of sensors, $S_{\max}$ is the circumference of the wheel.

It can be concluded from the above formula that at least four vibration sensors are installed on one side of the track. Acceleration sensor installation diagram is shown in Fig. 2.

The wheel axle position sensor is extracted to obtain the time t_{2i} and t_{3i} of the wheel passing through G_2 and G_3, where i is the wheel number. The effective detection time interval is $t_{2i} + 1/10\Delta t_i$ to $t_{3i} - 1/10\Delta t_i$. Therefore, the effective detection time interval is divided into four segments:

The first: $t_{2i} + 1/10\Delta t_i$ to $t_{2i} + 3/10\Delta t_i$, Sensor $L_1(R_1)$ output signal data is valid, and the rest of the signal data is not considered.
The second: $t_{2i} + 3/10\Delta t_i$ to $t_{2i} + 1/2\Delta t_i$, Sensor $L_2(R_2)$ output signal data is valid, and the rest of the signal data is not considered.
The third: $t_{2i} + 1/2\Delta t_i$ to $t_{2i} + 7/10\Delta t_i$, Sensor $L_3(R_3)$ output signal data is valid, and the rest of the signal data is not considered.
The forth: $t_{2i} + 7/10\Delta t_i$ to $t_{3i} - 1/10\Delta t_i$, Sensor $L_4(R_4)$ output signal data is valid, and the rest of the signal data is not considered.

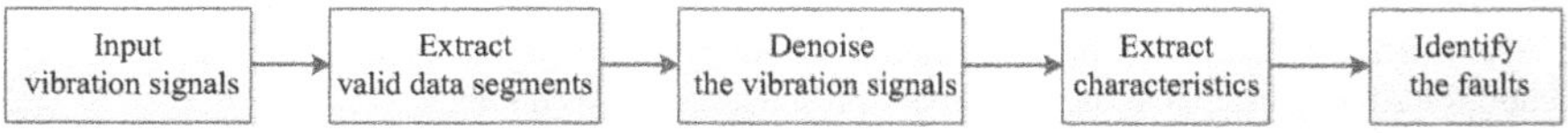

Fig. 1 Algorithm circuit of the detection system

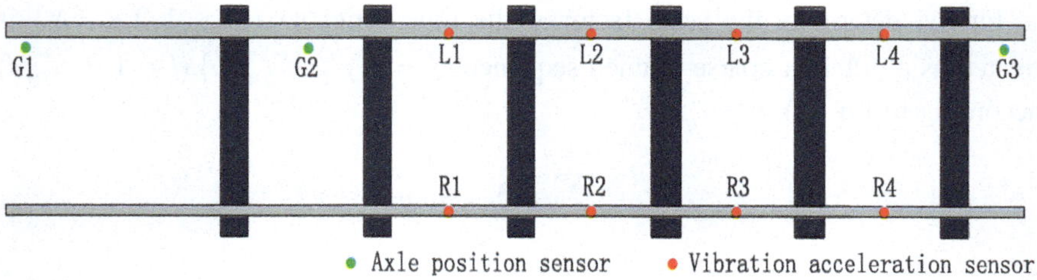

Fig. 2 Acceleration sensor installation diagram

2.2 Denoise the Vibration Signals

In this paper, the denoising algorithm is improved by the traditional wavelet threshold denoising [4] algorithm to denoise the vibration signals. The wavelet threshold denoising process is mainly divided into wavelet decomposition, threshold selection and signal reconstruction.

This paper adds the decomposition scale as the principle of threshold selection based on the general threshold principle, which is consistent with the phenomenon that the wavelet coefficient of noise decreases with the increase of the decomposition scale. The improved threshold selection principle [5] is

$$\lambda_j = \sigma\sqrt{2\log N}/\log(j+1) \tag{2}$$

where σ is the noise standard deviation in the signal, j is the scale of decomposition, N is the length of the signal.

The improved threshold function [6] combines soft and hard threshold functions with the expression:

$$\overline{W_{j,k}} = \begin{cases} \operatorname{sgn}(W_{j,k})\left(\left|W_{j,k}\right| - \dfrac{2}{1+e^{\frac{\left|W_{j,k}\right|-\lambda}{N}}}\lambda\right), & \left|W_{j,k}\right| \geq \lambda \\ 0, & \left|W_{j,k}\right| < \lambda \end{cases} \tag{3}$$

where $W_{j,k}$ is the wavelet coefficient, $\overline{W_{j,k}}$ is the wavelet coefficients obtained by the threshold function, λ is the threshold, $\operatorname{sgn}(*)$ is a symbolic function.

2.3 Extract Characteristics

In practical engineering, the characteristic information contained in the vibration signal is not only reflected on a single scale, but also on other scales, there may be important features related to the system state. Therefore, improved multi-scale permutation entropy (IMPE) [7] is introduced as the fault feature.

Firstly, determine the scale factor s, the time series $\{x(i), i = 1, 2, \ldots, N\}$ is treated as a different coarse-grained sequence $z_i^{(s)} = \{y_{i,1}^{(s)}, y_{i,2}^{(s)}, \ldots\}$ $(i = 1, 2, \ldots, s)$ according to Eq. (4):

$$y_{i,j}^{(s)} = \frac{\sum_{f=0}^{s-1} x_{f+i+s(j-1)}}{s} \tag{4}$$

Secondly, the entropy values are, respectively, calculated for each coarse-grained sequence corresponding to each scale, and then, the average value is calculated as the permutation entropy value at this scale.

The permutation entropy [8] is defined as follows.

There is a time series $\{x(i), i = 1, 2, \ldots, N\}$, reconstruct its phase space, to get the following time series:

$$\left.\begin{array}{c} X(1) = \{x(1), x(1+\lambda), \ldots, x(1+(m-1)\lambda)\} \\ \vdots \\ X(i) = \{x(i), x(i+\lambda), \ldots, x(i+(m-1)\lambda)\} \\ \vdots \\ X(N-(m-1)\lambda) = \{x(N-(m-1)\lambda), \\ x(N-(m-2)\lambda), \ldots, x(N)\} \end{array}\right\} \tag{5}$$

where m is the embedded dimension, λ is the delay time, $X(i)$ is the reconstructed vectors.

Rearrange the m components of $X(i)$ in ascending order [9], which is

$$X(i) = \{x(i+(j_1-1)\lambda) \le x(i+(j_2-1)\lambda) \le \cdots \le x(i+(j_m-1)\lambda)\} \tag{6}$$

If $X(i)$ has equal components, i.e.,

$$x(i+(j_p-1)\lambda) = x(i+(j_q-1)\lambda) \tag{7}$$

The size of j value is used as the basis for sorting, i.e., when $j_p \le j_q$:

$$x(i+(j_p-1)\lambda) \le x(i+(j_q-1)\lambda) \tag{8}$$

Therefore, for any reconstruction vector in the above phase space, a set of symbols reflecting the order of its component size can be obtained:

$$S(g) = \{j_1, j_2, \ldots, j_m\} \tag{9}$$

Calculate the probability P_g $(g = 1, 2, 3, \ldots, k)$ of each symbol sequence appearing. According to the form of information entropy, it is possible to give the permutation entropy definition of the time series $\{x(i), i = 1, 2, \ldots, N\}$:

$$H_p = -\sum_{g=1}^{k} P_g \ln P_g, \quad \sum_{g=1}^{k} P_g = 1 \tag{10}$$

According to Eq. 10, when $P_g = 1/m!$, $H_p(m)$ reach the maximum value $\ln(m!)$. Normalize the permutation entropy $H_p(m)$ with $\ln(m!)$, i.e.,

$$H_p(m) = \frac{H_p(m)}{\ln(m!)} \tag{11}$$

where H_p is the complexity and randomness of the time series.

2.4 *Identify the Faults*

Using the extreme learning machine (ELM) [10] algorithm, a large number of samples are selected, each of which includes the IMPE of the vibration signal and the label of the wheel type. The samples are trained to obtain an ELM classification model, input the test samples into the classification model to get the classification result.

3 Experimental Analysis

3.1 *Improved Wavelet Threshold Denoising Analysis*

In order to verify the superiority of the improved wavelet threshold denoising algorithm, analyze on-site measured vibration signals. The measured vibration signal spectrum is shown in Fig. 3.

It can be seen from Fig. 3 that due to the influence of noise, the fault characteristic information has been difficult to extract accurately, so it is necessary to denoise the vibration signals.

The vibration signals after denoising are shown in Figs. 4 and 5.

From the two Figures, it can be clearly seen that the impact frequency is 19.1 Hz. According to the speed of the train is 60 km/h, the wheel diameter is 810 mm, the wheel circumference is 2.54 m, and the wheel rotation frequency can be calculated to be 6.56 Hz. 19.1 Hz is basically consistent with the theoretical fault frequency of 19.68 Hz of the third-order global non-roundness. In comparison, the filtering effect of the improved wavelet threshold denoising is better, and the high-frequency noise is more obviously suppressed.

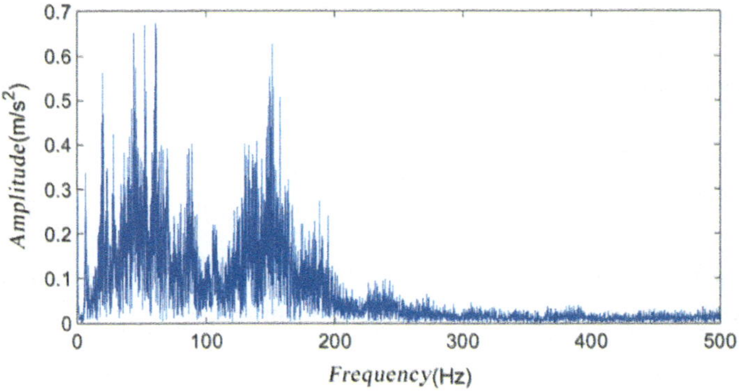

Fig. 3 Spectrogram of measured vibration signal

Fig. 4 Spectrogram of traditional wavelet threshold denoising

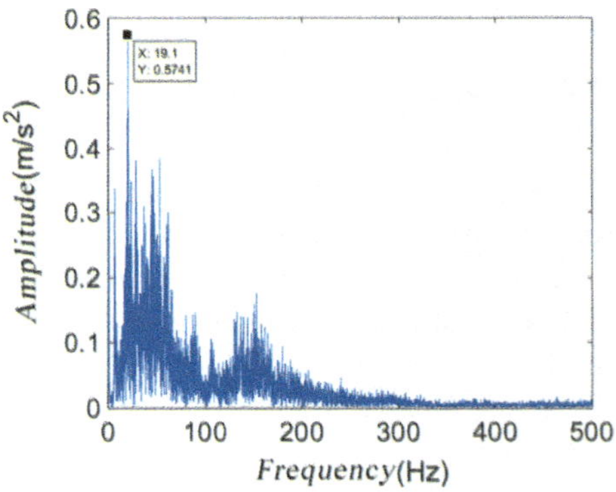

Fig. 5 Spectrogram of improved wavelet threshold denoising

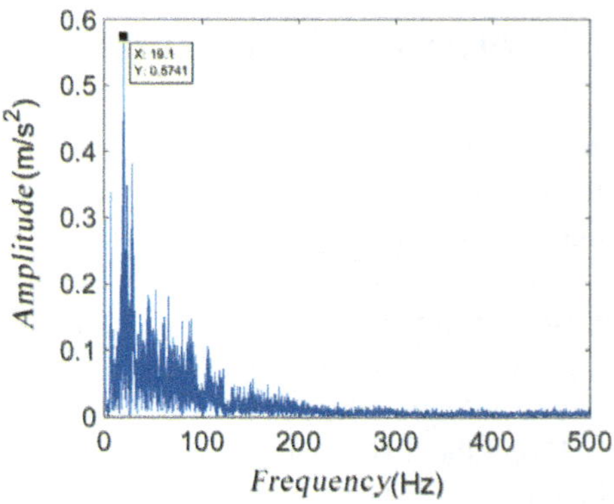

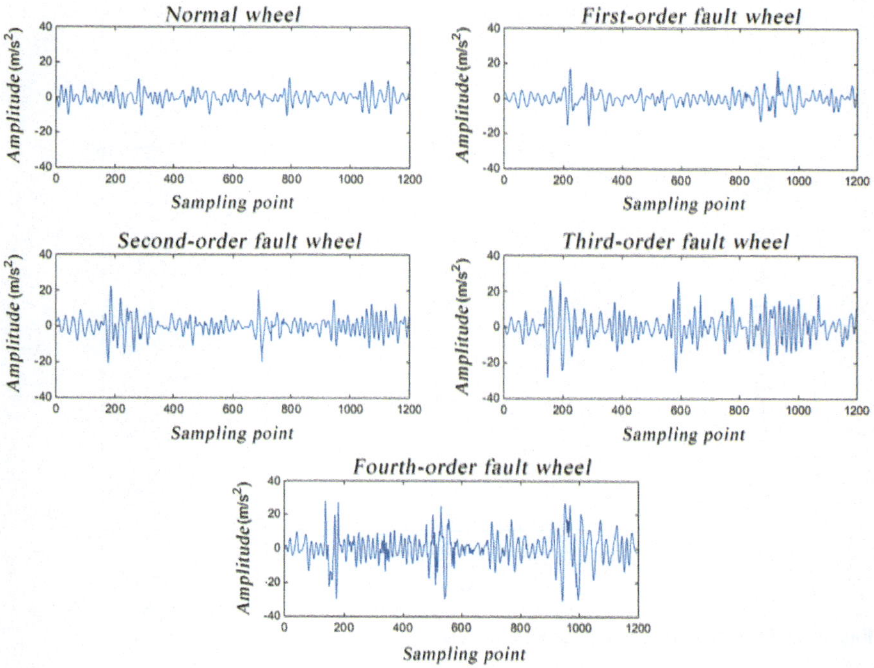

Fig. 6 Time-domain waveforms of five types of wheels

3.2 IMPE Fault Feature Extraction Analysis

The measured vibration signals of the normal wheel, the first-order fault wheel, the second-order fault wheel, the third-order fault wheel and the fourth-order non-circular wheel are extracted. The time-domain waveform is shown in Fig. 6.

Take the vibration signals corresponding to the above five types of wheels and take 40 groups, calculate their mean value of multi-scale permutation entropy.

When $m = 4$, $\lambda = 1$, $s = 8$. The final result is shown in Fig. 7.

With the increase of the scale factor, the value of permutation entropy gradually becomes fixed, which indicates that the time series contains multi-scale time mode information, and the IMPE algorithm can accurately extract the fault features.

3.3 Fault Identification

The IMPE value of the measured vibration signals of the eight wheels at the scene is extracted as the feature vector. The IMPE value of each wheel vibration signal is as shown in Table 1.

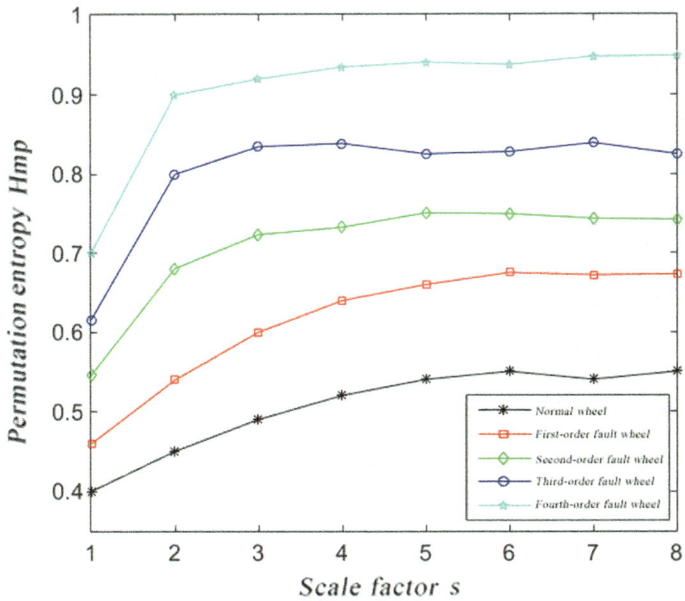

Fig. 7 IMPE of five types of measured signals

Table 1 IMPE value of vibration signal

Scale factors	No. 1	No. 2	No. 3	No. 4	No. 5	No. 6	No. 7	No. 8
1	0.3946	0.4187	0.4061	0.3829	0.4567	0.4652	0.4215	0.4014
2	0.4582	0.4491	0.4670	0.4427	0.5282	0.5306	0.4754	0.4673
3	0.4867	0.4928	0.4872	0.4763	0.5742	0.5978	0.5013	0.4960
4	0.5021	0.5184	0.5128	0.5078	0.6241	0.6283	0.5220	0.5175
5	0.5347	0.5431	0.5446	0.5392	0.6389	0.6377	0.5469	0.5343
6	0.5427	0.5465	0.5483	0.5429	0.6490	0.6525	0.5516	0.5427
7	0.5374	0.5381	0.5249	0.5456	0.6497	0.6508	0.5422	0.5449
8	0.5462	0.5398	0.5515	0.5480	0.6673	0.6742	0.5543	0.5396

Input the above IMPE values to the trained ELM model. The fault recognition result is shown in Fig. 8.

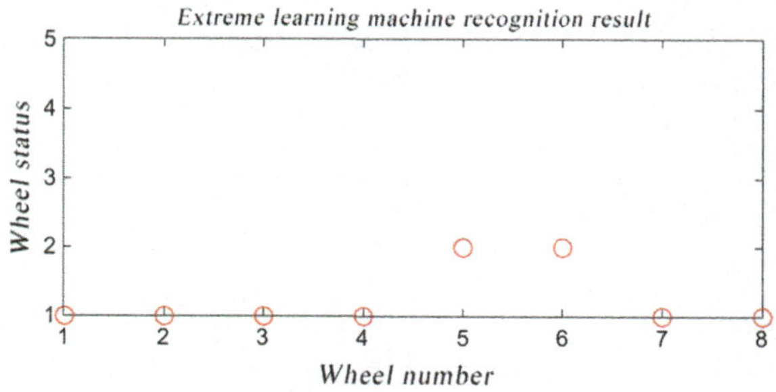

Fig. 8 Extreme learning machine fault recognition result

4 Summary

The improved wavelet threshold can effectively remove the noise interference in the vibration signal and retain useful information. IMPE is used as the characteristic value of the vibration signal. Experimental analysis shows that using IMPE construct feature vector can reflect the wheel fault feature information well.

The vibration data of the field is used to verify the wheel global non-round fault detection algorithm proposed in this paper meet the actual engineering requirements.

References

1. Yu XL (2014) Fault diagnosis of urban rail vehicle running system based on information fusion. Nanjing University of Science and Technology (in Chinese)
2. Wang XL (2017) Research on the detection method of urban rail train wheel circle based on improved Wigner-Ville time-frequency analysis. Nanjing University of Science and Technology (in Chinese)
3. Wang W, Zeng J, Luo R (2009) Research status of train wheels not smooth. Foreign Railway Veh 46(01):39–43 (in Chinese)
4. Donoho DL (1995) De-noising by soft-thresholding. IEEE Trans Inf Theory 41(3):613–627
5. Wu YJ, Gao X (2013) Application of improved wavelet threshold method in denoising of logging curves. J Comput Syst 22(3): 182–185+152 (in Chinese)
6. Chen YG, Luo BQ (2010) Heart sound signal denoising based on a new wavelet threshold function. J Comput Simul 27(11):319–323 (in Chinese)
7. Azami H, Escudero J (2016) Improved multiscale permutation entropy for biomedical signal analysis: interpretation and application to electroencephalogram recording. Biomed Signal Process Control 23:28–41
8. Xu ML (2013) Fault diagnosis of rolling bearing based on wavelet denoising and empirical mode decomposition. Harbin Institute of Technology (in Chinese)
9. Ke D-G et al (2006) Easily calculable complexity measure for finite time series. Nonlinear Science
10. Yu C (2015) Prediction and control of ship roll based on extreme learning machine. Dalian Maritime University (in Chinese)

Rolling Bearings Fault Diagnosis Method Based on EWT Approximate Entropy and FCM Clustering

Lin Bai, ChunLin Zhu, Zhen Ye and Meng Hui

Abstract In this paper, a novel method for rolling bearings fault diagnosis based on empirical wavelet transform with approximate entropy and fuzzy c-means clustering is proposed. The method can identify different types of rolling bearing faults. Firstly, the original vibration signals of rolling bearing are decomposed by empirical wavelet transform to obtain several amplitude and frequency modulation components that have physical meaning. Then, with correlation analysis, the first three amplitude–frequency modulation components most relevant to the original vibration signal are selected and their approximate entropy is calculated as the eigenvector. Finally, the constructed eigenvector matrix is used as the feature for fuzzy c-means clustering to realize the fault diagnosis of rolling bearing. The experimental results show that compared with the fault diagnosis method based on empirical mode decomposition or approximate entropy, the proposed rolling bearing fault diagnosis method has a better fault recognition effect.

Keywords EWT · Bearing fault diagnosis · Approximate entropy · FCM

1 Introduction

The vibration signal processing algorithms commonly used in mechanical fault diagnosis mainly include time domain, frequency domain, and time–frequency domain methods [1–3]. Traditional methods mainly study how to extract and analyze fault vibration signals validly. However, the vibration signals detected by sensors are too weak, unstable, and often disturbed by large noise [3, 4]. Empirical mode decomposition (EMD) can decompose vibration signals adaptively into some narrow-band transient oscillation components, which called intrinsic mode functions (IMFs). Usually, EMD method decomposes vibration signals into several IMF, selects appropriate IMFs, and then uses Hilbert transform to reconstruct the

L. Bai (✉) · C. Zhu · Z. Ye · M. Hui
Chang'An University, Middle Section of Nan'er Huan Road, Xi'an, Shaanxi Province, China
e-mail: Linbai@chd.edu.cn

Y. Qin et al. (eds.), *Proceedings of the 4th International Conference on Electrical and Information Technologies for Rail Transportation (EITRT) 2019*, Lecture Notes in Electrical Engineering 639, https://doi.org/10.1007/978-981-15-2866-8_7

envelope spectrum [5, 6]. Unfortunately, there are some problems for fault diagnosis, such as end effect, pattern mixing, even lack of mathematical basis of constraints, which limit its application [7].

Empirical wavelet transform (EWT) based on Fourier spectrum segmentation proposed in [8]. After each EWT, series of amplitude–frequency modulations (AM-FMs) will be obtained. Some AM-FMs related to fault information, others just contain useless information; therefore, these useful AM-FM signals should be selected reasonably [9].

Approximate entropy (ApEn) is a time series complexity measurement method based on edge probability distribution statistics [10, 11]. The mechanical vibration signals can be quantified by approximate entropy, and the purpose of feature quantity extraction can be achieved by means of the complexity information contained in it. One of the most widely used algorithms in fuzzy clustering is fuzzy c-means clustering (FCM) [12, 13]. Based on the similarity between different feature data samples, FCM can classify the similarity of sample objects into the same class by iteration optimization of the objective function, thus realizing the classification of data.

A novel method combining ApEn of EWT with FCM clustering to realize rolling bearing fault diagnosis is proposed in this paper. The approximate entropy of the AM-FM components obtained by the decomposition of mechanical vibration by EWT, which is used as the eigenvectors and put into the FCM clustering classifier to recognition variety of rolling bearing malfunctions. The validity of the method proved by analyzing the data of the rolling bearing fault test.

The rest of this paper arranged as follows. The basic principle of this method introduced in the second section. The third section carries out experiments on different types of bearing faults and verifies the correctness of the method with experimental data. Finally, some discussions and conclusions provided in the section fourth.

2 Principle of Fault Diagnosis Methodology

2.1 Empirical Wavelet Transform

The EWT method is an adaptive signal decomposition method based on Fourier spectrum segmentation. Firstly, the Fourier support in $[0, \pi]$ is detected and the corresponding wavelet is established based on the support; then, the input signal is filtered by the obtained filter banks and different components are obtained. As shown in Fig. 1, the Fourier support $\cup_{n=1}^{N} \Lambda_n = [0, \pi]$ divided into N consecutive segments $\Lambda_n = [\omega_{n-1}, \omega_n]$, where $\omega \in [0, \pi]$. The transition phase defined with each ω_n as the center and the width $2\tau_n$, shown as shadow region in Fig. 1.

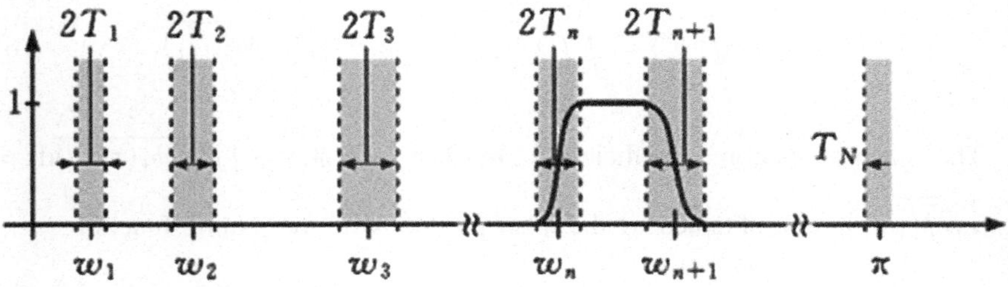

Fig. 1 Partitioning of the Fourier axis

After determining the segmentation Λ_n, the empirical wavelets defined as a group of band pass filters on each $\wedge_n$. Therefore, $\forall n > 0$, empirical scaling function and empirical wavelets constructed as Eqs. (1) and (2), respectively.

$$\hat{\phi}_n(\omega) = \begin{cases} 1 & \text{if } |\omega| \leq \omega_n - \tau_n \\ \cos\left[\frac{\pi}{2}\beta\left(\frac{1}{2\tau_n}(|\omega| - \omega_n + \tau_n)\right)\right] & \text{if } \omega_n - \tau_n \leq |\omega| \leq \omega_n + \tau_n \\ 0 & \text{otherwise} \end{cases} \tag{1}$$

$$\hat{\psi}_n(\omega) = \begin{cases} 1 & \text{if } \omega_n + \tau_{n \leq |\omega|} \leq \omega_{n+1} - \tau_{n+1} \\ \cos\left[\frac{\pi}{2}\beta\left(\frac{1}{2\tau_{n+1}}(|\omega| - \omega_{n+1} + \tau_{n+1})\right)\right] & \text{if } \omega_{n+1} - \tau_{n+1} \leq |\omega| \leq \omega_{n+1} + \tau_{n+1} \\ \sin\left[\frac{\pi}{2}\beta\left(\frac{1}{2\tau}(|\omega| - \omega_n + \tau_n)\right)\right] & \text{if } \omega_n - \tau_n \leq |\omega| \leq \omega_n + \tau_{n+1} \\ 0 & \text{otherwise} \end{cases} \tag{2}$$

The $\beta(\chi)$ in Eqs. (1) and (2) defined as Eq. (3)

$$\beta(\chi) = \begin{cases} 0 & \text{if } \chi \leq 0 \\ \beta(\chi) + \beta(1-\chi) = 1 & \forall \chi \in [0, 1] \\ 0 & \text{if } \chi \geq 1 \end{cases} \tag{3}$$

The $\beta(\chi)$ in literatures usually chosen as follows:

$$\beta(\chi) = \chi^4(35 - 84\chi + 70\chi^2 - 20\chi^3) \tag{4}$$

where,

$$\tau_n = \gamma\omega_n, \gamma < \min_n\left(\frac{\omega_{n+1} - \omega_n}{\omega_{n+1} - \omega_n}\right) \tag{5}$$

The details of coefficients obtained by taking the inverse of the convolution operation between f and wavelet function ψ_n as follows:

$$W_f(\mathrm{n,t}) = \langle f, \psi_n \rangle = \int f(\tau)\overline{\hat{\psi}_n(\tau - \mathrm{t})}\mathrm{d}\tau = \left(\hat{f}(\omega)\overline{\hat{\psi}_n(\omega)}\right)^{\vee} \tag{6}$$

The approximation coefficients $W_f(0,\mathrm{t}) = \langle f, \phi_1 \rangle = \int f(\tau)\overline{\phi_1(\tau - t)}\mathrm{d}\tau = \left(\hat{f}(\omega)\overline{\widehat{\phi_1}(\omega)}\right)^{\vee}$ can obtained by the inner product with the scaling function:

$$W_f(0,\mathrm{t}) = \langle f, \phi_1 \rangle = \int f(\tau)\overline{\phi_1(\tau - \mathrm{t})}\mathrm{d}\tau = \left(\hat{f}(\omega)\overline{\widehat{\phi_1}(\omega)}\right)^{\vee} \tag{7}$$

Using Eqs. (1) and (2), the construction signal obtained by fallow equation:

$$\begin{aligned} f(t) &= W_f(0,t) * \phi_1(t) + \sum_{n=1}^{N} W_f(n,t) * \psi_n(\mathrm{t}) \\ &= \left(\widehat{W}_f(0,\omega)\widehat{\phi}_1(\omega)\right) + \sum_{n=1}^{N} \widehat{W}_f(n,\omega)\widehat{\psi}_n(\omega))^{\vee} \end{aligned} \tag{8}$$

Derivation from the above formulas, the empirical mode f_k provided by

$$f_0(t) = W_f(0,t) * f_1(t); f_k(k,t) = W_f(k,t) * \psi_k(t) \tag{9}$$

After EWT, Hilbert transform performed on each empirical mode to obtain meaningful AM-FM components, and then the Hilbert spectrum can obtained.

2.2 *Approximate Entropy*

ApEn is a non-negative number representing the complexity of time series, which reflects the overall characteristics of fault signals from the perspective of measuring a time series complexity. Assuming the time series $\{\mathrm{x(i)}\}$ with a data length N, the m-dimensional space reconstruction executed, the reconstructed ith vector can expressed as follow functions:

$$X(i) = [x(i), x(i+1), \ldots, x(i+m-1)],\ i = 1, 2, \ldots, N - m + 1 \tag{10}$$

Usually $m = 2$ is the preselected mode dimension. The distance between $X(i)$ and $X(j)$ is defined as $d[X(i), X(j)]$. It means the maximum difference between two corresponding elements.

$$d[X(i), X(j)] = \max_{k=0,\ldots,m-1} \{|x(i+k) - x(j+k)|\} \tag{11}$$

Based on tolerance r ($r > 0$), to every i, calculate the ratio of $d[X(i), X(j)] < r$ to the total distance $N - m$, taken as $C_i^m(\mathrm{r})$. Take logarithm of $C_i^m(\mathrm{r})$, count the mean of all i ($i = 1, \ldots, N - m + 1$), expressed as $\Phi^m(\mathrm{r})$.

$$\Phi^m(r) = \frac{1}{N - m + 1} \sum_{i=1}^{N-m+1} \ln C_i^m(r) \tag{12}$$

Frist raising the m by 1, become $m + 1$ dimension, and then repeating above steps to obtain $C_i^{m+1}(\mathrm{r})$ and $\Phi^{m+1}(\mathrm{r})$, finally the ApEn of this sequence definition as:

$$\mathrm{ApEn}(m, r) = \lim_{N \to \infty} \left[\Phi^m(r) - \Phi^{m+1}(r)\right] \tag{13}$$

When the sequence length is N (a finite value), an estimate of ApEn can obtained by above steps, as follow equation:

$$\mathrm{ApEn}(m, r, N) = \Phi^m(r) - \Phi^{m+1}(r) \tag{14}$$

2.3 Fuzzy C-Mean Clustering

FCM is an iterative optimization algorithm that minimizes the cost function as denote:

$$\mathrm{J} = \sum_{k=1}^{n} \sum_{i=1}^{c} \mu_{ik}^{l} \|\chi_k - v_i\|^2 \tag{15}$$

The number of data points expressed by n, c is the clusters quantity, χ_k is kth data point, v_i is the ith cluster center μ_{ik} is the degree of membership of the kth data in the ith cluster, and l is a constant greater than one (typically $l = 2$). The grade of membership μ_{ik} defined by:

$$\mu_{ik} = \frac{1}{\sum_{j=1}^{c} \left(\frac{\|\chi_k - v_i\|}{\|\chi_k - v_j\|}\right)^{2/(m-1)}} \tag{16}$$

Beginning with the desired clusters number c and the initial guess of each cluster center v_i, $i = 1, 2, 3, \ldots, c$, the FCM converges to the solution for v_i representing either a local minimum or a saddle point cost function. The FCM method uses fuzzy partitioning, so that each point can belong to multiple clusters with membership between zero and one. FCM includes predefined parameters such as the weighting exponent m and clusters number c. Clustering evaluation index adopts classification coefficient (PC) and classification entropy (CE) as significant parameters to determine the effect of clustering. Rolling bearings fault diagnosis method based on EWT-ApEn-FCM clustering described as follows:

Step 1: The fault signals decomposed into a group of AM-FM components by EWT method.
Step 2: Selection of useful AM-FM components based on correlation analysis of original vibration signals.
Step 3: The first three AM-FM components with the largest correlation coefficient with the original vibration signal selected and their ApEn calculated as the eigenvector.
Step 4: The constructed eigenvectors matrix put into FCM as features to implement rolling bearings fault detection and recognition.

3 Experimental Analysis and Results

The vibration signals used in this paper provided by the bearing data center of Case Western Reserve University [14]. The motor power of the bearing fault testbed is about 1494 W and the speed is 1730 r/min. The fault arranged at a single point in rolling bearing (6205-2RS JEM SKF) by EDM technology. The diameter of the fault is 0.1778 mm and the depth is 0.2974 mm. When the sampling frequency is 12 kHz, the vibration signal time domain waveforms of rolling element fault (BB), inner ring fault (IR), outer ring fault (OR), and normal (NORM) are shown in Fig. 2. The EWT used to decompose the original signal. The original vibration signal of bearing inner ring fault decomposed by EWT and the AM-FM component shown in Fig. 3.

Cross-correlation coefficient represents the relationship between sequences. The larger the value, the greater the relationship between sequences. Here, we can use to determine whether the components contain the original signal information. The correlation coefficients of the first three components and the corresponding original signal are all greater than 0.1 and larger than other components. Therefore, it can be considered that the first three components can best reflect the characteristics of original signal.

The sampling frequency is 1.2 kHz, take the data length of 2400, that is each time the data of 0.2 s is analyzed. Thirty sets of data for each of the four types of bearing signals (BB Fault, IR Fault, OR Fault, and NORM states) are taken as samples. The approximate entropy of the first three AM-FM components of all samples is obtained, and a 120×3 data matrix is constructed. The data is taken as the eigenvector matrix of clustering analysis and analyzed by FCM clustering. In order to verify the superiority of EWT method, EMD method is used to decompose the original fault signals. The clustering results are shown in Figs. 4 and 5. In the process of experiment, the clustering centers quantity $k = 4$, $\varepsilon = 0.0001$.

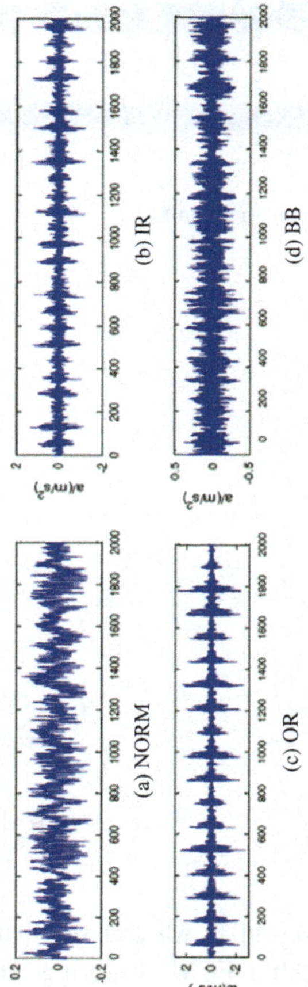

Fig. 2 The time domain signal of different rolling bearing faults

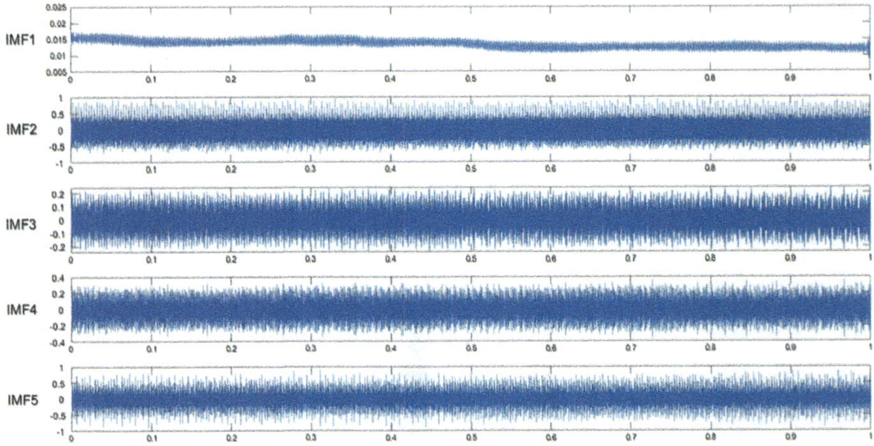

Fig. 3 EWT processing results of inner race fault

Fig. 4 Three-dimensional clustering diagram of EWT approximate entropy

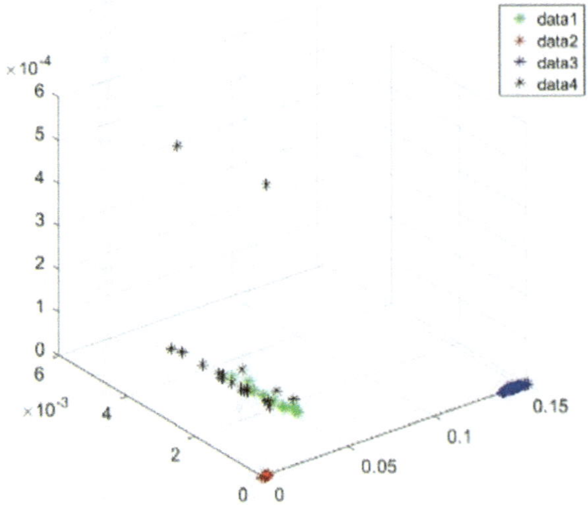

From the analysis of Figs. 4 and 5, we can see that EWT-ApEn has obvious advantages over EMD-ApEn. After FCM clustering, all eigenvalue data samples can be distributed to different clustering centers according to the different type of fault because of the similarity of the same type of signal samples. The clustering effect of two methods tested with parameters PC and CE. The experiment results are shown in Table 1. From the experimental results in Table 1, it can be seen that the PC value of EWT-ApEn is greater than EMD-ApEn, which is closest to 1. At the same time, CE value of EWT-ApEn is less than EMD-ApEn, most close to 0. This is because EWT-ApEn method does not have mode aliasing, and it has good adaptability to bearing fault signal. Compared with the EWT-ApEn method,

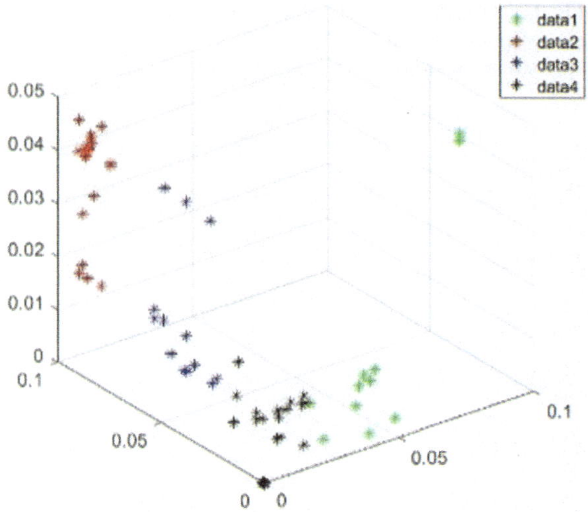

Fig. 5 Three-dimensional clustering diagram of EMD approximate entropy

Table 1 Evaluation of fault clustering effect using different clustering methods

Experimental methods	Clustering evaluation index	
	Classification coefficient	Classification entropy
EWT-ApEn	0.9672	0.0531
EMD-ApEn	0.8317	0.3651

EMD-ApEn has a poor clustering effect. The PC value of EMD-ApEn method is smaller than EWT-ApEn, because it has modal aliasing and the signal decomposition is not ideal.

In order to further verify the validity of this method for bearing fault diagnosis, experiments were carried out to verify the inner race signals with different degrees of damage. In this experiment, there are three types: normal, slight, and severe injuries. The damage diameter of minor faults is 0.1778 mm and that of serious faults is 0.5334 mm. For different types of vibration signals, 30 groups of data are taken, each group of data length is 2400. According to the above steps, 90 groups of data are decomposed into EWT and EMD. The first three AM-FM components are selected as data sources, and then the approximate entropy is calculated. The approximate entropy obtained by two different decomposition methods is constructed into two 90×3 matrices, respectively. The results of FCM cluster analysis are shown in Fig. 6 and 7. In the process of experiment, the clustering centers number $K = 3$, $\varepsilon = 0.0001$.

Fig. 6 Three-dimensional clustering diagram of EWT approximate entropy

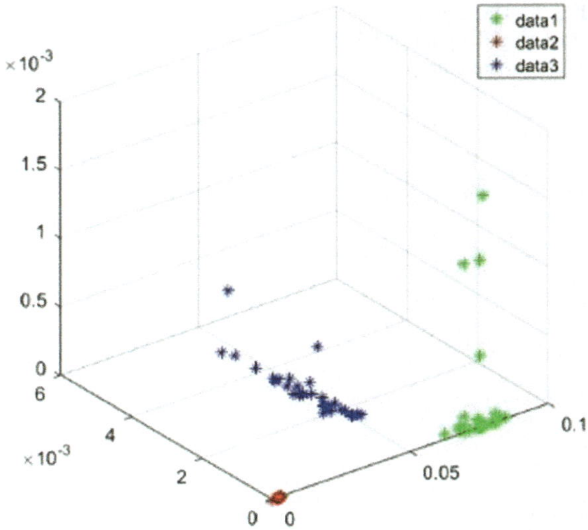

Fig. 7 Three-dimensional clustering diagram of EMD approximate entropy

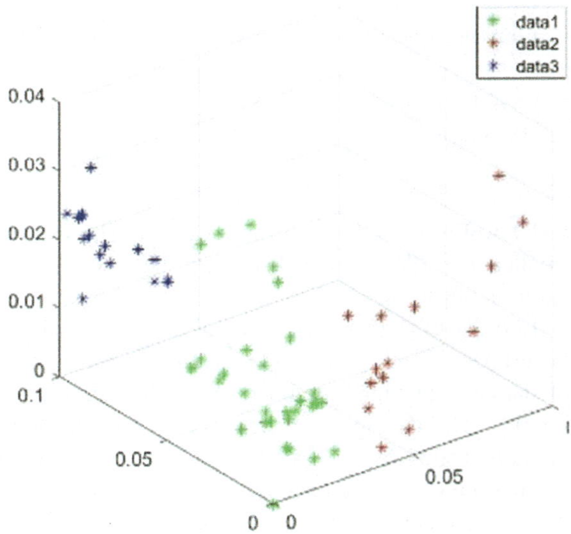

The EWT approximate entropy and FCM clustering method can effectively distinguish the fault types compared with the EMD approximate entropy and FCM clustering method for bearing vibration signals with different damage degrees. The fault clustering effectiveness of two methods was examined by parameters PC and CE. The experiment results are shown in Table 2. It can also be proved from the clustering test index that the FCM clustering method based on EWT approximate entropy is superior.

Table 2 Evaluation of fault clustering effect using different clustering methods

Experimental methods	Clustering evaluation index	
	Classification coefficient	Classification entropy
EWT-ApEn	0.9538	0.0473
EMD-ApEn	0.8935	0.4675

4 Conclusion

A novel method of bearing fault diagnosis based on EWT with approximate entropy and FCM clustering presented in this paper. In order to verify the validity of the fault diagnosis method, the analysis results of inner ring fault, outer ring fault, and rolling element fault cases are given in this paper. The experiment results show that the proposed method achieves a good clustering effect and can distinguish different states of bearings well. There is no overlap between different fault categories. It is an effective adaptive fault information extraction and classification method for fault diagnosis.

Acknowledgements This research was supported by the special foundation of basic scientific research of central colleges, Chang'an University, No. 300102328201 and No. 300102328203.

References

1. Hong S, Zhou Z, Zio E, Han K (2014) Condition assessment for the performance degradation of bearing based on a combinatorial feature extraction method. Digit Signal Proc 27:159–166
2. Li J, Li M, Zhang J (2017) Rolling bearing fault diagnosis based on time-delayed feed back monostable stochastic resonance and adaptive minimum entropy deconvolution. Sound Vib 401:139–151
3. Feng Z, Liang M, Chu F (2013) Recent advances in time–frequency analysis methods for machinery fault diagnosis: a review with application examples. Mech Syst Sig Process 38:165–205
4. Chen J, Li Z, Pan J, Chen G, Zi Y, Yuan J, Chen B, He Z (2016) Wavelet transform based on inner product in fault diagnosis of rotating machinery: a review. Mech Syst Sig Process 71:1–35
5. Feldman M, Braun S (2017) Nonlinear vibrating system identification via Hilbert decomposition. Mech Syst Sig Process 84:65–96
6. Lei Y, Lin J, He Z, Zuo MJ (2013) A review on empirical mode decomposition in fault diagnosis of rotating machinery. Mech Syst Sig Process 35:108–126
7. Zhongliang LV, Peiwen AN, Tang B, Zhang L (2013) The application of EEMD to fault diagnosis of rolling bearing. In: Advanced materials research, pp 765–767
8. Gilles J (2013) Empirical wavelet transform. IEEE Trans Sig Process 61(16):3999–4010
9. Cao H, Fan F, Zhou K, He Z (2016) Wheel-bearing fault diagnosis of trains using empirical wavelet transform. Measurement 82:439–449
10. Zhang J, Sun G, Li L (2013) Research on mechanical fault diagnosis based on LMD approximate entropy and FCM clustering. J Sci Instrum 34(3):714–720

11. Yan R, Gao RX (2007) Approximate entropy as a diagnostic tool for machine health monitoring. Mech Syst Sig Process 21(2):824–839
12. Lotfan S, Salehpour N, Adiban H, Mashroutechi A (2015) Bearing fault detection using fuzzy C-means and hybrid C-means-subtractive algorithms. In: 2015 IEEE international conference on fuzzy systems (FUZZ-IEEE). IEEE, Istanbul, Turkey, pp 1–7
13. Wang YY, Yang JG, Song BY (2011) Study on gasoline engine knock diagnosis based on wavelet transform and fuzzy c-means clustering. Chin Intern Combust Engine Eng 32(4): 56–59
14. Bearing Data Center. Case Western Reserve University. http://www.eecs.case.edu/laboratory/bearing

Research on Equilibrium Evacuation of Station Based on Improved Iterative Weighting Algorithm

Siyao Li, Jie Xu, Hui Zhang and Mengdi Liang

Abstract Evacuation efficiency is the key to ensuring safety in emergency situations. Combined with the connection structure of station facilities and pedestrian motion characteristics, an equilibrium evacuation model is constructed based on the reliability of evacuation time and a penalty rule. The model is solved by an improved iterative weighting algorithm. To verify the model validity, we simulate the evacuation process of Guomao Station and select evacuation time as the evaluation index. The simulation results show that the equilibrium evacuation model uses shorter evacuation time.

Keywords Subway station · Passenger flow distribution · Equilibrium evacuation model · Improved iterative weighting algorithm

1 Introduction

Subway stations are often crowded with people, and evacuation efficiency is the key to ensuring safety and maximizing evacuation capability in case of emergency. The equilibrium evacuation of passengers on a network will substantially decrease their evacuation time [1]. In stations, some circulating facilities, such as escalators, staircases, corners between passages, are bottlenecks that increase the swings,

S. Li · J. Xu (✉) · H. Zhang · M. Liang
State Key Laboratory of Rail Control and Safety, Beijing Jiaotong University, Beijing, China
e-mail: jxu1@bjtu.edu.cn

S. Li · J. Xu · H. Zhang · M. Liang
School of Transportation and Traffic, Beijing Jiaotong University, Beijing, China

S. Li · J. Xu · H. Zhang · M. Liang
Beijing Research Center of Urban Traffic Information Sensing and Service Technologies, Beijing, China

S. Li · J. Xu · H. Zhang · M. Liang
Research and Development Center of Transport Industry of Technologies and Equipment of Urban Rail Operation Safety Management, MOT, PRC, Beijing, China

Y. Qin et al. (eds.), *Proceedings of the 4th International Conference on Electrical and Information Technologies for Rail Transportation (EITRT) 2019*, Lecture Notes in Electrical Engineering 639, https://doi.org/10.1007/978-981-15-2866-8_8

arching and clogging of pedestrians [2, 3]. These bottlenecks will prolong the evacuation process [4]. Based on the macroscopic characteristics of pedestrian evacuation, mathematical models were established to study the relationship among pedestrian speed, pedestrian density and pedestrian volume [5, 6]. A mathematical model was established to compare the time difference between the use of escalators and the use of stairs based on a binomial logit model [7, 8]. To study the route choice strategy during evacuation process, the probability selection of all origin-destination (OD) pairs in the station was investigated by establishing routes of the station circulation facility network [9, 10]. However, the equilibrium evacuation is related to individual evacuation behavior and appropriate assignment of pedestrians to every evacuation route [1]. The complex structure of station network is rarely addressed by pre-studies of pedestrian equilibrium evacuation and most studies ignore the reliability of evacuation time.

2 Problem Description

2.1 Definition of Equilibrium Evacuation

We define equilibrium evacuation as an efficient evacuation method. This method can reduce the crowdedness of overcrowded routes and appropriately assign excessive passengers to relatively sparse routes.

2.2 Factors that Influence Emergency Evacuation Time

Evacuation capacity of a single facility

The circulation facilities include stairs, escalators, and passages. We collected the data of passenger walking behavior on these facilities during peak hours of weekdays. The evacuation capacity of the circulation facilities is described as follows: the circulation capacity of an escalator is 120 p/m/min, the circulation capacity of a stairway (ascending) is 70 p/m/min, the circulation capacity of a stairway (descending) is 80 p/m/min and that of a passageway is 92 p/m/min.

Evacuation capacity of a facility junction

Subway stations have two typical junctions according to the passenger flow direction, i.e., a diverging facility and merging facility (see Fig. 1).

(1) Capacity of a diverging junction

According to the blocked probability of passengers at the prior and target facilities, we will calculate the passenger volume and evacuation capacity of the junction.

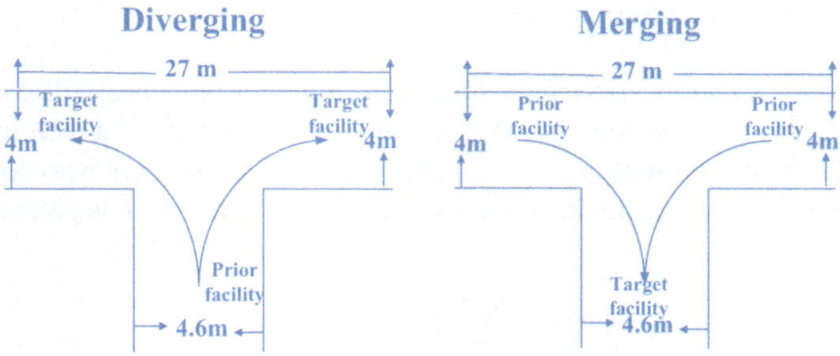

Fig. 1 **a** Diverging facility junction. **b** Merging facility junction

The number of gathering passengers at the exit of prior facility from time *0* to *T* (*T* is the total evacuation time) is calculated as

$$N_{f1} = \int_{0}^{T} f_f'(t) B_f'(t) \mathrm{d}t \tag{1}$$

where f_f' is the passenger volume of the prior facility, and B_f' is the width of the prior facility.

The number of passengers who pass through the entrances of the target facilities from *0* to *T* is calculated as

$$N_{f2} = \int_{0}^{T_0} f_f'(t) B_f'(t) \mathrm{d}t + \sum_{j=1}^{m} \int_{T_0}^{T} f_{fj}(t) B_{fj}(t) \mathrm{d}t \tag{2}$$

where T_0 is the time of passengers who are blocked, $f_{fj}(t)$ is the passenger volume of branch passage *j*, and $B_{fj}(t)$ is the width of branch passage *j*.

The volume of blocked passengers at the diverging facility is calculated as

$$N_{fs} = N_{f1} - N_{f2} = \int_{T_0}^{T} f_f'(t) B_f'(t) \mathrm{d}t - \sum_{j=1}^{m} \int_{T_0}^{T} f_{fj}(t) B_{fj}(t) \mathrm{d}t \tag{3}$$

when $N_{fs} = 0$, the number of passengers who pass through the junction will be equal to the number of passengers who gather at prior facility, and the utilization rates of the facility will be maximum. When $N_{fs} > 0$, traffic congestion will occur at the connection of facilities, and the effective evacuation capacity of the facility is determined by the capacity of target facilities.

(2) Capacity of a merging junction

For the merging facility, passenger flow from different branches will converge at the same exiting passage, and the effective evacuation capacity of a merging facility can be calculated according to Eq. (8). The number of passengers who converge to exits of each branch passage of prior facilities from time *0* to T is calculated as

$$N_{h1} = \sum_{i=1}^{n} \int_{0}^{T} f'_{hi}(t) B'_{hi}(t) \mathrm{d}t \tag{4}$$

where f'_{hi} is the passenger from branch i of the prior facilities, and B'_{hi} is the width of branch i of the prior facilities.

The number of passengers who pass through the entrance of the target facility from 0 to T is calculated as

$$N_{h2} = \int_{T_0}^{T} f_h(t) B_h(t) \mathrm{d}t + \sum_{i=1}^{n} \int_{0}^{T_0} f'_{hi}(t) B'_{hi}(t) \mathrm{d}t \tag{5}$$

where $f_h(t)$ is the passenger volume of the target facility, and $B_h(t)$ is the width of the target facility.

The number of blocked passengers of the merging facility is calculated as follows:

$$N_{hs} = N_{h1} - N_{h2} = \sum_{i=1}^{n} \int_{T_0}^{T} f'_{hi}(t) B'_{hi}(t) \mathrm{d}t - \int_{T_0}^{T} f_h(t) B_h(t) \mathrm{d}t \tag{6}$$

when $N_{hs} = 0$, the maximum utilization rates of the facilities will be achieved. When $N_{hs} > 0$, traffic congestion will occur in the junction of the facilities, and the effective evacuation capacity of merging facility is determined by the target facility.

Construction of station network

The construction of station network G = (V, E, w, c) is based on the topology structure of circulation facilities (see Fig. 2).

In Fig. 2, T, L, Z, and A-B denote the passage, stair, platform and four exits, respectively. The node set V consists of start nodes, intermediate nodes and destination nodes. In this paper, we consider that the station platforms are start nodes, the escalator and passage are intermediate nodes and the exits of station are destination nodes. The edge set Z is composed of directed and undirected edges, and it represents the connection relation of the facilities in the station. The dynamic travel time w of the circulation facility is determined by its length and passenger density; c denotes the evacuation capacity of the circulation facility.

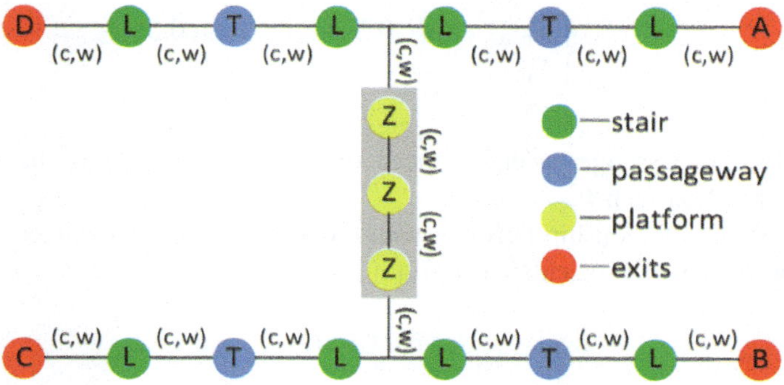

Fig. 2 Evacuation network diagram

2.3 *Route Impedance*

In this paper, "route" denotes the sequence circulation facilities from the origin node O to the destination node D, "segment" denotes a part of the route from an intermediate node r to another node s, and "path" denotes a single circulation facility. Their route impedances are calculated based on the evacuation time and its reliability. The reliability of evacuation time is the probability that passengers will successfully evacuate from the origin node to safe areas within the prescribed time. The evacuation time reliability of route k is calculated as

$$R_k = \begin{cases} \frac{t_{\max} - t_k}{t_{\max} - t_0}, & t_k \in [t_0, t_{\max}] \\ 0, & t_k \notin [t_0, t_{\max}] \end{cases} \tag{7}$$

where t_0 is the passenger initial evacuation time through route k, which is calculated when passengers evacuate at free-flow speed; $t_{\max}$ is the maximum allowable evacuation time through route k; and t_k is the prevailing travel time of passengers through route k, which is determined by the distance of route k and the path crowdedness.

The path crowdedness is denoted as c_i and is calculated as

$$c_i = \frac{N_i}{C_i} \tag{8}$$

where N_i is the number of passengers who evacuate through path i, and C_i is the maximum capacity of path i and is calculated as

$$C_i = \frac{A \times B}{\pi r^2} \tag{9}$$

$$v_i = \begin{cases} V_{\max} \cdot \exp(-0.5 \times c_i), & c_i > 0.5 \\ V_{\max}, & c_i \leq 0.5 \end{cases} \tag{10}$$

where v_i is the passenger average speed on path i, and $V_{\max}$ is the passenger free-flow speed on path i.

Based on the relationship between path crowdedness and passenger speed, the travel time of a path is calculated by Eq. (11),

$$t_i = l_i / v_i(c_i) \tag{11}$$

where t_i is the evacuation time on path i, and l_i is the length of path i.

The total evacuation times of routes that consist of several paths can be calculated by Eq. (12),

$$t_k^{rs} = f(N_k) = \sum_i \sigma_{pi} t_{pi}(N_k) + \sum_i \sigma_{si} t_{si}(N_k), \forall k \in W_{rs} \tag{12}$$

when σ_{pi} and σ_{si} are equal to 1, path i belongs to route k; when σ_{pi} and σ_{si} are equal to 0, path i does not belong to route k, and t_{pi} and t_{si} represent the evacuation time of passenger flow traversing passages and the evacuation time of passenger flow traversing stairs, respectively.

The total time impedance of route k from node r to node s in the evacuation network T_k^{rs} is calculated as

$$T_k^{rs} = (1 - \rho) t_k^{rs} + \lambda \rho (1 - R_k) \tag{13}$$

where ρ is a criterion coefficient that indicates the passenger preference, which is considered to be the reliability of the travel time of route k, the value range is 0–1, and λ is a transfer coefficient between the reliability of time and the reliability of travel time.

3 Model

3.1 Modeling

When the population density is high, the walking directions of occupants are similar to vehicle operation on a road: They line up in an orderly manner and walk in straight lines. Assume that the passengers exhibit rational behavior during the evacuation process, based on the equilibrium distribution model proposed by Beckman (Beckmann M, McGuire CB 1956). An equilibrium evacuation model that considers the travel time reliability of evacuation routes is established and shown in Eq. (14).

$$\min Z = \sum_k T_k^{rs}(K_k) + C$$

$$\text{s.t.}\begin{cases} t_k^{rs} = f(N_k) \\ \sum_k \sum_r \sum_s N_k^{rs} = N \\ N_k^{rs} \le C_k^{rs} \\ N_k^{rs} \ge 0 \\ x_k^{(0)} = 0 \end{cases} \tag{14–19}$$

Equation (14) is the optimal objective function, where K_k denotes the route set and C is a penalty factor. When $\sum_k T_k^{rs}(K_k) \le 360\text{s}$, $C = 0$; or $C = M$, where M is an infinite value. The objective function is used to minimize the time impedance of the total evacuation routes, that is, to minimize the evacuation time and crowdedness of evacuation routes. Equation (15) indicates that the travel time impedance of route k between point pair (r, s) is the sum of travel time of each passage and stair. Equation (16) ensures the volume of start node and the volume of stop node are conserved. Equation (17) is the quantity constraint condition of each path, which indicates that the number of evacuation passengers is less than the capacity of routes. Equation (18) indicates the volume of each segment is not negative, and the initial state of station network is expressed by Eq. (19).

3.2 Algorithm

The passenger equilibrium evacuation model is solved with an improved iterative weighting algorithm method. The steps are described as follows:

Step 1: Construct the evacuation network and identify all feasible routes. Set the passenger volume of all facilities on the network to 0, that is, $x_i^0 = 0, f_k^{rs(0)} = 0$, $n = 0$, where x_i^n is the passenger volume on facility i of iteration n, and $f_k^{rs(n)}$ is the volume of passengers on route k at iteration n;
Step 2: n = n + 1; the free-flow walking time of each segment in the evacuation network is calculated as $t_a^n = t_a\left(X_a^{n-1}\right), \forall a$;
Step 3: Sum the travel time of each segment to obtain the travel time of the routes. Based on the travel time, search the shortest route k_n;
Step 4: Based on the time reliability of each route in the evacuation network, the time impedance of each route T_{kn} is obtained. Search the optimal route with the minimum time impedance;

Step 5: Based on the route impedance calculated in step 4, assign passengers to the shortest route between all OD pairs. The additional volume of passenger flow of each segment is obtained:$F_a^n, a \in k_n$;

Step 6: Calculate the passenger volume on segments in the current train operation cycle according to the equation $X_a^n = (1 - 1/n)X_a^{n-1} + F_a^n/\text{n}, \forall a$,, based on the weighted coefficient;

Step 7: Repeat steps 3 and 4. When the time impedance of route k satisfies the inequality $|T_{kn} - T_{kn-1}| \leq \varepsilon$, where $\varepsilon > 0$ is a sufficiently small number, the iteration stops and the final distribution results for the volume of passenger flow are obtained. Otherwise, return to Step 2.

4 Case Study

4.1 Passenger Flow Distribution

Guomao station is a transfer station and is located on Line 1 and Line 10 of Beijing subway. The station has 7 exits, i.e., exits A, B, C, D, E, F and G. Based on the layout and circulation capacity of the facilities, the evacuation network of the station is constructed and shown in Fig. 3.

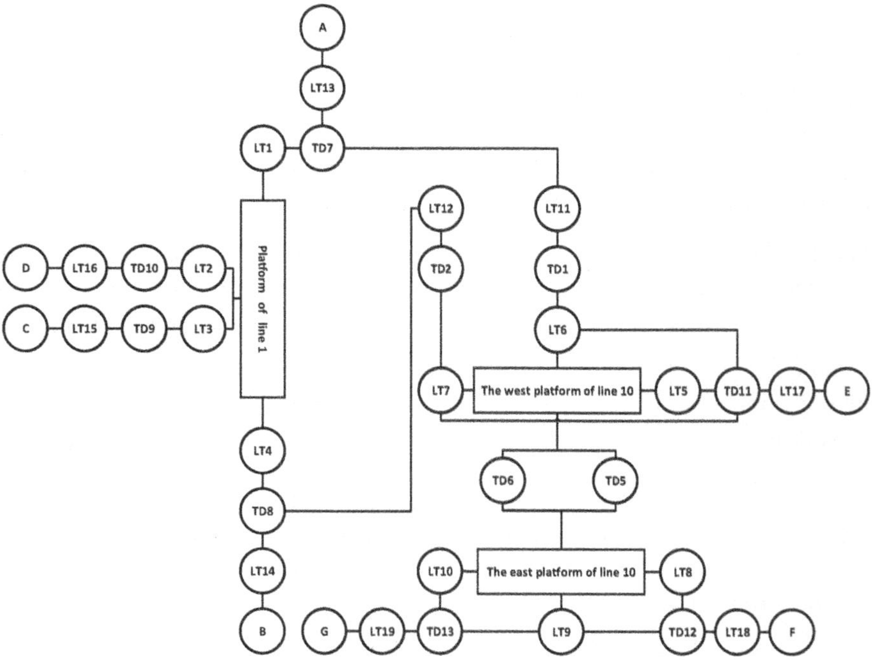

Fig. 3 Topology structure of the Guomao station network

According to the density-speed function, the free-flow speed of passengers on the passage under emergency evacuation is 1.94 m/s and that on ascending stairs is 1.02 m/s. Assume that an accident occurred on the platform of Line 1. Only one train dwells at the platform of Line l. The passengers on the three platforms need to be evacuated. In the most adverse condition, 704 initial passengers exist on the platform of Line 1, and the train capacity is 1840. The initial passengers of Line 10 on each side of platform are 620. The specific accident scenario is discussed as follows:

An emergency occurred on the platform of Line 1. With the first evacuation strategy, the criterion coefficient in the time impedance function is 0, which indicates that passengers select routes according to evacuation time only. The evacuation routes and distribution ratio are shown in Table 1.

With the second evacuation strategy, the criterion coefficient in the time impedance function is 0.5, which indicates that passengers will select routes after considering the evacuation time and the reliability of time. The evacuation routes and distribution ratio are shown in Table 2.

Table 1 Evacuation routes and passenger flow distribution

Route No.	Routes	Travel time at free-flow speed (s)	Volume (peds)	Distribution ratio (%)
1	1-LT1-TD7-LT13-A	59.36	534	20.98
2	1-LT1-TD10-LT16-D	57.31	338	13.28
3	1-LT2-ZT1-TD7-LT13-A	85.36	191	7.52
4	1-LT2-ZT1-TD10-LT16-D	83.31	227	8.91
5	1-LT2-LT11-TD1-ZT10-TD11-LT17-E	202.03	0	0
6	1-LT3-ZT1-TD8-LT14-B	84.33	212	8.34
7	1-LT3-ZT1-TD9-LT15-C	85.36	189	7.46
8	1-LT3-LT12-TD2-ZT10-TD11-LT17-E	202.07	0	0
9	1-LT4-TD8-LT14-B	58.33	327	12.84
10	1-LT4-TD9-LT15-C	59.36	526	20.67
11	10-LT5-TD11-LT17-E	67.96	229	18.43
12	10-LT6-ZT10-TD11-LT17-E	88.96	74	5.94
13	10-LT6-TD4-TD13-LT19-G	74.65	150	12.14
14	10-LT7-ZT10-TD11-LT17-E	109.19	21	1.71
15	10-LT7-TD4-TD13-LT19-G	74.65	133	10.73
16	10-LT8-TD12-LT18-F	65.89	234	18.92
17	10-LT9-ZT10-TD12-LT18-F	86.89	90	7.25
18	10-LT9-ZT10-TD13-LT19-G	87.92	70	5.62
19	10-LT10-TD13-LT19-G	66.92	239	19.26

Table 2 Passenger flow distribution results

No.	Volume (peds)	Distribution ratio (%)	No.	Volume (peds)	Distribution ratio (%)
1	479	17.82	11	236	19.05
2	337	13.24	12	68	5.46
3	208	8.18	13	148	11.97
4	191	7.51	14	40	3.24
5	75	3.82	15	135	10.88
6	204	8.01	16	224	18.03
7	194	7.63	17	81	6.54
8	55	3.15	18	76	6.09
9	320	12.06	19	232	18.74
10	483	18.47			

4.2 *Simulation and Discussion*

To verify the validity of the model, *AnyLogic* 8.2.3 is used to simulate the evacuation process of the two evacuation strategies. The passengers are assumed to be uniformly distributed on the platform when they begin to evacuate.

A contrast analysis of evacuation time and distribution of passenger density changes with different strategies indicates that the evacuation equilibrium will increase the evacuation efficiency. Compared with Fig. 4a, b, when passengers select routes based on travel time only, the passenger density near LT1 and LT4 is relatively high. If passengers select evacuation routes based on the evacuation time and the reliability of time, the number of passengers who evacuate through the transfer passage connected to the station concourse of Line 10 will increase, and the passenger density will decrease near Exit A, Exit B, Exit C, and Exit D.

In the case of two different distribution ratios of passenger flow, the evacuation times are 323.7 and 310.8 s. Therefore, the passenger flow density at the evacuation bottleneck and the evacuation time can be reduced considering the time reliability.

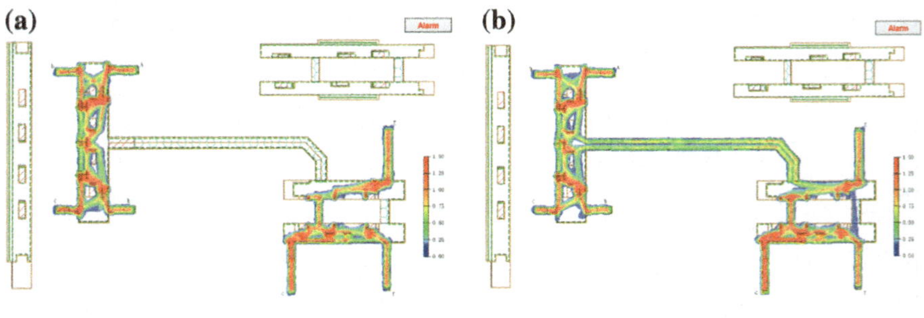

Fig. 4 **a** Passenger distribution based on travel time.
b Passenger distribution based on the reliability of travel time

The social force model is a built-in simulation model of AnyLogic, and the simulation process is determined by the model. The simulation results show that passengers will transfer from overcrowded evacuation routes to alternate routes in an emergency evacuation. From the perspective of social force model, the congestion degree of overcrowded routes can be alleviated after considering the reliability of evacuation time.

Therefore, the reduction in passenger density on overcrowded routes indicates that the force between passengers, the repulsion between passengers and obstacles in the social force model will decrease, which enables a smoother passenger evacuation process and a decreasing trend in the total evacuation time.

5 Conclusion

In this paper, to promote passenger evacuation efficiency in emergency conditions in a subway station, an equilibrium evacuation model is established to balance the distribution of passengers and reduce evacuation time. The key to assigning passengers to balanced evacuation routes is that the evacuation model is constructed based on the evacuation time and time reliability. To increase travel time reliability, we have to lessen the passenger density on crowded routes and transfer overcrowded passengers on the shortest routes to spare routes. According to the relationship between density and speed, the transfer of passengers will accelerate during the evacuation process to ensure a reduction in evacuation time. We simulate the evacuation process of Guomao subway station, and the result shows that evacuation time can decrease from 2.5 to 3.7%. With some evacuation route disruption, the number of available routes will decrease, and some overcrowded passengers will not transfer to alternate routes, and the total actual travel time undergoes minimal fluctuation. Therefore, the evacuation equilibrium model is a more effective passenger allocation method.

Acknowledgements The authors gratefully acknowledge the support provided by China National "13th Five-Year" key research project "Safety assurance technology of urban rail system" (Grant No. 2016YFB1200402).

References

1. Wu Y, Xu J, Jia L, Qin Y (2018) Estimation of emergency evacuation capacity for subway stations. J Transp Saf Secur 10:586–601. https://doi.org/10.1080/19439962.2017.1357059
2. Almeida JE, Rosseti RJF, Coelho AL (2013) Crowd simulation modeling applied to emergency and evacuation simulations using multi-agent systems. In: DSIE'11—6th Doctoral symposium on informatics engineering, pp 93–104
3. Sun L, Gong Q, Yao L (2018) A dynamic time warping algorithm based analysis of pedestrian shockwaves at bottleneck 2018:1–7

4. Jin JG, Tang LC, Sun L, Lee DH (2014) Enhancing metro network resilience via localized integration with bus services. Transp Res Part E Logist Transp Rev 63:17–30. https://doi.org/10.1016/j.tre.2014.01.002
5. Ronchi E, Fridolf K, Frantzich H, Nilsson D, Walter AL, Modig H (2017) A tunnel evacuation experiment on movement speed and exit choice in smoke. Fire Saf J 1–11. https://doi.org/10.1016/j.firesaf.2017.06.002
6. Thi N, Anh N, Daniel ZJ, Du NH, Drogoul A, An VD (2014) A hybrid macro-micro pedestrians evacuation model to speed up simulation in road networks. https://doi.org/10.1007/978-3-642-27216-5
7. Cheung CY, Lam HKW (1998) Pedestrain route choices between escalator and stairway in MTR stations. J Transp Eng 124:277–285
8. Lam WHK, Cheung C (2000) Pedestrian speed/flow relationships for walking facilities in Hong Kong. J Transp Eng 2:343–349
9. Sun L, Jin JG, Lee DH, Axhausen KW, Erath A (2014) Demand-driven timetable design for metro services. Transp Res Part C Emerg Technol 46:284–299. https://doi.org/10.1016/j.trc.2014.06.003
10. Zhu KJ, Shi Q (2016) Experimental study on choice behavior of pedestrians during building evacuation. Procedia Eng 135:206–215. https://doi.org/10.1016/j.proeng.2016.01.110

Fault Diagnosis of MVB Based on Random Forest and Ensemble Pruning

Zhaozhao Li, Lide Wang, Ping Shen, Hui Song and Xiaomin Du

Abstract Due to the complex working environment and lack of network management protocol, it is difficult to comprehensively monitor the operating states and diagnose the faults of the Multifunction Vehicle Bus (MVB). In this paper, an MVB fault diagnostic method based on physical waveform features and ensemble pruning has been proposed. Firstly, MVB waveforms of the bus administrator node (BA) in normal and fault conditions are sampled by an MVB analyzer based on high-speed A/D sampling technology. Network features are extracted from the waveforms, and a random forest (RF) classifier has been trained to classify different MVB faults. An ensemble pruning method based on diversity index and the k-mean algorithm has been proposed to reduce the number of decision trees and improve the ensemble performance. The experimental results show that the proposed feature extraction method and ensemble pruning classifier can recognize the MVB faults correctly, reduce the number of base classifiers, and improve the accuracy of the subforest compared with the original RF ensemble.

Keywords Fault diagnosis · Multiple vehicle bus · Ensemble pruning

1 Introduction

MVB is the central nervous of a train which transmits control and monitoring data [1]. However, the distributed MVB system and hostile operating environment make it difficult to diagnose MVB faults in practice. Many MVB faults will lead to impedance mismatch and cause signal reflection of the network. Different physical waveform features can characterize different fault patterns. Based on this fact, it is reasonable to change the MVB fault diagnosis into a classification problem.

Ensemble learning has been proven to have a better performance than a single classifier [2]. However, the higher performance of the ensemble classifier is

Z. Li (✉) · L. Wang · P. Shen · H. Song · X. Du
Beijing Jiaotong University, Beijing 100044, China
e-mail: jore_lee@163.com

Y. Qin et al. (eds.), *Proceedings of the 4th International Conference on Electrical and Information Technologies for Rail Transportation (EITRT) 2019*, Lecture Notes in Electrical Engineering 639, https://doi.org/10.1007/978-981-15-2866-8_9

achieved at the price of heavy storage and computational resources requirements. Random forest (RF) is one of the typical ensemble methods [3]. Generally, many trees in RF are useless to improve the ensemble accuracy but occupy too much computing and storage resources. Ensemble pruning aims to get rid of those useless trees and get a subset while maintaining or even improving the accuracy of the ensemble. Successful ensemble pruning method depends on how to select good and diverse base classifiers from the original ensemble [4]. In traditional ordering pruning, the subset is initialized to an empty set and the base classifiers which can optimize the diversity index of the subset are added one by one [5]. But the diversity index is hard to choose and no existing index can apply to all datasets or classification algorithms. Genetic algorithm (GA) has also been proposed to select proper decision trees in the ensemble pruning process by Adnan et al. [6]. But in their work, both test and ensemble pruning operations are on the same test set which will lead to overfitting and cannot prove effective on a new dataset.

In this paper, the k-mean algorithm has been introduced to cluster similar trees into groups and parts of the trees in each group are selected into subforest based on diversity index interrater agreement κ. A testbed has been built and different fault features are extracted from the MVB waveforms of the master node. The experimental results show that different MVB faults can be diagnosed by a trained RF classifier correctly. And the proposed ensemble pruning method can reduce the number of decision trees and improve accuracy at the same time.

2 MVB Fault Diagnosis System

Typical MVB faults include packet loss, node shutdown fault, network disconnection, impedance mismatch, and so on. In this paper, we mainly focus on impedance mismatch. Many MVB faults can come down to impedance mismatch problem.

Terminating resistor mismatch. It is suggested that the characteristic impedance of the MVB cable is 120 Ω and each end of the network needs to terminate with a 120 Ω terminator [1]. Inappropriate maintenance or vibration may lead to this kind of fault.

Parallel impedance or serial impedance mismatch. Network cables, cable connectors, and network card electron components can be considered as durable components in the poor operating environment. Improper maintenance can also change the electrical properties of those network infrastructures. For example, the DB9 connector pins will suffer from corrosion, wear and then result in the serial impedance increase of the cable. Bad contact of DB9 connector under vibration will also lead to this kind of fault.

Impedance mismatch faults will lead to the waveform distortion of MVB as shown in Fig. 1. Therefore, we have presented an MVB fault diagnostic system as shown in Fig. 2. An MVB analyzer with high-speed A/D sample function has been

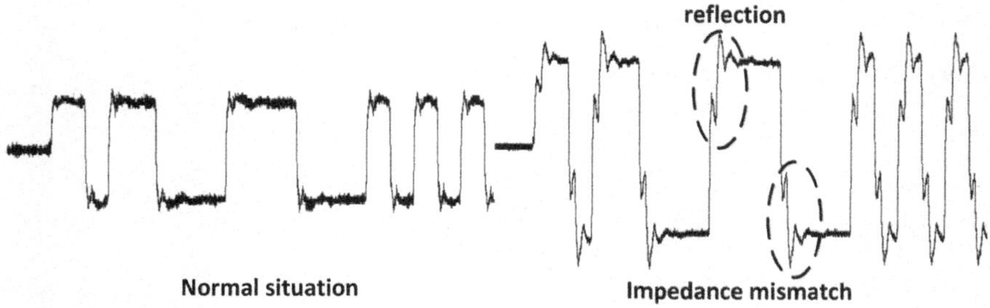

Fig. 1 Waveforms of MVB in the normal and terminating fault

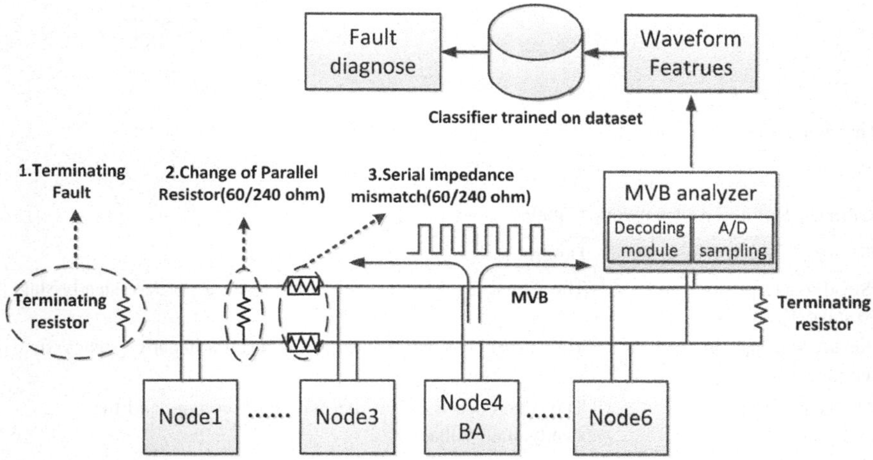

Fig. 2 MVB fault diagnostic framework

used to sample the waveforms of BA node in both normal and fault states. Network features are extracted from the waveforms and MVB faults' diagnostic model is trained and tested based on the extracted dataset.

2.1 *Feature Extraction*

Feature extraction is the base of the MVB fault diagnosis. Reasonably extracted features can effectively represent the states of the network. Impedance mismatch faults will affect network waveforms notably [7]. Based on the analog acquisition of the MVB analyzer, waveform parameters are extracted from each bit of the start

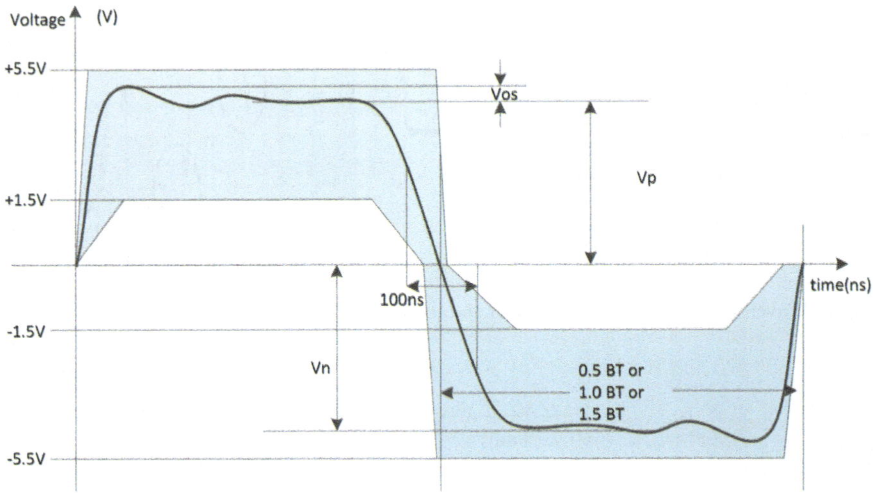

Fig. 3 Features extraction

Table 1 Definition of network features

Features	Description
Steady-state positive voltage V_p	The signal level when waveforms reach a positive steady-state
Steady-state negative voltage V_n	The signal level when waveforms reach a negative steady-state
Overshoot V_{os}	The difference between the maximum voltage and the steady-state voltage
Slew rate S	The slope of the waveform at zero crossing point
Rising slope K_p	The slope when the frame signal rises from 10% of V_p to 90% of V_p
Falling slope K_n	The slope when the frame signal descends from 10% of V_n to 90% of V_n

delimiter and the network features are the averages of these parameters. Figure 3 shows the feature extraction from MVB waveforms. These physical waveform features are listed in Table 1.

2.2 *Diversity Measures and Ensemble Pruning Method*

Random forest is a bagging ensemble method consisting of different decision trees trained on the different training sets. When splitting a node during the construction of the decision tree, the chosen split feature is no longer the best split among all features but the best split among a random subset of the features. The randomness leads to the diversity of the base classifiers and improves the accuracy of the ensemble [3]. However, as the increase of the ensemble size, the computing and storage resource increases. Besides, it has been proven that not all base classifiers are helpful to increase the performance of the ensemble [4]. It is meaningful to get rid of the redundant base classifiers from the ensemble and find a subensemble which can maintain or even increase the performance of the model. This process is called ensemble pruning.

In this paper, the dataset of MVB has been divided into three parts: training set, validation set, and testing set. The training set is used to train the base classifiers, namely the RF model. The validation set is used for ensemble pruning. When predicted by the RF model, each output of the trees of the validation set is converted into an oracle output (a binary type output: "1" means the sample is classified correctly and "0" means misclassified). The testing set is used to verify the performance of the pruned subforest.

The key to the success of ensemble learning or ensemble pruning is that we should build a set of diverse and good classifiers. Diversity measures will be helpful to find proper base classifiers [8]. Interrater agreement measurement κ is a diversity measure defined in Formula (1). It is related to the average individual classification accuracy p_{av} defined in Formula (2) and average disagreement measure Dis_{av} defined in Formula (3).

$$\kappa = 1 - \frac{1}{2p_{\text{av}}(1 - p_{\text{av}})}\text{Dis}_{\text{av}} \tag{1}$$

$$p_{\text{av}} = \frac{1}{NL}\sum_{j=1}^{N}\sum_{i=1}^{L} y_{j,i} \tag{2}$$

$$\text{Dis}_{\text{av}} = \frac{2}{L(L-1)}\sum_{i=1}^{L-1}\sum_{k=i+1}^{L} \text{Dis}_{i,k} \tag{3}$$

$$\text{Dis}_{i,k} = \frac{N_{01} + N_{10}}{N_{00} + N_{10} + N_{01} + N_{11}} \tag{4}$$

N is the size of the validation set $D = \{x_1, x_2, x_3,\ldots, x_N\}$, L is the size of RF, and $y_{j,i}$ is the oracle output of the ith decision tree on sample x_j. $\text{Dis}_{i,k}$ is the disagreement measure for the ith and kth tree as defined in Formula (4). For a team of L classifiers, the averaged $\text{Dis}_{i,k}$ over all pairs of classifiers is defined as (3). N_{00}, N_{10}, N_{01}, and N_{11} are elements of the contingency table of classifier C_i and C_k as shown in Table 2. For example, N_{11} is the number of samples that are classified correctly by both C_i and C_k. $N = N_{00} + N_{10} + N_{01} + N_{11}$.

Table 2 Contingency table of the relationship between a pair of classifiers C_i and C_k

	C_i correct(1)	C_i wrong(0)
C_k correct(1)	N_{11}	N_{10}
C_k wrong(0)	N_{01}	N_{00}

From the definition of interrater agreement measurement κ, we can see that it not only focuses on the diversity of the ensemble but also the prediction performance of the entire model. The higher the value is, the less is the diversity of the ensemble.

Based on the oracle output of the validation set, a K-mean method is used to cluster similar trees into M groups. Trees are more similar to each other in the same group and more diverse with each other in different groups. To maintain the performance of the ensemble, parts of the trees in each group are selected into the subforest based on interrater agreement measurement κ.

The K-mean ensemble pruning process is shown in Fig. 4. The number of trees in each cluster is Num(i). An in-cluster-pruning operation is conducted as follow:

Step 1. For each cluster, we select the most accurate tree C_{acc} firstly.
Step 2. Find the tree C_k in the rest of the cluster which can make the κ value of the selected subset F_i $\{C_{\text{acc}}, C_k\}$ lowest. In this way, we can get the proper trees to make the subset F_i most diverse.
Step 3. Repeat the second step and the trees are put into F_i one by one until the size of F_i reaches a certain value V.
Step 4. Finally, we can get a subforest F which is the set of F_i. $F = \{F_1, F_2, \ldots, F_i, \ldots, F_{\text{sub}}\}$.

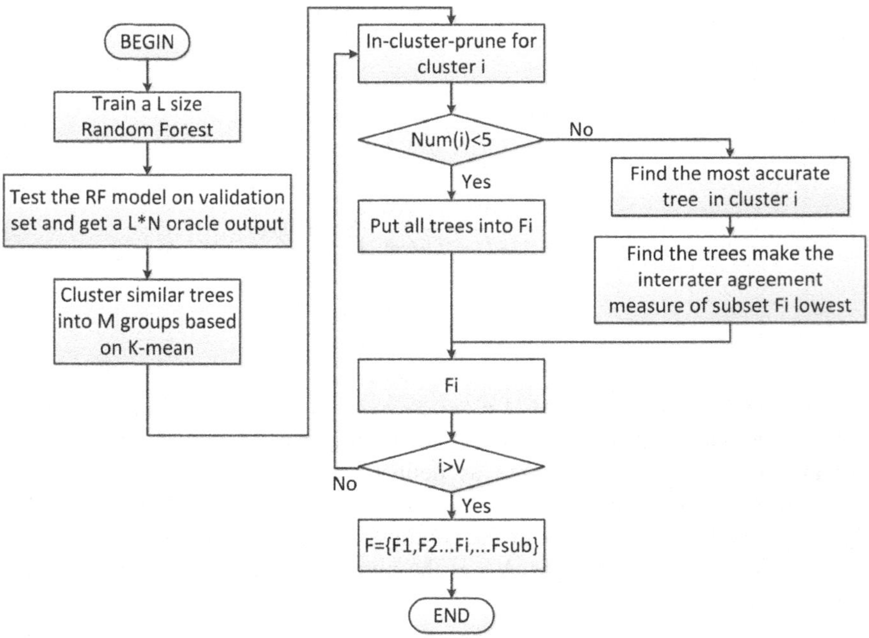

Fig. 4 K-mean ensemble pruning process

3 Experiments

A testbed has been built in the laboratory environment to test the proposed method. The testbed includes 1 MVB bus administrator node (BA) and 5 slave nodes. Besides, a test board is used to change the impedance of the network and an MVB analyzer is used to sample the waveforms of the BA node. In the experiment, we have conducted 6 operating conditions including 5 fault conditions and 1 normal condition as shown in Fig. 2. These faults are loss of the left terminator (Fault 1), serial impedance changes into 60 Ω (Fault 2), serial impedance changes into 240 Ω (Fault 3), parallel impedance changes into 60 Ω (Fault 4), and parallel impedance changes into 240 Ω (Fault 5) (Fig. 5).

For each operating condition, we have sampled 1000 frames and formed a 6000 size dataset including 6 classes. RF size is 200 and in the subset selection of K-mean pruning, we set V equal to 0.6*Num(i). M is equal to 10. In the training phase, the ensemble pruning and testing process have been conducted for 10 times as shown in Figs. 6, 7, and 8. The dataset is divided into a training set (3000), a validation set (1500), and a testing set (1500) randomly each time. Kmean-ka means the proposed pruning method based on the k-mean cluster and κ diversity measure. The ordering method is similar to the mentioned in-cluster-pruning process above. The difference is that it is based on the whole RF without any clustering operation. Besides, the selection index of Ordering-e is the classification error of the subforest and Ordering-c is complementariness of the subforest. Formula (5) is the definition of the classification error in Ordering-e. $F_i(x_i)$ is the oracle output of subset F_i on sample x_i. Formula (6) is the definition of complementariness. $C_k(x_i)$ is the oracle output of the candidate tree C_k. Complementariness is an index that how many samples' oracle outputs are different between the selection subset F_i and the candidate tree C_k.

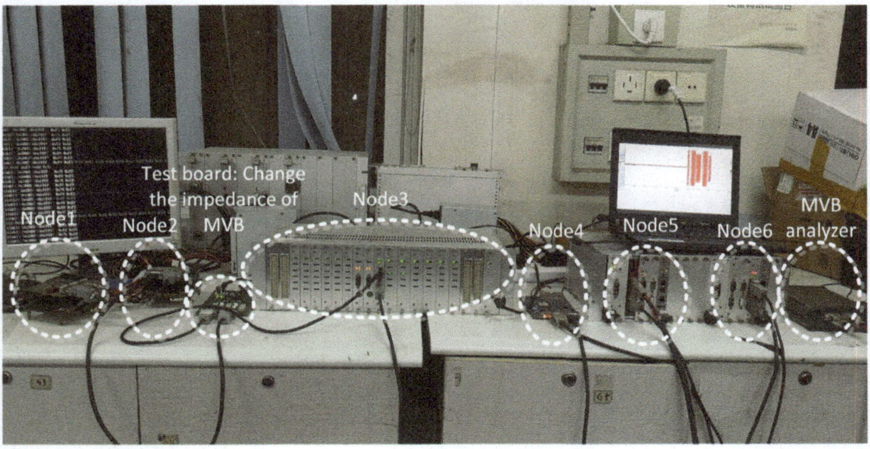

Fig. 5 Testbed

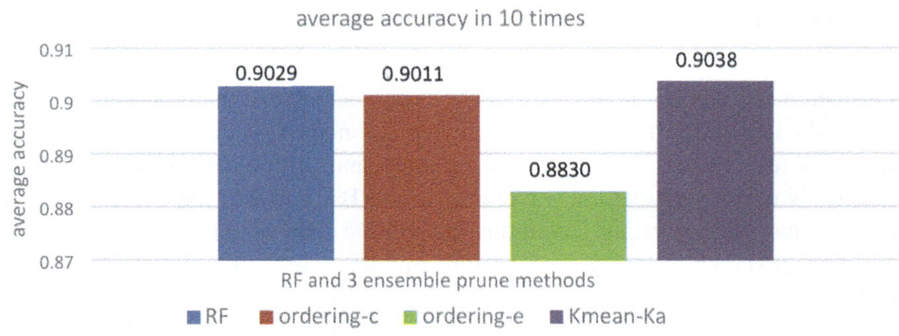

Fig. 6 Average accuracy

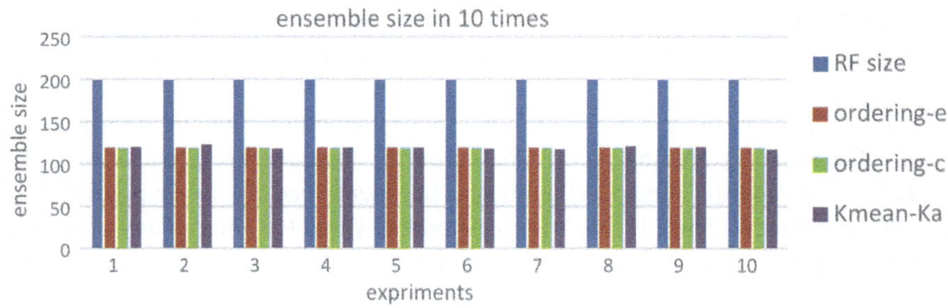

Fig. 7 Ensemble size

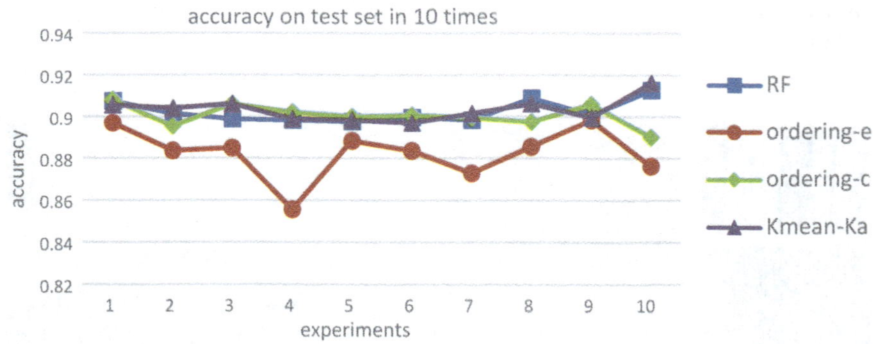

Fig. 8 Accuracy in 10 times

$$E = 1 - \frac{1}{N}\sum_{i=1}^{N}(y_i \neq F_i(x_i)) \tag{5}$$

$$C = \sum_{i=1}^{N} (C_k(x_i) \neq F_i(x_i)) \tag{6}$$

From Fig. 6, we can see that the classification accuracy is about 90.38% on average. The network feature extraction and MVB faults diagnosis methods are verified. Figures 6, 7, and 8 show that the Kmean-ka method can significantly reduce the ensemble size of RF while slightly improve the mean accuracy. The tree number of the Kmean-ka method is about 60% of the original RF. Both Ordering-e and Ordering-c methods can reduce the RF size but the accuracy of the subforest cannot be maintained.

The interrater agreement measure κ not only focuses on the diversity of the whole system but also the average accuracy of base classifiers. Through the k-mean cluster, similar decision trees are grouped into the same cluster. Based on the parameter κ, complementary decision trees are selected from each cluster, resulting in the performance improvement of the whole subforest. Ordering-e pruning will lead to overfitting on the validation set so the generalization performance of the corresponding subforest is poor. Ordering-c pruning only focuses on complementariness but neglects the average accuracy of the base classifiers. Although the selected trees in Ordering-c are more complementary, they are less accurate as well. Hence, the performance of the subforest is restricted by those less accurate trees.

4 Conclusion

In this paper, we proposed an MVB network feature extraction method from physical waveforms. The MVB impedance mismatch faults diagnosis is converted to a classification problem. A Kmean-ka ensemble pruning method has also been proposed to reduce the ensemble size. Test results show that the feature extraction and fault diagnosis method can recognize different MVB impedance mismatch faults correctly. Compared with RF and traditional ordering pruning methods, the proposed ensemble pruning method cannot only reduce the ensemble size significantly but also improve the average accuracy of the model.

Acknowledgements This work is supported by the Beijing Municipal Natural Science Foundation under Grant L171009.

References

1. Kirrmann H, Zuber PA (2003) The IEC/IEEE train communication network. Micro IEEE 21 (2):81–92
2. Zhou ZH (2012) Ensemble methods—foundations and algorithms, 1st edn. Chapman & Hall/ CRC

3. Cutler A, Cutler DR, Stevens JR (2004) Random forests. Mach Learn 45(1):157–176
4. Zhou ZH, Wu J, Wei T (2002) Ensembling neural networks: many could be better than all. Artif Intell 137(1):239–263
5. Martınez-Munoz G, Suárez A (2004) Aggregation ordering in bagging. In: Proceedings of the IASTED international conference on artificial intelligence and applications. Acta Press, Innsbruck Astria, pp 258–263
6. Adnan MN, Islam MZ (2016) Optimizing the number of trees in a decision forest to discover a subforest with high ensemble accuracy using a genetic algorithm. Knowl-Based Syst 110: 86–97
7. Lei Y, Djurdjanovic D, Barajas L et al (2011) DeviceNet network health monitoring using physical layer parameters. J Intell Manuf 22(2):289–299
8. Sun B, Wang JD, Chen HY, Wang YT (2014) Diversity measures in ensemble learning source. Kongzhi yu Juece/Control Decis 29(3):385–395

Denoising Method of Train Vibration Signal Based on Improved Wavelet Threshold

Zihao Wang, Xinhai Liu and Zongyi Xing

Abstract In view of the wheel state monitoring, the wheel vibration signal is often used to reflect the wheel state information, but the original vibration signal contains a lot of noise. In order to extract the effective wheel information, it is necessary to reduce the noise of the original vibration signal. The problems existing in the traditional wavelet threshold denoising method include unreasonable threshold setting, lack of continuity of hard threshold function and fixed deviation of wavelet coefficients processed by soft threshold function. In this paper, the traditional wavelet threshold denoising method is improved from threshold selection and threshold function, and an improved wavelet threshold denoising method is proposed. This method is used to reduce the noise of the measured wheelset vibration signal and compared with the traditional wavelet denoising method from evaluation indexes and graphical results. The results show that the improved wavelet threshold denoising method has better noise reduction effect.

Keywords Wheel vibration signal · Wavelet threshold denoising · Threshold selection · Threshold function

1 Introduction

As an important part of urban rail transit vehicles, the wheels carry the weight of the entire vehicle. When the train is running at high speed, due to the continuous impact and collision between the wheel and the track, the wheel will be damaged, causing the impact vibration of the wheel–rail system and even derailment of the train. Therefore, it is necessary to identify wheel faults effectively. As a common failure mode, wheel out-of-roundness is the difference between the maximum and minimum diameters of the wheel tread cross-section [1]. For the monitoring of wheel out-of-roundness, vibration signal analysis method is one of the most commonly

Z. Wang · X. Liu · Z. Xing (✉)
Nanjing University of Science and Technology, Nanjing 210014, China
e-mail: xingzongyi@163.com

Y. Qin et al. (eds.), *Proceedings of the 4th International Conference on Electrical and Information Technologies for Rail Transportation (EITRT) 2019*, Lecture Notes in Electrical Engineering 639, https://doi.org/10.1007/978-981-15-2866-8_10

used real-time monitoring methods. However, due to the influence of working environment and other factors, original vibration signal contains a large amount of noise. Therefore, in order to effectively extract the fault information in the vibration signal, it is important to reduce the noise of original vibration signal.

Since the noise signals in the vibration signal are mostly non-stationary signals, time-frequency analysis processing techniques are required. Wavelet analysis has better time-domain resolution and frequency-domain resolution, which can meet the requirements of time-frequency analysis [2]. The essence of wavelet denoising is to deal with the wavelet coefficients obtained after decomposition, so as to complete the noise reduction [3]. And the method of wavelet threshold denoising is most widely used in engineering. The key points of wavelet threshold denoising are the selection of threshold and threshold function. At present, there are four principles for threshold selection, including the general threshold principle, the stein unbiased likelihood estimation threshold principle, the heuristic threshold principle and the minimax threshold principle [4]. For the selection of threshold, if the threshold is set too high, it is easy to filter out some of the useful signals and cause distortion when the signal is reconstructed; if the threshold is set too low, it is easy to mistake some noise for useful signals and affect the effect of noise reduction [5]. There are two threshold functions, including hard threshold functions and soft threshold functions. For the threshold function, the hard threshold function does not have continuity, which causes a large mean square error of the signal. In contrast, the soft threshold function has the better overall stability. However, when the absolute value of the wavelet coefficients is much larger than the threshold value, there will always be a constant deviation between the wavelet coefficients before and after the processing, which will affect the noise reduction effect [6].

In order to solve the above problems, this paper proposes an improved wavelet threshold denoising method from the aspects of threshold and threshold function. It compares the traditional wavelet threshold denoising method and the improved wavelet threshold denoising method based on measured data, which proves the validity and practicability of the improved wavelet threshold denoising method proposed in this paper.

2 Improved Wavelet Threshold Denoising

A one-dimensional noisy signal can be represented as [7]:

$$S(t) = f(t) + \sigma e(t), \quad t = 0, 1, \ldots, n-1 \tag{1}$$

where $f(t)$ is a useful signal, $e(t)$ is noise, $S(t)$ is the original signal containing noise, σ is the noise standard deviation in the signal.

2.1 Wavelet Threshold Denoising Principle

The basic principle of wavelet threshold denoising is that after wavelet decomposition of the original signal containing noise, the useful signal is decomposed into components with large wavelet coefficients, and the noise portion is decomposed into components with small wavelet coefficients [8]. According to this characteristic, the noise can be effectively filtered out from the original signal by comparing wavelet coefficients and thresholds.

As shown in Fig. 1, the wavelet threshold denoising process is mainly composed of three steps:

- Wavelet decomposition of the original signal: determining the appropriate wavelet basis function and wavelet decomposition layer *N*, and the noise signal is carried out by *N* Layer wavelet decomposition;
- Threshold processing of wavelet coefficients for each layer: setting a reasonable threshold, and the wavelet coefficients of each layer are processed by threshold function;
- Signal reconstruction: After the signal reconstruction of the wavelet coefficients obtained by the threshold processing, the wavelet threshold denoising process is completed.

In the process of wavelet threshold denoising, the most important steps are the selection of the threshold and the determination of the threshold function. Aiming at the problems of traditional threshold selection principle and threshold function, this paper improves the threshold selection principle and threshold function.

2.2 Improved Threshold Selection Principle

The traditional threshold selection principles cannot guarantee the reasonableness of the threshold. In addition, as the decomposition scale increases, the wavelet coefficient of noise gradually decreases [9]. However, the traditional threshold selection principles do not take into account the influence of decomposition scale on threshold selection. Therefore, based on the general threshold principle, this paper proposes an improved threshold selection principle:

$$\lambda_j = \sigma\sqrt{2\log N}\Big/\log(j+1) \tag{2}$$

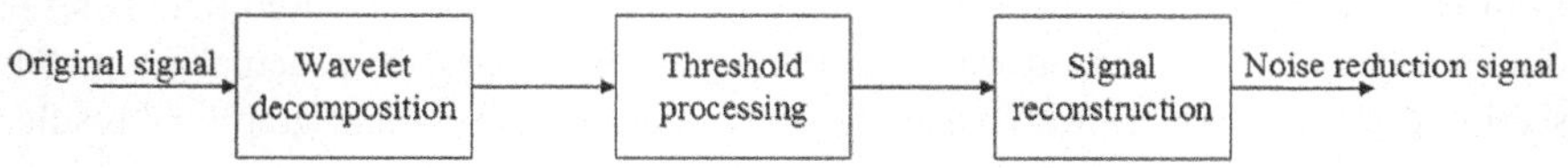

Fig. 1 Wavelet threshold denoising flowchart

where λ_j is threshold of wavelet decomposition layer j, N is the length of the signal, σ is the noise standard deviation in the signal.

The improved threshold selection principle reflects the influence of the reserved signal length and the noise standard deviation on the threshold selection and also adds the decomposition scale. Therefore, threshold decreases with the increase of the decomposition scale, just as the wavelet coefficient of the noise decreases with the increase of the decomposition scale, and the deviation caused by the unreasonable threshold is also solved.

2.3 Improved Threshold Function

In view of the problems of hard and soft threshold functions, this paper proposes an improved threshold function, which should be designed to meet the following basic principles:

- The threshold function has continuity to prevent additional oscillations;
- The threshold function has high-order conductivity, which is convenient for calculation;
- The threshold function is as simple as possible, and it is convenient to adjust by parameters;
- The asymptote of the threshold function is $\overline{W_{j,k}} = W_{j,k}$, and it requires a good degree of approximation.

Based on the above design principles, the improved threshold function proposed in this paper is

$$\overline{W_{j,k}} = \begin{cases} \operatorname{sgn}(W_{j,k})\left(\left|W_{j,k}\right| - \dfrac{2}{1+e^{\frac{\left|W_{j,k}\right|-\lambda}{N}}}\lambda\right), & \left|W_{j,k}\right| \geq \lambda \\ 0, & \left|W_{j,k}\right| < \lambda \end{cases} \tag{3}$$

where $\overline{W_{j,k}}$ is the wavelet coefficients processed by the threshold function, $W_{j,k}$ is the wavelet coefficients before processing, λ is the threshold, $\operatorname{sgn}(*)$ is the symbolic function.

3 Experiment Analysis

In order to further verify the effect of the improved wavelet threshold denoising in practical applications, based on vibration signal measured from actual train, this section does research from two aspects of evaluation indexes and graphical results.

3.1 Evaluation Index

In order to quantitatively evaluate the performance of the noise reduction algorithm, signal-to-noise ratio (SNR) and root mean square error (RMSE) are usually selected. The larger the value of the SNR, the smaller the value of RMSE, indicating that the noise reduction effect is better.

The SNR is

$$\mathrm{SNR} = 10\log_{10}\left[\frac{\sum_{n=1}^{N} x^2(n)}{\sum_{n=1}^{N} [X(n) - x(n)]^2}\right] \tag{4}$$

The RMSE is

$$\mathrm{RMSE} = \sqrt{\frac{1}{N}\sum_{n=1}^{N} [X(n) - x(n)]^2} \tag{5}$$

where $x(n)$ is an original vibration signal, $X(n)$ is vibration signal after denoising.

3.2 Results and Analysis

This section takes the vibration signal measured from a certain metro line as an example, and its time-domain waveform is as shown in Fig. 2.

Then, a spectrum diagram of the measured vibration signal is obtained by fast Fourier transform as shown in Fig. 3.

As shown in Figs. 2 and 3, the measured vibration signal is seriously affected by external interference, and the effective part of vibration signal has been submerged in the noise. Therefore, it is necessary to reduce the noise of original vibration signal in order to extract the characteristic information effectively.

In this section, traditional wavelet threshold denoising and improved wavelet threshold denoising are used to compare the denoising effect. For traditional

Fig. 2 Measured vibration signal waveform

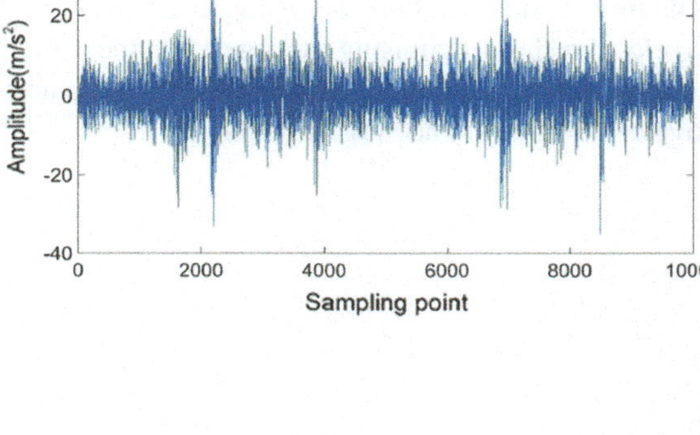

Fig. 3 Spectrogram of measured vibration signal

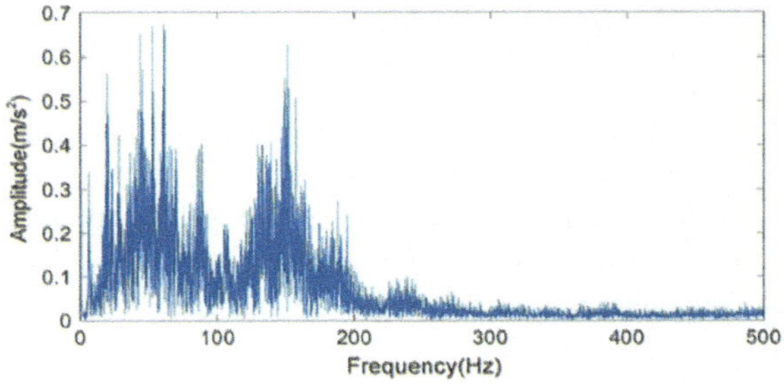

Fig. 4 Vibration signal processed by traditional wavelet threshold denoising

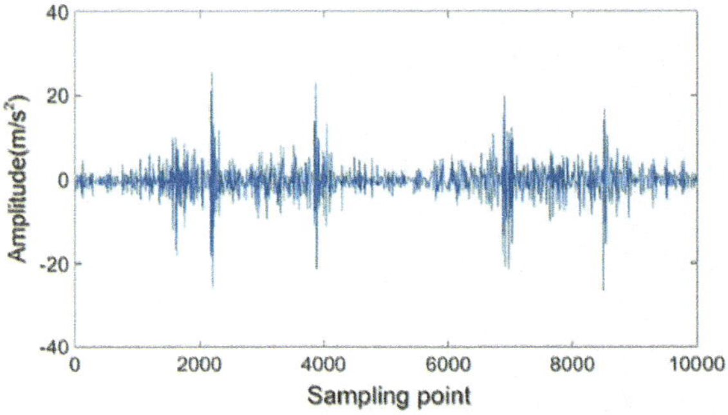

wavelet threshold denoising, the parameters are: the stein unbiased likelihood estimation threshold principle and soft threshold function, sym8 wavelet basis function, four-layer wavelet decomposition [10, 11]. For improved wavelet threshold denoising, the parameters are: the improved threshold selection principle and the improved threshold function, sym8 wavelet basis function, four-layer wavelet decomposition. After the two types of denoising processing, the vibration signal is shown in Figs. 4 and 5. Compared with Fig. 2, the noise of the original signal is obviously suppressed after denoising processing. In order to compare the effects of the two noise reduction methods, the spectrum of the processed signal is obtained by fast Fourier transform as shown in Figs. 6 and 7. The evaluation results of two denoising method are shown in Table 1.

By comparing Figs. 6 and 7, we can know that the improved wavelet threshold denoising has a more significant suppression effect in the high-frequency noise portion than the traditional wavelet threshold denoising. Table 1 presents the evaluation results of two denoising methods. For the value of SNR, improved wavelet threshold denoising is larger; for the value of RMSE, improved wavelet threshold denoising is smaller, indicating that the noise reduction effect of improved wavelet threshold denoising is better.

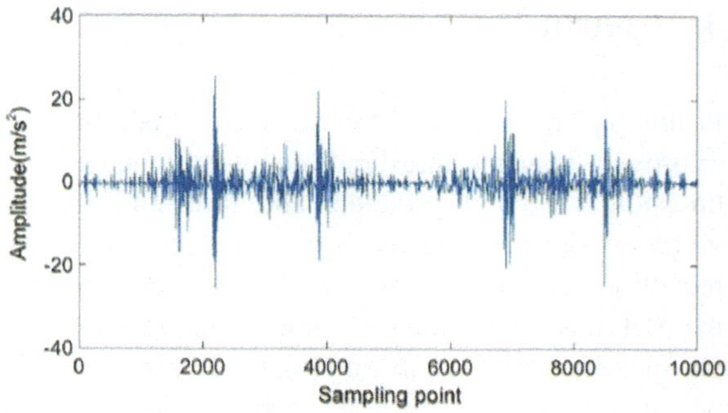

Fig. 5 Vibration signal processed by improved wavelet threshold denoising

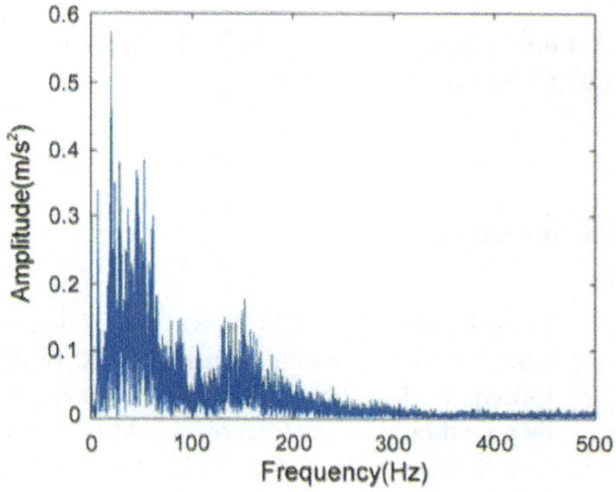

Fig. 6 Spectrogram of noise reduction signal for traditional wavelet threshold denoising

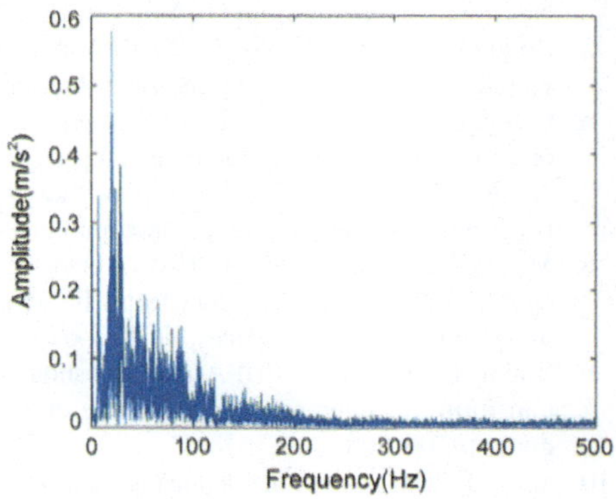

Fig. 7 Spectrogram of noise reduction signal for improved wavelet threshold denoising

Table 1 Evaluation results of two denoising methods

Evaluation index	Traditional wavelet threshold denoising	Improved wavelet threshold denoising
SNR	12.7266	13.1199
RMSE	0.3703	0.3539

4 Conclusion

In this paper, in view of the defects of traditional wavelet threshold denoising, an improved wavelet threshold denoising method is proposed by improving the threshold selection principle and threshold function. At the same time, using vibration signals collected from a certain line to compare the effects of two noise reduction methods. The results show that the improved wavelet threshold denoising method has more obvious noise suppression effect. Therefore, the improved wavelet threshold denoising method has good effectiveness and practicability, which ensures the effective extraction of the characteristic information in the vibration signal.

Acknowledgements This work is supported by National Key R&D Program of China (2016YFB1200402).

References

1. Hong L, Xinhong L (2015) Review of detecting methods of out of roundness of locomotive wheel tread. Automobile Parts 1:19–20 (in Chinese)
2. Lidong W, Kai Z, Liangrun W (2015) Improved wavelet threshold algorithm application in motor vibration signal denoising. Meas Control Technol Instrum 41(5):77–80 (in Chinese)
3. Yue S, Liyong W, Tao C (2015) Engine vibration signal denoising based on wavelet packet new threshold quantization. Mech Eng Autom 6:1–2 (in Chinese)
4. Xiangxin Z, Xiaomin W, Yang Y (2014) De-noising of high-speed turnout vibration signals based on wavelet threshold. J Vibr Shock 33(23):200–206 (in Chinese)
5. Bingren W, Yanxia Y, Wei C (2008) Application of wavelet threshold de-noising technique in vibration signals processing. Noise Vibr Control 28(6):9–12 (in Chinese)
6. Guangjun L, Jing Z, Ci L (2016) Signal process technology of train vibration. In: Proceedings of 35th Chinese control conference, pp 10187–10192
7. Amin S, Eni DW (2017) Analysis of the vibration signal detection for rail train arrival. In: IEEE international conference on communication, networks and satellite, pp 85–87
8. Maleki YK, Khajavi MN (2016) Over hang slant cracked rotor vibration signal processing based on discrete wavelet transform. In: Proceedings of 13th international multi-conference on system, signals & devices, pp 672–676
9. Yating L, Jianli C (2010) A new threshold function for signal denoising based on wavelet transform. In: International conference on measuring technology and mechatronics automation, pp 200–203
10. Ying Q (2018) A wavelet denoising method based on improved threshold and autocorrelation. In: Proceedings of 30th Chinese control and decision conference, pp 4058–4063
11. Huimin C, Ruimei Z (2012) Improved threshold denoising method based on wavelet transform. Phys Procedia 33(1):1354–1359

Identifying Critical Component Set of High-Speed Train System Based on Topological Integrated Importance Analysis

Huiru Zhang, Limin Jia, Li Wang and Mingming Wang

Abstract The aim of this study is to introduce a critical component set identification method based on topological integrated importance and to illustrate the application of the proposed measure to high-speed train system. Firstly, a multi-layer-multi-granularity network model is proposed including virtual and actual nodes in view of equipment and other closely related factors, such as the environment, human and management. In addition, for convenience of calculation, the fusion rules of the links of different layers are designed, which merge different connections of the same node. Then, a topological integrated importance ranking algorithm by combining complex networks with actual function and risk characteristics is presented to identify critical components. Finally, a typical high-speed train system is provided for illustration of method purposes, and the critical component set is obtained.

Keywords High-speed train system · Network model · Topological integrated importance

1 Introduction

The high-speed train is an extremely complex system consisting of a large number of components with different structures and functions [1, 2]. Not only the train itself but other closely related factors, such as the environment, human and management,

H. Zhang · L. Jia · L. Wang (✉) · M. Wang
State Key Laboratory of Rail Traffic Control and Safety, Beijing Jiaotong University, Beijing 100044, China
e-mail: wangli@bjtu.edu.cn

H. Zhang · L. Wang · M. Wang
School of Traffic and Transportation, Beijing Jiaotong University, Beijing, China

L. Jia
Beijing Engineering Research Center of Urban Traffic Information Intelligent Sensing and Service Technologies, Beijing 100044, China

Y. Qin et al. (eds.), *Proceedings of the 4th International Conference on Electrical and Information Technologies for Rail Transportation (EITRT) 2019*, Lecture Notes in Electrical Engineering 639, https://doi.org/10.1007/978-981-15-2866-8_11

are significant in the process of train operation as well [3]. Therefore, the high-speed train system is a broad concept in this study, which mainly refers to equipment, but also includes other relevant factors. The identification of system-level critical components facilitates safe, reliable and high-speed operation of high-speed train system.

Reliability importance indices are widely used in existing research to identify critical components. Birnbaum importance is one of the most widely used importance measures in reliability theory, and it is a sensitivity analysis in component reliability [4]. Criticality Importance (CI) is used to measure the impact of critical component failures on system failures and is primarily applied to identify critical components in the fault tree that cause top events to occur [5].

Most research on complex network involves two main steps: the representation of the problem as a complex network and then the analysis of the topological features [6]. For the distributed and complex electromechanical system, a penetrable visibility graph method is generated, which combines the phase space reconstruction [7].

As the size and complexity of the high-speed train system increase, it makes the system-level critical components identification become infeasible since large numbers of reliability tests and statistical data are needed. However, the methods based on basic complex networks focus only on the topological features of the network and are limited from the point of view of the physical analysis [8].

This paper introduces a novel importance measure, the so-called topological integrated importance, which is a meaningful extension of the identification of critical component in complex network. The rest of this paper is structured as follows: Sect. 2 describes a multi-layer-multi-granularity network model and details the rule of fusion. Section 3 explains the topological integrated importance ranking algorithm. A case study is used to verify the effectiveness of the network model and importance ranking algorithm in Sect. 4. Finally, conclusions are drawn in Sect. 5.

2 Methodology

2.1 *The Network Model*

The high-speed train system is divided into four systems: equipment system, environment system, human system and management system. Equipment system is a complex mechatronic system, and relationships can be presented in many ways. Some researchers describe the relationships as material, energy and information dependency [9]. The multi-layer-multi-granularity network $S = \{S^x | x \in \{1,2,3,4\}\}$ is built, where S is the set of entire system, and x represents the above four systems in turn. Equations (1) and (2) are network expression of the equipment system and other three systems separately.

$$\begin{cases} S^1 = \{s_1^1, s_2^1, \ldots, s_i^1, \ldots\} \\ s_i^1 = \left\{G_{i,u}^1 | u \in \{\alpha, \beta, \gamma\}\right\} = \left\{V_i^1, E_{i,u}^1 | u \in \{\alpha, \beta, \gamma\}\right\} \end{cases} \tag{1}$$

$$\begin{cases} S^x = \{s_1^x, s_2^x, \ldots, s_i^x, \ldots\} \\ s_i^x = \{V_i^x, E_i^x\} \end{cases}, \quad x \in \{2, 3, 4\} \tag{2}$$

where s_i^x is the set of subsystem i of system x; $G_{i,u}^1$ is the different layer of subsystem i of equipment system; V_i^x is the set of nodes; E_i^x and $E_{i,u}^1$ are the sets of links.

$$\begin{cases} V_i^x = \left\{v_i^{\text{vir},x}, V_i^{\text{real},x}\right\} \\ V_i^{\text{real},x} = \left\{v_{i,1}^{\text{real},x}, v_{i,2}^{\text{real},x}, \ldots\right\} \end{cases}, \quad x \in \{1, 2, 3, 4\} \tag{3}$$

where $v_i^{\text{vir},x}$ is the virtual node of subsystem i of system x; $V_i^{\text{real},x}$ is the set of real node, and $v_{i,s}^{\text{real},x}$ represents the real node s. Note that nodes in different layers of the same subsystem are identical.

$$\begin{cases} E_i^x = \left\{E_i^{\text{affi},x}, E_i^{\text{act},x}\right\}, \quad x \in \{1, 2, 3, 4\} \\ E_i^{\text{affi},x} = \left\{e_{v_i^{\text{vir},x}, v_{i,1}^{\text{real},x}}, \ldots, e_{v_i^{\text{vir},x}, v_{i,s}^{\text{real},x}}\right\}, \quad x \in \{1, 2, 3, 4\} \\ E_i^{\text{act},x} = \left\{e_{v_{i,1}^{\text{real},x}, v_{i,2}^{\text{real},x}}, \ldots, e_{v_{i,s}^{\text{real},x}, v_{i,t}^{\text{real},x}}\right\}, \quad x \in \{2, 3, 4\} \end{cases} \tag{4}$$

where $E_i^{\text{affi},x}$ and $E_i^{\text{act},x}$ are the sets of links representing affiliation and action relationship, respectively; $e_{v_i^{\text{vir},x}, v_{i,s}^{\text{real},x}}$ and $e_{v_{i,s}^{\text{real},x}, v_{i,t}^{\text{real},x}}$ are links between two nodes.

$$\begin{cases} E_i^{\text{act},1} = \left\{E_{i,\alpha}^{\text{act},1}, E_{i,\beta}^{\text{act},1}, E_{i,\gamma}^{\text{act},1}\right\} \\ E_{i,u}^{\text{act},1} = \left\{e_{v_{i,u,1}^{\text{real},1}, v_{i,u,2}^{\text{real},1}}, e_{v_{i,u,1}^{\text{real},1}, v_{i,u,3}^{\text{real},1}}, \ldots, e_{v_{i,u,s}^{real,1}, v_{i,u,t}^{real,1}} | u \in \{\alpha, \beta, \gamma\}\right\} \end{cases} \tag{5}$$

2.2 Fusion Rules of Links of Different Layers

The multi-layer complex network abstracted from the real system needs to be further simplified to calculate network features, and simplification by fusion rules is an effective method [10].

(1) If there are multiple connection relationships among three nodes (node s, o and t), the links can be fused into a same layer through the same node s (see Fig. 1a).
(2) If there are multiple connection relationships between two nodes (node s and t), the two links can be fused into one (see Fig. 1b).

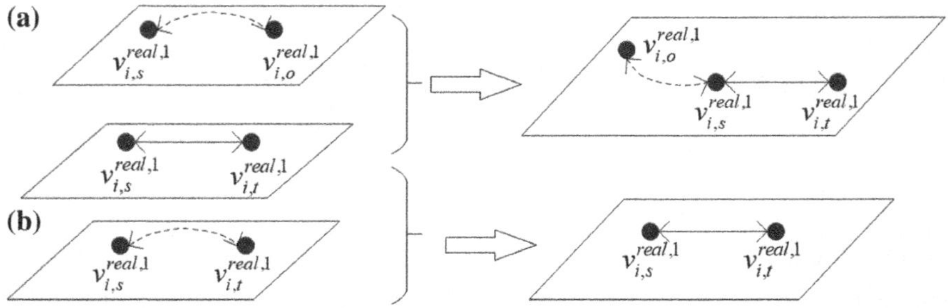

Fig. 1 Fusion rules of links

3 Algorithm

3.1 Selection and Calculation of Indices

Indices of Function Characteristics. The high-speed train system is such a complex system that each component has specific function. The most important components of this study are components that are closely related to train operation. Thus, the function indicator K_s^{func} is proposed based on the topology structure.

$$K_s^{\text{func}} = F\left(I_{deg}^{\text{func}}(s), I_{\text{close}}^{\text{func}}(s), I_{\text{ne}}^{\text{func}}(s), I_{\text{betw}}^{\text{func}}(s)\right)$$

$$\begin{cases} I_{deg}^{\text{func}}(s) = \omega_s * \sum_{t=1}^{m} a_{st} \\ I_{\text{close}}^{\text{func}}(s) = \omega_s * (m-1)/\sum_{t=1}^{m} d_{st} \\ I_{\text{ne}}^{\text{func}}(s) = \omega_s * 1/m(m-1)\sum_{s\neq t} 1/d_{st} \\ I_{\text{betw}}^{\text{func}}(s) = \omega_s * \sum_{a\neq s\neq b\in V} \sigma_{ab}(s)/\sigma_{ab} \end{cases} \quad (6)$$

where a_{st} is the value of the adjacency matrix; m is the total number of nodes; d_{st} is the shortest number of links between node s and t, and the value is equal to the longest path if two nodes are disconnected; $\sigma_{ab}(s)$ is the number of the shortest paths between node a and node b via node s; σ_{ab} is the number of the shortest paths between two nodes; ω_s represents the function importance of node s in complex network calculated commonly through expert evaluating method (see Table 1). On the basis of practical experience, experts rank the importance of each node, with

Table 1 Function importance standard

Score	[0, 0.3]	[0.4, 0.7]	[0.8, 1]
Standard	Less	Important	Very

scores ranging from 0.1 to 1. The larger the value of K_s^{func} is, the more important the function of the component corresponding to the node is in the process of train operation.

Indices of Risk Characteristics. Each component could be in a failure state during the train operation, while the possibility and impact of the failure may be quite different. Hence, proposing an indicator K_i^{risk} descripting the risk characteristic is significant.

$$K_s^{\text{risk}} = F\left(I_{deg}^{\text{risk}}(s), I_{\text{close}}^{\text{risk}}(s), I_{\text{ne}}^{\text{risk}}(s), I_{\text{betw}}^{\text{risk}}(s)\right)$$
$$\begin{cases} I_{\text{deg}}^{\text{risk}}(s) = p_s * l_s * \sum\limits_{t=1}^{m} a_{st} \\ I_{\text{close}}^{\text{risk}}(s) = (m-1) / \sum\limits_{t=1}^{m} d_{st}^{-} \\ I_{\text{ne}}^{\text{risk}}(s) = 1/m(m-1) \sum\limits_{s \neq t} 1/d_{st}^{-} \\ I_{\text{betw}}^{\text{risk}}(s) = \sum\limits_{a \neq s \neq b \in V} \sigma_{ab}^{-}(s)/\sigma_{ab}^{-} \end{cases} \tag{7}$$

where p_s is the coefficient representing the failure frequency obtained from the statistical data (see Table 2); l_s is the severity degree of impact to train operation when the node is in failure, which is got from the historical data and ranges from 0 to 100; q_{sa} is the impact on node a after node s fails; $d_{st}^{-} = p_s l_s q_{sa} + p_a l_a q_{ab} + \cdots + p_c l_c q_{ct}$ is the risk shortest path between two nodes. The larger the value of K_s^{risk}, the more severe the impact on train operation when the component corresponding to the node fails.

3.2 Integrated Importance Measure

Indices measuring the function and risk characteristics of the complex network are selected and calculated. Next, an integrated importance ranking algorithm will be introduced to measure the topological integrated importance.

The normative decision matrix is calculated as follows.

$$z_{sx} = I_{sx} / \sqrt{\sum\nolimits_{s=1}^{m} I_{sx}^2} \tag{8}$$

$$y_{sx} = \mu_x * z_{sx} \tag{9}$$

Table 2 Failure frequency grading standard

Score	1	2	3
Frequency	High	Medium	Low

where I_{sx} is the xth index of node s; z_{sx} is the normative decision matrix; μ_x is the weight of index x, and y_{sx} is the weighted normative matrix. The nodes are then ranked based on the proximity of the indices.

$$\begin{cases} y_x^* = \left(\max(I_{deg}), \max(I_{\text{close}}), \min(I_{\text{ne}}), \max(I_{\text{betw}})\right) \\ y_x^0 = \left(\min(I_{deg}), \min(I_{\text{close}}), \max(I_{\text{ne}}), \min(I_{\text{betw}})\right) \end{cases} \tag{10}$$

$$\begin{cases} d_s^* = \sqrt{\sum_{x=1}^{n} \left(y_{sx} - y_x^*\right)^2} \\ d_s^0 = \sqrt{\sum_{x=1}^{n} \left(y_{sx} - y_x^0\right)^2} \end{cases} \tag{11}$$

$$C_s^* = d_s^0 / \left(d_s^* + d_s^0\right) \tag{12}$$

where y_x^* is the optimal solution; y_x^0 is the worst solution; d_s^* and d_s^0 are the distance between index s and two solutions separately; C_s^* is the proximity of the index s. The larger the value of C_s^*, the more important a given node s is in the complex network.

4 Case Study

4.1 Data and Parameters

Considering that many components are so small that they are not suitable to be studied as separate nodes with specific function in this paper and some components have little to do with the safety of train operation, the nodes selected in the network do not correspond exactly to the real world. Finally, the multi-layer-multi-granularity network consists of 172 nodes, of which eight are virtual nodes and 164 are actual nodes (see Table 3).

Table 3 Virtual and actual nodes of the whole network (only partial actual nodes are listed)

Type	Label	Component	Label	Component
Virtual nodes	PANTOGRAPH	Pantograph subsystem	SIGNAL	Signal subsystem
	TRACTION	Traction subsystem	INFRASTRU	Infrastructure
	AIRBRAKE	Air and brake subsystem	DISPATCH	Dispatch subsystem
	BOGIE	Bogie subsystem	PASERVICE	Passenger service
Actual nodes	1	Carbody	7	Dropper
	3	Pillar	……	……
	4	Insulator	……	……
	5	Location device	……	……
	6	Messenger wire	172	Earthquake

Note that, virtual and actual nodes are encoded together, so the label of actual nodes is not continuous, and a missing number represents a virtual node.

4.2 Results

The fused network of the high-speed train system is shown in Fig. 2, with 172 nodes and 491 links. The node is colored according to its outdegree, which reflects the total amount of activities of the node in this network. The outdegree of virtual nodes is relatively large compared with other actual nodes because virtual nodes have affiliation connections with other actual nodes.

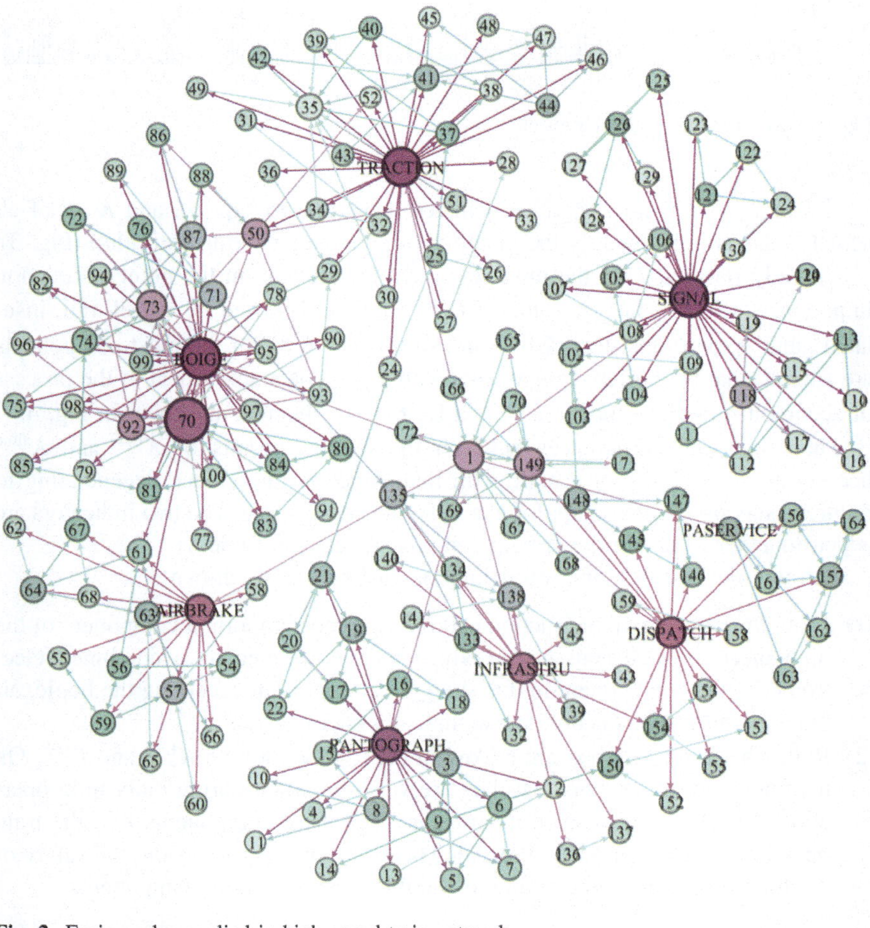

Fig. 2 Fusion rules applied in high-speed train network

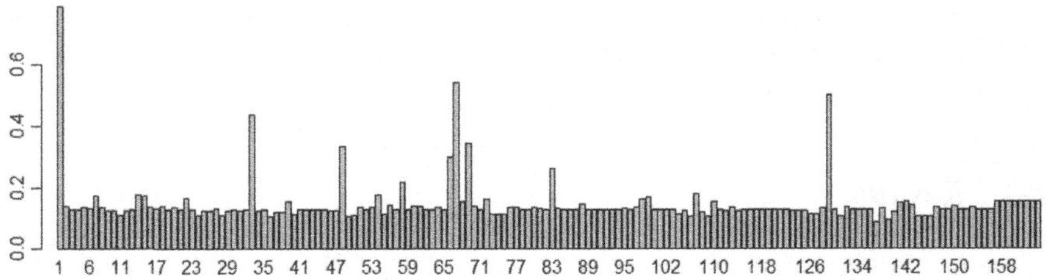

Fig. 3 Function importance ranking results

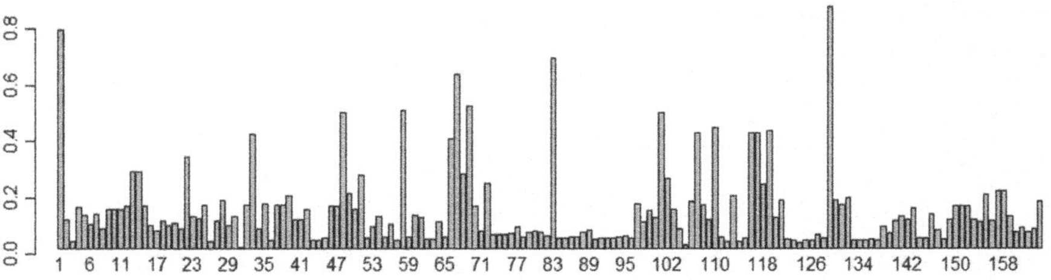

Fig. 4 Risk importance ranking results

$I_{\circ}^{\text{func}}$, $I_{\text{close}}^{\text{func}}$, $I_{\text{ne}}^{\text{func}}$ and $I_{\text{betw}}^{\text{func}}$ are calculated according to Eq. (6), and K^{func} of all actual nodes are obtained by the integrated importance ranking algorithm (Fig. 3).

Carbody (labeled 1) is the most important component on the point of function importance, with the largest value of K_1^{func} (0.778), followed by wheel, rail, insulated gate bipolar transistor (IGBT) and so on. The K^{func} of the most components are similar, that is, the few components with high scores are critical in the process of accomplishing train function. Similarly, K^{risk} is obtained as shown in Fig. 4.

There are some differences between the ranking results of K^{risk} and K^{func} because the probability of a node with high function importance degenerating to failure is supposed to be small, for example, the bogie frame. The two indicators are added together to obtain the critical component set (see Table 4).

Through the research, the following conclusions can be drawn:

(1) From the perspective of function and risk, the most important component of the high-speed train system is carbody, which is defined as a combined electromechanical component in this study, i.e., a rail train excepting the bogie, air brake, pantograph, traction and signal subsystem.
(2) Rail, wheel and axle box are ranked in the top five in both K^{func} and K^{risk}. On the one hand, this means that they are very important during daily train operation. On the other hand, it means that once these components fail, the train operation will be seriously affected. Therefore, in order to ensure the safety of the train operation, more attention should be paid to these components.

Table 4 Nodes in top 20 of the integrated importance

Order	Name	K^{func}	K^{risk}	K	Order	Name	K^{func}	K^{risk}	K
1	Carbody	0.78	0.78	1.55	11	BTM	0.18	0.41	0.59
2	Rail	0.52	0.86	1.37	12	Speed and distance measurement device	0.16	0.43	0.58
3	Wheel	0.56	0.62	1.18	13	Lock device	0.13	0.42	0.55
4	Gearbox	0.27	0.67	0.94	14	Switch rail	0.13	0.41	0.54
5	Axle box	0.35	0.5	0.85	15	Moveable point rail	0.13	0.41	0.54
6	IGBT	0.44	0.41	0.85	16	Main breaker	0.17	0.33	0.49
7	Motor bearing	0.34	0.48	0.82	17	Carbon slide	0.18	0.27	0.45
8	Brake cylinder	0.22	0.49	0.71	18	Axle box bearing	0.16	0.26	0.42
9	Bogie frame	0.3	0.39	0.69	19	Neutral section	0.13	0.27	0.4
10	Matching transformer	0.13	0.48	0.61	20	Main air compressor	0.14	0.26	0.4

(3) In Table 4, the top-ranked nodes are the critical components that mainly affect the topology structure, train operation and failure consequence, that is, the set of critical components we hope to obtain.

5 Conclusion

This study aims to present a novel critical component set identification method applying to high-speed train system. A multi-layer-multi-granularity network model is proposed to describe the system, and some fusion rules of the links are designed, too. Based on the indices of the network characteristics, such as degree, closeness centrality, efficiency and betweenness, the topological integrated importance ranking algorithm is proposed both considering function and risk factors. A typical high-speed train system is taken as an example, and the results are reasonable because the characteristics of the complex network in different dimensions are considered. The critical component set obtained allows the decision-makers to improve the reliability during the design stage and to allocate maintenance resources more reasonably during the operational phase.

Acknowledgements This study is funded by the National Key Research and Development Program of China (2016YFB1200401) and the National Natural Science Foundation of China (71701010).

References

1. He H (2007) The Innovative Chinese High-speed Railway Technology. Eng Sci 9(9):4–18 (in Chinses)
2. He H (2007) The Innovative Chinese high-speed railway technology. Eng Sci 9(10):4–18 (in Chinses)
3. Wang M, Wang L, Xu X, Qin Y, Qin L (2019) Genetic algorithm-based particle swarm optimization approach to reschedule high-speed railway timetables: a case study in China. J Adv Transp 13–16
4. Wang W, Loman J, Vassiliou P (2004) Reliability importance of components in a complex system. In: Annual symposium reliability and maintainability, pp 6–11
5. Espiritu J, Coit D, Prakash U (2007) Component criticality importance measures for the power industry. Electr Power Syst Res 77(5–6):407–420
6. Lin S, Wang Y, Jia L, Zhang H, Li Y (2018) Intuitionistic mechanism for weak components identification method of complex electromechanical system. J Intell Fuzzy Syst 34(1):583–598
7. Wang R, Gao J, Gao Z, Gao X, Jiang H (2016) Complex network theory-based condition recognition of electromechanical system in process industry. Sci China Technol 59(4):604–617
8. Zio E, Golea L (2012) Analyzing the topological, electrical and reliability characteristics of a power transmission system for identifying its critical elements. Reliab Eng Syst Saf 101:67–74
9. Wang Y, Bi L, Lin S, Li M, Shi H (2017) A complex network-based importance measure for mechatronics systems. Phys A Stat Mech Appl 466:180–198
10. Lin S (2018) Method of system reliability and safety evaluation for high-speed train based on topology. Beijing Jiaotong University, Beijing, pp 25–40 (in Chinses)

Bayesian Network-Based Reliability Analysis of Traction Transmission System

Chaoyue Dong, Lei Wang, Ming Li, Lei Zhang and Yanqin Wang

Abstract Traction transmission system is one of the important components of the electric multiple unit (EMU), and its malfunction could lead to the train late even abnormal parking. Reliability analysis can guide us to scientifically formulate the maintenance cycle, select the maintenance content and method of the electric multiple unit and reasonably determine the service life of the unit components so that we can timely replace and repair them. Through reliability analysis, the safe and reliable operation of the EMU can be ensured. Therefore, it is extremely important to study the reliability of the traction drive system. This paper takes the CRH3 multiple unit as an example and builds Bayesian networks of traction transmission system. Using the results of the Bayesian network to get reliability index of the system, identify the weak links and do reliability allocation to the system. The results show that Bayesian network is a convenient and reliable method to evaluate the reliability of traction drive system.

Keywords BN · Traction transmission system · Reliability analysis · Fault tree

C. Dong · L. Wang (✉)
School of Electrical Engineering, Beijing Engineering Research Center of Electric Rail Transportation, Beijing Jiaotong University, Beijing 100044, China
e-mail: leiwang@bjtu.edu.cn

C. Dong
e-mail: 17121422@bjtu.edu.cn

M. Li · L. Zhang · Y. Wang
CRRC Tangshan Locomotive & Rolling Stock Co. LTD, Tangshan, China
e-mail: sjc-liming.ts@crrcgc.cc

L. Zhang
e-mail: sjc-zhanglei.ts@crrcgc.cc

Y. Wang
e-mail: wangyanqin@crrcgc.cc

Y. Qin et al. (eds.), *Proceedings of the 4th International Conference on Electrical and Information Technologies for Rail Transportation (EITRT) 2019*, Lecture Notes in Electrical Engineering 639, https://doi.org/10.1007/978-981-15-2866-8_12

1 Introduction

China has independently developed many different types of electric multiple unit (EMU), but no matter which type of EMU cannot do without traction drive system. The reliability of traction drive system directly affects the operation of the entire train, so it is necessary to analyze the reliability of the traction drive system. This paper takes the traction drive system of CRH3 multiple unit as an example and uses Bayesian network to analyze its reliability.

The traction transmission system of CRH3 multiple unit is divided into two mutually symmetrical traction units [1]. The main circuit of each traction unit is mainly composed of high-voltage electrical equipment, traction transformer (MT), two traction converters (TC), eight traction motors and eight sets of traction transmission devices [2]. The two traction units of the CRH3 multiple unit can isolate the faulty unit when a traction unit fails and working with 50% traction. Similarly, when one of the four traction converters fails, it can also run with 75% traction [1, 2]. In this paper, when the reliability modeling of traction drive system is carried out, the situation that traction force is damaged will be classified into the fault situation.

The Bayesian network (BN) consists of a directed acyclic graph (DAG) and conditional probability tables (CPT) [3]. A BN with known DAG and CPT can perform bidirectional reasoning calculation. Top-down reasoning, also called causal reasoning, refers to the process in which the probability of leaf node is obtained by hierarchical downward reasoning under the condition of known prior probability of root nodes. Bottom-up reasoning, also called diagnostic reasoning, refers to the process in which the posterior probability of each root node is finally obtained by hierarchical upward reasoning under the condition of known leaf nodes (known results) [4, 5].

Compared with other reliability evaluation methods such as fault tree evaluation, BN has the following advantages:

1. It can express the causal relationship between faults more easily and clearly
2. It can describe the polymorphism of an event
3. It has the ability to describe non-deterministic logical relationships between events
4. It can achieve bidirectional reasoning to calculate more available indicators
5. It can be extended to dynamic Bayesian networks.

With a specific transformation rule shown in Fig. 1 (state 0 represents fault, and state 1 represents work), Bayesian networks can be obtained by transforming fault trees [6, 7]. This paper uses Bayesian analysis software GeNIE to construct and analyze the Bayesian network.

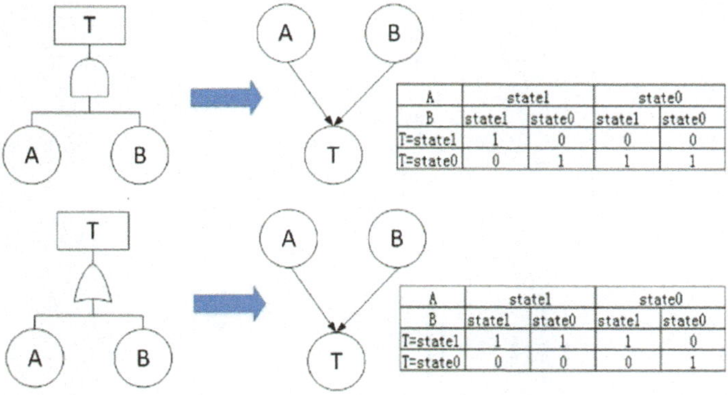

Fig. 1 Logic gate transformation rules

2 Modeling of Traction Drive System Based on BN

The components of the traction drive system, such as traction converters and traction transformers, are composed of more basic components. If these basic components are placed in the overall BN, then the BN will be very huge, so this paper constructs a Bayesian network with a modular structure.

The auxiliary system is an indispensable part of the EMU. The cooling device and control unit in the traction drive system are the loads of the auxiliary system, and the auxiliary system is direct powered by the intermediate DC of TC [1, 8]. So, there is coupling between the traction drive system and the auxiliary system, and it is not easy to decouple. So, the influence of the auxiliary system on the traction drive system should be considered when modeling the traction drive system.

The modeling process can be divided into the following steps

1. The basic components of the traction drive system are divided into modules according to their functions in the system. For example, the transformer current transformer, the traction transformer and the traction transformer cooling system are divided into one module, which is named as traction transformer.
2. Build a BN for each module and get its failure rate.
3. Analyze the composition structure and main circuit of the traction drive system and construct the reliability block diagram.
4. Obtain fault tree of traction transmission system by reliability block diagram.
5. Construct BN of the system from the fault tree, the events represented by the root node of the obtained BN are those modules divided in the first step.

From the above process, we can get the fault tree model and BN of the traction drive system as shown in Figs. 2 and 3. We believe that when four or more of the six single auxiliary converters of the auxiliary system are in an inoperable state,

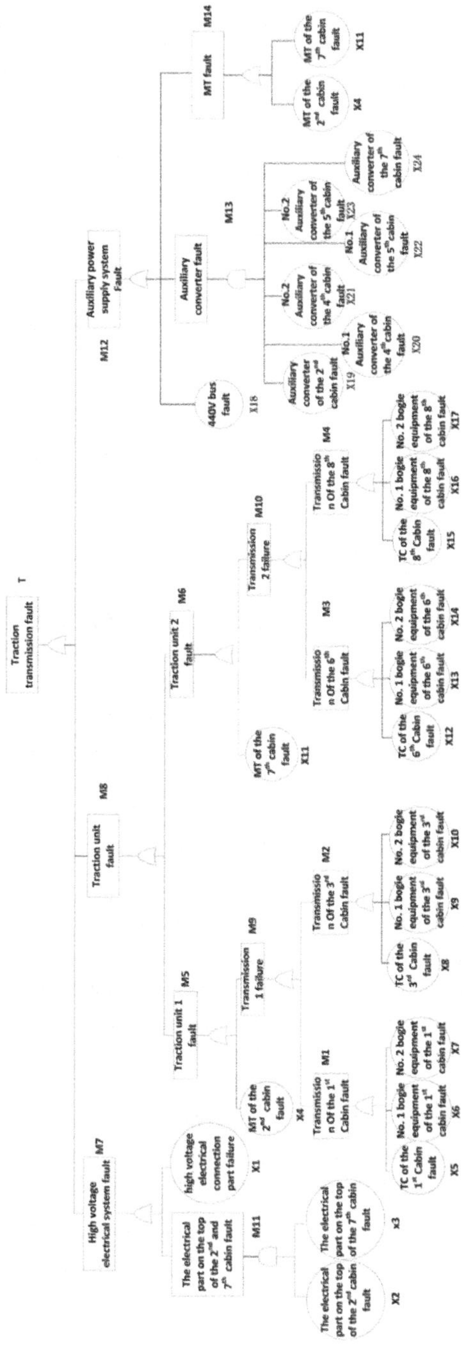

Fig. 2 Fault tree model of traction drive system

M20	M14	M15	M16	M17	M18	M19
1	1	1	1	1	1	1
1	1	1	1	1	1	0
1	1	1	1	1	0	0
1	1	1	1	0	0	0
...				...		
0	1	1	0	0	0	0
0	1	0	0	0	0	0
...				...		
0	0	0	0	0	0	0

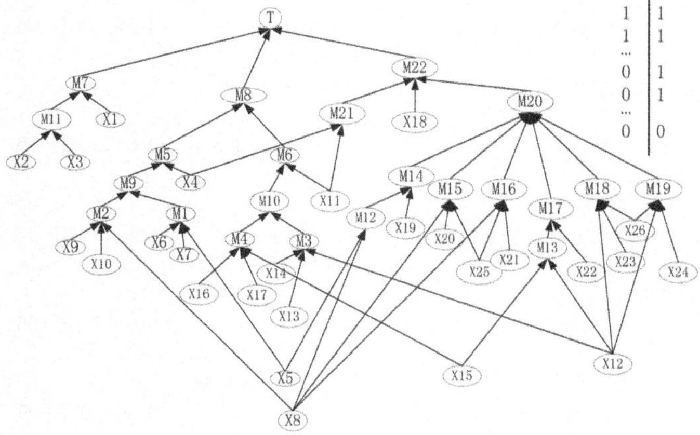

Fig. 3 BN of traction drive system

the traction drive system will be affected. Therefore, the logical relationship between the node M20 and the nodes M14–M19 is non-deterministic, which is difficult to represent with the fault tree, but the CPT of the BN can perfectly describe this logical relationship. The specific representation is shown in the table in Fig. 3. (Table 1 represents the node is in a normal state, and 0 represents a fault state).

The prior probability of the root node in the Bayesian network of the traction drive system is shown in Table 1.

3 Reliability Analysis

3.1 The Diagnostic Reasoning of Traction Drive System

In the BN that we built, the traction drive system is set to fault. Through the diagnosis reasoning, we can get the inference results as shown in Figs. 4 and 5.

It can be seen from Fig. 4 that among the three main components of the traction drive system (the high-voltage electrical system, the traction unit and the auxiliary system), the traction unit has the largest posterior probability value. This is because when the traction drive system is in working state, the components in the traction unit have no redundancy, and all components are connected in series, so the failure rate is high, while the components in the other two systems have redundancy, and the components are connected in parallel, so the failure rate is much lower than the rate of traction unit.

Table 1 Prior probability of root node

Node of BN	Events in the fault tree	Failure rate (times per km)	Node of BN	Events in the fault tree	Failure rate (times per km)
X1	High-voltage electrical connection part failure	5.984032E−07	X14	No. 2 bogie equipment of the sixth cabin fault	1.258280E−06
X2	The electrical part on the top of the second cabin fault	4.263092E−06	X15	TC of the eighth Cabin fault	1.632364E−06
X3	The electrical part on the top of the seventh cabin fault	3.283264E−06	X16	No. 1 bogie equipment of the eighth cabin fault	9.594384E−07
X4	MT of the second cabin fault	4.041788E−07	X17	No. 2 bogie equipment of the eighth cabin fault	1.017230E−06
X5	TC of the first Cabin fault	1.338941E−06	X18	440 V bus fault	7.396001E−07
X6	No. 1 bogie equipment of the first cabin fault	1.100995E−06	X19	Auxiliary converter of the second cabin fault	6.664405E−07
X7	No. 2 bogie equipment of the first cabin fault	1.022353E−06	X20	No. 1 Auxiliary converter of the fourth cabin fault	8.079972E−07
X8	TC of the third Cabin fault	1.600551E−06	X21	No. 2 Auxiliary converter of the fourth cabin fault	9.389479E−07
X9	No. 1 bogie equipment of the third cabin fault	1.080144E−06	X22	Auxiliary converter of the seventh cabin fault	7.502057E−07
X10	No. 2 bogie equipment of the third cabin fault	8.127593E−07	X23	No. 1 Auxiliary converter of the fifth cabin fault	8.496996E−07
X11	MT of the seventh cabin fault	6.715643E−07	X24	No. 2 Auxiliary converter of the fifth cabin fault	8.025141E−07
X12	TC of the sixth cabin fault	1.925714E−06	X25	Jumper cable between the third and fourth cabin fault	1.060580E−08
X13	No. 1 bogie equipment of the sixth cabin fault	1.100474E−06	X26	Jumper cable between the fifth and sixth cabin fault	1.060580E−08

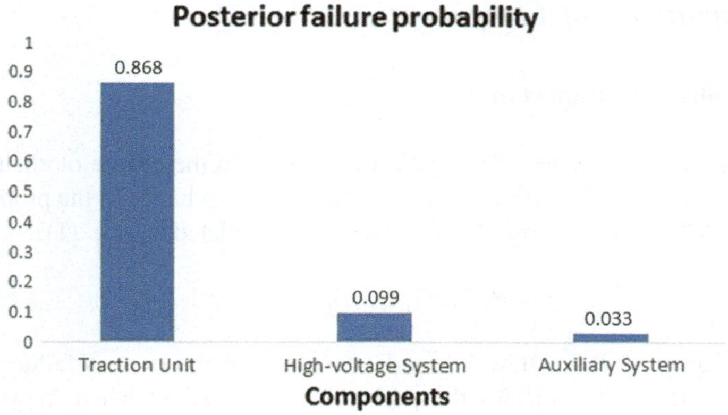

Fig. 4 First diagnosis reasoning result of traction drive system

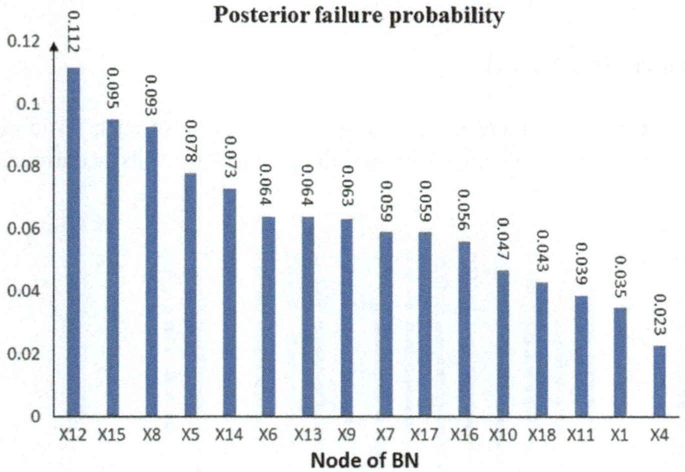

Fig. 5 Second diagnosis reasoning result of traction drive system

The results in Fig. 5 represent that the weakest link in the traction unit is the traction converter followed by the bogie part (including the traction motor). Therefore, the daily maintenance and protection of the traction converter and bogie should be strengthened to improve the reliability of the traction drive system.

3.2 Importance of Components

3.2.1 Probability Importance

The probability importance of the basic event refers to the degree of change in the probability of occurrence of the top event caused by the change in the probability of occurrence of the basic event. Expressed in I^{pr}, calculated by Eq. (1)

$$I_{T=1}^{pr}(X_i) = P(T = 1|X_i = 1) - P(T = 1|X_i = 0) \quad (1)$$

In the equation, $P(T = 1|X_i = 1)$ indicates the probability of T failure when X_i fails, $P(T = 1|X_i = 0)$ indicates the probability of T failure when X_i works, and $I_{T=1}^{pr}(X_i)$ indicates the probability importance of X_i, when top event T is out of order.

The critical importance of each component of the traction drive system is calculated as shown in Fig. 6.

3.2.2 Critical Importance

The critical importance measures the importance criteria of each basic event from the double perspectives of sensitivity and the probability of its occurrence, and the

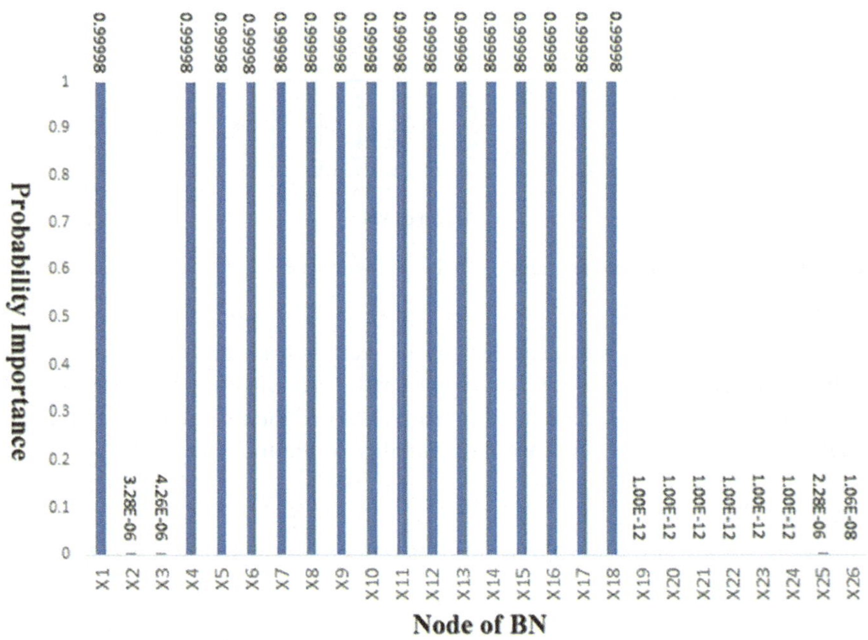

Fig. 6 Probability importance of each component

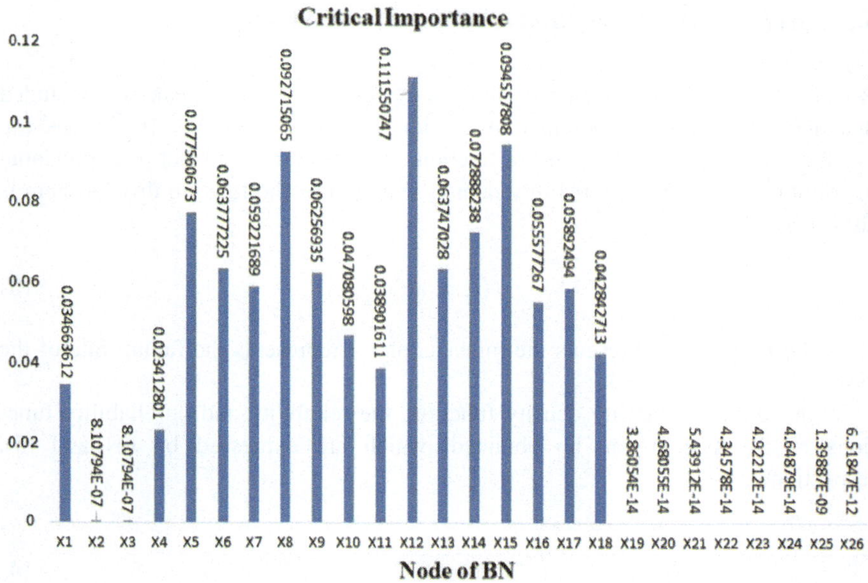

Fig. 7 Critical importance of each component

difference between it and the probability importance is that the degree of difficulty of the event's own probability changes is considered. It can be obtained by Eq. (2). The critical importance of each component of the traction drive system is calculated as shown in Fig. 7.

$$I_{T=1}^{\mathrm{Cr}}(X_i) = \frac{p(X_i = 1)}{p(T = 1)} * I_{T=1}^{\mathrm{Pr}}(X_i) \tag{2}$$

In (2), $I_{T=1}^{\mathrm{Cr}}(X_i)$ represents the critical importance of X_i, $P(X_i = 1)$ represents the failure rate of X_i, $P(T = 1)$ represents the failure rate of T, and $I_{T=1}^{\mathrm{pr}}(X_i)$ represents the probability importance of X_i, when top event T is out of order.

According to the analysis results as shown in Fig. 6, it is known that the components that have the greatest influence on the failure probability of the system from the perspective of the importance of the basic components are those without redundant design. And the results of Fig. 7 are that the basic component of those components that is the easiest to reduce the failure rate and can greatly improve the reliability of the system is the TC module. So, in order to improve the reliability of traction drive system, we should focus on improving the reliability of traction converter module.

3.3 Reliability of Traction Transmission System

Based on causal inference of the obtained Bayesian network, it can be concluded that the failure rate of traction transmission system is 1.6539×10^{-5} times/km, assuming that the reliability of the traction drive system is subject to exponential distribution, the failure probability density function of the traction drive system is shown by (3):

$$f(t) = \lambda e^{-\lambda t} \tag{3}$$

In the formula, t represents the mileage, and λ represents the failure rate of the system.

With failure probability density function, the reliability and unreliability functions of the system can be obtained, which are expressed by (4) and (5), respectively.

$$R(t) = 1 - \int_0^t f(t)\mathrm{d}t = e^{-\lambda t} = e^{-1.8338591 \times 10^{-5} t} \tag{4}$$

$$F(t) = 1 - R(t) = 1 - e^{-1.8338591 \times 10^{-5} t} \tag{5}$$

With the failure probability density function, the mean time between failures (MTBF) can also be got, which means the average operating mileage between two adjacent faults of a traction drive system.

$$\mathrm{MTBF} = \int_0^\infty tf(t)\mathrm{d}t = \int_0^\infty t\lambda e^{-\lambda t}\mathrm{d}t = \frac{1}{\lambda} \tag{6}$$

In (6), t represents the mileage, and λ represents the failure rate of the system.

From the equation, it can be known that the average fault interval of the traction drive system is 60,463 km.

The reliability curve of the traction drive system obtained by (4) is shown in line1 in Fig. 8. The line2, line3 and line4 in Fig. 8 separately show the reliability curve of traction drive system when one TC, one bogie part, one MT's failure rate reduced by the same magnitude. It can be shown from Fig. 8 that reducing the failure rate of the traction converter can greatly improve the reliability of the traction drive system, which is the same as the analysis result of Sect. 3.2.2.

As can be seen from the figure, the reliability of the traction drive system decreases rapidly with the increase of the running mileage, and the reliability at 100,000 km is very low. This is because the maintenance factor of the device is not considered in the analysis. This phenomenon quantitatively shows the importance

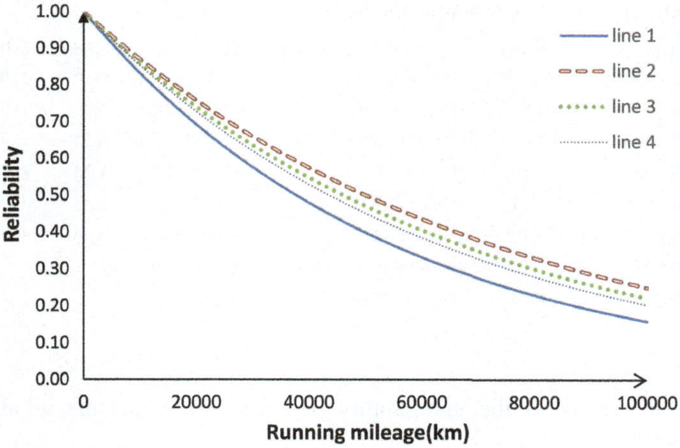

Fig. 8 Curve of traction drive system reliability

of routine maintenance of the system. If the system is not regularly maintained and the fault is repaired immediately, the traction drive system will become extremely unreliable.

3.4 Reliability Distribution

Through the posterior probability of the root node in the BN, the reliability distribution of all the basic components of the system can be achieved, which is an important part of the reliability design. If the reliability of the traction drive system is 50% when the train is operated for 100,000 km without considering the maintenance factor, reliability distribution is performed.

1. Calculate the sum of the posterior probabilities of all root nodes when the traction drive system fails by Bayesian inference

$$L = \sum_{i=1}^{26} L_i = 1.003014005 \tag{7}$$

L_i: the posterior probabilities of root node X_i.

2. Calculate the unreliability that each of the basic components should be assigned according to the design requirements

$$F_i = (1 - R_t)L_i/L \tag{8}$$

Table 2 Reliability of each component after reliability distribution

Nodeof BN	Allocated reliability	Node of BN	Allocated reliability	Node of BN	Allocated reliability	Node of BN	Allocated reliability
X1	0.982552587	X8	0.953639730	X15	0.952642735	X22	0.999999626
X2	0.999997471	X9	0.968594656	X16	0.972084139	X23	0.999999576
X3	0.999997959	X10	0.976570616	X17	0.970588646	X24	0.999999600
X4	0.988534557	X11	0.980558596	X18	0.978564606	X25	0.999999795
X5	0.961117193	X12	0.944168277	X19	0.999999668	X26	0.999999795
X6	0.968096158	X13	0.968096158	X20	0.999999597		
X7	0.970588646	X14	0.963609681	X21	0.999999532		

In (8), F_i represents the unreliability of X_i, R_t represents the reliability that required.

3. According to the unreliability, the allocated reliability of each basic event in the system is calculated by (9). The results are shown in Table 2

$$R_i = 1 - F_i$$

R_i: The allocated reliability of X_i.

4. According to the reliability calculation formulas of the series system and the parallel system shown in (9) [9] and Eq. (10), after the reliability distribution, the 100,000 km reliability of the system is 0.587453296908622 > 0.5, that means the distribution results meet the reliability design requirements.

$$R_s = \prod_{i=1}^{n} R_i \quad \text{(series system)} \tag{9}$$

$$R_s = 1 - \prod_{i=1}^{n} (1 - R_i) \quad \text{(parallel system)} \tag{10}$$

In the equation, R_i is the reliability of the basic component, and R_s is the reliability of series system or parallel system.

4 Conclusion

In this paper, a BN is constructed for the traction drive system of CRH3 multiple unit. By constructing Bayesian network, the failure rate and reliability of the system, importance of each component can be evaluated. The system can also be assigned reliability based on inference results of BN. Thereby, the key components that need to be paid more attention to during the maintenance and maintenance of the system are obtained, and the purpose of increasing the reliability of the traction drive system is achieved.

Acknowledgements This work is partially supported by the National Key Research and Development Program (2017YFB1200802-01).

References

1. Xiuguo C (2013) Study on reliability of electrical system of CRH3 emu. Beijing Jiaotong University (in Chinese)
2. Yongfeng S, Yang L, Jiebo L et al (2013) Reliability modeling and index allocation for traction transmission system of CRH3C emu. Railw Locomotive Car 33(5) (in Chinese)
3. Darwiche A (2010) Bayesian networks. Commun ACM 53(12):80–90
4. Marquez D, Neil M, Fenton N (2010) Improved reliability modeling using Bayesian networks and dynamic discretization. Reliab Eng Syst Saf 95(4):412–425
5. Alyson GW, Aparna VH (2007) Bayesian networks for multilevel system reliability. Reliab Eng Syst Saf 92(10):1413–1420
6. Shuting Z (2018) Reliability analysis of subway train braking system based on bayesian network. Beijing Jiaotong University (in Chinese)
7. Sousen L, Chingmiao C (2015) Bayesian-network-based fall risk evaluation of steel construction projects by fault tree transformation. J Civ Eng Manage 21(3):334–342
8. Jianqiang L, HuaiLong G, Huiqian D et al (2015) Reliability study of CRH_3 type emu auxiliary power supply system. J Railw 2015(11):44–51 (in Chinese)
9. Yisan C (2014) Reliability modeling and safety evaluation of high speed train traction transmission system. In: The 9th China intelligent transportation annual conference (in Chinese)

Mutation-Based Test Case Generation of Train-Ground Transmission Function for CTCS

Yujie Sun, Kaicheng Li and Lei Yuan

Abstract Safety tests are used to verify the safety functions of Chinese Train Control System (CTCS). The structural complexity and interface diversity of the system bring some challenges to generate test cases including possible failure modes for specific safety functions. In this paper, we propose a method of generating test cases for the train-ground transmission function based on model-based mutation testing. The timed automata model of CTCS is established. Then, mutation operators are designed according to the types of the failure modes which may occur in the process of train-ground information transmission. By applying these mutation operators, the mutants are obtained. Finally, test cases are generated via conformance checks between the mutants and the original model. The application of this method is illustrated by the level conversion scene. The result shows that this method can automatically generate 208 safety test cases covering five failure modes for the scene.

Keywords CTCS · Timed automata · Mutation · Test case

1 Introduction

As a typical safety-critical system, Chinese Train Control System (CTCS) is the railway infrastructure to ensure trains' operation safety and improve the transportation efficiency. With the increase in the complexity of the train control system,

Y. Sun · K. Li
National Engineering Research Center for Rail Transportation Operation and Control System, Beijing Jiaotong University, Beijing 100044, China
e-mail: 17120270@bjtu.edu.cn

K. Li
e-mail: kchli@bjtu.edu.cn

L. Yuan (✉)
State Key Laboratory of Rail Traffic Control and Safety, Beijing Jiaotong University, Beijing 100044, China
e-mail: lyuan@bjtu.edu.cn

Y. Qin et al. (eds.), *Proceedings of the 4th International Conference on Electrical and Information Technologies for Rail Transportation (EITRT) 2019*, Lecture Notes in Electrical Engineering 639, https://doi.org/10.1007/978-981-15-2866-8_13

the reliability, safety, and maintainability of the system are put forward higher requirements. Safety test is an effective way to verify the safety functions of the train control system and to confirm the consistency between the safety requirements and the real system. Test cases play an important role in safety tests, which can guide the execution of tests. Therefore, a key challenge is how to generate feasible test cases including faults to test whether the train control system meets the safety requirements under certain fault conditions.

With the wide application of formal methods, it is an effective way to generate test cases by applying model-based testing. Model-based testing (MBT) [1] is used for automatically generating test cases based on formal models and can avoid the subjectivity of manually generated test cases. Model-based mutation testing [2] is a special type of model-based testing, which combines ideas from mutation testing [3] and model-based testing. For the past few years, it has been applied to several formalisms, including probabilistic finite state machines [4], UML models [5], Simulink models [6], and timed automata models [7].

In this paper, we utilize timed automata models and model-based mutation testing to generate safety test cases of train-ground transmission function for CTCS. The remainder of the paper is organized as follows. Section 2 presents a description of the train-ground transmission function of CTCS. In Sect. 3, the method of test case generation for the safety function is introduced based on the model-based mutation testing theory. We explain the application of the method with the level conversion scene in Sect. 4 and conclude in Sect. 5.

2 Train-Ground Transmission Function

CTCS is composed of ground equipment and on-board equipment. By receiving route information, movement authority (MA), and other information provided by ground equipment, on-board equipment can calculate the speed monitoring curve in real time, so as to generate the current limiting speed and supervise the safety operation of trains. Therefore, the correct realization of train-ground information transmission function is crucial to the safe operation of on-board equipment.

At present, the train control systems applied to the high-speed railway mainly include the CTCS-3 train control system and the CTCS-2 train control system. CTCS-3 train control system uses the Global System for Mobile Communications-Railway (GSM-R) to realize two-way transmission of train-ground information. In addition, the system uses balises to transmit position correction information, level conversion command, and other information to on-board equipment. CTCS-2 train control system transmits MA, line parameters, temporary speed restriction, positioning information, and other control information to on-board equipment through balises and track circuits. A simple model of the train-ground transmission function for CTCS is shown in Fig. 1.

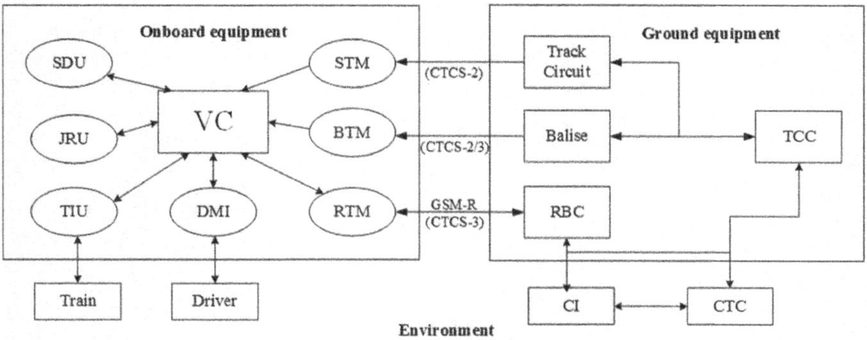

Fig. 1 We present a simplified model of the train-ground transmission function. At CTCS-3 level, the manner of train-ground information interaction is through GSM-R and balises, while at CTCS-2 level, the on-board equipment interacts with ground equipment via track circuits and balises

3 Model-Based Mutation Testing

Model-based mutation testing is a fault-based testing technique. In model-based mutation testing, the test case generation is guided by the altered versions of the original model under the action of certain mutation operators.

As shown in Fig. 2, the process of model-based mutation testing for train-ground transmission function comprises three main stages: test model construction, mutant generation, and test case generation. To begin with, the model of CTCS in specific scene is established based on timed automata theory and related specification. Then, the types of failure modes that may lead to the failure of the function are obtained by Failure Mode and Effect Analysis (FMEA). The corresponding mutation operators are designed, and mutants are generated by applying these mutation operators. Finally, the original model is compared to the mutants via conformance checks. If a mutant conforms to the original model, the mutant is treated as an equivalent mutant and discarded. Otherwise, the conformance check produces a timed trace that can be converted into a test case. The stages are described in detail as follows.

3.1 Timed Automata

The timed automata theory [8] is an extension of the automata theory for the purpose of specializing in real-time systems, which adds time factors to the traditional finite state machine. CTCS is a typical real-time system, and the information interaction between the on-board equipment and the ground equipment has very high requirements on real-time performance. Therefore, we use timed automata theory to construct test models of the train control system.

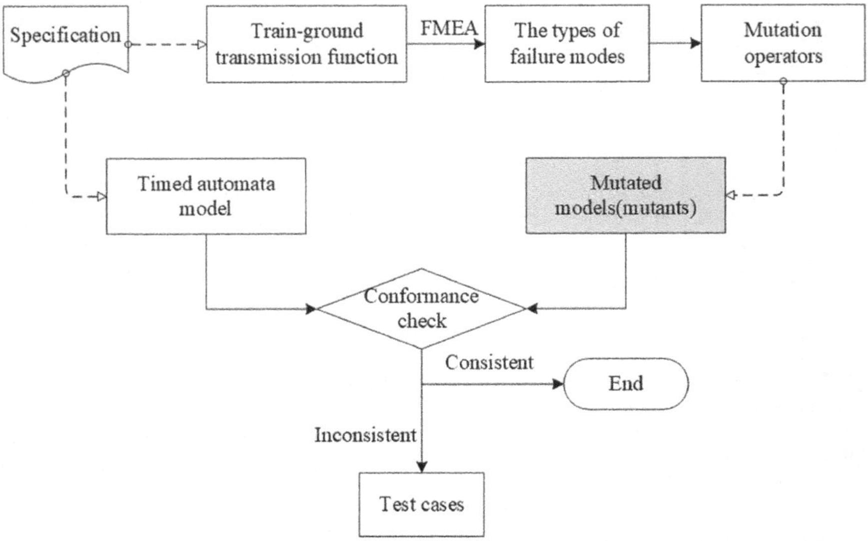

Fig. 2 Framework of test case generation for train-ground transmission function

A timed automaton A is a tuple $(Q, \hat{q}, \Sigma, X, I, G, T)$, where Q is a finite set of locations and $\hat{q} \in Q$ is the initial location; Σ is a finite set of observable actions; X is a finite set of clock variables; I: $Q \rightarrow LI$ is a mapping from locations to location invariants, where each location invariant $li \in LI$ is a conjunction of constraints of the form *true*, $x < n$ or $x \leq n$, with $x \in X$ and $n \in \mathbb{N}$[1]; G is a set of transition guards, in which each guard is a conjunction of constraints of the form $x \sim n$, where $x \in X$, $\sim \in \{\langle, \leq, =, \geq, \rangle\}$, and $n \in \mathbb{N}$; $T \subseteq Q \times \Sigma \times G \times P(X) \times Q$ is a finite set of transitions of the form $(q, \alpha, g, \delta, q')$, where q, $q' \in Q$ are the source and the target locations, $\alpha \in \Sigma$ is the transition action, $g \in G$ is the transition guard, and $\delta \subseteq X$ is the subset of clocks to be reset. A more comprehensive introduction can be found in [8].

3.2 Mutation Operator

FMEA is a systematic analysis method to identify potential failure modes [9]. In this paper, FMEA is used to analyze the failure modes of train-ground transmission function according to the structure and working principle of the train control system. The failure modes of the function can provide object support for the design of mutation operators. Taking the wireless transmission mode of CTCS-3 train control

[1]$\mathbb{N}$ is the set of natural numbers.

system as an example, the failure modes of train-ground transmission function are mainly divided into the following five types: error message, message loss, repeated message, communication timeout, and message timing error. Table 1 gives a detailed description of these failure modes.

Based on the semantics of the timed automata, we can change the synchronization channel of the original transition to model an error message or a repeated message; to model the message loss, we can change the target location of the original transition or the clock constraint in the guard of a transition; the other two time-related failure modes can be implemented by changing the clock reset or the guard of the original transition. Hence, the following four mutation operators for the five kinds of failure modes are designed.

- Change Action (μ_{ca}): replaces the synchronization channel on the transition in the timed automata model A with different output channels. This mutation can result in the mutant M_{ca}: $(Q, \hat{q}, \Sigma, X, I, G, (T\backslash\{t\}) \cup t_M)$, where t is the transition in A represented by $(q, \alpha, g, \delta, q')$, t_M is the mutated transition in M_{ca} represented by $(q, \alpha_M, g, \delta, q')$, $\alpha_M \in \Sigma$, and $\alpha_M \neq \alpha$.
- Change Guard (μ_{cg}): replaces the clock constraint in the guard of a transition in the original model A by replacing every equality or inequality sign appearing in the guard by another one. This mutation can result in the mutant M_{cg}: $(Q, \hat{q}, \Sigma, X, I, G, (T\backslash\{t\}) \cup t_M$, where t is the transition in A represented by $(q, \alpha, g, \delta, q')$, t_M is the mutated transition in M_{cg} represented by $(q, \alpha, g_M, \delta, q')$, o_i is the equality or inequality sign in the original guard, while o_i^m is the equality

Table 1 Description of failure modes for wireless transmission function

No.	Failure mode	Description	Characterization in timed automata
1	Error message	The message cannot be decoded or the variable in the message is invalid or incorrect	Incorrect synchronization signal
2	Message loss	The desired wireless message was not sent/received	No expected signal/other synchronization signal received
3	Repeated message	The equipment sends or receives repeated wireless messages	Repeated synchronization signals
4	Communication timeout	The timestamp of the latest received message is greater than *T_NVCONTACT*[a] with the current time of the equipment	Clock reset or violation of clock constraint in guards and no process to continue
5	Message timing error	The time stamp of the latest received message is older than the time stamp of the previous message	Clock reset or violation of clock constraint in guards

[a]*T_NVCONTACT* represents the system's acceptable downtime for train-ground communication, which is generally 7–20 s

or inequality sign in the mutated guard, $o_i, o_i^m \in \{<, \leq, =, \geq, >\}$ and $o_i \neq o_i^m$; for any $i \in I$, $g = c_i o_i d_i$, $g_M = c_i o_i^m d_i$, where c_i is a clock variable and d_i is a constant.

- Invert Reset (μ_{ir}): reverses the clock reset of a transition in the original model A. If the clock was originally reset, the reset is removed; otherwise a reset will be added. This mutation can result in the mutant M_{ir}: $(Q, \hat{q}, \Sigma, X, I, G, (T \backslash \{t\}) \cup t_M)$, where t is the transition in A denoted by $(q, \alpha, g, \delta, q')$, t_M is the mutated transition in M_{ir} denoted by $(q, \alpha, g, \delta_M, q')$; if δ exists, $\delta_M = \varnothing$ where $\varnothing$ stands for the empty set; if $\delta = \varnothing$, $\delta_M \neq \varnothing$ and $\delta_M \in X$.
- Self Loop (μ_{sl}): changes a single output transition in the original model A by replacing the target location of the transition by the source location. This mutation can result in the mutant $M_{\mathrm{sl}} : (Q, \hat{q}, \Sigma, X, I, G, (T \backslash \{t\}) \cup t_M)$, where t is the transition in A denoted by $(q, \alpha, g, \delta, q')$, t_M is the mutated transition in M_{sl} denoted by $(q, \alpha, g, \delta, q'_M)$, $q'_M \in Q$, and $q'_M \neq q'$.

3.3 Conformance Relation

To determine the consistency of the original model and the mutant, the timed input–output conformance (tioco) relation is proposed in [10]. Given a TA *A*, a mutant M, and a timed trace σ, *L*(*A*) is the set of timed traces induced by all runs of *A*.

The tioco is defined as:

$$M\,\mathrm{tioco}\,A\;\mathit{iff}\;\forall\,\sigma \in L(M), \sigma \in L(A)\mathrm{or}\;M\,\mathrm{tioco}\,A \Leftrightarrow L(M) \subseteq L(A) \tag{1}$$

Therefore, the set of observable traces of M is included in the set of observable traces of A, if there is a conformance relation between them. If the mutant is inconsistent with the original model, the mutation introduces a timed trace which is not in the original model, and the trace can be turned into a test case.

It is very appropriate to generate test cases for train control systems using conformance checks, since there is an implicit fault propagation process within conformance checks, and test cases are generated only when an observable error output is detected. This helps improve the feasibility of actual test execution to some extent.

4 Test Case Generation of Train-Ground Transmission Function

4.1 Test Model Construction

Level conversion is a typical operation scene in CTCS which performs the conversion through the interaction of the on-board equipment with specific balises and

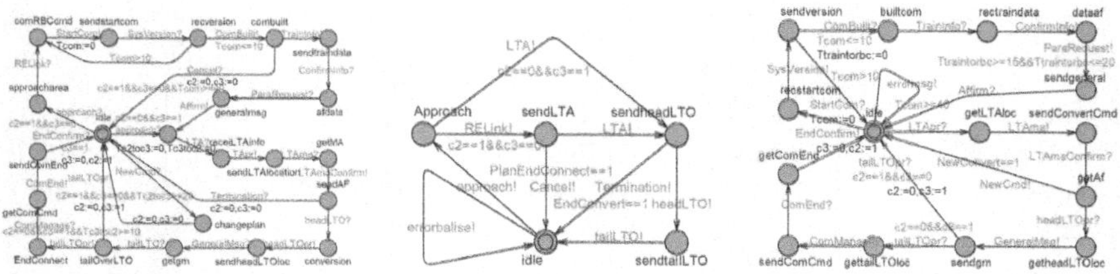

(a) The train model (b) The balise model (c) The RBC model

Fig. 3 We present the TA models of the level conversion scene. The three models of train, balise, and RBC constitute the timed automata network of the scene

RBC. The failure in the train-ground interaction process may cause the level conversion unsuccessful, which may affect the operation efficiency and even endanger the operation safety. Therefore, we illustrate our approach with the conversion scene, which includes the two-way conversion between CTCS-2 level and CTCS-3 levels.

In the conversion process from CTCS-2 to CTCS-3, when the train passes RE (the RBC connection balise), the train calls RBC according to connection information provided by the balise. If the on-board equipment successfully communicates with the RBC, the train sends the position report to the RBC. When the front end of the train passes the LTA (the conversion notice balise), the train obtains MA and the conversion command from the RBC. If the front end of the train passes LTO (the conversion execution balise), the train will automatically turn into the CTCS-3 level. The conversion from CTCS-3 to CTCS-2 is similar to the process described above. UPPAAL [11] is a tool for modeling, simulation, and verification of real-time systems. As shown in Fig. 3, the timed automata models of the train, RBC and the balise, are constructed by UPPAAL tool, respectively.

4.2 *Mutant Model*

The mutant models are generated by applying mutation operators to the test models. Each operator is applied to every transition or location, and only one operator is applied at a time. Here we select a model fragment of the RBC in which the train connects with RBC to illustrate the forms of the mutant models.

The reference model TA A is shown in Fig. 4 (TA A). It represents that "when RBC receives *StartCom* (the communication session start message) sent by the train, RBC sets the clock *Tcom* to 0 and sends *SysVersion* (the system version message) to the train, and establishes communication with the train within 10 time units after receiving *ComBuilt* (the communication session establishment message) sent by the train." According to the predefined four mutation operators, the mutant models are shown in Fig. 4 ($M1$–$M4$).

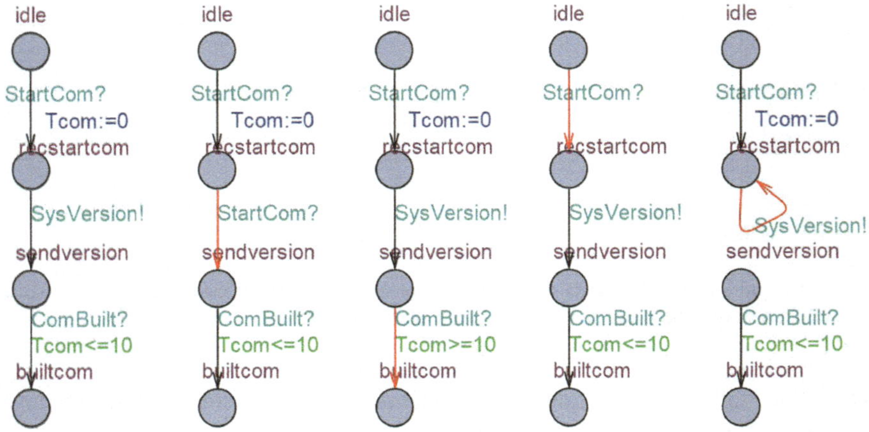

Fig. 4 Reference model TA *A* and the four mutants *M*1–*M*4

4.3 Test Case Generation

MoMuT::TA [12] is a test case generation tool for timed automata models by applying model-based mutation testing. It provides tioco conformance checks between the test model and the mutants and can generate test cases automatically. Taking the on-board equipment as the system under test (SUT), we use MoMuT::TA to process the balise model and RBC model to complete the test case generation by applying the mutation operators we defined.

The number of mutants and the number of test cases generated are shown in Table 2. It can be seen that there are 318 mutants and 208 abstract test cases. These test cases contain fault injection conditions that guide the execution of safety tests. For example, the content of a concrete test case which is extracted from one of the abstract test cases is "*after*(0.0) → *StartCom* → *after*(1.0) → *errormsg* → *fail*." This means "after zero delay, when RBC receives *StartCom*, RBC sends *errormsg* instead of *SysVersion* to the train and the system reaches a fail state after one delay." By simulating this wireless message error in the test, we can judge whether the corresponding actions of the train conform to the specification and safety requirements.

Table 2 Mutants and test cases of level conversion scene

Mutation operator	Mutant number	Test case number
Change action	250	167
Change guard	12	7
Invert reset	37	15
Self loop	19	19
Total	318	208

5 Conclusion

In this paper, we propose a method of generating test cases for CTCS based on model-based mutation testing. Firstly, the models of CTCS are constructed based on the timed automata theory. Then, four mutation operators are designed according to the characteristics of the failure modes. Finally, test cases are generated via conformance checks. A case study of the level conversion scene is presented. The result shows that this method not only makes the process of test case generation automatic, but also can generate test cases containing fault conditions, which can provide guidance for the execution of safety tests and ensure the feasibility of test execution to some extent.

Acknowledgements This work was supported by the National Natural Science Foundation of China (No. U1734210) and by National Engineering Laboratory for Urban Rail Transit Communication and Operation Control.

References

1. Apfelbaum L, Doyle J (1997) Model based testing. In: Software quality week conference, pp 296–300
2. Aichernig BK, Jöbstl E, Tiran S (2015) Model-based mutation testing via symbolic refinement checking. Sci Comput Program 97:383–404
3. Jia Y, Harman M (2011) An analysis and survey of the development of mutation testing. IEEE Trans Softw Eng 37(5):649–678
4. Hierons RM, Merayo MG (2007) Mutation testing from probabilistic finite state machines. In: Testing: academic and industrial conference practice and research techniques-Mutation (TAICPART-Mutation 2007). IEEE, pp 141–150
5. Schlick R, Herzner W, Jöbstl E (2011) Fault-based generation of test cases from UML-models–approach and some experiences. In: International conference on computer safety, reliability, and security. Springer, Berlin, Heidelberg, pp 270–283
6. Brillout A, He N, Mazzucchi M et al (2009) Mutation-based test case generation for simulink models. In: International symposium on formal methods for components and objects. Springer, Berlin, Heidelberg, pp 208–227
7. Aichernig BK, Lorber F, Ničković D (2013) Time for mutants—model-based mutation testing with timed automata. In: International conference on tests and proofs. Springer, Berlin, Heidelberg, pp 20–38
8. Alur R, Dill DL (1994) A theory of timed automata. Theoret Comput Sci 126(2):183–235
9. Stamatis DH (2003) Failure mode and effect analysis: FMEA from theory to execution. ASQ Quality press
10. Krichen M, Tripakis S (2009) Conformance testing for real-time systems. Formal Methods Syst Des 34(3):238–304
11. Larsen KG, Pettersson P, Yi W (1997) UPPAAL in a nutshell. Int J Softw Tools Technol Transfer (STTT) 1(1):134–152
12. Fellner A, Krenn W, Schlick R et al (2019) Model-based, Mutation-driven test-case generation via heuristic-guided branching search. ACM Trans Embedded Comput Syst (TECS) 18(1):4

Urban Rail Train Wheel Fault Diagnosis Based on Improved EEMD

Ning Fu, Kaijie Qian and Zongyi Xing

Abstract In order to accurately identify the types of wheel faults in urban rail trains, a method based on improved ensemble empirical mode decomposition (EEMD) and Hilbert transform is proposed. The improved EEMD decomposition of the acquired vibration signal obtains several intrinsic mode functions (IMFs), and the Hilbert transform is performed on the IMF component containing the main information components, and judged the type of failure of the train wheels according to the fault characteristic frequency of the Hilbert spectrum. The experimental results show that the method can be used to identify the fault types of urban rail train wheels effectively and accurately.

Keywords Wheel · EEMD · Fault diagnosis · Hilbert transform

1 Introduction

The out-of-round failure of the wheel is one of the main faults of the wheel, which has a great influence on the running safety of the train; it not only reduces the comfort of the train, but also causes damage to the wheels and the track. Therefore, in order to operate the train safely, it is necessary to perform on-line detection of the working state of the wheel [1].

Various feature extraction methods have been widely used in rolling bearing fault diagnosis [2, 3]. Such as wavelet transform (WT), Wigner–Ville distribution (WVD), morphological filter (MF), and sparse decomposition. Although these methods have been widely used in fault detection in bearings, they have their own

N. Fu · K. Qian
School of Mechanical Engineering, Nanjing University of Science and Technology, Nanjing 210094, China

Z. Xing (✉)
School of Automation, Nanjing University of Science and Technology, Nanjing 210094, China
e-mail: xingzongyi@163.com

Y. Qin et al. (eds.), *Proceedings of the 4th International Conference on Electrical and Information Technologies for Rail Transportation (EITRT) 2019*, Lecture Notes in Electrical Engineering 639, https://doi.org/10.1007/978-981-15-2866-8_14

unique drawbacks. For example, WT is one of the most advanced methods of analysis, but it requires the prior selection of wavelet bases and mother wavelets [4, 5]. The WVD is affected by cross-interference items [6, 7]. However, these methods are not suitable for dealing with complex vibration signals that reflect nonlinear and non-stationary characteristics, while empirical mode decomposition (EMD) is not limited by this adaptive capability [8]. EMD can decompose a complex signal into a series of eigenmode functions (IMFs) with amplitude and frequency modulation characteristics [9], but the end effect and mode mixing of EMD may cause IMFs to lose its specific physical meaning.

Ensemble empirical mode decomposition (EEMD), a strengthened approach of EMD, is proposed by Wu and Huang [10] and it is able to eliminate the troubles of the EMD end effects and mode mixing by adding a finite white series to the analysis signal. EEMD has obtained a lot of attentions in the field of rotating machinery fault diagnosis [11–13]. Cheng et al. combined EEMD and kernel principal component analysis (KPCA) to identify the extent of planetary gear failure damage [11]. Wang et al. applied EEMD to improved tunable Q-factor wavelet transform (TQWT) to well extract fault feature of rolling element bearings [12]. Žvokelj et al. proposed a multivariate analysis method based on EEMD and independent component analysis (ICA) for rolling element bearing fault detection and diagnosis [13].

2 EEMD Decomposition

Huang [14] proposed an empirical mode decomposition (EMD) for time–frequency processing methods for analyzing nonlinear non-stationary signal sequences. The purpose of this method is to decompose the original signal into a series of IMF components (intrinsic). The modal function, which lays the foundation for the distraction of fault features. The eigenmode component needs to satisfy two conditions: the function extremum point is equal to the function zero point or at most one phase difference; the local extrema of the function form an envelope mean of zero. The EMD method has defects such as modal aliasing. The modal aliasing means that signals of different scales or frequencies appear in the same IMF, or signals of the same scale or frequency are decomposed into a plurality of different IMF components. The factors causing modal aliasing mainly include intermittent signals, pulse signals, and intermittent interference.

The overall mean empirical mode decomposition (EEMD), EMD decomposition by adding different white noise to the original signal [15], the results of multiple decompositions are averaged to obtain the final IMF. After white noise is added to the original signal, the statistical characteristics of the white noise frequency are uniformly distributed to eliminate the intermittent phenomenon in the original signal, thereby suppressing the modal aliasing problem effectively.

EEMD solves the modal aliasing problem by adding white noise in each decomposition process. However, since the decomposition depends on the added white noise amplitude and the number of additions, if the parameter selection is inappropriate, the pseudo component is likely to occur. Aiming at the modal aliasing problem of EMD decomposition and the pseudo-component problem of EEMD, the paper improves the EEMD, and proposes an improved overall mean empirical mode decomposition. The overall average empirical mode decomposition method based on information entropy. Through each decomposition process, white noise with the same amplitude and opposite sign is added. In the process of decomposition, an abnormal signal with large randomness is detected by setting the information entropy threshold, and it is separated from the original signal, and then obtained. The remaining signals are EMD decomposed.

3 Information Entropy Algorithms

Shannon uses the methods of mathematical statistics and probability theory to introduce the definition of entropy into information theory, and the concept of information entropy was created [16].

Information entropy is an index that characterizes the state of the overall information of the source. It is used to describe the uncertainty of the source information and the randomness of the event, and solve the problem of information quantification that has long plagued humans. The size of information entropy is closely related to the degree of order of the system. The more ordered the system is, the smaller the information entropy value is. Conversely, the more chaotic the system is, the larger the information entropy value is. Information entropy can reflect the characteristics of the system operation in essence, and the operating state of the system can be judged by comparing the size of the entropy.

Suppose the source is a discrete random variable $X = \{x_1, x_2, \ldots, x_n\}$, different states x_i correspond to different states, and the corresponding probability of occurrence of each state is $P = \{p_1, p_2, \ldots, p_n\}$, then the information uncertainty of each state can be expressed as:

$$I(x_i) = -\log_a p_i, \sum_{i=1}^{n} p_i = 1 \quad (3.1)$$

The logarithmic base a is generally taken e. Shannon defines the mathematical expectation as information entropy, so the information entropy of the source can be expressed as:

$$H(X) = E[I(X)] = -\sum_{i=1}^{n} p_i \ln p_i \quad (3.2)$$

As we can see from the above definition, information entropy has the following characteristics:

(1) When there is one and only one state whose probability of occurrence is not 0, the probability of occurrence of the state must be 1, and the information entropy of the system is $H(X) = 0$, which indicates that the state of the system is stable;
(2) When the probability of occurrence of each state in the source is the same, the information entropy of the system is the $H(X) = \ln(n)$, and the uncertainty of the system state is the largest at this time;
(3) When the order of appearance probability corresponding to a state changes, the information entropy does not change;
(4) The range of information entropy is $0 \leq H(X) \leq 1$, which reflects the complexity and randomness of the time series. The larger the value $H(X)$ is, the higher the complexity of the time series and the stronger randomness is. Conversely, the smaller the value $H(X)$ is, the lower the complexity of the time series and the weaker the randomness is.

4 Wheel Fault Diagnosis Based on Improved EEMD Method and Hilbert Marginal Spectrum

For the signal $X(t)$, the improved EEMD decomposes the urban train wheel vibration signal into multiple IMF components, performs a Hilbert transform on the IMF component to obtain the instantaneous frequency and instantaneous amplitude of each component, and superimposes them together to obtain Hilbert Spectrum. The method decomposition steps are as follows.

Step 1: In the original signal, add a white noise sum with a mean of zero $m_i(t)$ and $-m_i(t)$, namely:

$$X_i^+(t) = X(t) + a_i m_i(t) \tag{4.1}$$

$$X_i^-(t) = X(t) - a_i m_i(t) \tag{4.2}$$

$m_i(t)$ indicates the added white noise signal, a_i indicates the magnitude of the added noise signal, $i = 1, 2, \ldots, N_e$, N_e indicates the addition of a white noise logarithm. EMD decomposition is performed on $X_i^+(t)$ and $X_i^-(t)$, respectively, to obtain a first-order IMF component sequence, $\{K_{i1}^+(t)\}$ and $\{K_{i1}^-(t)\}$ $(i = 1, 2, \ldots, N_e)$. Integrate the average component obtained:

$$K_1(t) = \frac{1}{2N} \sum_{i=1}^{N_e} \left[K_{i1}^+(t) + K_{i1}^-(t)\right] \tag{4.3}$$

Check if it is an abnormal signal. If the entropy value is larger than the signal that is considered to be an abnormal signal, it is considered to be a stationary signal. After many experiments, it is more appropriate to take 0.55–0.6.

Step 2: If $K_1(t)$ is an abnormal signal, continue with step (1) until the IMF component $K_p(t)$ is not an abnormal signal.
Step 3: Separating the decomposed first component from the original signal, namely:

$$r(t) = S(t) - \sum_{i=1}^{p-1} K_i(t) \tag{4.4}$$

Step 4: EMD decomposition is performed on the remaining signals $r(t)$, and all the obtained IMF components are arranged in a high frequency to a low frequency.
Step 5: For each intrinsic mode function $K_i(t)$, perform a Hilbert transform to obtain:

$$\widetilde{K}_i(t) = \frac{P}{\pi} \int_{-\infty}^{+\infty} \frac{K_i(\tau)}{t-\tau} \mathrm{d}\tau \tag{4.5}$$

where P is the main value of Cauchy, and the structure is parsed by $K_i(t)$ and $\widetilde{K}_i(t)$:

$$z_i = K_i(t) + j\widetilde{K}_i(t) = b_i(t)\mathrm{e}^{j\varphi_i(t)} \tag{4.6}$$

Step 6: The amplitude function can be obtained by analyzing the signal:

$$b_i(t) = \sqrt{K_i^2(t) + \widetilde{K}_i^2(t)} \tag{4.7}$$

The phase function can also be obtained:

$$\varphi_i(t) = \arctan \frac{\widetilde{K}_i(t)}{K_i(t)} \tag{4.8}$$

Step 7: Further obtain the instantaneous frequency:

$$f_i(t) = \frac{\mathrm{d}\phi_i(t)}{\mathrm{d}t} \tag{4.9}$$

The residual term is a monotonic function, and there is no frequency, so it is negligible for the spectrum of the signal to be analyzed. After ignoring the residual component, the signal to be processed can be expressed as:

$$x(t) = R\left(\sum_{i=1}^{n} b_i(t)e^{j\varphi_i(t)}\right)$$
$$x(t) = R\left(\sum_{i=1}^{n} b_i(t)e^{j\int \omega_i(t)\mathrm{d}t}\right) = R\left(\sum_{i=1}^{n} b_i(t)e^{j\omega_i(t)}\right) \tag{4.10}$$

where R represents the real part of the signal, whereby the Hilbert amplitude spectrum can be expressed as:

$$H(\omega, t) = \mathrm{Re}\sum_{i=1}^{n} b_i(t)e^{j\int \omega_i(t)\mathrm{d}t} \tag{4.11}$$

It is not difficult to see from Hilbert's amplitude mathematical expression that the spectrum is the instantaneous frequency and its amplitude expressed from the time domain. The Hilbert spectrum is integrated over time to obtain the marginal spectrum. The definition is as follows:

$$h(\omega) = \int_{0}^{S} H(\omega, t)\mathrm{d}t \tag{4.12}$$

where S is the length of the original signal. The marginal spectrum reflects the magnitude and variation of the instantaneous frequency at all frequencies.

5 Simulation Signal Analysis

To illustrate the superiority of the improved EEMD method, it was used to simulate the decomposition of the signal and compared it to the EMD method. This article uses a mixture of three simulated signals.

$$S(t) = S1 + S2 + S3 \tag{5.1}$$

Among them, $S1$ is a high-frequency sinusoidal signal, the expression is $S1 = 2\sin(2\pi 80t)$; $S2$ is a low-frequency sinusoidal signal, the expression is $S2 = 3\sin(2\pi 10t)$; $S3$ is an intermittent white noise signal; the signal sampling frequency is 2000 Hz, and the sampling time is 1s. The time domain waveforms of the mixed simulation signal and its component signals are shown in Figs. 1, 2 and 3.

The simulation signals $S(t)$ are decomposed using the EMD method and the improved EEMD method, respectively. It can be seen that after the EMD method is decomposed, due to the influence of intermittent noise, the modal aliasing phenomenon appears in the high-frequency signal. The components obtained by the

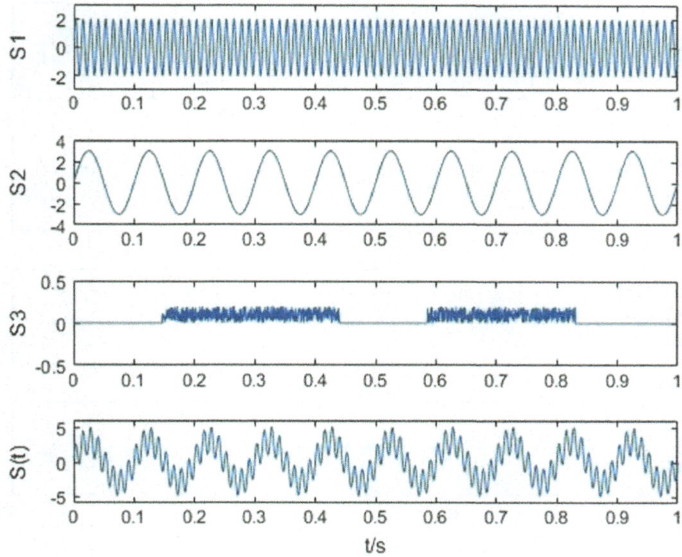

Fig. 1 Mixed simulation signal $S(t)$ and its components

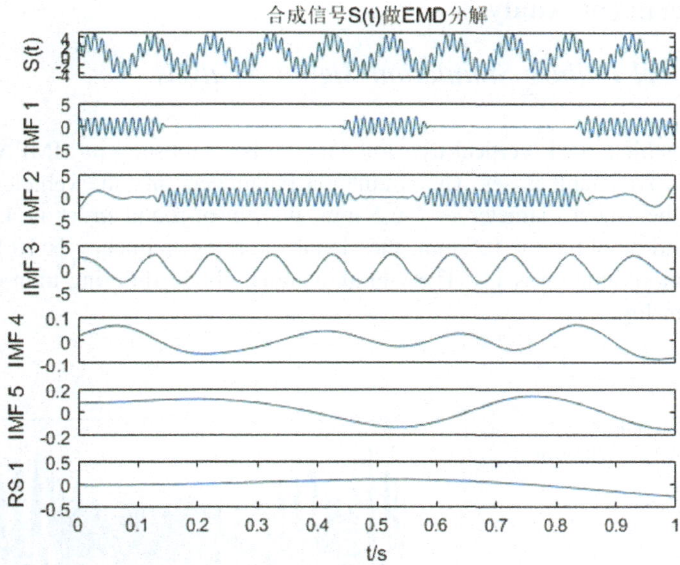

Fig. 2 Simulation signal for EMD decomposition

improved EEMD method are not affected by intermittent noise, and there is no modal aliasing. Two different signal components are decomposed from the mixed signal.

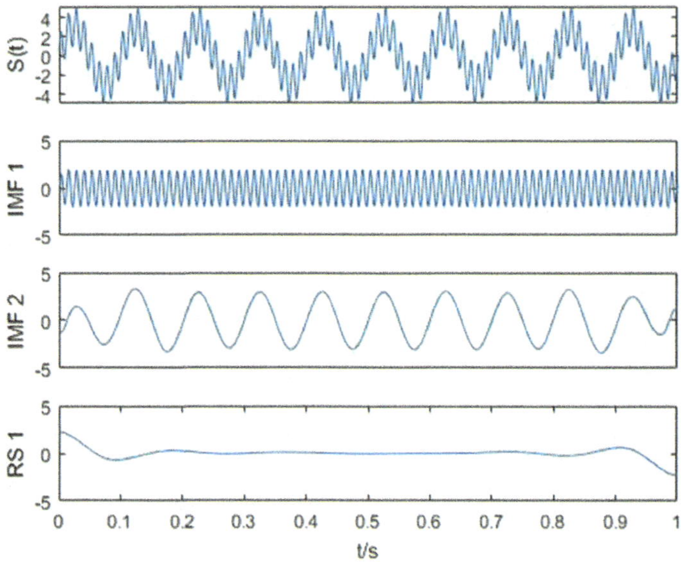

Fig. 3 Improved EEMD decomposition of simulated signals

6 Experiment Analysis

6.1 Model Failure Simulation Signal Analysis

The train vehicle-track vertical dynamics model is established in SIMPACK, and the out-of-round fault is set. The simulation parameters are: the vehicle speed is 60 km/h, the wheel diameter is 746.5 mm, the out-of-round order is 4, and the out-of-round amplitude is 0.5 mm, the signal sampling frequency is 10 kHz, and the simulation duration is 1 s. The vibration signal obtained during train operation is shown in Fig. 4.

Fig. 4 Model failure simulation signal with noise

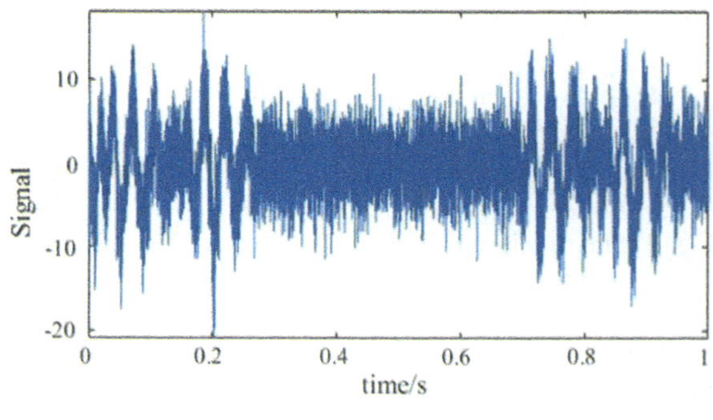

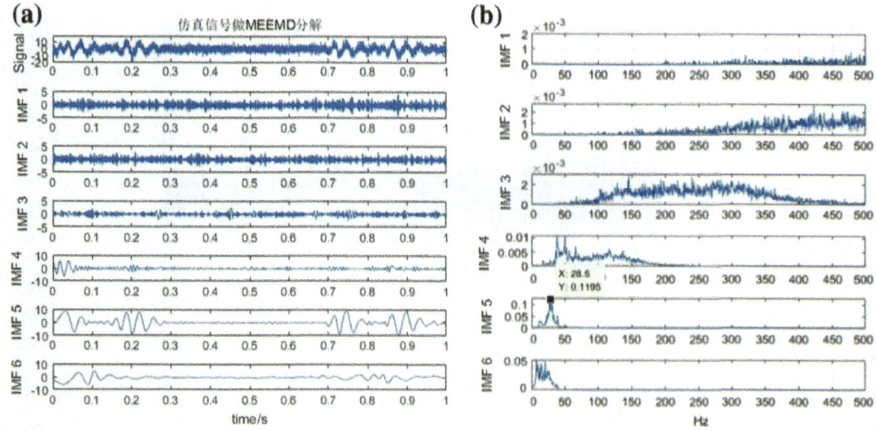

Fig. 5 Outcome of model failure simulation signal. (Left) Improved EEMD decomposition of signals. (Right) Hilbert marginal spectrum

The improved EEMD decomposition is performed on the noisy signal. The Hilbert transform is performed on the decomposed IMF to obtain the Hilbert marginal spectrum. They are showed in Fig. 5.

It can be seen from the figure that the main frequency of the impact of the IMF5 is 28.5 Hz, and the amplitude is the largest, and according to the vehicle speed of 60 km/h, the wheel diameter is 746.5 mm, the wheel circumference is 2.344 m, and the rotational frequency of the wheel is 7.11 Hz, then the theoretical failure frequency of the 4th order wheel out-of-round is 28.44 Hz. This is close to the main frequency of the impact of the IMF5.

6.2 *Measured Vibration Signal Analysis*

In this section, the on-site measured orbital vibration signal of the Guangzhou Metro Company's train number 8788 passing through the acquisition system is used to improve the EEMD decomposition to verify the engineering feasibility of the proposed method. The sampling frequency of the signal is 10 kHz, the train running speed is 65 km/h, and the wheel diameter is 800 mm (Fig. 6).

The improved EEMD decomposition is performed on the Measured vibration signal. The Hilbert transform is performed on the decomposed IMFs to obtain the Hilbert marginal spectrum. They are showed in Fig. 7.

It can be seen from the figure that the main frequency of the impact of the IMF6 is 14.5 Hz, and the amplitude is the largest, and according to the vehicle speed of 65 km/h, the wheel diameter is 800 mm, the wheel circumference is 2.512 m, and the wheel rotation frequency is 7.2 Hz, then the second order, the theoretical fault frequency of the wheel is not rounded at 14.4 Hz. This is close to the impact

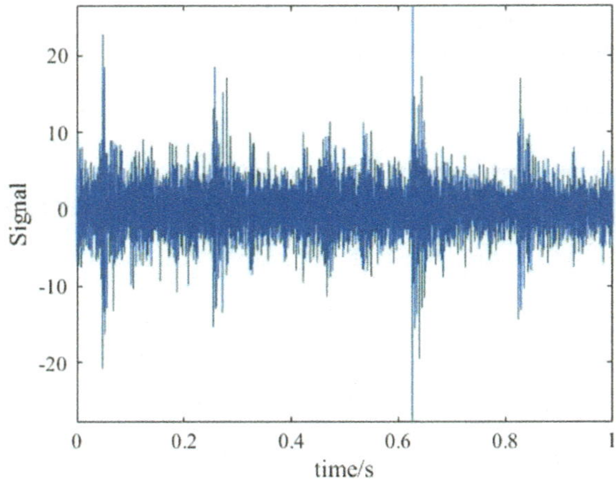

Fig. 6 Measured vibration signal

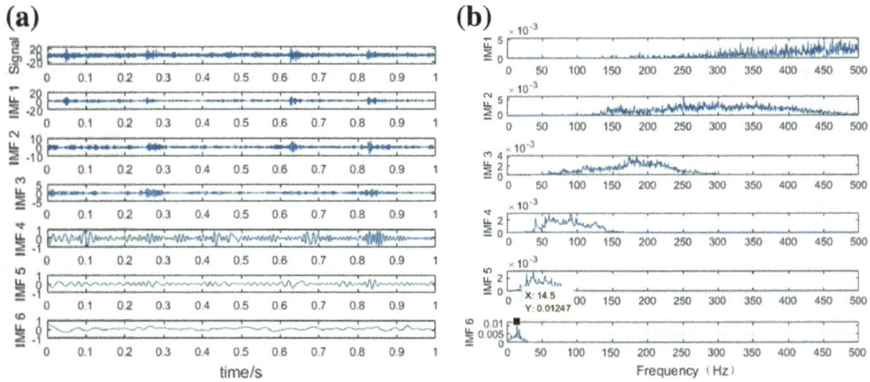

Fig. 7 Outcome of measured signal in Guangzhou. (Left) Improved EEMD decomposition of signals. (Right) Hilbert marginal spectrum

frequency of the IMF6. Therefore, it can be considered that after the improved EEMD decomposition, the obtained IMF component correctly extracts the characteristic frequency in the original signal, which provides conditions for further signal analysis processing.

7 Conclusion

Aiming at the modal aliasing problem of EMD decomposition and the pseudo-component problem of EEMD, the paper improves the EEMD and proposes an improved overall mean empirical mode decomposition. The overall average empirical mode decomposition method based on information entropy, hybrid simulation. The decomposition results of the signal show that the improved EEMD method can solve the modal aliasing and pseudo-component problems, and can reduce the calculation time and improve the adaptability. The analysis results of the model fault signal and measured vibration signal show that the method studied in this paper can well judge the type of wheel non-circular fault.

Acknowledgements This work is supported by National Key R&D Program of China (2016YFB1200401).

References

1. Sharma SC, Harsha SP (2014) EPFM Analysis of subsurface crack beneath a wheel flat using dynamic condition. Procedia Mater Sci 6:43–60
2. Zhang Q, Gao J, Dong H et al (2018) WPD and DE/BBO-RBFNN for solution of rolling bearing fault diagnosis. Neurocomputing 312:27–33
3. Shao H, Jiang H, Wang F et al (2017) Rolling bearing fault diagnosis using adaptive deep belief network with dual-tree complex wavelet packet. ISA Trans 69:187–201 (in Chinese)
4. Shao R, Hu W, Wang Y et al (2014) The fault feature extraction and classification of gear using principal component analysis and kernel principal component analysis based on the wavelet packet transform. Measurement 54(6):118–132 (in Chinese)
5. Liu Z, He Z, Guo W et al (2016) A hybrid fault diagnosis method based on second generation wavelet de-noising and local mean decomposition for rotating machinery. ISA Trans 61:211–220
6. Pachori RB, Nishad A (2016) Cross-terms reduction in the Wigner-Ville distribution using tunable-Q wavelet transform. Sig Process 120:288–304 (in Chinese)
7. Liu Y, Zhang J, Ma L (2016) A fault diagnosis approach for diesel engines based on self-adaptive WVD, improved FCBF and PECOC-RVM. Neurocomputing 177:600–611
8. Wang L, Liu Z, Miao Q et al (2018) Time-frequency analysis based on ensemble local mean decomposition and fast Kurtogram for rotating machinery fault diagnosis. Mech Syst Sig Process 103:60–75
9. Kopsinis Y, McLaughlin S (2009) Development of EMD-based denoising methods inspired by wavelet thresholding. IEEE Trans Sig Process 57:1351–1362
10. Wu Z, Huang NE et al (2009) Ensemble empirical mode decomposition: a noise assisted data analysis method. Adv Adaptive Data Anal 1:1–41 (in Chinese)
11. Cheng G, Chen X, Li H et al (2016) Study on planetary gear fault diagnosis based on entropy feature fusion of ensemble empirical mode decomposition. Measurement 91:140–154
12. Wang H, Chen J, Dong G (2014) Feature extraction of rolling bearing's early weak fault based on EEMD and tunable Q-factor wavelet transform. Mech Syst Sig Process 48:103–119
13. Žvokelj M, Zupan S, Prebil I (2016) EEMD-based multiscale ICA method for slewing bearing fault detection and diagnosis. J Sound Vibr 370:394–423
14. Huang NE, Wu Z (2009) A review on Hilbert–Huang transform: method and its applications to geophysical studies. Adv Adaptive Data Anal 1:1–23
15. Cheng Y, Zhang Z, Zhang H, Qi Y (2018) Application of improved EMD algorithm in post-explosion shock wave processing. Electron Meas Technol (23) (in Chinese)
16. Shannon CEA (2001) A mathematical theory of communication. Acm Sigmobile Mob Comput Commun Rev 5(1):3–55

A New Method for Fault Location of Metro Traction Network Based on Transient Ratio Method

Teng Li, Yongjun Zhou, Mingli Wu, Xiaojie You and Yong Mu

Abstract When the rail short-circuit fault occurs in the traction network of the railway, especially when the traction network is short-circuited through a large transition resistance or the fault occurs near the substation, the existing fault location methods often have larger error. Compared with traveling wave method and artificial intelligence-based fault location method, the two-terminal synchronous fault analysis method is more suitable for fast and accurate short-circuit fault location of metro traction network. Steady-state ratio method is derived from the steady-state of fault analysis method. However, due to the influence of post-fault protection tripping, the synchronous acquisition of steady-state values at both ends cannot be guaranteed. A new fault location method based on transient ratio method is proposed to solve the problem of low accuracy of existing fault location methods for metro traction network. Compared with the traditional fault analysis method and the steady-state ratio method, this method does not need the synchronization of two-terminal data and is not affected by the protection tripping after the fault occurs. Fault location can be realized by using the ratio of initial short-circuit current on both sides of fault point. The simulation results by MATLAB/Simulink show that this method is less affected by the inaccurate estimation of transition resistance and traction network parameters. The method is simple and accurate and has a certain application prospect.

Keywords Short circuit of metro traction network · Feeder protection · Transient ratio method · Fault location

T. Li (✉) · Y. Zhou · M. Wu · X. You
School of Electrical Engineering, Beijing Jiaotong University, Beijing, China
e-mail: liteng@bjtu.edu.cn

Y. Mu
State Grid Hebei Power Co., Ltd. Tangshan Power Supply Company, Hebei, China

Y. Qin et al. (eds.), *Proceedings of the 4th International Conference on Electrical and Information Technologies for Rail Transportation (EITRT) 2019*, Lecture Notes in Electrical Engineering 639, https://doi.org/10.1007/978-981-15-2866-8_15

1 Introduction

As one form of the urban rail transit, the metro is becoming more important for modern cities. In order to reduce the personal casualties and economic losses, it is necessary to find out the fault point quickly and accurately after the fault occurs in the metro traction network.

The fault location methods of the metro traction network mainly include impedance method, traveling wave (TW) method and artificial intelligence (AI) method [1–4]. According to the location of the acquired data, the fault location methods can be divided into single-end and double-end measurement methods [5]. At present, the double-end impedance method is more accurate than the single-end method for fault location of the Metro DC traction network since it is less affected by factors such as the large transition resistance, inaccurate estimation errors of line parameters. It is also more suitable than the TW and AI methods for its simplicity and stability. It can be further divided into steady-state current ratio method and fault analysis method [6]. However, the acquired value of the steady-state current is affected by the tripping time when there is the short-circuit fault; the inconsistency of the tripping time at two ends causes distance measurement errors. So the fault analysis method is more often used currently for fault location.

This paper presents a new short-circuit location method for rail transit feeder, given the name of "transient ratio method". This method only needs to judge the sudden change time of the fault according to the increment of the current, and then calculate the fault distance by the ratio of the initial current values of two ends after the fault. Finally, we verify its accuracy by the simulation.

2 Principle of Fault Analysis Method

In the case of short-circuit fault, 96% of the fault current is provided by four traction substations on both sides of the fault point [7], and the two traction substations closest to the short-circuit point have the greatest impact on the short-circuit current. Therefore, when analyzing the principle of fault location, only the short-circuit fault model of two traction stations adjacent to the fault point is considered. It is assumed that both the catenary and the traveling track are uniformly distributed, that is, the resistance and inductance per unit length are constant. At the same time, the rectifier units supplied on both sides of the traction network are regarded as DC voltage sources with certain internal resistance when performing the principle analysis [8]. Then the transient process of traction network short circuit can be described by the model as shown in Fig. 1 (left).

Considering that catenaries, i.e. the contact lines are divided into upstream line and downstream line, it is necessary to obtain the equivalent circuit to simplify the two stream lines to the single line when there is the contact line to track short-circuit fault. Figure 1 (right) shows the simplified circuit after the Δ-Y transformation of the circuit in Fig. 1 (left). As a result, the equivalent unit resistance r_0 and unit inductance l_0 of the transformed catenary are half of the original ones.

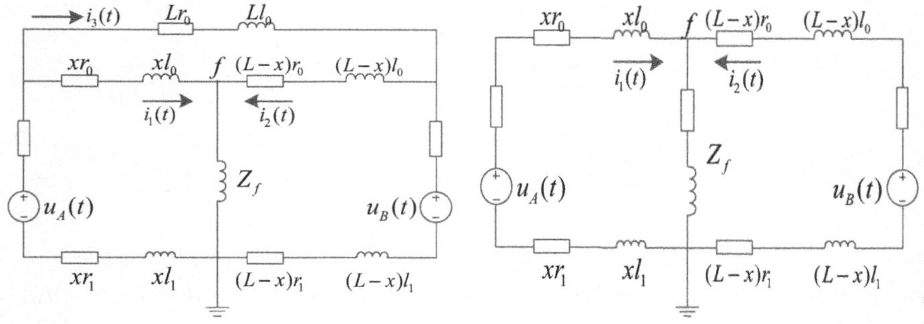

Fig. 1 Transient circuit diagram of short-circuit fault in metro. (Left) Schematic when one contact line is short-circuited with the track. (Right) Equivalent diagram after the Δ-Y transform

When the traction network is short-circuited, due to the influence of equivalent inductance of traction network, the current waveforms on both sides of the fault point are similar to the exponential change in the transient process. The transition impedance of the short-circuited point is expressed in Z_f. The unit resistance and inductance of the running track are r_1 and l_1. The feeder currents at both ends are $i_{1(t)}$ and $i_{2(t)}$, the feeder voltages are $u_{A(t)}$ and $u_{B(t)}$, The distance between the two traction stations is L, and the distance between the short-circuit point and the data acquisition point of the left traction station is x.

According to KVL, the time domain differential equation after the fault is:

$$u_A(t) = i_1(t)(r_0 + r_1)x + (l_0 + l_1)x\frac{\mathrm{d}i_1(t)}{\mathrm{d}t} + (i_1(t) + i_2(t))Z_f \tag{1}$$

$$u_B(t) = i_2(t)(r_0 + r_1)(L - x) + (l_0 + l_1)(L - x)\frac{\mathrm{d}i_2(t)}{\mathrm{d}t} + (i_1(t) + i_2(t))Z_f \tag{2}$$

By combining the above equations, the expression of short-circuit fault distance x can be obtained as follows:

$$x = -\frac{u_B(t) - u_A(t) + L(l_0 + l_1)\frac{\mathrm{d}i_2(t)}{\mathrm{d}t} + L(r_0 + r_1)i_2(t)}{(r_0 + r_1)(i_1(t) + i_2(t)) + (l_0 + l_1)\left(\frac{\mathrm{d}i_1(t)}{\mathrm{d}t} + \frac{\mathrm{d}i_2(t)}{\mathrm{d}t}\right)} \tag{3}$$

3 Fault Location Principle Based on Transient Ratio Method

When there is no transition resistance in the short-circuited branch, the traction network on both sides of the fault point can be equivalent to a R-L series circuit excited by DC power source [9, 10], which is shown in Fig. 2 (left). The transient short-circuited model with a transition resistances is shown in Fig. 2 (right).

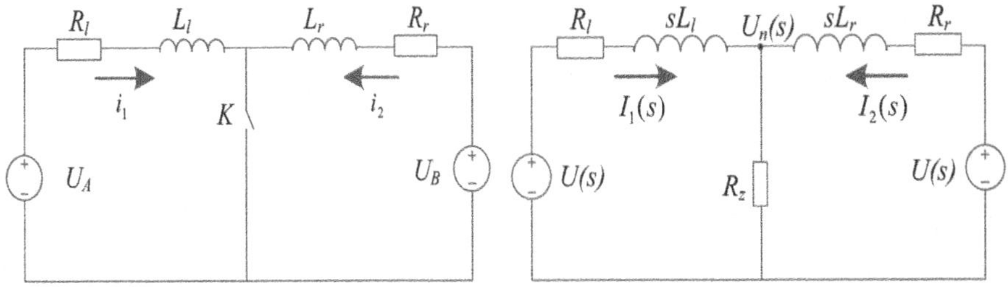

Fig. 2 Short-circuit transient model of traction network. (Left) Metallic short circuit. (Right) Impedance short circuit

As shown in Fig. 2 (left), when a metal short circuit occurs in the traction network, if the short-circuit current on the left side of the fault point is i_1, the short-circuit voltage is U_A, and the transient equivalent resistance and inductance are R_l and L_l, respectively, the results can be obtained from the KVL equation.

$$L_l \frac{di_1}{dt} + R_l i_1 = U_A \tag{4}$$

The fault current i_1 is obtained as follows:

$$i_1 = \frac{U_A}{R_l}\left(1 - e^{-\frac{R_l}{L_l}t}\right) \tag{5}$$

Similarly, it can be found that the right short-circuit current i_2 is:

$$i_2 = \frac{U_B}{R_r}\left(1 - e^{-\frac{R_r}{L_r}t}\right) \tag{6}$$

Here, U_B is equal to U_A. It can be seen from the formula that the resistance of traction network mainly affects the peak value and steady value of transient current, and the inductance mainly affects the rise rate of fault. At the time 0^+, we can see that the current ratio of both ends is:

$$\frac{i_1(0^+)}{i_2(0^+)} = \frac{R_r}{R_l} \approx \frac{L - x}{L} \tag{7}$$

In the formula, L is the distance between the two rectifiers and x is the distance between the fault point and the left rectifier.

As shown in Fig. 2 (right), when resistance short circuit occurs in traction network, in order to facilitate the analysis of transient circuit model in the presence of transition resistance, the time domain circuit diagram is transformed into frequency domain equivalent model.

According to the circuit, the formula of node voltage is listed as follows:

$$\left(\frac{1}{R_l + sL_l} + \frac{1}{R_z} + \frac{1}{R_r + sL_r}\right) \cdot U_n(s) = \frac{U(s)}{R_l + sL_l} + \frac{U(s)}{R_r + sL_r} \tag{8}$$

According to KVL, the short-circuit current of the left rectifier is:

$$I_1(s) = \frac{U(s) - U_n(s)}{R_l + sL_l} \tag{9}$$

Then the response transfer function of the short-circuit current network of the left rectifier is as follows:

$$H_1(s) = \frac{I_1(s)}{U(s)} = \frac{(R_r + sL_r)}{(L_l L_r)s^2 + (R_z L_r + R_l L_r + R_r L_l + R_z L_l)s + (R_r R_z + R_l R_r + R_z R_l)} \tag{10}$$

By further transforming the transfer function, the coefficients before the highest order of denominator and molecule in the transfer function are extracted, and the function gain $K_l = L_r / L_l L_r = 1/L_l$ is obtained. Similarly, the function gain $K_r = 1/L_r$ of the right rectifier can be obtained. In order to find out the relationship between transfer function molecule and denominator, different short-circuit fault distances are set respectively. The root locus of transfer function is plotted by using MATLAB programming. It is found that no matter where the fault distance is set, there always exists a pair of zero and pole, which can be canceled with each other. Therefore, by further simplifying the formula, the transfer function of the network can be obtained as follows:

$$H_1(s) = \frac{1/L_l}{s + \alpha} \tag{11}$$

There, $\alpha = \mathrm{R}_z + R_l + R_z \cdot (x/L - x)$, when the input is step function, i.e. $U(s) = 1/s$, By inversely transforming the transfer function into a time function in time domain through Laplace, the response equation of short circuit current in time domain is obtained as follows:

$$i_1(t) = \frac{1}{\alpha L_l} \mathrm{e}^{-\alpha t} \tag{12}$$

$$i_2(t) = \frac{1}{\alpha L_r} \mathrm{e}^{-\alpha t} \tag{13}$$

Therefore, the short-circuit fault distance formula can be obtained according to the ratio of the initial transient value of short-circuit current on both sides of traction station at the time of short-circuit fault occurrence:

$$\frac{i_1(0^+)}{i_2(0^+)} = \frac{1/L_l}{1/L_r} = \frac{L-x}{x} \tag{14}$$

From the above analysis, it can be seen that the fault distance can be determined by the transient ratio of short-circuit currents on both sides of the fault point at the time of fault occurrence, whether the traction network is metal short-circuited or resistance short-circuited.

4 Simulation Verification

As shown in Fig. 3, short-circuit fault simulation model based on MATLAB/Simulink is built. The rectifier is obtained by parallel connection of 12-pulse rectifier transformers. The short-circuit simulation parameters are as follows [11]: catenary resistance is 0.028 Ω/km, inductance is 3.424 mH/km; track resistance is 0.023 Ω/km, the inductance is 1.78 mH/km; transition resistance between short-circuit point catenary and track is 0.01 Ω; The distance between two adjacent rectifiers is 4 km and the start time of short-circuit fault is 0.1 s. Different short-circuit locations are set to run the simulation models separately so as to get the current waveforms of different distances.

When the metal short-circuit fault occurs at 100 m away from the traction station A, the short-circuit current waveforms measured from both ends A and B of the traction station are shown in Fig. 4. It can be seen that after the short-circuit fault occurs, the current at both ends increases exponentially to the steady-state value.

Figure 5 shows the waveform of the current ratio of the two traction stations when the resistance short-circuit fault occurs at a distance of 100 m from the traction station A. It can be seen from the figure that at the time of short-circuit fault, the current ratio will have an obvious pulse. When the pulse value is collected in time, the fault distance can be judged accurately according to the formula of transient ratio method.

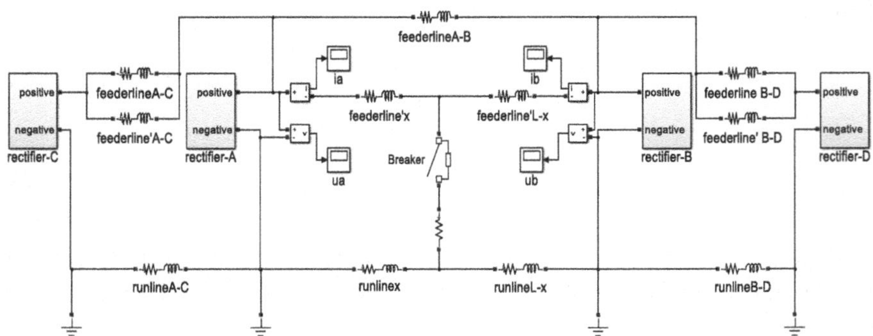

Fig. 3 Short-circuit simulation of traction network

Figure 6 shows the waveform of fault distance obtained by fault analysis method at the fault distance of 100 m at the fault time 0.1 s. When the short-circuit fault occurs near the rectifier, the current difference between the rectifiers on both sides is large, so the noise interference to the fault location waveform is relatively large.

For the problems of fault analysis method, the low-pass filtering method is used to reduce the location error. The definition of "post-filter" is to filter the waveform of fault distance x directly, and the definition of "pre-filter" is to filter the short-circuit current and voltage at both ends, and then to calculate the fault location by fault analysis method. Figure 7 shows the range finding results using "post-filter" at different cut-off frequencies. Considering the reduction of filtering overshoot and filtering effect, the cut-off frequency of 20 Hz is selected to filter x, and the average value of the stable data in the range of 0.45–0.5 s is selected as the final ranging result.

After a comprehensive analysis of the above fault location methods, MATLAB/Simulink simulation was carried out when metal short circuit occurs, and the accuracy of different fault location methods is obtained as shown in Table 1.

From the results of Table 1, it can be seen that no matter where the fault occurs, the accuracy of transient ratio method is the highest. Traction network parameters are used in two-terminal fault location based on fault analysis method and transient ratio method. Because of skin effect during the transient process of short circuit, the resistance and inductance parameters of the track will change a little; In addition, the actual values of traction network resistance and inductance parameters are also affected by factors such as soil conditions under the line, so there is a certain error between the actual parameters of traction network and the assumed unit impedance, which will also have a certain impact on the ranging accuracy. Therefore, fault location is carried out by assuming that the traction network parameters are increased or decreased with the same proportion, so as to study the influence of the uncertainty of traction network parameters on the results of fault location. The comparison results are shown in Fig. 8.

Fig. 4 Metallic short-circuit current waveform

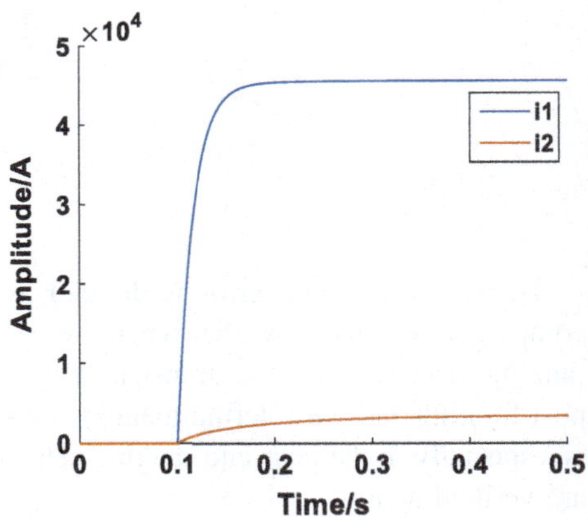

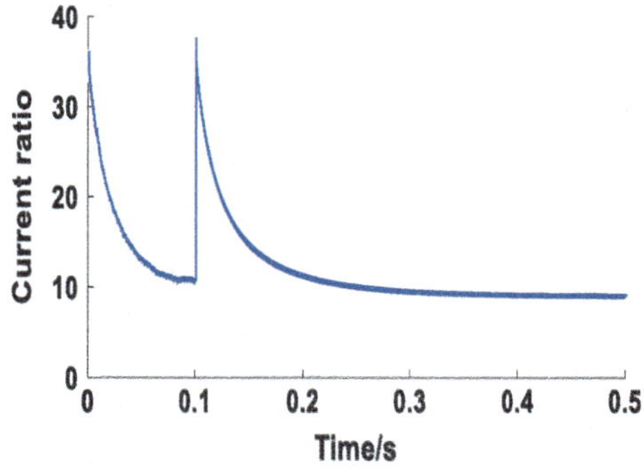

Fig. 5 Transient current ratio

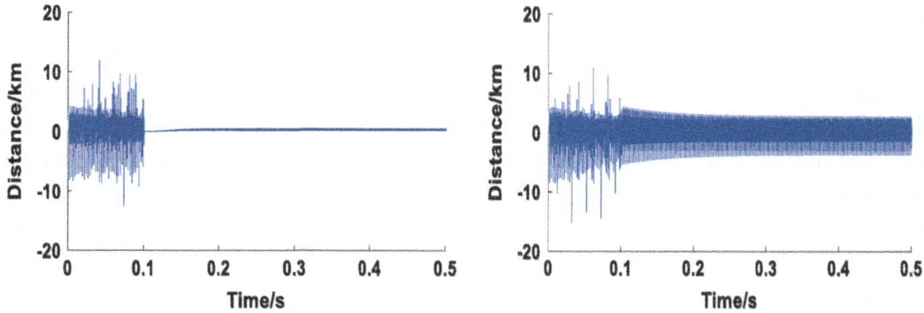

Fig. 6 Fault distance simulation diagram under different transition resistance. (Left) The transition resistance is 0.01 Ω. (Right) The transition resistance is 1000 Ω

Fig. 7 The waveform diagram after filtering the fault distance

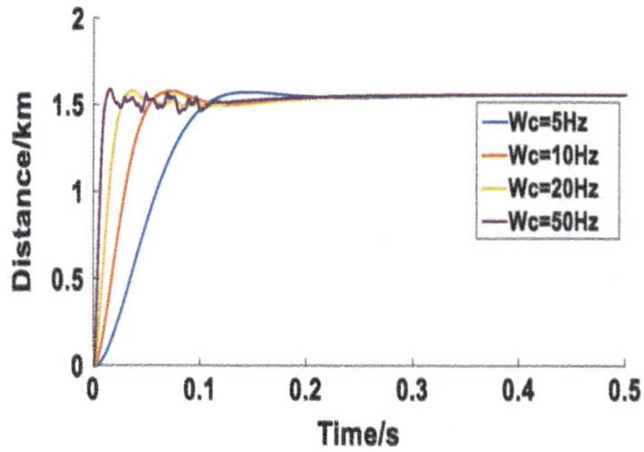

There, the absolute error is defined as AE and the relative error is RE. The comparison results show that when the parameters of traction network change, the ranging accuracy of the transient ratio method (red line) is better than the post-filtering and pre-filtering methods no matter how large the transition resistance is, especially in the near-end ranging. The accuracy and universality of this method are verified again.

Table 1 Comparison of errors of several ranging methods

The fault distance (m)	Methods	The measured distance (m)	Absolute error (m)	The relative error (%)
50	Post-filtering	84.7	34.7	69.4
	Prior-filtering	55.8	5.8	11.6
	Transient ratio	51.5	1.5	3
100	Post-filtering	149.0	49	49
	Prior-filtering	106.0	6.0	6
	Transient ratio	101.6	1.6	1.6
500	Post-filtering	514.6	14.6	2.92
	Prior-filtering	506.5	6.5	1.3
	Transient ratio	501.3	1.3	0.26
1500	Post-filtering	1470.1	29.9	1.9
	Prior-filtering	1502.6	2.6	0.173
	Transient ratio	1500.5	0.5	0.03

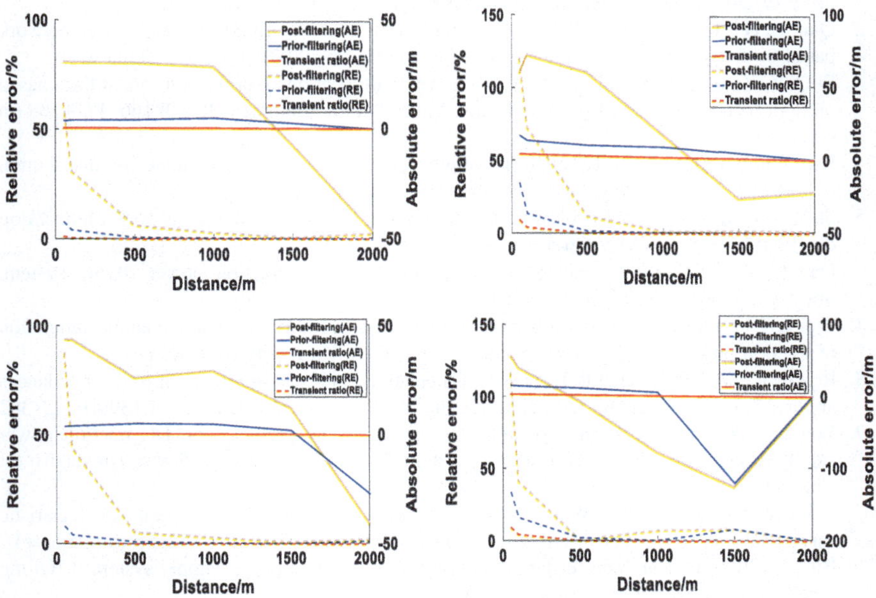

Fig. 8 (Top) Increase resistance and inductance by 50% at the same time. (Left) Metallic short circuit. (Right) High impedance short circuit. (Bottom) Resistance decreases by 50% and inductance increases by 50%. (Left) Metallic short circuit. (Right) High impedance short circuit

5 Conclusion

For the inaccuracy of short-circuit fault location of metro feeder, this paper proposes a transient ratio method, which uses the ratio of the initial value of current faults of traction stations at both ends to locate the fault. Compared with the traditional steady-state ratio method, this method overcomes the problem that the steady-state value cannot be obtained due to protection tripping, or that the ranging error occurs at different time of double-end tripping. Compared with the fault analysis method, it avoids the noise interference caused by the large current difference between the two terminals when short circuit occurs and is also less affected by the inaccurate parameters of the traction network. The principle of this method is simple, no matter whether it is metallic short circuit or resistance short circuit. It has high sensitivity and small ranging error, so it is a useful ranging method.

Acknowledgements This research was supported by the Fundamental Research Funds for the Central Universities (2017JBM062, 2017JBZ101).

References

1. Ren Qing W, Liu Song C, Li Y et al (2018) Simulation research on short-circuit fault of metro drive system. Electric Drive for Locomotives 6:89–92 (in Chinese)
2. Qian L, Hai Lang H (2018) Research on fault location of electrified railway traction network based on C-type traveling wave method. Electric Railw 29(03):14–18 (in Chinese)
3. Si Qi Z, Yong Li L, Xiao Long C (2018) Novel fault location algorithm for mixed lines based on fault section fast identification. Electric Power Autom Equipment 38(10):166–171+183 (in Chinese)
4. Kui Yin H (2018) Two Terminal Traveling wave fault location technique for distribution hybrid line. Telecom Power Technol 35(6):48–50 (in Chinese)
5. Saha MM, Izykowski J, Rosolowski E (2010) Fault location on power networks. https://doi.org/10.1007/978-1-84882-886-5
6. Dan C (2012) Study of fault location method for metro traction power supply system. Southwest JiaoTong University (in Chinese)
7. Wei L (2009) Research on optimization of urban railway train control and dynamic simulation of traction power supply system. Southwest JiaoTong University (in Chinese)
8. Rown JC, Allan J, Meilitt B (1992) Calculation and measurement of rail impedances applicable to remote circuit fault currents. Electric Power Appl IEEE Proc B 139(4):295–302
9. Wei K (2005) Investigation into fault analysis and some problems on DC traction power supply system for urban railway transit system. North China Electric Power University (in Chinese)
10. Liang Wei L, Qun Zhan L, Wei L (2007) Simulation and application of external characteristic carve of 24-pulse rectifier in urban rail transit. Urban Mass Transit 10:52–55 (in Chinese)
11. Bing L (2009) Research on feeder protection for DC traction power supply system. J Beijing JiaoTong Univ 33(5):65–68 (in Chinese)

Multi-level Criticality Analysis of Metro Vehicle Bogie Based on Fuzzy Theory

Xiuqi Wang, Yong Qin, Guidong Tao, Yong Fu and Meng Ye

Abstract This paper presents a fuzzy-based methodology using triangular fuzzy numbers to determine the weight of influencing factors and construct decision matrices in order to reduce the subjective uncertainty associated with influencing factors and decision-makers in traditional criticality analysis. A multi-level evaluation system is established to realize the hierarchical criticality quantitative analysis for failure modes of components and the system. The bogie system as a case example is used to demonstrate the proposed method and the quantitative criticality index of the bogie system can be obtained by using the multi-level fuzzy comprehensive evaluation method. The results show that the proposed method can be used effectively for quantitative criticality analysis of the systems.

Keywords Bogie · Criticality analysis · Fuzzy theory · Analytic hierarchy process (AHP) · Multi-level evaluation

1 Introduction

Criticality analysis is a ranking procedure for possible failure mode as per the combined influence of occurrence (O), severity (S), and detection (D). It can intuitively obtain the criticality of failure modes and present proposals to improve weak parts and links of the system. The bogie system is the key subsystem of the

X. Wang · Y. Qin · Y. Fu · M. Ye
State Key Laboratory of Rail Traffic Control and Safety, Beijing Jiaotong University, Beijing 100044, China

Y. Qin (✉)
Beijing Engineering Research Center of Urban Traffic Information Intelligent Sensing and Service Technologies, Beijing Jiaotong University, Beijing 100044, China
e-mail: yqin@bjtu.edu.cn

G. Tao
Qingdao Sifang Locomotive Co., Ltd., Qingdao, Shandong 266111, China

Y. Qin et al. (eds.), *Proceedings of the 4th International Conference on Electrical and Information Technologies for Rail Transportation (EITRT) 2019*, Lecture Notes in Electrical Engineering 639, https://doi.org/10.1007/978-981-15-2866-8_16

metro train. Its safety has a direct effect on the operation of the whole train. Therefore, the criticality analysis of the bogie is significant.

CA typically uses a Risk Priority Number (RPN) for ranking the failure modes. RPN is commonly calculated as the product of O, S, and D of the failure modes. Though it is used widely and operated easily, there are deficiencies of conventional RPN: (1) Different evaluation of O, S and D may lead to identical RPN values even if their risk implications are totally different; (2) there is great subjective uncertainty on expertise to assess the degree of criticality [1, 2].

In order to solve the above problems, some researchers attempted to use the fuzzy theory and establish the fuzzy comprehensive evaluation system to get more scientific and reasonable results: According to this method, Li et al. [3] obtained a more practical criticality sequence of freight trains; Li et al. [4] used the fuzzy number to obtain the weight of influencing factors, reducing the influence of subjective factors, and finally got the criticality ranking of failure modes on subway; Lin et al. [5] got the failure modes ranking of a circuit board in maglev train through fuzzy comprehensive evaluation system, which declines the subjective uncertainty of expert scoring.

However, these methods only evaluate the criticality of a single failure mode. In practical application, people also need the criticality of crucial components, equipment, and even the whole system, to optimize the components and equipment and improve the reliability of the system.

Therefore, this paper establishes a multi-level fuzzy comprehensive evaluation model and separates bogie into three levels: system, equipment, and component. First, fuzzy AHP is used to determine the weight of O, S and D followed by fuzzy comprehensive evaluation to obtain the criticality of each failure mode and then build the evaluation model of components based on the failure modes. Finally, the criticality of the system can be obtained step by step.

2 Fuzzy Comprehensive Evaluation

Fuzzy comprehensive evaluation refers to handle the various data and analyze factors influencing the evaluated object. Its basic idea is the use of the maximum membership principle and the fuzzy linear transformation theory [6, 7]. The specific steps are as follows.

(1) Establish influencing factors set

Choose occurrence, severity, and detection as influencing factors and constitute the influencing factors set.

$$U^1 = (u_1, u_2, u_3) \tag{1}$$

where u_1 is occurrence (O), u_2 is severity (S), and u_3 is detection (D).

(2) Establish the assessment set

Each influencing factor can be divided into five levels, and the classification criteria of different factors are according to Table 1.

$$V = \{1, 3, 5, 7, 9\} \tag{2}$$

where v_i represents the level that needs to be evaluated for each factor.

(3) Calculate the weight of influencing factors by FAHP

In conventional AHP, the ranking is basically done with the opinion of people but does not consider the uncertainty in judgment due to natural language. The evaluation of decision-maker has a significant effect on their preferences. This indirectly affects the results of AHP. To circumvent these, we integrate fuzzy theory with AHP to optimize the uncertainty. FAHP relies on pairwise comparison with fuzzy triangular numbers ($c_{ij} = (l_{ij}, m_{ij}, n_{ij})$). The process is as follows.

First, use fuzzy linguistic term sets to assign the priority of one criterion over that of another one and construct a pairwise comparison matrix. To achieve this, the linguistic terms will be converted to fuzzy numbers shown in Table 2.

$$C = \begin{bmatrix} 0 & c_{12} & c_{13} \\ c_{21} & 0 & c_{23} \\ c_{31} & c_{32} & 0 \end{bmatrix} \tag{3}$$

where C is the pairwise comparison matrix.

Second, aggregate the opinions of decision-makers and produce pairwise comparison matrices by Formula (4).

Table 1 Evaluation levels of influencing factors

Influencing factors	Evaluation levels				
	1	3	5	7	9
u_1(O)	Very low	Low	Medium	High	Very high
u_2(S)	Very low	Low	Medium	High	Very high
u_3(D)	Very high	High	Medium	Low	Very low

Table 2 Triangular fuzzy numbers with linguistic terms (Ambiguity is 0.2)

Linguistic terms	Fuzzy triangular numbers	Reciprocal fuzzy triangular numbers
Slightly high	(0, 0, 0.1)	(0.9, 1, 1)
Fairly high	(0.1, 0.3, 0.5)	(0.5, 0.7, 0.9)
High	(0.3, 0.5, 0.7)	(0.3, 0.5, 0.7)
Obviously high	(0.5, 0.7, 0.9)	(0.1, 0.3, 0.5)
Highly high	(0.9, 1, 1)	(0, 0, 0.1)

$$C = [c_{ij}]_{3\times3} = \left[\frac{1}{k}\left(c_{ij}^1 + c_{ij}^2 + \cdots + c_{ij}^k\right)\right]_{3\times3} \quad (4)$$

where $c_{ij} = (l_{ij}, m_{ij}, n_{ij})$ is the aggregated fuzzy comparison matrix and k is the number of decision-makers.

Third, calculate the fuzzy weight matrix $W^1 = (w_1^1, w_1^2, w_1^3)$ by using Formula (5).

$$w_i^1 = (w_{il}^1, w_{im}^1, w_{in}^1) = \left(\frac{\sum_{j=1}^3 l_{ij}}{\sum_{i=1}^3 \sum_{j=1}^3 l_{ij}}, \frac{2\sum_{j=1}^3 m_{ij}}{6}, \frac{\sum_{j=1}^3 n_{ij}}{\sum_{i=1}^3 \sum_{j=1}^3 n_{ij}}\right) \quad (5)$$

where w_i^1 denotes the fuzzy weights, $i = 1, 2, 3$.

Fourth, perform the defuzzification process. Convert the fuzzy weights that are obtained into crisp numbers. Formula (6) can be used for this procedure [8].

$$w_i^1 = \frac{w_{il}^1 + 2w_{im}^1 + w_{in}^1}{4} \quad (6)$$

Finally, normalize the crisp weights by Formula (6) as follows:

$$w_i^{1n} = w_i^1 / \sum_{i=1}^3 w_i^1 \quad (7)$$

(4) Construct the fuzzy decision matrix

In order to reduce the uncertainty, we deliberately transform the traditional precise values to five levels after normalizing the ratings. Fuzzy linguistic variables are: very low, low, medium, high, and very high. A triangular fuzzy number has been used to represent the five-level fuzzy linguistic variables.

The fuzzy decision matrix of failure mode f can be established by setting a suitable membership function, which reduces the subjectivity of the evaluation partly.

For a five-level evaluation set, the commonly used membership diagram is as follows: The abscissa expresses the fuzzy level of influencing factors, which increases in turn from 0 to 10, the ordinate represents the membership degree.

Taking the occurrence as an example, if the probability range of a failure mode is 1–5, the triangle is drawn with (1, 0), (3, 1), (5, 0) as vertices in Fig. 1a. The ordinate of the intersection on the graph constitutes the fuzzy evaluation set of the occurrence $X_1^f = [0.33 \quad 1 \quad 0.66 \quad 0 \quad 0]$.

After normalization, the matrix X^f is transformed into $X^{f'}$.

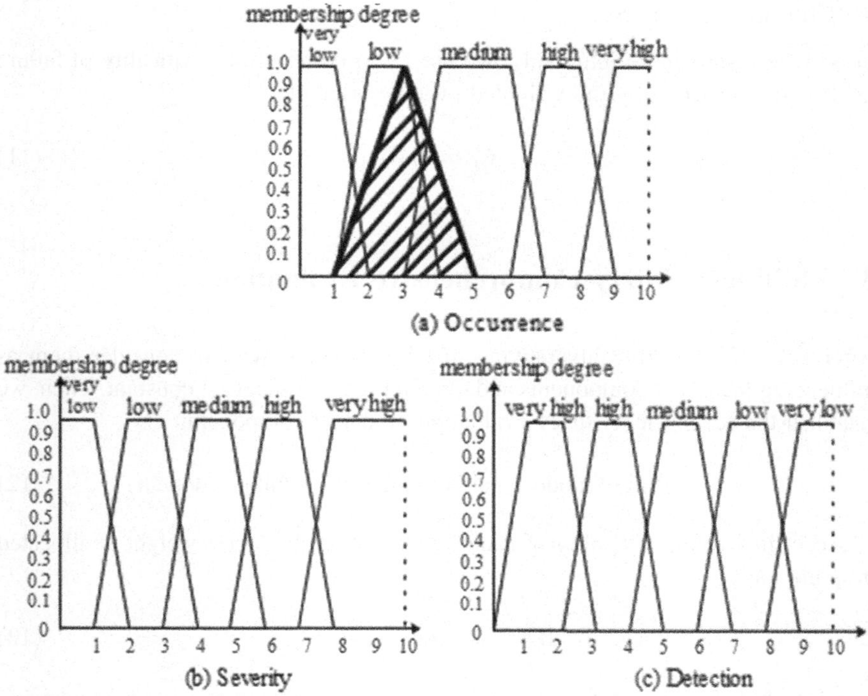

Fig. 1 Membership function

$$X^f = \begin{bmatrix} X_1^f \\ X_2^f \\ X_3^f \end{bmatrix} = \begin{bmatrix} x_{11}^f & x_{12}^f & x_{13}^f & x_{14}^f & x_{15}^f \\ x_{21}^f & x_{22}^f & x_{23}^f & x_{24}^f & x_{25}^f \\ x_{31}^f & x_{32}^f & x_{33}^f & x_{34}^f & x_{35}^f \end{bmatrix} \quad (8)$$

$$X^{f'} = \left[x_{ij}^{f'}\right]_{3\times 5} = \left[x_{ij}^f / \sum_{j=1}^{5} x_{ij}^f\right]_{3\times 5} \quad (9)$$

(5) Establish the evaluation model

Based on the normalized fuzzy decision matrix and influencing factors weights, the fuzzy comprehensive evaluation of failure mode f is as follows.

$$B_f = W^{1n} \cdot X^{f'} = [b_{f1}, b_{f2}, b_{f3}, b_{f4}, b_{f5}] \quad (10)$$

(6) Criticality sequences

To see the results more intuitively, we use C_f to represent the criticality of failure mode f to the system by the weighted average principle.

$$C_f = B_f V^T \tag{11}$$

3 Multi-level Fuzzy Comprehensive Evaluation

After finishing the criticality ranking of failure modes, we can consider them as influencing factors of components and remain the assessment set constant. Then, we establish the second-level fuzzy evaluation model of component.

$$U^2 = (f\text{ailure mode } 1, f\text{ailure mode } 2, \ldots, \text{failure mode } n) \tag{12}$$

According to the criticality of each failure mode, the factor weight is allocated proportionally.

$$W^2 = (w_1, w_2, \ldots, w_n) \tag{13}$$

After that, we repeat steps 4 through 6 of fuzzy comprehensive evaluation to acquire the criticality rankings of components. Then, the third-level fuzzy evaluation model of equipment can be established by the same method.

4 Case Verification and Result Analysis

In this case, we study of a certain metro bogie system to verify the correctness and feasibility of the model. The bogie system is divided into nine kinds of equipment by its function: frame, wheelset, axle box, primary suspension devices, secondary suspension devices, driving devices, brake devices, central traction connection devices, and auxiliary devices. Each type of equipment can be further divided into different components, and finally the three-level structure, namely "system-equipment-component," is constructed. Then, we can calculate the criticality step by step.

We hire three decision-makers from metro company and let them evaluate three factors, including occurrence, severity, and detection. Then, the weight of influencing factors is applied by using Formulas (3)–(7).

$$W^{1n} = \left(w_1^{1n}, w_2^{1n}, w_3^{1n}\right) = (0.26, 0.48, 0.26)$$

After that, three decision-makers score the probability range of influencing factors about each failure mode, as shown in Table 3. By membership diagram and normalization, we obtain the fuzzy decision matrixes corresponding to each failure mode. Then using Formulas (10) and (11), criticality rankings of failure modes can be obtained, shown in Table 4.

On this basis, failure modes of each component are taken its influencing factors and the second-level evaluation model is established. According to the criticality of each failure mode, the weight of influencing factors is allocated. An example of axle is shown in Table 5.

After computing, criticality sequences of components can be acquired. Taking components as the influencing factors of equipment, the third-level evaluation model is established, and the rankings of criticality are calculated in Table 6, which are consistent with the actual situation. It shows that this method has certain application value and can provide a data basis for subsequent reliability allocation.

From the above results, we can see that we should focus on the frame, wheels, gearboxes, and motors. Therefore, we put forward the following suggestions: (1) Select materials with higher strength and excellent tolerance to make the frame, improve the existing welding process [9]; (2) pay attention to the matching degree between track and wheel, and improve the distribution of train braking force [10]; and (3) record the external performance of the motor in time when it is broken, so as to provide information for fault diagnosis in daily maintenance [11].

Table 3 Fuzzy assessments of influencing factors for component failure modes (section)

Component	Failure mode	Occurrence	Severity	Detection
Axle	1. Small crack	2–6	6–7	4–5
	2. Crack or severe crack	1–3	8–10	2–3
…	…	…	…	…

Table 4 Criticality rankings of failure modes (section)

Criticality rankings	Component	Failure mode	Criticality
1	Frame assembly	Small crack	6.59635
2	Wheel	Tread crack	6.46017
…	…	…	…
68	Temperature sensor	Unable to detect abnormal temperature	4.14056

Table 5 Weight distribution of influencing factors in axle

Component	Influence factor	Criticality	Weight distribution
Axle	1. Small crack	5.58137	0.525
	2. Crack or severe crack	5.04800	0.475

Table 6 Criticality rankings of equipment

Criticality rankings	Equipment	Criticality
1	Frame	3.119
2	Driving device	2.912
3	Wheelset	2.790
4	Basic brake device	2.739
5	Axle box	2.736
6	Central traction connection device	2.730
7	Primary spring suspension device	2.528
8	Secondary spring suspension device	2.389
9	Auxiliary device	1.981

5 Conclusion

In this paper, the fuzzy theory is applied to the bogie system, which reduces the subjective uncertainty of traditional criticality analysis and quantitatively expresses the criticality degree of each failure mode clearly. By building the multi-level fuzzy comprehensive evaluation model, we can not only acquire criticality of failure modes but also calculate the influence of each subsystem on its superior system and finally prioritize the criticality sequences of failure modes, components, and equipment. As a result, fuzzy mathematics could effectively solve the ambiguity and complexity phenomenon involved in hazard analysis, indicating this method could be widely used in the bogie system.

Acknowledgements The authors gratefully acknowledge the financial support for this research from the National Key Research and Development Program of China (2016YFB1200505-014), High-level Talents Training Program of Ministry of Transportation of China (No. I18I00010), and Major Program of National Natural Science Foundation of China (No.61833002): Fundamental Theory and Methods in Operational Risk Assessment and Control of High Speed Train.

References

1. Shi W, Gu M, Li N et al (2017) Application of improved risk priority number ranking method in reliability analysis of punching machine. Manuf Technol Mach Tool 2:48–52 (in Chinese)
2. Chen Z, Fu G, Zhao Y (2011) Improved analysis method of risk priority number. J Beijing Univ Aeronaut Astronaut 11:1395–1399 (in Chinese)
3. Li Y, Zhao W (2009) Criticality analysis method for the failure mode effects of railway freight car. China Railw Sci 30(3):103–108 (in Chinese)
4. Li G, Tan N, Zhang J (2012) Criticality analysis of subway train equipment based on improved analytical hierarchy process. J Electron Meas Instrum 26(6):503–507 (in Chinese)
5. Lin G, Guo Y, Wang S (2017) Application of fuzzy criticality evaluation method in failure mode effects and criticality analysis for circuit board in maglev train. Mod Electron Tech 16:1–4, 9 (in Chinese)

6. Zhao Y, Jiao J, ZhaoT d (2015) Risk assessment method based on fuzzy logic. Syst Eng Electron 08:1825–1831 (in Chinese)
7. Liu R, Zhang J, Han X (2008) Risk identification based on triangular fuzzy number for construction project. J Harbin Inst Technol 10:1617–1620 (in Chinese)
8. Su SB, Huang RH (2006) Attribute hierarchical mode based on triangular fuzzy number. Syst Eng Theor Pract 12:115–119 (in Chinese)
9. Wang J (2004) Research on crack problem of bogie frame of Shanghai metro vehicle. Beijing Jiaotong University, Beijing (in Chinese)
10. Wu B (2011) Discussion of causes to cracking in wheel treads of urban rail vehicles. Rolling Stock 11:37–40+48 (in Chinese)
11. Liu H (2010) Development of on-line fault diagnosis system for motor bearing of railway vehicle bogie. Taiyuan University of Science and Technology, Taiyuan (in Chinese)

Research of Turnout Fault Diagnosis Method Based on Qualitative Trend Analysis

Yuanyuan Zhou, Yulin Han and Zongyi Xing

Abstract The turnout is connected to different tracks and usually installed between two or more strands, not only is responsible for switching track line but also ensures operation safety. It is of great practical significance to study the fault diagnosis methods. The fault diagnosis method based on qualitative trend analysis is to use the interval semi-division method to extract the trend of different state signals, and match the diagnostic signal with it to obtain the result. The experimental results show that the presented method has good accuracy.

Keywords Turnout · Qualitative trend analysis · Fault diagnosis · Section half discerption

1 Introduction

As the throat of the railway line, the state of the turnout directly affects the safe operation of the train and the smooth operation of the railway line [1]. However, with the increasing railway lines in China, the failure rate of turnout equipment is also increasing. Therefore, it is important to study the diagnosis method of turnout fault. Based on qualitative trend analysis, the fault diagnosis method of turnout is first analyzed. Then this essay introduces the qualitative trend analysis and fault diagnosis process, focuses on the three aspects of trend language, trend extraction technology, and trend matching technology, and gives the basic framework. Finally, method validation and results analysis are performed.

Y. Zhou · Y. Han · Z. Xing (✉)
School of Automation, Nanjing University of Science and Techonology, Nanjing, China

Y. Qin et al. (eds.), *Proceedings of the 4th International Conference on Electrical and Information Technologies for Rail Transportation (EITRT) 2019*, Lecture Notes in Electrical Engineering 639, https://doi.org/10.1007/978-981-15-2866-8_17

2 Turnout Failure Mode Analysis

This paper investigates the ZD6-type switch (electric switch machine) installed on the Guangzhou subway line [2]. According to the on-site maintenance data and the latest version of the "Railway Signal Maintenance Rules" and "Guangzhou Metro Line Maintenance Regulations," four types of fault modes with the highest frequency appearing on the line are summarized, which are faults for unlocking difficulties, faults for switching step resistance, faults for switching sawtooth resistance, and faults for locking difficulties [3].

3 Qualitative Trend Analysis

The idea of qualitative trend analysis (QTA) is to transform quantitative process trend data into qualitative trend segments and describe each trend segment in symbolic language [4]. The fault diagnosis method using qualitative trend analysis technology usually includes the following three parts [5]: (1) language describing the trend; (2) trend extraction; and (3) trend matching.

3.1 Primitive Trend Description Language

In the primitive trend description language, the primitive is the smallest component of the trend description language, and any process signal can be decomposed into a primitive sequence. As shown in Fig. 1, there are a total of 7 primitives.

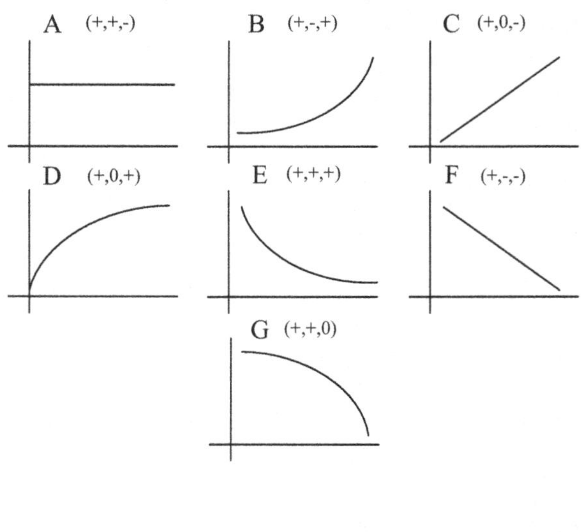

Fig. 1 Primitive diagram

The first- and second-order differential symbols of the primitive can be obtained by pre-differentiating the signal, as follows.

$$\frac{dy_i}{dx} = \frac{1}{\Delta_i}(y_i - y_{i-1}) \tag{1}$$

$$\frac{d^2y_i}{dx^2} = \frac{1}{\Delta_i^2}(y_i - 2y_{i-1} + y_{i-2}) \tag{2}$$

Each primitive includes the sign of the signal, the sign of the first-order differential of the signal, and the sign of the second-order differential, so the primitive can reflect the positive and negative, rising and falling and concave and convex conditions of the signal during the time period of the primitive, which also ensures the sequence of the primitive. It is possible to retain almost all important trend information in the process signal.

3.2 Trend Extraction Based on Interval Semi-division

Wavelet Denoising: Wavelet threshold denoising is applied to the signal before trending the signal. Any process signal can be expressed by Eq. (3).

$$y = f(t) + e(t) \tag{3}$$

where $y(t)$ is the noisy signal, $f(t)$ is an unknown original signal, and $e(t)$ is independent and obeys a Gaussian normal distribution of noise estimates. Wavelet decomposition of $y(t)$ yields Eq. (4).

$$y(t) \approx \sum_k s_{J,k}\phi_{J,k}(t) + \sum_{j=1}^{j=J} d_{j,k}\psi_{j,k}(t) \tag{4}$$

where J is the scale ordinal, k is the coefficient of the j layer, $s_{J,k}$ is the smoothing factor, and $d_{J,k}, \ldots, d_{J,1}$ is the detail coefficient. The functions $\phi_{J,k}(t)$ and $\psi_{j,k}(t)$ are approximate wavelet functions by scaling and translating the wavelet basis functions.

Interval Semi-Division Method: A time series function $y(t)$ can be approximated by a polynomial with exponent n, and likewise $y(t)$ can be approximated by a set of piecewise monomodal polynomial $p_i(t)$ over a unimodal region U_i with an exponent not exceeding n. That is $y(t) = \{p_1(t), p_2(t), p_3(t), \ldots, p_M(t)\}$. Thus, the segmentation of the time series function can be transformed into determining the start and end positions of the interval and the polynomial within the interval.

The algorithm part of the interval semi-division method has two aspects: (1) Determine the sequence of a polynomial or a unimodal region by the interval semi-division method; (2) Based on the sign of the first-order differential and the second-order differential of the polynomial p_i, the elementary assignment is performed on the single-peak region U_i. We perform a least squares curve fit on y, so coefficient matrix $\hat{\beta} = \left[\hat{\beta}_0, \hat{\beta}_1, \ldots, \hat{\beta}_n\right]$, polynomial matrix $\mathrm{p} = [p_0, p_1, \ldots, p_n]$, and fitting result $\hat{y}$ can be got, which are as below

$$\hat{y} = T\beta, p_i(t) = \sum_{k=0}^{k=n\leq 2} \hat{\beta}_k t^k \tag{5}$$

where $\hat{y} = T\beta, p_i(t) = \sum_{k=0}^{k=n\leq 2} \hat{\beta}_k t^k$. In order to avoid erroneously defining T, each single-peak interval is normalized. The significance of the fitting error is represented by $\varepsilon_{\text{fit}}^2$ and is used to compare with the noise estimate σ_{noise}^2.

3.3 *Trend Matching Technology Based on Fuzzy Logic*

The matching is divided into two processes, namely primitive matching and trend matching.

Primitive Matching: The degree of similarity between different primitives is different. Dash [6, 7] proposed a method of using the primitive similarity matrix to reflect the degree of approximation between primitives. The calculation results are shown in Table 1, where $S_{P_1P_2}$ represents the similarity between P_1 and P_2, $S_{P_1P_2}$ has a value range of [0, 1]. As can be seen from Fig. 1, some of the primitives such as B and G have completely opposite trends, so the similarity between these primitives is zero.

Trend Matching: After the trend is extracted, the sensor signal is transformed into a set of trend sequences that describe the trend. The trend matching is to compare the trend sequence in the real-time signal with the fault trend rule in the knowledge base, and calculate the similarity of the real-time signal trend sequence and fault trend [8].

Table 1 Primitive similarity matrix

$S_{P_1P_2}$	A	B	C	D	E	F	G
A	1	0	0.25	0	0	0.25	0
B	0	1	0.75	0.5	0	0	0
C	0.25	0.75	1	0.75	0	0	0
D	0	0.5	0	1	0	0	0
E	0	0	0	0	1	0	0.5
F	0.25	0	0	0	0.75	1	0.75
G	0	0	0	0	0.5	0.75	1

The real-time signal trend sequence placed in the same timeline and the fault trend rule in the knowledge base can be distinguished and matched from three aspects: (1) the order of the primitive *a*; (2) the duration of the primitive *b*; and (3) the size of the change in primitive *c*. The degree of matching between the real-time signal trend sequence and the knowledge base's fault trend rule can be defined as:

$$SI = \sum_{i=1}^{i=R} S_{P_i P_i^*} \frac{\Delta t_{ui}}{\mathrm{T}_u} \tag{6}$$

where a represents the similarity between the primitives *b* and *c*, *d* is the total length of the interval of the new partition, and *e* is the length of the *i*th interval of the new partition.

4 Case Analysis

Ten groups are randomly selected from the 30 sets of current operation signals in each state. A total of 50 groups are used to establish fault diagnosis rules for five states, and the remaining 100 sets of current operation signals are used for trend matching and final fault diagnosis.

4.1 Establishing a Fault Knowledge Rule Base

The following is a failure-free trend rule as an example to illustrate the process of establishing a rule. Firstly, the signal curve is filtered, and then the filtered curve is subjected to trend extraction based on interval half-division. As is shown in Fig. 2b, the interval semi-division method divides the set of curves into 10 trend intervals, and the trend sequence of the curve is {*B*[0.168], *C*[0.291], *D*[0.381], *G*[0.616], *E*[1.344], *C*[1.792], *A*[2.846], *C*[3.148], *G*[3.372], *E*[3.686]}. (The number in square brackets after the primitive is the time value at the end of the interval.) By analogy, the trend series of 10 sets of trouble-free operation curves are summarized, and the approximate common part of the trend sequence is BDGEAGE. The schematic diagram is shown in Fig. 2a, b.

The trend rules of the four faults are formulated in a similar manner to the no-fault trend rules. The schematic diagram is shown in Fig. 2c–f.

According to a series of 40 trends with the four faults, four fault trend rules are established which are summarized in Table 2.

It can be seen from Table 2 that the trend rules for unlocking difficult faults and no faults are basically the same, and the intervals of each trend interval are slightly different. The main feature of the unlocking fault operation curve is that the

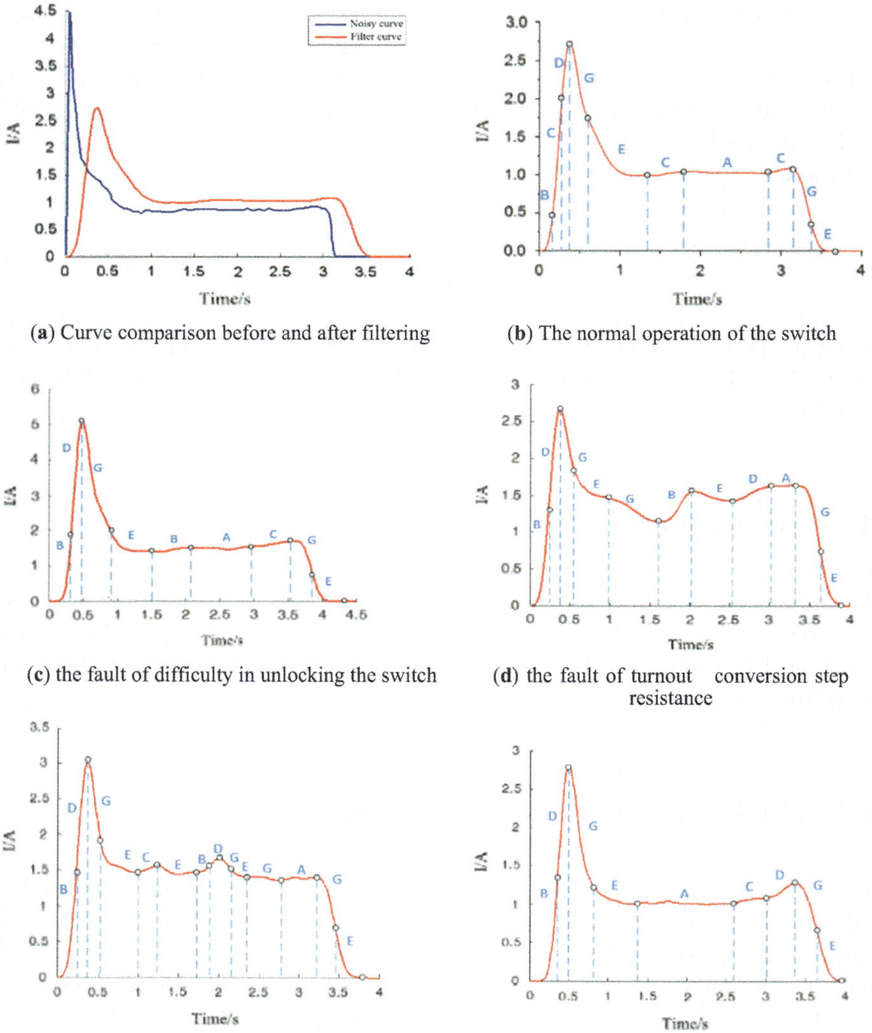

(**a**) Curve comparison before and after filtering

(**b**) The normal operation of the switch

(**c**) the fault of difficulty in unlocking the switch

(**d**) the fault of turnout conversion step resistance

(**e**) the fault of Switching sawtooth resistance

(**f**) the fault of difficulty in locking the switch

Fig. 2 Schematic diagram of the normal operation of the switch and the trend extraction of the fault current curve

unlocked portion of the curve has significantly increased amplitude compared to the unlocked portion of other state running curves, so in order to distinguish it from the non-faulty state, the matching degree between the trend sequence and the fault trend rule is calculated. The formula is modified to:

Table 2 Summary of failure trend rules

Number	Fault type	Trend rule sequence
0	Trouble free	BDGEAGE
1	Hard to unlock	BDGEAGE
2	Conversion step resistance	BDGEABAGE
3	Conversion sawtooth resistance	BDGEABEAGE
4	Difficulty in locking	BDGEACDGE

$$\mathrm{SI} = \frac{1}{T_u}\sum_{i=1}^{i=R} s_{P_iP_i^*}\frac{\Delta t_{ui}|\Delta d_{ui}|}{|\Delta d_{ui^*}|} \tag{7}$$

where Δd_{ui} is the amplitude difference between the first and last points of the *i*th trend interval of the diagnosed operation curve, and Δd_{ui} is the amplitude difference between the first and last points of the *i*th trend interval in the fault trend rule.

4.2 Fault Trend Matching

A group is randomly selected from the remaining 100 sets of operating current curves, taking this group as an example. Firstly, the running current curve of the group is filtered; secondly, the filtered curve is subjected to interval half-division trend extraction, and the extracted trend sequences are {*B*[0.355], *D*[0.494], *G*[0.816], *E*[1.372], *A*[2.593], *C*[3.005], *D*[3.383], *G*[3.656], *E*[3.905]}; finally, the obtained trend sequence is trend-matched with the normal and four fault trend rules, and the degree of matching of the trend sequence with respect to the five state trend rules is calculated.

As shown in Fig. 3, firstly, the curve corresponding to the faultless trend rule and the curve corresponding to the trend sequence are placed in the same time axis, and the action time of the shorter action curve is the upper limit of the interval, and the upper limit of the interval in Fig. 3, T_u is 3.83 s, and then the trend interval is redivided by integrating the original interval points of the two curves. The red point in the figure is the dividing point of the new partitioning interval, so the randomly selected running current curve is redivided into 12 intervals, namely $R = 12$.

After the trend segment resegmentation is completed, the trend primitive similarity sequence can be obtained as {1, 0.5, 0, 1, 0.5, 1, 0, 1, 0.25, 0, 0.5, 1}, and the random extraction curve is resegmented. The sequence of amplitude differences of the interval is {0.07, 1.24, 1.47, 0.67, 0.89, 0.10, 0.06, 0.01, 0.07, 0.22, 0.62, 0.6}, and the sequence of amplitude differences of the resegmentation interval of the non-fault trend rule is {1.35, 1.14, 0.59, 0.34, 0.37, 0.02, 0.02, 0.03, 0.02, 0.7, 0.52, 0.01}. Finally, the above parameters are substituted into Formula (7), and the

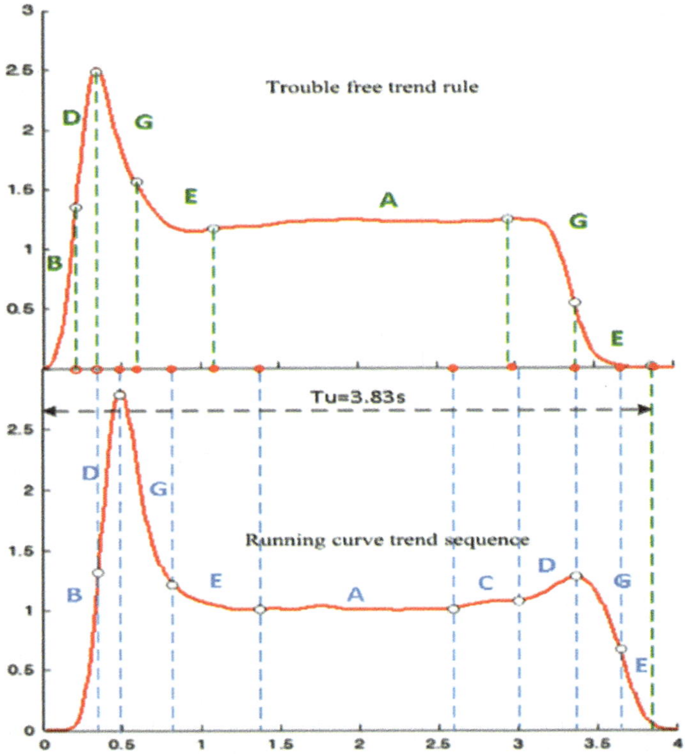

Fig. 3 Schematic diagram of trend matching calculation

matching degree between the randomly selected current running curve and the no-fault trend rule is 0.622.

Then the trend sequence is matched with the trend of the other four faults, and the matching degree of the curve calculated with the trend rule of the five states is 0.622, 0.5844, 0.5352, 0.6425, and 0.7950, respectively. The fault contained in the running current curve is a difficult locking fault.

5 Conclusion

In order to avoid the influence of random fluctuations and irrelevant trends in the signal curve on the fault diagnosis method of the switch, this paper introduces a method based on qualitative trend analysis to diagnose the switch fault and verify the method. Firstly, the fault type, qualitative trend analysis and fault diagnosis process based on qualitative trend analysis are briefly introduced. Secondly, it focuses on the three aspects of trend language, trend extraction technology and trend matching technology, and gives the basic framework of fault diagnosis based on qualitative trend analysis. Finally, the method of verifying the method with current running signal is described in detail.

Acknowledgements This work is supported by National Key R&D Program of China (2017YFB 1201102).

References

1. Zhang X (2010) Evaluation of Beijing Nanji Company's high-speed heavy-duty railway ballast project. Harbin Institute of Technology, Harbin (in Chinese)
2. Yao L (2014) Principle and analysis of ZD6 ballast current curve monitoring. Enterp Technol Dev 5:79–79 (in Chinese)
3. He Y (2014) Research on fault diagnosis method for high-speed railway turnout. Beijing Jiaotong University, Beijing (in Chinese)
4. Zhang H (2009) Application of qualitative trend analysis in fault diagnosis. Beijing University of Chemical Technology, Beijing (in Chinese)
5. Qing Y, Mo X, Wu S (2012) Overview of fault diagnosis technology and development trends. China Storage Transp 11:124–127 (in Chinese)
6. Song Z, Ningbo, Zhang X (2015) Research on fault diagnosis of nonlinear systems based on qualitative trend analysis. Comput Eng Appl 51(8):250–254 (in Chinese)
7. Dash S, Rengaswamy R, Venkatasubramanian V (2003) Fuzzy-logic based trend classification for fault diagnosis of chemical processes. Comput Chem Eng 27(3):347–362
8. Xue Y (2012) Research on the implementation method of turnout equipment fault diagnosis expert system. Beijing Jiaotong University, Beijing (in Chinese)

Recognition of Signal Fault Curves Based on Dynamic Time Warping for Rail Transportation

Shize Huang, Zaixin Wu, Fan Zhang, Kai Yu and Lingyu Yang

Abstract Signalling systems play a major role in railway reliability. However, microcomputer-based monitoring system (MMS), which monitors signal currents, simply raises an alarm when a signal has a failure but cannot recognize the exact reason of the failure. Therefore, we propose an intelligent diagnosis approach to help MMS to recognize faults automatically. First, the approach divides signal current curves collected by MMS into numerous sections with a same length of 600 s. Second, it utilizes dynamic time warping (DTW) to calculate similarities between reference curves and the 600 s-long curves and identify normal ones and a certain type of fault ones named as fluctuant curves. Third, our approach adopts three rules to further distinguish the rest into three types of fault curves. Finally, we conduct an experiment, and the results indicate that our approach can automatically diagnose signal fault curves with 100% accuracy.

Keywords Transportation · Engineering and technology · Transportation infrastructure · Roads · Signal · DTW · Fault recognition

---

S. Huang
Shanghai Key Laboratory of Rail Infra-Structure Durability and System Safety, Sichuan Railway Industry Investment Group Co., Ltd., Tongji University, 4800 Caoan Road, Shanghai, People's Republic of China
e-mail: hsz@tongji.edu.cn

Z. Wu
Sichuan Railway Industry Investment Group Co., Ltd., 4-1-1 Kitakaname, Hiratsuka, Kanagawa, 259-1292, Ja Building A, Liangjiang International, No. 535, Tianfu 1st Street, High-Tech Zone, Chengdu, People's Republic of China

F. Zhang · K. Yu
China Railway Second Hospital Engineering Group Co., Ltd., No. 3, Tongjin Road, Chengdu, Si-Chuan, People's Republic of China
e-mail: zhangfandyq@163.com

L. Yang (✉)
Key Laboratory of Road and Traffic Engineering of Ministry of Education, Tongji University, 4800 Caoan Road, Shanghai, People's Republic of China

Y. Qin et al. (eds.), *Proceedings of the 4th International Conference on Electrical and Information Technologies for Rail Transportation (EITRT) 2019*, Lecture Notes in Electrical Engineering 639, https://doi.org/10.1007/978-981-15-2866-8_18

1 Introduction

Demand for highly reliable railway transportation has been rising in many countries [1]. Maintaining safe and smooth traffic flows largely depend on signalling systems [2–7] as correct signals can avoid train overtaking and crashing or even serious accidents. A signalling system contains several devices, including a turnout, a signal, etc. All of them are monitored by a microcomputer-based monitoring system (MMS). MMS collects signal currents and raises alarms when signals are irregular. However, it cannot recognize reasons for those faults and has to depend on engineers inspecting the whole system to find out the problems. The manual inspection may cost extra time, labour and other resources, leading to train delay or even accidents. Further, human experiences are highly unstable, and a tiny misjudgement may cause a disaster. Therefore, fast and automatic identification of fault types is in a dire need.

Since MMS can present signal currents in chronological order in terms of curves in a current-time plane, fault diagnosis can be conducted by analysing those curves. Surprisingly, little attention has been devoted to this. But we could borrow similar approaches in other fields. Yan et al. collected landmines response curves and used back propagation (BP) neural network to detect landmines [8]. A. Gari et al. presented a method to estimate skew angle of printed documents based on Hough transform [9] X. Feng et al. utilized FCM to analyse actual power load curves and presented differences between various users [10]. Y. Liu analysed well logs with chaotic time series analysis [11]. X. Zhang introduced fuzzy neural network and support vector machines to diagnosis switch faults according to switch current curves [12]. H. Kim et al. proposed a dynamic time warping (DTW)-based method in order to tackle the duration-variation problem of railway point machine (RPM) movements [13].

Dynamic time warping (DTW) [14–18] can compare similarity between two time sequences, even the two sequences hold different nonlinear distortions [19]. DTW has been applied to many fields, including database index [20], handwriting recognition [21], engineering and biomedical information [22], gene regulatory network [23], computer vision [24] and so on. A variant of DTW is the extreme point warping technique [25] which selects points of the warped trace. Another modification includes the edit-distance-based string matching algorithm [26] which compares sequences of position external points between a test input and a set of reference signatures [27].

In this paper, we facilitate DTW to analyse signal current curves. First, we divide the curves into several parts with a same length of 600 s. Second, we utilize DTW to calculate the similarities between the parted curves and reference curves which include one normal curve and four types of fault curves. The DTW method only can distinguish a normal curve and a fault curve from the other three types of fault curves. Therefore, we adopt three rules to further determine the rest three faults and improve the diagnosis accuracy.

The rest of the paper is organized as follows. Section 2 introduces five types of signal current curves. Section 3 illustrates the method and principles of DTW. Section 4 presents numerical experiments and results of our approach. And we discuss the results and draw some conclusions in Sect. 5.

2 Signal Current Curve

Signal current curves can intuitively reflect the state of signals. A normal current curve is almost straight and stays at a standard value, as depicted in Fig. 1. Some peaks may emerge in a normal curve because of the instantaneous change of current when red lights turn into green or other colours. The standard value should not be zero, and the peaks should not last too long.

Typically, four types of curves are known as fault curves in practice, as shown in Figs. 2, 3, 4 and 5. Figure 2 depicts a curve, named as non-standard curve, whose values are lower than standard ones. Figure 3 plots a discontinuous curve which contains a sharp drop and a break. Figure 4 shows a fluctuant curve, which leads to a signal light brighter and darker at times. In Fig. 5, the current value falls suddenly and retains a low value, named as drop-off curve.

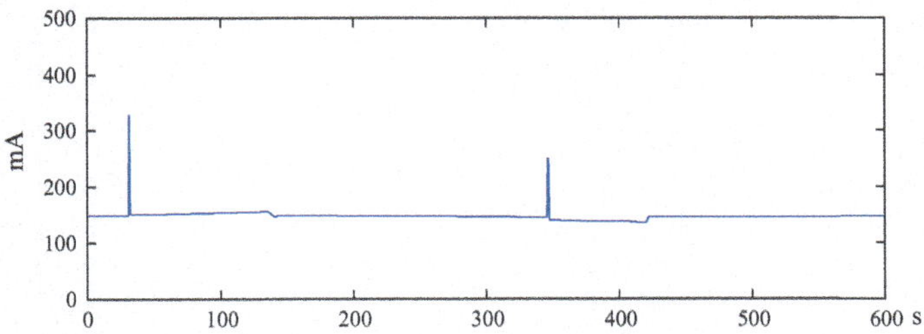

Fig. 1 Normal signal current curve

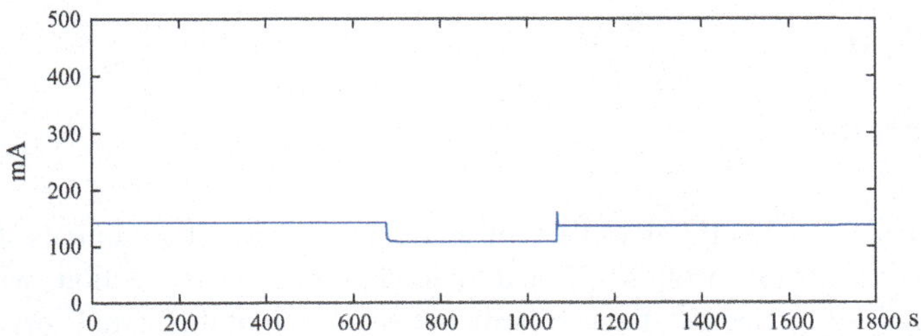

Fig. 2 Non-standard curve

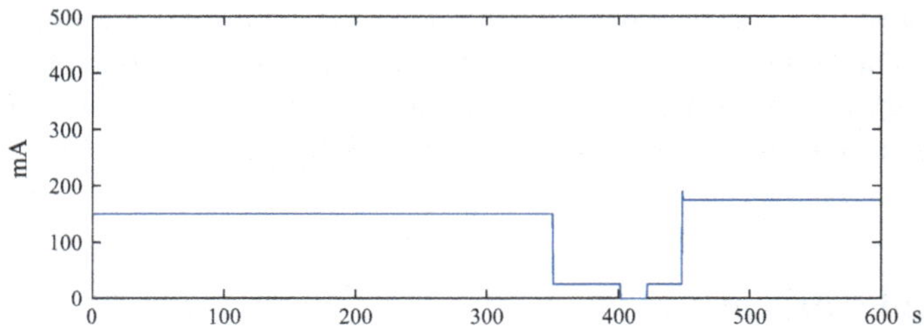

Fig. 3 Discontinuous curve

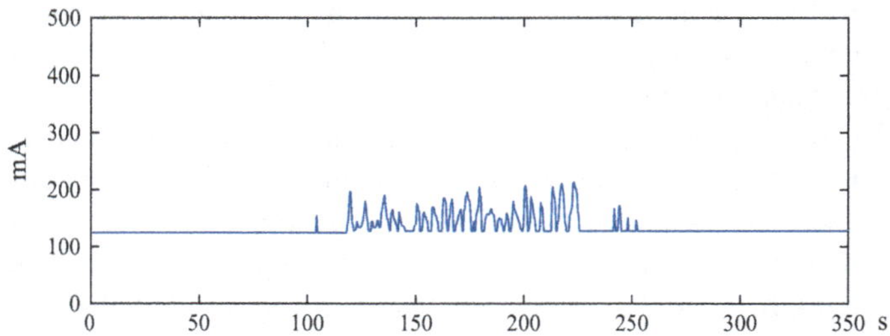

Fig. 4 Fluctuant curve

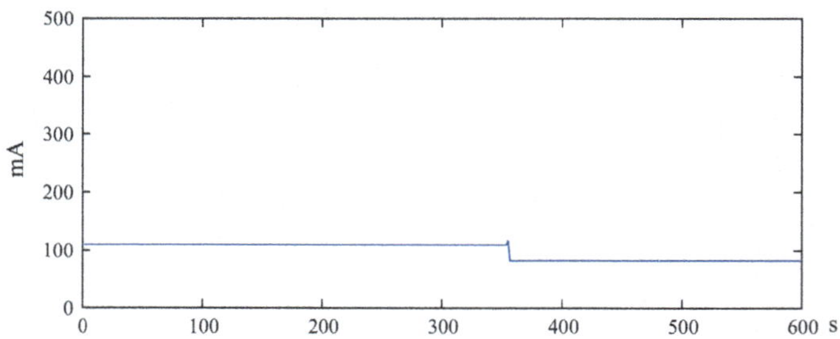

Fig. 5 Drop-off curve

3 Method

3.1 Structure

Figure 6 demonstrates the workflow of recognizing signal fault curves. First, current curves are gleaned from MMS and divided into numerous sections with a same time span of 600 s. Second, the similarity between each of the parted curves and the reference curve is calculated by DTW individually. And finally, the type of each curve is identified.

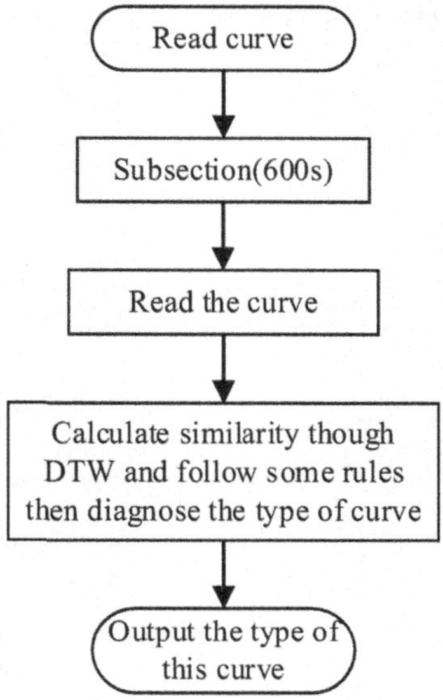

Fig. 6 Workflow of the recognition method

3.2 *Dynamic Time Warping*

Dynamic time warping (DTW) is employed to determine the similarity between a test curve and the reference curve. DTW is a method based on the idea of Dynamic Programming (DP) to calculate an optimal match between two given curves which may hold different lengths. DTW does not need pre-training, which saves much cost, but it still requires couples of template curves. DTW works in this way:

Suppose we are to find out an optimal path between a reference template *R* and a test template *T* by calculating the distance between *R* and *T*. The smaller the distance, the more similar *R* and *T*.

The optimal path is expressed as

$$m_i = \Phi(n_i) \tag{1}$$

where (n_i, m_i) is a pair of coordinates in a coordinate system where the horizontal axis represents the values of a test template while the vertical axis means the values of the reference template. Figure 7 depicts an optimal path.

The path passes points of $(n_1, m_1), L(n_i, m_i), L(n_N, m_M)$ where *N* is the number of the test template values and *M* the number of the reference template values. After the path passes the point of (n_{i-1}, m_{i-1}), the next point must be one of the three points shown as follows:

(1) $(n_i, m_i) = (n_i, m_{i-1})$;
(2) $(n_i, m_i) = (n_i, m_i)$;
(3) $(n_i, m_i) = (n_{i-1}, m_i)$.

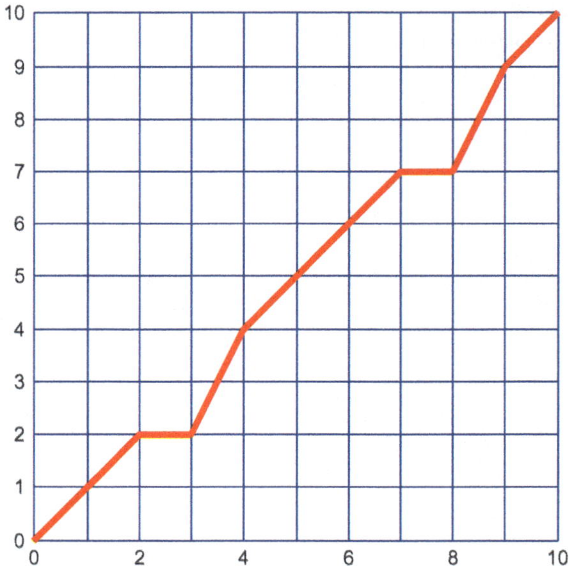

Fig. 7 Optimal path

According to the optimal path, the minimum distance of cumulative distance is obtained.

$$\sum_{n_i=1,m_i=\varphi(n_i)\in\eta}^{N} D[n_i, m_i] = \min \sum_{n_i=1,m_i=\varphi(n_i)\in\eta}^{N} D[n_i, m_i] \tag{2}$$

The total distance includes a cumulative distance and a frame matching distance. And the cumulative distance is

$$D[n_{i-1}, m_{i-1}] = \min\{D[n_{i-2}, m_{i-1}], D[n_{i-1}, m_{i-1}], D[n_{i-1}, m_{i-1}]\} \tag{3}$$

while the frame matching distance is $d[T(n_i), R(m_i)]$.

Thus, the total distance for time series T and R is

$$D[(n_i, m_{i)}] = d[T(n_i), R(m_i)] + D[(n_{i-1}, m_{i-1})] \tag{4}$$

4 Experiment and Results

We got current curve of a signal lasting a month from a MMS in Jinan Railway Bureau and divided it into numerous sections, each of which is 600 s long, given that a day-long curve is too large to analyse, and technically, 600 s can cover a complete version of any type of failure. So, we got 4320 curves.

And we enclosed the five typical curves, as mentioned in Sect. 2, into a template database, which is crucial to build a template database in order to calculate similarities.

DTW can almost recognize the normal curves and the fluctuant curves while cannot distinguish the other three types: the non-standard curves, the drop-off of curves and the discontinuous curves because the three all contain a process of sudden drop. Thus, we adopted three rules to further distinguish them, illustrated as follows, where abnormal point is a point whose value is larger than that of the next point.

(1) When the value of current is zero, the curve is determined as a discontinuous curve;
(2) When there are discontinuous abnormal points, the curve is determined as a non-standard curve;
(3) When there are three or more continuous abnormal points, the curve is determined as a drop-off curve.

The whole process is shown in Fig. 8. If a curve is determined neither a normal one nor a fluctuant one, the procedure will proceed to the next step. The curve is a discontinuous curve if the value of the current is zero. If not, the curve is a drop-off curve when there are three continuous abnormal points or more. Otherwise, the

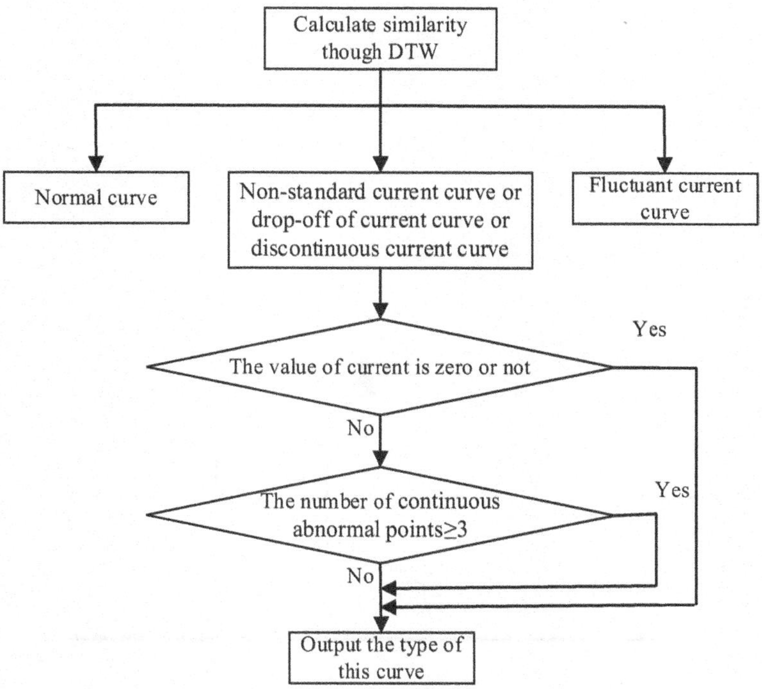

Fig. 8 Process of diagnosis

Table 1 Results of DTW diagnosis

	Normal curve	Non-standard current curve	Fluctuant current curve	Drop-off current curve	Discontinuous current curve
Quantity	4160	36	37	45	42
Accuracy (%)	100	63.16	100	71.11	80.95

curve is a non-standard curve. We conducted DTW on each of the 4320 curves, and the results are shown in Table 1.

The numbers of the non-standard curves, the drop-off curves and the discontinuous curves in Table 1 are not correct. Therefore, our approach continued to analyse the results by applying the three rules.

First, according to Rule 1, the approach identified 34 discontinuous curves, and the accuracy was 100%.

Second, the approach utilized Rule 2 to identify non-standard curves. Figure 9 illustrates an example, where "*" means an abnormal point occurs. As we can see, the curve has two continuous abnormal points and can be determined as a non-standard curve.

Further, the approach applied Rule 3, as shown in Fig. 10. There are five continuous abnormal points in the curve, and the curve can be determined as a drop-off curve.

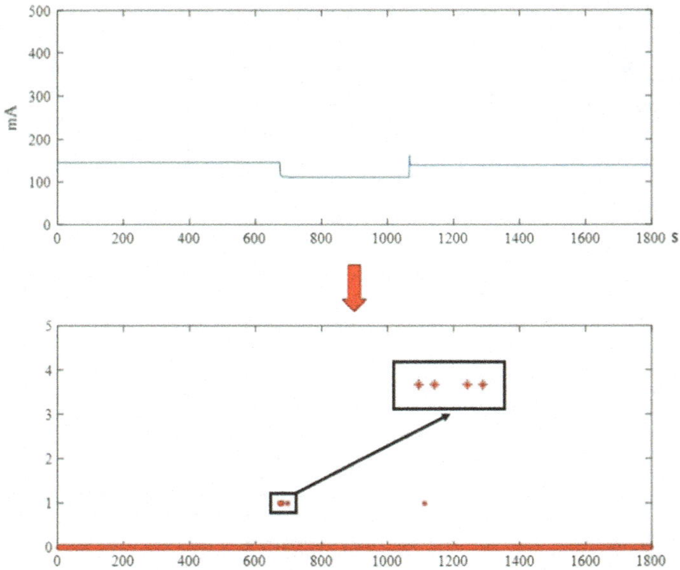

Fig. 9 Non-standard current curve and abnormal point curve

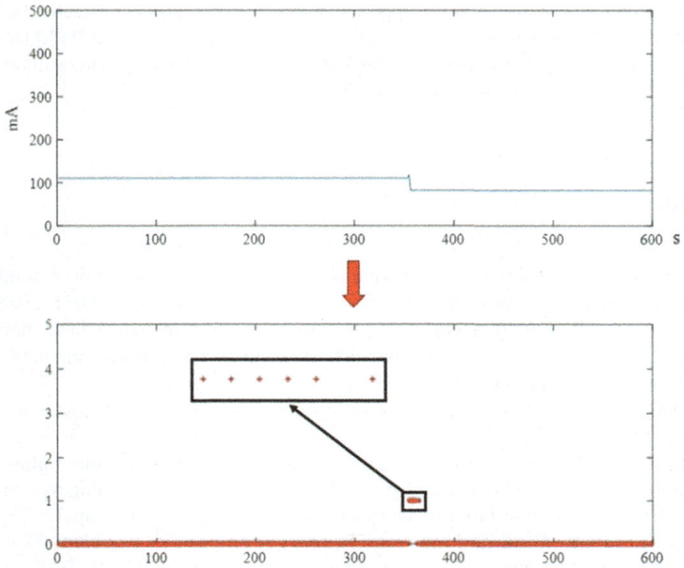

Fig. 10 Drop-off current curves

Table 2 Final result

	Normal curve	Non-standard current curve	Fluctuant current curve	Drop-off current curve	Discontinuous current curve
Quantity	4160	57	37	32	34
Accuracy (%)	100	100	100	100	100

Therefore, the final results of the fault diagnosis are shown in Table 2. It is clear that the accuracy is improved, and our approach can diagnose the types of signal current curves quickly and accurately.

5 Conclusion

In this study, we introduced a normal signal current curve and four abnormal curves and developed an approach to recognize types of signal current curves automatically. First, the approach segments the original curves into numerous 600 s-long-sub-curves. Then, DTW calculates the similarity between the input curve and the reference curve and determine whether it is a normal curve or a fluctuant curve. When the curve falls into other three types, the approach adopted three rules to identify whether it is a non-standard curve or a discontinuous curve or a drop-off curve.

Acknowledgements This research is supported by Sichuan science and technology program (2019YFG0040) and the National Key R&D Program of China (2016YFB1200402) and the National Natural Science Foundation of China (Grant No. 61703308). The authors gratefully acknowledge the invaluable contribution of the reviewers.

References

1. Xu Q et al (2017) Analysis and comparison of modular railway power conditioner for high-speed railway traction system. IEEE Trans Power Electron 32(8):6031–6048
2. Min YZ, Dang JW (2010) A dual-side comparison method of online fault monitoring for railway signal cable. In: 2010 International conference on information, networking and automation (ICINA), pp 486–489
3. Sibley HC (1957) Syncroscan: a high-speed control system for railway signaling. Electr Eng 76(1):38–43
4. Kunifuji T, Ito H, Saiki Y, Mori K (2011) A proposal of autonomous online expansion technology for real-time system and its application to railway signalling system. In: 2011 Tenth international symposium on autonomous decentralized systems, pp 73–80
5. Kunifuji T, Mori K, Miura T, Nishiyama J (2009) A proposal of flexible railway signalling system utilizing autonomous decentralized technology. In: 2009 29th IEEE international conference on distributed computing systems workshops, pp 300–305
6. Du X, Zou J, Wang Z (2015) Calculation of the impedance of a rail track with earth return for the high-speed railway signal circuit using finite-element method. IEEE Trans Magn 51(3):1–4
7. Pinkstone RA (1969) Developments in railway signal engineering. Stud Q J 40(158):75–80
8. Yan Y, Sun C, Zhou F, Liu C (2016) Curve recognition technology based on BP neural network and its application in landmine detection. China Meas Test 42(3):90–93
9. Gari A, Khaissidi G, Mrabti M, Chenouni D, El Yacoubi M (2017) Skew detection and correction based on Hough transform and Harris corners. In: 2017 International conference on wireless technologies, embedded and intelligent systems (WITS), Fez, pp 1–4
10. Feng X, Zhang T (2010) Research on electricity users classification technology based actual load curve. Electr Power Sci Eng 26(9):18–22
11. Liu Y (2005) Researches on recognition of well logs based on the chaotic time series analysis. Harbin Engineering University, pp 1–116
12. Zhang X (2016) The research and implementation of a method for recognizing switch fault current curve based on similarity. Lanzhou Jiaotong University, pp 1–64
13. Kim H, Sa J, Chung Y, Park D, Yoon S (2016) Fault diagnosis of railway point machines using dynamic time warping. Electron Lett 52(10):818–819
14. Li Y, Xue D, Forrister E, Lee G, Garner B, Kim Y (2016) Human activity classification based on dynamic time warping of an on-body creeping wave signal. IEEE Trans Antennas Propag 64(11):4901–4905
15. Meszlényi R, Peska L, Gál V, Vidnyánszky Z, Buza K (2016) Classification of fMRI data using dynamic time warping based functional connectivity analysis. In: 2016 24th European signal processing conference (EUSIPCO), pp 245–249
16. Han T, Liu X, Andy CCT (2017) Fault diagnosis of rolling element bearings based on multiscale dynamic time warping. Measurement 95:355–366
17. Sharma A, Sundaram S (2016) An enhanced contextual DTW based system for online signature verification using vector quantization. Pattern Recogn Lett 84:22–28
18. Hou W, Pan Q, Peng Q, He M (2017) A new method to analyze protein sequence similarity using dynamic time warping. Genomics 109(2):123–130
19. Meszlenyi RJ, Hermann P, Buza K, Gal V, Vidnyanszky Z (2017) Resting state fMRI functional connectivity analysis using dynamic time warping. Front Neurosci 11(75):1–17

20. Bing H, Trajcevski G, Scheuermann P, Wang X, Keogh E (2008) Querying and mining of time series data: experimental comparison of representations and distance measures. Proc VLDB Endowm 1(2):1542–1552
21. Vuori V, Laaksonen J, Oja E, Kanges J (2001) Experiments with adaptation strategies for a prototype-based recognition system for isolated handwritten characters. Int J Doc Anal Recogn 3(3):150–159
22. Forestier G, Lalys F, Riffaud L, Trelhu B, Jannin P (2012) Classification of surgical processing using time warping. J Biomed Inform 45(2):255–264
23. Lee C, Leu Y, Yang W (2012) Constructing gene regulatory networks from microarray data using GA/PSO with DTW. Appl Soft Comput 12(3):1115–1124
24. Hernández-Vela A, Ángel Bautista M, Perez-Sala X, Ponce-López V, Escalera S, Baró X, Pujol O, Angulo C (2014) Probability-based dynamic time warping and bag-of-visual-and-depth-words for human gesture recognition in RGB-D. Pattern Recogn 50:112–121
25. Feng H, Wah CC (2003) Online signature verification using a new extreme points warping technique. Pattern Recogn Lett 24(16):2943–2951
26. Gupta GK, Joyce RC (2007) Using position extrema points to capture shape in on-line handwritten signature verification. Pattern Recogn 40(10):2811–2817
27. Sharma A, Sundaram S (2017) On the exploration of information from the DTW cost matrix for online signature verification. IEEE Trans Cybern 99:1–14

Bionic Vision-Based Pantograph–Catenary Contact Point Detection Study in China High-Speed Railway

Zaixin Wu, Shize Huang, Liangliang Yu, Wei Chen and Lingyu Yang

Abstract The pantograph–catenary system of high-speed electric multiple units (EMU) is the only way to get electric power for high-speed trains. Over the past several years, video frames taken from traditional cameras have been used to analyze pantograph–catenary arcs. A traditional camera is built on mimicking human vision; thus, the natural phenomenon that an object will appear smaller if it is far from an observer but will become larger as it moves toward the observer will be reflected in the pictures that are taken. This is analogous to an implicit depth. Based on this observation, a bionic vision-based algorithm that utilizes the implicit depth is proposed in the present work to extract the touch point between the contact wire and pantograph slide under interference from the messenger wire. Experiments indicate that the proposed algorithm works quite well, with only marginal errors occurring, thus providing a strong base for future research activities.

Keywords Pantograph–catenary system · Arc detection · Image processing · Bionic vision · Railway safety

---

Z. Wu
Sichuan Railway Industry Investment Group Co., Ltd., 4-1-1 Kitakaname, Hiratsuka, Kanagawa, 259-1292, Ja Building A, Liangjiang International, No. 535, Tianfu 1st Street, High-Tech Zone, Chengdu, People's Republic of China

S. Huang
Shanghai Key Laboratory of Rail Infrastructure Durability and System Safety, Tongji University, 4800 Caoan Road, Shanghai, People's Republic of China
e-mail: hsz@tongji.edu.cn

L. Yu · W. Chen · L. Yang (✉)
Key Laboratory of Road and Traffic Engineering of Ministry of Education, Tongji University, 4800 Caoan Road, Shanghai, People's Republic of China
e-mail: 14weichen@tongji.edu.cn

Y. Qin et al. (eds.), *Proceedings of the 4th International Conference on Electrical and Information Technologies for Rail Transportation (EITRT) 2019*, Lecture Notes in Electrical Engineering 639, https://doi.org/10.1007/978-981-15-2866-8_19

1 Introduction

Pantograph–catenary systems are key components in electrified railways and play an essential role in powering electric locomotives [1]. To monitor pantograph–catenary system, China EMU have installed pantograph–catenary video surveillance systems (6C system) on all high-speed rails. While operating, the system produces massive pantograph–catenary monitoring images and videos, which allows the manual/automatic extraction of pantograph–catenary fault features through image processing [2].

While a train is running, power is transmitted through the point where the contact wire touches the pantograph slide, which is also known as the contact point. Correctly locating the contact point is the foundation of subsequent works such as arc detection, structure state detection, and running state analysis [3–6]. The contact point itself also becomes an important feature of the pantograph–catenary system. In a complex operating environment, quickly and correctly locating the contact point can not only improve the accuracy of fault detection but also provide key information that is required by other components of the system for train data analysis.

A camera is installed on the train's roof. Its position is fixed as is its view angle, as shown in Fig. 1. In the pantograph–catenary video, the contact wire is in a Z shape alignment. Together with the contact wire, a messenger wire exists and becomes a major source of interference. Thus, two contact points will be found in a single pantograph–catenary image, but only one is the true contact point. The other is false and is caused by the existence of messenger wire. Thus, automatically removing the noise from the image becomes a challenge.

Zhu and Feng et al. [7, 8] analyzed skateboard wear situations by extracting the edge positions of the pantograph slide from video image collected by CCD camera. However, they failed to get the pantograph–catenary position features and thus did not analyze runtime-specific wear points. Hulin and Schussler [9] studied the strategy of issuing warnings or dropping the pantograph via foreign object detection based on a train's operation route catenary video. Their studies did not distinguish between the contact wire and messenger wire at the foreign object's location; thus, they could not determine whether the foreign object is along a specific wire.

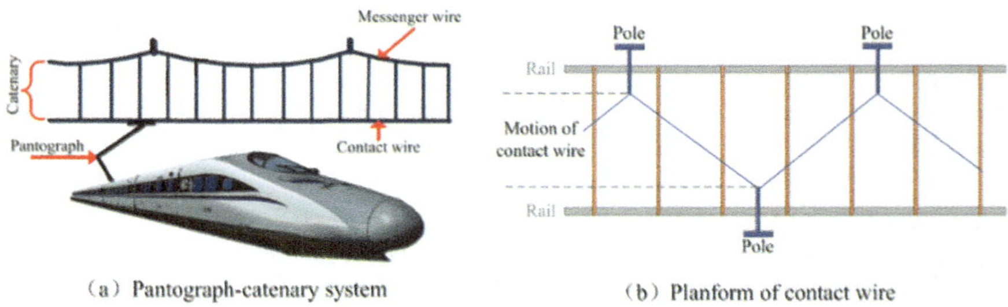

Fig. 1 Pantograph–catenary system and contact wire

This, in fact, harms the efficiency of their proposed strategy. Most of the existing pantograph–catenary video image processing methods extract either pantograph–catenary or catenary data for analysis or treat a pantograph–catenary system as a whole to use a fault image feature. The contact point between the pantograph and catenary is rarely located precisely to find if an arc occurs [10–12].

Based on image processing techniques, an original pantograph–catenary video is decomposed into frame by frame pictures. A traditional camera is built to emulate human vision such that an object will appear smaller and become gradually larger as it approaches the observer. Based on this observation, a bionic vision-based algorithm is proposed in this manuscript. Shortly, after preprocessing images, the appearance geometry of both the contact wire and the messenger wire, which actually share the same real geometry, will be extracted and compared. By noting the constant relative position along the camera, contact wire, and messenger wire and applying the vision philosophy, the contact wire position can be determined. The remaining contents are organized as follows. In Sect. 2, the proposed algorithm will be illustrated in detail. Conducted experiments are exhibited in Sect. 3. Conclusion and future research plans are in the end section.

2 Method

2.1 Image Preprocessing and Extraction

To obtain the contact point, the original image is pre-processed first to acquire a sub-image around the contact area. Since the pantograph is fixed onto the top of a train, the camera's location will change within a sub-area that can be determined beforehand. In addition, because the pantograph may shake while operating, the selected sub-range will be intentionally enlarged, so no useful information is missed. This would aid in the development of the algorithm to some extent. Figure 2 shows a flowchart of image preprocessing.

First, the original files are converted into image files because the pantograph monitoring system only provides videos. Next, the converted true color RGB images are transformed into grayscale images, with each pixel associated with a corresponding grayscale value. Next, a proper threshold value is determined such that in the derived grayscale images, pixels with gray values less than the threshold are marked in black, and the remaining pixels are marked in white. The final binary images are thus created. Based on a histogram, the Otsu threshold algorithm is used to determine the threshold value. The aim is to find the optimal threshold such that the grayscale value image is divided into two parts with maximal variance, and equivalently, maximal separation between the parts is achieved. When the entire image histogram is searched, only one threshold is determined to ensure that the image is separated into two sections, with target objects being black and the

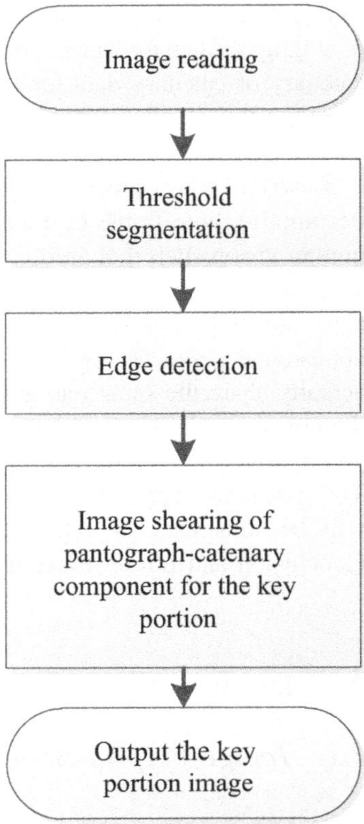

Fig. 2 Flowchart of image preprocessing

background being white, as shown in Fig. 8. Then, edges are detected using the selected Sobel differential operator:

$$\begin{bmatrix} -1 & 0 & 1 \\ -2 & 0 & 2 \\ -1 & 0 & 1 \end{bmatrix} \begin{bmatrix} -1 & 0 & 1 \\ -2 & 0 & 2 \\ -1 & 0 & 1 \end{bmatrix}$$

With all of the steps fulfilled, extracting the critical area is straightforward and easy once the pantograph arc coordinates are found and the image features are analyzed, which is of the few difficulties.

2.2 Localization Algorithm Based on Bionic Vision

An emerging imaging technology is based on the so-called RGB-D camera. Unlike traditional cameras, which only create RGB information, an RGB-D camera also produces depth information for each pixel [1, 13]. Using the additional depth information per pixel, computer algorithms are devised, so the 3D scenes that exist in the real world can automatically be reconstructed by using the videos taken. For instance, an RGB-D camera is applied to the automatic 3D modeling of indoor environments [14]. Up-to-date RGB-D cameras are yet not widely installed due to the following reason: (1) it is still a relatively new technology and needs time to become widespread and (2) technical limitations exist and should be conquered. For example, current RGB-D cameras have distance limitations (typically less than 5 m). Currently, in the Chinese high-speed railway pantograph system, traditional cameras are still installed to take videos. Thus, the algorithms and methodologies developed for RGB-D imaging cannot be applied directly.

Traditional cameras are known to be built by mimicking human vision, particularly camera lens that involve curvature to reflect eye lens' real shape. The curvature makes similar objects appear differently in an observer's vision, depending on the distance between them. Specifically, the object will appear smaller if it is far from the observer but will become larger as it moves toward the observer. In other words, if an object is imaged by a traditional camera, it is possible to tell the relative distance between the object and the camera, depending on the size of object's appearance in the photos. Certainly, the proposed methodology has limitations and is thus not a general purpose one. Fortunately, in the present study, the following observations make the proposed method feasible. First, in the Chinese high-speed railway pantograph system, the messenger wire and contact wire are composed of the same material and have the same geometry. Second, the video camera is installed on top of the high-speed train. From the viewpoint of video camera, the relative positions of the messenger wire and contact wire are kept the same, where the contact wire is closer to the camera than the messenger wire is.

2.3 Pantograph–Catenary Contact Point Extraction Algorithm

2.3.1 Positioning of Contact Wire

The contact wire and messenger wire sit on the top half of the extracted image. Because the messenger wire accompanies the contact wire, it becomes a source of severe interference when the contact wire is located. Distinguishing between the contact wire and the messenger wire is thus critical for positioning the contact point.

From the experimental video, the camera's position is fixed. The contact wire and messenger wire have obvious differences in terms of volume. The contact wire is closer to the camera; therefore, it appears to be thicker in the image. Correspondingly, the messenger wire is thinner. Thus, it is possible to design a computer algorithm that utilizes this feature to determine the position of the contact wire.

As seen in the extracted image, the two wires (contact wire and messenger wire) have four edge lines. The volume difference between the lines is reflected by measuring the line spacing of two adjacent edge lines. The wire of large spacing is deemed the contact wire. Because the four edge lines are straight lines, it is unnecessary that all of the points should be extracted from edge lines. Instead, only four points are extracted, which also speeds up the processing time.

The pixels representing image feature have the value 1, whereas the other pixels have the value 0. When determining the row's major orientation, the pixels in the image will be scanned every five rows. All pixels of value 1 will be treated as potential pixels forming the contact wire, messenger wire, or other important image features. When the edge lines span multiple rows, such an interval scan will not affect the position but will reduce the space consumption for data saving. All scanned pixels will be stored in a matrix P, inside which the contact point's coordinates are contained. A sample of the matrix can be found in Figs. 9, and 3 displays the flowchart.

The resulting matrix contains coordinates of the image features, including the contact wire and messenger wire. Thus, the functions of the contact wire and messenger wire can be determined through the coordinates. Due to the presence of erroneous data in the matrix, it is necessary to screen the raw data to ensure the precision of subsequent fitting. Since such a line fitting does not need many points, it is more suitable to delete the columns of invalid data and form a new matrix instead of trying to correct them. Figure 4 shows the flowchart.

After invalid data are removed and the matrix is reformed, the functions of the four edge lines can be derived and displayed through data fitting, as shown in Fig. 5.

Fig. 3 Flowchart for obtaining the edge feature matrix *P*

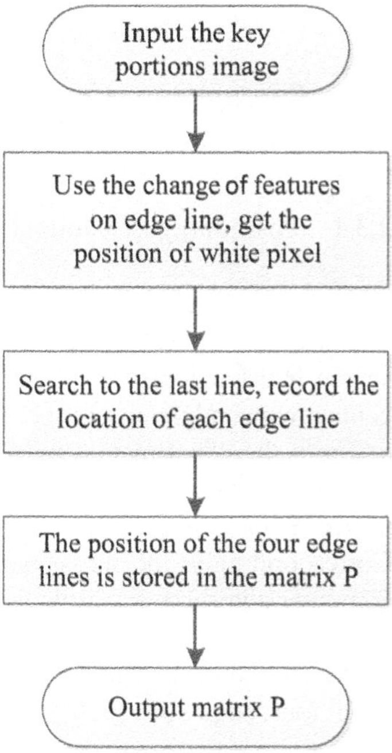

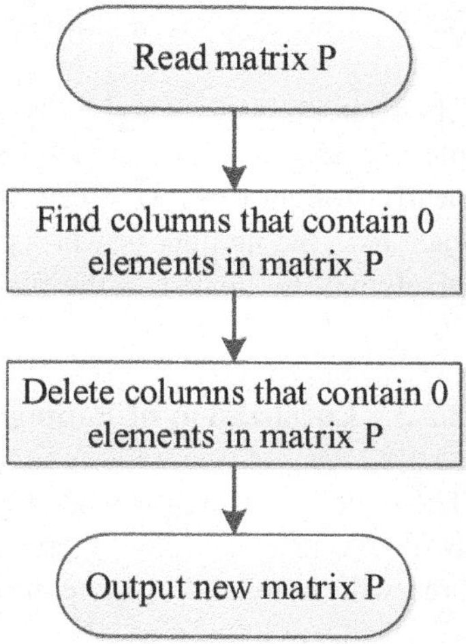

Fig. 4 Flowchart for renewing matrix *P*

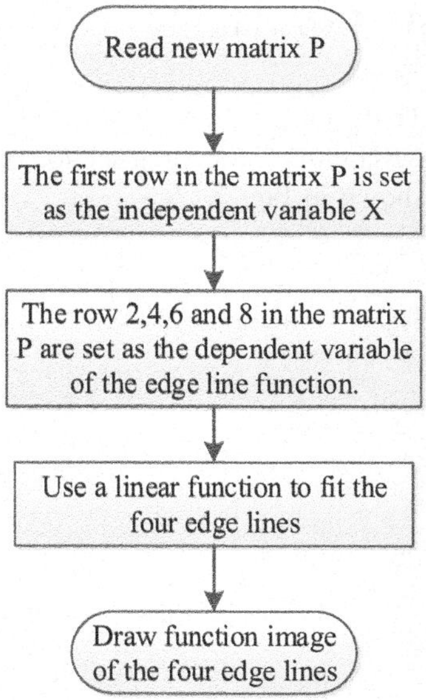

Fig. 5 Flowchart for linear fitting

After the functions of the four edge lines for the contact wire and the messenger wire are derived, the functions of the contact wire are determined by comparing the line spacing between the edge lines in the image. From the viewpoint of the camera, the contact wire is closer than the messenger wire. Thus, its geometry in the images taken should appear larger than the messenger wire. Thus, a larger separation distance, which is acquired by subtracting the two edge lines, will suggest that the contact wire exists.

2.3.2 Localization of Pantograph Slide

The pantograph shakes slightly while running, so its location will change slightly in the images. As the shape of the pantograph slide contains features such as being nearly straight (Fig. 8), it can be positioned via a straightforward way. That is, for each row, the number of white pixels will be counted. The row with the most white pixels will be treated as the function of pantograph slide.

2.3.3 Localization of Pantograph Slide

The intersection area can be obtained by solving the functions of the contact wire's two edge lines and pantograph slide simultaneously. The center of the intersection area will be treated as the contact point. The flowchart is shown in Fig. 6.

3 Experiments

To examine efficiency of the algorithm, the surveillance video of the Chinese EMU CRH380D Chongqing section in January 2016 is selected. A sample image frame is shown in Fig. 7.

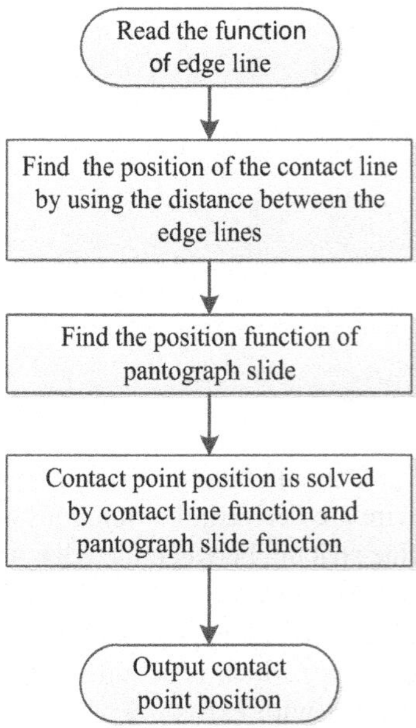

Fig. 6 Flowchart of contact point detection

Fig. 7 Original image

3.1 Image Preprocessing

According to the method exhibited in Sect. 2.1, Fig. 8 displays the final binary image obtained after pre-treatment, threshold segmentation, edge detection, and region segmentation.

3.2 Determination of Contact Line Position

Starting from a row of pixels in the image, white pixels are retrieved every five lines by using the edge line change feature. For each line, the retrieved information is stored in an orderly manner to form the matrix *P*. One example of the matrix *P* is shown in Fig. 9.

By closely examining the matrix, it is not difficult to find existence of erroneous data. For example, the data in rows 6 and 13 are incomplete. The key point information in lines 7 and 8 is missing. Rows 6 and 13 only contain coordinate information of three key points; after studying these, the existence of erroneous data is mainly due to four edge lines not being completely continuous after edge

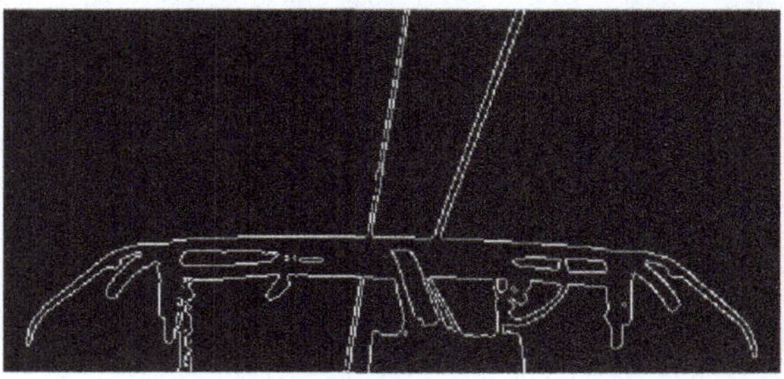

Fig. 8 Binary image

>> TS2

1	6	11	16	21	26	31	36	41	46	51	56	61	66	71
229	228	227	226	225	227	224	223	222	221	220	219	221	218	217
1	6	11	16	21	26	31	36	41	46	51	56	61	66	71
231	230	229	228	227	282	226	225	224	223	222	221	269	220	219
1	6	11	16	21	26	31	36	41	46	51	56	61	66	71
292	290	288	286	284	287	281	279	277	275	273	271	273	267	265
1	6	11	16	21	0	31	36	41	46	51	56	0	66	71
296	295	293	291	289	0	285	283	281	279	277	275	0	271	270

Fig. 9 Matrix of location

division. This further causes broken or intermittent edge lines when the data are retrieved because occasionally, only three or less coordinates can be read. One example of a broken edge line is shown in Fig. 10.

Due to the presence of erroneous information in the matrix, it is necessary to correct or reject erroneous data to ensure the correctness of the subsequent fitting operation. The straight-line fitting does not require many coordinates. Thus, deleting a row of incorrect information will not affect subsequent analyses, but a new matrix should be formed nonetheless. In the present work, the deletion strategy

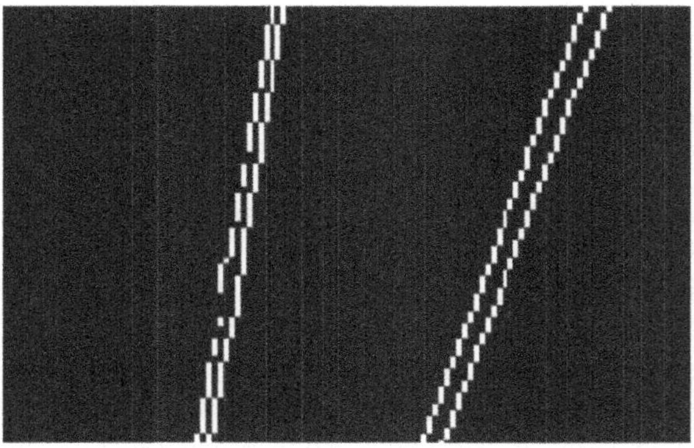

Fig. 10 Intermittent edge line

1	6	11	16	21	31	36	41	46	51	56	66	71
229	228	227	226	225	224	223	222	221	220	219	218	217
1	6	11	16	21	31	36	41	46	51	56	66	71
231	230	229	228	227	226	225	224	223	222	221	220	219
1	6	11	16	21	31	36	41	46	51	56	66	71
292	290	288	286	284	281	279	277	275	273	271	267	265
1	6	11	16	21	31	36	41	46	51	56	66	71
296	295	293	291	289	285	283	281	279	277	275	271	270

Fig. 11 New matrix of location

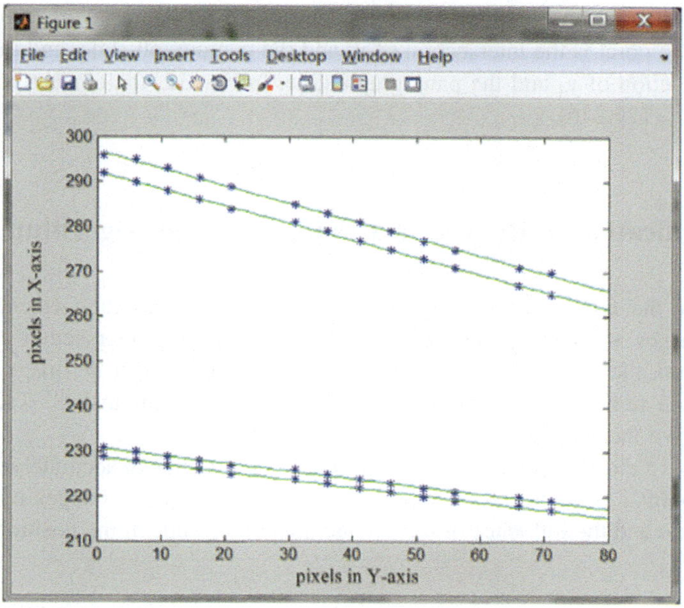

Fig. 12 Function of four edge lines

is used. Figure 11 shows the new matrix that is formed after rows of invalid data are removed. For the matrix (Fig. 11), performing fitting will give the four edge links' functions, as shown in Fig. 12.

3.3 *Localization of Pantograph Slide Position*

A pantograph slide shakes slightly while running, so its position varies with the image. Because a pantograph slide is nearly a straight line, it can be positioned by the head of pantograph. Every white pixel that is added and the line with the maximum value are the functions of the pantograph slide.

3.4 *Localization of the Contact Point*

The intersection can be obtained by solving the contact wire position function and the pantograph slide position function simultaneously. The position of contact wire is the intersection coordinate rounded to the nearest value. For example, in Fig. 12, the contact wire function can be confirmed as $y3 = 292.35 - 0.38 * x$ and $y4 = 297.01 - 0.39 * x$. The pantograph slide function is $L_{max} = 127$. Therefore, the final intersection coordinates can be determined by taking average of the two points. One point is the intersection of y_3 and the pantograph slide, and the other is the intersection of y_4 and the pantograph slide. In Fig. 12, the contact point is the coordinate (127,246).

4 Verification of the Contact Point Location Algorithm

To verify the contact point position algorithm, real pantograph–catenary video frame images are selected and tested. After obtaining the intersection coordinate information, the contact point is attached to the original grayscale image, which is drawn as a red circle of three-pixel radius to easily distinguish it. Table 1 and Fig. 13 give the outputs.

Figure 13 clearly indicates that the proposed algorithm can accurately locate the contact point's position. In some situations, due to irregular changes of the messenger wire and the influence of the surrounding environment, the position may not

Table 1 Calculation of contact point

Image	Edge function of contact line	Pantograph slide function	Contact point
1	$y_1 = 292.3514 - 0.3809x$ $y_2 = 297.0102 - 0.3888x$	L_max = 127	(127,246)
2	$y_1 = 305.2870 - 0.3304x$ $y_2 = 310.1083 - 0.3474x$	L_max = 130	(130,264)
3	$y_1 = 204 - 4.2833^{-16}x$ $y_2 = 208$	L_max = 128	(128,206)
4	$y_1 = 255.1122 - 0.2575x$ $y_2 = 259.4955 - 0.2609x$	L_max = 127	(128,224)
5	$y_1 = 317.6466 - 0.4239x$ $y_2 = 322.6466 - 0.4345x$	L_max = 129	(129,265)
6	$y_1 = 160.1015 - 0.1014x$ $y_2 = 164.6452 - 0.0906x$	L_max = 128	(128,176)
7	$y_1 = 136.1097 - 0.1999x$ $y_2 = 143.8945 - 0.1289x$	L_max = 128	(128,160)
8	$y_1 = 301.4 - 0.4x$ $y_2 = 309.3919 - 0.4180x$	L_max = 132	(132,253)
9	$y_1 = 140.0952 - 0.1714x$ $y_2 = 144.9124 - 0.1714x$	L_max = 129	(129,165)
10	$y_1 = 239.2 - 0.2x$ $y_2 = 244.4678 - 0.2179x$	L_max = 129	(129,215)

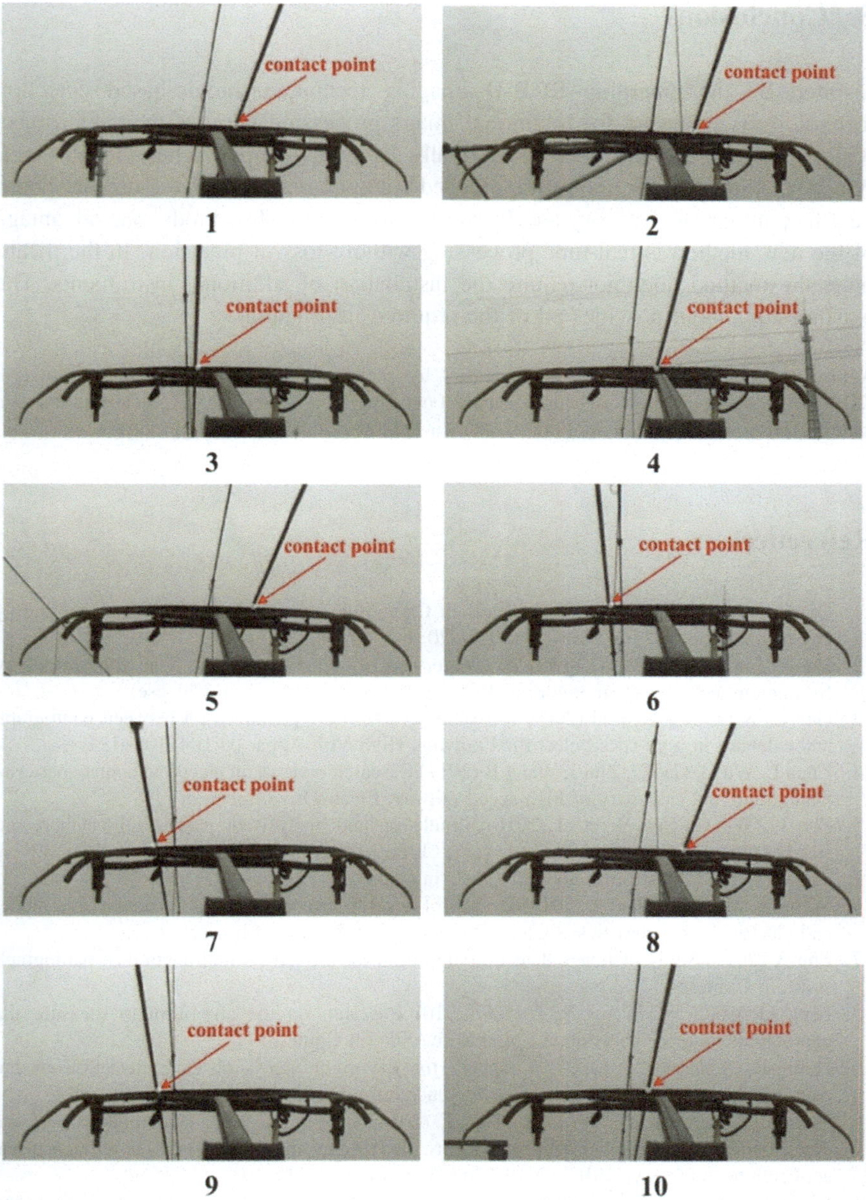

Fig. 13 Contact point

be very accurate. After the original data are screened and invalid points are removed, the position is still ensured to be the point predicted by the algorithm if it falls within a reasonable range around the real contact point in the image.

5 Conclusions

Inspired by the emerging RGB-D imaging technique, bionic-based detection technology is proposed for traditional image processing because the same object will appear smaller and become gradually larger as it approaches the observer. A new method is built upon the proposed technology to locate the contact point, i.e., the pantograph–catenary arc. In contrast to traditional methods, one advantage of the new method is real-time processing without loss of precision. In the meantime, the method does not require the installation of additional instruments. The existing equipment provides all of the required information.

Acknowledgements This work is supported by Sichuan science and technology program (2019YFG0040) and National Natural Science Foundation of China (61703308). The authors are grateful for the reviewer of initial drafts for their helpful comments and suggestions.

References

1. Qian Q, Gao S, He Z et al (2015) Study of China high-speed railway traction power supply key technology. J China Railw Soc 04:9–20 (in Chinese)
2. Han Z, Liu Z, Zhang G, Yang H (2013) Overview of non-contact image detection technology for pantograph-catenary monitoring. J China Rail Soc 06:40–47 (in Chinese)
3. Gao Z, Wu G, Lu W et al (2009) Research review of arc phenomenon between pantograph and catenary in high-speed electrified railway. High Volt Appar 03:104–108+127
4. Chen L, Wu G, Gao G, Zhu J, Wang B (2011) Research reviews on electrical contact between pantograph and cantenary of high-speed railway. Electr Drive Locomot
5. Zhu G, Wu G, Han W et al (2016) Simulation and analysis of pantograph-catenary arc steady-state characteristics during static lifting and lowering of high-speed railway pantograph. J China Railw Soc 02:42–47 (in Chinese)
6. Wang Y, Liu Z, Fan F et al (2013) Review of development of pantograph-catenary arc model and electrical characteristics. J China Railw Soc 08:35–43 (in Chinese)
7. Zhu X (2011) Study of image detection algorithm for the typical malfunction of pantograph slide (in Chinese)
8. Feng Q, Chen W, Wang Y, Yu G (2010) Research on the algorithm to measure the pantographic slipper abrasion. J China Railw Soc (in Chinese)
9. Hulin B, Schussler S (2007) Concepts for day-night stereo obstacle detection in the pantograph gauge. In: International conference on industrial informatics, pp 449–454
10. Yaman O, Karakose M, Aydin I, Akin E (2014) Image processing and model based arc detection in pantograph catenary systems. In: Signal processing and communications applications conference (SIU), 2014 22nd, Trabzon, pp 1934–1937
11. Wang Y, Liu Z, Mu X, Huang K, Wang H, Gao S, An extended Habedank's equation-based EMTP model of pantograph arcing considering pantograph-catenary interactions and train speeds. IEEE Trans Power Deliv 99:1
12. Ma L, Wang ZY, Gao XR, Wang L, Yang K (2009) Edge detection on pantograph slide image. In: 2nd International congress on image and signal processing, 2009, CISP'09, Tianjin, pp 1–3
13. Song S, Lichtenberg SP, Xiao J, SUN RGB-D: a RGB-D scene understanding benchmark suite
14. Henry P, Krainin M, Herbst E, Ren X, Fox D, RGB-D mapping: using Kinect-style depth cameras for dense 3D modeling of indoor environments

Application of Foreign Matter Detection in EEF Bogie by Measuring Technology of Linear-Structure Light

Zhaoqin Zhang, Yong Qin, Xiaoqing Cheng and Tangwen Yang

Abstract Non-contact detection is safe and efficient and less used in railway safety detection. This paper proposes a greatly new detection system that combines non-contact detection with railway safety detection. Based on the non-contact detection method—three-dimensional laser detection technology—it can detect foreign objects on the running gear of the train. The technology mainly includes two aspects: laser stripe center line extraction and point cloud processing. In order to improve the detection accuracy, the effect of laser stripe center line extraction must be valued. In this paper, some methods are compared to select the most suitable method for detecting foreign objects in the running gear of the train by the comparison of the stripe centerline extraction effect and the length of detection time. Besides, the paper, on the one hand, also introduces two subsystems of the detection system: hardware system and software system. On the other hand, the implementation principle and working steps of the detection system are also introduced in detail.

Keywords The running gear · Extraction method · Laser triangulation

1 Introduction

With the rapid development of railways in the past few decades, the speed and density of trains have been increased and the safety of trains has become the most important issue. The train running gear plays an important role in ensuring the flexibility of the vehicle during the process of train operation. At the same time, due to the influence of the environment, some foreign matter will adhere to the running

---

Z. Zhang · Y. Qin (✉) · X. Cheng
State Key Laboratory of Rail Traffic Control and Safety, Beijing Jiaotong University, Beijing, China
e-mail: 18125784@bjtu.edu.cn

T. Yang
School of Computer and Information Technology, Beijing Jiaotong University, Beijing, China

Y. Qin et al. (eds.), *Proceedings of the 4th International Conference on Electrical and Information Technologies for Rail Transportation (EITRT) 2019*, Lecture Notes in Electrical Engineering 639, https://doi.org/10.1007/978-981-15-2866-8_20

gear of the train. For example, in the cold weather, the low temperature can cause ice deposits on the train's running gear. It will affect the daily operation of the train and increase the risk possibility when the train is working. The non-contact detection method—linear-structure light measurement technology can be used to scan the train running gear to determine whether there is foreign matter on the running gear, which can greatly reduce the difficulty of the working night, increase the efficiency of inspection work, and reduce the time of inspection work.

For the method of measuring by linear-structure light, there are a lot of related researches at home and abroad. C. Alippi extracted and reconstructed the center of the light stripe through the neural network [1]; the German company GOM developed the Atos series of mobile optical scanners with an accuracy of 0.02 mm; Wang Hao proposed to use the gray gravity method in the normal direction of the light stripe and it greatly increased the extraction precision [7]; Zhu Lu explored the laser triangulation in depth and proposed how to choose a suitable method [8]. The system in this paper is based on the principle of visual imaging to establish a three-dimensional imaging system. After the calibration of camera and light plane, the light stripe is projected on the object by the laser; the industrial area array camera is used to obtain the light stripe information and the acquired information is stored on the computer. Then, we will use VS2013 and OpenCV library to extract the light stripe information and use the information to obtain the point cloud coordinates and rebuild 3D object model. By processing the data on the reconstructed 3D object model, the system can detect the foreign matter that attached on the running gear and obtain the corresponding foreign object information. The whole detection process includes the collection of image data, the extraction of laser stripe center line, and the processing of point cloud [2, 3]. Among, center line extraction and point cloud processing can directly affect the accuracy of detection and the efficiency of detection work. They are the technical difficulties and core parts of the system [4–6].

All in all, the measuring technology of linear-structure light has a wide range of applications, but there are few studies about the train safety. The train running gear plays an important role in the running process of the train. The detection of foreign objects on the EEF bogie of the train can greatly improve the safety of train operation, and it is important for people's lives and property.

2 Methods

This paper mainly introduces two parts: the centerline extraction and detection principle in the linear-structure light detection system. The light stripe centerline extraction introduces the comparison of various extraction methods. The detection principle introduces the laser triangulation in this system. The whole detection system is running under the hardware environment of Intel(R) Core(TM) i7-855U CPU @ 1.8 GHz 1.99 GHz; RAM 8G, operating system 64 bit.

2.1 Light Stripe Centerline Extraction

There are many influences and impurity when the system collects detection information. Among them, the centerline extraction of the laser stripe is greatly affected by the change of illumination and impurities. These will cause the shape of the laser stripe to change. It results in a decrease in the accuracy of laser detection. The laser extraction methods mainly include the extremum method, the gray gravity method, the refinement method, and the Steger method. In this paper, by the comparison of extracting efficiency and time from these extracting methods, the most efficient method is selected for the three-dimensional linear-structure light measurement system.

The extremum method

This method is based on the maximum of the laser image gray value on each row and the coordinate of the maximal gray value is defined as the position of the center point. These points will form a centerline. For the centerline extraction of the laser stripe, the extremum method has a good effect. However, this method is susceptible to impurity, which will result in inaccurate positioning center points.

The gray gravity method

The gray gravity method is derived from the solution of the position of the rigid body barycenter in physics by the formula:

$$x_{iG} = \sum_{j=1}^{n} I(i,j)x_{(i,j)} / \sum_{j=1}^{n} I(i,j) \tag{1}$$

The position of the gravity center on the i line in the image is x_{iG}, $i = 1, 2, 3, \ldots, m$, $I(i,j)$ represents the gray value in the pixel coordinates, and $x(i,j)$ represents the pixel coordinates. The gray gravity method is applied widely now, However, as same as the extremum method, the method is susceptible to impurity. Besides, the center is determined on the partial pixels of the light bar which has higher gray value. The distribution of laser stripes will be greatly changed in accuracy so the method has a poor robustness.

The refinement method

The refinement method can peel off the laser stripe layer by layer and remove some points from the original image. It ensures the original shape of the laser. Refinement method can be visually understood as the extraction of the skeleton. For example, the circle is refined to obtain the center of the circle, and the line is refined to obtain the median line of the line itself. After the laser stripe is processed by the refinement method, the line will just have a width of one pixel. However, the method has less calculation efficiency.

Steger method
Steger's method is based on the Hessian matrix to achieve accurate positioning of the sub-pixels in the center of the light stripe. Specifically, the method has two steps: the first step can obtain the normal direction of the light bar in the image; the second step can obtain the extreme point in the normal direction and these points will form a centerline. For any point (u, v) on the image, the Hessian matrix can be expressed as:

$$H(u,v) = \begin{bmatrix} \partial_{uu} & \partial_{vu} \\ \partial_{uv} & \partial_{vv} \end{bmatrix} \tag{2}$$

Among them, ∂_{uu} indicates the second-order derivation of the picture in the horizontal direction and other parameters are similar. The eigenvalues and eigenvectors of each matrix are obtained by calculation, and the eigenvector corresponding to the largest eigenvalue (n_u, n_v) is taken as the direction of the normal. The sub-pixel coordinates can be obtained by using the Taylor expansion on the normal vector. It is assumed that (u_0, v_0) are used as reference points. The corresponding sub-pixel coordinates is $(u_0 + kn_u, v_0 + kn_v)$, where:

$$k = \frac{n_u \partial_u + n_v \partial_v}{n_u^2 \partial_{uu} + 2n_u n_v \partial_{uv} + n_v^2 \partial_{vv}} \tag{3}$$

It is worth noting that at the time of $(kn_u, kn_v) < [-0.5, 0.5] \times [-0.5, 0.5]$, the reference point will be taken as the center point of the light stripe.

2.2 *Laser Triangulation*

In this part, the principle of direct laser triangulation is introduced. When the laser is vertically emitted, the surface of object and laser stripe is formed. Then, the diffuse reflection is generated on the surface of the object, and the light can be injected into the CCD camera for imaging. When the displacement occurs, the light stripe on the surface will also move. It can cause the light stripe to change in the camera and the light stripe's change is processed in the later stage. Finally, the computer will save the displacement information of the object.

As shown in Fig. 1, the principle of direct laser triangulation is carried out. The laser is emitted by the optical maser and vertically projected on the surface of the object to generate diffuse reflection. The reflected light is imaged on the photosensitive element through the camera lens. Assume that the intersection of the reference plane of the surface height of the measured object and the incident ray is a point called O. The angle between the incident ray and the camera plane is α. The angle between the camera plane and the photosensitive element is β. The distance between the incident point and the center of the lens is l. The distance between the

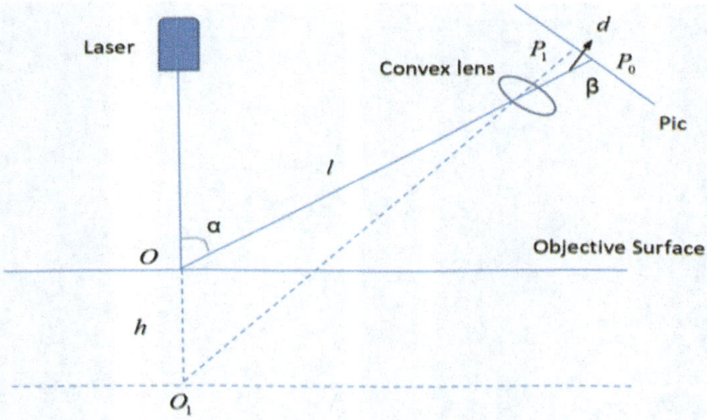

Fig. 1 Principle of direct laser triangulation

photosensitive plane and the center of the lens is d. The focal length of the lens is f. According to the imaging formula of the lens, it is necessary for ensuring a clear image:

$$\frac{1}{f} = \frac{1}{l} + \frac{1}{d} \tag{4}$$

Similarly, if the reflection point of the reflected light is O_1, the lens imaging formula is also satisfied. According to the derivation of the geometric formula, the result is as follow

$$h = \frac{ly\sin\beta}{d\sin\alpha - y\sin(\alpha+\beta)} \tag{5}$$

Among, the displacement distance between point O and point O_1 is h, y indicates the distance between point P_0 and point P_1, point P_0 indicates the imaging point of the point O on the photosensitive element, and P_1 indicates the imaging point of the point O_1 on the photosensitive element.

3 Results Comparison and Application

3.1 Comparison of Results

After these methods for the extraction of laser stripes, the following results can be obtained (Figs. 2, 3, 4 and 5).

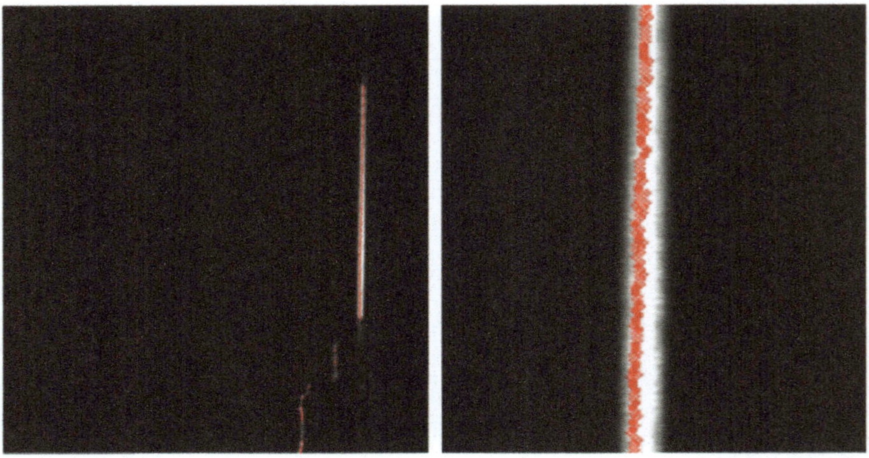

Fig. 2 Extremum method

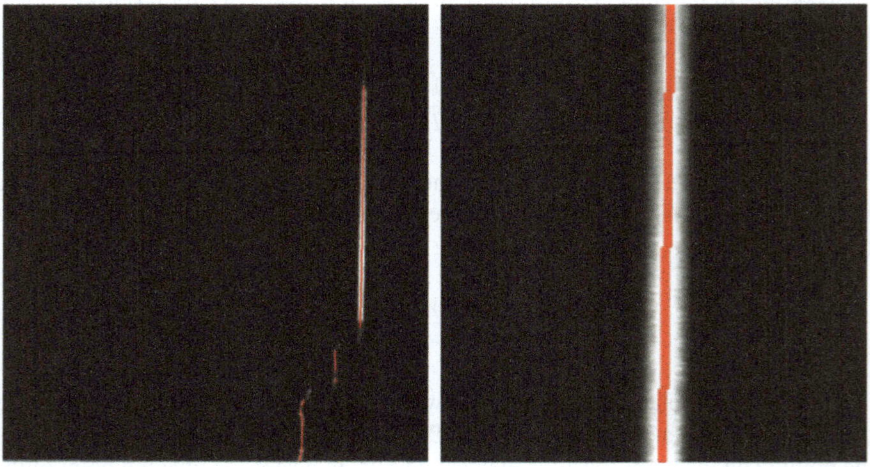

Fig. 3 Gray gravity method

After comparison, it can be seen that the Steger method and the gray gravity method have more better extraction effect than others. The extremum method and the refinement method are greatly affected by the impurity. The view of extraction time is been showed in Table 1.

As can be seen from the table, it takes more time than the other methods to extract the centerline of the image of one light stripe. The extraction time of the centerline greatly affects the working efficiency of the whole system. Therefore, the gray gravity method has a better effect and has a shorter time in the circumstances. So, it is supposed to be selected.

Fig. 4 Refinement method

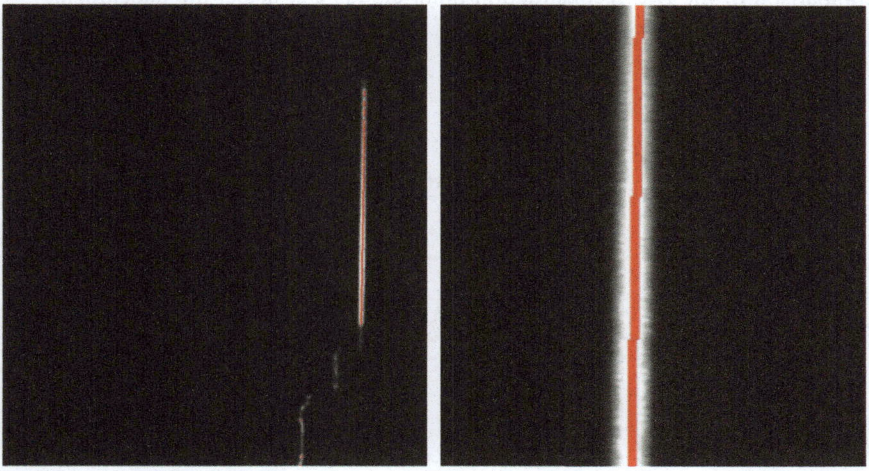

Fig. 5 Steger method

Table 1 Extraction time

Extraction method	Time (s)
Extremum method	3.446
Gray gravity method	3.555
Refinement method	4.846
Steger	3.7

3.2 Application

The entire system includes two parts: the hardware system and the software system. As shown in Fig. 6 [9], the hardware system includes experimental car, camera, laser, and trigger part; the software system includes camera calibration system, optical plane calibration system, and linear-structure light measurement system.

The specific technical process includes two parts: the acquisition process and the data processing process. The acquisition process is: when the train runs to the front device of the trigger, the camera starts to collect the laser stripe projected by the optical maser on the surface of train, and the picture data is saved to the computer. Until the train runs to the trigger-end device, the acquisition process is over. The data processing process is mainly processing the image data, including processing of the image data, extraction of the centerline of the laser stripe, and reconstruction of the point cloud.

The entire inspection technology needs to be collected at least twice. The first stage is the reconstruction of the base point cloud with no attachments. The second stage is the reconstruction of the point cloud of the object to be detected. The comparison of the two reconstruction point clouds can detect the attachment.

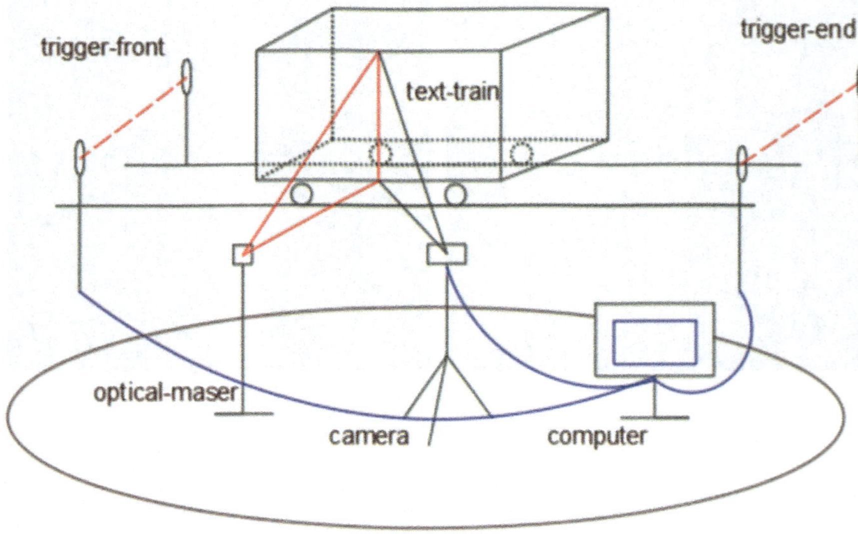

Fig. 6 Hardware system

4 Conclusion

In order to obtain the detection information of the attachment on the train running gear, this paper proposes a method based on linear-structure light detection. For the later stage of the detection process, the laser stripe centerline extraction method in the image data processing is compared with the various laser stripe centerlines. The paper selects the most suitable extraction method to improve the working efficiency of the system. In the later work, the improvement of the attachment volume detection, coloring, and other relating algorithms can be improved. That will make the attachment information more comprehensive and more visual. Besides, the post-processing of the point cloud can be improved to improve the efficiency of point cloud data processing.

Acknowledgements The paper was funded by the subject Modeling and Analysis Theory of Safety Behavior of Urban Rail Transit System (subject number: 2016YFB1200402). Thanks for the fund.

References

1. Alippi C, Casagrande E, Scotti F et al (2000) Composite realtime image processing for railways track profile measurement. IEEE Trans Instrum Meas 49(3):559–564
2. Ringbeck T, Hagebeuker B (2007) A 3D time of flight camera for object detection. Opt 3D Meas Tech 1–10
3. Griesser A, Koninckx TP, Van Gool L (2004) Adaptive real-time 3D acquisition and contour tracking within a multiple structured light system. In: Proceedings of the 12th pacific conference on computer graphical application, 2004, PG 2004, Proceedings 2004, pp 361–70
4. Liu Z, Sun J, Wang H et al (2011) Simple and fast rail wear measurement method based on structured light. Opt Lasers Eng 49(11):1343–1351
5. Gao W, Wang L, Hua Y (2009) Flexible calibration of a portable structured light system through surface plane. Acta Autom Sin 34(11):1358–1362 (in Chinese)
6. Lv Ghihua, Ghang Ghiyi (2012) Build 3D scanner system based on binocular stereo vision. J Comput 7(2):399–404
7. Hao W, Research and application of dynamic measuring method of rail profile based on structured light projection (in Chinese)
8. Lu Z, Research on 3D measurement system based on line-structure light (in Chinese)
9. Sun Y, Yang T, Qin Y (2019) Application research of 3D measurement technology of structure-light in foreign matter detection in train running gear. China Railw 01:122–126 (in Chinese)

Study on the Construction Rules and Sequencing Method of Train Operating Condition Sequence Based on Regenerative Energy

Yanhui Wang, Weifu Xia, Li Liu, Wanxiao Xiang and Shuai Lin

Abstract Obtaining an effective sorting method for train operating conditions is the premise for studying the optimization method of renewable energy utilization in the operation of urban rail trains. But there are few studies in this field. So that, in this paper, firstly, based on the given running time of the train, the construction rules of the running condition sequence of the interval train are given for the interval of different speed limiting types. Then, based on the train operating condition sequence solution set obtained by the construction rules, the regenerative energy utilized in the traction phase of each sequence and the regenerative energy utilized in the braking phase are used as ranking indices, and the interval number is used to express the uncertainty of the two ranking indices. The interval numbers of all solutions are obtained and sorted by the TOPSIS method based on entropy weight, and the sequence values of all train operating conditions corresponding to the interval are obtained. Finally, an example is given to illustrate the practical application of this sorting method.

Keywords Operating conditions sequence · Urban rail transit · Regenerative energy utilization · Train operation optimization

1 Introduction

Due to rising energy prices and environmental concerns, the energy efficiency of urban rail transit has attracted much attention from both researchers and practitioners in recent years. At present, the effective utilization of renewable energy is a common method in the energy saving operation of urban rail transit. Scholars [1, 2]

Y. Wang · W. Xia (✉) · L. Liu · W. Xiang · S. Lin
State Key Laboratory of Rail Traffic Control and Safety, Beijing Jiaotong University, Beijing 100044, China

School of Traffic and Transportation, Beijing Jiaotong University, Beijing 100044, China

Research and Development Center of Transport Industry of Technologies and Equipments of Urban Rail Operation Safety Management, MOT, PRC, Beijing, China

Y. Qin et al. (eds.), *Proceedings of the 4th International Conference on Electrical and Information Technologies for Rail Transportation (EITRT) 2019*, Lecture Notes in Electrical Engineering 639, https://doi.org/10.1007/978-981-15-2866-8_21

have confirmed that the regenerative energy in the train braking process is about 40% of the traction energy consumption, but only about 15–19% can be truly utilized and the rest is consumed by a rheostat in the form of thermal energy [3, 4]. Therefore, at present, many scholars [5–8] take improving the utilization rate of renewable energy as the goal of energy conservation.

The establishment of an efficient sequencing method for train operating conditions is a prerequisite for the establishment of an optimization model based on renewable energy train operation. However, there are few studies in this field. Therefore, the main work of this paper is to propose an efficient sequencing method for train operating conditions. The main steps are as follows. Firstly, for the different speed limiting types, the construction rules of train operating condition sequences are proposed. Then the calculation method of ranking index based on regenerative energy is put forward. Finally, the sequence of all train operating conditions corresponding to the same interval is sorted, and the sorting method is applied through an example.

2 Train Operation Sequence Sequencing Method Based on Regenerative Energy

2.1 Study on Train Operating Condition Sequence Rules Based on Different Speed Limits

In this paper, the change of the speed limits of the train running in the interval is a change of jump transition, which remains constant for a certain period and jumps to a larger or smaller speed limit at a certain moment. According to the variation times of interval speed limit, this paper mainly about the m-speed limit interval. The m-speed limit interval is one in which there are m-speed limits $\tilde{v}_k(k \in [1, m])$; therefore, this region can be divided into m-intervals of the single speed limit. Suppose the train is running at its initial speed $\tilde{v}_k$ within a certain subinterval $[\tilde{y}_{k-1}, \tilde{y}_k]$ with a speed limit of $\tilde{v}_k$, then we can get the following formulas:

$$v_{kn} = v_{(k+1)0} \tag{1}$$

$$v_{k0} \leq \min\{\tilde{v}_k, \tilde{v}_{k-1}\} \tag{2}$$

$$v_{kn} \leq \min\{\tilde{v}_{k+1}, \tilde{v}_k\} \tag{3}$$

Similar to the above in this paper, the existence of cruise condition mainly depends on whether the terminal velocity in the condition of using "maximum traction" and "maximum braking" exceeds the range $(0, \widetilde{v}_k]$, as shown in Eqs. (4) and (5).

$$\Delta\widetilde{y}_k \le \left(\widetilde{v}_k^2 - v_{k0}^2\right)/2a_{t\,\max} + \left(v_{k2}^2 - \widetilde{v}_k^2\right)/2a_{b\,\max} \tag{4}$$

$$\Delta\widetilde{y}_k < v_{k2}^2/2a_{t\,\max} - v_{k0}^2/2a_{b\,\max} \tag{5}$$

In addition, the minimum running time of the train in the sub-interval X should be the track of "maximum braking–(cruise)–maximum braking" with the brown line in the diagram, and the maximum running time corresponds to the track of "maximum braking–(cruise)–maximum braking" with the black line in the diagram.

The initial and terminal speeds of the train in the subregion are constant, and then the running process of the train in the subinterval under the conditions of given running distance and running time can be obtained, as shown in Eqs. (6) and (7).

The starting and ending speeds of the train in the adjacent subintervals and the running time of all the subintervals are constrained so as to obtain the running track of all the trains that meet the requirements in the whole interval, namely the sequence of the running conditions of the train in the interval.

$$\Delta\widetilde{y}_k = \sum_{i=1}^{n}\left(v_{ki} + v_{k(i-1)}\right)t_{ki}/2 = \sum_{j=1}^{m}\left(v'_{kj} + v'_{k(j-1)}\right)t'_{kj}/2 \tag{6}$$

$$t_k = \sum_{i=1}^{n} t_{ki} = \sum_{i=1}^{n}\left(v_{ki} - v_{k(i-1)}\right)/a_{ki} = \sum_{j=1}^{m} t'_{kj} = \sum_{j=1}^{m}\left(v'_{kj} - v'_{k(j-1)}\right)/a'_{kj} \tag{7}$$

$$\sum_{k=1}^{|\widetilde{V}|} t_k \in \left[t_r^{\min}, t_r^{\max}\right] \tag{8}$$

2.2 *Ranking Index Calculation*

For different train operation sequences in a certain interval, the energy consumed/generated by train traction/braking is different from the corresponding time. In order to avoid the error caused by the insufficient consideration of the above relation between energy and time, this paper mainly by the unit traction/braking time/energy consumption as to study the related parameters, calculate the two order index, namely the total train traction time can use his car E_t to regenerate and train braking time always produced by his car use of renewable energy E_b.

If there is a secondary traction condition in a train operating condition sequence, the energy consumption per unit time of a single traction is arranged in an array from small to large, and then the fifth element in the array is the energy consumption per unit traction time, as shown in Eq. (9).

$$\overline{E}_k^t = \tilde{E}_k^t / \tilde{t}_k^t \tag{9}$$

where $\tilde{E}_k^t$ is the train traction energy consumption corresponding to the fifth element in the array (J). $\tilde{t}_k^t$ is the train traction time corresponding to the fifth element in the array.

$\tilde{E}_t$ represents the regenerative braking energy of other trains used in the unit traction time of the train. The renewable energy of other vehicles used by the train in the total traction time is shown in Eq. (10)

$$E_t = \sum_{k=1}^{n} \tilde{t}_k^t \sum_{k=1}^{n} \min\left\{\overline{E}_k^t, \overline{E}_t\right\} / n = \sum_{k=1}^{n} \tilde{t}_k^t \sum_{k=1}^{n} \left[\overline{E}_t + \mu_k\left(\overline{E}_k^t - \overline{E}_t\right)\right] / n \tag{10}$$

In the case that the overlapping time of traction and braking between the train and other vehicles is unknown, $\tilde{E}_t$ is an uncertain value. If it is inserted into the array $\tilde{E}_1^t, \tilde{E}_2^t, \ldots, \tilde{E}_n^t$ according to the numerical value, there are $n+1$ possible cases ϕ. All cases can be expressed by Eq. (11)

$$\phi = \sum_{k=1}^{n} \mu_k = j (0 \le j \le n)$$

$$\begin{cases} \overline{E}_t < \overline{E}_1^t, & j = 0 \\ \overline{E}_j^t \le \overline{E}_t < \overline{E}_{j+1}^t, & j \in [1, n-1] \\ \overline{E}_t \ge \overline{E}_n^t, & j = n \end{cases} \tag{11}$$

The range of values E_t is also different according to different situations. Through calculation, we can get all the indicators of interval number $\tilde{E}_t(j)$. The probability of case $\phi = j(0 \le j \le n)$ can be thought of as inserting $\tilde{E}_j^t$ into the array of $\tilde{E}_1^t, \tilde{E}_2^t, \ldots, \tilde{E}_n^t$ which is probability behind or in front. Put all the cases after the indexes interval number $\widehat{E}_b$ can be expressed as Eq. (12)

$$\begin{aligned} E_b &= \sum_{k=1}^{j} \overline{E}_k^b \sum_{k=1}^{m} \tilde{t}_k^b / m + (m-j)\overline{E}_b \sum_{k=1}^{m} \tilde{t}_k^b / m \\ \widehat{E}_b &= \sum_{j=0}^{m} \widehat{E}_b(j) / p \end{aligned} \tag{12}$$

where $\bar{E}_t$ is the energy consumption per unit traction time, and it is the kth element in the array. $\bar{E}_k^b$ represents the regenerative braking energy of the train corresponding to the kth element in the array (J). $\bar{t}_k^b$ is the train braking time (s) corresponding to the kth element. $\tilde{E}_b(j)$ denotes that under the circumstance $\phi' = j$, the interval number (J) of regenerative energy E_b generated by the train and used by other vehicles within the total braking time. And, p is the number of $\widehat{E}_b(j) \neq 0$ within $m+1$ cases.

2.3 Sequencing Method Based on Regenerative Energy

The objects sorted in this section are V train operating condition sequences, namely $X_1, X_2, \ldots, X_v$, which are calculated by rules in Sect. 2.2. In this section, the more objective entropy weight method [9] is mainly used to allocate the weight of the index, and the distance between the index interval number and the ideal interval number is calculated by considering the expected value and the wide EW-type geometric distance [10]. The specific steps of sorting values corresponding to train operating condition sequences are shown as follows:

Step 1: Establish index matrix H, as shown below,

$$H = \left(h_{ij}\right)_{\upsilon \times 2} \tag{13}$$

where h_{ij} is the interval number of the jth index of the operating condition sequence X_i of the ith train.

Step 2: Normalize the index matrix H to obtain the matrix $\tilde{H}$,

$$\tilde{H} = \left(\tilde{h}_{ij}\right)_{\upsilon \times 2} \tag{14}$$

where $\tilde{h}_{ij}$ denotes the normalized interval number of the jth index of the ith train operating condition sequence.

Step 3: The entropy value of the jth index can be obtained by the following formula,

$$\varepsilon_j = \left(\varepsilon_j^- + \varepsilon_j^+\right)/2, \quad j = 1, 2 \tag{15}$$

where ε_j^- is the entropy of the lower bound array $\tilde{h}_{ij}^-(i = 1, 2, \ldots, v)$ denotes the entropy of the higher bound array $\tilde{h}_{ij}^-(i = 1, 2, \ldots, v)$.

Step 4: The entropy weight of the jth index can be obtained by the following formula

$$\omega_j = (1 - \varepsilon_j)/(2 - \varepsilon_1 - \varepsilon_2), \quad j = 1, 2 \tag{16}$$

Step 5: The index matrix H is normalized to obtain the matrix $\widehat{H}$

$$\widehat{H} = \left(\widehat{h}_{ij}\right)_{\upsilon \times 2} \tag{17}$$

Where $\widehat{h}_{ij}$ is the normalized interval number of the ith index of the operating condition sequence X_i of the ith train, and its upper and lower boundary values $\tilde{h}_{ij}^{+}, \tilde{h}_{ij}^{-}$ as follows:

$$\widehat{h}_{ij}^{+} = \left(h_{ij}^{+} - \min_i\left\{h_{ij}^{-}\right\}\right) / \left(\max_i\left\{h_{ij}^{+}\right\} - \min_i\left\{h_{ij}^{-}\right\}\right)$$
$$\widehat{h}_{ij}^{-} = \left(h_{ij}^{-} - \min_i\left\{h_{ij}^{-}\right\}\right) / \left(\max_i\left\{h_{ij}^{+}\right\} - \min_i\left\{h_{ij}^{-}\right\}\right)$$

Step 6: Construct the weighted normalized matrix Q, as follows,

$$Q = (q_{ij})_{\upsilon \times 2} \tag{18}$$

where q_{ij} is the weighted standardized interval number of the jth index of the ith train operating condition sequence X_i and $q_{ij} = \omega_j \widehat{h}_{ij} = \left[\omega_j \widehat{h}_{ij}^{-}, \omega_j \widehat{h}_{ij}^{+}\right]$.

Step 7: Determine the positive and negative ideal solutions and calculate the distance between them and the weighted normalized index. Adopt the EW-type geometric distance for calculation, and we can get the following formula when $j = 1, 2$.

$$d^{-}(q_{ij}) = \sqrt{\left[\Upsilon(q_{ij}) - \Upsilon([0,0])\right]^2 + \left[\Omega(q_{ij}) - \Omega([0,0])\right]^2 / 3} \tag{19}$$

$$d^{+}(q_{ij}) = \sqrt{\left[\Upsilon(q_{ij}) - \Upsilon([1,1])\right]^2 + \left[\Omega(q_{ij}) - \Omega([1,1])\right]^2 / 3} \tag{20}$$

where $\Upsilon(q_{ij})$ is the expected value of the interval number $q_{ij} = \left[q_{ij}^{-}, q_{ij}^{+}\right]$. $\Omega(q_{ij})$ denotes the width of the interval number $q_{ij} = \left[q_{ij}^{-}, q_{ij}^{+}\right]$.

Step 8: The closeness between the weighted standardized index and the positive ideal solution is calculated, and the specific calculation formula is shown below

$$\widetilde{d}_{ij} = d^-(q_{ij}) / \left[d^-(q_{ij}) + d^+(q_{ij})\right], \quad j = 1, 2 \tag{21}$$

Step 9: The closeness degree $\widetilde{d}_{i1}, \widetilde{d}_{i2}$ corresponding to the two indices is taken as the value of the horizontal and vertical coordinates to calculate the distance from the origin of the coordinates, that is, the sequencing value $O(i)$ of the train operating condition sequence

$$O(i) = \sqrt{\widetilde{d}_{i1}^2 + \widetilde{d}_{i2}^2} \tag{22}$$

On this basis, the sequencing of all train operating sequences can be obtained by taking the train operating sequence with larger sequencing value as the better choice.

3 Case Studying

The initial data of this section is from the Beijing subway Yizhuang line, which starts from Songjiazhuang station and ends at Yizhuang railway station based on the length of each interval and the line speed limit data.

The distribution of all solutions in the closeness degree space can be obtained as shown in Fig. 1. It can be clearly seen from the figure that the solution distribution area is relatively concentrated, and the solution area can be roughly divided into three parts except for a few outliers. Area 1: $x \in [0.2, 0.4], y \in [0.6, 0.8]$; Area 2:

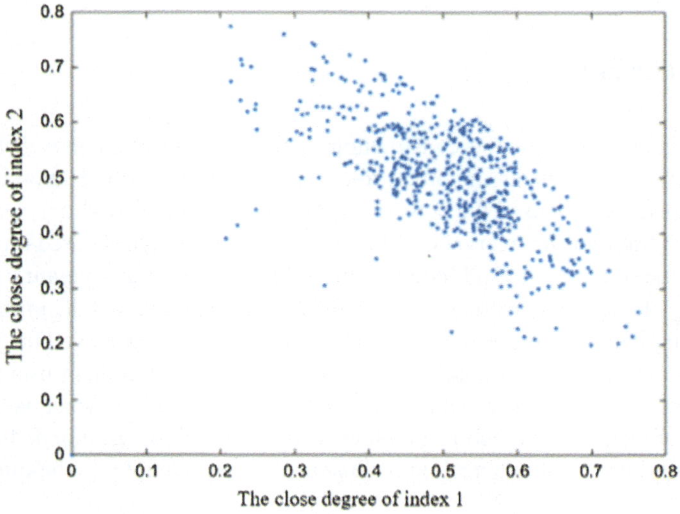

Fig. 1 Distribution of all the solutions

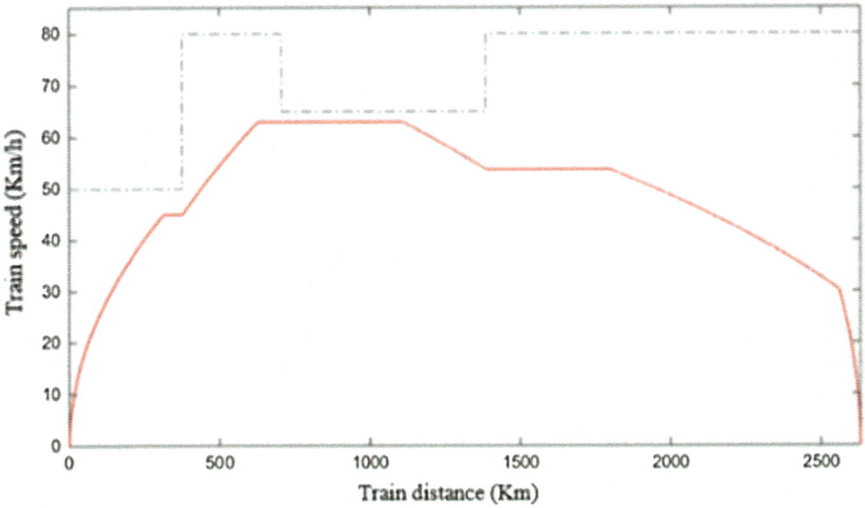

Fig. 2 Train operation condition sequence with maximum ranking value

$x, y \in [0.4, 0.6]$; Area 3: $x \in [0.6, 0.8], y \in [0.2, 0.4]$. Among them, the number of solutions in areas 1 and 3 is not significantly different, while the number of solutions in Area 2 is obviously higher than that in other two areas.

Of all the above solutions, choose the one with the largest ranking value $O = 0.8152$, the degree of closeness of the ranking index corresponding to the train operating condition sequence is $\bar{d}_{i1} = 0.6268, \bar{d}_{i2} = 0.6268$,

The interval running time is 186 s. Output its specific situation to get Fig. 2.

4 Conclusions

As a necessary work for the construction of train optimization model, train condition sequencing plays an important role in improving the optimization efficiency. In this paper, we proposed the construction rules of running condition sequence of the interval train for the interval of different speed limiting types based on the given running time of the train. And based on the solution set of train operating condition sequence obtained by constructing rules, the ranking indices, and the uncertainty of two ranking indices expressed by interval numbers, we have achieved the following results: firstly, the interval number of all solutions and sorted them using TOPSIS method based on entropy weight is obtained; secondly, the sequence values of all train operating conditions corresponding to the interval are acquired; finally, the practical application of this method is illustrated by an example. In the future study, this construction rule can also be used to obtain the solution set of train operating condition sequences of all intervals, and the database composed of all solution sets can contribute to the online optimization of train operation process in case of delay and shorten the optimization time.

Acknowledgements The authors gratefully acknowledge the support from "National Key R&D Program of China." (Grant No. 2017YFB1201105).

References

1. Gunselmann W (2005) Technologies for increased energy efficiency in railway systems. In: European conference on power electronics and applications. IEEE, 10 p
2. Söylemez MT, Açıkbaş S (2007) Parameters affecting braking energy recuperation rate in DC rail transit. In: ASME/IEEE joint rail conference & internal combustion engine spring technical conference. IEEE, pp 263–268
3. Li L (2016) Research on energy saving optimization of urban rail transit trains based on particle swarm optimization algorithm. Beijing Jiaotong University
4. Gupta SD, Tobin JK, Pavel L (2016) A two-step linear programming model for energy-efficient timetables in metro railway networks. Transp Res Part B 93:57–74
5. Su S, Tang T, Roberts C (2015) A cooperative train control model for energy saving. IEEE Trans Intell Transp Syst 16(2):622–631
6. Nasri A, Moghadam MF, Mokhtari H (2010) Timetable optimization for maximum usage of regenerative energy of braking in electrical railway systems. In: International symposium on power electronics electrical drives automation and motion. IEEE, pp 1218–1221
7. Aradi S, Becsi T, Gaspar P (2013) A predictive optimization method for energy-optimal speed profile generation for trains. In: IEEE, international symposium on computational intelligence and informatics. IEEE, pp 135–139
8. Huang Y, Yang L, Tang T et al (2017) Joint train scheduling optimization with service quality and energy efficiency in urban rail transit networks. Energy 138
9. Guo X (2012) Improvement of entropy method for determining index weights of interval numbers. Stat Decis Mak 17:32–34 (in Chinese)
10. Bao Y, Peng X, Zhao B (2013) Interval number distance based on expected value and width and its completeness. Fuzzy Syst Math 27(6):133–139 (in Chinese)

Fault Diagnosis for Levitation-Gap Sensor of Maglev Train Based on the Tracking Differentiator

Lili Xu, Jie Xu, Hongwei Gao and Yiwen Chen

Abstract To determine the additive fault and multiplicative fault of gap sensor, based on improved step-size sampling control of the optimal tracking differentiator, a fault diagnosis method for levitation-gap sensor of electromagnetic suspension (EMS) medium-speed maglev trains is proposed. The model of single electromagnet suspension system is established first. The tracking and differentiation effect of the classic tracking differentiator and the optimal tracking differentiator for gap sensor signal is obtained by Simulink. Based on the analysis of signal time domain features, the comparison of the effect is obtained. The result shows that the noise interference can be further reduced by tracking differentiator with sampling control. Meanwhile, the differentiator maintains good tracking performance. Diagnose the fault of the gap sensor with improved sampling control of the optimal tracking differentiator. Based on the analysis of amplitude variation of gap residual signal, the type of additive and multiplicative fault of gap sensor is distinguished.

Keywords Maglev train · Single electromagnet suspension system · Tracking differentiator · Levitation-gap sensor · Fault diagnosis

---

L. Xu · J. Xu (✉) · H. Gao · Y. Chen
State Key Laboratory of Rail Traffic Control and Safety, Beijing Jiaotong University, Beijing, China
e-mail: jxu1@bjtu.edu.cn

School of Transportation and Traffic, Beijing Jiaotong University, Beijing, China

Beijing Research Center of Urban Traffic Information Sensing and Service Technologies, Beijing, China

Research and Development Center of Transport Industry of Technologies and Equipments of Urban Rail Operation Safety Management, MOT, PRC, Beijing, China

Y. Qin et al. (eds.), *Proceedings of the 4th International Conference on Electrical and Information Technologies for Rail Transportation (EITRT) 2019*, Lecture Notes in Electrical Engineering 639, https://doi.org/10.1007/978-981-15-2866-8_22

1 Introduction

Electromagnetic suspension (EMS) medium-speed maglev train is a new type of rail transportation. It runs at high speed driven by a linear motor and controls the electromagnetic force between the electromagnet and the track to achieve the suspension and guidance of the train. Maglev trains are therefore quieter, smoother, and safer than other trains [1, 2]. As a key component of medium-speed maglev trains, the suspension control system consists of suspension controller, sensors (including acceleration sensors, gap sensors, and current sensors), choppers, and power supply. Levitation-gap signal is an important input of the suspension controller which is collected by the gap sensor. Practice verification shows that the gap sensor is prone to failure. If the sensor failure can not be detected and eliminated in time, it will affect the stable suspension of the bogie and even cause an accident because of falling or sucking [3]. Therefore, it is meaningful to study the fault diagnosis method of the gap sensor to improve the safety of the vehicle operation.

In the relevant fault diagnosis research, Long proposed a diagnosis algorithm which regards the electromagnet fault as input variation [4]. Wu proposed the fault detection methods about the main sections of gap sensor, including phase error detection, standard signal switch circuit, and voltage convert compare circuit [5]. Hu resolved some key problems of the gap sensor such as resonance and signal processing by selecting a special resonant circuit and taking other measurements [6]. At present, there are few studies on the additive and multiplicative fault diagnosis of the levitation-gap sensor. It is of great significance to carry out this work.

In this paper, the suspension behavior model of the single electromagnet suspension control system is developed considering the influence of the feedback control. Comparing the classical linear tracking differentiator with the new discrete of the sampling control tracking differentiator proves the later has more superiority. Further, a sampling control tracking differentiator with differential step size is proposed to improve its noise reduction ability. Diagnose the fault of the gap sensor with the improved tracking differentiator. The simulation proves that the method has good diagnostic effect.

2 Single Electromagnet Suspension System Model

The suspension bogie of the maglev train is symmetrically installed with two pairs of suspension modules [7]. Each suspension module has two electromagnets symmetrically in front and rear. Each electromagnet is controlled by a suspension controller, and an integrated sensor is installed on the electromagnet. In this paper, a single suspension unit consists of an electromagnet, a controller and a set of sensors is considered as a single electromagnet suspension system. Such model is generic and can well reflect the suspension behavior characteristics of the maglev train. In

Fig. 1 Structure diagram of single magnet suspension system

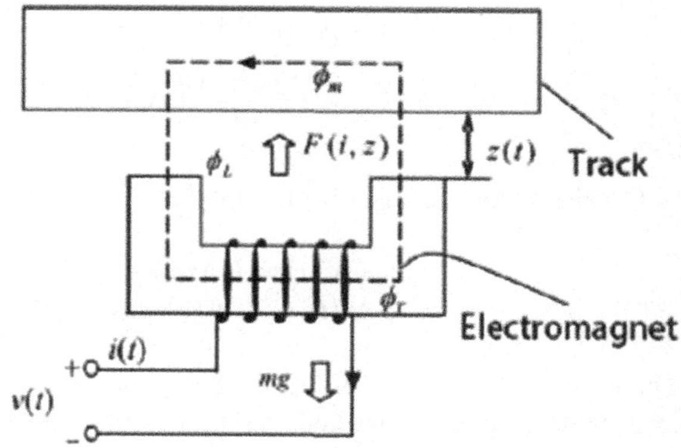

order to simplify the study of the problem, when establishing the physical model of the single electromagnet suspension system, it is assumed that the carriage and the bogie are rigidly connected, regardless of the rail deformation. For most of the laboratory-scale maglev train test systems, the track-fixed structure is much heavier than the test car, so this model can get accurate results. The structure diagram of the single electromagnet suspension system is shown in Fig. 1.

Assumptions:

(1) Ignore the magnetic flux leakage of winding;
(2) Ignore the reluctance in the solenoid stator and the guide rail;
(3) The positive direction of the speed is vertically downward.

The vertical dynamics model of single electromagnet suspension system can be completely determined by the following equations [8]:

$$\begin{cases} m\ddot{z}(t) = -F(i,z) + mg + f_d(t) \\ v(t) \approx Ri(t) + \frac{\mu_0 N^2 A}{2}\frac{\dot{i}(t)}{z(t)} - \frac{\mu_0 N^2 A i(t)}{2}\frac{\dot{z}(t)}{[z(t)]^2} \\ F(i,z) = \frac{\mu_0 N^2 A}{4}\left[\frac{i(t)}{z(t)}\right]^2 \\ mg = F_0(i_0, z_0) = \frac{\mu_0 N^2 A}{4}\left(\frac{i_0}{z_0}\right)^2 \end{cases} \quad (1)$$

where N denotes the electromagnet winding turns; $z(t)$ is the distance between electromagnet and lower surface of the track; A represents the polar area of the iron core;μ_0 is permeability of vacuum, and its value is $4\pi \times 10^7$; $i(t)$ is current in control coil; $v(t)$ is voltage applied across the control coil; m represents total weight carried by single electromagnet; F denotes the electromagnetism; i_0 is the stable current; and z_0 is the stable gap.

The nonlinear equations are linearized with binary Taylor expansion at steady-state operation point $i_0 - z_0$ as follows:

$$F(i, z) = F(i_0 + \Delta i, z_0 + \Delta z)$$
$$\approx F(i_0, z_0) + \frac{\mu_0 N^2 A}{4}\left(\Delta i \cdot \frac{2i_0}{z_0^2} + \Delta z \cdot \frac{-2i_0}{z_0^3}\right) \cdot F(i_0, z_0) \tag{2}$$

Get the following linearization model:

$$\begin{cases} m\Delta\ddot{z}(t) = k_z\Delta z(t) - k_i\Delta i(t) + \Delta f_d(t) \\ \Delta v(t) = R\Delta i(t) + L_0\Delta\dot{i}(t) - k_i\Delta\dot{z}(t) \end{cases} \tag{3}$$

where $\Delta z(t) = z(t) - z_0$ is the deviation between the vertical gap value of the suspension electromagnet and the steady-state gap value, $\Delta i(t) = i(t) - i_0$ is the deviation between current value and steady-state current value, $\Delta v(t) = v(t) - v_0$ is the deviation between the control voltage and the voltage of the system in steady state, $k_z = \frac{\mu_0 N^2 A i_0^2}{2z_0^3}$ is the coefficient of gap, $k_i = \frac{\mu_0 N^2 A i_0}{2z_0^2}$ is the coefficient of current, and $L_0 = \frac{\mu_0 N^2 A}{2z_0}$ is the coefficient of balance inductance.

Since the suspension system controls the train suspension by adjusting the input voltage, the control voltage value and the steady-state voltage value are equal, so $\Delta v(t) = 0$.

The suspension control system has three sensors: gap sensor, acceleration sensor, and current transformer sensor. Therefore, the gap deviation signal, vertical speed signal, and current deviation signal are selected as the system feedback signal to achieve stable closed-loop control. $X = (x_1, x_2, x_3) = (\Delta z, \Delta\dot{z}, \Delta i)^{\mathrm{T}}$ is the state variables. Take the control input variable as $u = \Delta v(t)$, and then obtain state space equations of (3) as follows:

$$\dot{X} = \begin{bmatrix} 0 & 1 & 0 \\ \frac{k_z}{m} & 0 & -\frac{k_i}{m} \\ 0 & \frac{k_i}{L_0} & \frac{-R}{L_0} \end{bmatrix} X + \begin{bmatrix} 0 \\ 0 \\ \frac{1}{L_0} \end{bmatrix} u \tag{4}$$

According to the suspension system data of the maglev train, the specific physical parameters of the single electromagnet suspension system are as follows:

$$m = 750\,\mathrm{kg}, \quad A = 0.021\,\mathrm{m}^2, \quad N = 324,$$
$$z_0 = 0.008\,\mathrm{m}, \quad i_0 = 26\,\mathrm{A}, \quad R = 0.5\,\Omega$$

According to the typical control scheme in actual engineering, gap deviation, vertical absolute speed, and current deviation are used as feedback variables to design the closed-loop controller. The feedback law is as follows:

$$u = K[k_1x_1 + k_2x_2 + k_3x_3] + \Delta v \tag{5}$$

where K is the amplifying multiple of the drive circuit, take $K = 320$, k_p is the feedback gain of gap deviation, k_v is the feedback gain of vertical absolute speed, and k_c is the feedback gain of current deviation.

In order to make the control system have ideal dynamic performance, take $[k_1 \quad k_2 \quad k_3] = [8000 \quad 400 \quad -0.12]$. After the discretization of the state space equation, the simulation study can be carried out. The stable closed-loop control system equation can be obtained as follows:

$$\dot{X} = \begin{bmatrix} 0 & 1 & 0 \\ \frac{k_z}{m} & 0 & -\frac{k_i}{m} \\ \frac{k_p}{L_0} & \frac{k_i + k_v}{L_0} & \frac{-R + k_c}{L_0} \end{bmatrix} X = \begin{bmatrix} 0 & 1 & 0 \\ 2438.40 & 0 & 0.75 \\ 46205.19 & 5560.26 & 3.58 \end{bmatrix} X \tag{6}$$

3 Performance Analysis of Fault Diagnosis Method

3.1 The Classical Linear Tracking Differentiator

It is necessary to differentiate the monitored signals in practical engineering applications. In classical theory, the discrete forms of the classical linear tracking differentiator are computed as follows:

$$\begin{cases} x_1(k+1) = x_1(k) + hx_2(k) \\ x_2(k+1) = x_2(k) + h(-w^2(x_1(k) - v(k)) - 2\varphi w x_2(k)) \end{cases} \tag{7}$$

where $v(k)$ is the input signal, $x_1(k+1)$ is the tracking signal of $v(k)$, $x_2(k+1)$ is the differential signal of $v(k)$, φ is the damping factor [9], and w is the turbo factor.

The equation shows that the tracking signals in the discrete form of the classical tracking differential are calculated by the tracking signal and the differential signal at the previous time point, which is due to a zero-order holder present in the conventional discrete form of the tracking differentiator. Therefore, the discrete form of the tracking differentiator output signal has phase delay of one sampling period at least, which also causes signal distortion to some extent, and it is easy to generate high-frequency vibration in the calculation.

3.2 The Optimal Tracking Differentiator

In order to avoid high-frequency vibration in the calculation and to reduce the phase delay of the tracking signal, an inertia link in the conventional discrete form can be eliminated by sampling control. Discrete form of second-order sampling control system [10] is computed as follows:

$$\begin{cases} x_1(k+1) = x_1(k) + hx_2(k) + \frac{1}{2}u(k)h^2 \\ x_1(k+1) = x_1(k) + hu(k), \quad |u| \le r \end{cases} \tag{8}$$

Assume $fc(x_1, x_2, r, h)$ is the optimal control integrated function that can be obtained from (8), and the new discrete form of the optimal tracking differentiator is computed as follows:

$$\begin{cases} x_1(k+1) = x_1(k) + hx_2(k) \\ x_1(k+1) = x_1(k) + hu(k) \\ u(k) = fc(x_1(k) - v(k), x_2(k), r, h) \end{cases} \tag{9}$$

$fc(x_1, x_2, r, h)$ is defined as follows:

$$\begin{cases} y = 2x_1 + hx_2 \\ a_0 = \frac{1}{2}\left(\sqrt{1 + \frac{4|y|}{h^2 r}} - 1\right) \\ a = \mathrm{fix}(a_0) + 1 \\ fc = -\mathrm{sat}\left(\frac{x_2}{h} + \frac{(a-1)r \cdot \mathrm{sign}(y)}{h} + \frac{y}{2ah^2}, r\right) \end{cases} \tag{10}$$

where $\mathrm{fix}(x)$ is the truncated rounding function, and $\mathrm{sat}(x, \delta)$ is the saturation function.

In order to verify the actual performance of the classical linear tracking differentiator and the sampling control differentiator, the Simulink simulation tool is used to simulate the single electromagnet suspension system and simulation step is 0.001 s. In practice, the maximum signal update frequency of the sensor exceeds 1500 Hz, which can achieve the requirements of output every 0.001 s and feedback control. When the turbo factor of the linear differentiator is 1000, the damping factor value is $\sqrt{3}/2$ the system has a good dynamic result [11]. After debugging, the turbo factor of the sampling control tracking differentiator has a good differential effect when it is 1000. Tracking and differentiation conditions of gap signal by classical differentiator and sampling control differentiator for maglev train from landing state to stable suspension period are shown in Figs. 2, 3, 4, and 5.

Fig. 2 Tracking effect of linear tracking differentiator

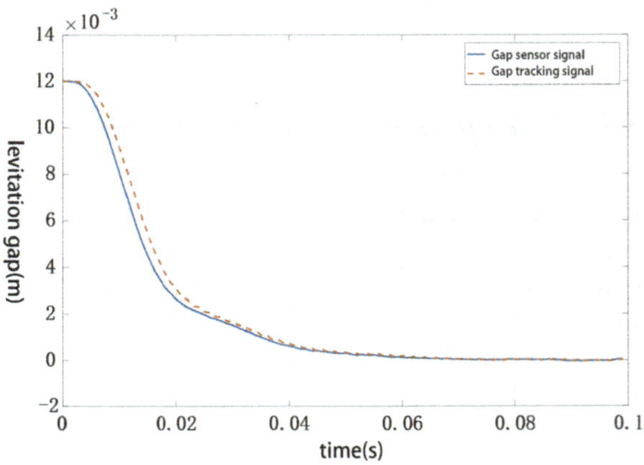

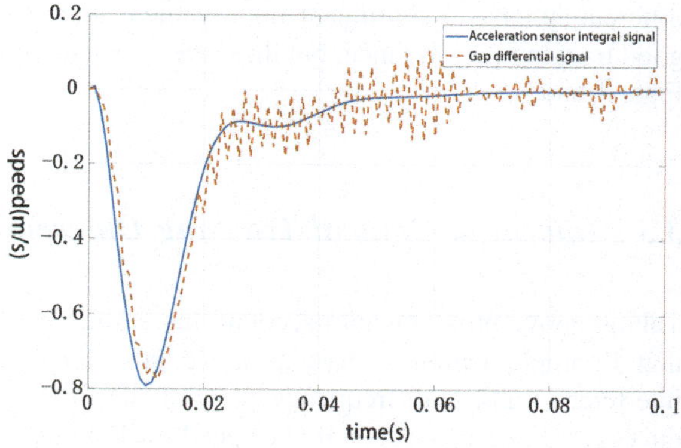

Fig. 3 Differential effect of linear tracking differentiator

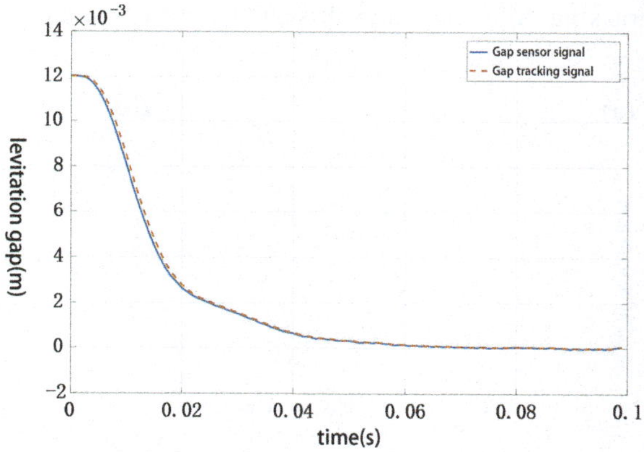

Fig. 4 Tracking effect of the speed tracking differentiator

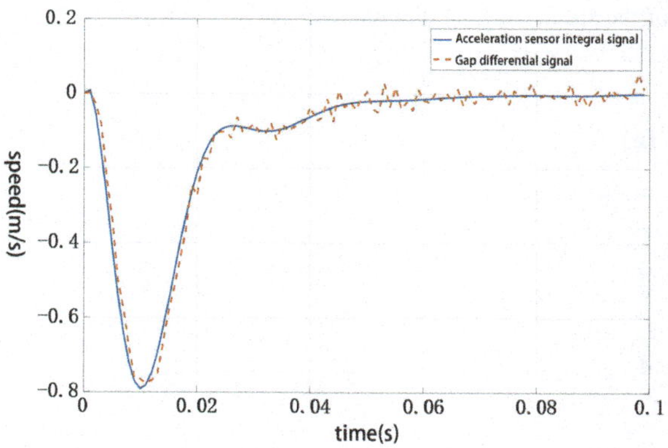

Fig. 5 Differential effect of the speed tracking differentiator

The classic tracking differentiator has better tracking effect, but the speed signal obtained by differentiating gap is seriously interfered by noise, as shown in Figs. 2 and 3. Compared with the linear tracking differentiator, the speed tracking differentiator has a significant improvement in the noise reduction while keeping good tracking performance, but there are still many noise interferences, as shown in Figs. 4 and 5.

3.3 Improved Optimal Tracking Differentiator

Taking above problems into account, increasing the differential time appropriately, that is, implementing a certain step delay, can be used to further reduce noise interference and high-frequency jitter of the differential signal. This paper introduces c_0 (a filtering factor) in the speed tracking differentiators and requires $c_0 \geq 1$. Then, change $fc(x_1, x_2, r, h)$ to $fc(x_1, x_2, r, c_0h)$ and take $c_0 = 2, c_0 = 3$, and the tracking differential signal is shown in Figs. 6 and 7.

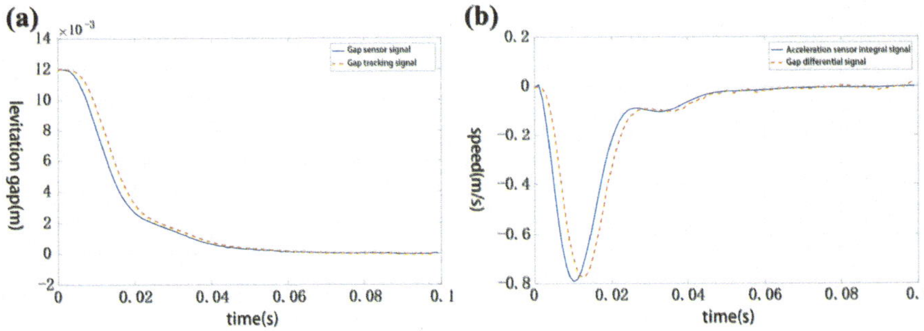

Fig. 6 **a** Tracking effect of the speed tracking differentiator $(c_0 = 2)$ and **b** the differential effect of the speed tracking differentiator $(c_0 = 2)$

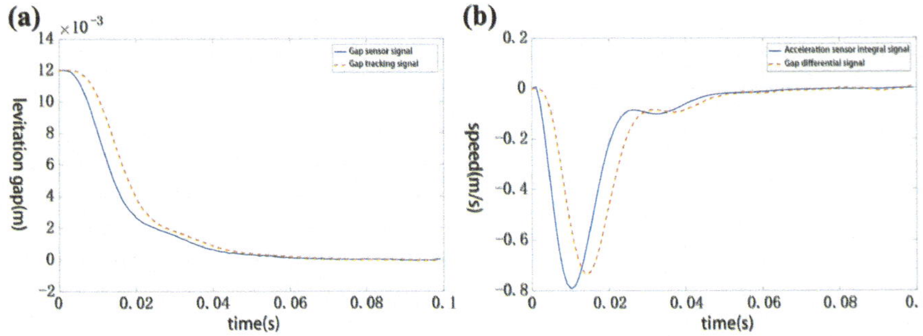

Fig. 7 **a** Tracking effect of the speed tracking differentiator $(c_0 = 3)$ and **b** the differential effect of the speed tracking differentiator $(c_0 = 3)$

After introducing a filtering factor, the tracking signal had a certain delay. However, removing the zero-order holder can still keep the tracking delay low. The tracking effect is better, as shown in Fig. 6. The reduction of signal distortion reduces the high-frequency jitter to a certain extent, and the noise reduction result is great, as shown in Fig. 7. Comparing Fig. 6 with Fig. 7, we can know that the high tracking delay caused the signal differential effect with a large gradient to decrease, and the anti-noise effect is not significantly improved. The filter factor value is not bigger is better. In this paper, the filtering factor is 2 for sensor diagnosis of single electromagnet suspension system.

4 Simulation Verification

For practical application, the gap sensor is considered to have ideal linear operating features under normal operating conditions, and the train's suspension gap range is 0.00–20.00 mm. Its displacement characteristic curve can be expressed as follows:

$$I = kd - d_0 \tag{11}$$

where I is the sensor output current, k is the fixed gain, d is the real gap, and d_0 is the reserved gap to ensure the sensor in linear interval. Usually, take $d_0 = 5\,\text{mm}$.

The simulation of train suspension behavior begins in the state of maglev train drop-off. After a period of stable suspension, the simulation joins the fault at $t = 2$ s. The way to simulate faults is divided into additive faults and multiplicative faults according to fault types.

(1) The additive fault of the gap sensor is regarded as the change of d_0. In the simulation, the feedback gap deviation signal increases by 0.2 mm. The residual signal figure of the gap sensor additive fault is as Fig. 8.
When an additive fault occurs in the gap sensor, the residual signal can significantly and quickly show the special characteristic signal of the fault. This is

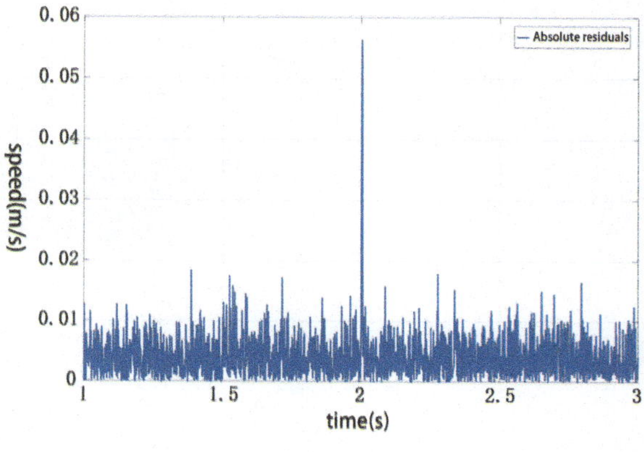

Fig. 8 Residual absolute value of speed of gap sensor with additive failure

due to the sudden change of the gap sensor causing a sharp change in amplitude in a short time and increased its differential value obviously. The fault also caused a sudden change in the acceleration signal, and the speed signal obtained by integral also changed obviously. However, the feedback control rate immediately stabilized the suspension system, means the gap and current generated a fixed offset, and the velocity went to zero. Finally, it produced a clear pulse signal on the residual, and it can be seen that the tracking differentiator has a better diagnostic result to the additive fault.
As shown in Fig. 8.

(2) The multiplicative fault of the gap sensor can be regarded as the change of k (the fixed gain which is generally considered to be 1) in the curve equation of sensor characteristic. Set one-half gain and double gain in fault simulation, respectively. The residual signal figures of the gap sensor multiplicative fault are as Figs. 9 and 10.

After the multiplicative faults of gap sensor occurs, residual signal of speed has obvious fault characteristics. Signal change has a positive correlation with the gain, and the fault characteristics react quickly, which can be reflected in the residual signal immediately, as shown in Figs. 9 and 10. The reason is that the tracking differentiator differentiates the gap signal, and the amplitude change of the gap signal can directly reflect on the speed signal obtained by differentiating.

Although the speed signal obtained by integrating the acceleration will change to some extent after the gap signal multiplicative fault occurs, compared with the change of differential speed causing by gain is small, which can be ignored in the actual diagnosis. Therefore, the variation of the residual amplitude can accurately diagnose the multiplicative fault of the gap sensor with different gain value. The tracking differentiator diagnostic method has a good diagnostic result for the multiplicative fault of the gap sensor.

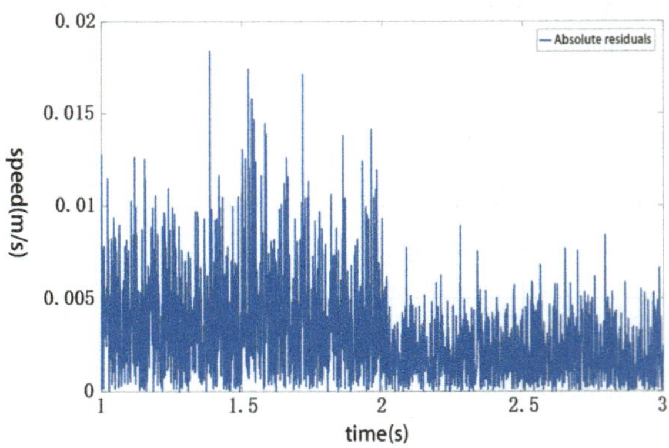

Fig. 9 Residual absolute value of speed of one-half multiplicative failure of gap sensor

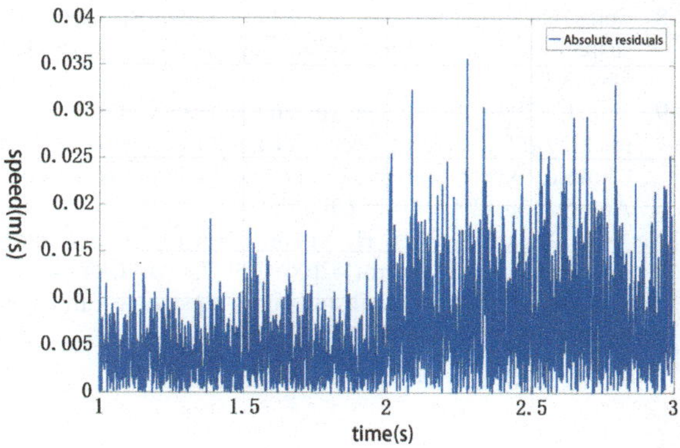

Fig. 10 Residual absolute value of speed of double multiplicative failure of gap sensor

5 Conclusion

This paper mainly studies the sampling control of the optimal tracking differentiator for the fault diagnosis of the gap sensor of single electromagnet suspension system. According to the deficiency of the classical linear tracking differentiator, a sampling control tracking differentiator with increasing differential step size is proposed. The tracking differentiator has better anti-noise effect while maintaining good tracking performance. The improved speed tracking differential is used to diagnose the gap sensor. The simulation results show that the method has a good diagnostic result to the multiplicative fault and the additive fault of the gap sensor.

Acknowledgements The authors gratefully acknowledge the support provided by the National Key R&D Program of China (2016YFB1200601).

References

1. Shang H (2004) Research on fault diagnosis technology of maglev train. National University of Defense Technology (in Chinese)
2. Jayawant BV, Sinha PK, Aylwin DC (1976) Feedback control systems for d.c. electromagnets in passenger-carrying vehicles. Int J Control 24(5):627–639
3. Yitao Z (2009) Fault detection of sensor in middle-low velocity maglev train. National University of Defense Technology (in Chinese)
4. Zhiqiang L, Yun L, Guang H (2010) Research on electromagnet fault diagnosis technology of suspension control system of maglev train. Control Decis 25(07):1004–1009 (in Chinese)
5. Jun W, Liang L (2009) Design on fault detection for levitation gap sensor of high speed maglev train. Comput Meas Control 17(10):1923–1924+1930 (in Chinese)
6. Hua H, Chuanrun Z, Dehui T, Guifen T (2006) Research of key technology of suspend gap sensor for EMS maglev. Instrum Tech Sensor 2006(09):6–7+62
7. Ying C (2007) Research on comprehensive fault evaluation of maglev train based on machine learning. National University of Defense Technology (in Chinese)

8. Song X (2007) Research on fault diagnosis and fault-tolerant control technology for suspension system of low speed maglev train. National University of Defense Technology (in Chinese)
9. Yun L (2008) Sensor active fault tolerant control of maglev suspension system based on tracking-differentiator. Technical committee on control theory. In: Proceedings of the 27th Chinese control conference. Technical committee on control theory. Chinese Association of Automation (2008/4) (in Chinese)
10. Xi L, Xiuxia S, Zhen H, Yukun L (2013) A new discrete-time form of optimal tracking differentiator. Inf Control 42(06):729–734 (in Chinese)
11. Jingqing H (2007) Auto disturbances rejection control technique. Front Sci (1):24–31. https://doi.org/10.3969/j.issn.1673-8128.2007.01.004

Research on Train Bogie Frame Service Performance Characterization Parameters Based on Dynamic Loads

Li Liu, Yanhui Wang and Kang Mei

Abstract The bogie frame, an important component of an urban rail train, plays a vital role in the safe operation of the trains. In this paper, a three-dimensional finite element numerical simulation analysis model is constructed. The mechanical deformation trend and modal parameters of the frame under different operating conditions are analyzed to obtain deformation and modal characteristics of the frame. Furthermore, the extraction and calculation methods of the service performance characterization parameters of the frame under different operating conditions are proposed based on the numerical simulation.

Keywords Bogie frame · Service performance · Finite element analysis · Characterization parameters

1 Introduction

As one of the most important components in the structure of a rail train, the bogie bears not only the weight of the train itself, but also various forces generated between the wheel set and the track. The frame plays an important role in the traction, braking and maintaining the stability of the train. Therefore, once it is damaged, it will have extremely serious consequences. This paper focuses on the analysis of the service performance of the bogie frame.

L. Liu · K. Mei
School of Traffic and Transportation, Beijing Jiaotong University, Beijing, China

L. Liu · Y. Wang (✉) · K. Mei
State Key Laboratory of Rail Traffic Control and Safety, Jiaotong University, Shanghai, China
e-mail: 18120979@bjtu.edu.cn

Y. Wang · K. Mei
Research and Development Center of Transport Industry of Technologies and Equipments of Urban Rail Operation Safety Management, MOT, PRC, Beijing, China

Y. Qin et al. (eds.), *Proceedings of the 4th International Conference on Electrical and Information Technologies for Rail Transportation (EITRT) 2019*, Lecture Notes in Electrical Engineering 639, https://doi.org/10.1007/978-981-15-2866-8_23

Bai et al. [1] of CRRC Tangshan Co., Ltd. tracked the performance of CRH3 high-speed train in Wuhan–Guangzhou Passenger Dedicated Line for two years, and through data analysis, the fault law, wheel wear law and dynamic performance evolution of the train were obtained. Based on the analysis of the service performance of high-speed trains, Lin et al. [2] proposed a real-time detection system for high-speed train service performance, developed a high-speed train bogie tracking test technology and got the service performance of high-speed trains.

2 Problem Statement

2.1 *Definition*

Service performance is the ability that the bogie frame makes the train run smoothly and safely under different service environments with the service status changes constantly. Take the bogie of the B-type subway train with 120 km/h speed as research object. Through the analysis of the bogie system fault data, the main reasons for the bogie frame failure are weld crack and bending deformation. It is caused by the deterioration of weld joint position, the deterioration of matrix material performance and the wear of matrix material.

2.2 *Analysis*

Based on the basic theoretical knowledge of material mechanics, we present three service performance characterization parameters of bogie frame, which are degradation degree, spectrum variation coefficient and structural stability coefficient. The formulas are shown in Table 1.

Table 1 Formulas for calculating characterization parameters

Parameter	Formula	Meaning
Degradation degree	$\mu = \frac{\bar{\sigma}}{\sigma_s}$	$\bar{\sigma}$: the Mises equivalent stress; σ_s: the allowable stress of the material
Spectral variation coefficient	$\varepsilon = \frac{(f_g - f_c)\cdot(f_g - f_z)}{f_g^2}$	f_g: the fundamental vibrational frequency of bogie frame; f_c: the vibration frequency of car body;f_z: the vibration frequency of bogie
Structural stability coefficient	$\delta = \frac{l_{T\max}}{L} + \frac{(l_G - 10)^2}{l^2}$	$l_{T\max}$: the maximum shape variable; L: the allowable elastic deformation value of the structure; l_G: the value of the rigid deformation of the structure; l: the total amount of displacement that occurs in the structure

The degradation degree reflects the strength properties of the material. When it goes from 0 to 1, means the material is getting worse and worse; when is larger than 1, the material structure has undergone yield deformation. The spectral variation coefficient reflects the relationship among the vibration frequency of the frame, the train body and the bogie; if it is bigger than 0, all three matches. And the greater the value, the greater the difference among the three. The structural stability coefficient reflects the effect of structural deformation on structural stability. The larger is the deformation, the larger is the value, and the structural stability is deteriorated.

3 Methodology

3.1 Modeling

In this paper, the characterization parameter analysis is carried out based on ANSYS software. The grid division diagram is shown in Fig. 1.

3.2 Model Analysis

In the process of train operation, the forces are complex. According to the relevant standard formulas [3, 4], the operating conditions are divided into four categories, as shown in Table 2. These four conditions have certain representativeness by considering the actual running environment of urban railway train comprehensively.

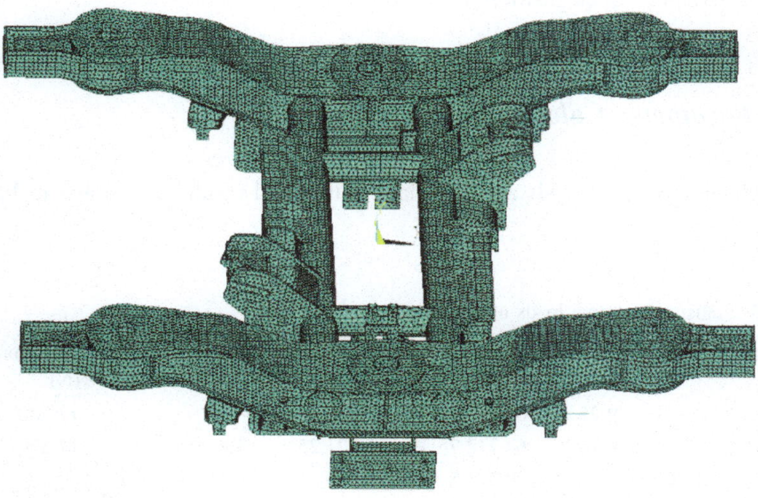

Fig. 1 Grid division of bogie frame model

Table 2 Bogie frame normal service conditions and loads

Conditions	Normal service condition		Abnormal service condition	
	1	2	3	4
k	–	–	1.4	2.0
Vertical load F_{z1}	$(1+\alpha-\beta)F_z$	$(1+\alpha-\beta)F_z$	F_{z1max}	F_{z1max}
Vertical load F_{z2}	$(1-\alpha-\beta)F_z$	$(1-\alpha-\beta)F_z$	F_{z2max}	F_{z2max}
Lateral load	F_x	F_x	F_{xmax}	F_{xmax}
Torsional load	–	F_n	F_{nmax}	F_{nmax}
Special load	F_g	F_g	–	–

4 Results

4.1 Model Calculation

Load analysis results are shown in Table 3.

We carried out finite element analysis of the bogie frame and used ANSYS for strength analysis and post-processing. The deformation cloud diagram and stress cloud diagram obtained are shown in Figs. 2 and 3, and strength analysis results are shown in Table 4.

Modal analysis. The first six modes of the bogie frame model are shown in Fig. 4. The analysis results are shown in Table 5.

According to the above modal analysis results, the fundamental vibration frequency of the bogie frame is 83.4 Hz. Moreover, the bogie frame appears the floating and sinking vibration pattern in the first order, indicating it has a strong ability to adapt to track distortion and irregularity.

4.2 Parameter Calculation

Through the above analysis and calculation, the final results are shown in Fig. 5.

Table 3 Statistics of frame loads under service conditions

Conditions	F_{z1} (N)	F_{z2} (N)	Lateral load (N)	Torsional load (N)	Gearbox load (N)
1	83,279.25	64,772.75	65,721.25	–	17,500
2	83,279.25	64,772.75	65,721.25	5962.34	17,500
3	129,545.5	129,545.5	106,220	7949.79	–
4	185,065	185,065	106,220	7949.79	–

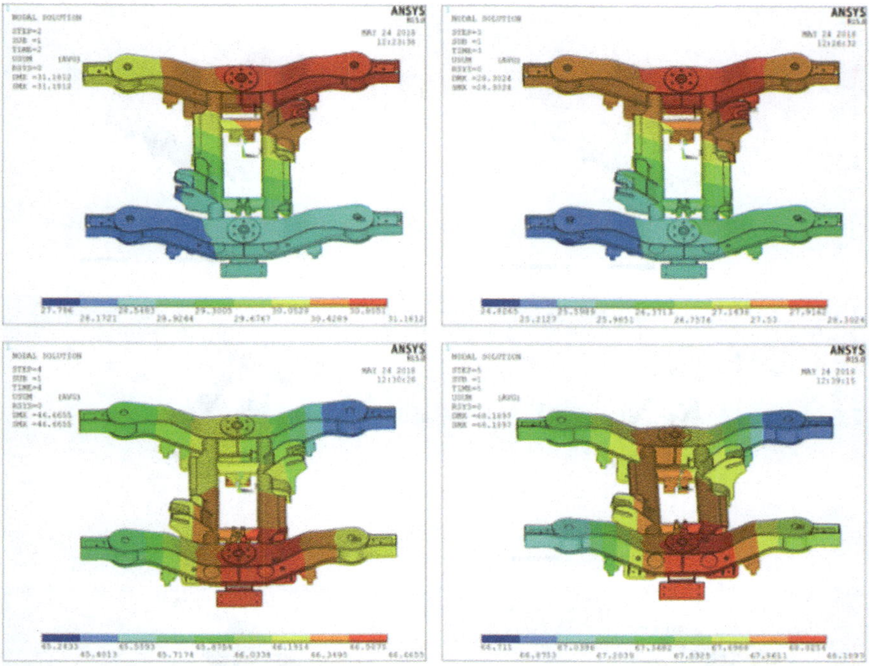

Fig. 2 Deformed cloud chart of bogie frame

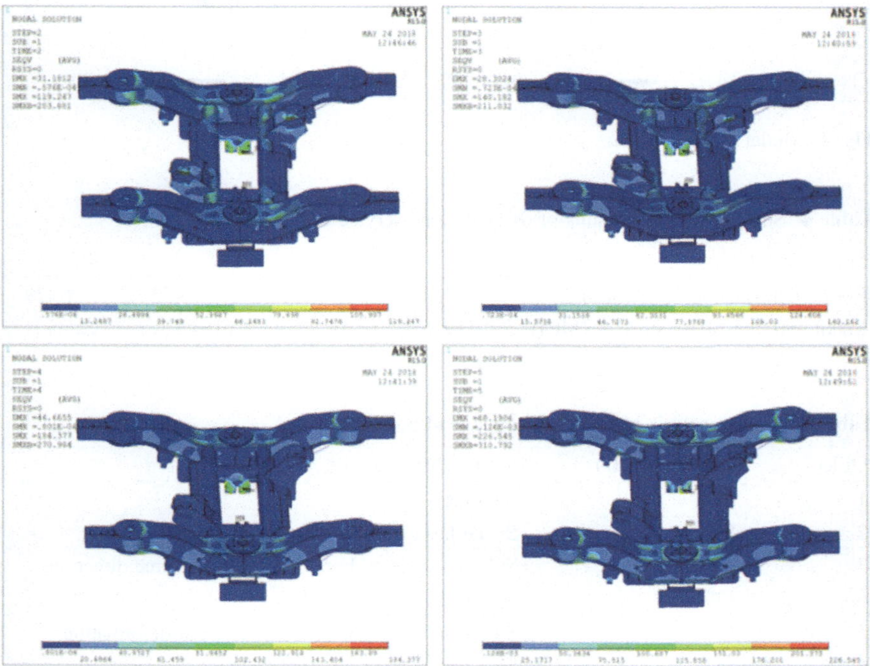

Fig. 3 Stress cloud chart of bogie frame

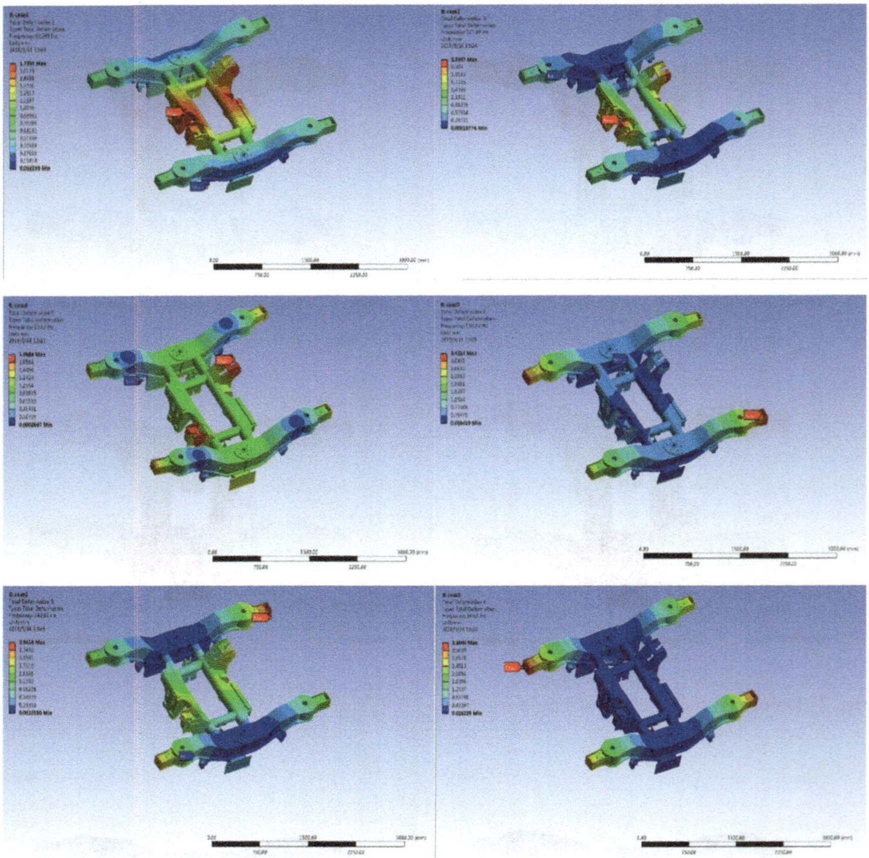

Fig. 4 Modal mode diagrams

Table 4 Strength analysis results of dynamic load service condition

Service condition	1	2	3	4
Elastic deformation $l_{T\max}$ (mm)	3.385	3.475	1.422	1.479
Maximum von Mises stress $\overline{\sigma}$ (MPa)	119.2	140.2	184.7	226.5

Table 5 Modal analysis results of dynamic load bogie frame

Order	Frequency (Hz)	Modal characteristics
1	83.4	Floating
2	122	Shake head
3	133.7	Side beams bend horizontally in the same direction
4	139.4	Anti-roll
5	142	Side beams horizontal figure of eight deformation
6	143	Side beams bend longitudinally

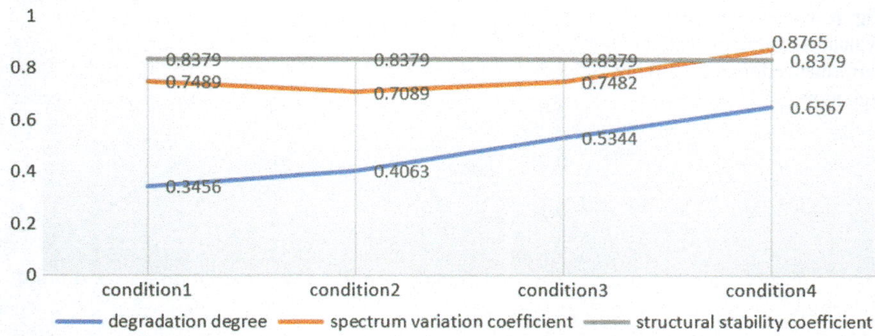

Fig. 5 Performance qualification parameters for structural service under dynamic loads

All values are less than 1, and the parameters are all in the normal range. And the modal analysis is independent of the load subjected to the structure and shows there is no resonance phenomenon. The trend of deterioration degree shows that the deterioration degree increases as the service environment becomes more severe and the load bearing becomes more complex in the service process. Besides, under the condition of abnormal service, the degree of deterioration is more serious. The structural stability coefficient always fluctuates, and the fluctuation range is small. The spectral variation coefficient remains unchanged under dynamic loads.

4.3 Assessment

The comprehensive evaluation of service performance under four operating conditions is realized by using characterization parameters.

Steps: Construct the decision matrix of evaluation object and evaluation index; standardized treatment of evaluation indicators; construct the evaluation index deviation matrix; the evaluation index judgment matrix of each evaluation object is constructed according to the idea of maximum deviation [5, 6]; weighting of evaluation indicators [7, 8]; calculate the value of multi-index fusion evaluation [9].

Considering the changes of service conditions, we use the nine-fractal scale to construct the decision matrix B in this paper. A decrease in the scale value from 9 to 1 indicates a gradual decrease in the importance of b_j over b_j, and the increase from 1/9, 1/7, 1/5 to 1/3 indicates that b_i is less important than b_j gradually.

$$B = \begin{bmatrix} b_{11} & b_{12} & \cdots & b_{1m} \\ b_{21} & b_{22} & \cdots & b_{2m} \\ \vdots & \vdots & \vdots & \vdots \\ b_{m1} & b_{m2} & \cdots & b_{mm} \end{bmatrix}$$

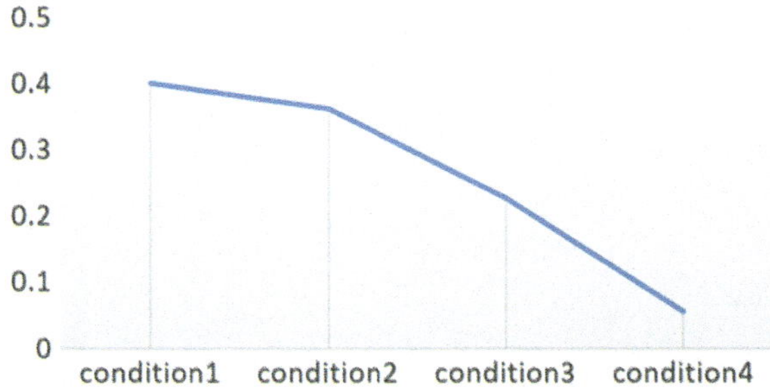

Fig. 6 Comprehensive evaluation value of service performance for each working condition

Finally, the multi-index fusion evaluation value vector of each evaluation object is obtained [9], that is, $Q = (0.4019, 0.3624, 0.2286, 0.0558)$, and the line diagram is shown in Fig. 6.

As can be seen from the figure, with the gradual deterioration of the service environment, the comprehensive evaluation value shows a downward trend, indicating that the service performance of bogie frame is getting worse and worse.

5 Conclusion

In this paper, the three-dimensional finite element numerical simulation analysis model of bogie frame is established for the bogie frame. Secondly, the deformation characteristics and modal changes of the frame under different service conditions are carried out. At the same time, the calculation of the characterization parameters based on numerical simulation is proposed. Finally, based on the dynamic load and the characterization parameters, the multi-index fusion evaluation was carried out, and the operating conditions with good service performance were obtained.

Acknowledgements The authors gratefully acknowledge the support from "the Fundamental Research Funds for the Central Universities" (Grant No. 2019YJS239).

References

1. Bai CY, An C, Li GM et al (2018) Research on service performance tracking of CRH3 type EMUs in Wuhan-Guangzhou passenger dedicated line. Collection 1-011 (in Chinese)
2. Lin HJ, Yi C, Zhang HW et al (2015) Research on service performance of high speed train. Chin Eng Sci 17(4):83–89 (in Chinese)
3. TB/T1335-1996, Hong SY (1996) Railway vehicle strength design and test identification specification. China Railway Publishing House (in Chinese)

4. BS EN 13749 (2005) Railway applications-methods of specifying structural requirements of bogie frames
5. Jin WL, Kim SH (2001) An integrated approach for interdependent information system project selection. Int J Project Manag 19(2):111–118
6. Liu C, Li K, Liu Z (2005) Medical image registration by maximization of combined mutual information and edge correlative deviation. In: International conference of the engineering in medicine & biology society, p 6379
7. Li H, Yu Y (2014) Research on the evaluation of expert scoring method in the competitiveness of high colleges and universities of Jiangxi province. In: International conference on information management, innovation management and industrial engineering. IEEE, pp 448–450
8. Guo JM, Prasetyo H, Su HS (2013) Image indexing using the color and bit pattern feature fusion. Academic Press, Inc., Cambridge
9. Wei XA (2015) The optimization decision of wind power prediction models based on multi-index fusion evaluation and case-based reasoning. Taiyuan University of Technology (in Chinese)

Maintenance Strategy Generation for Urban Railway Based on Improved Particle Swarm Optimization Algorithm

Zhichao He, Zhicheng Xia, Yanhui Wang and Lijie Li

Abstract At present, China's urban railway is faced with a large number of insufficient maintenance of failed components and surplus maintenance of normal components. It is urgent to carry out research on the generation of maintenance strategy of urban railway. As to satisfy the demand of transportation capacity in different periods, improve the safety of urban railway operation, and reduce the cost of urban railway maintenance. Particle swarm optimization is a stochastic global optimization method. It can be applied to find the optimal region in the complex search space by the interaction between particles. Particle swarm optimization algorithm is improved through position and particle variation similarity, which is presented in this paper. Based on this improved algorithm, a maintenance strategy generation for urban railway is proposed. The results from a case study show that the operation cost can be reduced by using the improved particle swarm optimization algorithm.

Keywords Particle swarm optimization algorithm · Maintenance strategy · Urban railway

Z. He · Y. Wang (✉) · L. Li
State Key Laboratory of Rail Traffic Control and Safety, Beijing Jiaotong University, Beijing 100044, China

Z. He · Y. Wang
School of Traffic and Transportation, Beijing Jiaotong University, Beijing 100044, China

Research and Development Center of Transport Industry of Technologies and Equipments of Urban Rail Operation Safety Management, MOT, PRC, Beijing, China

Z. Xia
Beijing Jinhong Xi-Dian Information Technology Corp., Beijing, China

Y. Qin et al. (eds.), *Proceedings of the 4th International Conference on Electrical and Information Technologies for Rail Transportation (EITRT) 2019*, Lecture Notes in Electrical Engineering 639, https://doi.org/10.1007/978-981-15-2866-8_24

1 Introduction

At present, Chinese urban railway adopts the operation and maintenance strategy of combining preventive periodic maintenance and corrective maintenance in operation. In a certain period of time, this maintenance strategy has played a positive role during the train operation. However, with the progress of science and technology, many disadvantages have been exposed, such as insufficient maintenance of failed parts and excessive maintenance of non-failed parts. If the maintenance surplus of non-failed parts persists persistently, it will greatly cause the waste of maintenance resources and the increase of maintenance costs. These two situations have seriously affected the high reliability, safety, efficiency, and low cost of the train operation.

Wim and Lennaert [1] focused on applying the proportional hazard model, with operational factors such as covariables, to generate reliability estimates for maintenance prediction. Joeri et al. [2] combined condition-based maintenance, preventive maintenance, and corrective maintenance states on a component and made maintenance strategy based on the actual state of the component. Sophie et al. [3] took into account instantaneous and periodic maintenance and established a visual maintenance model based on gamma process through non-uniform gamma process. Liu et al. [4] proposed a maintenance strategy formulation method based on interval for multi-component systems with potential faults. Chen et al. [5] studied a degradation system subjected to random environment and established an extended limit impact maintenance model. Li et al. [6] established a maintenance model of the three-state system based on state degradation and environmental impact. Li et al. [7] studied the state maintenance method considering the equipment defect maintenance model. Kim et al. [8] proposed a maintenance model with minimum maintenance and general maintenance, provided a method of modeling fault system, and effectively determined the optimal maintenance strategy.

To develop the maintenance strategy based on preventive periodic maintenance into the maintenance strategy based on preventive state maintenance is the key to optimize the maintenance strategy. Therefore, this paper optimized the generation of urban railway maintenance strategy based on the improved particle swarm algorithm.

The rest of this article is organized as follows: Sect. 2 introduces the algorithms involved in this paper. A case study is described in Sect. 3 to validate the proposed method. Section 4 is the conclusions.

2 Methodology

In this section, an improved particle swarm optimization algorithm is presented. A variation strategy is proposed from two aspects of position similarity and particle variation similarity. Considering external conditions and environmental requirements, the maintenance strategy of urban railway is realized.

2.1 *Improved Particle Swarm Optimization Algorithm*

2.1.1 Standard Particle Swarm Optimization Algorithm

In standard particle swarm optimization [9], every particle is a solution of the optimization problem to be solved in the solution space. All particles in the solution space have an optimized objective function, which is the fitness of each particle. In addition, each particle has a velocity that enables them to change the previous direction, which determines the position of the particle. Then, all the particles can pursue the current optimal particle trajectory in the population, as well as their historical optimal trajectory. Finally, all the particles can search the position in the solution space.

2.1.2 Improved Idea

(1) *Kent Function Mapping*

As the chaotic sequence generated by logistic mapping follows Chebyshev-type distribution, its probability density function has the characteristics of dense edges and sparse interior, which is not conducive to the efficiency and ability of search. The initial solution has a certain impact on the population. Therefore, the Kent function is selected to map the initial solution into chaos. It is randomized that the position j of the first particle is within 0 to 1. The formula of particle iteration is as follows:

$$x_{n+1} = \begin{cases} \frac{x_n}{0.2}, & 0 \le x_n \le 0.2 \\ \frac{1-x_n}{1-0.2}, & 0.2 \le x_n \le 0 \end{cases} \quad (1)$$

where x_n is initial value.

The initial positions of all the particles in the population can be obtained by mapping the random positions to the space of the original solution. The mapping formula is as follows:

$$p_{ij} = l_j + (u_j - l_j)\, x(i,j) \tag{2}$$

where p_{ij} is the position j of the particle i; l_j and u_j are parameters which are used to map the solution space; $x(i,j)$ is the initial parameter value, between 0 and 1.

(2) *Position Similarity and Fitness Similarity*

Fitness similarity can prevent the solutions of particles to be similar, which reduce the optimization ability of the algorithm. Fitness similarity can distinguish the particles of different solutions but similar fitness. As the number of iterations of algorithm increases, the positions of each particle in the population become more and more similar due to the guidance of the optimal solution. It is difficult for the standard particle swarm optimization algorithm to further optimize the current solution. The position similarity of particles in the population was calculated, and the calculation formula is as follows:

$$s(i) = \frac{\sum_{j=1}^{d} \left(p(i,j) - \overline{\mathrm{paver}(j)}\right)^2}{d} \tag{3}$$

where $s(i)$ is position similarity; $\overline{\mathrm{paver}(j)}$ is the average value of the position j; d is the number of variables.

The similarity of the particle position and fitness is different. Therefore, it is calculated again, and the calculation formula is as follows:

$$sf(i) = \sum_{j=1}^{n} \left(f(i) - \bar{f}\right)^2 \tag{4}$$

where $sf(i)$ is the fitness similarity; $f(i)$ is the fitness of the particle i; $\bar{f}$ is the average value of every particle.

Then, the coefficient of variation is

$$\mathrm{pm}(i) = a * s(i) * sf(i) \tag{5}$$

where $\mathrm{pm}(i)$ is mutation probability; a is mutation parameter.

We can find that the coefficient of variation is small when the positions of particles in the solution space are relatively close. When the fitness between two particles is relatively similar, the coefficient of variation is also small. The similarity is described from two aspects of particle position and fitness. The calculation formula of variation degree is as follows:

$$\begin{cases} \text{if rand} > \mathrm{pm}(i) p(i,j) = p(i,j) * \left(1 + \mu \dfrac{0.5 - \text{rand}}{\mathrm{sigmoid}(\mathrm{pm}(i))}\right) \\ \text{other}\, p(i,j) = p(i,j) \end{cases} \tag{6}$$

where μ is a tunable parameter.

This function is continuous, smooth, differentiable, and bounded.

2.2 *Maintenance Strategy Generation for Urban Railway Based on Safety*

When the maintenance resources are limited, it is necessary to give priority to railway safety, which can satisfy the transportation requirements. In other words, the railway safety can be maximized when the total cost of maintenance can be controlled and the transportation requirements can be satisfied, and the railway safety is higher than the minimum value.

(1) *Constraint condition*

In the time interval of $(t_b, t_e]$,

(1) The total maintenance cost C of the system is less than or equal to the limited maintenance cost constant C_0.

$$C = \sum_{i=1}^{n} \left[\text{INT}\left(\frac{t_e - t_b}{T_{\text{MI1}-i}}\right) \cdot C_{\text{MT1}-i} + \text{INT}\left(\frac{t_e - t_b}{T_{\text{MI2}-i}}\right) \cdot C_{\text{MT2}-i} + \text{INT}\left(\frac{t_e - t_b}{T_{\text{MI3}-i}}\right) \cdot C_{\text{MT3}-i} \right] \leq C_0 \tag{7}$$

where INT denotes rounding down.

(2) System safety S is greater than or equal to system safety limit value S_0.

$$\begin{aligned} S = S_G(t)\Big|_{t_b}^{t_e} &= \frac{1}{k_G}\sum_{\varepsilon=1}^{k_G}\int_{t_b}^{t_e} S_{G-\varepsilon}(t)\mathrm{d}t \\ &= \frac{1}{k_G}\sum_{\varepsilon=1}^{k_G}\int_{t_b}^{t_e} \left\{ \begin{array}{l} 1 - [1 - f_{5-\varepsilon}(T_{\text{MI1}-1}, T_{\text{MI2}-1}, T_{\text{MI3}-1}, \ldots, T_{\text{MI1}-q}, T_{\text{MI2}-q}, T_{\text{MI3}-q}, \ldots, \\ T_{\text{MI1}-m_{s-\varepsilon}}, T_{\text{MI2}-m_{s-\varepsilon}}, T_{\text{MI3}-m_{s-\varepsilon}}, t)] \times \sum_{w=1}^{\Lambda} X_{\varepsilon-w}\omega_{\varepsilon-w} \end{array} \right\} \mathrm{d}t \geq S_0 \end{aligned} \tag{8}$$

(3) After the integration and optimization, the maintenance interval of each component is greater than or equal to the minimum kilometer $T_{\min}$.

$$\min(T_{\text{MI1}-1}, T_{\text{MI2}-1}, \ldots, T_{\text{MI3}-m}) \geq T_{\min} \tag{9}$$

(4) After the integration and optimization, the absolute value of the maintenance interval difference of each component is greater than or equal to the minimum kilometer $T_{\min}$.

$$\min\left|T_{\mathrm{MI}-u}-T_{\mathrm{MI}-v}\right|\geq T_{\min} \tag{10}$$

(2) *Objective function*

In the time interval of $(t_b,\ t_e]$, let the objective function of the system safety be maximized.

$$S=S_G(t)\Big|_{t_b}^{t_e}=\frac{1}{k_G}\sum_{\varepsilon=1}^{k_G}\int_{t_b}^{t_e}S_{G-\varepsilon}(t)\mathrm{d}t$$
$$=\frac{1}{k_G}\sum_{\varepsilon=1}^{k_G}\int_{t_b}^{t_e}\left\{\begin{array}{l}1-\left[1-f_{5-\varepsilon}\left(T_{\mathrm{MI1}-1},T_{\mathrm{MI2}-1},T_{\mathrm{MI3}-1},\ldots,T_{\mathrm{MI1}-q},T_{\mathrm{MI2}-q},T_{\mathrm{MI3}-q},\ldots,\right.\right.\\ \left.\left.T_{\mathrm{MI1}-m_{s-\varepsilon}},T_{\mathrm{MI2}-m_{s-\varepsilon}},T_{\mathrm{MI3}-m_{s-\varepsilon}},t\right)\right]\times\sum_{w=1}^{\Lambda}X_{\varepsilon-w}\omega_{\varepsilon-w}\end{array}\right\}\mathrm{d}t \tag{11}$$

where $T_{\mathrm{MI1}-1},T_{\mathrm{MI2}-1},T_{\mathrm{MI3}-1},\ldots,T_{\mathrm{MI1}-i},T_{\mathrm{MI2}-i},\ T_{\mathrm{MI3}-i},\ldots,T_{\mathrm{MI1}-m_{s-\varepsilon}},T_{\mathrm{MI2}-m_{s-\varepsilon}},T_{\mathrm{MI3}-m_{s-\varepsilon}}$ are $3m_{s-\varepsilon}$ independent variables, which are about the maintenance interval; S is objective function; $f_{5-\varepsilon}$ is risk probability estimation function; everything else is constant.

The maintenance strategy oriented to safety of urban railway optimization calculation can be described as follows:

$$\begin{array}{l}\mathrm{Max}\,S(\vec{x})\\ \text{s.t.}\begin{cases}C(\vec{x})\leq C_0\\ x_i\geq T_{\min},\quad i=1,2,3,\ldots,3m\\ \min\left|x_i-x_j\right|\geq T_{\min},i\neq j\quad\text{and}\quad i,j=1,2,3,\ldots,3m\\ S(\vec{x})\geq S_0\end{cases}\end{array} \tag{12}$$

3 Case Study

According to the maintenance procedure, within the operating cycle of the time interval (0, 600000 km), $T_{\min}=2000\,\mathrm{km}$, $C_0=2{,}909{,}133\$$, $S_0=570{,}000\,\mathrm{km}$ the limited operation, and maintenance cost constant, and the system safety S is greater than or equal to the system safety limit value S_0 = 570,000 km and carries out the operation and maintenance strategy optimization calculation for security assurance:

(1) *Objective function*

$$
\begin{aligned}
S &= \int_{0}^{600000} S_{G\text{-I-II}}(t)\mathrm{d}t = \int_{0}^{600000} \left\{1 - F_{\text{I-II}}(t) \times \left[\frac{0.41F_{\text{I}}(t)}{F_{\text{I}}(t)+F_{\text{II}}(t)} + \frac{0.55F_{\text{II}}(t)}{F_{\text{I}}(t)+F_{\text{II}}(t)}\right]\right\}\mathrm{d}t \\
&= \int_{0}^{600000} \left\{1 - \frac{F_{\text{I-II}}(t)}{F_{\text{I}}(t)+F_{\text{II}}(t)} \times [0.41F_{\text{I}}(t) + 0.55F_{\text{II}}(t)]\right\}\mathrm{d}t \\
&= \int_{0}^{600000} \left\{\begin{array}{l} 1 - \dfrac{-f_{5\text{-I-II}}(T_{\text{MI1}-1},\ldots,T_{\text{MI1}-17},T_{\text{MI2}-17},T_{\text{MI3}-17},\ldots,T_{\text{MI3}-22},t)}{f_{5\text{-I}}(T_{\text{MI1}-1},\ldots,T_{\text{MI3}-22},t)+f_{5\text{-II}}(T_{\text{MI1}-1},\ldots,T_{\text{MI3}-17},t)} \\ \times\, [0.41f_{5\text{-I}}(T_{\text{MI1}-1},\ldots,T_{\text{MI3}-22},t) + 0.55f_{5\text{-II}}(T_{\text{MI1}-1},\ldots,T_{\text{MI3}-17},t)] \end{array}\right\}\mathrm{d}t
\end{aligned}
\tag{13}
$$

(2) *Constraint condition*

$$
\begin{aligned}
C = \sum_{j=1}^{11} & \left[\mathrm{INT}\left(\frac{600{,}000}{T_{\text{MI1}_j}}\right)\cdot C_{\text{MT1}_j} + \mathrm{INT}\left(\frac{600{,}000}{T_{\text{MI2}_j}}\right)\cdot C_{\text{MT2}_j}\right. \\
& \left. + \mathrm{INT}\left(\frac{600{,}000}{T_{\text{MI3}_j}}\right)\cdot C_{\text{MT3}_j}\right] \le 2000
\end{aligned}
\tag{14}
$$

$$
\begin{aligned}
S &= \int_{0}^{600{,}000} S_{G\text{-I-II}}(t)\mathrm{d}t \\
&= \int_{0}^{600{,}000} \left\{1 - F_{\text{I-II}}(t) \times \left[\frac{0.41F_{\text{I}}(t)}{F_{I}(t)+F_{\text{II}}(t)} + \frac{0.55F_{\text{II}}(t)}{F_{I}(t)+F_{\text{II}}(t)}\right]\right\}\mathrm{d}t \\
&= \int_{0}^{600{,}000} \left\{1 - \frac{F_{\text{I-II}}(t)}{F_{\text{I}}(t)+F_{\text{II}}(t)} \times [0.41F_{I}(t) + 0.55F_{II}(t)]\right\}\mathrm{d}t \\
&= \int_{0}^{600{,}000} \left\{\begin{array}{l} 1 - \dfrac{-f_{5\text{-I-II}}(T_{\text{MI1}-1},\ldots,T_{\text{MI1}-17},T_{\text{MI2}-17},T_{\text{MI3}-17},\ldots,T_{\text{MI3}-22},t)}{f_{5\text{-I}}(T_{\text{MI1}-1},\ldots,T_{\text{MI3}-22},t)+f_{5\text{-II}}(T_{\text{MI1}-1},\ldots,T_{\text{MI3}-17},t)} \\ \times\, [0.41f_{5\text{-I}}(T_{\text{MI1}-1},\ldots,T_{\text{MI3}-22},t) + 0.55f_{5\text{-II}}(T_{\text{MI1}-1},\ldots,T_{\text{MI3}-17},t)] \end{array}\right\}\mathrm{d}t \\
&\ge 570{,}000
\end{aligned}
\tag{15}
$$

$$\min(T_{\text{MI1}-1}, T_{\text{MI2}-1}, \ldots, T_{\text{MI3}-31}) \ge 2000 \tag{16}$$

$$\min|T_{\text{MI}-u} - T_{\text{MI}-v}| \ge 2000 \tag{17}$$

According to the objective function and constraint conditions, the optimization calculation carried out in this section can be described as follows:

Table 1 Optimized post-operational strategy for safety priority (unit: km)

Component	z_1	z_2	z_3	z_4	z_5	z_6	z_7	z_8	z_9	z_{10}	z_{11}
T_{MI1_j}	8E+03	6E+03	1E+04	4E+03	6E+03	6E+03	4E+03	4E+03	4E+03	1E+04	6E+03
T_{MI2_j}	4E+03	1E+04	8E+03	1E+04	1E+04	1E+04	6E+03	6E+03	6E+03	4E+03	1E+04
T_{MI3_j}	8E+03	1E+04	1E+04	1E+04	1E+04	1E+04	1E+04	1E+04	1E+04	2E+04	2E+04

$$\text{Max}\, S(\vec{x})$$
$$\text{s.t.}\begin{cases} C(\vec{x}) \leq 2000 \\ x_i \geq 2000, \quad i = 1, 2, 3, \ldots, 33 \\ \min|x_i - x_j| \geq 2000, \quad i \neq j \,\text{and}\;\; i, j = 1, 2, 3, \ldots, 33 \\ S(\vec{x}) \geq 570,000 \\ x_i/2000 \in N^* \end{cases} \tag{18}$$

where N^* is a set of positive integers.

Based on improved particle swarm optimization algorithm, the maximum value of objective function is optimized. In Table 1, the maintenance strategy for urban railway can be obtained.

According to the optimized maintenance strategy, the railway safety operating kilometrage increased from 494,600 to 580,100 km within the operation cycle (0, 600,000 km). At this time, the total system maintenance cost was 29,091$.

4 Conclusion

In this paper, an improved particle swarm optimization algorithm was proposed. The algorithm is improved from two aspects based on the existing research results. A particle position locating strategy has been provided by position similarity and fitness similarity. Then, the maintenance strategy generation is completed by improved particle swarm optimization algorithm. The model is based on staged fault data. If the fault data can be more sufficient, the research results obtained will be more representative.

Acknowledgements The authors gratefully acknowledge the support from "the Fundamental Research Funds for the Central Universities" (Grant No. 2017JBZ103).

References

1. Wim JCV, Lennaert WMDB (2018) Predictive maintenance for aircraft components using proportional hazard models. J Ind Inf Integr 12:23–30
2. Joeri P, Robert NB, Marc RL (2018) A hybrid condition-based maintenance policy for continuously monitored components with two degradation thresholds. Eur J Oper Res 268:515–532
3. Sophie M, Castro IT (2019) Stochastic comparisons of imperfect maintenance models for a gamma deteriorating system. Eur J Oper Res 273:237–248
4. Liu B, Yeh RH, Xie M et al (2017) Maintenance scheduling for multicomponent systems with hidden failures. IEEE Trans Reliab 66(4):1280–1292
5. Chen JY, Li ZH (2017) An extended extreme shock maintenance model for a deteriorating system. Reliab Eng Syst Saf 93(8):1123–1129

6. Li Y, Ma XB, Zhao Y (2017) A condition-based maintenance model for a three-state system subject to degradation and environmental shocks. Comput Ind Eng 105:210–222
7. Li YY, Han XS, Xu B, Wang YP (2015) A condition-based maintenance approach to an optimal maintenance strategy considering equipment imperfect maintenance model. In: 2015 5th International conference on electric utility deregulation and restructuring and power technologies. IEEE, Changsha, pp 11.26–11.29
8. Kim MJ, Makis V (2010) A maintenance model with minimal and general repair. IMA J Manag Math 21(4):385–393
9. Kennedy J, Eberhart R (1995) Particle swarm optimization. In: Proceedings of IEEE international conference on neural networks, pp 1942–1945

Passenger-Flow-Status Assessment Method of Urban Rail Network Based on GA-WNN

Chenyang Zhao, Yanhui Wang and Man Li

Abstract Along with the rapid development of urban transportation in recent years, the urban rail network has gradually formed. In the context of networked operations, mass passenger flow brings challenges to security management. This paper presents a passenger flow index system for urban rail network to assess passenger-flow-status, from three aspects of the passenger flow in urban rail network, including the capacity, the connection state and the transferability. Based on the index system, wavelet neural network with the genetic algorithm (GA-WNN) is used to form the assessment method of passenger-flow-status in urban rail network. The advantages of the assessment method are verified by a case study.

Keywords Passenger flow · Status assessment · Index system · GA-WNN

1 Introduction

Many large cities in China are committed to the development of rail transit. The urban rail transit road network has been gradually formed [1]. This has brought great challenges to the safety management of the urban rail transit.

C. Zhao (✉) · Y. Wang · M. Li
State Key Laboratory of Rail Traffic Control and Safety,
Beijing Jiaotong University, Beijing, China
e-mail: 18120960@bjtu.edu.cn

Y. Wang
e-mail: wangyanhui@bjtu.edu.cn

M. Li
e-mail: 9213@bjtu.edu.cn

C. Zhao
School of Traffic and Transportation, Beijing Jiaotong University, Beijing, China

Y. Wang · M. Li
Research and Development Center of Transport Industry of Technologies
and Equipments of Urban Rail Operation Safety Management, MOT, Beijing, China

Y. Qin et al. (eds.), *Proceedings of the 4th International Conference on Electrical and Information Technologies for Rail Transportation (EITRT) 2019*, Lecture Notes in Electrical Engineering 639, https://doi.org/10.1007/978-981-15-2866-8_25

At present, scholars are studying the operation safety of urban rail transit. However, many studies have focused on the safety assessment of a specific subsystem of the urban rail transit, such as a vehicle system [2]. In addition, WNN has strong learning ability, and it has certain adaptability to its application in passenger flow state evaluation [3].

Therefore, this paper presents a study on the passenger-flow-status assessment method of urban rail network. The rest of the paper is structured as follows. The definitions and calculation methods of passenger flow index of urban rail network are introduced in Sect. 2. Section 3 introduces the method of passenger-flow-status assessment based on GA-WNN. Based on the previous method, Sect. 4 presents a case study and describes the results. Finally, the conclusions of this paper are presented in Sect. 5.

2 Passenger Flow Index System of Urban Rail Network

From the three aspects of the passenger flow in urban rail network, we selected the corresponding index of passenger-flow-status assessment and constructed the passenger flow index system in urban rail network [4–6]. The calculation methods of the three indices are introduced in this section.

2.1 The Accordance Degree of Design Capacity and Operation (AD)

This index is the probability that the road network transport capacity can meet the demand level of a certain passenger flow.

$$\mathrm{AD} = \frac{r(s)}{m(s) * 2} \tag{1}$$

where AD is the accordance degree of design capacity and operation; $r(s)$ is the number of intervals with the load rate exceeding 80% in the urban rail network; $m(s)$ is the number of urban rail network intervals.

2.2 Connectivity Index (CD)

This index is the ratio of the road network degree at the current time to the degree of normal condition.

$$\mathrm{CD}(s) = \frac{\sum_{i=1}^{m}\sum_{j=1}^{n}(d(s_{ij})) - \sum_{i=1}^{m}\sum_{j=1}^{n}(\Delta d(s_{ij}))}{\sum_{i=1}^{m}\sum_{j=1}^{n}(d(s_{ij}))} \tag{2}$$

where $\mathrm{CD}(s)$ is the connectivity index of urban rail network; $d(s_{ij})$ is the degree of each point in the urban rail network; $\Delta d(s_{ij})$ is the change value of each point degree in the urban rail network.

2.3 The Accordance Degree of Design Transfer Capacity and Operation (TC)

This index is the weighted average of design transfer capacity and operation.

$$\mathrm{TC}_{(t,t+\Delta t)}(s) = \sum_{s_{ij}\in E} \varsigma(s_{ij}) \frac{\mathrm{tc}^{\max}(s_{ij})}{\mathrm{tc}_{(t,t+\Delta t)}(s_{ij})} \tag{3}$$

$$\varsigma(s_{ij}) = \frac{k(s_{ij})}{\sum_{s_{ij}\in E} k(s_{ij})} \tag{4}$$

where $\mathrm{TC}_{(t,t+\Delta t)}(s)$ is the accordance degree of urban rail network transferability; E is the set of transfer stations; the weight coefficient $(\varsigma(s_{ij}))$ of transfer station (s_{ij}) is obtained by Delphi method; $\mathrm{tc}_{(t,t+\Delta t)}(s_{ij})$ is the transfer passenger flow; $\mathrm{tc}^{\max}(s_{ij})$ is the maximum transfer quantity; n is the number of transfer stations in the road network.

3 Method of Passenger-Flow-Status Assessment

In order to improve operational safety management, we need to assess the passenger-flow-status. Based on the index system constructed in Sect. 2, the state of passenger flow is predicted and assessed.

3.1 Min-Max Normalization

The passenger flow indices for urban rail network have a different physical meaning, so there are differences in dimension. In order to facilitate the assessment process, this paper uses the method of min-max normalization to normalize the data [7].

$$r_i = \frac{x_i - \min_{1 \leq j \leq n}\{x_j\}}{\max_{1 \leq j \leq n}\{x_j\} - \min_{1 \leq j \leq n}\{x_j\}} \tag{5}$$

where r_i is the ith standardized processed data; $\{x_j\}(j = 1, 2, \ldots, n)$ is the raw data set that needs to be standardized; x_i is the ith raw data that needs to be standardized.

3.2 Assessment Method of Passenger-Flow-Status in Urban Rail Network

This paper uses GA-WNN to evaluate the passenger-flow-status of urban rail network [8]. The genetic algorithm is used to train the initial connection weight and the extension scale of the wavelet neural network.

Wavelet neural network

The basic structure of wavelet neural network for assessment of passenger-flow-status in urban rail network is plotted in Fig. 1.

The predicted value of passenger-flow-status assessment is defined as:

$$\widehat{Y} = \sum_{i=1}^{N} W_i \varphi\left(\frac{t_i - b_i}{a_i}\right) \tag{6}$$

where $\widehat{Y}$ is the predicted value of passenger-flow-status; W_i is the weight of the output of the ith wavelet basis unit; a_i and b_i are the scale factor and displacement factor of the wavelet function; t_i is the input of the ith wavelet basis.

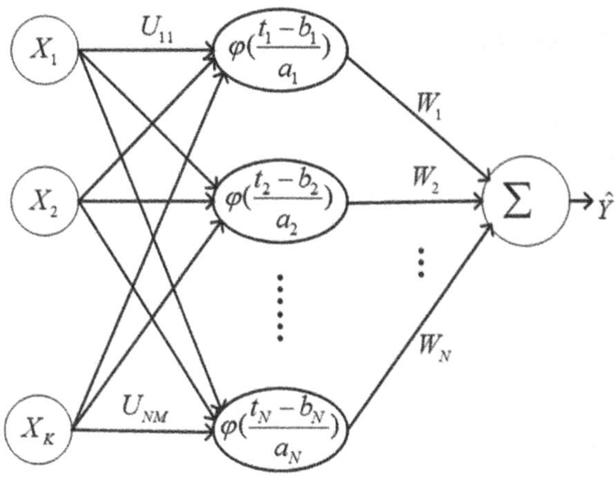

Fig. 1 Structure of wavelet neural network

The input of the ith wavelet basis is:

$$t_i = \sum_{j=1}^{N} U_{ij} X_j \tag{7}$$

where U_{ij} is the weight of the ith wavelet basis unit connected to the jth input X_j; Q is training sample set $(X_{k1}, X_{k2}, \ldots, X_{kM}, Y_k)(k = 1, 2, \ldots, Q)$. We choose the network parameters U_{ij}, a_i, b_i, W_i through network learning and training. And the number of wavelet bases N can be determined through self-tuning algorithms of hidden layer node.

Optimization is performed using a minimum mean squared error function, and it is defined as:

$$E_Q = \frac{1}{Q} \sum_{k=1}^{Q} \left[Y_k - \widehat{Y_k} \right]^2 \tag{8}$$

where Y_k is ideal output for the kth sample; $\widehat{Y_k}$ is the output of network for the kth sample.

Parameter optimization based on genetic algorithm

The method we chose is an adaptive genetic algorithm. We determine the initial population p, which contains W_i, U_{ij}, a_i and b_i, and initialize the code for the population p.

Individuals in each population contain Q genes. The chromosome structure of the individual in GA-WNN is shown in Fig. 2, where H is the number of hidden layer nodes and I is the number of input layer nodes.

The fitness function is used to calculate the fitness of each individual, and it is defined as:

$$f = \frac{1}{1 + E_Q} \tag{9}$$

Selection operation

After calculating the individual fitness values of each individual and sorting them, the network parameter individuals are selected according to roulette wheel selection [9]. The selection probability is defined as:

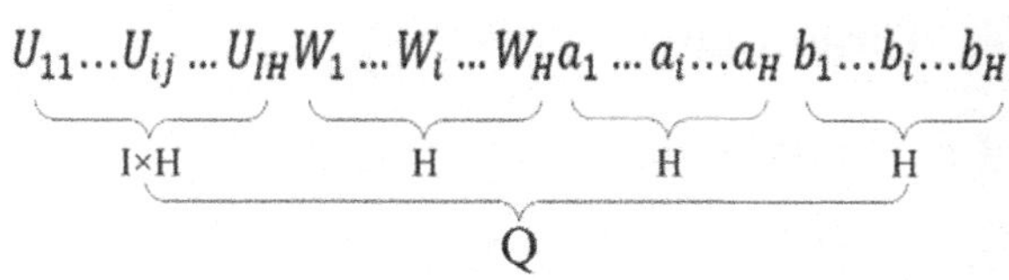

Fig. 2 Chromosome structure of the individual

$$p_s = \frac{f(n)}{\sum_{n=1}^{N} f(n)} \tag{10}$$

where $n(n = 1, 2, \ldots, N)$ is the population size.

Crossover operation

The new individuals G'_A and G'_B are created by arithmetic crossing individual G_A and G_B. The cross-probability p_c is defined as:

$$p_c = \begin{cases} \frac{k_1(f_{\max} - f)}{f_{\max} - f_{\text{avg}}}, & f \geq f_{\text{avg}} \\ k_2, & f < f_{\text{avg}} \end{cases} \tag{11}$$

where f is the larger fitness value of the two individuals to cross; $f_{\max}$ is the maximum fitness value in the population; f_{avg} is the average fitness value in the population; k_1 and k_2 are the adjustment coefficients, and $k_1 < k_2$.

Mutation operation

New individuals are mutated by means of uniform mutation. The mutation probability p_m is defined as:

$$p_m = \begin{cases} \frac{k_3(f_{\max} - f')}{f_{\max} - f_{\text{avg}}}, & f' \geq f_{\text{avg}} \\ k_4, & f' < f_{\text{avg}} \end{cases} \tag{12}$$

where f is the fitness of the individual to be mutated; k_3 and k_4 are the adjustment coefficients, $k_3 < k_4$.

New individuals are inserted into the population p, and their fitness values are calculated. If the fitness value meets the requirements, the genetic process ends; otherwise, the crossover and mutation operations continue until the end condition is reached.

We decode the best individuals in the final population to get W_i, U_{ij}, a_i and b_i. U_{ij} are the first $I \times H$ code values in the optimal individual; W_i are the next H code values (after the $(I \times H)$ code); a_i are the next H code values (after the $(I \times H + H)$ code); b_i are the last H code values.

Based on the initial value of the connection weight and the extension scale obtained by genetic algorithm, the final parameters of the GA-WNN are trained by the gradient descent method of wavelet neural network.

4 Case Study

4.1 Simulation

The 70-day data of urban rail transit operation is selected, and the calculation results of each index value are obtained according to the method in Sect. 2 (Table 1).

Table 1 Operation sample data of the urban rail network

NUM	AD	CD	TC	Y_k
1	0.783	1	0.1414	86
2	0.729	0.9996	0.1515	89
3	0.697	0.9995	0.2024	86
4	0.729	0.9992	0.1477	91
…	…	…	…	…

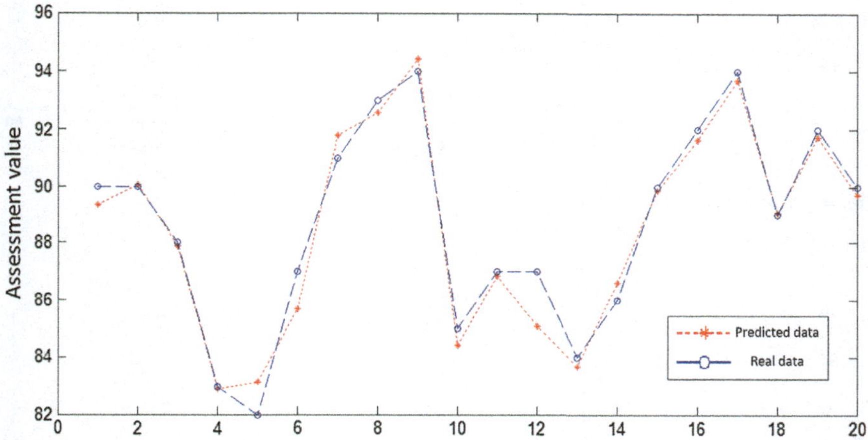

Fig. 3 Output of training value and the ideal value under GA-WNN

We choose three indices as the inputs of the GA-WNN and the assessment results as the output. That is, the network has three input nodes and one output node.

In this paper, the first 50 sets of data are used as training samples, and the last 20 sets of data are used as test samples. The result of simulation is shown in Fig. 3, and the error curve is shown in Fig. 4.

After 89 iterations, the GA-WNN converges. In the 20th iteration, the error accuracy has reached about 0.004, and the convergence speed is fast.

4.2 Comparison with the Results of Back-Propagation Neural Network

The number of hidden nodes and the value of error we set are the same as the assessment model based on GA-WNN. The result of simulation is shown in Fig. 5, and the error curve is shown in Fig. 6.

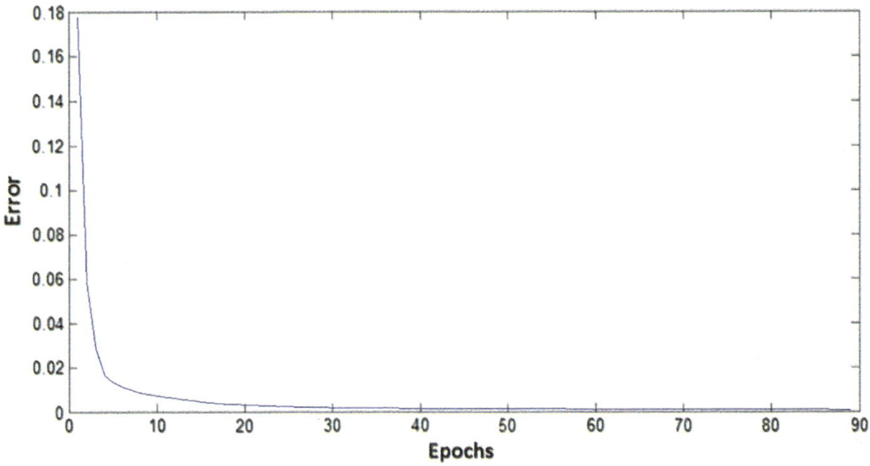

Fig. 4 Error curve of GA-WNN

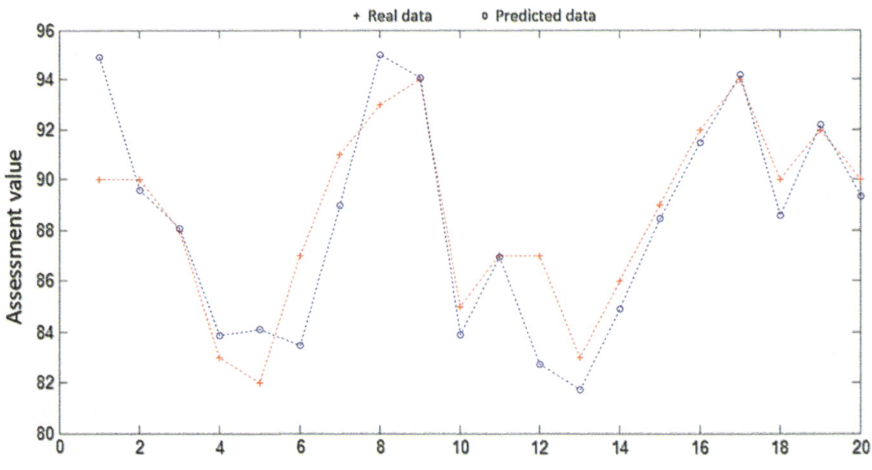

Fig. 5 Output of training value and the ideal value under BP neural network

Through the above results, it is found that the fitness of wavelet neural network is better than BP neural network. After 2959 epochs, the BP neural network converges, and the convergence speed is lower than the method of GA-WNN. Therefore, it is suitable to select GA-WNN as the assessment method of passenger-flow-status in urban rail network.

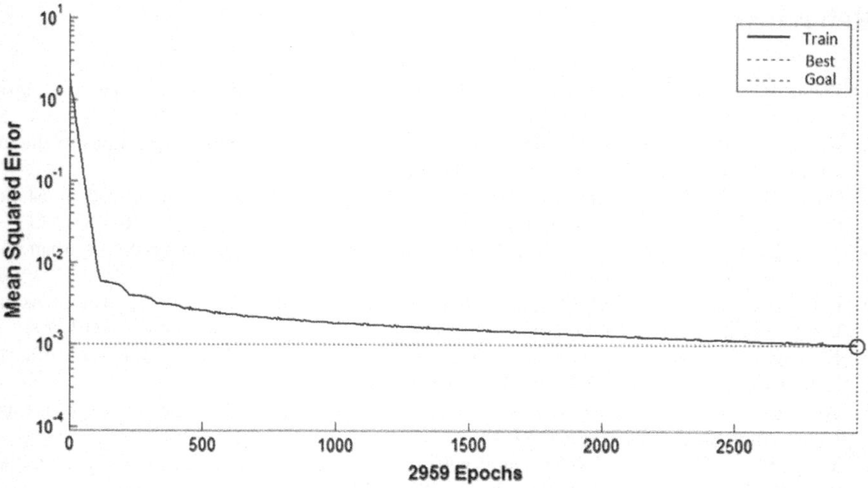

Fig. 6 Error curve of BP neural network

5 Conclusions

This paper involves theoretical research, establishment of index system, method of status assessment and case study.

Firstly, from the operational point of view, we present the index system for passenger-flow-status assessment. The calculation methods of passenger flow index in urban rail network are introduced in detail. Then, we improve the wavelet neural network using genetic algorithm. Genetic algorithm can improve the convergence speed and network learning accuracy of wavelet neural network. Based on the GA-WNN method, the assessment method of passenger-flow-status in urban rail network is constructed. Finally, we present a case study. The results demonstrate the advantages of the assessment method we use.

There are still some shortcomings in this paper: (1) When calculating transfer station weights, this paper relied on Delphi method, which has a certain impact on the final evaluation results; (2) the index system we built needs to be modified and improved further; (3) when the dimension of the input index increases, the convergence speed of the GA-WNN is greatly reduced.

Acknowledgements This work was supported by Key Research and Development Project of Safety Assurance Technology of Urban Rail System under China's National 13th Five-Year Plan, Award Number 2016YFB1200402.

References

1. Zhang JH, Zhao MW, Liu HK et al (2013) Networked characteristics of the urban rail transit networks. Physica A 392(6):1538–1546
2. Hua P, Xin JY, Lu BH (2018) Case study of safety assessment system for rail transit vehicle. J Dalian Jiaotong Univ 39(4):47–82 (in Chinese)
3. Wang XX, Xu LH (2018) Wavelet-based short-term forecasting with improved threshold recognition for urban expressway traffic conditions. IET Intel Transp Syst 12(6):463–473
4. Zhou ZF, Kang YZ, Zhou JH (2018) Study on comprehensive evaluation system of urban rail transit transfer stations. Transp Sci Technol 6:114–117 (in Chinese)
5. Liu J, Lu H, Ma H et al (2017) Network vulnerability analysis of rail transit plans in Beijng-Tianjin-Hebei region considering connectivity reliability. Sustainability 9(8):1479
6. Xiao XM, Jia LM (2016) Evaluation model for connectivity reliability of urban rail transit network. China Railway Sci 37(1):132–137 (in Chinese)
7. Polini C, Ulrich B, Vasconcelos WV, Villarreal RH (2019) Indices of normalization of ideals. J Pure Appl Algebra 223(9):3681–3694
8. Yang HJ, Hu X (2016) Wavelet neural network with improved genetic algorithm for traffic flow time series prediction. Optik 127(19):8103–8110
9. Cerf R (2017) The quasispecies regime for the simple genetic algorithm with roulette wheel selection. Adv Appl Probab 49(3):903–926

K-Step Fault Propagation Method for High-Speed Train System

Yanhui Wang, Pengfei Sun, Shuai Lin, Boyan Hao and Mingtao Wu

Abstract The research on the fault propagation path has become a hot spot in the field of the operation of the high-speed train system. First, the failure propagation network of the high-speed train system is constructed based on network theory and system structure. Second, the K-step fault propagation method is proposed to study the impact of the system topology on the fault propagation according to the failure propagation network. Finally, the proposed method is applied to analyse the process of fault propagation for the bogie system. The results show that the paths of the fault propagation for bogie system obtained by our method agree with the ones in reality. This work may be helpful for the maintenance and the optimal design of high-speed train system, which inspires the following works concerning the reliability and the safety.

Keywords High-speed train system · Fault propagation · Bogie system · Fault path

1 Introduction

High-speed railway is becoming one of the most promising large-scale transportation modes. High-speed train system is the core system to complete the safe and efficient operation of high-speed railway system. Once the high-speed train system fails, especially some catastrophic failures, it will lead to huge economic

Y. Wang · P. Sun (✉) · S. Lin · B. Hao · M. Wu
State Key Laboratory of Rail Traffic Control and Safety, Beijing Jiaotong University, 100044 Beijing, China
e-mail: 17120967@bjtu.edu.cn

School of Traffic and Transportation, Beijing Jiaotong University, 100044 Beijing, China

Research and Development Center of Transport Industry of Technologies and Equipments of Urban Rail Operation Safety Management, MOT, Beijing, China

Y. Qin et al. (eds.), *Proceedings of the 4th International Conference on Electrical and Information Technologies for Rail Transportation (EITRT) 2019*, Lecture Notes in Electrical Engineering 639, https://doi.org/10.1007/978-981-15-2866-8_26

losses and casualties. The failure of components and their corresponding connections in high-speed train system will propagate through the related components, which will produce a series of chain reactions, eventually leading to the system breakdown. Therefore, controlling fault propagation is very important to improve the security of the system.

Complex network has advantages in representing the hierarchy, propagation and correlation of the system, so it can be used to analyse the fault propagation process of complex systems.

Liu [1] analysed the statistical characteristics of the system's topological structure using small-world network theory, proposed a fault propagation model based on small-world clustering and used Dijkstra's algorithm to get the fault propagation path. Wang [2] described three kinds of cellular automata models, carried out a series of experiments on coupled networks and single-layer networks and gave the fault propagation process of unprotected schemes. Olami [3] and others put forward OPA model to describe the frequency and propagation of power grid faults. Dobson [4] proposed a cascade fault model in order to better understand the distribution of fault frequency and scale changes in the process of load increase. Wang [5] proposed a successive fault model and studied the successive faults in complex networks on this basis. Xu [6] proposed the temporal causal graphs (TCG) model from the bond graph (BG) model of CRH5 traction system. Guru [7] simulates the concurrent and distributed behaviour of SOA-based systems under fault conditions through Coloured Petri Nets (CPN). Hu [8] studied the fault propagation behaviour of process system. A root cause reasoning framework based on dynamic Bayesian network is proposed to deal with abnormal situations.

Because of the physical and functional connection of components, the influence factors of fault propagation are no longer determined by simple component characteristics. This paper studies the influence of network topological characteristics on fault propagation in the system.

2 Research on Propagation Path Based on Step-by-step Fault Propagation Principle

2.1 Construction Method of Fault Propagation Network

High-speed train system is a multi-energy coupling system which integrates mechanical, electrical and hydraulic components. Its essence is to combine components into a highly coupled complex network through the transmission of energy, electricity and information. Components are defined as follows.

Definition 1 Components in high-speed train system are an autonomous entity that undertakes specific functions and has certain mechanical, electrical or information attributes.

According to the structural characteristics of high-speed train system, the connection modes between components are divided into three types: mechanical connection, electrical connection and informational connection. Connection modes are defined as follows.

Definition 2 Mechanical connection is defined as the fasteners are used to connect the components, including bolt connection, stud connection, screw connection, thread fastening, pin connection, riveting, welding and gluing, etc.

Definition 3 Electrical connection is defined as the connections of conductors in different ways, such as electrical connectors, wires and cables, wire fixtures and wire protection devices.

Definition 4 Informational connection is defined as the component transmits the command or state information from one end to the other through the transmission medium and is received by the other side.

Taking components as nodes and mechanical connection, electrical connection and informational connection as edges, the system mechanical, electrical and informational networks $G_m = \{V_m, E_m\}$, $G_e = \{V_e, E_e\}$, $G_i = \{V_i, E_i\}$ are constructed successively, and finally, we integrated the networks into the system fault propagation network $G = \{V, E\}$.

2.2 *K-Step Fault Propagation Algorithm*

In the high-speed train system, component fault propagation does not follow a fixed path. That is to say, when some nodes in the network have poor performance (such as performance requirements are not up to standard and faults), the performance of the nodes associated with them will deteriorate gradually or fail directly. This effect will spread through the network and gradually spread to other non-neighbour nodes, causing chain reaction of high-speed train system. The K-step fault propagation algorithm is used to describe the propagation process.

2.2.1 Hypothesis

- Hypothesis 1: Fault propagation is transitive. For fault node A, there is $A \rightarrow B$; if node B fails, there is $B \rightarrow C$. If both nodes A and B fail in the process of fault propagation, there must be $A \rightarrow B \rightarrow C$. This is the logical basis of fault propagation relationship.

- Hypothesis 2: Component failure is a probability value. During the process from normal to failure, any component will experience multiple states. 1 represents the failure of the component, 0 represents the normal of the component, and other values represent the transition of the component from normal state to failure state, that is the failure probability value of the component.

2.2.2 Process of K-Step Fault Propagation

The fault propagation network is $G = \{V, E\}$, where the set of nodes is $V = \{V_1, V_2, \ldots, V\}_n$, and the set of fault edges is $E = \{E_1, E_2, \ldots, E_m\}$. The distance value of each edge is set to 1. The failure propagation probability matrix of graph G is $P = [w_{ij}]_{n \times n}$, where w_{ij} is the probability that faults propagate from V_i to V_j, and the value of w_{ij} is between 0 and 1.

The initial one-step probability matrix of fault propagation is $P(1)$, which means that the fault propagation is only one step. $P(2), \ldots, P(k)$ means that the fault has passed through two steps, …, k steps. The probability matrix of fault propagation after k-step propagation corresponds to the length of fault propagation paths of 1, 2, …, k. $P(k) = P(k-1)P(1)$; that is, the probability of a fault path with length k equals the product of the probability of a path with length $k - 1$ and the probability of another path with length 1. So $P(2) = P(1)P(1)$, $P(k) = P(1)^k, 1 \le k \le n$. Some nodes need at least k steps to arrive, while others will arrive earlier than k steps. After k-step propagation, with the increase of k, the element value in $P(k)$ decreases by an order of magnitude. When the propagation probability is less than 10^{-8}, that is to say $\prod_k P(w_{ij})^k < 10^{-8}$, the node is considered to be in a safe state. This condition is regarded as the constraint condition of whether the fault can continue to propagate.

2.2.3 K-Step Fault Propagation Algorithm

Step 1: In the first step of diffusion, $V_{i0}^{(0)}$ is set as the fault source and marked as the fault state, initially making $F_0 = \{V_{i0}\}$. In the fault edge of F_0 node, the edge with higher propagation probability is chosen to propagate first. If there are multiple edges with the same weight, the propagation will be carried out at the same time. It might be as well as assume the next fault point is $V_{j1}^{(1)}, V_{j2}^{(1)} \ldots, V_{jr}^{(1)}, r \ge 1$, $F = F_0 \cup F_1$, the point in F_1 is marked as the fault state and the probability matrix of fault propagation is $P(1)$. The probability value is checked whether the constraint condition is satisfied or not. If it is satisfied to go to the second step, record the propagation path and add it to the fault edge set $\tilde{E}$. If it is not satisfied, then select the edge with high probability to propagate and repeat the above process.

Step 2: In the second step of diffusion, the maximum probability of fault propagation is selected from the edge of the fault point set F. The nodes propagated are recorded as $V_{k1}^{(2)}, V_{k2}^{(2)} \ldots, V_{ks}^{(2)}, s \geq 1$, which is the fault point set F_2, and they are marked as fault states, so that $F = F_0 \cup F_1 \cup F_2$. The fault propagation probability matrix $P(2)$ is updated to verify whether the probability value satisfies the constraints. If satisfied, the fault edge is added to $\tilde{E}$ and transferred to the next round; if not, other paths are checked according to probability value.

Step 3: Repeat the above steps, after $k(k \leq n)$ step diffusion, the fault propagation probability matrix $P(k)$ reaches the boundary value of the constraint condition, then stops propagation, and gets the final fault propagation path; if the node is spread to a node, it stops propagation.

The steps of K-step fault propagation algorithm in high-speed train system are shown Fig. 1.

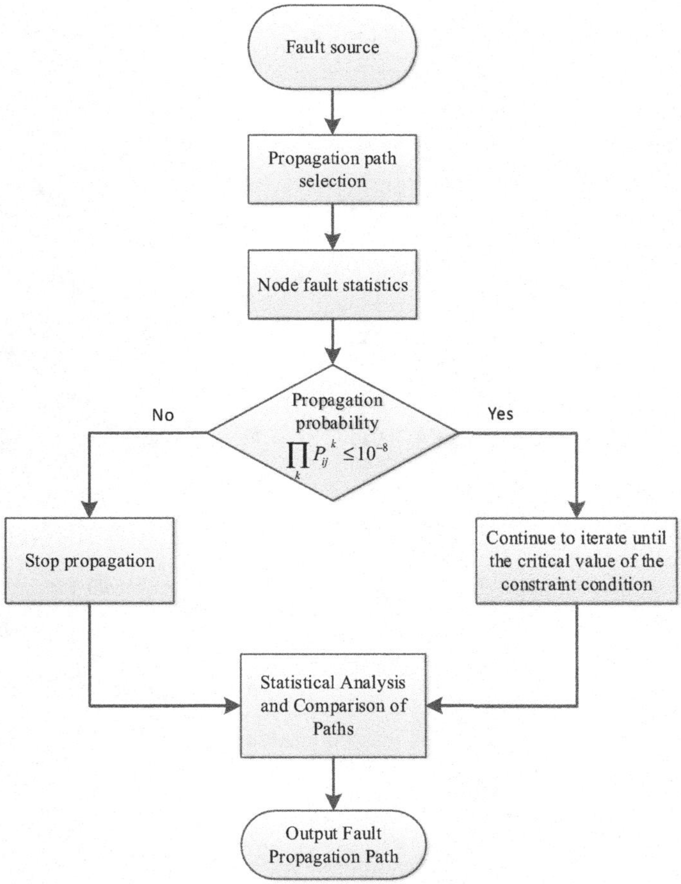

Fig. 1 Steps of K-step fault propagation algorithm

3 Research on Fault Propagation Path of Bogie System

3.1 Fault Propagation Network of Bogie System

According to the method proposed in Sect. 2.2.1, fault propagation network of bogie system $G = \{V, E\}$ is established as shown in Fig. 2 (Table 1).

3.2 Fault Propagation Path of Bogie System

It is assumed that the faults in the bogie system of high-speed train are brake clamp faults. According to the principle of K-step fault propagation algorithm proposed in Sect. 2.2.2, the fault propagation path is studied.

The fault source of high-speed train bogie system:

$$F_0 = \{v_4\}$$

Fig. 2 Fault propagation network of bogie system

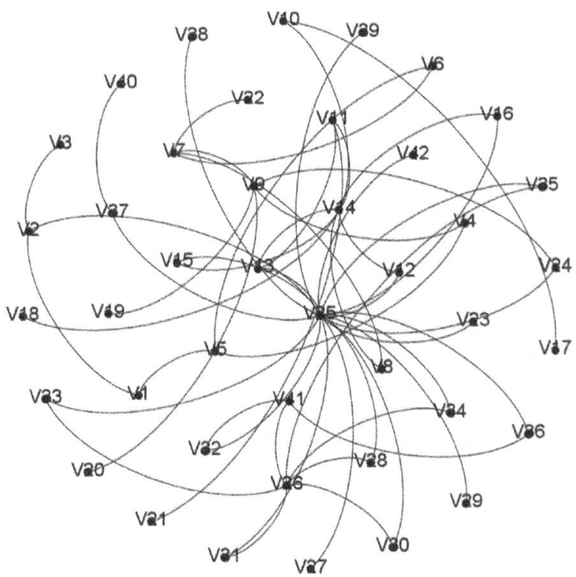

Table 1 Components of bogie system

Node	Component name	Node	Component name	Node	Component name
v_1	Brake pad	v_{15}	A series of vertical shock absorbers	v_{29}	Height control stem assembly
v_2	Brake cylinder	v_{16}	Stop structure (first series)	v_{30}	Anti-snake shock absorber
v_3	Brake clamp	v_{17}	Axle-end speed sensor	v_{31}	Air spring control valve set
v_4	Coupling	v_{18}	Axle-end temperature sensor	v_{32}	Secondary transverse shock absorber
v_5	Roulette	v_{19}	Gearbox shaft temperature sensor	v_{33}	Composition of anti-roll bar
v_6	Wheel	v_{20}	Gearbox speed sensor	v_{34}	Stop structure (second series)
v_7	Axle	v_{21}	Speed sensor of traction motor	v_{35}	Secondary vertical shock absorber
v_8	Axle box bearing	v_{22}	Speed gear	v_{36}	Traction rod device
v_9	Main body of gear box	v_{23}	Transverse acceleration sensor	v_{37}	Tread cleaning device
v_{10}	Axle box cover	v_{24}	Gear box boom	v_{38}	Flange lubrication device
v_{11}	Rubber pad	v_{25}	Framework composition	v_{39}	Sanding device
v_{12}	Location node	v_{26}	Composition of connecting pillow beam	v_{40}	Tread cleaning grinder
v_{13}	Axle box spring	v_{27}	Traction motor shock absorber	v_{41}	Traction centre pin and traction beam
v_{14}	Rotary arm shaft box	v_{28}	Empty spring composition	v_{42}	Stone sweeper

Probability matrix of fault propagation:

$$P(1) = \begin{bmatrix} 0 & 0.156 & 0 & 0 & 0.039 & \cdots & 0 & 0 & 0 & 0 \\ 0.0142 & 0 & 0.174 & 0 & 0.178 & \cdots & 0 & 0 & 0 & 0 \\ 0 & 0.036 & 0 & 0.145 & 0 & \cdots & 0 & 0 & 0 & 0 \\ 0 & 0 & 0.157 & 0 & 0 & \cdots & 0 & 0 & 0 & 0 \\ 0.057 & 0.072 & 0 & 0 & 0 & \cdots & 0 & 0 & 0 & 0 \\ \vdots & \vdots & \vdots & \vdots & \vdots & \cdots & \vdots & \vdots & \vdots & \vdots \\ 0 & 0 & 0 & 0 & 0 & \cdots & 0 & 0 & 0 & 0 \\ 0 & 0 & 0 & 0 & 0 & \cdots & 0 & 0 & 0 & 0 \\ 0 & 0 & 0 & 0 & 0 & \cdots & 0 & 0 & 0 & 0 \\ 0 & 0 & 0 & 0 & 0 & \cdots & 0 & 0 & 0 & 0 \end{bmatrix}$$

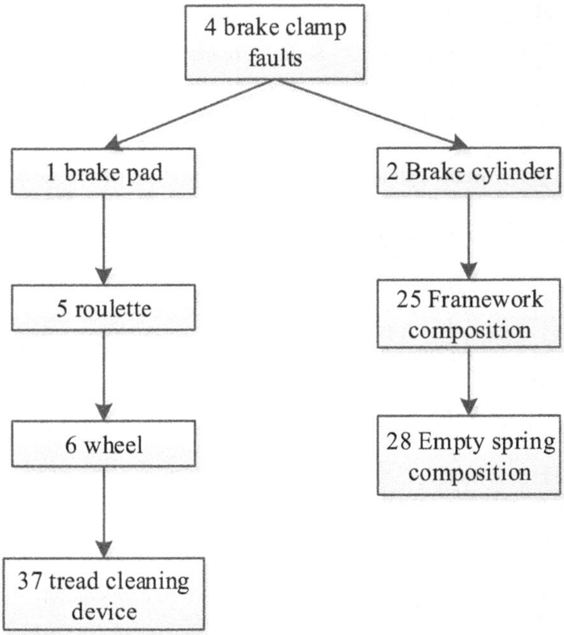

Fig. 3 Fault propagation path

After four steps of fault propagation, the fault propagation stops because the constraint condition is reached. In Fig. 3, fault propagation path can be obtained.

4 Conclusion

Based on the fault propagation network model, K-step fault propagation method is proposed, and the fault propagation path in bogie system is analysed. The path of the fault propagation obtained by our method agrees with the ones in reality. The result of our method will be helpful for the maintenance and the optimal design of high-speed train system, which inspires the following works concerning the reliability and the safety.

Acknowledgements The authors gratefully acknowledge the support from "the Fundamental Research Funds for the Central Universities" (Grant No. 2017JBZ103).

References

1. Liu X, An S (2014) Failure propagation analysis of aircraft engine systems based on complex network. Procedia Eng 80:506–521
2. Wang Z, Yao H, Yang G (2015) Failure risk propagation and protection schemes in coupled systems. Chaos, Solitons Fractals 80:62–75
3. Olami Z, Feder HJS, Christensen K (1992) Correlation functions in the fully frustrated 2D XY model. Phys Rev Lett 68:1224
4. Dobson I, Carreras BA, Newman DE (2003) A probabilistic loading-dependent model of cascading failure and possible implications for balckouts. In: 36th Annual Hawaii international conference on system sciences
5. Wang XF, Xu J (2004) Cascading failures in coupled map lattices. Phys Rev E 70:056–113
6. Xu X, Lu NY, Yong JQ, Jiang B (2016) Fault propagation analysis of IGBT fault in CRH5 traction system based on signed directed graph. In: 2016 Prognostics and system health management conference (PHM), Chengdu
7. Guru PB, Ratneshwer (2018) Dependency-based fault diagnosis approach for SOA-based systems using Colored Petri Nets. J King Saud Univ Comput Inf Sci
8. Hu JQ, Zhang LB, Cai ZS, Wang Y, Wang AQ (2015) Fault propagation behavior study and root cause reasoning with dynamic Bayesian network based framework. Process Saf Environ Prot 97:25

Forecast of Passenger Flow Under the Interruption of Urban Rail Transit Operation

Hongwei Gao, Jie Xu, Siyao Li and Lili Xu

Abstract When the subway is interrupted for a long time with a wide impact range during the operation of urban rail transit, a lot of passengers will be stranded in the station. In this case, buses will be needed to evacuate the passengers to ensure the safety and the normal travel of passengers. Therefore, effectively predicting the volume of stranded passengers at each station is highly important to improve the efficiency of the passenger evacuation. This paper mainly describes the forecast of passenger flow in the disruption of urban rail transit operation. In this paper, we focus on transport organization under interruptions in emergency and according to the adjustment of traffic schemes determine the section to be connected. We use historical passenger flow data to estimate the stranded passenger flow at the interruption stations and in the trains, and according to the passenger destination summarize all the passenger flows. So, the bus company can arrange the connection plan based on the passenger flow data. In order to illustrate the calculation process of the algorithm, we take the Guangzhou Metro Line 1 as an example to carry out simulation calculation.

Keywords Subway · Interruption in emergency · Emergent passenger transportation · Passenger flow forecast · Public transport evacuation

H. Gao · J. Xu (✉) · S. Li · L. Xu
State Key Laboratory of Rail Control and Safety, Beijing Jiaotong University, Beijing, China
e-mail: jxu1@bjtu.edu.cn

School of Transportation and Traffic, Beijing Jiaotong University, Beijing, China

Beijing Research Center of Urban Traffic Information Sensing and Service Technologies, Beijing, China

Research and Development Center of Transport Industry of Technologies and Equipment of Urban Rail Operation Safety Management, MOT, PRC, Beijing, China

Y. Qin et al. (eds.), *Proceedings of the 4th International Conference on Electrical and Information Technologies for Rail Transportation (EITRT) 2019*, Lecture Notes in Electrical Engineering 639, https://doi.org/10.1007/978-981-15-2866-8_27

1 Introduction

As a crucial part of urban rail transit project research, passenger flow forecast is the basis of rail transit design, construction, operation and maintenance [1]. In the process of subway operation, once a serious accident occurs, it will cause a long-term operation interruption and a large number of passengers will be delayed in the station [3]. Overcrowded passengers in subway systems may endanger safety of passengers and security of subway's operation [2]. To prevent overcrowdedness from trains and platforms during peak hours, the subway company need to make use of buses in time to evacuate the stranded passenger, to substitute rail transit to provide transport services [4, 5]. Before the buses start to evacuate the crowd, we need to analyze and predict the passenger flow in the subway network [6]. Therefore, effective prediction on stranded passengers of some stations during certain periods is important, especially during peak hours, which prevents injuries from stampede, falling down rail and so on [7, 8].

2 Analysis of Stranded Passenger Flow

2.1 Adjustment of Traffic Schemes After Interruption of Train Operation

Some unavoidable accidents may occur under the long-term operation of rail transit. Vehicles, communication signals, electrical and mechanical power supply and other equipment may go out of order. In addition, there are other accidents such as people and vehicles conflicts, fires, train derailment, train overturning, foreign body intrusion and so on. In the face of various emergencies, especially major accidents, the relevant subway traffic dispatchers must make scientific and reasonable adjustments to the traffic dispatch as soon as possible and formulate relevant and targeted solutions for different abnormal situations, to ensure the safety of subway driving and minimize risks and losses (Fig. 1).

When the above-mentioned disaster occurs, the accident will cause the train to run out, affecting people's normal travel. Rail trains will not be able to operate as planned after accident, the subway company usually take "short routing operation and long routing operation (in normal operating sections) and public transport evacuation (in interrupted operating sections)" to maintain the basic service level of the urban public transport system.

According to the adjusted driving plan, it can be known that in the short routing operation or the alternating operation of long or short routing mode, the passenger route of the relevant stations can form a complete closed loop, so the volumes of passenger in these stations are not considered. However, in the stations where we adopt the "bidirectional operation on single track" operation mode or the stations

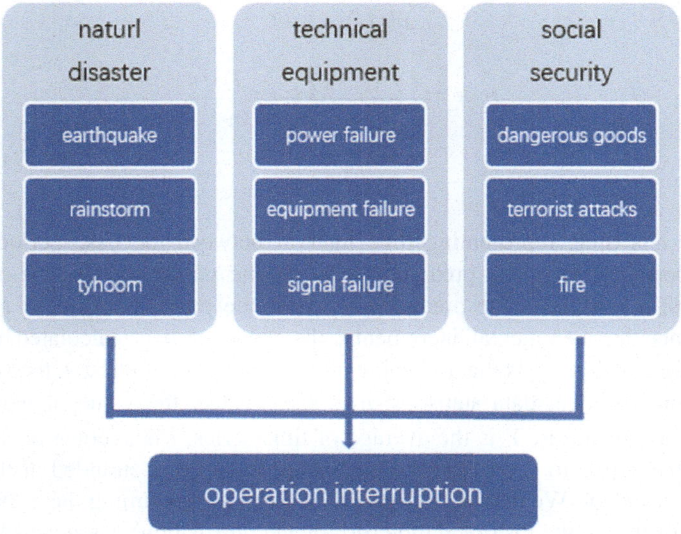

Fig. 1 Interruptions are classified by causes

completely interrupted due to serious accidents, the route of passenger travel in these stations cannot form a closed loop, so we have to use the buses to evacuate this part of the passenger flow. Therefore, we only predict the passenger flow of the interrupted section, where both up and down directions are not open to traffic or only one direction can be opened. The passengers in those stations which adopted short routing operation or alternating operation of long or short routing running mode are not analyzed.

2.2 *Analysis and Forecast of Waiting Passengers in the Station*

We can identify stations with passengers stranded according to the driving plan. The following is the passenger flow forecasting process of the station to be connected. In China, automatic fare collection (AFC) is applied in all of urban rail transit systems, and an electronic ticketing system will record some of travel information of passengers. The inbound and outbound time of passengers and origin and destination (O/D) stations of passengers can be derived from AFC. According to the current situation, it is difficult to obtain passenger data in time; hence, based on historical passenger flow data, we use the time series analysis method to predict passenger flow.

$$Y(t) = a + bt \tag{1}$$

$$b = \frac{\sum t * Y_t - (\sum t \sum Y_t)/n}{\sum t^2 - (\sum t)^2/n} \tag{2}$$

$$a = \overline{Y} - b\overline{t} \tag{3}$$

in which: t is time, representing time interval between the base period and the forecast period; $Y(t)$ is the predicted value of the passenger flow, a and b are parameters; a represents the predicted value of the passenger flow at $t = 0$, and b represents the incremental increments, the value of b is calculated using the least-squares method; Y_t is the true value of the time series of phase t, the value of Y_t is based on historical data statistics; n is stage value, the value of n is defined according to our needs; $\overline{Y}$ is the average of time series; $\overline{t}$ is average of t.

The next step is to determine the number of passengers stranded at the station during the accident. We record the time of the accident (assuming the accident time is T_1); according to the historical O/D (origin and destination) passenger flow table, we can calculate the volume of passenger flow from the connection site to the downstream site within a period of time after the accident, and volume is $Y(t)$. Combined with train schedules, we should define the departure time of the last shuttle bus at each connecting station before the accident occurs, and the time is recorded as T_2. Then the passenger flow at each station during this time period is:

$$Y = Y(t) * (T_1 - T_2)/T \tag{4}$$

in which: T is time, representing time interval between the base period and the forecast period; Y is the volume of stranded passenger.

2.3 Analysis and Forecast of Passengers on the Train

The line will be out of service when the accident occurs, and there may be trains stopped in the interruption section. The passengers on this train will be forced to get off and be staying at a certain station. These passengers will be the main evacuation passengers. When analyzing the number of passengers staying on the train, we mainly base on historical hourly cross-sectional passenger flow. The principle of data processing is referred to Sect. 2.2. We adopt the linear trend prediction method to calculate the passengers in an hour based on the average data of multi-day data. The number of hourly section passenger flow is denoted as S.

Next, calculate the passengers on the train. According to train schedule, we obtain the number of trains in an hour during the accident and recorded the number as d. The number of passengers on the mth train is,

$$Y(m) = S/d \tag{5}$$

After calculating the number of passengers on the train, it is also necessary to consider the passenger flow direction. Then we need to calculate origin and destination (O/D) distribution ratio. O/D distribution ratio of passenger flow is the ratio of outbound passengers of station j account for passengers who enter into station s for taking the mth train, which is denoted as $B_s|_m^j$ and computed as follows,

$$B_m|_s^j = P_{s_m j}^{\text{out}} \Big/ P_{s_m}^{\text{in}} \tag{6}$$

in which, $P_{s_m}^{\text{in}}$ is inbound passenger flow of station s during the mth train operation cycle, $P_{s_m j}^{\text{out}}$ is outbound passenger flows of station j, who enter station s during the mth train operation cycle. $P_{s_m}^{\text{in}}$ and $P_{s_m j}^{out}$ are attained by adding up key value of smart card, which will be collected from AFC.

$$Y(j) = Y(m) * B_m|_s^j \tag{7}$$

2.4 *Total Passenger Flow to Be Transported*

Both the volume of stranded passengers in stations and the trains calculated in the previous section have considered the passengers' direction. Therefore, we can summarize the volume by passenger destination. The summary results are computed as follows,

$$Y(j') = Y + Y(j) \tag{8}$$

3 Case Study

We take the accident case of Guangzhou Metro Line 1 as an example to illustrate. There are 16 stations on Line 1. If an accident happened from the Fangcun (FC) Station of Line 1 to the Gongyuanqian (GYQ) Station, the train will not be able to pass this line. The subway company adjust the driving plan according to the site situation: short routing operation from Xilang (XL) Station to Fangcun (FC) Station; short routing operation from Guangzhoudong (GZD) Station to Gongyuanqian (GYQ) Station; bidirectional operation on single track from Fangcun (FC) Station to Gongyuanqian (GYQ) Station. Therefore, the section that needs to be connected is the upward direction from Fangcun Station to the Gongyuanqian Station. We need to analyze the passenger stranded from Fangcun Station to the Gongyuanqian Station.

Table 1 Direction of passengers stranded in the stations

OD	FC	HS	CSL	CJC	XMK
HS	33	–	–	–	–
CSL	26	5	–	–	–
CJC	67	13	20	–	–
XMK	44	9	8	4	–
GYQ	35	6	16	15	10
NJS	31	4	14	14	16
LSLY	21	3	13	19	21
DSK	10	4	4	10	7
YJ	7	3	2	6	8
TYXL	7	4	5	10	5
TYZX	11	2	5	11	5
GZDZ	21	3	15	27	24
SUM	313	56	102	116	96

The stranded passengers in stations

The total numbers of inbound passengers at each station are estimated by using the transaction data accessing from AFC. The data set does not include weekend and holiday days. In order to ensure the prediction is as accurate as possible, this paper selects the O/D traffic flow of passenger with the granularity of 15 min. Using the time series analysis method, we calculate the volume of inbound passengers in 15 min. Inbound passengers at stations along the individual route are calculated based on the transaction data of AFC and tabulated in Table 1. In Table 1, line heading indicates passenger stranded sites, and the column heading indicates the stations the passenger is going to.

The stranded passengers in the train

We get the train schedule in line 1 as Table 2 shows and then calculate the hourly cross-sectional passenger flow in Table 3 based on the historical data. According to the OD distribution ratio, the arrival passengers in each train are estimated.

The final calculation of the passengers in the train is shown in Table 4, line heading indicates passenger stranded sites and the column heading indicates the stations the passenger is going to.

Summary of all stranded passengers

We add up the number of these two passengers, and the result is the passenger flow to be transported by buses. The final result is shown in Table 5, line heading indicates the interrupted stations, and passengers are stranded at these stations. The

Table 2 The train schedule in line 1

NJS	8:03:33	8:06:21	8:09:09	8:11:57	8:14:45	8:17:26	8:20:21	8:23:09	8:25:57
	8:02:00	8:04:48	8:07:36	8:10:24	8:13:12	8:15:53	8:18:48	8:21:36	8:24:24
GYQ	8:01:05	8:03:53	8:06:41	8:09:29	8:12:17	8:14:58	8:17:53	8:20:41	8:23:29
	7:59:49	8:02:37	8:05:25	8:08:13	8:11:01	8:13:42	8:16:37	8:19:25	8:22:13
XMK	7:59:17	8:02:05	8:04:53	8:07:41	8:10:29	8:13:10	8:16:05	8:18:53	8:21:41
	7:58:02	8:00:50	8:03:38	8:06:26	8:09:14	8:11:55	8:14:50	8:17:38	8:20:26
CJC	7:57:30	8:00:18	8:03:06	8:05:54	8:08:42	8:11:23	8:14:18	8:17:06	8:19:54
	7:55:47	7:58:35	8:01:23	8:04:11	8:06:59	8:09:40	8:12:35	8:15:23	8:18:11
CSL	7:55:13	7:58:01	8:00:49	8:03:37	8:06:25	8:09:06	8:12:01	8:14:49	8:17:37
	7:53:47	7:56:35	7:59:23	8:02:11	8:04:59	8:07:40	8:10:35	8:13:23	8:16:11
HS	7:53:13	7:56:01	7:58:49	8:01:37	8:04:25	8:07:06	8:10:01	8:12:49	8:15:37
	7:51:24	7:54:12	7:57:00	7:59:48	8:02:36	8:05:17	8:08:12	8:11:00	8:13:48
FC	7:50:49	7:53:37	7:56:25	7:59:13	8:02:01	8:04:42	8:07:37	8:10:25	8:13:13
	7:49:06	7:51:54	7:54:42	7:57:30	8:00:18	8:02:59	8:05:54	8:08:42	8:11:30
HDW	7:48:33	7:51:21	7:54:09	7:56:57	7:59:45	8:02:26	8:05:21	8:08:09	8:10:57
	7:47:19	7:50:07	7:52:55	7:55:43	7:58:31	8:01:12	8:04:07	8:06:55	8:09:43
KK	7:46:25	7:49:45	7:52:01	7:55:21	7:57:37	8:00:39	8:03:34	8:06:01	8:09:10
	7:44:41	7:47:40	7:50:17	7:53:16	7:55:53	7:58:34	8:01:29	8:04:17	8:07:05
XL						7:57:14			
DEP						▲			
TRAINID	0106	1404	1504	1604	0206	2502	1704	1804	0406
Running interval	7:44:41	0:02:59	0:02:37	0:02:59	0:02:37	0:02:41	0:02:55	0:02:48	0:02:48

Table 3 Passenger flow distribution results

Line	1	1	1	1	1	1	1
Starting site		HDW	FC	HS	CSL	CJC	XMK
terminal site	...	FC	HS	CSL	CJC	XMK	GYQ
5:00-6:00		1969	2613	2667	2843	3051	3275
6:00-7:00		8158	10,943	11,060	12,331	13,765	15,079
7:00-8:00		12,218	17,105	17,678	20,217	23,529	25,714
8:00-9:00		9162	12,199	12,616	13,574	15,356	16,904
9:00-10:00		7977	10,260	10,652	10,997	12,301	13,768
…		6677	8872	9363	10,056	11,107	12,680

column heading indicates the passenger's destination stations. The Numbers in the table have the following meanings, such as 33, it said that among the stranded passengers in Fangcun station, there are 33 people were going to Huangshan station.

Table 4 The calculation of the passengers in the train

OD	HS	CSL	CJC
HS	–	–	–
CSL	67	–	–
CJC	120	190	–
XMK	119	158	154
GYQ	112	140	148
NJS	105	127	139
LSLY	96	114	128
DSK	86	100	112
YJ	79	91	101
TYXL	73	83	93
TYZX	68	77	87
GZDZ	67	78	90
SUM	992	1156	1052

Table 5 The final calculation of the passengers

OD	FC	HS	CSL	CJC	XMK
HS	33	–	–	–	–
CSL	26	72	–	–	–
CJC	67	133	210	–	–
XMK	44	128	166	158	–
GYQ	35	118	156	163	10
NJS	31	109	141	153	16
LSLY	21	99	127	147	21
DSK	10	90	104	122	7
YJ	7	82	93	107	8
TYXL	7	77	88	103	5
TYZX	11	70	82	98	5
GZDZ	21	70	93	117	24
SUM	313	1048	1258	1168	96

4 Conclusion

The subway is the backbone of urban passenger traffic, once the subway line is interrupted, it will greatly affect traffic and normal life of residents. The subway company should do daily maintenance to avoid accidents and failures and maximize the advantages of rail transportation. We must make timely and accurate adjustments based on the accident situation and initiate the bus emergency procedures, so it is necessary to make predictions for the connection of passenger flow to provide data support for the emergency connection of the bus. We should not only ensure the travel experience of the passengers but also avoid the waste of public transportation vehicles.

Acknowledgements The authors gratefully acknowledge the support provided by China National "13th Five-Year" key research project "Safety assurance technology of urban rail system" (Grant No. 2016YFB1200402).

References

1. C Yang (2018) Analysis of train operation adjustment scheme under abnormal condition of rail transit operation. Constr Eng Technol Des J 23:397 (in Chinese)
2. Shi Q, He N, Ling X (2014) Analysis on the complexity and controllability of passenger flow forecast of urban rail transit. In: Transformation and innovation of urban transportation development mode–2011 annual meeting and the 25th symposium of urban transportation planning in China (in Chinese)
3. Zhao S (2015) Research on the organization and scheduling strategy of urban rail transit. Lanzhou Jiaotong University (in Chinese)
4. Zheng X (2010) Discussion on related problems of emergency bus connection in subway. J Mod Urban Rail Transit 04:18–20 (in Chinese)
5. Xiao G (2013) Research on urban passenger transport emergency passenger flow and distributing based on sudden interrupt condition. Southwest Jiaotong University (in Chinese)
6. Chen X (2008) Research on passenger flow prediction model of guangzhou metro. Chinese High-Tech Enterp J 16:3–4. https://doi.org/10.3969/j.issn.1009-2374.2008.16.006 (in Chinese)
7. Chen C, Liu G (2015) Discussion on passenger flow calculation method of urban rail transit section. J Transp Technol Econ 17(04):43–46 (in Chinese)
8. Liu J (2015) Study of bus emergency scheduling method under urban rail transit emergency. Beijing Jiaotong University (in Chinese)

Research on PHM Evaluation of Point Machine with State Transition Variable Weight

Minzhe Wu and Zhipeng Li

Abstract In order to better reflect the continuity of the failure process of the point machine and the whole process from normal to failure, this paper proposes a health machine evaluation plan based on ResNet and Markov state transfer contingency model. ResNet is used to evaluate the failure probability of the real-time curve, and the Markov state transition model is built according to the different operating states of the equipment to comprehensively judge the health of the switch. This paper uses the actual operational data of 173 sets of equipment in China to verify the algorithm. The results show that the algorithm described in this paper can reflect the continuity and complexity of the health of the switch and provide effective guidance for the state maintenance of the switch.

Keywords Point machine health assessment · ResNet · Markov state transition contingency model

1 Introduction

The point machine is one of the important basic equipments in the railway system, and its health status directly affects the train operation safety and railway transportation efficiency. Traditional post-repair and scheduled maintenance are resource-intensive and inefficient, so it is important to assess the health of the switch.

In recent years, research on the fault diagnosis of ballast equipment has been increasing [1]. Ardakani et al. used several point machines as an example for PHM

M. Wu · Z. Li (✉)
Department of Information and Communication Engineering, Tongji University, Shanghai, China
e-mail: lizhipeng@tongji.edu.cn

M. Wu · Z. Li
The Center of National Railway Intelligent Transportation System Engineering and Technology, Beijing, China

Y. Qin et al. (eds.), *Proceedings of the 4th International Conference on Electrical and Information Technologies for Rail Transportation (EITRT) 2019*, Lecture Notes in Electrical Engineering 639, https://doi.org/10.1007/978-981-15-2866-8_28

analysis [2]. Yuming et al. used the fuzzy synthesis method to evaluate the health of the switch [3], and the algorithm model could not be adaptive to the change weight of the device in real time. In order to better reflect the gradual change and continuity of equipment failure process and the whole process of equipment from normal to failure, a health assessment method of switch equipment based on state transition contingency is proposed in this paper.

2 Health Model of Point Machine

2.1 Health Definition of Point Machine

According to the reliability maintenance (RCM) theory [4], this paper defines the health (HI) to reflect the ability of the point machine to perform the specified function within the specified time and under the specified conditions. The health indicators at time t of the device are represented by $x_1(t), x_2(t), \ldots, x_n(t)$ as follows:

$$\mathrm{HI} = F\{x_1(t), x_2(t), \ldots, x_n(t)\} \tag{1}$$

In the formula, F is a health function, and the range of HI is (0–100).

2.2 Selection of Health Evaluation Index of Point Machine

Considering the characteristics of the point machine and the actual operation and maintenance of the site, select the operating period, fault record, curve failure probability and operating environment to reflect the health status of the point machine.

(1) Operating period and maintenance record

This paper will reflect the failure rate distribution of the equipment life cycle according to the famous bathtub model [5]. At the same time, when a device fails multiple times during its maintenance cycle, the algorithm reduces its health score according to the law of increasing exponentiality.

(2) Failure probability of real-time electrical characteristic curve

As shown in Fig. 1a, the normal operation is divided into four stages: start, unlock, conversion and lock, and there will be corresponding reactions on the working characteristic curve. When the performance is degraded, the frictional force increases during the action. As shown in Fig. 1b, the action curve is normal in the early stage, and the curve suddenly rises during the locking process. The high probability is that the turnout action is not in place, which is caused by the blockage in the locking process.

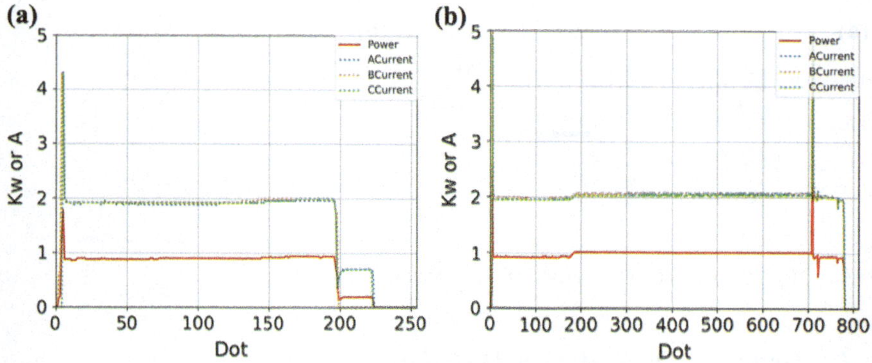

Fig. 1 Point machine electrical characteristic curve

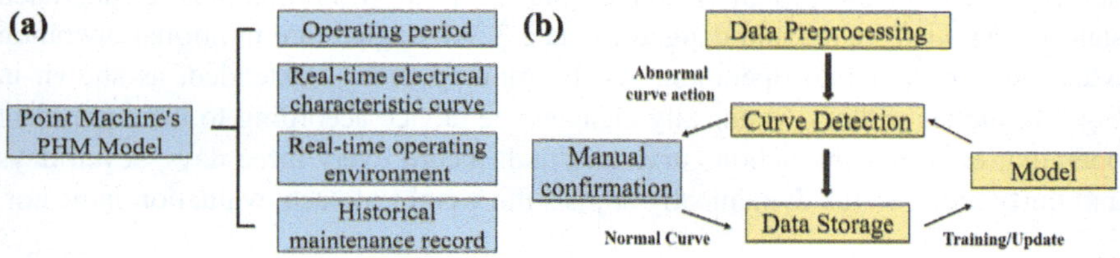

Fig. 2 Health composition and algorithm architecture

In this paper, the fault probability of the curve is used as one of the indicators to comprehensively evaluate the health of the equipment.

(3) Real-time environment of point machine

Deviation from the rated environment of the equipment operating environment will have a negative impact on equipment performance [6]. In this paper, based on the deviation of the real-time environment from the rated environment, the real-time operating environment score is obtained.

In summary, the health assessment indicators of the switch equipment are shown in Fig. 2a.

3 Dynamic Degradation Distribution Based on Markov State Transition

Part of the health assessment indicators is highlighted when the state of the device deteriorates. In order to overcome the shortcomings of traditional fuzzy synthesis method which cannot reflect the nonlinear characteristics of the system dynamically, a dynamic weight model based on Markov state transition is introduced [7].

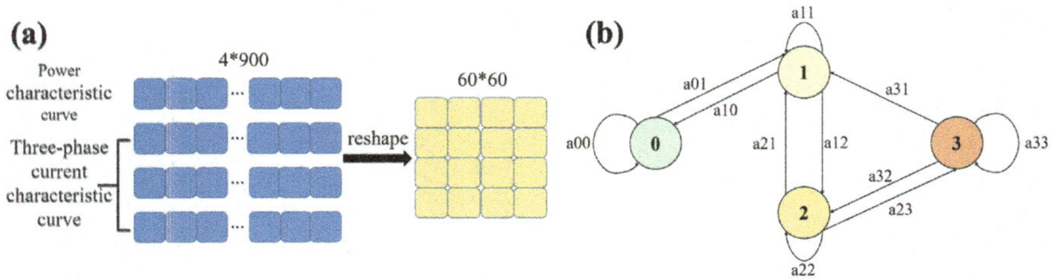

Fig. 3 Health composition and algorithm architecture

3.1 Dynamic Degradation Distribution

Through the analysis of the failure development of the point machine. The degraded state of the switch is divided into four states: normal state 0, degraded state 1, degraded state 2 and dangerous state 3. All devices are in normal operation when they are put into operation. As the running time is extended, as shown in Fig. 3b, the algorithm dynamically changes the device according to the equipment operating environment, action curve and fault record every three days, seven days and thirty days. Status dynamically adjusts the weight of each evaluation indicator.

3.2 Calculating the Index Weights of Each State Based on the Order Relationship Analysis Method

This paper defines three indicators $x_1(t), x_2(t)$ and $x_3(t)$, represents operation, fault record, real-time curve fault probability and real-time operation environment, respectively, and w_1, w_2 and w_3 are the weights of the three indicators. The ratio of the importance of x_{n-1} and x_n is w_{n-1}/w_n, denoted r_n, where $r_n \in [1, 4]$. The higher the importance of x_{n-1} relative to x_n, the larger the value of r_n.

The above r_n must satisfy the following constraints:

(1) If the indicator set $\{x_1, x_2, \ldots, x_n\}$ has an order relationship $x_1 > x_2 > \cdots > x_n$, then r_n and r_{n-1} satisfy:

$$r_{n-1} > \frac{1}{r_n}, n = m, m-1, m-1, \ldots, 3, 2 \tag{2}$$

(2) When r_n satisfies the condition in the above equation, the weight w_m is calculated as

Table 1 Individual state health weight

Operating status	Operating period and environment (ProMain)	Real-time curve (RealEnvir)	Maintenance record (DayCurve)
Normal status	0.4355	0.2016	0.3629
Degraded state 1	0.4330	0.3093	0.2577
Degraded state 2	0.3211	0.4495	0.2294
Dangerous state 3	0.2069	0.6207	0.1724

$$w_m = \left(1 + \sum_{n=2}^{m} \prod_{i=n}^{m} r_i\right)^{-1} \tag{3}$$

$$w_{n-1} = r_n w_n, k = m, m-1, \ldots, 3, 2 \tag{4}$$

In the normal state, the attribute x_1, x_2, x_3 has an order relationship $x_1 > x_3 > x_2$, and the calculation process is as follows:

$$r_2 = \frac{w_1}{w_3} = 1.2, r_3 = \frac{w_3}{w_2} = 1.8, r_2 * r_3 = 2.16, r_3 = 1.8, r_2 * r_3 + r_3 = 3.96$$

$$w_2 = (1 + 3.96)^{-1} = 0.2016$$
$$w_3 = w_2 r_3 = 0.2016 * 1.8 = 0.3629$$
$$w_1 = w_3 r_2 = 0.3629 * 1.2 = 0.4355$$

Similarly, other state weights are available. The results are as follows (Table 1).

3.3 Real-Time Curve Recognition Based on ResNet

The architecture of real-time curve recognition system based on ResNet [8] is shown in Fig. 2b, and it is divided into three parts: data preprocessing, model training and curve recognition. The data preprocessing converts the electrical characteristic curve into a pseudo image format that the algorithm can recognize and then uses the neural network to train the classification model. Finally, the fault curve is identified according to the model.

The actual operating time of the switch characteristic curve is about 8–10 s, which is sampled at a sampling interval of 25 points per second, and about 250 points are obtained. The duration of the abnormal action may reach 30 s (Fig. 1b, which is preprocessed as shown in Fig. 3a.

After the training is completed, the algorithm can obtain the probability of failure of the curve whenever the new curve is input.

4 Engineering Case Analysis

In order to verify the effectiveness of the proposed algorithm, this chapter uses 173 equipment operating data from eight stations across China for case analysis, with a time span of 117 days. The data of a total of 158 point machines in seven stations are used as a training set, and the data of 15 devices in the remaining station are used as a detection set.

Based on ResNet, the electrical characteristic curve recognition algorithm of the point machine evaluates the fault probability of the real-time action curve of the test set as shown in the following Table 2.

Secondly, the equipment of the test set was selected for the health monitoring of the historical monitoring data with a time span of 117 days. The evaluation results are as follows:

In Fig. 4a, the dotted line indicates the score curve of each indicator, and the solid red line indicates the comprehensive health score curve of the device. The health of the equipment is obviously lower than the health warning value of the equipment in about 40 days. The operation status of the equipment must be checked and analyzed. The statistics of the abnormal curve of the equipment are shown in Fig. 4b. When the health of the equipment is abnormal, the equipment has multiple abnormal curves for many days and must be repaired. This situation is in good agreement with the health score.

Figure 5a is the evaluation result according to the traditional fuzzy synthesis method, and Fig. 5b is the evaluation result obtained by the algorithm. Under the condition that the real-time curve index of the point machine is normal, the algorithm can better reflect the impact of the real-time environment, the operation period and the fault record on the health.

5 Conclusion

The results of the algorithm verification on the actual data set indicate the rationality of the evaluation method described in this paper. This algorithm can effectively reduce the number of inspections of equipment by operation and maintenance personnel and eliminate equipment safety hazards in time. After expansion, this model can also be applied to the evaluation of other systems other than point machines, which is of great significance for equipment health management.

Table 2 Abnormal curve occurrence date and probability of failure

Action day	34	36	36	36	36	38	39	43	43	50	50	74	81
Probability (%)	100	100	100	100	100	100	76	100	100	100	99	100	100

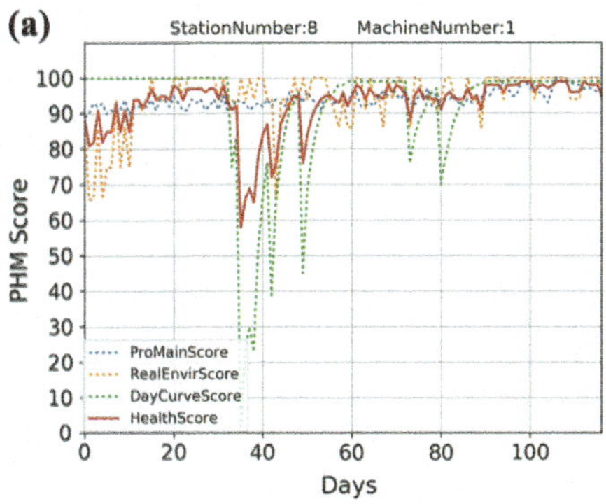

Fig. 4 Point machine health and fault statistics

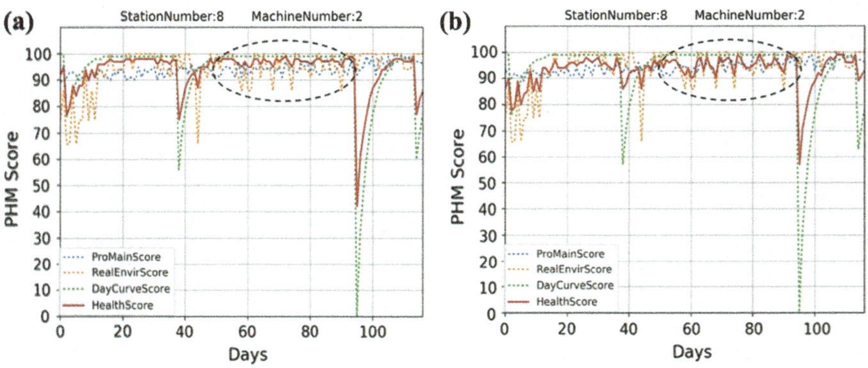

Fig. 5 Comparison of health assessment algorithms

Acknowledgements This work is by the Natural Science Foundation of China under Grant Nos. 61773290 and 71571107 and the Opening Foundation of The Center of National Railway Intelligent Transportation System Engineering and Technology under Grant No. RITS2018KF04.

References

1. Yang X, Li X, Ning B et al (2016) A survey on energy efficient train operation for urban rail transit. IEEE Trans Intell Transp Syst 17(1):2–13
2. Ardakani HD, Lucas C, Siegel D et al (2012) PHM for railway system—a case study on the health assessment of the point machines. In: 2012 IEEE conference on prognostics and health management, Denver, CO, pp 1–5. https://doi.org/10.1109/icphm.2012.6299533
3. Fu Y, Liu B, Song S (2017) Research on health assessment methods for switch machine based on fuzzy comprehensive evaluation. J Railw Sci Eng 05:192–198 (in Chinese)

4. Fischer K, Besnard F, Bertling L (2017) Reliability-centered maintenance for wind turbines based on statistical analysis and practical experience. IEEE Trans Energy Convers 27(1):184–195. https://doi.org/10.1109/TEC.2011.2176129
5. Xie M, Tang Y, Goh TN (2002) A modified Weibull extension with bathtub-shaped failure rate function. Reliab Eng Syst Saf 76(3):279–285. ISSN 0951-8320
6. Wang G, Xu T, Tang T et al (2017) A Bayesian network model for prediction of weather-related failures in railway turnout systems. Expert Syst Appl 69(1):247–256
7. Byon E, Perez E, Ding Y (2011) Simulation of wind farm operations and maintenance using discrete event system specification. Simulation 87(12):1093–1117
8. He K, Zhang X, Ren S et al (2016) Identity Mappings in Deep Residual Networks. In: Computer Vision–ECCV 2016, pp 630—645

An Effective Intrusion Detection System Based on Network Traffic and Packets for CBTC

Xueqian Chen, Bing Bu and Xuetao Yang

Abstract With the application of advanced network and computer technologies, communication-based train control (CBTC) systems are facing increasingly serious security risks. Intrusion detection can help detect attacks of CBTC systems and avoid major accidents. The traditional intrusion detection systems (IDS) do not consider the characteristics of CBTC systems, so they cannot be applied to CBTC systems directly. In this paper, we analyze the characteristics of network data of CBTC systems and propose an IDS based on network traffic and packets to detect typical attacks of CBTC systems, such as the denial of service (DoS) and data tempering attacks. The self-organizing maps (SOM) neural networks are used to improve the density-based spatial clustering of applications with noise (DBscan) method since DBscan only can detect anomalies offline with low detection rate. By testing on a simulation platform of CBTC systems, it is verified that the designed IDS is suitable for CBTC systems for its great detection performance and real-time performance.

Keywords CBTC · Intrusion detection · Network traffic · Network packets

1 Introduction

Urban rail transit is becoming an important mode of urban transportation due to its advantages of punctuality, high speed and high transport capacity. CBTC system is a new generation of train operation control system in urban rail transit. Advanced computer, communication and network technologies are adopted in CBTC systems to improve the automation and information level of the system, which introduces many serious security risks to the system at the same time.

X. Chen (✉) · B. Bu
Beijing Jiaotong University (BJTU), Beijing, China
e-mail: 18120211@bjtu.edu.cn

X. Yang
Beijing Traffic Control Technology Company, Beijing, China

Y. Qin et al. (eds.), *Proceedings of the 4th International Conference on Electrical and Information Technologies for Rail Transportation (EITRT) 2019*, Lecture Notes in Electrical Engineering 639, https://doi.org/10.1007/978-981-15-2866-8_29

Intrusion detection is efficient to detect external intrusion and internal illegal operations. Considering the characteristics of CBTC systems, such as fail-safe mechanism, the redundant equipment and networks, the traditional intrusion detection methods in IT fields cannot be applied to CBTC systems directly. At present, the research on intrusion detection of CBTC systems is in the early stage.

Yu [1] proposed an improved nonparametric cumulative sum (CUSUM) detection algorithm to detect anomalies in CBTC systems. However, this method cannot detect which attack is happening. So, it is difficult for managers to analyze and defend intrusions. According to different intrusion types in urban rail transit systems, Gao [2] designed corresponding snort rules to detect known attacks in the rule base. However, if a new attack occurs, it cannot be detected. Luo [3] designed a wireless intrusion detection algorithm based on the train states. He used static detection method and finite state machine to detect whether the state of wireless terminal was abnormal. Chen [4] proposed a joint detection method based on residual chi-square and Pearson coefficient to improve the detection performance of replay attacks in urban rail transit. These methods only can detect partial types of attacks.

In this paper, we analyze the impact of typical cyberattacks on the CBTC system combining with the characteristics of CBTC systems and propose an IDS based on network traffic and packets, which can not only detect typical attacks in CBTC systems but also new attacks. The previous research does not have a data set dedicated to training and testing the IDS for CBTC systems. Therefore, we analyze the communication protocol, data format and payload of the systems to choose appropriate data features and generate a data set on a simulation platform of CBTC systems. Due to the low detection rate of DBscan method, we use SOM neural networks to improve DBscan method for better detection performance. The traffic detection model is combined to improve the real-time performance of the designed IDS simultaneously.

The rest of the paper is organized as follows. Section 2 presents the impact of cyberattacks on CBTC systems. In the next section, we introduce the proposed IDS and the detection model. Section 4 presents the experimental results of the designed IDS. We summarize the paper and put forward the future research plan aiming at the limitations of the existing research in the last section.

2 Security Threat Analysis of CBTC Systems

2.1 A Typical CBTC System

The architecture of CBTC systems is shown in Fig. 1. CBTC is mainly composed of wayside equipment, on-board equipment and data communication system (DCS). The automatic train supervision (ATS), the zone controller (ZC), the computer interlocking (CI) and the database storage unit (DSU) are wayside equipment. The

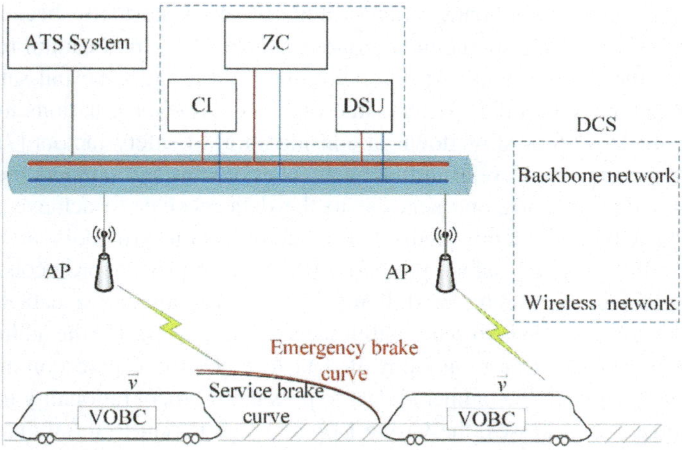

Fig. 1 Architecture of a typical CBTC system

on-board equipment mainly includes vehicle on-board component (VOBC) and some auxiliary equipment. DCS consists of the backbone network and the wireless network [5].

According to the operation instructions of ATS, the status of signal equipment provided by CI and the static speed limit information stored in DSU, ZC synthesizes the train status information, obtains the movement authority (MA) of the train through calculation, and sends it to the train through wireless network. After receiving MA, VOBC calculates the protection curve of train operation in real time according to the information of train position, speed and MA and controls the train operation according to the curve.

In order to ensure the reliability and security of the CBTC system, the fail-safe mechanism and the redundant structure of equipment and networks are adopted [6]. When the CBTC system is attacked by the cyber intruder, these characteristics can play a certain defense ability to avoid some accidents.

2.2 *The Impact of Cyberattacks on CBTC Systems*

The data tampering attacks, DoS attacks, probing, remote to local (R2L) and user to root (U2R) attacks are main attacks occur in CBTC systems. An attacker usually uses U2R and R2L attacks to obtain computer permissions. The probing attacks are implemented to obtain the information of key devices through port scanning, so as to perform DoS attacks or data tampering attacks.

DoS attacks typically exhaust the resources of the target computer or network and cause the service interrupted. Taking the vehicle-ground communication of CBTC systems as an example, if a DOS attack is implemented on the on-board wireless unit, another on-board wireless unit can work normally because of the redundant design. When the redundant one is attacked, the train cannot receive the MA sent by the ZC normally. At this point, the train triggers the fail-safe mechanism for emergency braking. At the same time, ZC issues instructions to the subsequent trains to stop or slow down in consideration of safety factors [7].

Data tampering attacks refer to that an attacker obtains and tampers the network transmission data illegally and sends it to the data receiver to achieve a specific purpose. In CBTC systems, dedicated communication protocols are adopted, including railway signal safety protocol (RSSP) and private protocols of some signal manufacturers, making it difficult for attackers to realize data tampering attacks. The attacker needs to scan within the system network for the address of ZC and the train, collect a large amount of data to crack the secure communication protocol and guarantee the tampered data to pass the protocol rationality inspection. Once the attacker successes, the wrong MA message is sent to the train. The train mistakenly thinks there is no car ahead, leading to the crash eventually.

In general, due to the fail-safe mechanism of CBTC systems, a DoS attack affects the train operation efficiency of CBTC systems but does not cause safety accidents. The implementation of data tampering attacks is difficult and costly, but once it is realized, it is likely to cause accidents and affect the safety of train operation.

3 The Intrusion Detection System Modeling

3.1 Framework of the Intrusion Detection System

The architecture of the IDS proposed is shown in Fig. 2, which is mainly composed of the traffic detection model and the packet detection model. Firstly, the traffic detection model preprocesses the network data and uses auto regression (AR) algorithm to detect the abnormal flow. Since the traffic detection model cannot detect some types of attacks, such as data tempering attacks, the packet detection model is used for further detection. At the same time, the abnormal flow usually contains a large number of packets. By sampling, the computational resource consumption of packet detection model can be reduced, and the real-time performance of the IDS can be improved. The packet detection model preprocesses the sampled abnormal traffic and normal traffic and uses the improved DBScan detection method to divide results into three categories: attacks, normal and abnormal. The abnormal category represents new attacks. At last, detection results are sent to the reporter.

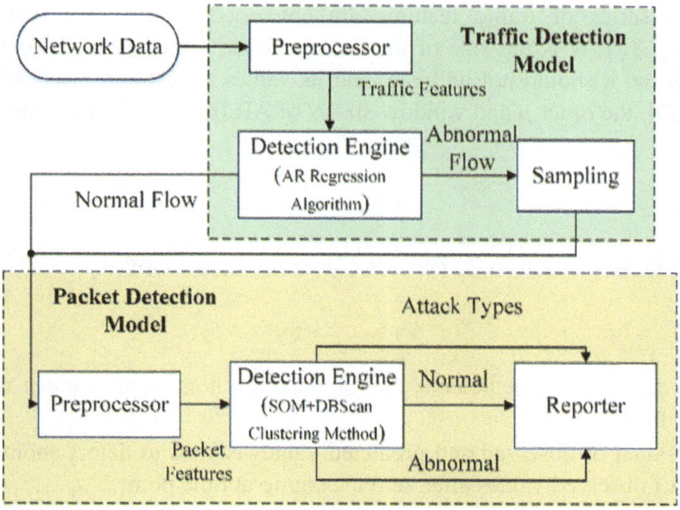

Fig. 2 Hybrid IDS based on traffic and packet features

3.2 *The Traffic Detection Model*

The traffic detection model is shown in Fig. 2, and the preprocessor is used to extract traffic features from network data. Considering the common features of traditional IT traffic detection and the effectiveness of detecting attacks in CBTC systems, three basic statistical features are adopted as follows.

1. Average traffic packet count. The average number of packets in traffic over a period of time.
2. Average traffic byte count. The average number of bytes of traffic over a period of time.
3. Average packet size. The average number of bytes per packet in traffic over a period of time.

Since traffic from different protocols has different statistical features, we divide the traffic according to the protocol types of packets in CBTC systems. Due to the periodic communication between devices of CBTC systems, the traffic flow is small and stable. Considering that the flow may increase in the event of attacks, the collection time interval of the traffic should be moderate, which is set as 1 min here.

AR regression algorithm is chosen in the detection engine because it is useful for smooth traffic modeling. With simple calculation and fast speed, AR regression algorithm can meet the real-time requirements of the IDS. AR regression algorithm uses a series of observed values to obtain the predicted value to compare with the observed value at the current moment [8]. If the residual exceeds a certain threshold, it indicates an anomaly.

First, a series of traffic features are converted into a time series $\{X\} = \{X_1, X_2, \ldots, X_N\}$, N is the size of the time window, which is used to smooth the time series, so it should not be large, and its values are 20 here. According to the literature [9], the order p and window size N of AR (p) model satisfy the constraint conditions:

$$0 \leq p \leq 0.1N \tag{1}$$

So, we make p equal to 2. The AR (2) model is as follows:

$$x_t = \psi_1 x_{t-1} + \psi_2 x_{t-2} + \varepsilon_t \tag{2}$$

where ψ_1 and ψ_2 are coefficients, ε_t is a Gaussian noise with a mean of 0 and a variance of σ_ε^2.

The residual of observed and predicted values is used to detect anomalies. The sequence of observed values after zero averaging at time point $\ldots t_{n-1}, t_n, t_{n+1} \ldots$ is predicted by the AR (2) model to obtain the residual series $\{\ldots, e_{t+1}, e_{t+2}, e_{t+3}, \ldots\}$:

$$e_{t+i} = X_{t+i} - \widehat{Y_{t+i}} \tag{3}$$

where $\widehat{Y_{t+i}}$ is the predicted value. The decision function is defined as follows to determine whether the residual sequence is abnormal:

$$W_t(N+1) = \frac{e_{t+N+1}}{\hat{\zeta}} \tag{4}$$

where $\hat{\zeta} = \left[\left(e_{t+1}^2 + e_{t+2}^2 + \ldots + e_{t+N+1}^2\right)/(N+1)\right]^{1/2}$.

When the value of $W_t(N+1)$ is out of the range $\{-(\bar{\zeta} + k\sigma), \bar{\zeta} + k\sigma\}$, it means this traffic is abnormal. $\bar{\zeta}$ and σ are the mean and the standard deviation of the residual sequence. By sampling, $p\%$ of the abnormal flow is sent to the packet detection model with the normal traffic for further detection.

3.3 The Packet Detection Model

As shown in Fig. 3, the packet detection model consists of the preprocessor, the detection engine and the reporter. The preprocessor extracts features from each packet. In order to choose appropriate features to detect attacks, we analyze the communication protocol, data format and payload of the systems and convert the packet header and payload into the canonical form consisting of 20 features, as shown in Table 1. Most of the features are derived from the structure of the packets, while some features are chosen combined with the practical issues.

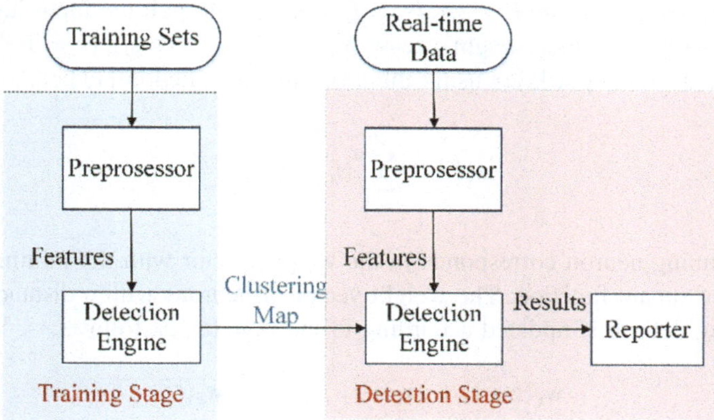

Fig. 3 Packet detection model

Table 1 Packet features

Classes	Features
IP header	time-stamp-diff, ip-headerlen, ip-len, is-home-src-mac, is-home-dest-mac, is-home-src-ip, is-home-dest-ip
TCP header	tcp-src-port, tcp-dest-port, tcp-seq, tcp-ack, tcp-offset
UDP header	udp-src-port, udp-dest-port
ICMP header	icmp-type, icmp-code, icmp-id, icmp-sequence
Payload	payload-type, payload-len

The first issue is to replace the source and destination IP address with is-home-src-ip and is-home-des-ip. In previous researches on IDS [9], source and destination IP addresses are used as features directly. It reduces the flexibility of neural network to detect new attacks and increases false positive rate (FPR) due to many attacks cheating IP addresses. Therefore, we extract the feature by determining whether the source/destination IP address belongs to the local network of the CBTC system. The next issue is to consider the effect of the time correlation between packets. We use the time difference between packets as a feature. The third issue is to select the type and length of the payload as features because different network service has different types and length of payload in the CBTC system.

DBScan is a data mining method that can cluster data with similar features into clusters of arbitrary shape and is used for IDS in the traditional IT field [10]. However, it is not applicable to big data sets and only can detect anomalies offline. Therefore, we use the SOM neural networks to improve DBScan in the detection engine.

First, the feature vector $\boldsymbol{F} = (f_1, f_2, \ldots f_{20})$ is sent through the input layer and is compared with the initial weight vector $\mathbf{w}_k = (w_1, w_2, \ldots w_{20}), (k = 1, 2 \ldots m)$ of m neurons in the output layer using the discriminant function [11]:

$$d(\boldsymbol{F}) = \sum_{i=1}^{20} (f_i - w_{k,i})^2 \tag{5}$$

The winning neuron corresponds to the weight vector with the minimum value of the discriminant function. The weight vector of neurons within distance D from the winning neuron is updated according to the function as follows:

$$\mathbf{w}_k(\mathrm{t}+1) = \mathbf{w}_k(\mathrm{t}) + \eta(t)[\boldsymbol{F} - \mathbf{w}_k(\mathrm{t})] \tag{6}$$

where $\eta(t)$ is the learning rate which generally decreases with time:

$$0 < \eta(t) \leq \eta(t-1) \leq 1 \tag{7}$$

Repeat the above operation to adjust the weight vector until $\eta(t)$ decays below a certain threshold θ. After the adjustment, the weight vector of neurons with similar patterns in the output layer is clustered using DBScan algorithm. Randomly select a weight vector $\mathbf{w}_k$, if there are at least n vectors within distance d from its neighborhood, add these vectors to a cluster. Otherwise, $\mathbf{w}_k$ is marked as a noise point. Do the above operation for other weight vectors in the cluster until the cluster cannot be extended. In order to obtain the next cluster, a random selection is made from the remaining weight vectors to continue the clustering. Since packets with similar features are divided into the same cluster, we label the clusters as the normal cluster, the attack cluster and the abnormal cluster in the training stage. When detecting real-time packets of CBTC systems, the results can be known by the label of the cluster which the winning neuron belongs to.

4 Experimental Results

4.1 Experimental Data Sets and Evaluation Indexes

We use some means of attacks to collect the network data of the CBTC system in the simulation platform. Then, we preprocess these network data by extracting features and converting them into the canonical form. The generated data sets are as shown in Table 2, for training and testing the detection performance of the IDS for CBTC systems. The training data set includes normal packets and four types of attack packets. Since the data tampering attack is difficult to implement in CBTC systems, so we add it into the test data set as a new attack.

There is an imbalance between the number of various types of packets in the training data set collected originally. For example, there are 547,208 packets of DoS

Table 2 Training and test data sets

Classes	Training data set	Training data set (sampling)	Test data set
Normal	183,617	18,361	40,904
U2R	4820	9640	1205
R2L	10,712	21,424	3431
Probing	39,645	39,645	9911
DoS	547,208	54,720	136,802
Data tampering	/	/	573
Amount	786,002	143,610	192,826

attacks in the training set, while there are only 4820 packets of U2R attacks. We sample normal and DoS attack packets by 10% and oversample U2R and R2L attack packets by replicating. In this way, U2R and R2L attacks can get more training in the SOM neural networks, so as to improve the detection performance. At the same time, the total number of packets in the training set is reduced, thus shortening the training time.

We use the accuracy rate, false positive rate (FPR) and false negative rate (FNR) as evaluation indexes to evaluate the detection performance of the IDS. The packet average detection time is used to evaluate the real-time performance of the IDS. The meaning of each index is as follows.

1. Accuracy rate: the ratio of all correctly classified samples to total test set samples.
2. FPR: the ratio of normal samples detected as an attack to the total normal samples.
3. FNR: the ratio of attack samples that have not been correctly detected to the total attack samples.
4. Detection time: the ratio of the total detection time to the total number of packets in the data set.

4.2 Evaluation Results

We evaluate the detection performance of the designed IDS by calculating evaluation indexes and compare it with the traditional DBScan clustering method and the packet detection model to prove the increase in the detection performance and real-time performance of our IDS. The evaluation results are shown in Table 3.

The results show that the packet detection model has the highest accuracy rate and the lowest FRR and FNR. Compared with the DBScan method, its detection performance is significantly improved. The designed IDS also has the accuracy rate as high as 96.94%. In terms of real-time performance, the average detection time of the packet detection model is long. In practical application, if there are a large

Table 3 Evaluation results

	Accuracy (%)	FPR (%)	FNR (%)	Detection time (us/packet)
DBScan	88.16	14.43	11.14	/
The packet detection model	97.49	2.21	2.58	18,300
The traffic detection model	90.35	2.36	10.85	450
The designed IDS	96.94	2.23	4.56	1020

number of attack packets, it is difficult for the packet detection model to detect all packets, resulting in the reduction of detection performance. Combined with the traffic detection model, the detection time is obviously shortened. Since the DBScan method cannot continuously detect packets, its real-time performance is not considered here. Considering the detection performance and real-time performance, the designed IDS combining the traffic detection model and the packet detection model is more suitable for CBTC systems.

5 Conclusions

In this paper, we analyzed the characteristics of network data of CBTC systems and proposed an IDS based on network traffic and packets to detect typical attacks of CBTC systems, such as the denial of service (DoS) and data tempering attacks. Experiments showed that the designed IDS was suitable for CBTC systems for its great detection performance and real-time performance. However, our IDS still had a little classification errors. In the future, we can improve the algorithm used in the IDS to enhance the detection performance.

Acknowledgements This paper was supported by grants from the National Natural Science Foundation of China (No. 61790575, 61603031), Beijing Natural Science Foundation (No. L181004), projects (No. RCS2018K008) and Beijing Laboratory for Urban Mass Transit.

References

1. Yu SY (2017) Research on information security detection technology of urban rail transit train control system
2. Gao CC (2017) Design of intrusion detection system for urban rail transit industrial control system
3. Luo YH (2017) Research on DCS network information security of urban rail transit and implementation of network monitoring system
4. Chen JM (2018) Study on information security risk monitoring approach of rail transit signal system

5. Bu B, Yu FR, Tang T (2014) Performance improvements of communication-based train control (CBTC) systems with unreliable wireless networks. Wirele Netw 20(1):53–71
6. Zhu L, Yu FR, Ning B (2012) Cross-layer handoff Design in mimo-enabled WLANs for communication-based train control (CBTC) systems. IEEE J Sel Areas Commun 30(4): 719–728
7. Bao ZT (2017) Research on active defense of information security risk of train control system
8. Lu W, Ghorbani AA (2003) Network anomaly detection based on wavelet analysis. EURASIP J Adv Signal Process 2009(1):837601
9. Mowla N, Doh I, Chae K (2018) Evolving neural network intrusion detection system for MCPS. In: International conference on advanced communication technology. IEEE
10. Chen Z, Li YF (2011) Anomaly detection based on enhanced DBScan algorithm. Procedia Eng 15(1):178–182
11. Yin H (2008) The self-organising maps: background, theories, extensions and applications. J Jiamusi Educ Inst

Safety Assessment of Bogie System Based on Intuitionistic Fuzzy Set and Complex Network Theory

Shuai Lin, Yanhui Wang and Limin Jia

Abstract The bogie system, which is the critical component of high-speed train (HST), plays an important and irreplaceable role in the bearing, traction and braking of the HST. This paper mainly presents a new method to assess bogie system safety. First, in order to reduce the effect of subjective factors and uncertainty of historical failure data, hesitant interval-valued intuitionistic fuzzy set (HIVIFS) is introduced to reflect failure propagation mechanism and fault consequences of components in bogie system. Second, from the perspective of fault propagation, bogie system safety measurement is put forward to assess system state through integration of the failure consequences of multi fault paths according to hesitant interval-valued intuitionistic fuzzy Choquet integral (HIVIFCI). Results show that the proposed model is capable of (i) achieving a more accurate result for safety assessment of bogie system and (ii) identifying all possibility bogie system's fault propagation paths. The achievements have important theoretical value and practical significance and could provide a basis for making maintenance strategies and reducing economic losses.

Keywords Bogie system · System safety · Fuzzy set · Failure consequence

1 Introduction

Bogie system, which is critical part of HST, plays an important and irreplaceable role in the running, bearing, traction and braking of HST. What is more, its safety directly affects the operational reliability and safety of the HSTs. For example,

S. Lin (✉) · Y. Wang · L. Jia
State Key Laboratory of Rail Traffic Control and Safety, Beijing Jiaotong University, Beijing 100044, China
e-mail: linshuai@bjtu.edu.cn

S. Lin · Y. Wang · L. Jia
Research and Development Center of Transport Industry of Technologies and Equipments of Urban Rail Operation Safety Management, MOT, PRC, Beijing, China

Y. Qin et al. (eds.), *Proceedings of the 4th International Conference on Electrical and Information Technologies for Rail Transportation (EITRT) 2019*, Lecture Notes in Electrical Engineering 639, https://doi.org/10.1007/978-981-15-2866-8_30

a crack in the bogie system of a Shinkansen in Japan was observed (see Fig. 1) [1] in 2017. If a break occurs during the running process, it may result in major accidents such as the derailment of the Shinkansen [2]. It explains that bogie system safety assessment is of great significance to ensure the efficient and safe operation of HST.

Recently, a larger number of methodologies have been put forward to assess system safety for the bogie system, including fault tree analysis (FTA) [1], event tree analysis (ETA) [2], Markov chain [3], hazard and operability study (HAZOP) [4], failure mode and effects analysis (FMEA) [5], Bayesian network (BN) [6], fuzzy set theory [7] and common cause analysis (CCA) [8]. Through the summarization and analysis of the above methods, we find that generally there are two steps to assess system safety. The first step is to construct function structure model to determine the causality relationship among failures, and the second step is to quantify system risk or hazard to evaluate bogie system safety in combination with data set and function structure model.

Existing approaches have yielded some results; however, several problems are still to be resolved, especially when these models are applied to the bogie system. For example, (1) these approaches are more or less impacted by activities of humans or other defects; therefore, the effectiveness of the method in bogie system is limited. (2) The uncertainty of large-scale fault data is not considered into assessing system safety. (3) Several approaches concentrate on integrating multi-indices, which excessively depend on failure data, but ignore the influence of inherent topology that determines the nature of bogie system safety.

In view of the shortcomings of the above methods, a novel method for assessing bogie system safety is presented based on interval-valued fuzzy set and complex network theory.

Fig. 1 Crack in bogie system of the Shinkansen

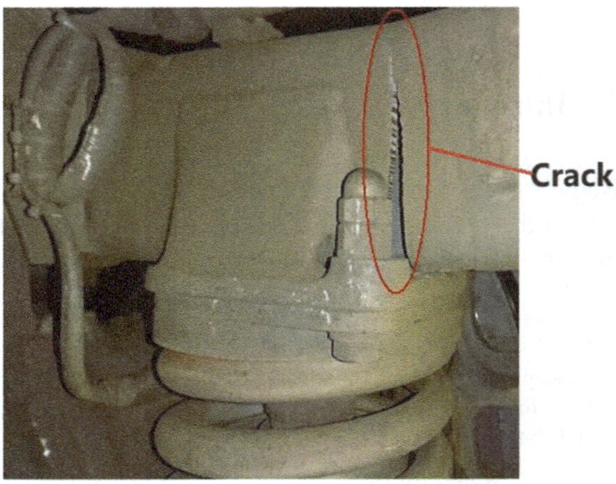

2 Background

2.1 Topological Network Model of Bogie System

Bogie system usually is divided into 33 components according to minimum maintenance unit. Components are connected by electricity, machinery or information relationships. The topological network (see Fig. 2) is constructed [11], where components are abstracted as points/nodes, and their connections are represented as edges/links.

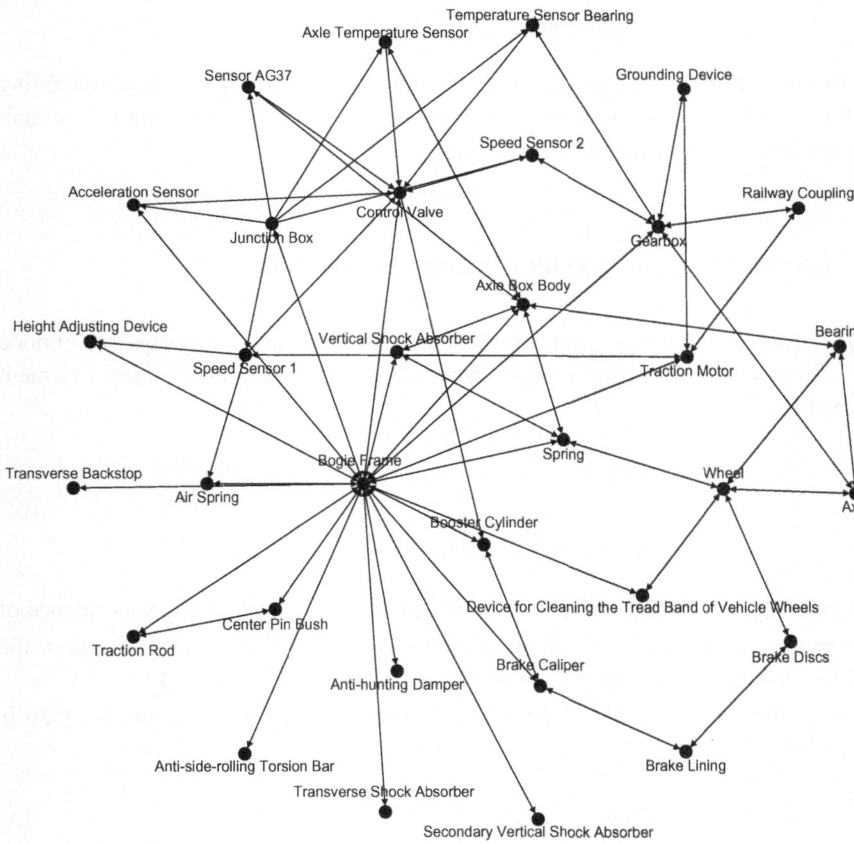

Fig. 2 Topological network of bogie system

2.2 Fault Propagation Model of Bogie System

According to literature [11], the fault propagation model is constructed

$$\begin{aligned}
&M_k = [M(i)]_k = M_{k-1}(i) + \left\{w_p P_k(i) + w_s T_k(i)\right\} \otimes F_k(i) \\
&P_k(i) = [\mathbf{V}, \mathbf{A}((\mathbf{R_{k-1}}(\mathbf{i})\mathbf{p}(\mathbf{E})) * \mathbf{p}(\mathbf{V}))]^T \\
&T_k(i) = [\mathbf{A}(\mathbf{R_k}(\mathbf{i})), \mathbf{I}] \\
&F_k(i) = \max_j \left(FM^j_{D(R_{k-1}(i)) \times D(R_k(j))} \left(fm_{D(R_{k-1}(i)),u} \right) \right) \\
&R_k = \begin{cases} M_k, & k = 0 \\ M_k - M_{k-1}, & k = 1, 2, \ldots \end{cases} \\
&i = 1, \ldots, \|R_{k-1}\|_0, j = 1, \ldots, \|R_k\|_0
\end{aligned} \tag{1}$$

In view of the limited space, Eq. (1) is not introduced in detail here. According to Eq. (1), the set $\tilde{V}_t(n)$ of nodes, which are in the nth path, and the number of fault paths l also can be calculated.

3 Methodology of System Safety Assessment

Due to the uncertainty and subjectivity of fault data, the failure consequence of node v_t is firstly described by hesitant interval-valued intuitionistic fuzzy element (HIVIFE).

$$\begin{aligned}
fc^n_{v_t} &= \left(\left[\mathrm{FCQ}^-_{v_t}, \mathrm{FCQ}^+_{v_t} \right], \left[1 - \mathrm{FCQ}^+_{v_t}, 1 - \mathrm{FCQ}^-_{v_t} \right] \right) \\
&= \left(\left[\gamma^-_{v_t}, \gamma^+_{v_t} \right], \left[\eta^-_{v_t}, \eta^+_{v_t} \right] \right), v_t \in \tilde{V}_t(n)
\end{aligned} \tag{2}$$

where $\mathrm{FCQ}^-_{v_t}$ and $\mathrm{FCQ}^+_{v_t}$ are the slightest and the most serious fault consequence of the node v_t, respectively. To facilitate subsequent illustration and description, the unified denotation of symbols is used to character $\mathrm{FCQ}^-_{v_t}$ and $\mathrm{FCQ}^+_{v_t}$.

Then, the set of HIVIFE for fault consequences of all nodes in the nth path is given by

$$fc^n = \left\{ fc^n_{v_{(1)}}, \cdots, fc^n_{v_{(mn)}} \right\}, n = 1, \cdots, l; v_{(m_n)} \in \tilde{V}_t(n) \tag{3}$$

where $v_{(m_n)}$ is the mth node in the nth path.

Then, the fault consequence's HIVIFE of the nth path is calculated as

$$\begin{aligned}
\mathrm{FCS}_n^{\mathrm{path}} &= (C)\int\left\{\mathrm{CI}_{v_{(1)}}(t)\times fc_{v_{(1)}}^n,\ldots,\mathrm{CI}_{v_{(m_n)}}(t)\times fc_{v_{(m_n)}}^n\right\}\mathrm{d}g_n\\
&=\left\{\left(\begin{array}{l}\left[\prod_{i=1}^{m_n}\left(\mathrm{CI}_{v_{(i)}}(t)\times\gamma_{v_{(i)}}^-\right)^{\mu(E_{v(i)})-\mu(E_{v(i-1)})},\prod_{i=1}^{m_n}\left(\mathrm{CI}_{v_{(i)}}\times\gamma_{v_{(i)}}^+\right)^{\mu(E_{v(i)})-\mu(E_{v(i-1)})}\right],\\
\left[1-\prod_{i=1}^{m_n}\left(1-\mathrm{CI}_{v_{(i)}}\times\gamma_{v_{(i)}}^-\right)^{\mu(E_{v(i)})-\mu(E_{v(i-1)})},1-\prod_{i=1}^{m_n}\left(1-\mathrm{CI}_{v_{(i)}}\times\gamma_{v_{(i)}}^+\right)^{\mu(E_{v(i)})-\mu(E_{v(i-1)})}\right]\end{array}\right)\middle|1\leq i\leq m_n\right\}\\
&=\left\{\left[\theta_{n,\mathrm{path}}^-,\theta_{n,\mathrm{path}}^+\right],\left[\vartheta_{n,\mathrm{path}}^-,\vartheta_{n,\mathrm{path}}^+\right]\middle|1\leq i\leq m_n\right\};n=1,\ldots,l
\end{aligned}\tag{4}$$

where μ is the parameter.

If the bogie system is in operation for a long time, multiple propagation paths may happen at the same time. The failure consequences of paths will determine bogie system safety together. Indeed, these failure paths are not independent of each other. Therefore, fault consequence's HIVIFE of the bogie system by integrating the fault consequences of paths is computed as

$$\begin{aligned}
\mathrm{SFC} &= (C)\int\left\{\mathrm{OP}_1\times\mathrm{FCS}_1^{\mathrm{path}},\ldots,\mathrm{OP}_l\times\mathrm{FCS}_l^{\mathrm{path}}\right\}\mathrm{d}g\\
&=\left\{\left(\begin{array}{l}\left[\prod_{i=1}^{l}\left(\mathrm{OP}_{(i)}\times\theta_{i,\mathrm{path}}^-\right)^{\mu(\alpha_{(i)})-\mu(\alpha_{(i-1)})},\prod_{i=1}^{l}\left(\mathrm{OP}_{(i)}\times\theta_{i,\mathrm{path}}^+\right)^{\mu(\alpha_{(i)})-\mu(\alpha_{(i-1)})}\right],\\
\left[1-\prod_{i=1}^{l}\left(1-\mathrm{OP}_{(i)}\times\vartheta_{i,\mathrm{path}}^-\right)^{\mu(\alpha_{(i)})-\mu(\alpha_{(i-1)})},1-\prod_{i=1}^{l}\left(1-\mathrm{OP}_{(i)}\times\vartheta_{i,\mathrm{path}}^+\right)^{\mu(\alpha_{(i)})-\mu(\alpha_{(i-1)})}\right]\end{array}\right)\right\}\\
&=\left\{\left[\bar{\varphi}_i^-,\bar{\varphi}_i^+\right],\left[\bar{\eta}_i^-,\bar{\eta}_i^+\right]\right\}
\end{aligned}\tag{5}$$

where $\mathrm{OP}_{(i)}$ is the parameter of the ith path and could be written as

$$\mathrm{OP}_{(i)}=\frac{\sum_{j=1}^{N_{G_{\mathrm{path,i}}}}\mathrm{BC}_j/\bar{N}_{G_{\mathrm{path},i}}}{\sum_{j=1}^{N}\mathrm{BC}_j/N}\tag{6}$$

CC_j represents the BC of the node v_j. $G_{\mathrm{path},i}$ is a subgraph, which removes the links and the points in the ith path. $\bar{N}_{G_{\mathrm{path},i}}$ represents the number of nodes in the $G_{\mathrm{path},i}$.

Next, the system safety (SS) measurement is computed as below

$$\begin{aligned}
&\mathrm{SS}=\frac{d(\mathrm{SFC},\alpha^-)}{d(\mathrm{SFC},\alpha^+)+d(\mathrm{SFC},\alpha^-)}\\
&d(\mathrm{SFC},\alpha^+)=\sqrt[q]{\frac{1}{4n(t+1)^q}\sum_{i=1}^{n}\left\{\begin{array}{l}\left|t\left(\bar{\varphi}_i^- -\max\varphi_{\alpha_{ij}}^-\right)-\left(\bar{\eta}_i^- -\min\eta_{\alpha_{ij}}^-\right)\right|^q+\left|t\left(\bar{\eta}_i^- -\min\eta_{\alpha_{ij}}^-\right)-\left(\bar{\varphi}_i^- -\max\varphi_{\alpha_{ij}}^-\right)\right|^q\\
+\left|t\left(\bar{\varphi}_i^+ -\max\varphi_{\alpha_{ij}}^+\right)-\left(\bar{\eta}_i^+ -\min\eta_{\alpha_{ij}}^+\right)\right|^q+\left|t\left(\bar{\varphi}_i^+ -\min\varphi_{\alpha_{ij}}^+\right)-\left(\bar{\eta}_i^+ -\max\eta_{\alpha_{ij}}^+\right)\right|^q\end{array}\right\}}\\
&d(\mathrm{SFC},\alpha^-)=\sqrt[q]{\frac{1}{4n(t+1)^q}\sum_{i=1}^{n}\left\{\begin{array}{l}\left|t\left(\bar{\varphi}_i^- -\min\varphi_{\alpha_{ij}}^-\right)-\left(\bar{\eta}_i^- -\max\eta_{\alpha_{ij}}^-\right)\right|^q+\left|t\left(\bar{\eta}_i^- -\max\eta_{\alpha_{ij}}^-\right)-\left(\bar{\varphi}_i^- -\min\varphi_{\alpha_{ij}}^-\right)\right|^q\\
+\left|t\left(\bar{\varphi}_i^+ -\min\varphi_{\alpha_{ij}}^+\right)-\left(\bar{\eta}_i^+ -\max\eta_{\alpha_{ij}}^+\right)\right|^q+\left|t\left(\bar{\varphi}_i^+ -\max\varphi_{\alpha_{ij}}^+\right)-\left(\bar{\eta}_i^+ -\min\eta_{\alpha_{ij}}^+\right)\right|^q\end{array}\right\}}\\
&t=2,3,4,\ldots;q=1,2,3,4,\ldots
\end{aligned}\tag{7}$$

where $d(\text{SFC}, \alpha), \alpha = \alpha^{+}$ or α^{-} is the distance between SFC and hesitant interval-valued intuitionistic fuzzy ideal solution. If $\alpha = \alpha^{+}$, we called positive-ideal, otherwise negative-ideal.

$$\begin{aligned} \tilde{h}^{+} &= \left\{\left(\left[\max\varphi_{\alpha_{ij}}^{-}, \max\varphi_{\alpha_{ij}}^{+}\right], \left[\min\eta_{\alpha_{ij}}^{-}, \min\eta_{\alpha_{ij}}^{+}\right] \middle| \alpha_{ij} \in B\right), \left(\left[\min\varphi_{\alpha_{ij}}^{-}, \min\varphi_{\alpha_{ij}}^{+}\right], \left[\max\eta_{\alpha_{ij}}^{-}, \max\eta_{\alpha_{ij}}^{+}\right] \middle| \alpha_{ij} \in C\right)\right\} \\ \tilde{h}^{-} &= \left\{\left(\left[\min\varphi_{\alpha_{ij}}^{-}, \min\varphi_{\alpha_{ij}}^{+}\right], \left[\max\eta_{\alpha_{ij}}^{-}, \max\eta_{\alpha_{ij}}^{+}\right] \middle| \alpha_{ij} \in B\right), \left(\left[\max\varphi_{\alpha_{ij}}^{-}, \max\varphi_{\alpha_{ij}}^{+}\right], \left[\min\eta_{\alpha_{ij}}^{-}, \min\eta_{\alpha_{ij}}^{+}\right] \middle| \alpha_{ij} \in C\right)\right\} \end{aligned} \tag{8}$$

where B and C are the parameters.

When SS > Threshold, we consider that the system state is safe; Otherwise, we can deem that the system state is unsafe.

The basic idea of bogie system safety assessment is plotted in Fig. 3.

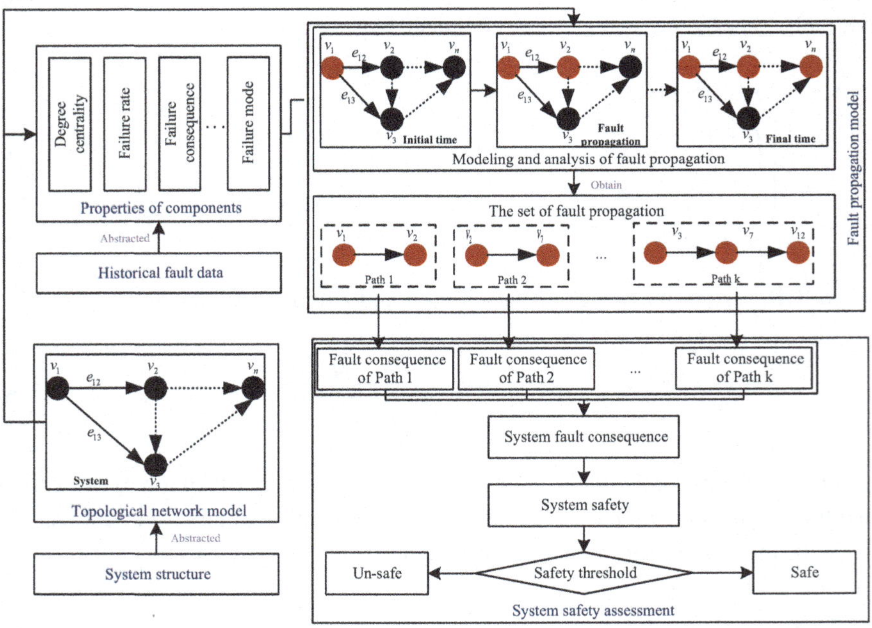

Fig. 3 Idea of bogie system safety assessment

Table 1 Failure consequences

HIVIFE	System State I	System State II	System State III
Failure consequences for nodes	$fv^1_{v_2} = ([0.033, 0.03], [0.967, 0.97])$ $fv^1_{v_3} = ([0.033, 0.045], [0.967, 0.955])$ $fv^1_{v_4} = ([0.033, 0.03], [0.967, 0.97])$	$fv^2_{v_2} = ([0.033, 0.03], [0.967, 0.97])$ $fv^2_{v_5} = ([0.033, 0.03], [0.967, 0.97])$	$fv^1_{v_2} = ([0.033, 0.03], [0.967, 0.97])$ $fv^1_{v_3} = ([0.033, 0.045], [0.967, 0.955])$ $fv^1_{v_4} = ([0.033, 0.03], [0.967, 0.97])$ $fv^2_{v_2} = ([0.033, 0.03], [0.967, 0.97])$ $fv^2_{v_5} = ([0.033, 0.03], [0.967, 0.97])$
Failure consequences for paths	$\mathrm{FCS}^{\mathrm{path}}_{2-3-4} = ([0 \quad 0.3147], [0.3397 \quad 0.3456])$	$\mathrm{FCS}^{\mathrm{path}}_{2-5} = ([0.2290 \quad 0.0899], [0.3383 \quad 0.3428])$	$\mathrm{FCS}^{\mathrm{path}}_{2-3-4} = ([0 \quad 0.3147], [0.3397 \quad 0.3456])$ $\mathrm{FCS}^{\mathrm{path}}_{2-5} = ([0.2290 \quad 0.0899], [0.3383 \quad 0.3428])$
Failure consequences for system	$\mathrm{SFC}^{\mathrm{path}}_{2-3-4} = ([0 \quad 0.3147], [0.3397 \quad 0.3456])$	$\mathrm{SFC}^{\mathrm{path}}_{2-5} = ([0.2290 \quad 0.0899], [0.3383 \quad 0.3428])$	$\mathrm{SFC}^{\mathrm{path}}_{\mathrm{both}} = ([0.2845 \quad 0.1136], [0.2989 \quad 0.3030])$

4 Case Study

Based on Fig. 3, failure consequences' HIVIFEs of the nodes [Eq. (1)], failure paths [Eq. (2)] and the holistic bogie system [Eq. (3)] are calculated and listed in Table 1, respectively.

Consider that the safety threshold is 0.5. On the basis of Table 2 and Eq. (6), Table 2 shows bogie system safety of different system states.

In order to verify the validity of the proposed model, the bogie system safety of statistical analysis for multiple trains of the same batch, which run on the same line, is compared with the proposed method. Figure 4 compares the occurrence probability of different bogie system states according to the statistical analysis of historical failure data. It can be seen that the statistical results are found to coincide well with the results of the proposed method. It also should be noted that the bogie system safety of statistical analysis reflects the average safety level of bogie system based on a huge amount of data set, which contains the failure information of multiple trains of the same batch. Although these trains run on the same line, uncertainties, such as the operating environment, are still unavoidable. Therefore, the statistical results are still slightly different from the modelling results.

Figure 5 shows the evaluation results of bogie system safety by maintenance experts under different system states. When the bogie system is in state III, more than 90% of experts evaluate that bogie system is unsafe. It can be seen that the expert's evaluation results are consistent with those of the model.

Table 2 Bogie system safety

State	Occurrence probability	SS	Safety level
State I	0.6113	0.5001	Safe
State II	0.3684	0.4801	Safe
State III	0.0204	0.1405	Unsafe

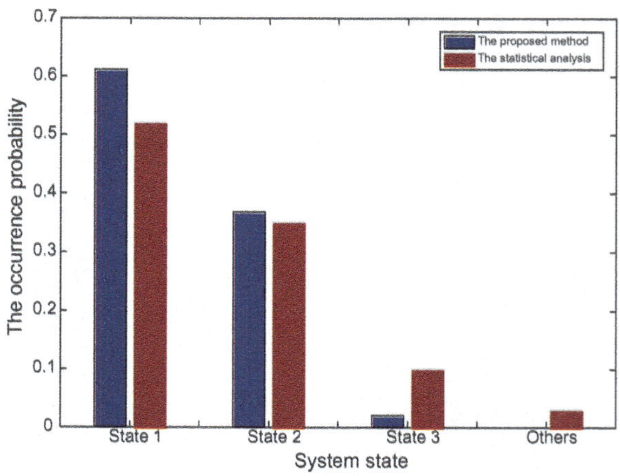

Fig. 4 Occurrence probability of bogie system state

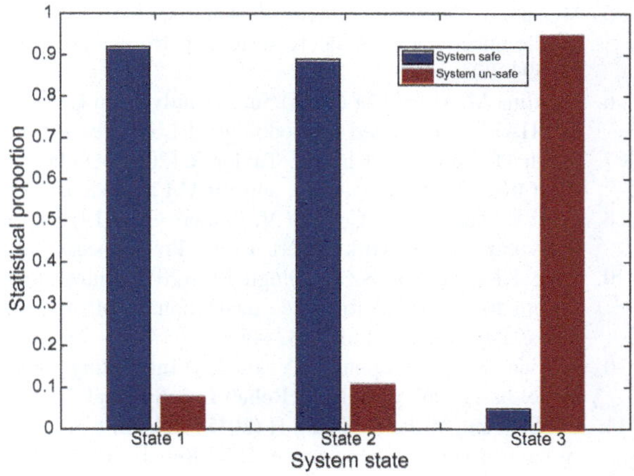

Fig. 5 Bogie system safety by statistical analysis

5 Conclusions

From the point of view of the bogie system fault propagation mechanism, this paper studies the feasibility of application system safety assessment model. From the view of practical application, our method is a universal approach to evaluate system safety and could be applied to the arbitrary complex electromechanical system. In the first place, it can easily find out all possible failure propagation paths and rapidly locate fault nodes, consequently providing valuable references to managers, operators as well as maintenance engineers. Secondly, this model can be revised continuously according to local conditions and the understanding of historical fault data.

Acknowledgements The authors gratefully acknowledge the support from "the Fundamental Research Funds for the Central Universities" (Grant No. 2017JBZ103).

References

1. Lu YH, Xiang PL, Dong P (2018) Analysis of the effects of vibration modes on fatigue damage in high-speed train bogie frames. Eng Fail Anal 89:222–241
2. Lu Y, Zheng H, Zeng J, Chen T, Wu P (2019) Fatigue life reliability evaluation in a high-speed train bogie frame using accelerated life and numerical test. Reliability Engineering & System Safety 188:221–232
3. Rao KD, Gopika V, Rao VS, Kushwaha HS, Verma AK, Srividya A (2009) Dy-namic fault tree analysis using Monte Carlo simulation in probabilistic safety assessment. Reliability Engineering & System Safety 94(4):872–883
4. Ferdous R, Khan F, Sadiq R, Amyotte P, Veitch B (2011) Fault and event tree analyses for process systems risk analysis: uncertainty handling formulations. Risk Anal 31(1):86–107

5. Mechri W, Simon C, Bicking F, Othman KB (2013) Fuzzy multiphase Markov chains to handle uncertainties in safety systems performance assessment. J Loss Prev Process Ind 26 (4):594–604
6. Giardina M, Morale M (2015) Safety study of an LNG regasification plant using an FMECA and HAZOP integrated methodology. J Loss Prev Process Ind 35:35–45
7. Smith D, Veitch B, Khan F, Taylor R (2017) Understanding industrial safety: compar-ing Fault tree, Bayesian network, and FRAM approaches. J Loss Prev Process Ind 45:88–101
8. Zarei E, Khakzad N, Cozzani V, Reniers G (2019) Safety analysis of process systems using fuzzy bayesian network (FBN). J Loss Prev Process Ind 57:7–16
9. Tang KHD, Dawal SZM, Olugu EU (2018) Integrating fuzzy expert system and scoring system for safety performance evaluation of offshore oil and gas platforms in Ma-laysia. J Loss Prev Process Ind 56:32–45
10. Sharvia S, Papadopoulos Y (2015) Integrating model checking with HiP-HOPS in model-based safety analysis. Reliab Eng Syst Saf 135:64–80
11. Lin S, Wang Y, Jia L, Zhang H (2018) Reliability assessment of complex electrome-chanical systems: a network perspective. Qual Reliab Eng Int 34(5):772–790

Correction Scheme and Error Analysis of Compound Fault Location in Direct Power Supply Mode of High-Speed Railway Traction Network

Guosong Lin, Nanling Chen, Wei Quan and Xuanzhi Feng

Abstract Autotransformer (AT) all-paralleling feeding systems have been widely used in high-speed railways in China. At present, short circuit faults in traction networks are mainly divided into three types, such as T-R, F-R and T-F. When permanent faults occur on traction power supply network, all parallel AT feeding systems are changed into direct power supply mode. If the fault is a compound fault, i.e., T-R-F fault, the traditional fault location based on reactance-distance method will have a large error. Formula of measuring reactance in a compound fault was deduced and fault location errors of a compound fault in direct power supply mode were analyzed in this paper. And then, a new correction scheme was proposed under the existing conditions. MATLAB/Simulink simulation results show that there is almost no error in the correction results under different transition resistance and fault distance.

Keywords High-speed railway · Fault location · Compound fault · Reactance method

1 Introduction

With the rapid development of high-speed railway in China, the up and down AT all-paralleling feeding system has been widely used in the traction power supply system. The parallel position is at traction substation, AT station and sectioning post. In AT feeding system, the fault types include T-R(between T-line and rail), F-R(between F-line and rail), T-F(between T-line and F-line). If the AT station and sectioning post are cut off, this feeding system changes into direct power supply mode. If a fault occurs at this power supply mode, fault location system will use reactance-distance method [1] to locate the fault position. In practical application, reactance-distance table method is used. Normally, the fault location method only

G. Lin (✉) · N. Chen · W. Quan · X. Feng
School of Electrical Engineering, Southwest Jiaotong University, Chengdu 611731, China
e-mail: linguosong@swjtu.cn

Y. Qin et al. (eds.), *Proceedings of the 4th International Conference on Electrical and Information Technologies for Rail Transportation (EITRT) 2019*, Lecture Notes in Electrical Engineering 639, https://doi.org/10.1007/978-981-15-2866-8_31

considers the pure T-R, F-R, T-F short circuit faults [2]. When a T-F short circuit fault with grounding which is a compound fault occurs, the previous fault location principle may lead to great error.

The traction network is always in a harsh working environment. Ensuring the normal power supply of the traction network is the prerequisite, which will ensure the safety of the railway transportation. Therefore, when a short circuit fault occurs, it is crucial to determine the fault position quickly and accurately [3–5]. In this paper, the reason for inaccurate fault location for a compound fault by using reactance-distance method is analyzed and a reasonable correction scheme is proposed in this paper.

2 Operation of Protection and Fault Location in a Short Circuit Fault

2.1 Relay Protection Operation Behavior

As shown in Fig. 1, a short circuit fault occurred at 14:36 on November 16, 2018, in Beijing–Guangzhou high-speed railway. The feeder circuit breakers 211 at down line and 212 at up line are at traction substation of tripped for distance protection. After up and down line of power supply arm is powered off, the circuit breakers 271, 272 at AT station and 273, 274 at sectioning post tripped for low voltage protection in succession [6]. Two seconds later, circuit breaker 211 reclosed successfully and circuit breaker 212 failed for post-protection acceleration. Finally, the upside of power supply arm from substation to sectioning station is powered off. The down line was AT power supply mode and up line was direct power supply mode in power supply network. At 14:39, the dispatcher failed to power to up line by switching on circuit breaker 212 in the substation. During the troubleshooting, the maintenance personnel found that the post-porcelain insulator of F wire in circuit breaker 274 GIS high voltage cabinet at sectioning post was broken down. It caused the short circuit fault between T-F and the rail.

Fig. 1 The schematic of the power supply arm in high-speed railway

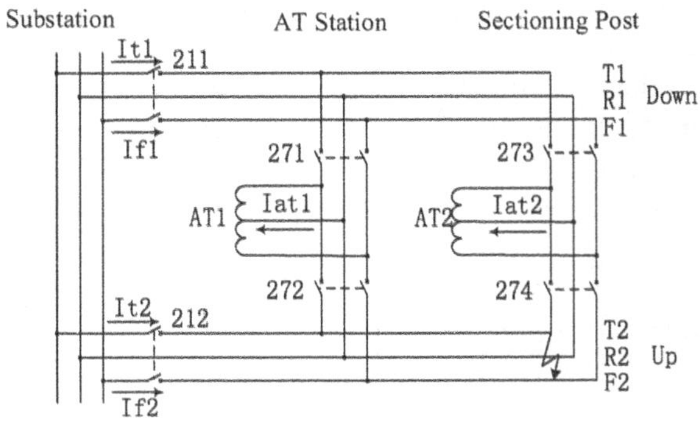

2.2 Analysis of Fault Location

During this fault, the first location result by using AT neutral current ratios method [7] is correct. At the moment of circuit breaker 212 reclosed, the up line traction network is in direct power supply mode because the up line and down line were separated at AT station and sectioning post. Therefore, the fault location measuring device adopted reactance-distance method to locate fault position. The ranging error is 10.69 km.

If a pure T-F short circuit fault occurred, the T-line current and the F-line current of one side in substation should be nearly equal, and the existing judgment of T-F fault location method in direct power supply mode is also based on this characteristic. However, this fault is extremely rare, and it is not a pure T-R, F-R or T-F fault, but a compound fault of T-F short circuit with grounding. Besides, the fault location manufacturers have no consideration for failure judgment and corresponding calculation formula for this kind of compound fault. Therefore, we need to propose a correction scheme for this kind of compound fault, which will solve the problem fundamentally and reduce or eliminate the ranging error.

3 Analysis of Compound Fault Location Error and Correction Scheme

3.1 Analysis of Fault Location Error

The high-speed railway is generally designed for the viaduct and adopted the slag-free track. At the same time, the integrated grounding wire is arranged in parallel with the rails. To simplify the analysis and ignore the track-to-ground leakage, the integrated grounding wire and the rail can be equivalent to one conductor R.

The circuit diagram of compound fault in single-line direct power supply mode of traction network is shown in Fig. 2. R_T, R_R, R_F represent the transition resistances, respectively, between T-line, F-line and R rails, and $z_T, z_R, z_F, z_{TR}, z_{FR}, z_{TF}$ represent unit self-impedance and unit mutual impedance, respectively.

Based on Fig. 2, the following formula can be obtained:

$$\begin{cases} \dot{U}_T = \dot{I}_T z_T l + \dot{I}_F z_{TF} l - \dot{I}_R z_{TR} l + \dot{I}_T R_T + \dot{I}_R (z_R l + R_R) - \dot{I}_T z_{TR} l - \dot{I}_F z_{FR} l \\ \dot{U}_F = \dot{I}_F z_F l + \dot{I}_T z_{TF} l - \dot{I}_R z_{FR} l + \dot{I}_F R_F + \dot{I}_R (z_R l + R_R) - \dot{I}_T z_{TR} l - \dot{I}_F z_{FR} l \\ \dot{I}_R = \dot{I}_T + \dot{I}_F \end{cases} \quad (1)$$

where $\dot{U}_T, \dot{U}_F$ are the voltages of up and down wingding in secondary side, respectively; $\dot{I}_T, \dot{I}_F$ and $\dot{I}_R$ are the currents of T-line, F-line and Rail, respectively; l is the fault distance from substation to fault position.

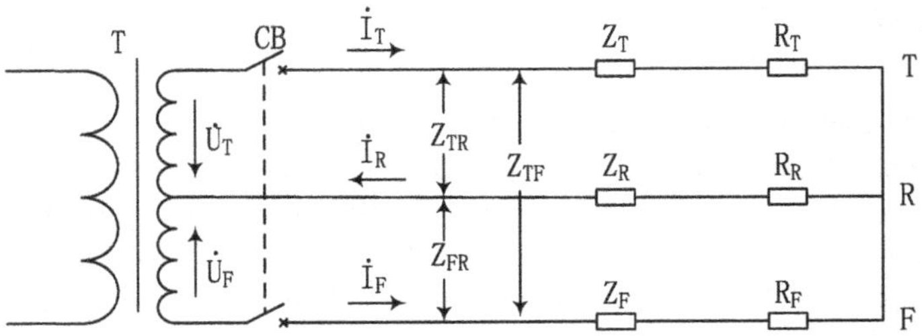

Fig. 2 Circuit diagram of compound fault in single-line direct power supply mode

Equation (2) can be derived from Eq. (1):

$$\begin{aligned}\dot{U}_T - \dot{U}_F &= \dot{I}_T z_T l - \dot{I}_F z_F l + \dot{I}_F z_{TF} l - \dot{I}_T z_{TF} l - \dot{I}_R z_{TR} l + \dot{I}_R z_{FR} l + \dot{I}_T R_T - \dot{I}_F R_F \\ &= \dot{I}_T (z_T l + R_T - z_{TF} l) + \dot{I}_F (z_{TF} l - z_F l - R_F) + \dot{I}_R (z_{FR} l - z_{TR} l) \\ &= \dot{I}_T (z_T l + R_T - z_{TF} l + z_{FR} l - z_{TR} l) - \dot{I}_F (z_F l + R_F - z_{TF} l - z_{FR} l + z_{TR} l)\end{aligned} \quad (2)$$

There are currents in the T, R and F lines when the compound fault occurs, so the fault impedance calculation can be divided into three circumstances:

If $\dot{I}_T > \dot{I}_F$, and $\dot{I}_R > \dot{I}_F$, according to the first line of Eq. (1), Eq. (3) can be derived:

$$\begin{aligned}\frac{\dot{U}_T}{\dot{I}_T} &= z_T l + \frac{\dot{I}_F}{\dot{I}_T} z_{TF} l - \frac{\dot{I}_R}{\dot{I}_T} z_{TR} l + R_T + \frac{\dot{I}_R}{\dot{I}_T}(z_R l + R_R) - z_{TR} l - \frac{\dot{I}_F}{\dot{I}_T} z_{FR} l \\ &= z_T l + z_R l - 2 z_{TR} l + R_T + R_R + \frac{\dot{I}_F}{\dot{I}_T}(z_R l + R_R - z_{TR} l + z_{TF} l - z_{FR} l)\end{aligned} \quad (3)$$

If $\dot{I}_F > \dot{I}_T$, and $\dot{I}_R > \dot{I}_T$, according to the second line of Eq. (1), Eq. (4) can be derived:

$$\begin{aligned}\frac{\dot{U}_F}{\dot{I}_F} &= z_F l + \dot{I}_T z_{TF} l - \dot{I}_R z_{FR} l + R_F + \dot{I}_R (z_R l + R_R) - \dot{I}_T z_{TR} l - z_{FR} l \\ &= z_F l + z_R l - 2 z_{FR} l + R_F + R_R + \frac{\dot{I}_T}{\dot{I}_F}(z_R l + R_R - z_{TR} l + z_{TF} l - z_{FR} l)\end{aligned} \quad (4)$$

If $\dot{I}_T > \dot{I}_R$, and $\dot{I}_F > \dot{I}_R$, according to Eq. (2), Eq. (5) can be derived:

$$\begin{aligned}\frac{\dot{U}_{\mathrm{T}}-\dot{U}_{\mathrm{F}}}{\dot{I}_{\mathrm{T}}} &= zl_{\mathrm{T}}+\frac{\dot{I}_{\mathrm{F}}}{\dot{I}_{\mathrm{T}}}(z_{\mathrm{TF}}l-z_{\mathrm{F}}l-\mathrm{R}_{\mathrm{F}}) \\ &\quad+\frac{\dot{I}_{\mathrm{R}}}{\dot{I}_{\mathrm{T}}}(z_{\mathrm{FR}}\mathrm{l}-z_{\mathrm{TR}}\mathrm{l})+\mathrm{R}_{\mathrm{T}}-z_{\mathrm{TF}}l \\ &= z_{\mathrm{T}}l+z_{\mathrm{F}}l-2z_{\mathrm{TF}}l+\mathrm{R}_{\mathrm{T}}+\mathrm{R}_{\mathrm{F}} \\ &\quad+\frac{\dot{I}_{\mathrm{R}}}{\dot{I}_{\mathrm{T}}}(z_{\mathrm{TF}}l+z_{\mathrm{F}}l-\mathrm{R}_{\mathrm{F}}+z_{\mathrm{FR}}l-z_{\mathrm{TR}}l)\end{aligned} \tag{5}$$

In particular, if $\dot{I}_{\mathrm{F}}=0$, $\dot{I}_{\mathrm{T}}=\dot{I}_{\mathrm{R}}$, Eq. (3), indicates that a pure T-R short circuit fault occurs. If $\dot{I}_{\mathrm{T}}=0$, $\dot{I}_{\mathrm{F}}=\dot{I}_{\mathrm{R}}$, Eq. (4) indicates that a pure F-R short circuit fault occurs. If $\dot{I}_{\mathrm{R}}=0$, $\dot{I}_{\mathrm{T}}=-\dot{I}_{\mathrm{F}}$, Eq. (5) indicates that a pure T-F short circuit fault occurs. Under these circumstances, the fault location by using reactance-distance method for T-R, F-R or T-F faults is not affected by transition resistance. On the contrary, according to Eqs. (3), (4) and (5), the fault location accuracy is affected by transition resistance and the relationship between the currents of T-line and F-line when a compound fault occurs. That is why the fault location results are inaccurate in this fault analysis.

3.2 Correction Scheme of Fault Location

In the direct power supply mode, the T-R, F-R and T-F short circuit faults can be directly measured by the reactance-distance table configured in the protection device or fault location device. But for a compound fault, the application of the calculation formula can be determined by the larger of the three currents of the T-line, the R-line and the F-line. When the current of F-line is the smallest, which is called the compound T-R fault temporarily, Eq. (3) is used to calculate the impedance; when the current of T-line is the smallest, which is called the compound F-R fault temporarily, Eq. (4) is used to calculate the impedance; when the current of R-line is the smallest, which is called the compound T-F fault temporarily, Eq. (5) is used to calculate the impedance.

Taking the compound T-F fault as an example, according to Eq. (5), the following equation can be derived:

$$\begin{aligned}&\tfrac{\dot{U}_{\mathrm{T}}-\dot{U}_{\mathrm{F}}}{\dot{I}_{\mathrm{T}}}-\tfrac{\dot{I}_{\mathrm{R}}}{\dot{I}_{\mathrm{T}}}(z_{\mathrm{TF}}\mathrm{l}-z_{\mathrm{F}}l-\mathrm{R}_{\mathrm{F}}+z_{\mathrm{FR}}l-z_{\mathrm{TR}}l) \\ &=z_{\mathrm{T}}l+z_{\mathrm{F}}l-2z_{\mathrm{TF}}l+\mathrm{R}_{\mathrm{T}}+\mathrm{R}_{\mathrm{F}}\end{aligned} \tag{6}$$

Equation (6) shows that the measured short circuit reactance value in compound fault minus the compensation value $\Delta X=\frac{\dot{I}_{\mathrm{R}}}{\dot{I}_{\mathrm{T}}}(z_{\mathrm{TF}}l-z_{\mathrm{F}}l-\mathrm{R}_{\mathrm{F}}+z_{\mathrm{FR}}l-z_{\mathrm{TR}}l)$ equals the short circuit reactance in the pure T-F fault. And, the distance correction of the compound fault is realized according to the reactance-distance table. It can also be seen that the compensation effect is affected by the transition resistance R_F.

According to Eqs. (3), (4) and (5), when $\dot{I}_T$ and $\dot{I}_F$ are 180° out of phase or $\dot{I}_T$ and $\dot{I}_R$ are 180° out of phase, the calculated reactance is not affected by the transition resistance. Similarly, it can be perfectly corrected by Eq. (6) when metallic fault occurs.

4 Simulation

According to Figs. 2 and 3, a simulation model based on MATLAB/Simulink in single-line direct power supply mode of 2 × 25 kV power supply system is built. The total length of the line is 25 km. The main components are as follows: system power: 220 kV, 50 Hz; the capacity of traction transformer: 40MVA, 220/27.5/27.5 kV; the unit impedance matrix of traction power supply network is shown as Eq. (7) [8].

$$z = \begin{bmatrix} & T & R & F \\ T & 0.145+0.55i & 0.05+0.314i & 0.05+0.413i \\ R & 0.05+0.314i & 0.084+0.407i & 0.05+0.314i \\ F & 0.05+0.413i & 0.05+0.314i & 0.145+0.734i \end{bmatrix} \Omega/\text{km} \quad (7)$$

The magnitude and direction of the currents of T-line, R-line and F-line are affected by three transition resistances. The setting value of the resistance in the distance protection is set according to the maximum load of the traction power supply system. The resistance setting value of distance protection in the feeder relay avoids to cover the load impedance. By changing the value of three transition

Fig. 3 Reactance-distance curve 1 with all transition resistance equal to 0 in compound fault

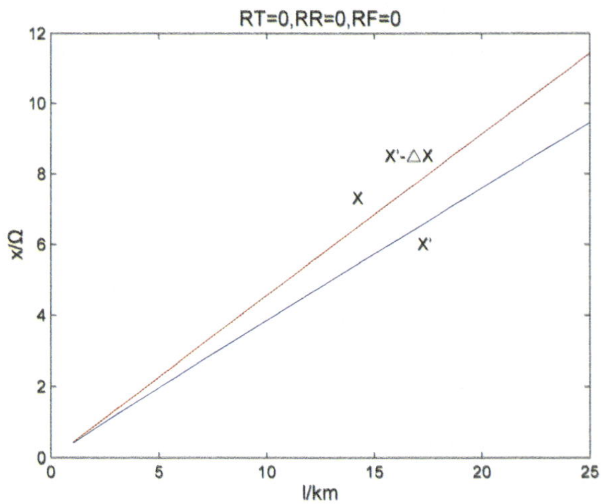

resistances from 0 Ω to resistance setting value and fault distance from 0 to 25 km, many simulations of compound faults are accomplished.

Equation (6) was used as an example of the correction scheme. As is shown in simulation, in the direct power supply mode, when the three transition resistances are all 0, the correction effect based on Eq. (6) is shown in Fig. 3. Under different transition resistances conditions, the compensation effect is shown in Figs. 4 and 5. In these figures, X is the measured reactance of pure T-F fault and X' is the measured reactance without compensation of compound fault. ΔX is the compensation component.

Figures 3, 4 and 5 show that the X' minus the compensation component ΔX will overlap X. These simulations verify that the compensation method is feasible and accurate.

Figure 3 shows when the compound fault occurs and the transition resistance R_T, R_R and R_F are 0 in the single-line direct feeding mode, the reactance-distance curve, which is compensated by the given correction scheme, is perfectly coincident with the measured reactance-distance curve of the T-F fault. It indicates that the derivation calculation is consistent with the simulation result.

Figures 4 and 5 show the compensation effect of compensation component ΔX are perfect, even though transition resistance R_T, R_R and R_F exist. Transition resistance only causes the change of ΔX.

According to statistics of the traction power supply network faults from China Railway Corporation, the value of the short circuit transition resistance is small, nearly zero, in most cases; therefore, this correction scheme proposed in this paper is completely in compliance with the requirements.

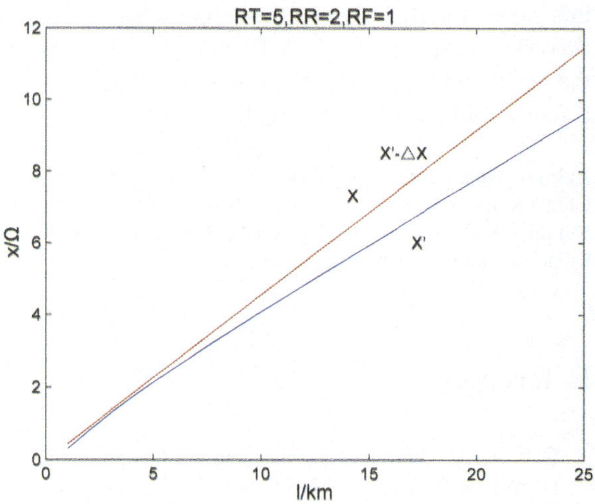

Fig. 4 Reactance-distance curve 2 with compensated transition resistance in compound fault

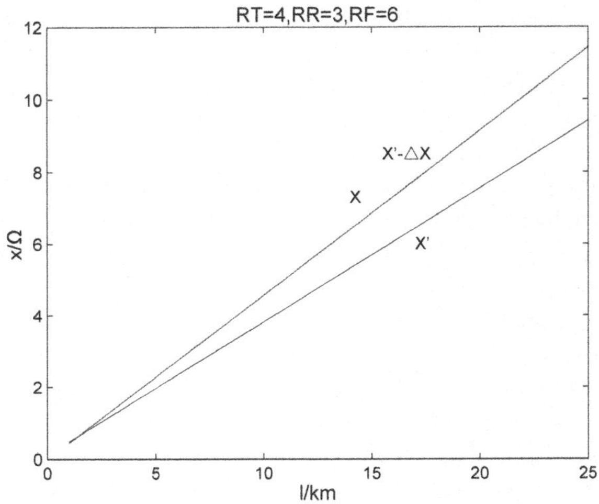

Fig. 5 Reactance-distance curve 3 with compensated transition resistance in compound fault

5 Conclusion

In this paper, large-ranging error in a short circuit fault in high-speed railway is taken as the entry point. The compound fault location error of T-F short circuit with grounding is analyzed comprehensively, and the formula of measured reactance in direct feeding mode is derived, and a new correction scheme for compound fault location is proposed. MATLAB/Simulink simulation results showed that when a T-R-F compound fault occurs on single-line direct power supply arm in high-speed railway, despite the transition resistance is 0 or not, the compensation method in this paper is perfect. In practical application, we just set the reactance-distance table according to the pure T-R, F-R and T-F fault. If a compound fault occurs, based on the measuring reactance and compensation component, the original reactance-distance table also can be used.

Acknowledgements This work is supported by the Research and Development Plan Project of China Railway Corporation (No. 2017J001-F) and Open Project Fund (NEEC-2018-B09) of National Rail Transportation Electrification and Automation Engineering Technology Research Center of Southwest Jiaotong University.

References

1. Guosong L, Qunzhan L, Xiaochuan C (2005) Summary of fault location of electric railway power supply traction network. In: 20,000 km Academic conference of electric railway in China 05
2. Yu Z (2018) Research on fault location method for traction network in special connection mode. Railway Stand Des 62(09):140–144

3. Guosong L, Jinfei M (2019) Fault location scheme for electric railway based on reactance measured at section post. J Railway Eng Soc 234(3):79–83
4. Bidong M, Rui L, Guangning A (2019) Research and correction of fault location method for highspeed railway traction power supply system. Electr Railway 30(01):20–23
5. Cheng X (2016) Judgment on the fault distance accuracy of catenary tripping of Nanjing-Hangzhou high speed railway. Railway Technol Shanghai (02):119–120+122
6. Liebin X (2006) Protection and control scheme of feeder in parallel autotransformer pow-supply systems. Autom Electr Power Syst 22:73–76
7. Bidong M, Rui L (2017) Discussion on defects of reactance approach method in terminal parallel direct supply system. In: China railway society electrification committee 2017 annual meeting and new technology seminar 03
8. Guosong L (2010) Study on novel protection and location schemes for traction power supply system. Southwest Jiaotong University Dr: 113

Ground Optimization Study on the Reconnected CRH3 EMUs Considering Double-Pantograph Off-line Arcing

Li Cao, Ying Wang, Ying Lu, Xiaoqiang Chen and Yanlong Shen

Abstract The grounding system of electric multiple units (EMUs) is an indispensable part of the traction power supply system. In this paper, the ground optimization study is carried out on the reconnected CRH3 EMUs considering double-pantograph off-line arcing. Firstly, each train body is rearranged based on the 8-unit CRH3 train. Secondly, the grounding system of the reconnected CRH3 EMUs is analyzed. Then, the actual electric parameters of China Lanzhou–Xinjiang line are determined to establish a reasonable reconnected CRH3 train body model. Finally, the optimization scheme of reconnected CRH3 train body grounding is proposed considering the actual double-pantograph arcing condition.

Keywords Reconnected electric multiple units (EMUs) · Double-pantograph off-line arcing · Train body · Ground optimization

1 Introduction

With the rapid development of high-speed railway in China, when an electric multiple units (EMUs) train cannot meet the transport requirements, it will make the reconnected EMUs running again to increase the transport capacity [1]. However, the operation condition of the 16-units train body of the reconnected EMUs is more complex. At the same time, the running double-pantograph of the reconnected EMUs flow quality will become unstable, and it will also bring about the more frequent pantograph–catenary off-line arcing [2]. The reliable grounding of running train bodies [3] is related to the onboard train equipment operation performance, so it is necessary to analyze in depth the reconnected CRH3 EMUs ground issue considering double-pantograph off-line arcing.

L. Cao · Y. Wang (✉) · Y. Lu · X. Chen · Y. Shen
School of Automation & Electrical Engineering, Rail Transit Electrical Automation Engineering Laboratory of Gansu Province, Lanzhou Jiaotong University, No. 88 West Anning Road, Lanzhou, China
e-mail: wangying01@mail.lzjtu.cn

Y. Qin et al. (eds.), *Proceedings of the 4th International Conference on Electrical and Information Technologies for Rail Transportation (EITRT) 2019*, Lecture Notes in Electrical Engineering 639, https://doi.org/10.1007/978-981-15-2866-8_32

The causes of appearing train body overvoltage about Japanese ac locomotive are analyzed in [4]. The propagation characteristics of the bow lifting moment affected by the overvoltage of the pantograph head and the overvoltage of the car body were studied in [5]. The generation mechanism of operating overvoltage is studied, and the characteristics of overvoltage amplitude range, duration, propagation rule and so on are analyzed in [6]. The overvoltage characteristics of vehicle body surge, and the influence of grounding resistance was studied in [7, 8]. In [9], the influence of different grounding modes on the surge overvoltage of the rising bow based on the grounding mode of a certain type of EMUs was studied.

Generally, many researches on the train body voltage are implemented, but the study object is mainly limited to the ordinary EMUs, and the analysis on a large number of reconnected EMUs has not been carried out. In this paper, based on the reconnected CRH3 EMUs of China Lanzhou–Xinjiang line, the optimization of the reconnected EMUs body grounding technology is done. Firstly, the grounding model of reconnected CRH3 EMUs under normal working conditions is established. Secondly, the working grounding and protective grounding mechanisms are analyzed in detail. Thirdly, on the basis of normal working conditions, the reconnected CRH3 EMUs grounding model is considered when the double-pantograph is off-line, and the laws of the changes of train body voltage with the train body grounding parameters are compared and studied. Finally, the optimization method of train body grounding for reconnected CRH3 EMUs considering the double-pantograph off-line is given.

2 Train Body Ground Modeling

Train body and grounding diagram of reconnected CRH3 EMUs are depicted in Fig. 1. In Fig. 1, the reconnected CRH3 EMUs have 16 train bodies, which are composed of two groups, and each group has a total of 8 train bodies. The pantograph is located on the roof of locomotive no. 2, 7, 10 and locomotive no. 15. In actual operation, each group usually adopts the single-pantograph to receive flow. The pantograph is generally located on the body of locomotive no. 2 and locomotive no. 10. In Fig. 1, when the reconnected CRH3 EMUs are in normal operation, the double-pantograph takes the flow from the contact line. There are two main kinds of train body ground methods, i.e., working ground and protection ground. Ground layout of reconnected CRH3 train body is depicted in Fig. 1. On the primary side of the onboard transformer, the traction current is brushed into the ground through the EMUs working ground, axle ground terminal box and ground carbon. In addition to working grounding, protective grounding is provided on some train bodies. In addition, the high-voltage cable on the roof can also induce induction voltage to the train body, which will also cause the formation of the reconnected CRH3 EMUs train body grounding circulation. In the operation process of reconnected CRH3 EMUs, due to the influence of pantograph–catenary vibration, when the double-pantograph located in train body 2 and train body 10

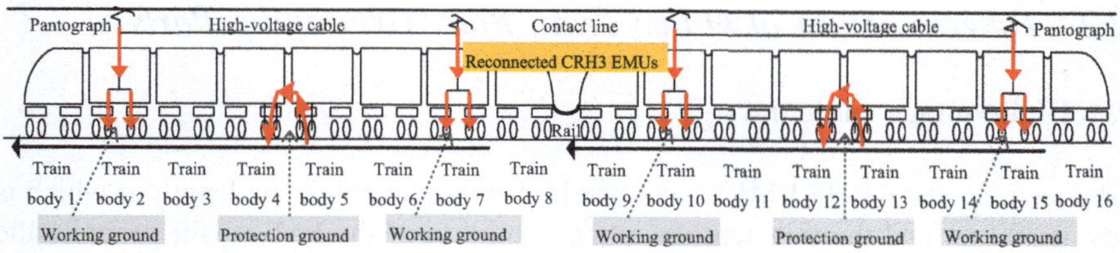

Fig. 1 Train body ground diagram of reconnected CRH3 EMUs

slides along the contact line to take the flow, off-line arcing phenomenon often occurs. Due to the reconnection of two CRH3 trains, it is more complicated to solve the problems of train body circulation and arcing discharge. Therefore, it is necessary to optimize the reconnected CRH3 EMUs ground layout. In order to analyze the reconnected CRH3 EMUs train body overvoltage under the pantograph arcing, corresponding coupled model should be established.

As shown in Fig. 1, it is necessary to establish the corresponding CRH3 EMUs reconnection train body ground model. The model mainly consists of the pantograph-arcing part, the high-voltage cable of the EMUs roof, the body of the EMUs, the working ground and the protective grounding system of EMUs.

2.1 *Determination of Pantograph Off-Line Arcing*

In order to study the electrical characteristics of train body caused by pantograph arcing, we have established and extended a suitable model of Habedank pantograph arc [3]. In view of the fact that the electrical characteristics of pantograph arc have both the phenomenon of arc current crossing the zero point and the characteristics of large current, it is pointed out that the pantograph arc can be simulated by the black box arc model combined by Mayr and Cassie in series [2]. The extended model can be expressed as:

$$\begin{cases} \frac{1}{g} = \frac{1}{g_m} + \frac{1}{g_c} \\ \frac{dg_m}{dt} = \frac{1}{\tau_0 g^{\alpha}} \left[\frac{i^2}{kg^{\beta}(1.535 \times 10^{-4} v^2 - 0.0505v + 5.842)} - g_m \right] \\ \frac{dg_c}{dt} = \frac{1}{\tau_0 g^{\alpha}} \left[\frac{i^2}{(2.3025 \times 10^{-3} v^2 - 0.7575v + 87.63)^2 g_c} - g_c \right] \end{cases} \quad (1)$$

In formula (1), g is the instantaneous arc conductance; i is the arc current; g_m and g_c are instantaneous conductance of Mayr and Cassie parts in formula (1); α is a constant; τ_0 is the initial time of the pantograph arcing; v is the actual train speed; β is the arc-related parameter. Under the high speed condition, the parameter settings are as follows: v = 200 km/h, k = 2000 W·Ω·cm-1, β = 0.9, τ_0 = 1.5 μs, α = 0.17 and $g_m = g_c = g_0 = 0.8$ s.

2.2 Determination of Other Train Body Grounding Parts

(1) **High-voltage cable part**

The reconnected CRH3 EMUs have two high-voltage cables, the length of which is about 200 m, and the transient operation condition of the EMUs will generate the train body overvoltage. When the voltage frequency reaches a range of several megahertz, the phase difference caused by cable length is extremely large. Considering the symmetry, roof cable of train bodies 3 and 14 is the same and so on. So, the roof cable needs to adopt a distributed parameter model, and the specific values of inductance, resistance and capacitance are shown in the following Table 1.

(2) **Train body part**

Figure 2 is the schematic diagram of the train body, which mainly includes three parts: connecting line impedance between train bodies, car body impedance, axle-to-rail impedance. The connection resistance of the train body is fixed, and two train bodies are connected in series.

Table 1 Reconnected Lanzhou–Xinjiang CRH3 EMUs train body parameters

Train body parts	Resistance value	Inductance value (mH)	Capacitance value (uF)
Roof cable of train bodies 3 and14	0.278 mΩ	0.0026	0.0127
Roof cable of train bodies 4 and 13	0.348 mΩ	0.0032	0.008
Roof cable of train bodies 5 and12	0.348 mΩ	0.0032	0.008
Roof cable of train bodies 6 and11	0.348 mΩ	0.0032	0.008
Roof cable of train bodies 7 and10	0.625 mΩ	0.0058	0.0184
Connection between train bodies	6.4 mΩ	0	0
Train body	0.47 mΩ	0.23	0.0002
Part between axil to train bottom	1 mΩ	0	0
Rail	0.01 Ω	0.0064	0.0009
Part between axil and rail	1 mΩ	0	0
Grounding brush	0.01 Ω	0	0
Traction line	0.165 Ω	10.8	0.0184

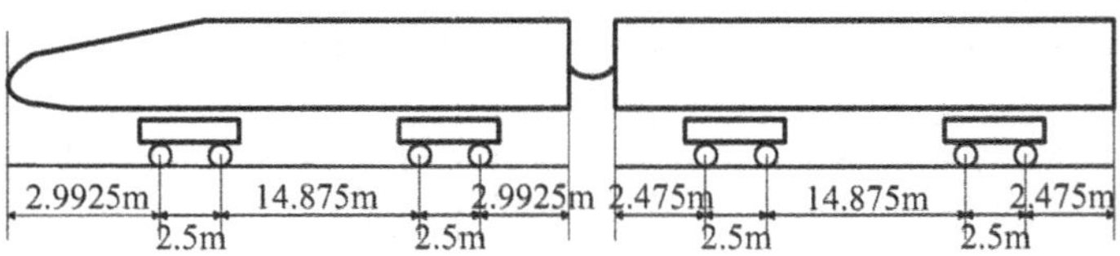

Fig. 2 Schematic diagram of train body

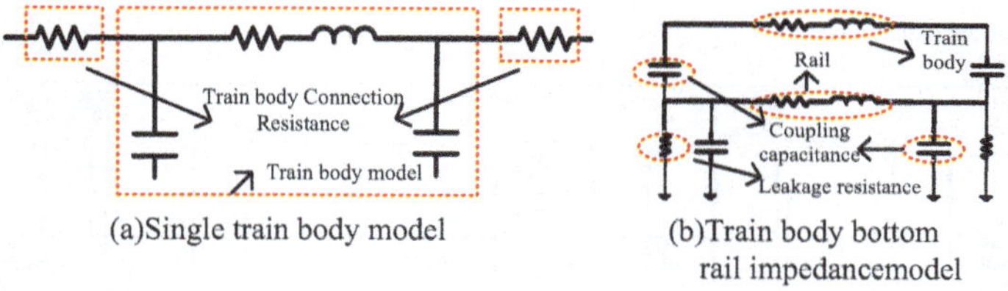

Fig. 3 Train body part model

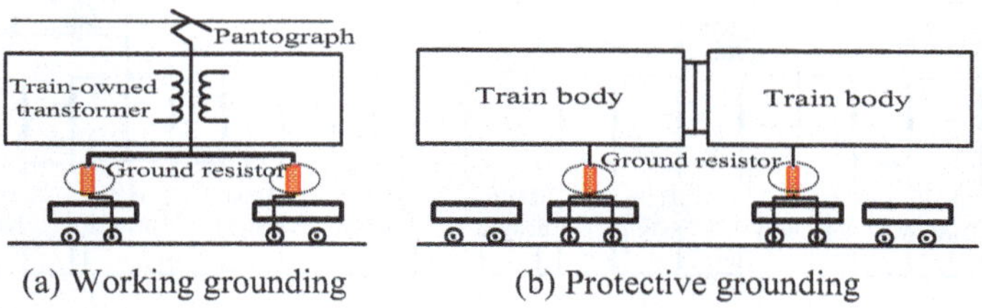

Fig. 4 Schematic diagram of train body grounding

The train body model also adopts the distributed parameter model, which can be divided into multiple parts in series. The single train body model is shown in Fig. 3a. The inductance of train body includes self-inductance and external inductance. The effect of capacitance should be considered when the pantographs are off-line. In addition, considering the connection between train body and rail, the rail impedance between each connection point can be taken into account in the model of car body, as shown in Fig. 3b.

(3) **Train body grounding part**

Figure 4 is the grounding diagram of reconnected CRH3 EMUs, including working grounding and protective grounding.

In Fig. 4a, the working grounding is connected to the rail by a ground carbon brush at the train shaft. For the reconnected CRH3 EMUs, its working grounding is set at train bodies 2, 7, 10 and 15. The protective grounding design in Fig. 4b is installed on the bogie axle 1 and axle 2 of two adjacent trailers, that is, the intermediate train is selected to install protective grounding. For the reconnected CRH3 EMUs, its protective grounding can also refer to the form or grounding layout of 4, 5, 12 and 13 train bodies.

Based on the reconnected CRH3 EMUs operated by Lanzhou–Xinjiang line as the object, the electrical parameters of each aforementioned model part are obtained through the analysis, and the calculations are as shown in Table 1.

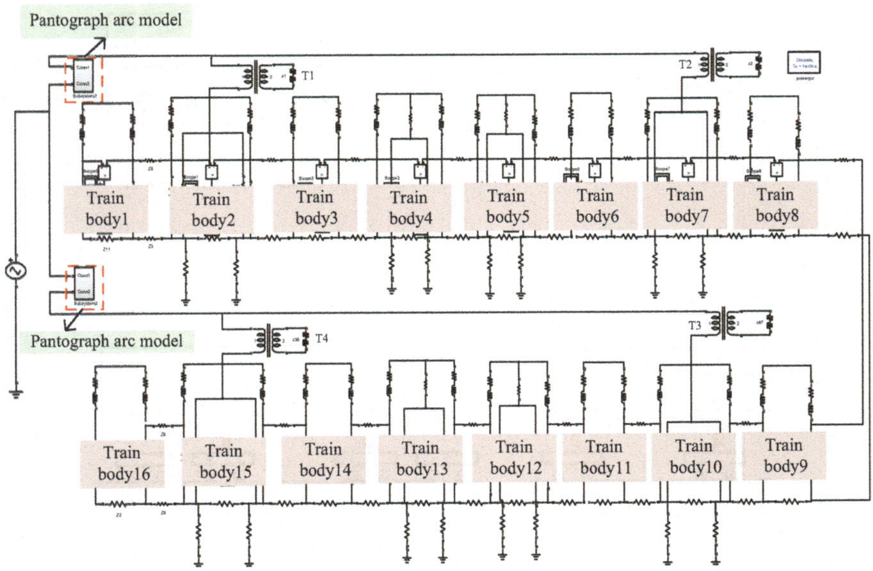

Fig. 5 Model of reconnected train body grounding considering double-pantograph off-line

2.3 *Establishment of Reconnected Train Body Grounding Model*

As depicted in Fig. 5, based on the former analysis and calculation, the reconnected CRH3 train body grounding model considering double-pantograph arcing can be gained by MATLAB/Simulink platform. In Fig. 5, the reconnected EMUs adopt autotransformer supply system of electric traction, the transformer capacity is 3060kVA, the original auxiliary side voltage ratio is 25 kV:1.5 kV, the ac voltage source is used to simulate the contact network voltage and the value is set to 25 kV. In Fig. 5, T1, T2, T3 and T4, respectively, represent the four traction transformers. The layout is the same as that of the reconnected CRH3 EMUs, and the double-pantograph off-line arcing model is set at the pantograph of train body 2 and train body 10, respectively. The body impedance is evenly distributed on both sides of the body to facilitate the connection between the protective ground and the body, and the connection between each body is through the connection resistance. Working ground is set at 2, 7, 10 and 15 train bodies, and protective grounds are set at 4, 5, 12 and 13 train bodies.

3 Analysis of Reconnected CRH3 Train Body Voltage

As shown in Fig. 6, when reconnected CRH3 EMUs run normally, the train body potential is relatively flat and the amplitude is low. As the off-line arc occurs at different positions of the train roof, the potential of the train body rises greatly.

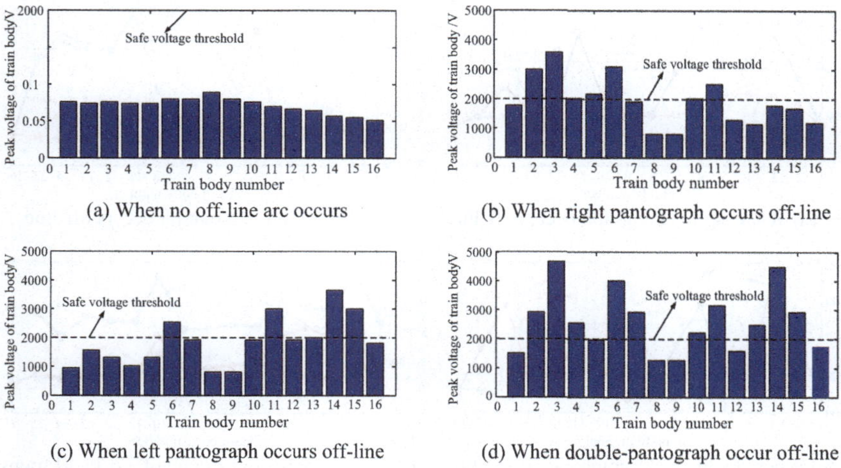

Fig. 6 Model of reconnected train body grounding considering double-pantograph off-line

With off-line arcs occurring at the left pantograph, the voltages of train 6, 11, 14 and 15 exceed 2 kV and the voltage of the left train rises more obviously. When the off-line arc both occur at the left and right pantograph simultaneously, the train body voltage is about 1.5 times of left or right pantograph where off-line arc occurs alone.

4 Ground Optimization Measures on Reconnected CRH3 EMUs

Figure 7 demonstrates the different ground optimization measures on the reconnected CRH3 EMUs considering the double-pantograph off-line arcing. Figure 7a shows the comparison of peak voltages between each train body before optimization and after setting protective grounding (3, 6 trains and 11 and 14 trains) when off-line arc occurs at the left pantograph. Figure 7b shows the comparison of peak voltages between each train body before optimization and after setting protective grounding (3, 6 trains and 11 and 14 trains) when off-line arc occurs at the right pantograph. Figure 7c shows the comparison of waveforms for peak voltages between each train body before optimization and after setting protective grounding (3, 6 trains and 11 and 14 trains) when off-line arc both occur at the left and right pantograph. Figure 7d shows the comparison of the peak voltages for each train body when the left or right pantograph respectively generates off-line arcs and the peak voltage of each train body when the two pantographs simultaneously generate

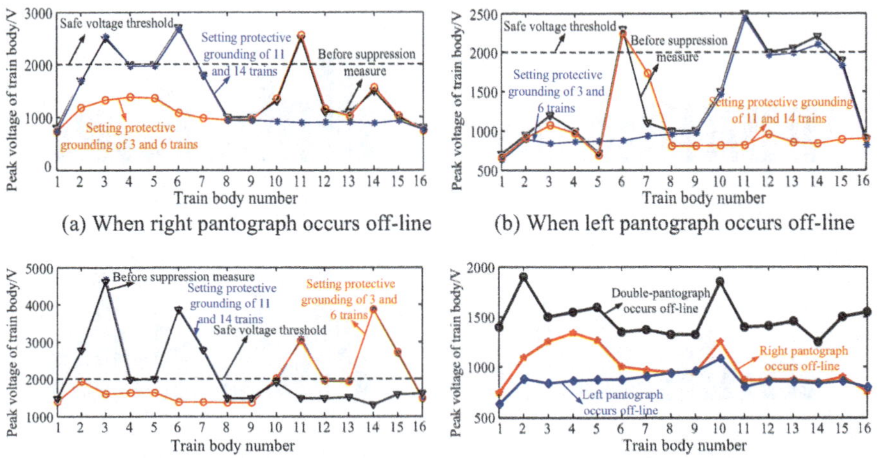

(a) When right pantograph occurs off-line (b) When left pantograph occurs off-line

(c) When double-pantograph occurs off-line (d) Setting protective gound of 3,6,11,14 trains

Fig. 7 Different ground suppression measures on the reconnected CRH3 EMUs

off-line arcs after setting the protective grounding at 3, 6, 11 and 14 trains. After optimization, according to the above analysis, the voltage of each train body is significantly decreased, and the peak voltage of each train body is relatively flat, and the potential difference between train bodies is obviously reduced.

5 Conclusions

In this paper, the reconnected CRH3 train body model is completely established in China Lanzhou–Xinjiang line, and the double-pantograph off-line arcing of train roof is taken into account in the model. Firstly, the grounding model of the reconnected CRH3 EMUs train is established, and the working grounding and protective grounding mechanisms of the train body under the condition of reconnection are analyzed. Secondly, under normal working conditions, the train body grounding model is added when the double-pantograph off-line arcing occurs. The variation rules of the train body voltage with the train body grounding parameters are studied. Finally, the optimal reconnection CRH3 EMUs grounding optimization scheme is obtained.

Acknowledgements This work was supported by the Natural Science Foundation of Gansu Province Science and Technology Department (18JR3RA111), the National Natural Science Foundation of China (51767013), the Science and Technology Research and Development Plan of China Railway Corporation (2017J012-A), the Natural Science Foundation of Gansu Province Education Department (2017A-020) and the Program for Excellent Team of Scientific Research in Lanzhou Jiaotong University (201701).

References

1. Yang L, Changxian L, Long C (2018) Time-delay prediction and compensation of EMU coupling network control system. J Railway Sci Eng 15(12):3044–3049
2. Ying W, Zhigang L, Xiuqing M et al (2016) An extended Habedank's equation-based EMTP model of pantograph arcing considering pantograph-catenary interactions and train speeds. IEEE Trans Power Delivery 31(3):1186–1194
3. Ying W, Fengyang G, Xiuqing M et al (2017) Analysis of EMUs vehicle body voltage caused by pantograph-catenary off-line arc. In: EITRT 2017, lecture notes in electrical engineering, 20–22 Oct, Changsha, China
4. Hatsukade S, Fujino K, Kanno S (2013) Suppressing over-voltage on carbodies of a high speed EMU. In: Proceeding of 2013 15th European conference on power electronics and applications, Lille, France, 2–6 Sept 2013, pp 1–9
5. Baojiang C, Yongbao S, Wenhan X et al (2018) Simulation analysis of surge voltage of rising pantograph for high-speed electric multiple unit based on PSCAD. High Volt Eng 44(8):2746–2753 (in Chinese)
6. Yong D (2016) Analysis of optimal body grounding technology in case of electric multiple unit passing neutral section device. Proc CSU-EPSA 28(5):1–8 (in Chinese)
7. Huiqian D, Liquan S (2014) Design of protective grounding for CRH380B(TX) EMU Train. Railway Locomot Car 34(4):51–54 (in Chinese)
8. Ke H, Zhigang L, Ying W et al (2016) Analysis on railcar's body over-voltage distributional characteristics considering operating conditions of high-speed railway station. J China Railway Soc 38(9):38–45 (in Chinese)
9. Shi X, Yinghong W, Jinbao Z et al (2013) Impedance characteristics of EMU MR-139 earthing resistor. J Beijing Jiaotong Univ 37(6):39–44 (in Chinese)

An Intrusion Detection Method for Railway Based on Fast Feature Extraction and Matching of UAV Camera

Ling Guan, Xiaofeng Li and Limin Jia

Abstract With the rapid increase of railway mileage, especially in complex geomorphic environments, it is necessary to use unmanned aerial vehicle (UAV) to automated monitor railway environmental security. This paper presents a railway intrusion detection method based on fast feature extraction and matching. Firstly, a low-degree polynomial detector (ALP) feature-based template database indexed by geolocation is established. Secondly, ALP descriptors are extracted in the region of interest (ROI) from the detection system for the sequences of railway images by the onboard camera of the UAV captured. Finally, ALP descriptors are matched between those from the real-time images and those from the feature database in the same geolocation and check whether there is an intrusion according to the matching ratio. The experiment results show that the proposed method can detect the invader effectively due to the obvious decrease of matching ratio and the successful detection ratio can up to 96%.

Keywords Railway intrusion detection · Feature template database · Track detection · ALP descriptor · Feature matching

1 Introduction

With the rapid development of railway transportation, especially on the complex landform area, it is especially important for railway intrusion detection to guarantee safe operation. Traditional detection systems, which arranging the monitor cameras or using fences to cover along the track line, are costly and require regular main-

L. Guan (✉) · X. Li
School of Traffic and Transportation, Beijing Jiaotong University,
Beijing 100044, China
e-mail: 18114020@bjtu.edu.cn

L. Jia
State Key Laboratory of Rail Traffic Control and Safety, School of Traffic
and Transportation, Beijing Jiaotong University, Beijing 100044, China

Y. Qin et al. (eds.), *Proceedings of the 4th International Conference on Electrical and Information Technologies for Rail Transportation (EITRT) 2019*, Lecture Notes in Electrical Engineering 639, https://doi.org/10.1007/978-981-15-2866-8_33

tenance [1]. Some researchers studied to use intelligent computer vision method to automate detect intrusion in order to reduce the hidden danger caused by staff's negligence [2]. Other researchers tried to use train mounted forward-viewing camera to detect intrusion [3, 4]. However, these methods cannot perform accurately.

In recent years, UAVs are rapidly expanding worldwide and are used in various areas. Some railroads are beginning to use UAVs to inspect track conditions [5]. The use of UAVs can offer direct benefits for routine inspection activities compared to tradition detection methods, such as early detecting obstacles on the track, improving work efficiency and decreasing the costs [6]. Therefore, it is necessary to research the intelligent detection methods based on UAVs to achieve the routine automated inspection process.

This paper presents a fast visual feature extraction and matching method based on ALP descriptors to detect railway intrusion for UAVs, which only employ the monocular camera, embedded computer platform and the GPS system.

2 Design of Intrusion Detection Algorithm

The UAV-based intrusion detection system is shown in Fig. 1. Actually, the UAV can fly 20–30 min, and the flight altitude is about 15 m. The length of the detection region is about 15–20 m which depends on the viewing angle and lens of the onboard camera. The UAV can fly automatically along the railway track according to the track location in the images and geolocations to achieve the beyond visual line of sight operations (BVLOS) for long distance surveying, which means that it does not require continuous manual control [6].

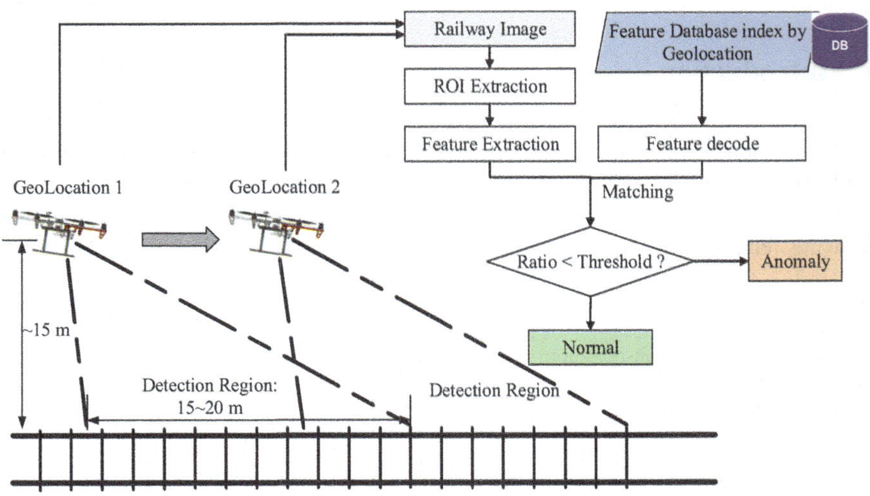

Fig. 1 Railway intrusion detection system

Firstly, this paper establishes a feature template database for a section of the track line. ALP descriptors are extracted and encoded to the feature database from collected railway images under the safe railway environment. Every feature descriptor is indexed by corresponding geolocation.

Secondly, the intrusion detection progress mainly consists of three steps:

1. ROI extraction. In order to remove the interference of complicated background and reduce the range of feature extraction, the specific region around the railway track is segmented.
2. ALP features extraction. This paper extracts ALP features in the ROI. We can get a 32-dimensional ALP feature vector.
3. Features matching. Matching extracted features with these from the database in the same geolocation and computing the matching ratio. This paper sets a threshold to decide whether there is the intrusion on the track.

2.1 ROI Extraction

The gradient direction shall be consistent on the edge of the track due to the special shape characteristics [7]. In this paper, the image orientation field method based on the Sobel operator is used to locate the track. The method is described below.

Firstly, calculating each pixel coordinates (i, j) horizontal gradient $\partial_x (i, j)$ and vertical gradient $\partial_y (i, j)$, respectively, in the image using the 3×3 Sobel gradient operator. Secondly, calculating each pixel horizontal squared gradient $\partial_{s,x} (i, j)$ and vertical squared gradient $\partial_{s,y} (i, j)$. As shown in Eq. 1.

$$\begin{cases} \partial_{s,x}(i,j) = \partial_x^2(i,j) - \partial_y^2(i,j) \\ \partial_{s,y}(i,j) = 2\partial_x(i,j)\partial_y(i,j) \end{cases} \tag{1}$$

Then, calculating the average horizontal squared gradient $\Delta_x (i, j)$ and average vertical squared gradient $\Delta_y (i, j)$ of each fixed slide window $w \times w$ with each pixel as the center:

$$\begin{cases} \Delta_x(i,j) = \sum_w \partial_{s,x}(i,j) \\ \Delta_y(i,j) = \sum_w \partial_{s,y}(i,j) \end{cases} \tag{2}$$

Finally, calculating each slide window average orientation $\varphi(i, j)$:

$$\varphi(i,j) = \begin{cases} \frac{1}{2}\arctan\frac{\Delta_y(i,j)}{\Delta_x(i,j)} + \frac{\pi}{2}, \Delta_x(i,j) < 0 \\ \frac{1}{2}\arctan\frac{\Delta_y(i,j)}{\Delta_x(i,j)} + \pi, \Delta_x(i,j) > 0, \Delta_y(i,j) < 0 \\ \frac{1}{2}\arctan\frac{\Delta_y(i,j)}{\Delta_x(i,j)}, \Delta_x(i,j) > 0, \Delta_y(i,j) > 0 \end{cases} \tag{3}$$

After obtaining the orientation field image, threshold segmentation, morphological corrosion method and Canny edge detection will be applied to locate the railway position. The railway image processing results are shown in Fig. 2.

According to the rules of the railway clearance, the distances between each pixel and its nearest track pixel are calculated, and the areas whose distance is less than or equal to the track width are segmented. The extracted scope of railway clearance is shown in Fig. 3.

2.2 ALP Features' Extraction

ALP feature descriptor is proposed by Telecom Italia in 2013 [8]. Comparing to the classical SIFT descriptors [9], ALP performs better on the localization of keypoints and running speed [10]. Therefore, ALP was adopted by Compact Descriptors for Visual Search (CDVS) in MPEG international standard as the key algorithm for visual search.

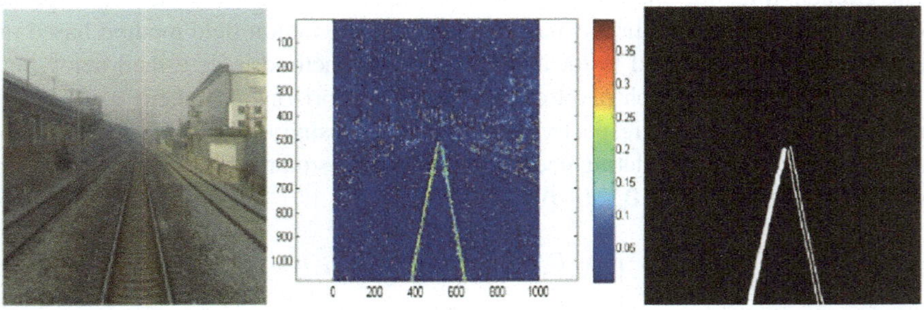

(a) Raw railway image (b) Orientation field image (c) Location of the track

Fig. 2 Extracting the position of the track

Fig. 3 Region of railway clearance

Similar to the SIFT descriptor, the image is processed in a scale-space pyramid which is obtained by Gaussian blur with different scale factors σ in order to calculate interest points. The key of the ALP algorithm is using a polynomial to approximate the scale-space function for each pixel (x, y) in the image:

$$p(x, y, \sigma) = \alpha_3(x, y)\sigma^3 + \alpha_2(x, y)\sigma^2 + \alpha_1(x, y)\sigma + \alpha_0(x, y) \quad (4)$$

where $\alpha_0(x, y) \sim \alpha_3(x, y)$ are obtained by computing weighted sums of the octave images. The result of extracted ALP keypoints is shown in Fig. 4.

After extracting keypoints from image pyramid, local feature selection and compression are applied to produce the accurate feature vector and encode feature descriptor.

2.3 Feature Matching

Ideally, the extracted features should be highly similar to the features from the template database at the same geolocation. Therefore, the matching ratio should be at a very high level.

This paper computes the distance ratio r of each keypoint based on the L2 distance and k-nearest neighbor (kNN) method to determine the keypoint matches. Then, the matched keypoint pair is filtered by a ratio threshold. Finally, the Random Sample Consensus (RANSAC) algorithm is used to reject the wrong matching pairs.

The ALP keypoints matching result is shown in Fig. 5.

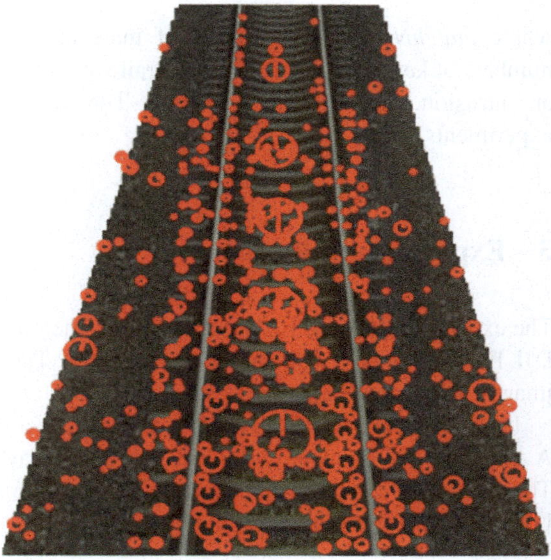

Fig. 4 Results of ALP descriptors in the railway image

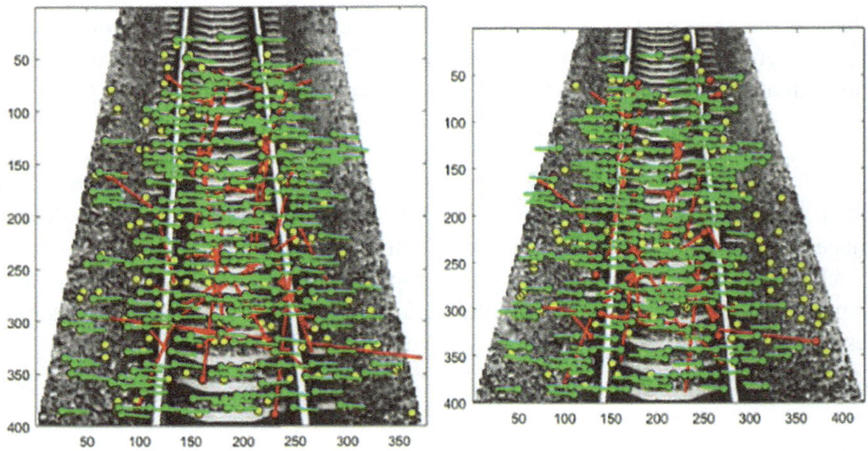

(a) realtime image (b) template image

Fig. 5 Results of ALP keypoints matching after RANSAC (green lines denote the successful matching, while red lines denote the wrong matching)

2.4 *Intrusion Detection*

After the feature matching between the real-time railway image and template feature, this paper sets a threshold value *R* to determine whether there is an intrusion.

Assuming a successful matching ratio:

$$sratio = matchNum/ALLNum \tag{5}$$

where *matchNum* is the number of matching keypoints pairs, *ALLNum* is the numbers of keypoints in the real-time railway image. If $sratio > R$, there should exit an intrusion, otherwise it is normal. The value of *R* is determined by many experiments.

3 Experimental Results

The experimental scene is shown in Fig. 6. The railway images are collected by the DJI F450 equipped with the HERO camera. The resolution of all experimental images is 640 × 480.

In the experiment, the UAV tries to maintain a stable hover in a fixed position. A random environment produced by wind may cause UAV to be unstable. Therefore, this condition can simulate the unstable dynamic flying status when the UAV detects intrusion at specific geolocation.

Fig. 6 Experimental scene

The detection accuracy is defined as *Acc*:

$$Acc = SuccessNum/TotalNum \quad (6)$$

where *SuccessNum* is the number of correctly detected. *TotalNum* is the total number of samples.

There are three video sequences in the experiment (shown in Fig. 7): normal, invader stays on the track and invader crosses the track. The matching ratio curves of each frame are shown in Fig. 8. The value of *R* is 0.45 by the red line as also shown in Fig. 8.

From the experiment results, it is easy to know that when the invader gets into the center of the railway track, the matching rate has an obvious decrease. In summary, the experimental results under various circumstances are shown in Table 1.

From the above experimental results, it can be seen that in the case of normal circumstances, the matching ratio is approximately stable at about 0.55 and is higher than the threshold *r*. For invader stays on or crosses the track, the matching ratio is reduced to about 0.33, and the detection accuracy is above 96% and 98%. The results show the effectiveness of the proposed method.

(a) normal circumstances (b) invader stays on the track (c) invader crosses the track

Fig. 7 Different experimental scenes

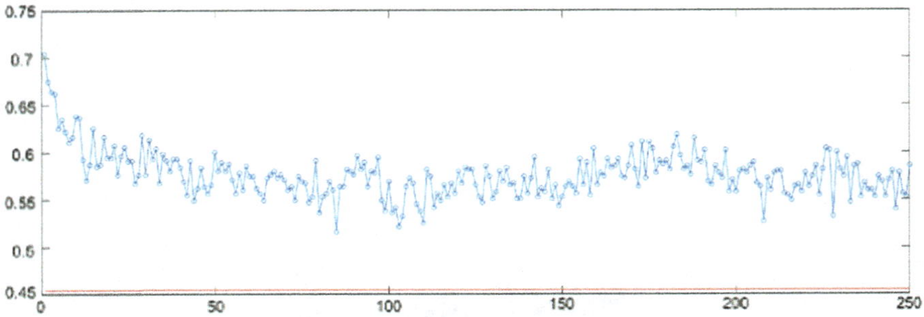

(a) normal circumstances

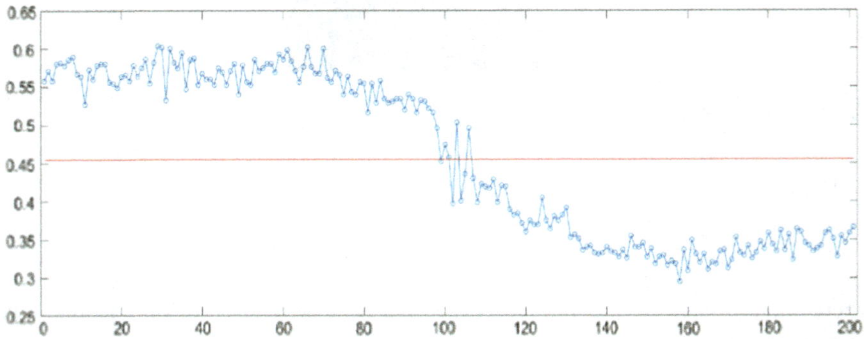

(b) invader stays on the track

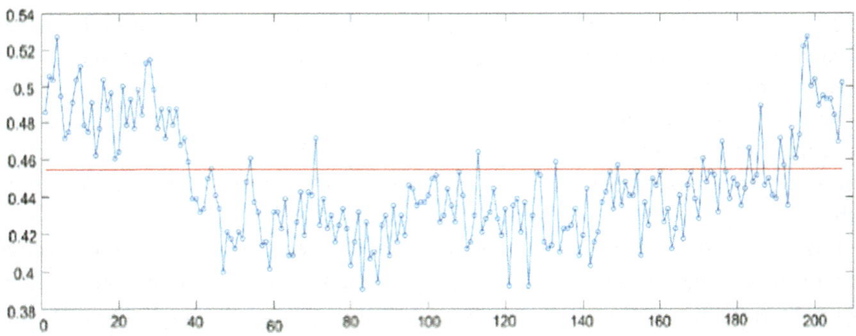

(c) invader crosses the track

Fig. 8 Curve of matching ratio

Table 1 Experiment results of difference invading circumstances

Scenes	Number	Average match ratio	Misclassified number	Average success ratio (%)
Normal	250	0.5766	0	100.0
Single invader-1	100	0.3500	2	98.0
Single invader-2	130	0.3300	5	96.1

4 Conclusion

This paper proposed a new intrusion detection method for the UAV based on ALP descriptors extraction and matching. The proposed method overcomes the defects of the traditional intrusion detection system, which can only be equipped with a fixed location. The method also greatly improves the automation of railway safety system. The experimental results show that the proposed method can identify intrusion effectively.

Acknowledgements This work is supported by the National Key Research and Development Program of China (No. 2016YFB1200402).

References

1. Liu Q, Qin Y et al (2017) Intrusion detection for high-speed railway perimeter obstacle. In: Proceedings of the 3rd international conference on electrical and information technologies for rail transportation (EITRT 2017), pp 465–473
2. Wang Q, Liang X, Liu Y et al (2014) Visual detection method for the invasion of slowly changing foreign matters to railway lines. China Railway Sci 35(3):137–143 (in Chinese)
3. Rodriguez LAF, Uribe LAF, Bonilla JFV (2012) Obstacle detection over rails using hough transform. In: 2012 XVll symposium of image, signal processing, and artificial vision (STSIVA), pp 317–322
4. Mukojima H, Deguchi D, Kawanishi Y et al (2016) Moving camera background-subtraction for obstacle detection on railway rails. In: 2016 IEEE international conference on image processing (ICIP), pp 3967–3971
5. US (2018) Department of transportation. In: Unmanned aircraft system applications in international railroads
6. Flammini F, Naddei R, Pragliola C, Smarra G (2016) Towards automated drone surveillance in railways: state-of-the-art and future directions. In: International conference on advanced concepts for intelligent vision systems, pp 336–348
7. Guan P, Gu X, Zhang L (2007) Automatic railroad detection approach based on image processing. Comput Eng 33(19):207–212 (in Chinese)
8. ISO/IEC JTC1 SC29 WG11 output document N12202. Compact Descriptors for Visual Search: Evaluation Framework. MPEG (2011)
9. Lowe DG (2004) Distinctive image features from scale-invariant keypoints. Int J Comput Vis 60(2):91–110
10. Chandrasekhar V, Chen DM et al (2010) Comparison of local feature descriptors for mobile visual search. In: IEEE international conference on image processing (ICIP), pp 3885–3888

Research on Controllability of Risk Chain Network in Urban Rail Transit System

Yucheng Hao, Yuanzhou Yang, Yanhui Wang and Limin Jia

Abstract In reality, the risk in urban rail transit system usually happens. In this paper, we construct the risk chain network in urban rail transit system, where the different factors are considered in detail. Moreover, in order to describe the characteristic of the risk chain network, the propagation risk is proposed. Based on these, the controllability of risk chain network is calculated in four kinds of systems in urban rail transit system. In addition, we analyze the key node, which has a serious impact on the controllability. Our paper may be useful to control the risk in urban rail transit system and improve the safety of urban rail transit.

Keywords Controllability · Complex network · Urban rail transit system

1 Introduction

Urban rail transit (URT) has become the preferred mode of transportation to solve urban traffic congestion due to its outstanding characteristics and advantages, such as large volume, fast, punctual, comfortable, energy-saving and environmental protection. The operational risk factors of urban rail transit can be divided into

Y. Hao · Y. Wang · L. Jia (✉)
State Key Laboratory of Rail Traffic Control and Safety, Beijing Jiaotong University, Beijing 100044, China
e-mail: 18114021@bjtu.edu.cn

School of Traffic and Transportation, Beijing Jiaotong University, Beijing 100044, China

Y. Hao · Y. Wang · L. Jia
Research and Development Center of Transport Industry of Technologies and Equipments of Urban Rail Operation Safety Management, MOT, PRC, Beijing, China

Y. Yang
Department of Transport Services, Ministry of Transport of the People's Republic of China, Beijing, China

Y. Qin et al. (eds.), *Proceedings of the 4th International Conference on Electrical and Information Technologies for Rail Transportation (EITRT) 2019*, Lecture Notes in Electrical Engineering 639, https://doi.org/10.1007/978-981-15-2866-8_34

human factors, equipment and facilities factors, environmental factors and management factors from the perspective of "man-machine-environment-managemnt". In the operation system of urban rail transit, the occurrence of staff irregularities or equipment failures will cause train delays or forced interruptions, which will affect the normal travel of citizens and also have a greater impact on the entire urban traffic network.

In the aspect of urban rail safety, Li [1] combined the key equipment status, passenger flow status and environmental status and constructed a comprehensive evaluation index system for urban rail transit safety at three different levels: station, line and road network. Jia et al. [2] analyzed the whole system of urban rail transit and its subsystems and on this basis, built a Bayesian network of the urban rail operation safety risk prediction system. It is found that the main factors affect the urban rail operation system under the forward and reverse reasoning. Zhang et al. [3] established a neural network model of subway operation safety based on BP neural network and used Nanjing Metro as an example to carry out risk assessment, which verified the theoretical selection of evaluation indicators and the feasibility of the evaluation model. On the basis of risk theory and system engineering method, Zhou et al. [4] established a quantitative risk analysis model on the operational safety assessment of the existing lines on maps.

In terms of the controllability of complex networks, Liu et al. [5] transformed the controllability of network structure into the problem of finding the minimum set of driving nodes in networks. They found that the minimum number of driving nodes is determined by the degree distribution of networks and other important characteristics. Yuan et al. [6] put forward the theory of strict controllability, which can be used to solve the most of the control problems encountered in the real network, so that we can more clearly understand the internal relationship between network structure and controllability. What is more, the problem concerning controllability was also studied [7, 8].

To assess the controllability of the risk in urban rail transit system, we built the risk chain network in urban rail transit system considering comprehensively the different factors. According to the characteristic of the risk chain network, the propagation risk is proposed, whose value is measured by the relative size of the driver nodes. Finally, we obtain the controllability of risk chain network in urban rail transit system and the key nodes.

2 Methodology

2.1 Controllability Model in Networks

For a linear time dynamics system, the state of the node is denoted by the following equation, which dose not consider the internal relationship:

$$\dot{x}(t) = Ax(t) + Bu(t) \tag{1}$$

where in the case of time is the state of the system with nodes. The adjacency matrix is defined by the interaction strength of nodes in a network which represents the input matrix that is used to describe how the signals connect with nodes in a network. Therefore, the Eq. (1) can be also denoted as follows. Thus, when the system is controllable, the controllability matrix

$$C = (B, AB, A^2B, \ldots, A^{N-1}B) \tag{2}$$

has full rank. This rule is called Kalmans controllability rank condition. This can represent the dimension of the controllable subspace for the system. To fully control the network, it is needed to choose right so that the matrix is a full rank matrix. In most of real networks, we hardly get weight of each edge in adjacency matrix, so it is hard to get the computation result in large-scale networks. Therefore, we solve the problem by the structural controllability. In this model, nonzero weights in B and $u(t)$ are chosen to achieve the full rank of C. In this case, the system can be used to be controllable for most of the combination of the weight edge, apart from the pathological case [5]. The exact controllability of network is dependent on the maximum multiplicity of eigenvalue. However, the weight of the edge is hardly obtained in reality. In the consideration of the computational efficiency, the method of structural controllability is adopted based on the maximum matching. Therefore, the controllability of risk chain network in urban rail transit system is measured by the relative size of the driver nodes, which is denoted as:

$$R = \frac{N_{\mathrm{d}}}{N} \tag{3}$$

where N_{d} is the number of the driver nodes. Obviously, the bigger R, the weaker the controllability.

2.2 *Construction of Risk Chain Network*

The most basic element of the network is the edge and the node, so when constructing the urban rail transit risk network, we mainly consider these two aspects. From the above risk analysis, we can know that the equipment factor is the main reason to cause accidents in the operation story of urban rail transit, so this paper will start from the equipment system and divide it into five systems: electromechanical system, vehicle system, track line system and track circuit system. The risk in each system is regarded as the node of the risk chain network. Referring to the topological network of each device system, if the occurrence of risk factor A causes the occurrence of risk factor B, an edge from node A to node B will be added to the network.

2.3 *Risk Propagation*

The equipment subsystem of the urban rail transit system studied in this paper will cause risk because of an accident at a certain node. Assuming that when a node factor occurs in a risk network, the node is removed from the network and the starting edge of the node is removed. If a node lacks any edge arriving it, the risk of the node will also occur. Risk propagation will continue to iterate until no risk occurs. There is a case as follows in Fig. 1.

3 Case Study

3.1 *Controllability of Risk Chain Networks*

First, according to the above construction rules, this paper establishes risk chain network in urban rail transit system and analyzes the controllability of static network and driver nodes. In the light of the controllability theory of complex network, the number of driving nodes and specific driving nodes of vehicle risk network, rail line risk network, data propagation subsystem risk network, rail circuit system risk network and electromechanical risk network are calculated as in Table 1.

Based on the above calculation results, we find that the controllability of the track circuit system is the highest. This shows that for this system, more driving nodes are needed, so the risk factors are difficult to control. On the contrary, for the

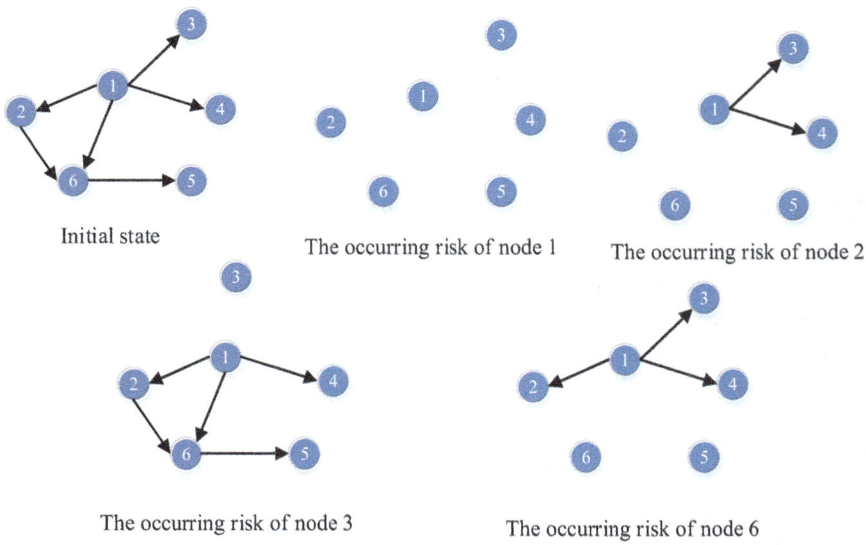

Fig. 1 Case of the risk propagation

Table 1 Values of parameters

System	Nodes	Edges	Diver nodes	R
Electromechanical system	65	82	38	0.58
Train system	42	48	22	0.52
Track line system	21	22	12	0.57
Track circuit system	15	20	9	0.6

vehicle system, its controllability is stronger than other systems. This shows that the network can be controlled by applying a small amount of signals in the system. That is to say, the risk control of the system is easier.

3.2 *Risk Chain Network Controllability Considering Risk Propagation*

In this part, we consider the risk propagation mechanism in the risk chain network. In order to measure the impact of each node's risk on network controllability, it is assumed that each node has a risk, and the risk is propagated to calculate the *R* of each node. The simulation results in the four systems are shown in Figs. 2, 3, 4 and 5.

In Fig. 2, we can find that the occurring risk of node 7 will lead to the bigger *R*, which means that node 7 plays an important role in the controllability of risk chain network in the electromechanical system. Additionally, when the risk of most of the nodes occurs, the diver nodes needed is not too much and *R* is equal to 0.52. From Fig. 3, it is clear that in the case of the occurring risk of nodes 9, 23 and 38, the

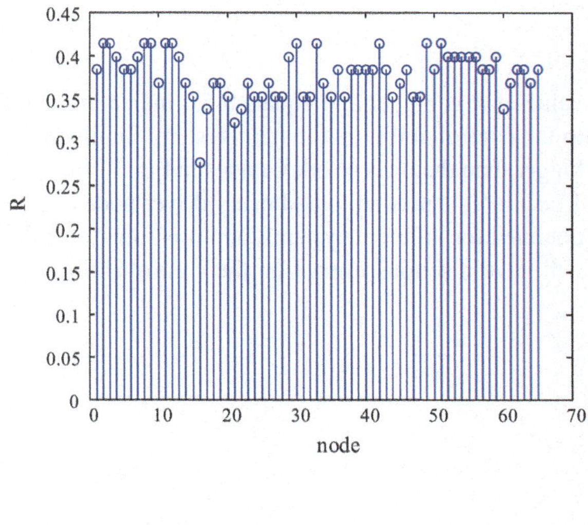

Fig. 2 Controllability of risk chain network considering the propagation risk in the electromechanical system

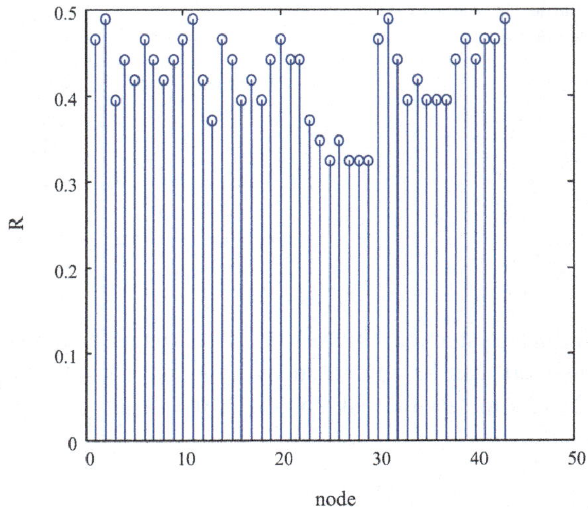

Fig. 3 Controllability of risk chain network considering the propagation risk in the train system

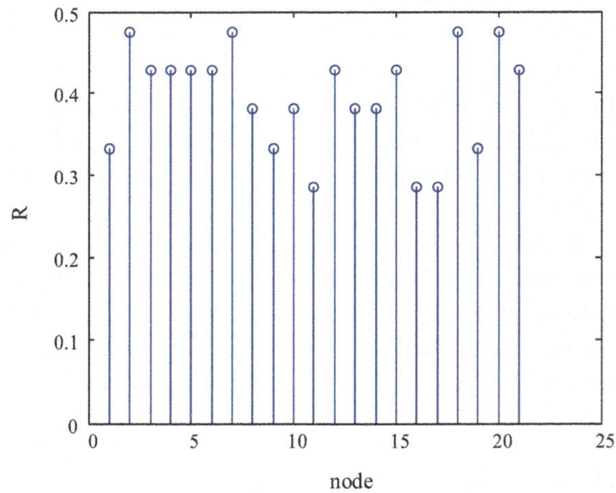

Fig. 4 Controllability of risk chain network considering the propagation risk in the track railway line system

value of R is bigger and equals 0.57, indicating that these three nodes are very key to the controllability in the train system. Similarly, R of the node 21 is significantly bigger than the ones of other nodes in Fig. 4. Figure 5 shows that if the risk of node 1 occurs, it is needed to add many driver nodes. All in all, the simulation results are obtained for different systems, and it is found that the key node whose risk happens will have a serious impact on the controllability.

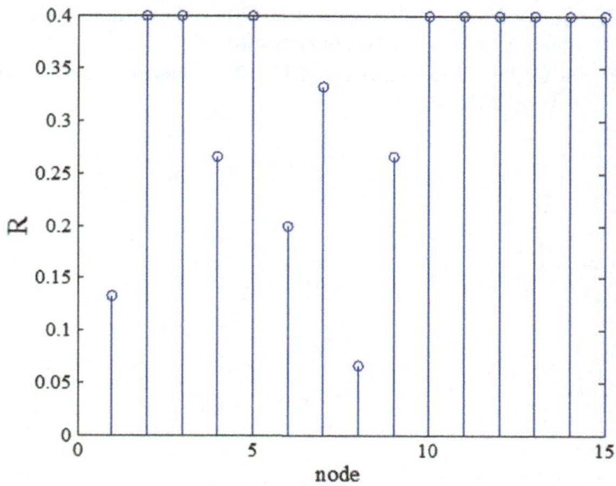

Fig. 5 Controllability of risk chain network considering the propagation risk in the track circuit line system

4 Conclusion

In this study, the risk chain network in urban rail transit system is built, and in this model, we consider comprehensively the different factors. Based on the characteristic of the risk chain network in reality, we put forward the propagation risk. On the basis of these, the simulation on the controllability of risk chain network is carried out in urban rail transit system with different systems. In addition, in the light of the results mentioned above, we obtain the components having a serious impact on the controllability. This work may be helpful for controlling the propagation of the risk in urban rail transit.

Acknowledgements The authors gratefully acknowledge the support from "the Fundamental Research Funds for the Central Universities" (Grant No. 2017JBZ103).

References

1. Li M (2012) Study on comprehensive evaluation model of urban rail transit operation safety [Ph.D.]. Beijing Jiaotong University, Beijing (in Chinese)
2. Jia C (2014) Research and application of safety risk prediction in urban rail transit operation areas [Ph.D.]. Tianjin University of Technology and Education, Tianjin, (in Chinese)
3. Zhenggui Z (2013) Use of BP neural network in metro operation safety evaluation. Mod Urban Transit 28(2):15–21 (in Chinese)
4. Zhou Y, Wang T, Dai B (2013) Research on quantitative classification method of risk in operating subway line. China Saf Sci J 23(02):103–108 (in Chinese)
5. Liu Y, Siotine J, Barabasi A-L (2011) Controllability of complex networks. Nature 473(7346): 167–173
6. Yuan Z, Zhao C, Di Z et al (2013) Exact controllability of complex networks. Nat commun 4:2447

7. Nie S, Wang X, Zhang H et al (2014) Robustness of controllability for networks based on edge-attack. PLoS ONE 9(2):e89066
8. Pu C, Pei W, Michaelson A (2012) Robustness analysis of network controllability. Phys A 391(18):4420–4425

Fault Diagnosis of Rolling Element Bearing for the Traction System of High-Speed Train Based on Wavelet Segmented Threshold De-Noising and HHT

Tao Xu, Xilian Wang and Zhipeng Li

Abstract Conventional wavelet threshold methods de-noise the original signal without considering the energy distribution of signal and noise varying with decomposition layer, which reduces the efficiency and adaptability of de-noising. A novel wavelet segmented threshold method is proposed in this paper to overcome this shortcoming. This method implements a more detailed threshold processing by piecewise nonlinear threshold function, in which crucial parameter is determined adaptively by the energy distribution on the wavelet coefficients in higher decomposition layer. The simulation results indicate that the proposed method has enhancement on the signal-to-noise ratio (SNR) and reduction on the root-mean-square error (RMSE) than conventional methods. The Hilbert–Huang transformation (HHT), an effective method for analyzing nonlinear and non-stationary signal and envelope spectrum analysis are applied to the de-noised signals of the faulty bearing, and the fault characteristic frequencies are extracted and the fault types are judged effectively.

Keywords Rolling bearing · Fault diagnosis · Wavelet de-noising · Threshold function · Hilbert–Huang transform

---

T. Xu · X. Wang (✉) · Z. Li
School of Electrical Engineering, Beijing Jiaotong University, Beijing, China
e-mail: xlwang1@bjtu.edu.cn

T. Xu
e-mail: 17121518@bjtu.edu.cn

Z. Li
e-mail: 18126122@bjtu.edu.cn

Y. Qin et al. (eds.), *Proceedings of the 4th International Conference on Electrical and Information Technologies for Rail Transportation (EITRT) 2019*, Lecture Notes in Electrical Engineering 639, https://doi.org/10.1007/978-981-15-2866-8_35

1 Introduction

Rolling element bearing fault is one of the commonest types of rotating machinery failure [1]. Therefore, the fault diagnosis of rolling element bearing for the traction system of high-speed train is so important to high-speed railway safety. The railway bearings work in so tough condition with frequent temperature and load changes [2] that the vibration signals of bearings are nonlinear and non-stationary, and the SNR is very low. How to extract the weak fault characteristic information from strong noise is a challenge.

Traditional signal de-noising methods by frequency domain analysis and linear filtering are validated to perform poorly when signal and noise overlap on multiple frequencies. Hence, the wavelet transform (WT) is applied to signal de-noising because of its time–frequency domain analysis ability and simplicity [3]. Wavelet threshold de-noising method [4] proposed by Donoho achieves good de-noising effect, while the effective information is occasionally filtered out while de-noising because of the poor adaptability resulting from global threshold and conventional threshold function. To overcome this defect, the threshold of each decomposition layer in Ref. [5] is determined according to the autocorrelation of the signal to realize the threshold automation. The correction coefficient is added to soft threshold function in Ref. [6] to reduce the constant deviation of soft threshold function and obtain better de-noising effect. A new wavelet threshold function with first-order continuous derivability is proposed in Ref. [7] to smooth transition between the noise coefficients and the signal coefficients, and the threshold of each decomposition layer depends on the respective wavelet coefficients. A lot of research has been done in the meantime. A novel wavelet segmented threshold method is proposed in this paper based on the above research. It implements more detailed threshold processing by piecewise nonlinear threshold function on the wavelet coefficients in higher decomposition layer. The crucial parameter in the threshold function is adaptively determined by the energy distribution of signal and noise varying with decomposition layer, which filters out the noise to the maximum extent.

The HHT [8] proposed by Huang is an effective method for representing the local characteristics of signals, especially suitable for the analysis of nonlinear and non-stationary signals. It is used to extract the effective fault characteristic information from the de-noised signal in this paper.

2 Wavelet Segmented Threshold De-noising and HHT Basic Principles

2.1 Traditional Wavelet Threshold de-Noising Theory

The process of traditional wavelet threshold de-noising includes selecting the wavelet basis and the decomposition scale, threshold processing of wavelet coefficients in each layer and reconstructing the de-noised signal [9]. Traditional wavelet threshold functions including hard and soft threshold functions [4] are showed in the following equation, respectively.

$$\omega_{j,k}^{\wedge} = \begin{cases} 0 & |\omega_{j,k}| < \lambda \\ \omega_{j,k} & |\omega_{j,k}| \geq \lambda \end{cases} \tag{1}$$

Here, $\omega_{j,k}$ is the wavelet coefficient, $\omega_{j,k}^{\wedge}$ is the wavelet coefficient processed, λ is the given threshold and sgn is the sign function. In hard threshold method, $\omega_{j,k}^{\wedge}$ is discontinuous in $\pm\lambda$, which makes the reconstructed signals oscillate. The reconstructed signals de-noised by soft threshold method are more continuous but have a constant deviation for subtracting λ from the wavelet coefficients larger than λ.

$$\omega_{j,k}^{\wedge} = \begin{cases} 0 & |\omega_{j,k}| < \lambda \\ \operatorname{sgn}(\omega_{j,k})\left(|\omega_{j,k}| - \lambda\right) & |\omega_{j,k}| \geq \lambda \end{cases} \tag{2}$$

2.2 *Wavelet Segmented Threshold De-noising Theory*

The energy distribution of signal varies with the decomposition layers [9]. It is ignored in many optimized threshold functions proposed based on the traditional threshold functions to affect the de-noising effect. An improved wavelet segmented threshold de-noising algorithm is proposed in this paper to overcome this shortcoming. This threshold function is continuously differentiable in the first order and between the hard and soft threshold function. Considering the ratio of effective signal energy increases with the decomposition layers, this proposed method deals more specifically with wavelet coefficients in higher decomposition layer. The mathematical expressions of the threshold function are as follows:

$$\begin{aligned} j = 1, \omega_{j,k}^{\wedge}(\omega_{j,k}, \lambda_{11}, m) \\ = \begin{cases} 0 & |\omega_{j,k}| < \lambda_{11} \\ \omega_{j,k} - \frac{1}{2^{1-1}}\left(\frac{\lambda_{11}^{m}}{|\omega_{j,k}|^{m-1}}\right) & |\omega_{j,k}| \geq \lambda_{11} \end{cases} \end{aligned} \tag{3}$$

$$\begin{aligned} j = 2, \omega_{j,k}^{\wedge}(\omega_{j,k}, \lambda_{22}, \lambda_{21}, m) \\ = \begin{cases} 0 & |\omega_{j,k}| < \lambda_{21} \\ 2^{\frac{1}{2-1}}\operatorname{sgn}(\omega_{j,k}) \frac{\lambda_{21}\left(|\omega_{j,k}| - \lambda_{21}\right)^{m}}{(\lambda_{22} - \lambda_{21})^{m}} & \lambda_{21} \leq |\omega_{j,k}| < \lambda_{22} \\ \omega_{j,k} - 2^{\frac{1}{2-1}}\operatorname{sgn}(\omega_{j,k}) \frac{\lambda_{22}^{m}}{|\omega_{j,k}|^{m-1}} & |\omega_{j,k}| \geq \lambda_{22} \end{cases} \end{aligned} \tag{4}$$

$$j = 3, \omega_{j,k}^{\wedge}(\omega_{j,k}, \lambda_{33}, \lambda_{32}, \lambda_{32}, m) = \begin{cases} 0 & |\omega_{j,k}| < \lambda_{31} \\ 2^{\frac{1}{2-1}} \mathrm{sgn}(\omega_{j,k}) \frac{\lambda_{32}(|\omega_{j,k}| - \lambda_{31})^m}{(\lambda_{32} - \lambda_{31})^m} & \lambda_{31} \leq |\omega_{j,k}| < \lambda_{32} \\ \omega_{j,k} - 2^{\frac{1}{2-1}} \mathrm{sgn}(\omega_{j,k}) \frac{\lambda_{32}^m}{|\omega_{j,k}|^{m-1}} & \lambda_{32} \leq |\omega_{j,k}| < \lambda_{33} \\ \omega_{j,k} - 2^{\frac{1}{3-1}} \mathrm{sgn}(\omega_{j,k}) \frac{\lambda_{33}^m}{|\omega_{j,k}|^{m-1}} & |\omega_{j,k}| \geq \lambda_{33} \end{cases} \tag{5}$$

$$\lambda_{jj} = \delta_j \sqrt{\ln N_j}, \lambda_{ji} = \alpha_{ji} \lambda_{jj} (i = 1, 2, \ldots, j) \tag{6}$$

$$\delta_j = \mathrm{median}(|\omega_{jk}|)/0.6745 \tag{7}$$

Here, m is the real number and larger than $1, j$ is the decomposition layer and δ_j is the estimated variance of wavelet coefficients. The threshold function is continuously differentiable in the first order. The following equation satisfies when $j = 3$.

$$\left. \frac{\partial\left(2^{\frac{1}{2-1}} \mathrm{sgn}(\omega_{j,k}) \frac{\lambda_{32}(|\omega_{j,k}| - \lambda_{31})^m}{(\lambda_{32} - \lambda_{31})^m}\right)}{\partial(\omega_{j,k})} \right|_{\omega_{j,k} = \lambda_{32}} = \left. \frac{\partial\left(\omega_{j,k} - 2^{\frac{1}{2-1}} \mathrm{sgn}(\omega_{j,k}) \frac{\lambda_{32}^m}{|\omega_{j,k}|^{m-1}}\right)}{\partial(\omega_{j,k})} \right|_{\omega_{j,k} = \lambda_{32}} \tag{8}$$

$$\left. \frac{\partial\left(\omega_{j,k} - 2^{\frac{1}{2-1}} \mathrm{sgn}(\omega_{j,k}) \frac{\lambda_{32}^m}{|\omega_{j,k}|^{m-1}}\right)}{\partial(\omega_{j,k})} \right|_{\omega_{j,k} = \lambda_{33}} = \left. \frac{\partial\left(\omega_{j,k} - 2^{\frac{1}{3-1}} \mathrm{sgn}(\omega_{j,k}) \frac{\lambda_{33}^m}{|\omega_{j,k}|^{m-1}}\right)}{\partial(\omega_{j,k})} \right|_{\omega_{j,k} = \lambda_{33}} \tag{9}$$

Hence:

$$\begin{cases} \alpha_{31} = \left(\frac{1}{m+1}\right) \left(\frac{1}{2}\right)^{\frac{1}{m}} \\ \alpha_{32} = \left(\frac{1}{2}\right)^{\frac{1}{m}} \end{cases} \tag{10}$$

It can be seen that α_{ji} is determined by m, which means the segmented thresholds change with m and j. Figure 1 is a comparison between the proposed threshold

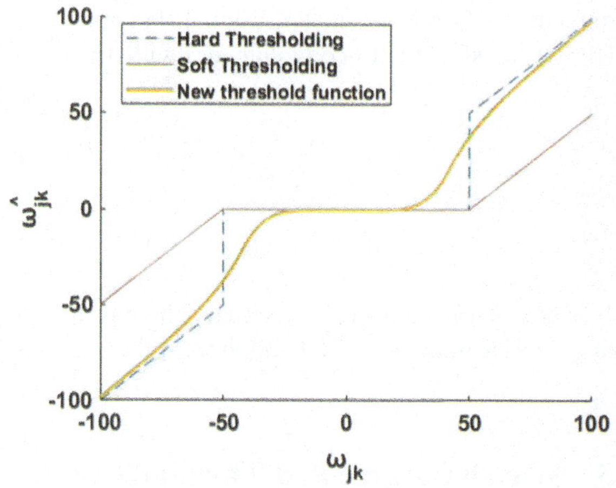

Fig. 1 Comparison of hard and soft threshold functions and new threshold functions

function and the traditional threshold functions. The wavelet segmented threshold function merges both the characteristics of the soft and hard threshold functions.

The noise account for a larger proportion for the wavelet coefficients in smaller layer, m is expected to be larger so that the threshold function of this layer can filter out more noise. The calculation formula of m is as follows:

$$m = 1 + \left[12 \frac{E_{nj}}{E_{dj}} \right] \tag{11}$$

$$\frac{E_{nj}}{E_{dj}} = \left(\frac{1}{2} \right)^{j} \tag{12}$$

Here, $[\cdot]$ is the integral function, E_{nj} is the energy of noise in j layer, E_{dj} is the energy of signal in layer j. When $j = 1$, $m = 13$, which is the maximum value.

2.3 *HHT Principles*

HHT is the process of Hilbert transform after EMD [8]. EMD decomposes the original signal into a linear combination of intrinsic mode functions (IMF) $c_i(t)$ $(i = 1, 2 \ldots n)$ and one remainder $r_n(t)$ by a series of iterations.

$$X(t) = \sum_{i=1}^{n} c_i(t) + r_n(t) \tag{13}$$

HHT can demodulate the low-frequency fault characteristic signal attached to the high-frequency carrier signal. Its basic principle is to transform the original time domain signal into an analytic signal. For a given continuous time signal $x(t)$, its Hilbert transform is to convolute $x(t)$ with $1/(\pi t)$, as follows:

$$\hat{x}(t) = x(t) * \frac{1}{\pi t} = \frac{1}{\pi}\int_{-\infty}^{+\infty} \frac{x(\tau)}{t-\tau} d\tau \tag{14}$$

$$a(t) = \sqrt{x(t)^2 + \hat{x}(t)^2} \tag{15}$$

Here, $a(t)$ is the envelope signal. The envelope spectrum of $a(t)$ is obtained by fast Fourier transform (FFT) applied to the envelope signal.

3 Wavelet Segmented Threshold De-noising Simulation Test

Simulation tests are implemented in the Bumps signal with 10 dB of white Gauss noise based on hard threshold method, soft threshold method, the method proposed in Ref. [7] and the method proposed in this paper to compare and verify the superiority of new method. Hard and soft threshold methods adopt sym8 wavelet for three-layer wavelet decomposition. Db5 wavelet for five-layer is adopted in the method in Ref. [7] and sym4 wavelet for four-layer is adopted in new method.

Figure 2 indicates Bumps original signal and Fig. 3 indicates Bumps noisy signal. Figure 4 indicates that hard threshold method can retain the peak mutation part of signal better, but the oscillation is obvious when signal approaches zero. Soft threshold method performs poorly for the filtering of the peak mutation part from Fig. 5 The method proposed in Ref. [7] shown in Fig. 6 has progress for the peak mutation signal reserving better, but the oscillation is obvious when signal approaches zero and the spike signal is generated. The proposed wavelet segmented threshold method is shown in Fig. 7 retains the peak mutation signal well, and the waveform is smoother than other methods. In Table 1, the indexes of SNR and RMSE that can measure the effectiveness of de-noising are calculated and compared. The wavelet segmented threshold method obtains the highest SNR and the lowest RMSE, which proves the superiority.

$$\mathrm{RMSE} = \sqrt{\frac{1}{n}\sum_{i=1}^{n} [x(i) - \hat{x}(i)]^2} \tag{16}$$

$$\mathrm{SNR} = 10\lg\left[\frac{\sum_{i=1}^{n} x^2(i)}{\sum_{i=1}^{n} [x(i) - \hat{x}(i)]^2}\right] \tag{17}$$

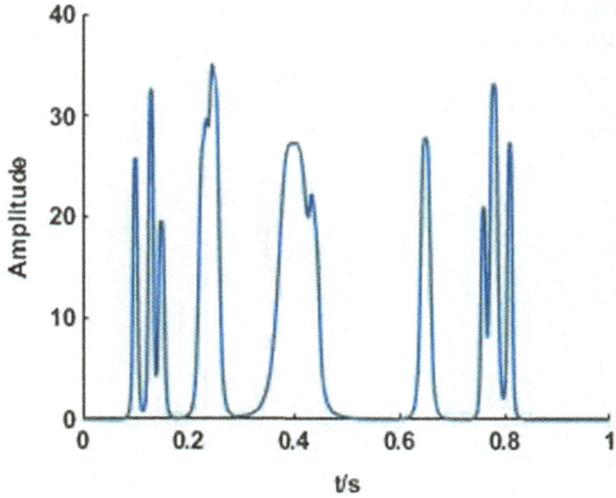

Fig. 2 Bumps original signal

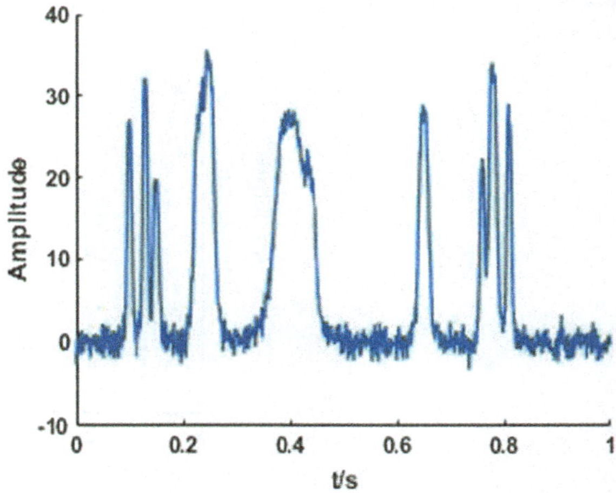

Fig. 3 Bumps noisy signal

Here, $x(i)$ is the original signal, $\hat{x}(i)$ is the de-noised signal and n is the signal length.

4 Rolling Element Bearing Fault Diagnosis Test

Experimental data of rolling element bearings are obtained from the Electrical Engineering Laboratory of Cast Western Reserve University, USA. The tested bearing type is 6205-2RS JEM SKF and the parameters are shown in Table 2. The experimental platform consists of a motor of two horsepower, a torque sensor, a power meter and an electronic control device. In the experiment, vibration signals

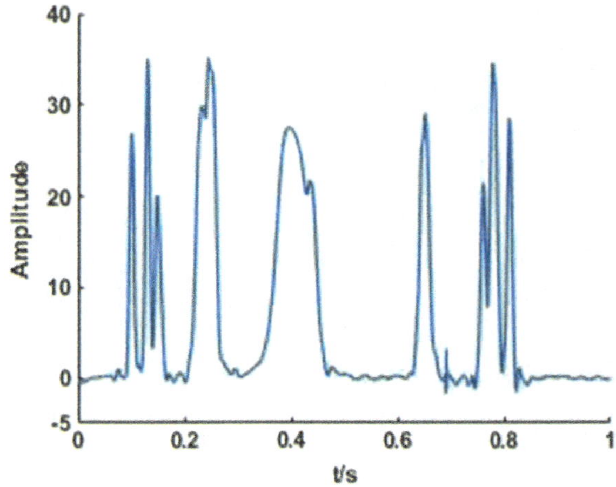

Fig. 4 Hard threshold de-noised signal

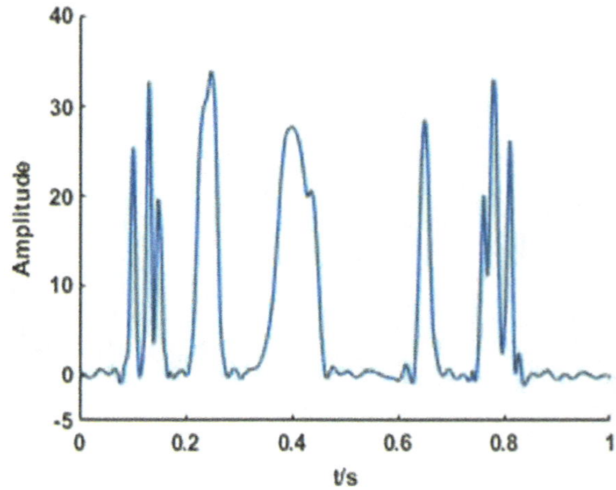

Fig. 5 Soft threshold de-noising

are collected by the acceleration sensor, which are installed on the motor housing through a magnetic base. The sampling frequency is 12 kHz. When the inner ring damage is 0.1778 μm and the motor load is 0, the vibration signal of up to 0.1 s is selected. The rotating speed frequency is 29.95 Hz and the inner ring failure frequency is 162.19 Hz.

Original signals are de-noised by the above method, respectively, and HHT is applied to de-noised signal. The IMF components of maximum kurtosis are selected for envelope spectrum analysis. Figures 8, 9, 10, 11 are the results. It can be clearly seen that the method proposed in this paper performs best for well filtering noise and identifying faulty frequency compared with the three other methods.

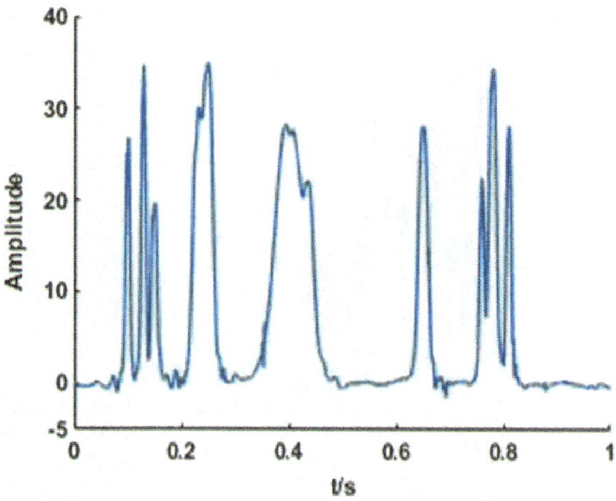

Fig. 6 De-noised signal based on the method in Ref. [7]

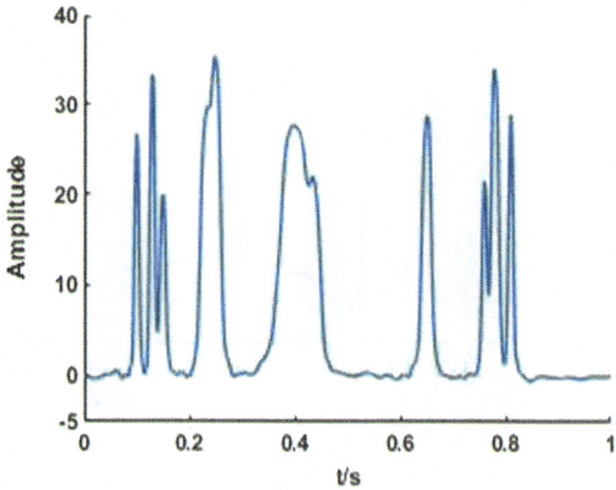

Fig. 7 De-noised signal based on the method in this paper

Table 1 Comparison of evaluation indexes of de-noising method

Index	Hard thresholding	Soft thresholding	Method in Ref. [7]	Font size and style
RMSE	0.9608	1.0525	0.9550	0.8915
SNR	49.9953	48.1715	50.1150	51.4914

Table 2 6205-2RS JEM SKF bearing parameters

Inner diameter (mm)	Outer diameter (mm)	Number of rollers	Roller diameter (mm)	Pitch diameter (mm)
25.0012	51.9989	9	7.9400	39.0398

Fig. 8 Envelope spectrum of de-noised signal based hard threshold method

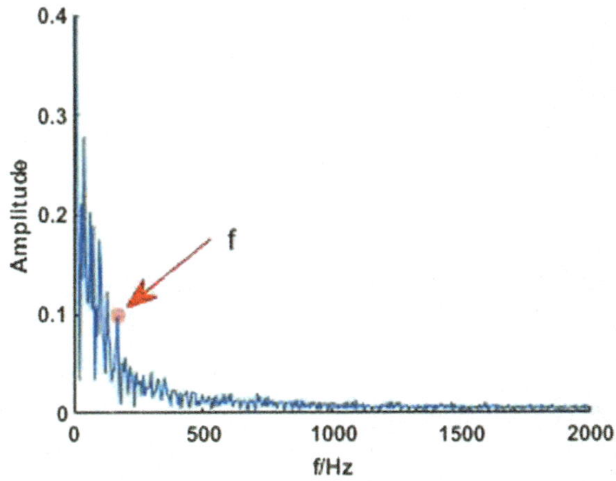

Fig. 9 Envelope spectrum of de-noised signal based on soft threshold method

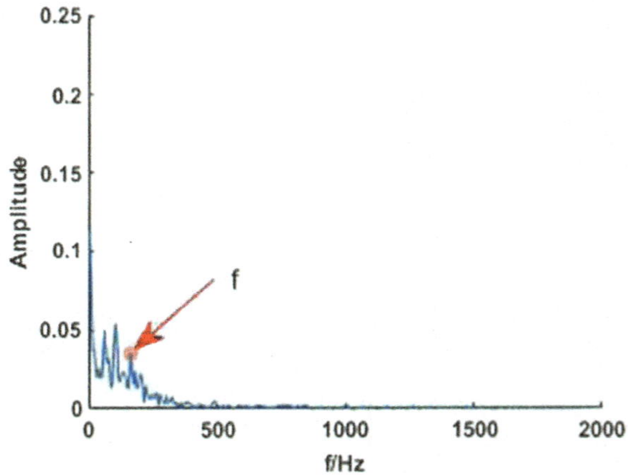

Fig. 10 Envelope spectrum of de-noised signal based on the method in Ref. [7]

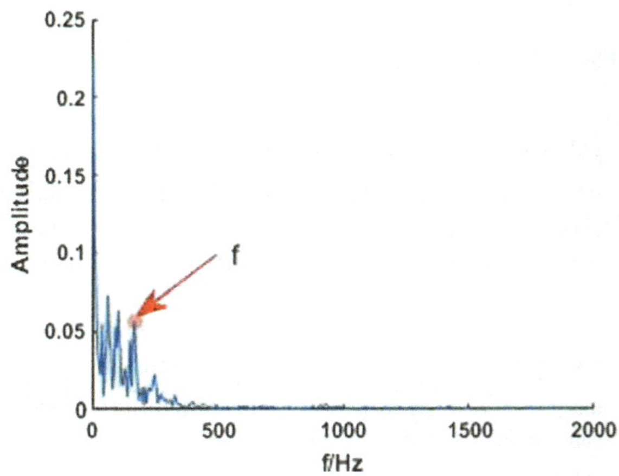

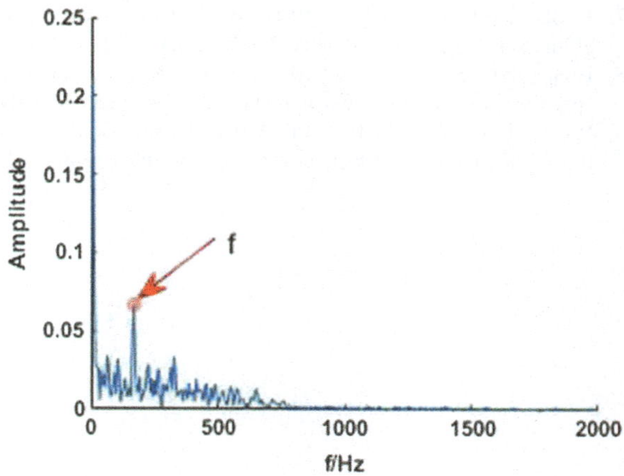

Fig. 11 Envelope spectrum of de-noised signal based on the method in this paper

5 Conclusion

The weak fault information of bearing in early stage is buried by strong noise and difficult to extract because of the complex work condition of the traction system of high-speed train. The conventional de-noising methods have drawbacks, such as the lack of adaptive ability to filter out the effective signal by mistake while de-noising. While the wavelet segmented threshold method proposed in this paper can retain fault feature information to the maximum extent while de-noising. By combining the proposed de-noising method with HHT, the faulty frequency information can be extracted and the fault gets diagnosed. The experimental results reveal the certain significance of the proposed method in the practical application.

References

1. Cheng Y, Zhou N, Zhang W (2018) Application of an improved minimum entropy deconvolution method for railway rolling element bearing fault diagnosis. J Sound Vib 425:53–69
2. Henao H, Kia SH, Capolino GA (2011) Torsional-vibration assessment and gear-fault diagnosis in railway traction system. IEEE Trans Ind Electron 58(5):1707–1717
3. Yi T, Li H, Zhao X, (2012) Noise smoothing for structural vibration test signals using an improved wavelet thresholding technique. Sensors 12(8):11205–11220
4. Donoho DL, (1995) De-noising by soft-thresholding. IEEE Trans Inf Theory 41(3):613–627
5. Zhang Y, Zhang D, Wang W (2009) Per-level threshold de-noising method using wavelet and its application in engine vibration analysis. J Vib Shock 28(8):57–60 (in Chinese)
6. Meng Z, Li S, (2013) Rolling bearing fault diagnosis based on improved wavelet threshold de-noising method and HHT. J Vib Shock 32(14):204–208, 214 (in Chinese)

7. Li SP, Zhou Y, Zhou YQ (2017) An adaptive wavelet shrinkage de-noising algorithm for low altitude flying acoustic targets. J Vib Shock 36(9):153–158, 230 (in Chinese)
8. Huang NE, Shen Z, Long SR (1998) The empirical mode decomposition and the Hilbert spectrum for nonlinear and non-stationary time series analysis. R Soc 454:903–995
9. Wu G, Wang C, Bao J (2014) A wavelet threshold de-noising algorithm based on adaptive threshold function. J Electron Inf Technol 36(6):1340–1347 (in Chinese)

Evacuation Capacity Estimation for Subway Station Based on Queueing Model

Hui Zhang, Jie Xu, Siyao Li and Yiwen Chen

Abstract Evacuation capacity is the foundation of the evacuation plan. Evacuation plans with accurate evacuation capacity can improve evacuation efficiency and decrease losses. To estimate evacuation capacity, the evacuation network is constructed firstly based on different queueing models and the relation between throughput and arrival rate is determined. When it comes to splitting nodes, the choice probability of the next target is calculated by travel time and the reliability of evacuation time. Next, we estimate the capacity of Jianguomen station of Beijing Subway as an example to validate the efficiency of the proposed model. Finally, density, saturation, and blocking probability of each facility are discussed. Results show that the maximum throughput of Jianguomen station is 855 peds/min, which can be approximately regarded as evacuation capacity.

Keywords Subway station · Evacuation capacity · Evacuation network construction · Choice probability · Queueing models

---

H. Zhang · J. Xu (✉) · S. Li · Y. Chen
State Key Laboratory of Rail Traffic Control and Safety, Beijing Jiaotong University, Beijing, China
e-mail: jxu1@bjtu.edu.cn

School of Transportation and Traffic, Beijing Jiaotong University, Beijing, China

J. Xu
Beijing Research Centre of Urban Traffic Information Sensing and Service Technologies, Beijing, China

Research and Development Center of Transport Industry of Technologies and Equipments of Urban Rail Operation Safety Management, MOT, Beijing, People's Republic of China

Y. Qin et al. (eds.), *Proceedings of the 4th International Conference on Electrical and Information Technologies for Rail Transportation (EITRT) 2019*, Lecture Notes in Electrical Engineering 639, https://doi.org/10.1007/978-981-15-2866-8_36

1 Introduction

Nowadays, more and more passengers prefer to commute by subway system. Urban rail transit stations are playing an increasingly important role in major cities [1]. Falling and trampling accidents will lead to injuries and fatalities. Subway station operators and managers are paying more attention to the evacuation capacity.

There are mainly two kinds of methods of estimating evacuation capacity for the subway station. One is the subjective assessment, and the other is the optimal evacuation models. The former method estimates evacuation capacity by picking up some indicators and building index system [2]. While this method cannot be applied in routine operations because of the great discrepancy between calculated results and actual evacuation flow [3], the latter method constructs evacuation network and obtains network throughput, evacuation time, and other indicators using the optimal evacuation model and specific algorithm [4].

Some of the previous researches on the crowd evacuation are based on queueing theory. Each facility is regarded as a queueing unit, and the units connect in series or in parallel to form the queueing network [5]. Cruz et al. [6] quantitatively describe the dynamic process of crowd formation, simulate the congestion and calculate the evacuation time. Gosavi and Smith [7] construct the queueing network with a server and exponential arrivals and service times, and analyze the routing decision to maximize network throughput. However, most literature regard all facilities in subway station as an M/G/c/c state-dependent queueing model, not considering the differences between facilities. The facilities with different location and attribution in the layout of subway station have different characteristics and capacities, which will influence the status of passenger flow and facility service. Thus, we formulate the M/G/c/N queueing model to describe the stairs and corridors connected to platform or hall, M/G/c/c model to illustrate other stairs and corridors, and M/M/1 model to depict gates.

In this paper, different queueing models are proposed considering different characteristics of facilities. Then evacuation network is constructed based on different queueing models. Moreover, evacuation bottlenecks are identified and network throughput is also estimated. Finally, a real case about Jianguomen station of Beijing Subway is implemented, which demonstrates the validity and efficiency of the proposed method.

2 Problem Statement

2.1 Network Construction

A graph $G = (V, E, T)$ is an evacuation network, where V is the set of nodes, E is the set of arcs, and T is the set of arc resistance. The nodes, arcs, and resistances represent the entrances and exits of facilities, the facilities, and the utilities of arcs, respectively.

A simple evacuation network is shown in Fig. 1. The origin *O* represents the location of passenger flow generation, and the destination node *D* denotes the location of passenger flow sink. Each facility of evacuation network is modeled as a queue service system, and the number of servers in one system is the effective facility capacity. When passengers walking side by side without interference, the width and length needed are *wd* and *le*, respectively. In this paper, *wd* and *le* are both 0.5 m. Then the effective facility capacity C_{eff} is computed as Eq. (1).

$$C_{\text{eff}} = \left\lfloor \frac{l}{\text{le}} \right\rfloor \cdot \left\lfloor \frac{W_{\text{eff}}}{\text{wd}} \right\rfloor = \left\lfloor \frac{l}{\text{le}} \right\rfloor \cdot \left\lfloor \frac{(w - 1.07)}{\text{wd}} \right\rfloor \tag{1}$$

Where, W_{eff} is the effective width of the facility, which equals the actual width w diffs 1.07 m [8], and ⌊ ⌋ stands for rounding the number down.

The maximum population density in walking facilities is 5.0 peds/m^2 [9]. Hence, the maximum facility capacity $C_{\max}$ can be described as,

$$C_{\max} = \lfloor 5.0 \times l \times w \rfloor \tag{2}$$

The evacuation network can be classified as three sub-networks, namely, series, merging, and splitting. The arrival rate as a significant parameter of the queueing model can be determined as follows.

1. Series. In Fig. 2a, the arrival rate of the target facility (the next facility passengers will occupy) equals the throughput of the prior facility. λ is the arrival rate, and θ is the throughput.
2. Merging. The throughput of *i*th prior facility is θ_{ui}, and the arrival rate of target facility is $\lambda_{\text{d}}, \lambda_d = \theta_{u1} + \theta_{u2} + \cdots + \theta_{un} = \sum_{i=1}^{n} \theta_{ui}$ (Fig. 2b).
3. Splitting. The chosen probabilities of each target are $p_{d1}, p_{d2} \ldots p_{dn}$. Thus, the arrival rates of each target facility are $p_{d1}\theta_u, p_{d2}\theta_u \ldots p_{dn}\theta_u$ (Fig. 2c).

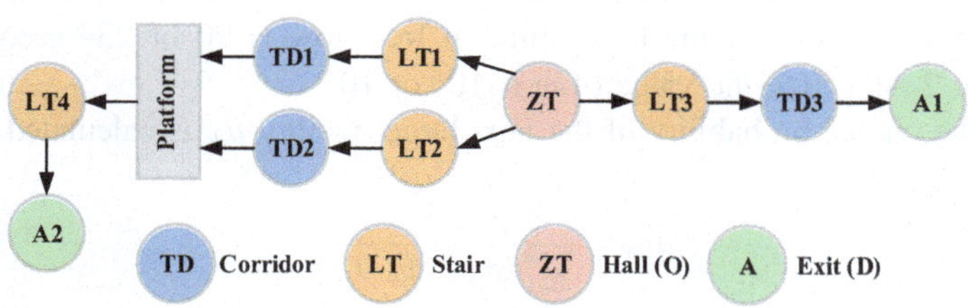

Fig. 1 A simple evacuation network

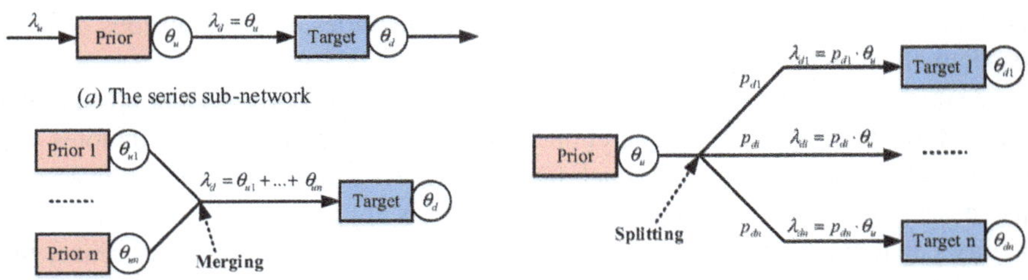

(a) The series sub-network

(b) The merging sub-network

(c) The splitting sub-network

Fig. 2 The topology of three sub-networks

2.2 Choice Probability

When it comes to splitting nodes, people generally prefer to choose the route with shorter travel time, which will easily cause congestion in shorter route. To avoid over-crowded, the resistance of a route not only depends on the real travel time, but the reliability of the travel time in free flow. Assume that the travel time of the route in free flow is t_{di}^{0}, and the maximum available evacuation time $t_{\max}$ is 5 minutes. The reliability of travel time R_{di} can be calculated as follow,

$$R_{di} = \begin{cases} \frac{t_{\max} - t_{di}^{0}}{t_{\max}} & t_{di}^{0} \in [0, t_{\max}] \\ 0, & \text{otherwise} \end{cases} \tag{3}$$

The average travel time of m occupants in the route is t_{di}^{c} and the route resistance T_{di} is

$$T_{di} = (1 - \sigma)t_{di}^{c} + \varphi\sigma(1 - R_{di}) \tag{4}$$

Where, σ is the preference of occupants choosing the route with higher travel time reliability, and $\sigma \in [0, 1]$. In this paper, we set $\sigma = 0.7$. φ is a coefficient to transform the reliability into travel time. The value of φ is related to the travel time of route. For example, if the travel time in free flow is 90 or 234 seconds, i.e., 9×10^{1} or 2.34×10^{2}, the value of φ is 10^{1} or 10^{2}.

Then the choice probability of the next target facility p_{di} is calculated as,

$$p_{di} = \frac{\exp(-\theta \cdot T_{di})}{\sum_{i}^{n} \exp(-\theta \cdot T_{di})} \tag{5}$$

Where, θ is a parameter measuring the familiarity of occupants to the evacuation network, and we set $\theta = 3.3$ [10].

3 Development of Models

A summary of the notation used in the proposed model is given in Table 1.

3.1 M/G/c/N Queueing Model

During the evacuation, the following passengers with higher speed will be hindered by the passengers with slower speed near exit [11]. We regard the stairs and corridors connected to the platform or hall as queueing system with a buffer and formulate the M/G/c/N queueing model. The buffer size is twice the number of servers. The related equations of the M/G/c/N queueing model are as follows. Firstly, the probability of m occupants in facility is described as,

$$p(m) = \begin{cases} p_0 \cdot \prod_{i=1}^{m} \frac{\lambda \cdot E(T_f)}{i \cdot f(i)} & 1 \le m < c \\ p_0 \cdot \frac{\left[\lambda \cdot E(T_f)\right]^m}{c!\left[c \cdot f(c)\right]^{m-c} \prod_{i=1}^{c} f(i)} & c \le m \le N \end{cases} \tag{6}$$

Table 1 Notation list

Parameters	Description
M	The queueing system with markovian arrival rates
G	The queueing system with generally distributed services time
c	c servers in the queuing system
N	$N = 2c$
l	Corridor length in meters
λ	Passenger arrival rate in ped/s
μ_m	The service rate when there are m occupants in facility
v_f	Average lone occupant walking speed in m/s
v_m	Average walking speed in facility for m occupants in m/s
$f(i)$	v_i/v_f, service rate of i passengers
$E(T_f)$	l/v_f, mean travel time for a lone occupant in facility in seconds
$E(T)$	Expected service time, including waiting time and servicing time
$p(m)$	Probability of m occupants in facility
p_0	Empty facility probability
p_b	Blocking probability
L	Mean number of occupants in the facility
θ	Throughput in ped/s
sa	Facility saturation

Based on the normalization theory, the empty facility probability is,

$$p_0 = \left[1 + \sum_{m=1}^{c-1} \prod_{i=1}^{m} \frac{\lambda \cdot E(T_f)}{i \cdot f(i)} + \sum_{m=c}^{N} \frac{[\lambda \cdot E(T_f)]^m}{c![c \cdot f(c)]^{m-c} \prod_{i=1}^{c} f(i)} \right]^{-1} \tag{7}$$

According to the Little's law, the blocking probability is,

$$p_b = \sum_{m=c}^{N} P(m) \tag{8}$$

The mean number of occupants in facility is,

$$L = \sum_{m=1}^{N} m \cdot p(m) \tag{9}$$

The throughput is,

$$\theta = \lambda \cdot (1 - p_b) \tag{10}$$

And the facility saturation is,

$$sa = \theta / c_{\max} \tag{11}$$

3.2 M/G/c/c Queueing Model

The capacity of escalator, stair, and corridor is limited, which is different from the platform and hall. There are c parallel servers in the M/G/c/c queueing model. When there are m occupants in the facility, the probability is

$$P(m) = \frac{[\lambda \cdot E(T_f)]^m \cdot p_0}{m! f(m) \cdot \cdots \cdot f(2) \cdot f(1)} = p_0 \cdot \prod_{i=1}^{m} \frac{\lambda \cdot E(T_f)}{i \cdot f(i)} \tag{12}$$

The empty facility probability is,

$$p_0 = \left[1 + \sum_{m=1}^{c} \prod_{i=1}^{m} \left(\frac{\lambda \cdot E(T_f)}{i \cdot f(i)} \right) \right]^{-1} \tag{13}$$

Some related indicators are calculated as follows.

$$p_b = p_0 \cdot \prod_{i=1}^{c} \frac{\lambda \cdot E(T_f)}{i \cdot f(i)} \quad (14)$$

$$L = \sum_{m=1}^{c} m \cdot p(m) \quad (15)$$

3.3 M/M/1 Queueing Model

To minimize the total evacuation time, all gates will be open to passengers in the evacuation. The gate can only pass one person at a time, and the service rate is constant. We assume that there are n gates in the queueing network, and the empty facility probability of each facility is,

$$p_0 = \mu/(\mu + \lambda) \quad (16)$$

When there is one occupant in facility, the probability is,

$$p_1 = \lambda/(\mu + \lambda) \quad (17)$$

The mean number of occupants in facilities is

$$L = n\lambda/(n\mu - \lambda) \quad (18)$$

3.4 Algorithm

The detailed algorithm is proposed as follows.

- Step 1: Initialize the arrival rates λ_1 and λ_2 as 6 ped/s and 4 ped/s, respectively. Free flow speed in corridors v_C^0 and in stairs v_S^0 and the geometric parameters of facilities are initialized.
- Step 2: Search for available paths.
- Step 3: Calculate the effective capacity and the maximum capacity of facilities.
- Step 4: Calculate the indicators of the queueing network.
  - Step 4.1: Update the number of occupants in each facility, and get the average walking speed in the ith facility for m occupants.
  - Step 4.2: Calculate the probability of m occupants in the facility, the empty facility probability, the blocking probability, throughput, and saturation.

Step 4.3: Calculate the probability of the ith target facility for splitting node p_{di} and get the arrival rate of the ith target facility.

Step 5: Increase arrival rate λ_2 by one. If $\lambda_2 > 9$, go to step 6; else, go to step 4.

Step 6: Increase the arrival rate λ_1 by one. If $\lambda_1 > 13$, end the algorithm; else, go to step 4.

4 A Case Study

4.1 Jianguomen Station

Jianguomen station is an interchange station of Lines 1 and 2, Beijing Subway. There are three exits (entrances) in the station. The evacuation source node is set in the platform midpoints. According to the building structure, Jianguomen station is divided into four parts (see Fig. 3). Considering the most disadvantaged situation, all passengers gather in the platform before evacuation.

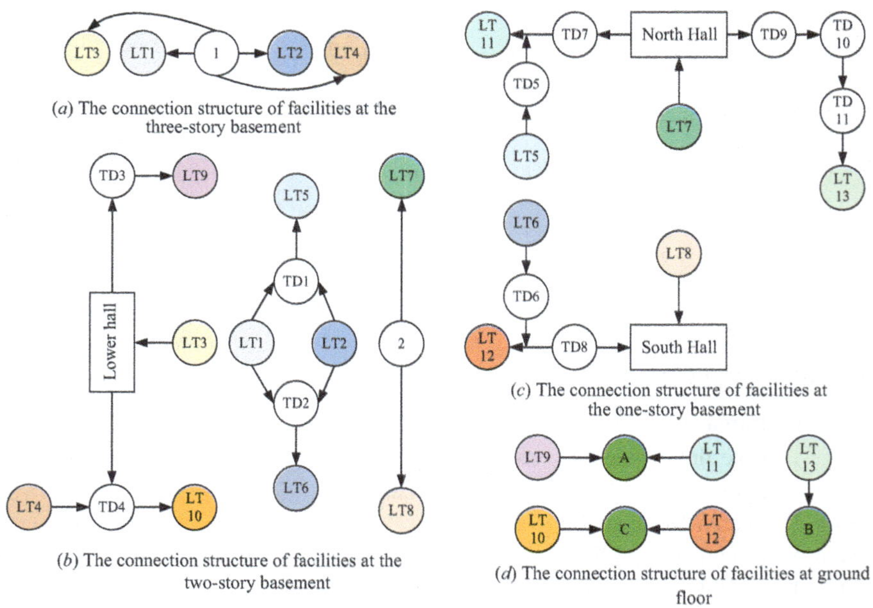

Fig. 3 The topology network of Jianguomen station

4.2 Discussion

The arrival rate of Line 1 ranges from 6 to 13 peds/s and that of Line 2 ranges from 4 to 9 peds/s. The relation between speed and density is referred to [12]. In Fig. 4, the annotate of each cell means the number of passengers completing evacuation in each scenario. When the arrival rate of Line 1 is 11 peds/s and that of Line 2 is 7 peds/s, the throughput reaches the maximum of 13.918 peds/s. The throughput of three exits is 5.227, 3.427, and 5.624 peds/s, respectively. The throughput of evacuation network is 835 peds/min, and 4175 people can be evacuated within the safe evacuation time.

When the throughput is maximum, we analyze the throughput, density, saturation, and blocking probability of each facility (see Fig. 6). In total, most blocking probabilities are not equal to zero. Half of the densities are more than 2 peds/m^2. Passenger flow is crowded, and the occupants may be stopped intermittently in the facility. The congestion of stair is higher than that of the corridor. The three stairs connected to the platform is more crowded because the stairs are necessary to evacuate no matter which route is chosen. Therefore, it is significant to arrange more staff guiding evacuation in crowded facilities to prevent accidents. The saturation of most facilities is above 0.5, and the mean is 0.7, which suggests that the evacuation network is fully utilized and meets the requirement of balanced evacuation.

Besides, we compare the expected time calculated by queueing models and evacuation time obtained by Anylogic (see Fig. 5). From the results, we can conclude that the actual evacuation time is not proportional to the length of the path, because the shorter route always attracts more passengers and easily causes congestion. The evacuation time of each path is less than 300 seconds and meets the standard.

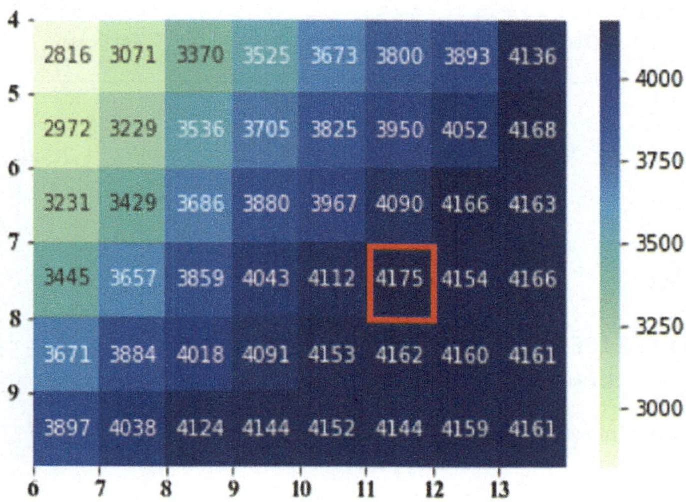

Fig. 4 Throughput of evacuation network in different scenarios

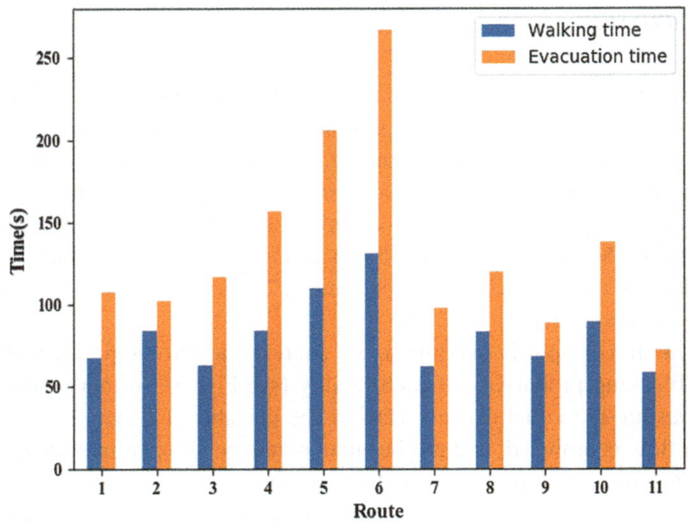

Fig. 5 Expected time and evacuation time of different routes

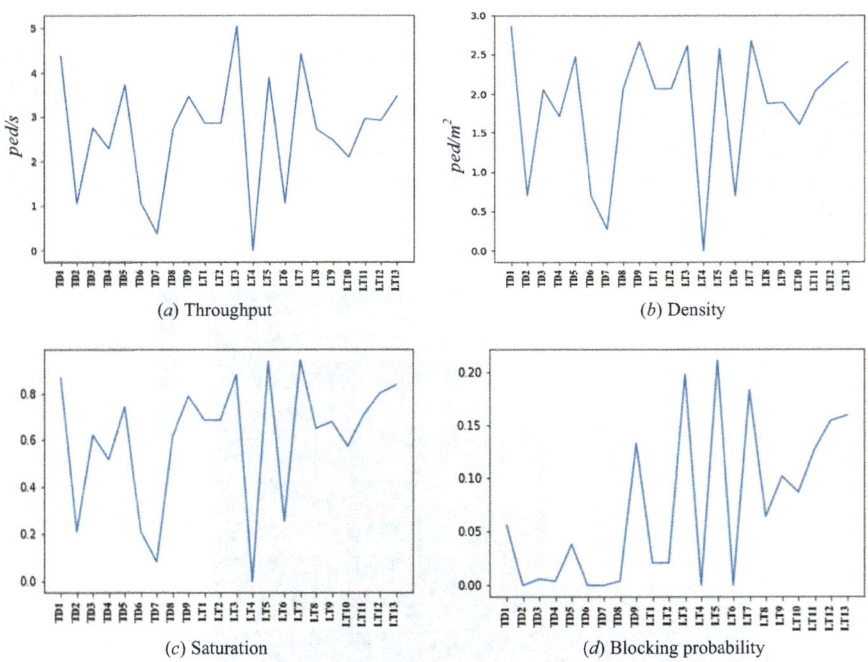

Fig. 6 Indicators of each facility

5 Conclusion

In this paper, the arrival rate and probability of route choice are determined, three different queueing models considering the different characteristics of facilities are proposed and the queueing network is constructed according to the layout of station. Moreover, an algorithm is proposed to solve this model. Next, the Jianguomen station of Beijing Subway is taken as an example to validate the proposed model. Because of the lower speed of passengers in stairs, the congestion of stairs is more serious than that of the corridors, and the staff should guide the evacuation in the crowded area. Generally speaking, the evacuation network is fully utilized and meets the requirement of balanced evacuation. The maximum throughput of Jianguomen station is 855 peds/min, and 4175 people can be evacuated within the safe evacuation time. The number of evacuees can be approximately regarded as the network throughput of the station.

Acknowledgements The authors gratefully acknowledge the support provided by China National "13th Five-Year" key research project "Safety assurance technology of urban rail system" (Grant No. 2016YFB1200402).

References

1. Shiwakoti N, Tay R, Stasinopoulos P, Woolley PJ (2017) Likely behaviours of passengers under emergency evacuation in train station. Saf Sci 91:40–48
2. Chen CLP, Liu Z (2018) Broad learning system: an effective and efficient incremental learning system without the need for deep architecture. IEEE Trans Neural Netw Learn Syst 29(1):10–24
3. Zhu G, Zhang J (2015) Evacuation capacity evaluation of urban rail transit station in fire. China Saf Sci J 25 (in Chinese)
4. Vermuyten H, Beliën J, De Boeck L (2016) A review of optimisation models for pedestrian evacuation and design problems. Saf Sci 2016(87):167–178
5. Gosavi HD, Smith JMG, Smith M (1995) Asymptotic bounds of throughput in series-parallel queueing networks. Comput Oper Research 22:1057–1073
6. Cruz FRB, Smith JMG, Queiroz DC (2005) Service and capacity allocation in M/G/c/c state-dependent queueing networks. Comput Oper Research 32:1545–1563
7. Gosavi HD, Smith JM (1997) An algorithm for sub-optimal routeing in series-parallel queueing networks. Int J Prod Res 35:1413–1430
8. Rahman K, Ghani NA, Kamil AA, Mustafa A (2013) Modelling pedestrian travel time and the design of facilities: a queuing approach. Plos One 8:1–11
9. Li F, Chen SK, Wang X (2014) Pedestrian evacuation modeling and simulation on metro platforms considering panic impacts. Procedia Soc Behav Sci 38(0):314–322
10. Osaragi T (2004) Modeling of pedestrian behavior and its applications to spatial evaluation. In: Proceedings of the third international joint conference on autonomous agents and multiagent systems, AAMAS 2004
11. Osorio C, Bierlaire M (2009) An analytic finite capacity queueing network model capturing the propagation of congestion and blocking. Eur J Oper Res 196:996–1007
12. Wu Y, Xu J, Jia L, Qin Y (2018) Estimation of emergency evacuation capacity for subway stations. J Transp Saf Secur 10:586–601

Estimate of Railway Line Capacity Under Adverse Operation Condition

Jie Xu, Mengdi Liang and Yanhui Wang

Abstract Amongst train operation factors that affect railway capacity, train speed is crucial. It fluctuates due to train operation equipment or facility impairments under adverse operation conditions. Firstly, the train operation models are transformed into Markov chain models (MCM). By solving the ergodic MCM, the actual speed of train can be estimated. Thus, the transfer probability matrix of train speed could be derived. According to the matrices, the transfer probability with fuzzy states using the conditional probability of the fuzzy event, the railway capacity can be estimated.

Keywords Available railway capacity · Markov chain model · Steady-state probability · Fuzzy set · Adverse operation condition

1 Introduction

When disturbance or disruption occurs, the train operation plan has to be modified [1]. The very important prerequisite condition for rescheduling of the plan is estimation of line capacity under adverse operation conditions [2]. The goal of railway line capacity analysis is to determine the maximum number of trains that can operate on a given railway infrastructure during a specific time period, under operational conditions [3]. However, the concept of railway line capacity cannot be easily defined or quantified [4]. Generally, it is assumed that the capacity is the outcome of a close interaction between different equipment and infrastructure during train operation [5]. In China, railway line capacity represents the maximal number of trains that can go through special line or section during a certain time period, such as 24 h [6]. Currently, railway managers and researchers usually utilize empirical or analytical methods to calculate line capacity [3, 7–9]. Optimizing the application of railway infrastructure is a complex and difficult task [4]; therefore,

J. Xu · M. Liang · Y. Wang (✉)
State Key Laboratory of Rail Control and Safety, Beijing Jiaotong University, Beijing, China
e-mail: jxu1@bjtu.edu.cn

Y. Qin et al. (eds.), *Proceedings of the 4th International Conference on Electrical and Information Technologies for Rail Transportation (EITRT) 2019*, Lecture Notes in Electrical Engineering 639, https://doi.org/10.1007/978-981-15-2866-8_37

numerous capacity studies have been performed in order to provide as many as possible route slots for railway operators, as well learn how much railway traffic can be supported by current railway network [4]. Railway line capacity differs under adverse operation conditions, since contingencies lead to conflicts with predefined timetables. The occurrence of impaired train, equipment, and infrastructure function gives rise to service irregularities.

As a closely related factor to capacity, train operation time calculation should consider the probabilistic aspects. And successively, it can contribute to analyzing the random chain of trains travelling through railway line [4]. This current work aims to focus on how railway line capacity fluctuates under adverse operation conditions, and hence provide some inputs for railway administrators in terms of scheduling or preliminary planning datum. For lack of commonly accepted measures of capacity calculation, it is necessary to investigate impact factors specifically under adverse operation condition. This paper proposed an approach to model the performance of train operation using SPNs under adverse condition, and interpret the steady-state probability of train operation scenario, which was isomorphic to Markov chain. Based on the transfer probability of train operation scenario and train speed, the flexible capacity of railway line capacity was estimated.

2 Line Capacity Fluctuation Under Adverse Condition

2.1 Control of Train Operation

The block and signaling systems manage the speed and headway of train. Trains running at operational speeds are separated by a headway interval, which is composed of different numbers of blocks due to different types of signaling system. Four-aspect signaling arrangements are most common for train control. The minimum headway time is calculated as following [10]:

$$t_I = [(4L_s + L_t)/v] \times 3.6 \tag{1}$$

where L_s is the length of the block (m), L_t is length of train (m), and v is the speed of the train (km/h).

As shown in (1), with a fixed block length, the faster trains run the more they go through the block. Capacity of the block or section within unit time can be calculated as following [10]:

$$c = (60 \times 60 - \tilde{t}_r - \tilde{t}_{zu})/t_I \tag{2}$$

where $\tilde{t}_r$ is the running time margin, and $\tilde{t}_{zu}$ is the added time.

Table 1 Speed limits in case of chinese train control system (CTCS) equipment failures

Failure facility	Signal failure	Track circuit failure	BTM failure	STM failure	Speed recorder failure
Speed limit (km/h)	20	60	20	20	40

Table 2 Definition of train operation scenarios corresponding speed limit

Speed limit (km/h)	300	250	200	160	110	60	40	20	0
Operation scenario	S_0	S_1	S_2	S_3	S_4	S_5	S_6	S_7	S_8

where S_0 is the train operation scenario with regular speed limit, S_i, (i = 1, 2, …, 8) are train operation scenarios subject to adverse operation conditions

2.2 Train Operation Scenarios

Adverse weather may cause damages and malfunctions of equipment and facilities, the malfunction and failures of equipment of block and signaling system will lead to corresponding train control modes. Nine train operation scenarios were defined to deal with these failures, such as signal failure, track circuit failure, balise transmission module (BTM) failure, specific transmission module (STM) failure, and train speed recorder failure. As shown in Table 1, each facility failure or operation scenario corresponds to a certain speed limit.

Based on these speed limits, train operation scenarios will be defined. Regardless of which type of failures, we considered scenarios with the same speed limit that they were homogenous. Corresponding to the change of speed limit, train operation scenarios under adverse condition were defined and shown in Table 2.

2.3 Capacity Fluctuating Under Adverse Operation Condition

The capacity calculation process will be divided into several scenarios. Assume the duration is divided into m intervals, and speeds of train are $v_1, v_2, \cdots, v_m$, accordingly. Fluctuation of available capacity under severe condition is defined. $\tilde{N}^1_{\text{mix}}$ is the normal capacity, and $\tilde{N}^2_{\text{mix}}, \cdots, \tilde{N}^{m-1}_{\text{mix}}$ and $\tilde{N}^m_{\text{mix}}$ denote the available capacity under adverse conditions. The set of section capacity C_m will be calculated based on Eqs. (1) and (2). The capacity can be described as a corresponding fuzzy set $\tilde{N}_m(m = 1, 2, \ldots)$. During $t \in [0, t_m]$, the available capacity of a given section is calculated as follows:

$$\tilde{N} = \sum_{m=0}^{m=M} \tilde{N}_{\text{mix}}^{m} \tag{3}$$

3 Model Development

3.1 Stochastic Petri Net Model

Petri net (PN) was proposed by Dr. Petri in 1962, assigning firing probability of transition with stochastic exponential distribution, stochastic Petri net (SPN) model will be built. The SPN is formulated as 6-tuple SPN = $(P, T, F, W, M_0, \lambda)$, where $P = \{p_1, p_2, \cdots, p_n\}$, $n > 0$, refers to the finite set of place; $T = \{t_1, t_2, \cdots, t_m\}$, $m > 0$, refers to the finite set of transition and satisfies $P \cap T = \emptyset$ and $P \cup T \neq \emptyset$; $F \subseteq (P \times T) \cup (T \times P)$ is the flow relation; W is a weigh function, weighting all arc lines. $\omega(p, t)$ or $\omega(t, p)$ denotes the weight of directed arc going from p to t or from t to p. $M \subseteq (P \rightarrow N)$ is the marking reflecting token distribution in all places in the SPN, where $N = \{0, 1, 2, \ldots\}$; forward markings decide the set of all possible states of the system with a given initial state; M_1 is the initial marking. $\Lambda = \{\lambda_1, \lambda_2, \cdots, \lambda_i\}$ denotes the set of average transition firing rates for the exponentially distributed transition firing times [11],

$$F_x(X) = P(X \leq x) = 1 - e^{-\lambda_i x}, \quad x \geq 0 \tag{4}$$

3.2 Scenario Transfer of Train Operation Using SPN

For safety concern, train driver will receive messages from dispatchers and adjusts speed to a lower predefined speed limit under adverse condition. Assuming that adjusting time of train is subject to exponential distribution, SPNs can depict the above-mentioned train operation processes as in Fig. 1. The adjusting time of train is considered as transition firing time of the SPN model as in Fig. 2.

Figure 2 shows the SPN model for train operation under adverse condition. In the model, we assume that a safety distance between two consecutive trains equals to minimum headway. The transfer of scenario sequences involves delay according to Fig. 2, and the mean adjusting time of train operation is computed according to the acceleration and difference between speed limits of the two train operation scenarios.

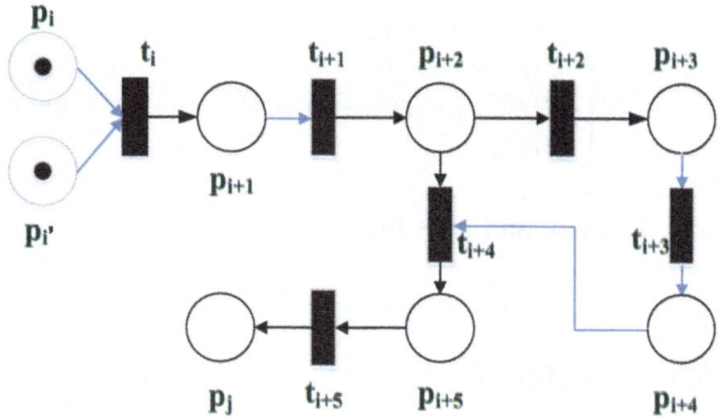

Fig. 1 SPN for train operation adjustment under adverse condition

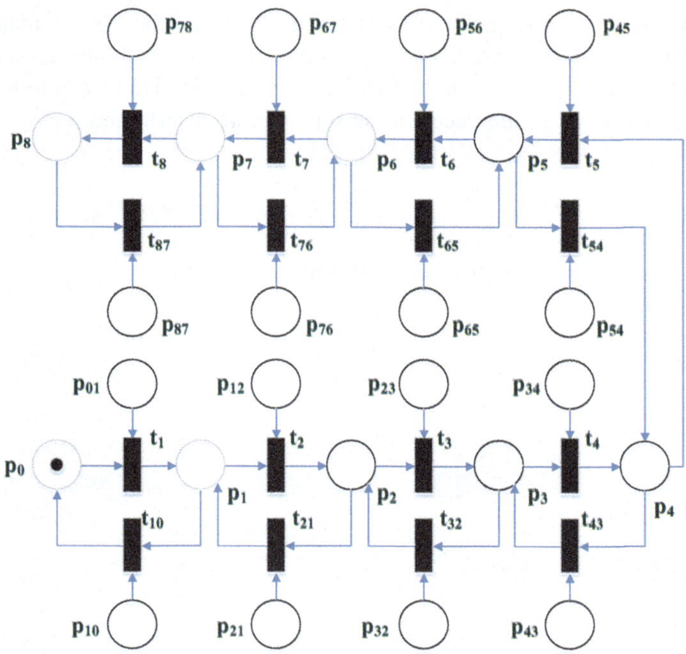

Fig. 2 Petri net model for operation scenario transfer

Enabling and simulation rules of transition [12]:

- Transition $t \in T$ is enabled, when and only when in respect of all $p \in P$, $\omega(p, t) = 1$, $M(p) > 0$;
- After the enabled transition t is stimulated by the marking M, a new marking M' is generated, in respect of $\forall p \in P$, is

$$M'(p) = \begin{cases} M(p) - \omega(p,t), & p \in^{\circ} t \\ M(p) + \omega(p,t), & p \in t^{\circ} \\ M(p) + \omega(t,p) - \omega(p,t), & p \in^{\circ} t \wedge p \in t^{\circ} \\ M(p), & \text{other} \end{cases} \tag{5}$$

where $^{\circ}t$ is the pre-set of p, and t° is the post-set of p.

3.3 Scenarios Transfer

When an SPN model is bounded, it is isomorphic to a bounded Markov chain (MC) [13]. Subsequently, the SPN model of train operation process under adverse operation condition will be considered as homogeneous MC, as shown in Fig. 3.

The directed arcs in Fig. 3 represent the transition of the PN for major events from one state to another. Train's acceleration/deceleration is assumed constant and no more than 0.8 m/s^2 according to ISO reference [14]. Train operation scenario transfers about $S_i \rightarrow S_j$, the deceleration time can be determined,

$$t_{ij} = \frac{|v_i - v_j|}{a} \tag{6}$$

The activation rate of transition of SPN is computed as,

$$\lambda_{ij} = \frac{1}{t_{ij}} \tag{7}$$

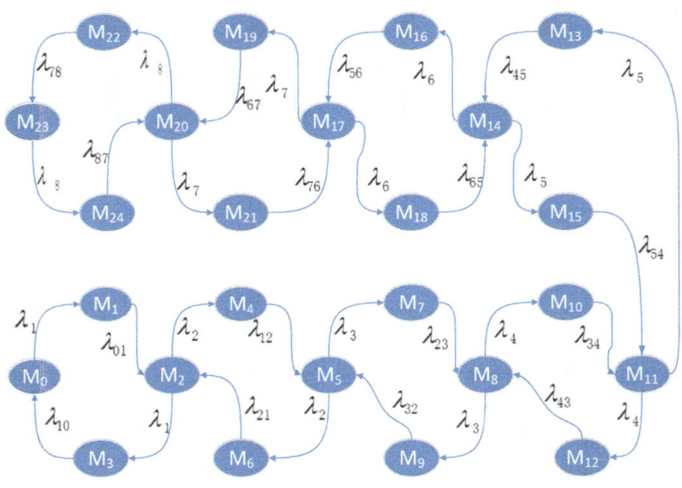

Fig. 3 MC for SPN model for scenario transfer

Define π_j as the probability of the system state M_i when SPN for train operation scenario is in the steady state. Also $\Pi = (\pi_0, \pi_1, \ldots, \pi_q)$ is probability matrix of steady state of MC, $A = (a_{ij})_{(q+1)\times(q+1)}$ denotes probability matrix of train operation scenario transfer, the system state probabilities expression can be obtained:

$$\begin{cases} \Pi A = 0 \\ \sum_{i=0}^{q} \pi_j = 1 \end{cases} \tag{8}$$

Solving the above linear equations, Π can be acquired, indicating the stable probability of possible states of train operation scenario. $P = [P(M_1), P(M_2), \ldots, P(M_n)]$ represents a row vector of the probability distribution of steady states. The busy probability of places indicator can be calculated as follows [15],

$$P(M(p) = i) = \sum_{j} P[M_j], M_j \in [M_0] > , M_j(p) = i \tag{9}$$

where $[M_0] >$ indicates the set of reachable marking from M_0.

3.4 Capacity Changing with Fuzzy States

Define states of operation condition as fuzzy partition, i.e., a set of fuzzy states $\tilde{A} = \{\tilde{A}_1, \tilde{A}_2, \ldots, \tilde{A}_n\}$ with corresponding membership function $\mu_{\tilde{A}} = \{\mu_{\tilde{A}_1}, \mu_{\tilde{A}_2}, \ldots, \mu_{\tilde{A}_n}\}$. $\sum_{i=1}^{n} \mu_{\tilde{A}_i}(M_t) = 1$ for all $M_t \in M$, and $0 < \mu_{\tilde{A}_i}(M_t) < 1$ for all $\tilde{A}_i \in A$, where n is a positive integer. The probability of fuzzy initial state $\tilde{P}(\tilde{A}_i) = \tilde{P}(X_0 = \tilde{A}_i)$ is calculated as,

$$\tilde{P}(\tilde{A}_i) = \sum_{s=0}^{N} p_s \mu_{\tilde{A}_i}(s) \tag{10}$$

in which p_s is the probability of speed that is determined by operation scenario, $\mu_{\tilde{A}_i}(s)$ is membership function of speed grade according to various speed on operation scenario. Although railway capacity is determined by train speed and headway, dispatchers estimate the capacity with a train speed, which is less than speed limit. To deal with fuzziness of the train speed, trichotomy method is used. Drivers associate the train speed interval with two states denoted $\tilde{A}_{\text{low}}$ and $\tilde{A}_{\text{high}}$ which correspond to "low interval" and "high interval," respectively. Correspond to the two intervals, we find the two fuzzy centers of train speed with fuzzy k-means clustering, i.e., $\tilde{v}^K_{\text{low}}$ and $\tilde{v}^K_{\text{high}}$. The intervals are described with triangular fuzzy numbers and interpreted as membership functions, respectively [16].

$$\tilde{v}_{\text{low}} = \begin{cases} 0 & \text{for } v < \min(v_{\text{low}}) \\ \frac{v - \min(v_{\text{low}})}{\tilde{v}_{\text{low}}^{K} - \min(v_{\text{low}})} & \text{for } \min(v_{\text{low}}) < v < \tilde{v}_{\text{low}}^{K} \\ \frac{\max(v_{\text{low}}) - v}{\max(v_{\text{low}}) - \tilde{v}_{\text{low}}^{K}} & \text{for } \tilde{v}_{\text{low}}^{K} < v < \max(v_{\text{low}}) \\ 0 & \text{for } v > \max(v_{\text{low}}) \end{cases} \tag{11}$$

and,

$$\tilde{v}_{\text{high}} = \begin{cases} 0 & \text{for } v < \min(v_{\text{high}}) \\ \frac{v - \min(v_{\text{high}})}{\tilde{v}_{\text{low}}^{K} - \min(v_{\text{high}})} & \text{for } \min(v_{\text{high}}) < v < \tilde{v}_{\text{high}}^{K} \\ \frac{\max(v_{\text{high}}) - v}{\max(v_{\text{high}}) - \tilde{v}_{\text{low}}^{K}} & \text{for } \tilde{v}_{\text{high}}^{K} < v < \max(v_{\text{high}}) \\ 0 & \text{for } v > \max(v_{\text{high}}) \end{cases} \tag{12}$$

Given the operation scenario initial state m, $m \in \{0, 1, \ldots, N\}$, the conditional probability of the fuzzy state $\tilde{A}_j$, with $j \in \{1, 2, \ldots, n\}$ is computed as follow,

$$\tilde{P}(\tilde{A}_i/m) = P\{\tilde{X}_1 = \tilde{A}_j/X_0 = m\} = \sum_{s=0}^{N} p_{ms}\mu_{\tilde{A}_j}(s) \tag{13}$$

in which p_{ms} is transition probability of train speed or operation scenario from state m to s. And the transition probability from fuzzy initial state $\tilde{A}_i$ to fuzzy final state $\tilde{A}_j$, $i, j \in \{1, \ldots, n\}$, is

$$\tilde{P}(\tilde{A}_j/\tilde{A}_i) = \sum_{m=0}^{N} P(\tilde{A}_j/m)\frac{p_m \mu_{\tilde{A}_i}(m)}{\tilde{P}(\tilde{A}_i)} \tag{14}$$

in which $P(\tilde{A}_i)$ is probability of fuzzy variable $(\tilde{A}_i)$ of line capacity, p_m is probability of train speed at state m. Given the transition probability of the initial state m, $(m \in \{0, \ldots, N\})$ to the fuzzy final state $\tilde{A}_j$, $(j \in \{1, \ldots, n\})$.

Then, define matrices Q_1,

$$Q_1 = \begin{bmatrix} \mu_{\tilde{A}_1}(0) & \mu_{\tilde{A}_2}(0) & \cdots & \mu_{\tilde{A}_n}(0) \\ \mu_{\tilde{A}_1}(1) & \mu_{\tilde{A}_2}(1) & \cdots & \mu_{\tilde{A}_n}(1) \\ \vdots & \vdots & \ddots & \vdots \\ \mu_{\tilde{A}_1}(N) & \mu_{\tilde{A}_2}(N) & \cdots & \mu_{\tilde{A}_n}(N) \end{bmatrix} \tag{15}$$

and Q_2,

$$Q_1 = \begin{bmatrix} \frac{p_0\mu_{\tilde{A}_1}(0)}{P(\tilde{A}_1)} & \frac{p_1\mu_{\tilde{A}_1}(1)}{P(\tilde{A}_1)} & \cdots & \frac{p_N\mu_{\tilde{A}_1}(N)}{p(\tilde{A}_1)} \\ \frac{p_0\mu_{\tilde{A}_2}(0)}{P(\tilde{A}_2)} & \frac{p_1\mu_{\tilde{A}_2}(1)}{P(\tilde{A}_2)} & \cdots & \frac{p_N\mu_{\tilde{A}_2}(N)}{P(\tilde{A}_2)} \\ \vdots & \vdots & \ddots & \vdots \\ \frac{p_0\mu_{\tilde{A}_n}(0)}{P(\tilde{A}_n)} & \frac{p_0\mu_{\tilde{A}_n}(0)}{P(\tilde{A}_n)} & \cdots & \frac{p_N\mu_{\tilde{A}_n}(0)}{P(\tilde{A}_N)} \end{bmatrix} \tag{16}$$

The matrix Q_1 contains the membership function values of the fuzzy partition which determines the fuzzy states of the system $\{\tilde{A}_1, \tilde{A}_2, \ldots, \tilde{A}_n\}$. With the matrices, one can derive:

$$\tilde{P} = Q_2 P Q_1 \tag{17}$$

$\tilde{P}$ is stochastic and given that the sum of each of its rows is 1, $\tilde{N}_0^{\text{mix}} = \tilde{c}_0$ denotes the available capacity under operation scenario M_0. The capacity of next operation scenario will be calculated as follow:

$$\tilde{N}_1 = \begin{cases} \tilde{c}_0, \; p(\tilde{A}_0/\tilde{A}_0) \\ \tilde{c}_1, \; p(\tilde{A}_1/\tilde{A}_0) \\ \vdots \\ \tilde{c}_8, \; p(\tilde{A}_8/\tilde{A}_0) \end{cases} \tag{18}$$

$\tilde{N}_l$ denotes the available capacity under train operation scenario M_l and be calculated as follow:

$$\tilde{N}_l = \begin{cases} \tilde{c}_0, \; p(\tilde{A}_0/\tilde{A}_{l-1}) \\ \tilde{c}_1, \; p(\tilde{A}_1/\tilde{A}_{l-1}) \\ \vdots \\ \tilde{c}_8, \; p(\tilde{A}_8/\tilde{A}_{l-1}) \end{cases} \tag{19}$$

The capacity of a given section under a given operation scenario is [7]:

$$N_{\text{sec}} = \sum \tilde{N}_i \cdot T_i + \sum \tilde{N}_{ij}^{tr} \frac{1}{\lambda_{ij}} \tag{20}$$

4 Example

4.1 Capacity Under Adverse Condition

Assume that adverse condition lasts 24 h, and train operates initially at a regular speed of 300 km/h. We discuss the railway section capacity under different adverse condition.

The train speeds are at “low interval” under all of adverse conditions. According to λ_{kj}, we can obtain the steady-state transition probability of train speed P = (0.0170, 0.0256, 0.0314, 0.0387, 0.0412, 0.0448, 0.0433, 0.0434, 0.0435, 0.0435, 0.0435, 0.0435, 0.0435, 0.0435, 0.0435, 0.0435, 0.0435, 0.0435, 0.0435, 0.0433, 0.0432, 0.0453, 0.0411, 0.0352, 0.0282), which is considered the steady-state probability of operation conditions.

Using Eq. (13), the transition probability matrix from state A_m to state A_s can be calculated based on P_m, Q_1 and Q_2. $\tilde{N}_i^{\text{low}}$ (i = 0, 1, …, 8) denotes the available capacity of “low interval” per unit time (one hour) under adverse condition S_i. Furthermore, we can calculate the probability of capacity of all operation conditions and tabulate in Table 3. Fuzzy numbers of dynamic capacity of railway section can be computed with Eq. (12), which is shown as follow,

Table 3 Conditional probability of capacity under various operation conditions

Conditional probability of capacity	0.53	0.41	0.06	0.09	0.54	0.34	0.04	0.09	0.54	0.34	0.04	
Capacity	$\tilde{c}_0$	$\tilde{c}_1$	$\tilde{c}_2$	$\tilde{c}_0$	$\tilde{c}_1$	$\tilde{c}_2$	$\tilde{c}_3$	$\tilde{c}_1$	$\tilde{c}_2$	$\tilde{c}_3$	$\tilde{c}_4$	
Operation scenario	S_0			S_1				S_2				
Conditional probability of capacity	0.09	0.54	0.34	0.04	0.09	0.54	0.34	0.04	0.09	0.54	0.34	0.04
Capacity	$\tilde{c}_2$	$\tilde{c}_3$	$\tilde{c}_4$	$\tilde{c}_5$	$\tilde{c}_3$	$\tilde{c}_4$	$\tilde{c}_5$	$\tilde{c}_6$	$\tilde{c}_4$	$\tilde{c}_5$	$\tilde{c}_5$	$\tilde{c}_6$
Operation scenario	S_3				S_4				S_5			
Conditional probability of capacity	0.09	0.54	0.34	0.04	0.1		0.54	0.34	0.09	0.91		
Capacity	$\tilde{c}_5$	$\tilde{c}_6$	$\tilde{c}_7$	$\tilde{c}_8$	$\tilde{c}_6$		$\tilde{c}_7$	$\tilde{c}_8$	$\tilde{c}_7$	$\tilde{c}_8$		
Operation scenario	S_6				S_7				S_8			

$$\tilde{N}_0^{\text{low}} = \begin{cases} \tilde{c}_0^{\text{low}}, 0.53 \\ \tilde{c}_1^{\text{low}}, 0.41 \\ \tilde{c}_2^{\text{low}}, 0.06 \\ \tilde{c}_3^{\text{low}}, 0 \\ \tilde{c}_4^{\text{low}}, 0 \\ \tilde{c}_5^{\text{low}}, 0 \\ \tilde{c}_6^{\text{low}}, 0 \\ \tilde{c}_7^{\text{low}}, 0 \\ \tilde{c}_8^{\text{low}}, 0 \end{cases} \Rightarrow \tilde{N}_0^{\text{low}} = \begin{cases} (14.16, 16.37, 18.57), 0.53 \\ (11.92, 14.24, 16.17), 0.41 \\ (9.19, 10.97, 12.46), 0.06 \end{cases} \quad (21)$$

where $\tilde{c}_i^{\text{low}}$ denotes fuzzy number of estimates of capacity at "low interval" under operation scenario Si. Hence, $\tilde{N}_0^{\text{low}} = (12.94, 15.17, 17.21)$. Similarly, we can compute the capacity under various conditions, i.e., $\tilde{N}_1^{\text{low}} = (11.13, 13.24, 15.03)$, $\tilde{N}_2^{\text{low}} = (8.69, 10.37, 11.78), \tilde{N}_3^{\text{low}} = (6.49, 7.74, 8.79), \tilde{N}_4^{\text{low}} = (4.13, 4.95, 5.65)$, $\tilde{N}_5^{\text{low}} = (2.16, 2.64, 3.05)$, $\tilde{N}_6^{\text{low}} = (1.02, 1.31, 1.55)$, $\tilde{N}_7^{\text{low}} = (0.28, 0.42, 0.52)$, and $\tilde{N}_8^{\text{low}} = (0.03, 0.04, 0.05)$. Assume that maintenance gap equals to zero and the adverse condition last 24 h, $N^{\text{low}} = \sum \tilde{N}_i^{\text{low}} = (124.97, 149.04, 169.75)$. Similarly, $\tilde{N}_i^{\text{high}} (I = 0, 1, 2, \ldots, 8)$ denotes the available capacity of "high interval" of adverse condition, consider the train speeds are at "high interval" under adverse conditions, $N^{\text{high}} = \sum \tilde{N}_i^{\text{high}} = (134.92, 155.46, 184.74)$.

The railway section has an expected capacity of 152, maximum capacity of 184, and minimum passable capacity of 125. The dispatcher would be able to make train operation plans or adjust predefined train schedule based on the expected capacity.

4.2 *Discussion*

In Fig. 4, assume that the worst operation scenarios are S_3 and S_8, respectively. Under the operation condition S_3, the available capacity is slightly affected. If the lasting duration of worst-case scenario increases, that is, λ_2 decreases, the available capacity decreases linearly. Under the operation scenario S_3, if the duration of worst-case scenario lasts one hour, the maximum available capacity is 369 for 24 h. If the duration of worst-case scenario last 23 h, the maximum available capacity will decrease 60.

Compare with the fluctuation of available capacities of various operation conditions in Fig. 5, the recovery duration will prolong and the train speeds have to slow down, which causes the train headway lengthening. Obviously, both the lasting duration of adverse operation condition and the severity of adverse condition will decrease the available capacity. Under the same condition, we recover the operation condition as soon as possible, and the available capacity will increase linearly.

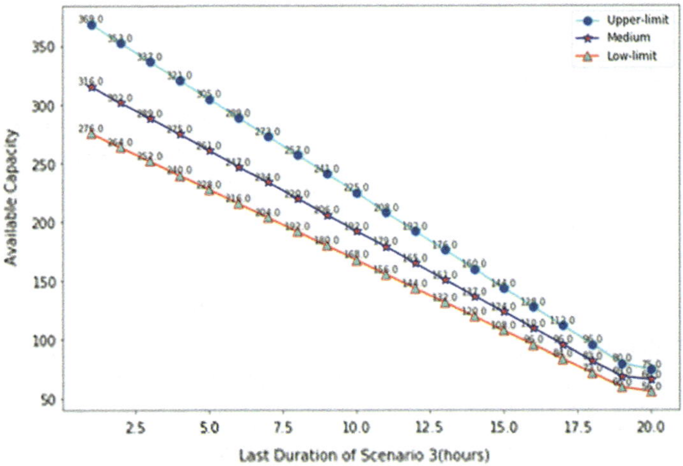

Fig. 4 Available capacity with the lasting duration of worst operation condition

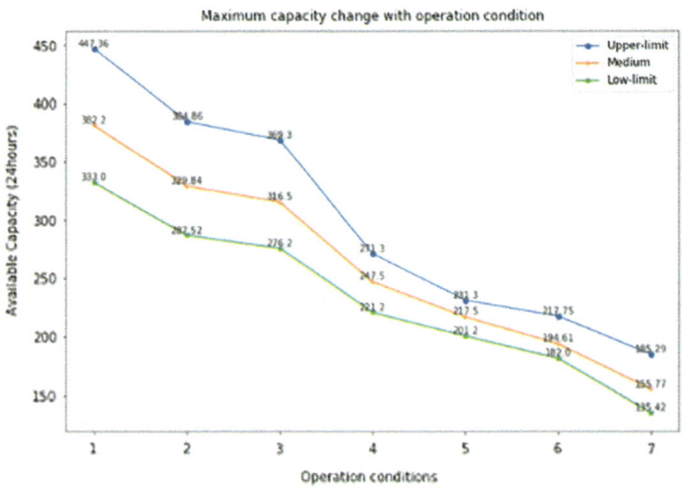

Fig. 5 Fluctuation of maximum capacity with operation condition

5 Conclusions

A novel approach for determining available capacity of railway sections was developed in this paper, based upon the SPN and fuzzy theory. Under adverse condition, the available capacity will be affected by numerous influencing factors, which are dynamic, fuzziness, and stochastic. The dominant factor is the train operation conditions, which are discrete and nondeterministic. SPN is suitable for analyzing and researching the transfer process of train operation. Based on homogeneous MCM of SPN, the steady probability of operation condition will be obtained. Consequently, lasting duration of operation condition was determined by solving an MCM. Then, we will obtain the range of the available capacity of worst-case conditions in general. This approach developed a generic stochastic and fuzzy measure to present various driving manners. It can quantify the available capacity further.

Determination of available capacity, however, depended on various conditions, e.g., it relied on but not merely the approximation of interference delays. For this reason, no simulation results or other empirical were given. Empirical proof of actual capacity through exact timetabling, on the other hand, could be further investigated.

Acknowledgements This study is supported by China National "13th Five-Year" key research project "Safety assurance technology of urban rail system" (Grant No. 2016YFB1200402).

References

1. Cacchiani V, Huisman D, Kidd M, Kroon L, Toth P, Veelenturf L et al (2013) An overview of recovery models and algorithms for real-time railway rescheduling, pp 1–36
2. Cuibus O, Letia T (2011) Genetic algorithm for transitions scheduling guidance. Acta Tech Napocensis. 52(3):14–18
3. Abril M, Barber F, Ingolotti L, Salido MA, Tormos P, Lova A (2008) An assessment of railway capacity. Transp Res Part E Logist Transp Rev 44(5):774–806
4. Mussone L, Wolfler Calvo R (2013) An analytical approach to calculate the capacity of a railway system. Eur J Oper Res 228(1):11–23
5. Sameni MK, Landex A, Preston J (2005) Developing the UIC 406 method for capacity analysis, pp 1–19
6. Meng XX-L, Jia L-ML, Qin Y, Xu J, Wang L. Calculation of railway transport capacity in an emergency based on markov process. J Beijing Inst Technol 21(1):7–10
7. Wang L, Qin Y, Xu J, Jia L. Capacity determination approach of railway section in speed restriction conditions. J Tongji Univ Natural Sci 42(6):880–6
8. Burdett RL, Kozan E (2006) Techniques for absolute capacity determination in railways. Transp Res Part B Methodol 40(8):616–632
9. Harrod S (2009) Capacity factors of a mixed speed railway network. Transp Res Part E Logist Transp Rev 45(5):830–841
10. Wang H, Schmid F, Chen L, Roberts C, Xu T (2013) A topology-based model for railway train control systems. IEEE Transcation Intell Transp Syst 14(2):819–827
11. Lin C, Marinescu DC (1988) Stochastic high-level petri nets and applications. IEEE Trans Comput 37(7):815–825
12. Shi C, Zhong M, Nong X, He L, Shi J, Feng G (2012) Modeling and safety strategy of passenger evacuation in a metro station in China. Saf Sci
13. Molloy MK (1982) Performance analysis using stochastic petri nets. IEEE Trans Comput C-31(9):913–7

14. ISO (2014) Mechanical vibration and shock—evaluation of human exposure to whole-body vibration—Part 1: general requirements
15. Shan S, Wang L, Li L (2012) Modeling of emergency response decision-making process using stochastic petri net : an e-service perspective, pp 363–76
16. Gani AN, Assarudeen SNM (2012) A new operation on triangular fuzzy number for solving fuzzy linear programming problem. Appl Math Sci 6(11):525–532

Maintenance Optimization Method Based on Service Status for Train Rolling Bearings

Meng Ye, Yong Qin, Hui Ma, Yong Fu and Xiaoqing Cheng

Abstract According to the current maintenance problems and the expectations of the operators, a fuzzy multi-objective maintenance model with the objectives of maximizing availability and minimizing cost under reliability constraint is developed, which is presented in this paper. Offline condition monitoring modes are discussed. An adjusting monitor interval method is further proposed. One kind of bearing condition data as a case study is used to verify the proposed model. The results from the case study show that the proposed model offers the best maintenance time and maintenance strategy based on the condition of the bearing is given, which has accomplished a whole process of the scientific proactive maintenance for the train rolling bearings.

Keywords Rolling bearing · Maintenance optimization · Fuzzy multi-objective

1 Introduction

As a key component of rail trains, rolling bearings are widely used in various mechanical parts of trains and are responsible for frequent transmission and heavy loads. Their operating status is directly related to driving safety.

For train rolling bearings, most scholars have studied maintenance methods based on reliability and risk, as well as condition-based maintenance [1]. Reference [2] established a semi-Markov maintenance optimization model, assuming that the degradation rate of each failure phase is the same, and the decision graph of dynamic threshold optimization is obtained. Reference [3] used Weibull proportional hazards

M. Ye · Y. Qin (✉) · H. Ma · Y. Fu · X. Cheng
State Key Laboratory of Rail Traffic Control and Safety, Beijing Jiaotong University, Beijing 100044, China
e-mail: yqin@bjtu.edu.cn

Y. Qin
Beijing Engineering Research Center of Urban Traffic Information Intelligent Sensing and Service Technologies, Beijing Jiaotong University, Beijing 100044, China

Y. Qin et al. (eds.), *Proceedings of the 4th International Conference on Electrical and Information Technologies for Rail Transportation (EITRT) 2019*, Lecture Notes in Electrical Engineering 639, https://doi.org/10.1007/978-981-15-2866-8_38

regression modeling theory to establish a functional relationship between the performance parameters and the service time, and the optimal maintenance decision threshold is determined with the minimizing maintenance cost and maximizing availability. Reference [4] established a condition space model based on the Gamma degradation process and used the service life prediction results to obtain the best maintenance time with the minimizing cost.

This paper studies multi-objective maintenance optimization model based on service status for train rolling bearings. Firstly, according to the actual problems, several maintenance optimization objectives are proposed, and the fuzzy multi-objective maintenance model is selected. Then, the proportional hazard model is used as the input of the maintenance optimization model. For offline monitoring, a method for dynamically adjusting monitoring interval is proposed. Finally, one kind of bearing condition data as a case study is used to verify the proposed model.

2 Maintenance Time Optimization Model

For bearings, this paper takes reliability as constraint, maximizing availability as main optimization objective, maintenance cost as secondary objective and maintenance opportunity as decision variable. Thus, on the basis of ensuring the safety and reliability of the train, the utilization rate of the vehicle is increased and the maintenance cost is reduced.

Firstly, the Weibull distribution is used as the basis failure rate function to obtain the proportional hazards model.

$$\begin{aligned} h(t,X) &= \left[\frac{\gamma}{\eta}\cdot\left(\frac{t}{\mu}\right)^{\gamma-1}\right]\cdot\exp(X\beta) \\ &= \left[\frac{\gamma}{\eta}\cdot\left(\frac{t}{\mu}\right)^{\gamma-1}\right]\cdot e^{\beta_1X_1(t)+\beta_2X_2(t)+\cdots+\beta_mX_m(t)} \end{aligned} \tag{1}$$

where $h(t,X)$ is a failure function with a covariate, related to time t and covariate X; $X(t) = [x_1(t), x_2(t), \ldots, x_m(t)]$, X is the covariate that affects the probability of system failure at time t, and m is the dimension of X; $\beta = [\beta_1, \beta_2, \ldots, \beta_m]$, β is the regression parameters vector; γ is shape parameter; η is the scale parameter.

Before applying the proportional hazard model, we need to determine the values of the three parameters in Eq. (1) and get the concrete form of the model. The parameters estimated by the MDPSO algorithm optimized by MATLAB are shown in Table 1.

Table 1 Proportional hazard model parameters

Parameter	γ	η	β_1	β_2	β_3
Value	16.08	423.73	5.58	4.40	−2.67

Taking the above parameters into the proportional hazard model, the proportional failure rate function for such bearings is

$$h(t,X) = \left[\frac{16.08}{423.73} \cdot \left(\frac{t}{423.73}\right)^{16.08-1}\right] \cdot e^{5.58X_1(t) + 4.40X_2(t) - 2.67X_3(t)} \tag{2}$$

Next, we analyze the maintenance time optimization with the maximizing availability. The condition-based maintenance includes preventive maintenance and corrective maintenance [5]. In preventive maintenance, the bearing will be disassembled and replaced, and the average preventive maintenance time is $\overline{T_p}$. If a fault occurs during operation, it needs to be repaired, and the average corrective maintenance time is $\overline{T_f}$. Combined with the proportional hazard model, the availability of train bearings is expressed as follows:

$$A(t) = \frac{\int_0^{\infty} R(t, X(t)\mathrm{d}t}{\int_0^{\infty} R\,[t, X(t)\mathrm{d}t + R(t, X(t)]\overline{T_p} + [1 - R(t, X(t))]\overline{T_f}} \tag{3}$$

According to the field experience, the minimizing reliability R^* of train running department is 0.8 or 0.9 [6], which is taken as the constraints of optimization. Considering that the conservation can prolong the residual life of component, preventive maintenance should be carried out when reliability is reduced to $(1-R^*)/4 + R^*$. Therefore, the average maintenance cost for a life cycle is

$$\bar{C}(T) = \frac{\bar{C}_{p1} + R[t, X(t)]\bar{C}_{p2} + [1 - R(t, X(t))]\bar{C}_f}{\int_0^{\infty} R[t, X(t))\mathrm{d}t]} \tag{4}$$

where $\bar{C}_{p1}$ is the conservation cost, $\bar{C}_{p2}$ is the replacement cost, and $\bar{C}_f$ is the repair cost.

For the multi-objective maintenance optimization, this paper uses the maximum and minimum fuzzy multi-objective decision-making method to weigh the optimization results of availability, maintenance cost and reliability constraints. There are i objective functions $f_1(X), f_2(X), \ldots, f_i(X)$, which maximize at $X \in D$, and D is a feasible solution set. In order to replace the optimal solution with satisfaction, we need to get the range of each objective function, that is, the maximum ideal expectation $f_i^*(X)$ and the minimum value $f_{i,\min}(X)$ under the condition. Using the linear membership function to express the decision-makers' requirements for objectives, then the membership function of the ith objective is

$$\mu_i(X) = \begin{cases} 1, & f_i(X) > f_i^* \\ \frac{[f_i(X) - f_{i,\min}]}{f_i^* - f_{i,\min}}, & f_{i,\min} \le f_i(X) \le f_i^* \\ 0, & f_i(X) < f_{i,\min} \end{cases} \tag{5}$$

When solving multi-objective problems, we try to make the objective function value larger than the maximum ideal expectation in the fuzzy relation, if it is not satisfied, the function value is allowed to be in the range $f_{i,\min} \leq f_i(X) \leq f_i^*$.

The weights of each objective are obtained by using fuzzy logic, and the scale is obtained by using the judgment method of the importance degree between the two objectives in the analytic hierarchy process [7]. The importance degree of the corresponding membership function μ_i is reflected by the triangular fuzzy number.

$$\max k,$$
$$s.t. \begin{cases} \mu_i(X) \geq k, i = 1, 2, \ldots, c, \quad X \in D \\ a_{ij} \leq \frac{\mu_i(X)}{\mu_j(X)} \leq b_{ij} \end{cases} \tag{6}$$

where a_{ij} and b_{ij} are the upper and lower limits of the triangular ambiguity, respectively, and k is the minimum value of the membership function. i and j are the objective function numbers.

3 Offline Monitoring Maintenance Strategy

The offline monitoring mode means that the condition monitoring of bearings will be obtained at a large interval, and there will be a gap period for monitoring. Therefore, the monitoring interval needs to be taken into account in the optimization. When the monitoring condition approaches the fault threshold, the frequency of condition monitoring is increased. Let Δt be the monitoring interval, and bring the last component monitoring time closest to the failure rate threshold into proportional hazard model, h_w^* is obtained. Where h_w^* is the judgment threshold of improving monitoring frequency. The strengthen monitoring curve $M_{\mathrm{E}}(t)$ is

$$M_E(t) = \ln\frac{\eta h_w^*}{\gamma} - (\gamma - 1)\ln\frac{t}{\eta} \tag{7}$$

In the application, if the actual failure rate is less than h_w^*, the monitoring is performed according to the normal period; if the failure rate is greater than h_w^* and less than h^*, the monitoring frequency is increased; when the failure rate is greater than h^*, the preventive maintenance is performed.

The size of the monitoring period can be determined by the probability of a sudden failure from the current monitoring time to the next monitoring time. Set $k\Delta t$ as the current monitoring time and $(k+1)\Delta t$ as the next monitoring time. $P_{f\Delta t}$ is the probability of a sudden failure during the $[k\Delta t, (k+1)\Delta t]$. Finding a satisfactory maximum Δt, it is the next monitoring interval, the specific process is shown in Fig. 1.

When PM is equal to 0, it means that the component is working normally; when PM is equal to 1, it means that the reliability of the component is degraded; when

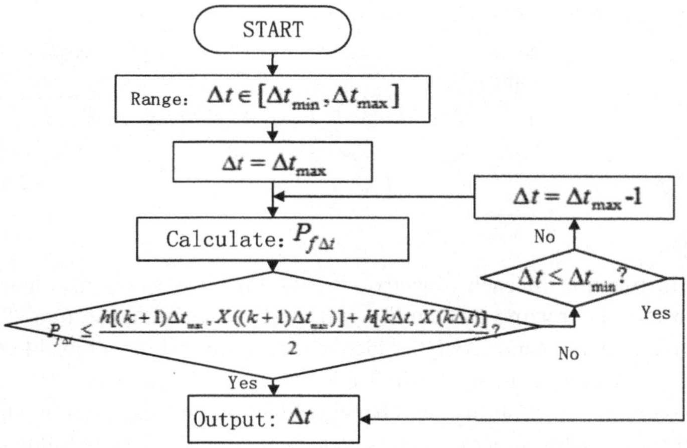

Fig. 1 Flow process of dynamically adjusting status monitor interval

Table 2 Decision-making basis of rolling bearings in offline mode

PM	Decision making	Maintenance strategy
0	$\gamma_1 Z_1(t) + \gamma_2 Z_2(t) \le \ln\frac{\eta h^*}{\beta} - (\beta - 1)\ln\frac{t}{\eta}$	Continue to monitor
1	$R[t, X(t)] = (1 - R*)/4 + R*$	Conservation
2	$\ln\frac{\eta h_w^*}{\gamma} - (\gamma - 1)\ln\frac{t}{\eta} \le \gamma_1 Z_1(t) + \gamma_2 Z_2(t) < \ln\frac{\eta h^*}{\gamma} - (\gamma - 1)\ln\frac{t}{\eta}$	Strengthen monitoring frequency
3	$\sum_{i=1}^{n} \beta_i X_i(t) \ge \ln\frac{\eta h^*}{\gamma} - (\gamma - 1)\ln\frac{t}{\eta}$	Preventive maintenance

PM is equal to 2, the monitoring frequency is strengthened; when PM is equal to 3, the component needs replacement. Table 2 shows the optimized component maintenance strategy.

4 Case Verification and Result Analysis

In this example, the bearing data provided by IMS is used for example verification. In Ref. [8], there are relevant parameters for vehicle equipment maintenance. Assuming that the average preventive maintenance time $\overline{T_p}$ is 1.5 h, and the average corrective maintenance $\overline{T_f}$ is 5 h, and the average conservation cost $\bar{C}_{p1}$ is 120 yuan, and the replacement cost $\bar{C}_{p2}$ is 320 yuan, and the repair cost $\bar{C}_f$ is 600 yuan, and the reliability threshold of the bearing is selected as 0.8.

When the fuzzy multi-objective decision-making method is used for optimization, we need to optimize separately with single objective, and then obtain the

Table 3 Optimal solution of each objective

Optimal target	Optimal solution (h)	f1 (%)	−f2 (yuan)
Max f1	$T_{\text{bestA}} = 221$	$f_1^* = f_1(T_{\text{bestA}}) = 99.75$	$-f_2(T_{\text{bestA}}) = -2.01$
Max f2	$T_{\text{bestC}} = 250$	$f_1(T_{\text{bestC}}) = 99.62$	$-f_2^* = f_2(T_{\text{bestC}}) = -1.88$
$\min[f_1(T_{\text{rul}}), f_1(T_{\cos t})]$		$f_{1,\min} = 99.62$	$-f_{2,\min} = -2.01$

membership function of each objective. According to the best maintenance time T_{bestA} with the maximizing availability and the best maintenance time T_{bestC} with the minimizing maintenance cost, the ideal expectation and the allowable minimum value of each objective are shown in Table 3.

The maximizing availability membership function and the minimizing maintenance cost membership function can be obtained from Eq. (5) as follows:

$$\mu_1(t) = \begin{cases} 1, & f_1(t) > 99.75\% \\ \frac{[f_1(t) - 99.62\%]}{99.75\% - 99.62\%}, & 99.62\% \le f_1(t) \le 99.75\%, \\ 0, & f_1(t) < 99.62\% \end{cases}$$

$$\mu_2(t) = \begin{cases} 1, & 1.88 < f_2^* \\ \frac{[2.01 - f_2(t)]}{2.01 - 1.88}, & 1.88 \le f_2(t) \le 2.01 \\ 0, & f_2(t) > 2.01 \end{cases}$$

By consulting on-site personnel, it is determined that the availability objective is slightly more important than the maintenance cost, and the corresponding triangular fuzzy number is (1, 3, 5). Finally, the optimal maintenance time interval T_m is 235 h, the maximizing availability is 99.72%, and the average maintenance cost is 1.9203 yuan.

For the offline monitoring system, verify with a failed bearing, assuming that the monitoring is performed at intervals of 24 h. Bringing condition data at time $T_m - \Delta t$ into the proportional hazards model, you can get the failure rate threshold h_w^* which is 0.0004. The obtained maintenance condition judgment curves and the status points of the bearing monitored at 24 h intervals are shown in Fig. 2.

It can be seen from the figure that the bearing should start to strengthen the monitoring frequency at 217 h. Assuming that the minimizing monitoring interval is 6 h, then $\Delta t \in [6, 24]$.When $P_{f\Delta t}$ is less than or equal to 0.0019, the maximum Δt is 18 h. Therefore, the next monitoring time interval should be 235 h. The condition point of strengthening monitoring is very close to maintenance. It can be judged that the condition-based is needed at 263 h. Finally, Table 4 gives specific maintenance strategies.

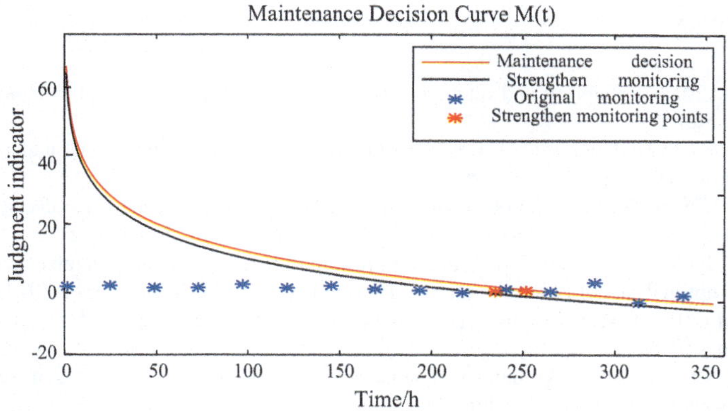

Fig. 2 Maintenance decision curve of bearing in offline monitor with adjusting monitor

Table 4 Decision-making basis of rolling bearing in offline mode

PM	Time (h)	Maintenance strategy
0	$t < 261$	Continue to monitor
1	$t = 227$	Conservation
2	$t = 217$	Strengthen monitoring, monitoring interval is 18 h
3	$t = 263$	Preventive maintenance

5 Conclusion

Aiming at the blindness of periodic preventive maintenance, the fuzzy multi-objective maintenance model with maximizing availability, minimizing maintenance cost and reliability constraint is established. In view of the particularity of monitoring interval in offline monitoring, the strategy of strengthening monitoring is adopted, and the specific method of dynamically adjusting the monitoring interval is proposed. Through the example verification, the optimization results of the maintenance strategy for the failed bearing are obtained.

Acknowledgements The authors gratefully acknowledge the financial support for this research from the National Key Research and Development Program of China (2016YFB1200402), High-level Talents Training Program of Ministry of Transportation of China (No. I18I00010), and Major Program of National Natural Science Foundation of China (No. 61833002): Fundamental theory and methods in operational risk assessment and control of high speed train.

References

1. Yan JH (2012) Reliability and intelligent maintenance. Harbin Institute of Technology Press (in Chinese)
2. Kallen MJ, van Noortwijk JM (2005) Optimal maintenance decisions under imperfect inspection. Reliab Eng Syst Saf 90:177–185
3. Zuo Hf, Zhang Hj, Qi X (2006) Condition based aero-engine maintenance decision method using proportional hazards model. J Aerosp Power 04:716–721 (in Chinese)
4. Gu YB, Jia Yx, Zhang Yb (2013) Study on residual life prediction and maintenance decision-making optimal model based on gamma deterioration process. Bearing 4:44–49 (in Chinese)
5. Ren LM (2013) Essential knowledge handbook for reliability engineers. China Standard Press (in Chinese)
6. Sun C, Xu YN (2015) Optimum maintenance circle model for rail transit vehicle's running gear. Urban Mass Transit (in Chinese)
7. Yang LL (2015) Research and application of key technologies for urban rail vehicle maintenance system. Nanjing University of Science and Technology (in Chinese)
8. Leng QJ (2011) Maintenance mode and suggestions for urban rail transit vehicles. Electr Drive Locomot 1:57–58 (in Chinese)

Design and Implementation of Online Inspection System for Linear Motor Air Gap

Lingqi Zhu, Zhaoyi Su, Jiawei Zhu and Zongyi Xing

Abstract In order to solve the problems of low detection efficiency and inaccurate detection data in the manual detection of the air gap of linear motor, an online air gap detection system based on a laser sensor is designed and developed. First of all, the hardware part of the online detection system is designed, and the control units including vehicle number recognition, axial position sensor, and PLC are described in detail, as well as the acquisition unit composed of laser displacement sensor, and then the software of the system is designed. Including data acquisition, data processing unit and web release, finally, the actual field operation test is carried out, the test results verify the accuracy and stability of the system and can meet the needs of operation and maintenance.

Keywords Linear motor · Air gap · Laser sensor · Online detection

1 Introduction

Linear motor metro is the most advanced urban rail transit mobile equipment in the world. It is famous for its linear motor traction technology [1]. Compared with the traditional rotary motor drive, the linear motor drive has strong climbing ability and good dynamic performance and has the advantages of low vibration noise and low construction cost [2]

The linear motor has been well applied in the subway drive, but the distance between the motor stator and the induction plate, that is, the air gap has become a hidden problem. If the gap between the air gaps is too large, the traction of the train is insufficient, and the energy consumption of the motor is increased. On the contrary, if the distance is too small, the motor will burn out, causing hidden

L. Zhu · J. Zhu · Z. Xing (✉)
School of Automation, Nanjing University of Science and Technology, Nanjing, China
e-mail: xingzongyi@163.com

Z. Su
Guangzhou Metro Group Co., Ltd., Guangzhou, China

L. Jia (ed.), *Proceedings of the 4th International Conference on Electrical and Information Technologies for Rail Transportation (EITRT) 2019*, Lecture Notes in Electrical Engineering 639, https://doi.org/10.1007/978-981-15-2866-8_39

dangers to the train and affecting the normal operation of the subway [3]. Therefore, the daily measurement and maintenance of the air gap are an important part of ensuring the normal operation of the train. Since the air gap measurement is a dynamic measurement and is affected by factors such as the length of the line, the line condition, and the electromagnetic field interference intensity of the linear motor, the actual measurement is difficult.

The traditional method of controlling the gap is to manually measure the height of the air gap and then adjust the height of the suspension beam. The method has a high working intensity, and the accuracy after measurement and adjustment has a considerable relationship with the level of experience of the staff, so the measurement data is not accurate enough, dynamic detection of air gaps is not possible, and unexpected events are unpredictable. Lv Liang scholar of National University of Defense Technology proposed an air gap detection method [4] for linear synchronous induction motor based on eddy current method. This method is to align the position of the linear motor cogging with two detection coils above the sleeper and pass the eddy current sensor. The equivalent inductance of the coil is detected, and the air gap height of the linear motor can be automatically measured by using the relationship between the inductance of the detecting coil and the air gap of the linear motor.

In order to ensure the safety of the train and provide scientific basis for motor maintenance, this paper proposes an air gap online detection system based on laser displacement sensor, which can achieve real-time measurement of air gap distance and provide real-time running status of trains in operation and maintenance department to ensure train safety run.

2 Overall System Design

The linear motor air gap online detection system includes a field control unit, a field data acquisition unit, a remote data transmission unit, a data analysis unit, a data storage, and a web publishing unit, and its system structure diagram is shown in Fig. 1. The on-site control unit is responsible for real-time control of the on-site data acquisition unit; the on-site data acquisition unit realizes the function of train air gap spacing acquisition; the remote data transmission unit transmits data remotely through TCP/IP protocol; the data analysis unit uses a specific algorithm for data processing. The data storage and web publishing unit saves the processed data to the server database and performs web publishing to realize data visualization.

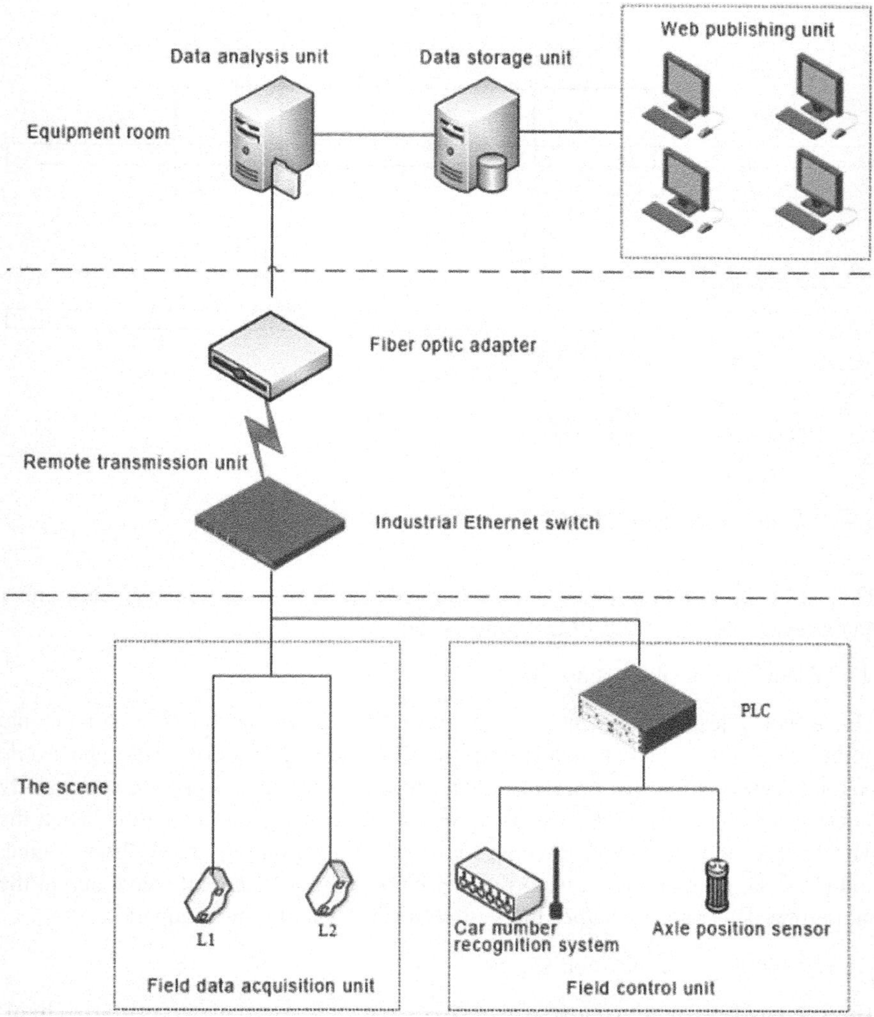

Fig. 1 System overall structure

3 System Hardware Design

The linear motor air gap online detection hardware system includes a field control unit and a field data acquisition unit. The equipment installation arrangement is shown in Fig. 2. The sensors along the train running direction are, respectively, the wheel axis position sensor, the car number identification antenna, and the laser distance measuring sensor.

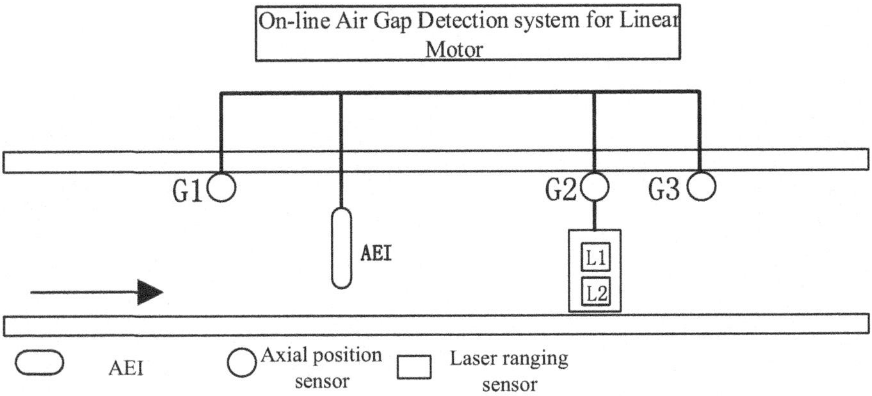

Fig. 2 Equipment installation layout

3.1 Field Control Unit

The on-site control unit mainly includes a wheel axle position sensor, a car number recognition system, and a PLC control system.

(1) Wheel axle position sensor

The wheel axle position sensor is an important part of the linear motor air gap online detection system to realize the train positioning function. The wheel axle position sensor uses the principle of electromagnetic induction to non-contactly detect the moment when the wheel passes over the axle position sensor. When the wheel passes the wheel axle position sensor, the sensor outputs an electrical signal, and the system marks the passing time by collecting the electrical signal and at the same time determines whether the system needs to start data acquisition.

(2) Car number identification system

The car number identification system uses radio frequency identification technology, which is an automatic identification technology based on the principle of radio frequency signal and space coupling. It has the characteristics of all-weather work, reliable operation, and strong adaptability to the environment [5].

The car number identification system mainly includes an electronic tag, a car number identification antenna, and a car number identification host. The identification information stored by the electronic tag through its internal circuit is used as the vehicle identification information, and the information stored in the electronic tag of each train is unique. The identification information in the electronic tag can be divided into vehicle attribute code, vehicle type, model, car number, manufacturer, and manufacturing time. The car number identification antenna is a key component for transmitting radio frequency signals and receiving car number signals. The car number identification host is a key component for receiving the car number information. The device transmits data with the PLC through the serial port communication mode and transmits the identified car number to the PLC.

(3) PLC control system

The system receives the signal outputted by the wheel axis position sensor through the PLC and realizes the transmission of the acquisition instruction, the triggering of the synchronization signal, and the reading and transmission of the vehicle number.

According to the system requirements, the PLC of the s7200-smart model produced by Siemens is selected as the core control module of the system field control unit. This PLC has different types of CPU modules with rich I/O points. The product configuration is flexible and easy to expand. At the same time, the CPU module is equipped with an Ethernet interface, which can communicate with a computer and easily networking [6].

3.2 Field Data Acquisition Unit

The on-site data acquisition unit realizes real-time data acquisition by the laser ranging sensor. The laser sensor can accurately measure the position and shape of the measured object without contact. It is mainly used for the measurement of precise geometric dimensions such as displacement, thickness, distance, and diameter [7]. The schematic diagram of laser sensor installation is shown in Fig. 3.

The laser displacement sensor uses the laser triangulation measurement principle for data measurement. The laser triangulation measurement principle is based on a simple geometric relationship [8]. Laser triangulation measurement is divided into direct type and oblique type, and direct type is suitable for measuring the surface of the object with good scattering, while oblique type measurement is suitable for

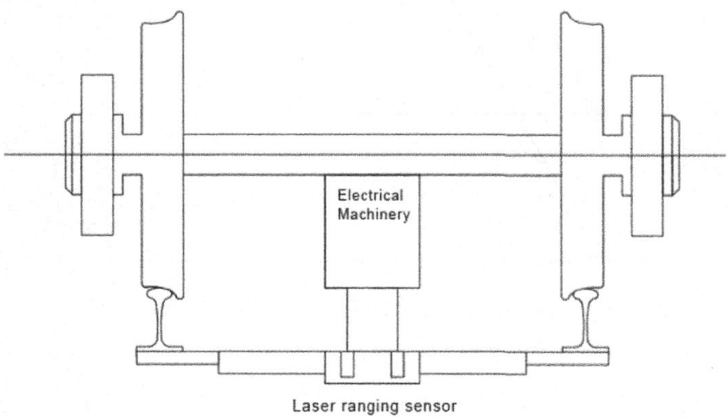

Fig. 3 Laser sensor installation diagram

measuring objects with smooth surfaces [9]. The project site uses the laser range sensor ild2300 series produced by the German rice bran company. The sensor is based on the direct laser triangulation method and has the characteristics of high sampling frequency, high measurement accuracy, and strong anti-interference ability.

Figure 4 shows the optical path diagram of the direct-type triangular displacement detection method. When the laser beam passing through the collimating lens is incident on the surface of the measuring target object, it is at an angle with the normal of the surface of the object, the optical axis of the receiving lens is placed at another angle with the normal of the surface of the measuring target object, and the reflection of the surface of the target object is measured. After the light or scattered light is received by the receiving lens, the filter is filtered and finally imaged by the photodetector [10, 11].

In Fig. 4, the laser optical axis is in line with the normal, and the angle between the reflected light OA and the normal is α. The angle between the photodetector and OA is β. The distance OP between the incident spot and the main plane of the imaging lens is b, the distance PA between the main plane of the imaging lens and

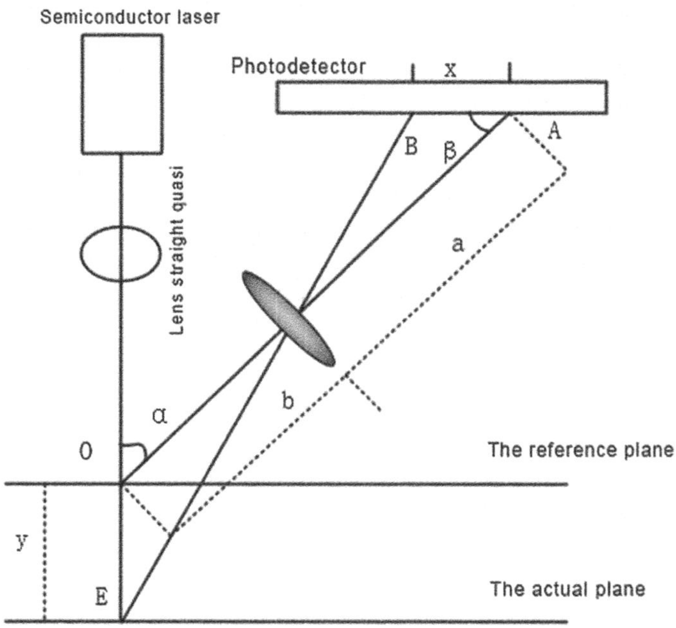

Fig. 4 Direct light triangle displacement detection optical path diagram

the photodetector is α, the surface distance of the measured object is y. The distance AB of the spot displacement due to the movement of the object to be measured is x. The focal length of the lens is over B point, the E point is the vertical line of the reflected light OA, and the vertical foot is C point and D point, respectively.

$$\frac{\overrightarrow{BC}}{\overrightarrow{ED}} = \frac{\overrightarrow{PB}}{\overrightarrow{PE}} = \frac{a - \overrightarrow{AC}}{b + \overrightarrow{OD}} \tag{1}$$

In Formula (1) $\vec{BC} = x \sin\beta$, $\vec{AC} = x\cos\beta$, $\vec{OD} = \vec{OE}\sin\beta$, also because $\vec{OE} = y$. So, taking the above formula into Eq. (2), you can get

$$\frac{x \sin\beta}{y\sin\alpha} = \frac{a - x\cos\beta}{b + y\cos\alpha} \tag{2}$$

Figure 4 shows that $u = b$ and $v = a$. According to the lens imaging formula $\frac{1}{u} + \frac{1}{v} = \frac{1}{f}$, available $\frac{1}{a} + \frac{1}{b} = \frac{1}{f}$.

Bring the above formula into Eq. 2 to get

$$y = \frac{x(b - f)\sin\beta}{f\sin\alpha \mp x\left(1 - \frac{f}{b}\right)\sin(\alpha + \beta)} \tag{3}$$

4 Software System Design

The software system is mainly composed of three parts: data acquisition unit, data processing unit, and web publishing unit. The system software function diagram is shown in Fig. 5. Data collection should include car number identification, data collection, data storage [12], and data transmission. Data processing software includes data reading, air gap calculation, hierarchical alarm, data management, and remote management. Web publishing software includes real-time data detection, historical reports, fault reports, and trend analysis.

5 Field Test and Analysis

In order to verify the accuracy and reliability of the system, on-site test was carried out at the installation site on May 4, 2018.

In order to obtain more accurate data, we select a train with the latest round as the dynamic detection object, select the 7th, 39th, and 73rd gear as the specific data

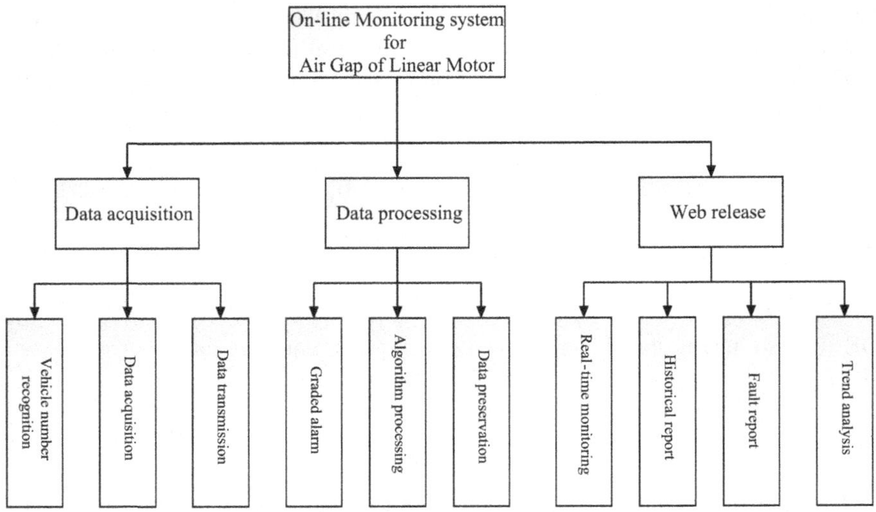

Fig. 5 System software function diagram

Table 1 Air gap detection data

Passing time	7th left tooth	7th right tooth	39th left tooth	39th right tooth	73rd left tooth	73rd right tooth
8:20:36	24.97	24.82	24.05	23.83	24.94	24.67
10:52:45	24.96	24.84	24.03	23.75	25.05	24.81
13:26:45	24.89	24.76	24.11	23.86	24.97	24.69
16:00:12	24.94	24.75	24.07	23.80	24.95	24.64
18:32:24	24.97	24.78	24.02	23.77	24.99	24.73

Table 2 Average value of air gap detection data

Passing time	7th left tooth	7th right tooth	39th left tooth	39th right tooth	73rd left tooth	73rd right tooth
Average value	24.946	24.79	24.056	23.802	24.98	24.708

analysis object, and repeat the measurement five times according to the system accuracy measurement requirements (Tables 1, 2 3, and 4).

It can be seen from Tables 1 and 3 that the maximum deviation of the data repeat accuracy of each analyzed tooth is 0.17 mm, which satisfies the requirement that the repeated measurement accuracy is less than 0.2 mm.

Table 3 Wedge detection data

Passing time	7th left tooth	7th right tooth	39th left tooth	39th right tooth	73rd left tooth	73rd right tooth
8:20:36	28.16	28.01	27.33	27.13	28.15	28.09
10:52:45	28.14	28.04	27.20	27.06	28.26	28.25
13:26:45	28.08	27.92	27.25	27.13	28.18	28.09
16:00:12	28.11	27.95	27.22	27.02	28.17	28.07
18:32:24	28.14	27.98	27.15	26.98	28.19	28.15

Table 4 Average value of wedge detection data

Passing time	7th left tooth	7th right tooth	39th left tooth	39th right tooth	73rd left tooth	73rd right tooth
Average value	28.126	27.98	27.23	27.064	28.19	28.13

6 Conclusion

This paper proposes a linear sensor online air gap detection system based on laser sensor, which can accurately and reliably detect the gap of the linear motor air gap and complete the data storage and data visualization functions. Field tests have shown that the measurement accuracy of the system fully meets the measurement error requirements, and the system measurement is superior to manual measurement. Therefore, the linear sensor online detection system based on laser sensor proposed in this paper can replace the manual measurement of workers, reduce the measurement labor intensity, provide scientific basis for the maintenance and adjustment of train linear motor, and has great market promotion prospects.

Acknowledgements This work is supported by National Key R&D Program of China (2017YFB1201201).

References

1. Genwu Z, Chuanfeng W (2012) Development and improvement of Japan's linear motor metro system. Urban Rapid Transit 25(1):19–22 (in Chinese)
2. Yongchun HAN (2015) Research on the relationship between air gap of linear motor and train energy consumption. Railw Locomot 35(6):69–72 (in Chinese)
3. Qingchao W, Jinghui X, Chuanqi Q et al (2017) Influence of air gap on dynamic response of linear motor metro system. J Traffic Transp Eng 6:10–18 (in Chinese)
4. Liang L, Shujiang F, Jun W (2005) Air gap detection of linear synchronous induction motor based on eddy current method. Ordnance Ind Autom 24(6):84–85 (in Chinese)
5. Yunfeng B (2014) R&D design and application of vehicle number identification and automatic weighing data recording system. China Coal, 2014(s1):346–350 (in Chinese)

6. Jun Y (2017) Design of temperature control system for large-scale basic environment simulation laboratory based on PLC. Nanjing University of Science and Technology (in Chinese)
7. Qiuhong Y (2014) Research on displacement measurement technology of one-dimensional laser triangulation. Harbin Institute of Technology (in Chinese)
8. Lee KC, Yang JS, Yu HH (2013) Development and evaluation of a petal thickness measuring device based on the dual laser triangulation method. Comput Electron Agric 99(9):85–92
9. Cigada A, Mancosu F, Manzoni S et al (2010) Laser-triangulation device for in-line measurement of road texture at medium and high speed. Mech Syst Signal Process 24(7):2225–2234
10. Su Y, Dinggen L, Zhaoliang Y (2012) Parameter optimization design and analysis of triangular laser displacement sensor. Electron Meas Technol 35(10):21–24 (in Chinese)
11. Ying D (2013) Research on digital contouring of wood segment by laser triangulation method. Beijing Forestry University (in Chinese)
12. Lee CH, Zheng YL (2015) Automatic SQL-to-NoSQL schema transformation over the MySQL and HBase databases. In: IEEE international conference on consumer electronics, IEEE, Taiwan, pp 426–427

Application Research of Escalators Status Monitor and Forecast Based on Vibration Analysis

Miaomiao Huo, Xin Li, Guanghong Wei and Chenghao Zhao

Abstract Escalators are special equipment. Escalators safety is closely related to the public safety. In 2011, a failure of an escalator at Beijing Zoo Metro Station, where some passengers were injured, resulted in a very serious incident. This highlighted the need to monitor and evaluate the condition of each escalator. Traditional escalator safety protection method was post-aware and did not provide a complete evaluation with respect to potential fatigue failure. This paper's principal objective was to reduce the risk of escalator hidden trouble such as fatigue failure by developing a suitable method in which to establish and continuously monitor the real condition. Monitor sensors were used to collect vibration data of the escalator main transmission components. These vibration data reflect the escalator real condition. Before the failure occurs, the forecasts can be provided by analysis of the changes of the real vibration data. These forecasts will be sent to a mechanic, who repairs the escalator, thereby preventing failure or accident. This technology has been applied to engineering practice on Beijing Subway's escalators, and the result is significant. One typical engineering practice case in this paper verified the vibration analysis method, which could perform escalator failure warning.

Keywords Escalator · Vibration analysis · Status monitor · Forecast

---

M. Huo (✉) · C. Zhao
Beijing Key Laboratory of Subway Operation Safety Technology, Beijing Subway Operation Technology Centre, The Suburban Railway Depot in Huojiaying Village, Dongxiaokou Town, Changping District, Beijing, China
e-mail: miaomiaohuo@163.com

X. Li
Beijing Rail and Transit Design and Research Institute CO., LTD., Beijing, China

G. Wei
Beijing Subway Electromechanical Branch, Beijing, China

Y. Qin et al. (eds.), *Proceedings of the 4th International Conference on Electrical and Information Technologies for Rail Transportation (EITRT) 2019*, Lecture Notes in Electrical Engineering 639, https://doi.org/10.1007/978-981-15-2866-8_40

1 Introduction

Escalators are special equipment. They are widely used in public places such as stations and shopping malls. They transport thousands of passengers every day. Escalators safety is closely related to public safety. Personal injury accidents caused by escalator failures happen sometimes. Various safety protection devices in the escalator prevent the incident and serious results. Most of them were electrical and mechanical switches, which stop the running when failure occurs. However, many serious accidents still happened occasionally. These protections are insufficient, because component defects could not forecast, such as the bolt breakage on the reducer shell. We have established an escalator status monitor system. Monitor sensors were used to collect vibration data of the escalator mechanical components, which reflect the real components running status. These forecasts are sent to mechanic from analysis of the vibration data that changes in the current condition. The engineering practice was carried out already and the result is significant [1]. One typical engineering practice case in this paper verified the vibration analysis method, which could perform escalator failure warning [2].

2 The Description of Monitor System

2.1 Principle

The escalator status monitor system consists of sensors, data collector, and data processor.

Vibration sensors were mounted on the drive machine to collect data [3]. We set a data collector and a processor on site. The data collector is responsible for analog-to-digital conversion quickly. The software analyzes the transmitted data to calculate extract fault feature. As shown in Fig. 1, these data from the sensors, processor, and fault diagnosis could be shown to the engineers through a remote network.

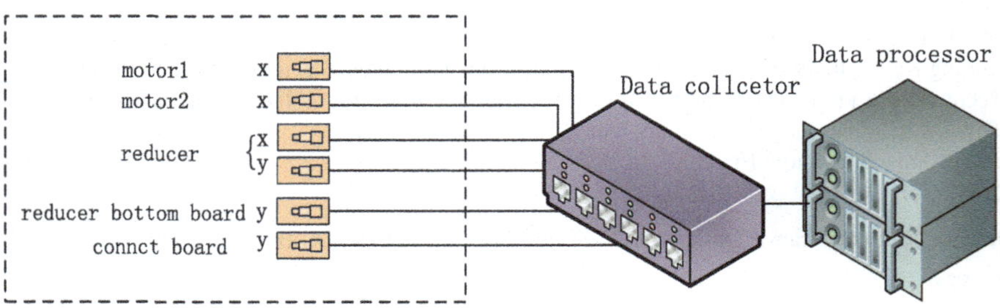

Fig. 1 Principle of escalator status monitor system

2.2 Sensors Positions

The drive machine is the most important component in an escalator, and it fails often. It consists of a motor, a reducer, a drive chain, a brake, and a flywheel. We mounted several vibrations acceleration sensors on motor shell, reducer shell, and drive machine fix plate. The sensors positions are shown in Table 1 and Figs. 2 and 3.

Table 1 Sensors position

Component	Position	Sensor direction	Object
Motor	Shell	Horizontal	Bearings
Reducer	Upper shell	Vertical	Gears and bearings
Reducer	Upper shell	Horizontal	Gears and bearings
Reducer	Bottom plate	Vertical	Bolts
Fix plate	Surface	Vertical	Bolts

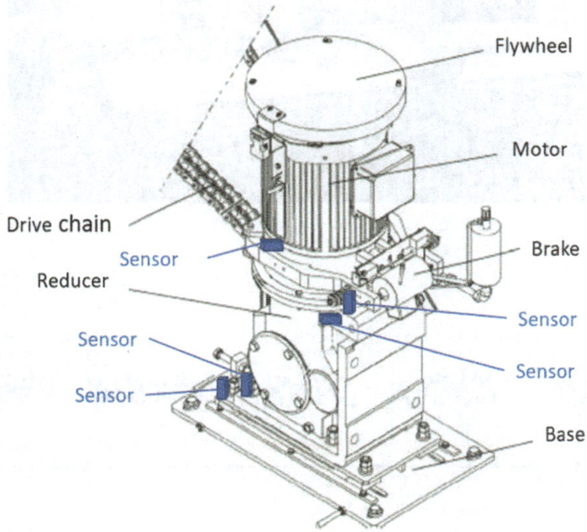

Fig. 2 Sensors position

Fig. 3 Sensors on motor shell (left), reducer upper shell (middle), reducer bottom plate and fix plate (right)

2.3 Monitor System

Escalators status monitor system is involved in Beijing Subway Internet of Things, short for IoT, covered elevators, escalators, and moving walks. In the IoT monitor center, all the escalators statuses and data could be shown and sent to the mechanics. The mechanics check and repair the equipment according to the analysis results and repair recommendations [4]. As shown in Fig. 4.

Fig. 4 Monitor center

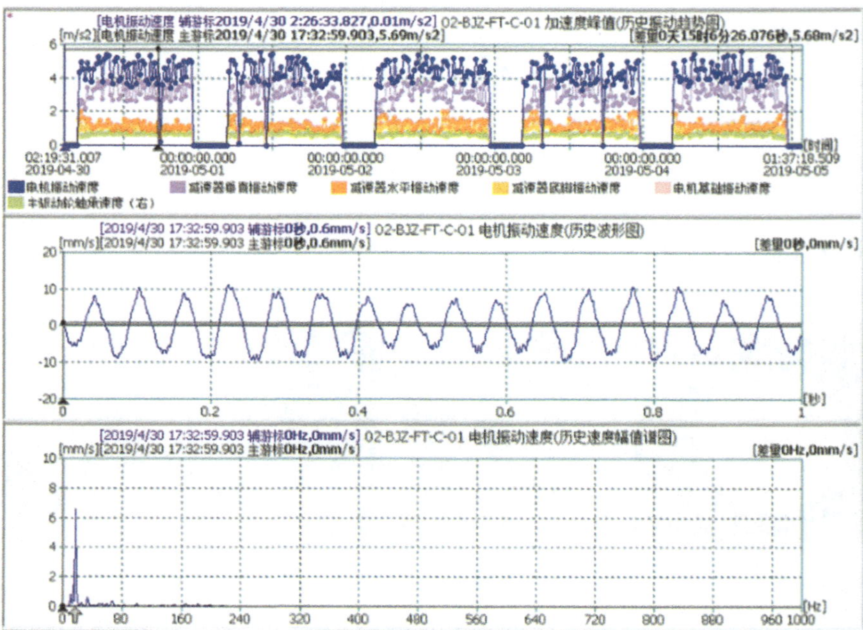

Fig. 5 Status monitor waveforms

2.4 *Diagnosis*

The rotating machinery diagnostic technology was adopted in the status monitor system, focusing on the gears and bearings. The faults of bearing inner/outer ring, roller, cage, run-off, lubrication, and gear broken tooth/abrasion, bolt fastening, chain abrasion could be analyzed and forecasted. As shown in Fig. 5, the vibration acceleration numerical value of the monitoring point and the corresponding spectrum can be displayed on the monitoring terminal. Technical experts analyze the data and waveforms collected through the network. They filter out abnormal data and waveforms by algorithm and establish a fault signature database and a maintenance suggestion library. Finally, the system gradually has the ability to diagnose intelligently.

3 A Typical Case Analysis

On November 16, 2018, several abnormal vibration waveforms of reducer bottom plate and fix plate were detected. The accelerations of the reducer bottom plate and fix plate were higher than the reducer upper shell by 1.0 m/s^2. Technical experts analyzed the waveforms and vibration data. It was considered that the bottom of the reducer is subjected to an external impact force. On site, it was found that the drive chain switch was tilted, and the press plate edges collided with the drive chain. The monitor data was restored to normal after adjusting the press plate position.

3.1 *Escalator Drive Principle*

This is a single-drive escalator driven by an end drive wheel. The drive machine is the power source. Power is transmitted from reducer to main drive shaft by a drive chain. The main drive shaft transmits the power to other drive wheels. Drive machine is bolted on the fix plate, as shown in Fig. 6.

3.2 *Normal Waveforms*

According to "vibration acceleration reflects the impact force," the order of acceleration values from large to small should be fix plate, reducer horizontal, reducer bottom plate, motor and reducer vertical. These are accurately reflected in the acceleration waveform, as shown in Fig. 7.

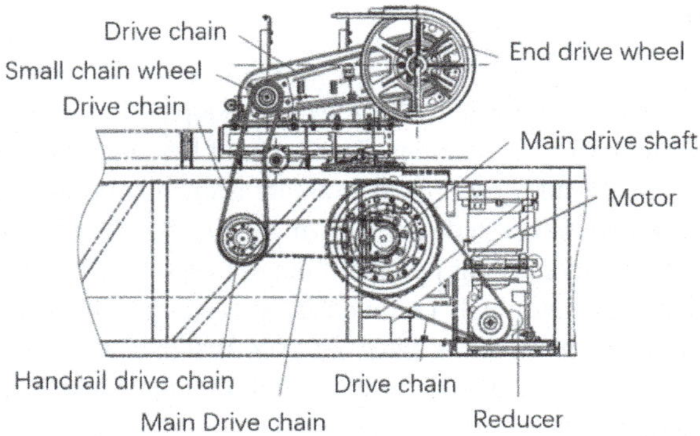

Fig. 6 Escalator drive principle

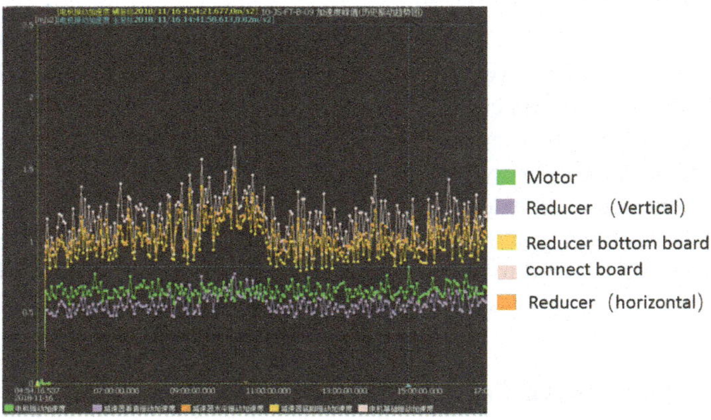

Fig. 7 Vibration acceleration waveform in normal

3.3 *Abnormal Waveforms*

A few days before November 16th, we found that the acceleration value of the fix plate changed from 1.2 to 1.7 m/s^2, the value of reducer bottom plate changed from 1.1 to 1.4 m/s^2. The other positions value did not change significantly, as shown in Fig. 8. That means the reducer bottom plate and the fix plate were subjected to an abnormal impact force. Maybe the anchor bolts were loose or the reducer axle was subjected to an abnormal impact.

3.4 *On-site Inspection and Adjustment*

The mechanic inspected the escalator. They heard obvious noise from the upper engine room when it was running. After shutdown and inspection, it was found that

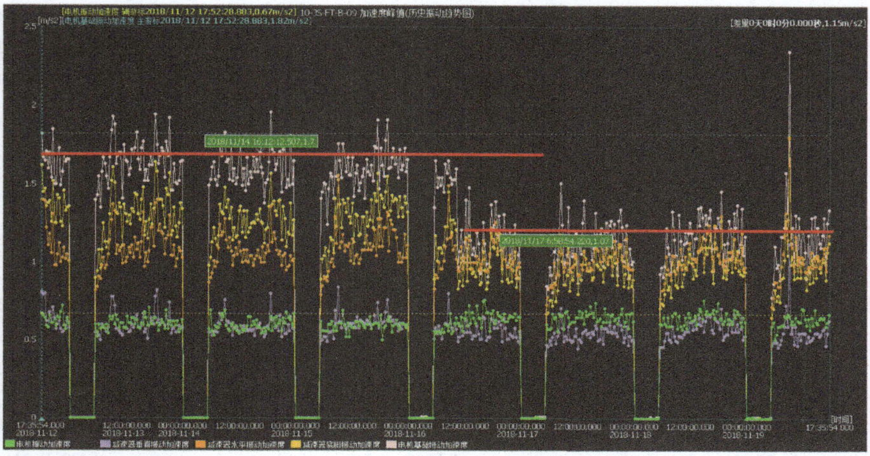

Fig. 8 Acceleration for several days before and after adjustment

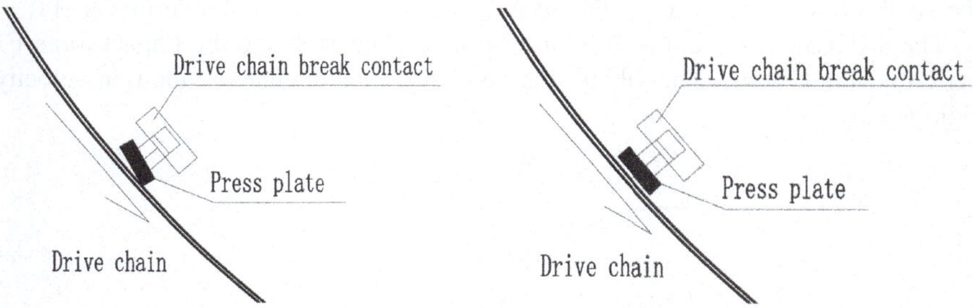

Fig. 9 Before (left) and after (right) adjustment

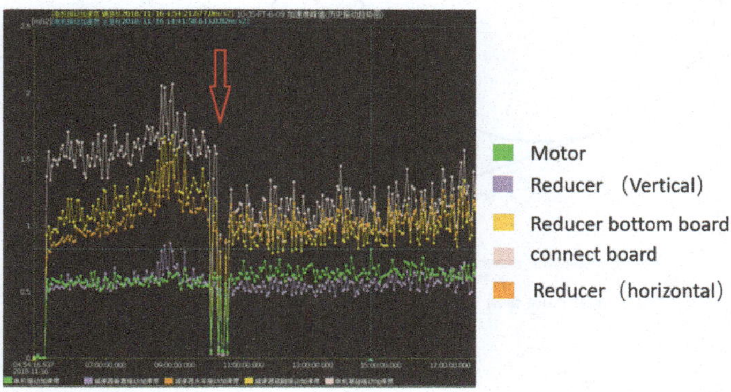

Fig. 10 Acceleration on November 16th

the drive chain break contact tilted. The noise came from the abnormal impact between the main drive chain and the press plate of break contact. After adjusting the press plate to its correct position, the noise was reduced and the monitor data was restored to normal, as shown in Figs. 9 and 10.

3.5 Failure Analysis

This up escalator has a rated speed of 0.65 m/s and a motor power of 37 kW. The press plate of the break contact is attached to the drive chain under the action of its own weight. When the chain breaks or sinks beyond a certain allowable range, the press plate causes the safety switch to operate and the motor is powered off. Because the press plate tilted, the chain received the impact of the press plate. If the impact is F, then F1 is the component of F along the chain movement, and F2 is the component perpendicular to the chain movement. The impact force of the chain on the reducer is F1', which is decomposed into F1x and F1y. F1y produced a vertical vibration acceleration to the reducer axle, which is transmitted from the reducer to the shell and is recognized by the acceleration sensor [5], as shown in Fig. 11.

The driving chain pitch is 38.1 mm, speed is 0.65 m/s, and the impact force F1 period is about 58 ms. The vibration acceleration cannot change vibration velocity significantly [6, 7].

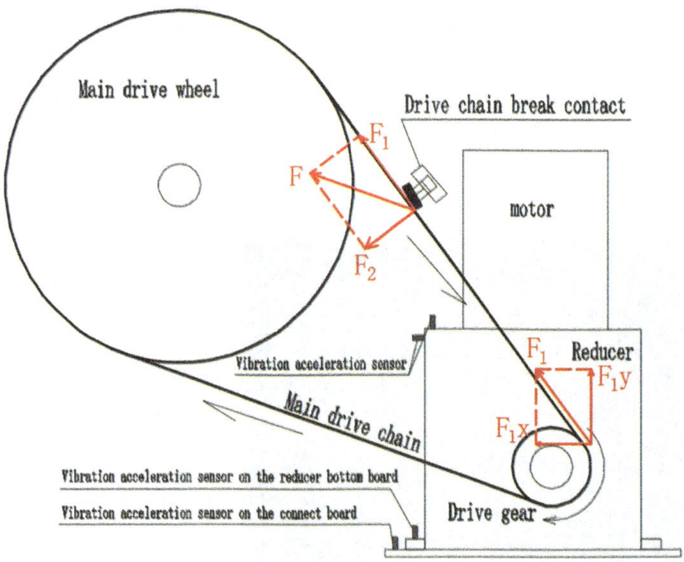

Fig. 11 Stress analysis chart

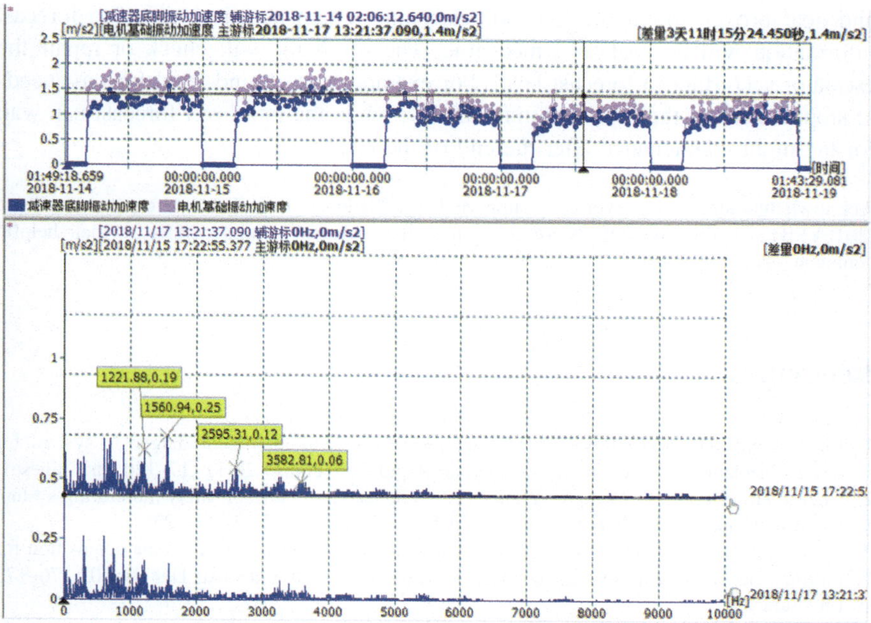

Fig. 12 Reducer bottom plate acceleration spectrum

Compare the frequency pattern of the normal and fault conditions of the reducer bottom plate, as shown in Fig. 12. As can be seen from the spectrogram, when the fault occurred, the drive chain was subjected to noise disturbances of 1200, 1500, 2500, 3500 Hz, and higher frequencies.

This typical case verifies that "the vibration speed reflects the energy, and the vibration acceleration reflects the impact force." When the vibration acceleration value of the reducer bottom plate and fix plate is significantly increased, the failure of the drive chain or axle can be initially determined [8].

4 Conclusion

Through the escalator status monitor and forecast system, vibration data of major components can be obtained in real time. The system helps to analyze and understand the running status of the components, and to provide forecast. Many cases already identified its efficiency and effectiveness. Technical experts are establishing a fault signature database and a maintenance suggestion library. Finally, the system gradually has the ability to diagnose intelligently.

By using this monitor system, the risk of escalator hidden trouble such as fatigue failure will be reduced. Many escalator failure information can be received before fails, such as the faults of bearing inner/outer ring, roller, cage, run-off, lubrication, and gear broken tooth/abrasion, bolt fastening, chain abrasion, etc. The forecast information will be sent to a mechanic. The mechanic will check or repair the escalator according to forecast level. For example, the second-level forecast needs to stop and repair the escalator immediately, while the first-level forecast can wait for the night maintenance after the end of the operation.

Acknowledgements This work is supported by National Key R&D Program of China (2016YFB1200402). The authors are grateful for the reviewer of initial drafts for their helpful comments and suggestions.

References

1. Li X, Peng JH (2015) Research of fault monitoring and intelligent diagnosis system for electro-mechanical device of subway. Urban Rapid Rail Transit 28(1):117–120 (in Chinese)
2. Chen J, Hong T, Liu G, Wang G, Zhou J (2013) Research of elevator safety monitoring system based con internet of things technology. J China Jiliang Univ 24(3):231–236
3. Li X, Li MX, Wang HY, Yan H (2017) Research and Application of the data acquisition for on-line state monitoring of escalator in urban rail transit station. Constr Des Proj 02:176–178 (in Chinese)
4. Li X, Mei Q, Huo MM, Wang HY, Jin MY (2016) Escalators of urban rail transit on-line monitoring and intelligent diagnosis system application functional design and research. Constr Des Proj 18:203–206 (in Chinese)
5. Wang N, Jiang DX, Han T, Xu HZ (2017) Dynamics of rub-impact of dual-rotor systems and the experimental verification based on vibration accelerations measurement. J Vib Shock 36 (14):71–76, 97
6. Lu B, Li Y, Wu X et al (2009) A review of recent advances in wind turbine condition monitoring and fault diagnosis. In: The power electronics and machines in wind applications, PEMWA (IEEE, 2009), pp 1–7
7. Zhou Q, Liu X, Zhao J, Shen H, Xiong X (2018) Fault diagnosis for rotating machinery based on 1D depth convolutional neural network. J Vib Shock (23):31–37
8. Tai XY, Ma H, Tan Z, Wen BC (2013) Dynamic characteristics of a continuous rotating beam model with a rubbing fault. J Vib Shock 32(18):43–48

Fault Simulation and Analysis of Escalator Step Chain

Lirui Zhao and Xiukun Wei

Abstract In this paper, the escalator step chain simulation model is established to simulate the operation of the step chain under four chain states, which are normal state, chain pitch elongation state, roller wearing and sprocket wheel wearing state, respectively. Then, the wavelet packet analysis method is used to analyze the lateral vibration acceleration data of the driven shaft in these states. The analysis results show that the time-domain energy of each component of the wavelet packet has different degrees of change after the step chain fault, which has great significance for the study of the fault law of the step chain.

Keywords Escalator step chain · SIMPACK simulation · Wavelet packet analysis

1 Introduction

With the development of science and technology, escalators have gradually entered people's lives and become an indispensable part. Now, the use of escalators in major cities is common and has produced enormous economic benefits. And it is widely loved by the public for its stability, safety and convenience. It is not only widely used in large shopping malls, supermarkets, airports, hotels, hospitals and other public places, but also plays a prominent role in the emerging urban rail system. As a transmission device of the escalator, the step chain is a key component, and it has an important position in the transmission system of the escalator. Once the ladder chain fails, it will directly affect the stability of the escalator and the safe operation of the escalator.

About the research in the field of chain drive, Ref. [1] introduces fault detection methods using penetrant inspection, magnetic particle inspection and ultrasonic flaw detection techniques. References [2, 3] build the normal sprocket shape and the

L. Zhao · X. Wei (✉)
State Key Laboratory of Rail Traffic Control and Safety,
Beijing Jiaotong University, Beijing, China
e-mail: xkwei@bjtu.edu.cn

Y. Qin et al. (eds.), *Proceedings of the 4th International Conference on Electrical and Information Technologies for Rail Transportation (EITRT) 2019*, Lecture Notes in Electrical Engineering 639, https://doi.org/10.1007/978-981-15-2866-8_41

circular gear shape dynamics model equation. In these two cases, the dynamic relationship between the components of the chain [4], the contact force between the roller and the sprocket wheel, the link force and the displacement of the link [5] is analyzed, which provides great technical support for the research of sprocket wheel failure. Reference [6], wearing elongation calculation model of the roller chain is established, and the relationship between the number of the roller chain collisions, material properties, impact energy and the cumulative plastic deformation of the chain plate is considered, which provides a reference for the calculation of the cumulative deformation of the chain under low stress and multiple impact loads. References [7, 8] establish the lateral and longitudinal vibration characteristic equation of the roller chain drive. The lateral and longitudinal motion laws and characteristics of the roller chain drive are studied under the influence of certain excitation. In this paper, the research on step chain fault based on Simpack simulation software is proposed to simulate the step chain under different degradation type of the step chain for the first time in China, which avoids the long time practicality experiments, saving amount of manpower, material and financial resources. The change rule of fault characteristics is summarized by analyzing the lateral vibration acceleration signal in driven shaft under normal and fault conditions, which is of great significance for fault detection and life prediction of the escalator step chain.

2 Structure and Main Failure Forms of Escalator Step Chain

As one of the important chain transmission systems of the escalator, the escalator step chain system mainly consists of roller chain, sprocket wheel and guide. The function of the guide rail is to support the load transferred by the step wheel and make the step chain run according to the established path. The probability of guide failure is very low. The main structure prone to failure is the inner chain plate, outer chain plate, pin shaft, sleeve, roller and sprocket, and the failure of these structures usually interact with each other. It is of great significance to study the failure of step chain structure to ensure the safety of escalator. The structure of the roller chain is shown in Fig. 1.

In the actual process, especially in the case of improper lubrication and improper protective measures, as a complex load transfer machinery with circular movement, the wear of the escalator step chain components is the main reason for the failure.

First, the inner chain plate and outer chain plate of the escalator step chain. In the process of running of the escalator step chain, vibration is inevitable, which makes the escalator step chain produce extension. At this time, the inner chain plate and the outer chain plate of the cascade chain rotate and rub with each other, and deform and elongate under the action of load.

Fig. 1 Roller chain structure

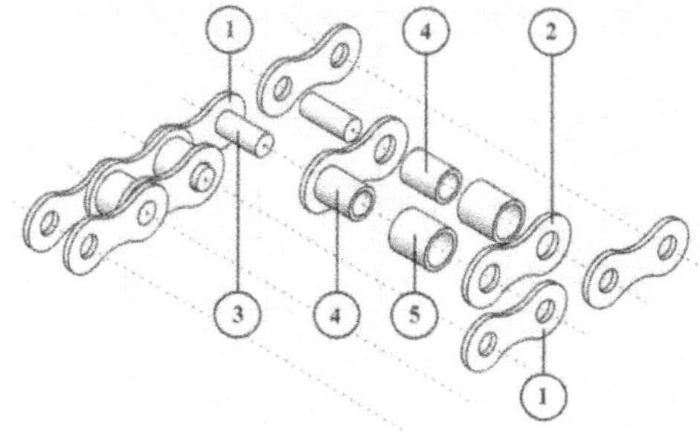

1-Outer Chain Plate 2-Inner Chain Plate 3-Pin Shaft 4-Sleeve 5-Roller

Next, the pin shaft and sleeve of the step chain. The operating environment of the escalator is complex. The rain and dust from the outside fall on the step chain, and mixed with the lubricant on the chain and get into the inside of the step chain structure forming the grinding media, which greatly enhances the wear effect between the pin shaft and the sleeve. The outer diameter of pin shaft is getting thinner and thinner, and the inner diameter of sleeve is getting bigger and bigger.

Finally, rollers and sprockets. The step chain is a transmission chain with low speed and heavy load, which is driven by sprocket wheel. The step chain keeps running through the continuous meshing between the sprockets wheel and the chain. When the step chain enters the sprocket teeth, there is a certain tangent angle between the step chain and the sprocket teeth. The friction and impact between the two make the roller and sprocket teeth worn or deformed.

3 Modeling and Simulation of Escalator Step Chain Fault

The simplified topological structure diagram of the step chain is designed, according to the structure of the step chain. And the entity model of the structure involved in step chain is established. Then, physical properties are set, and the connection among the components in the model is set according to the relationship among the components in the topological structure. After establishing the model, sensors are set to measure the motion change of the step chain. Then, online time integration is carried out to verify the correctness of the established model and then the model is optimized to achieve the optimal state. Different fault states of the step chain are simulated separately, and offline integration calculation is carried out to obtain the simulation data of the step chain. The simplified topological diagram of the step chain is drawn as shown in Fig. 2.

The escalator step chain model is shown in Fig. 3.

The normal state, elongation of chain pitch state, roller wearing state and sprocket wheel wearing state are simulated, respectively, and the outout is simulation data. The specific state is set as follows.

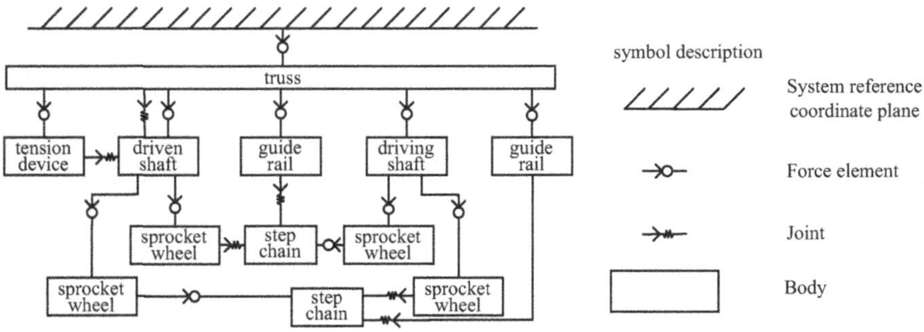

Fig. 2 Topology structure diagram of escalator step chain

Fig. 3 Escalator step chain model

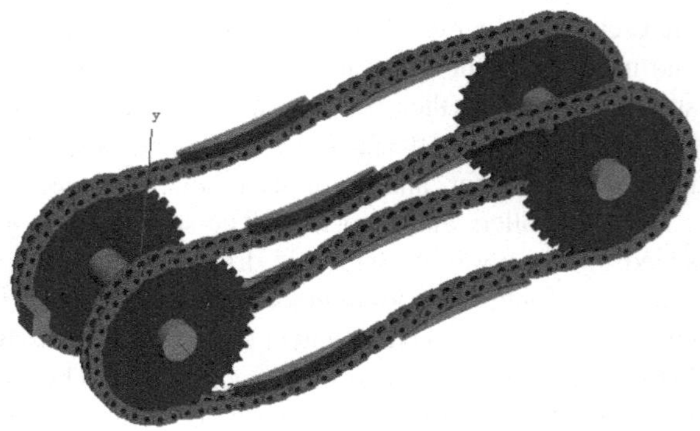

(1) Normal state

Normal state refers to the escalator step chain without any fault. The model of the escalator step chain is integrated offline directly and the required data is output.

(2) Elongation of chain pitch

When the step chain pitch is elongated, the length of the step chain link will be lengthened. The parameters of chain pitch in escalator step chain model are increased as the fault state of chain pitch elongation. The model is simulated and calculated, and the output is simulation data. The schematic diagram of chain pitch elongation is shown in Fig. 4. The dotted line represents the state of the step chain link before the chain pitch is elongated and the solid line represents the state of the step chain link after the chain pitch elongated.

(3) Wearing of the roller

After the roller wears, the outer diameter of the roller will become smaller. The roller outer diameter parameter in the step chain model is set smaller as the roller wearing failure state. Then, the output of the model is simulation data. The schematic diagram

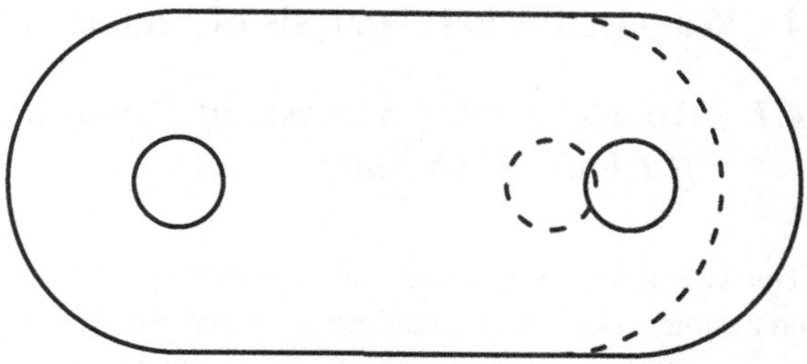

Fig. 4 Chain pitch elongation diagram

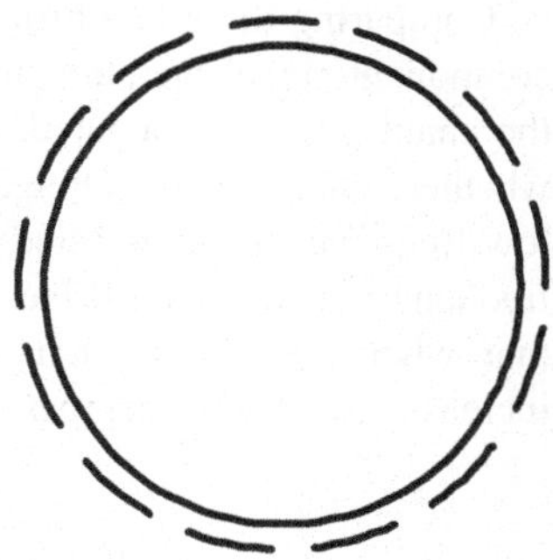

Fig. 5 Roller wear diagram

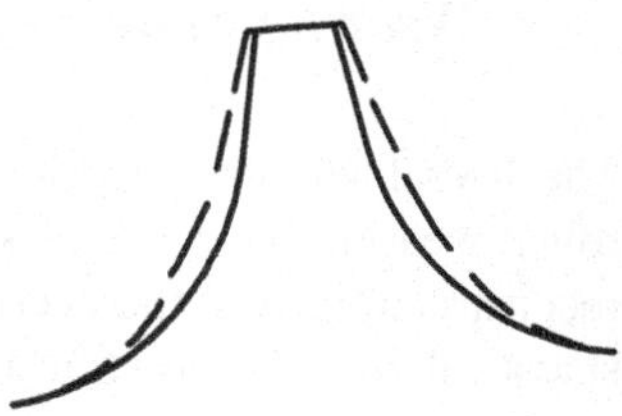

Fig. 6 Sprocket wheel wear diagram

of roller wearing is shown in Fig. 5, where the dashed line indicates the state of the step chain link before wearing and the solid line indicates the state of the step chain link after wearing.

(4) Wearing of the sprocket wheel

After the sprocket wheel wears, the roller seating radius and the roller seating angle become larger. The shape parameter of the sprocket wheel is modified in the step chain model as the sprocket wear failure state. Then, the outpot of the model is simulation data. The schematic diagram of the sprocket wheel wearing is shown in Fig. 6, where the dashed line indicates the state of the sprocket wheel before wearing and the solid line indicates the state the sprocket wheel after wearing.

4 Wavelet Packet Analysis of Signal in Fault State

4.1 Wavelet Packet Analysis of Signal in Elongation of Chain Pitch State

The lateral vibration acceleration signals of driven shaft under normal state and chain pitch elongation state are collected for wavelet decomposition.

The energy of each component is calculated for the layered signal, and the energy spectrum is shown in Fig. 7.

Comparing these two histograms (a) and (b), it observed that the energy of each component signal increases to varying degrees after the wavelet decomposition of the chain pitch is elongated. And the energy of the S30 component is the largest whether chain pitch elongation occurs or not, that is, the energy of the low-frequency band is relatively high, which is also matched with the meshing frequency of the chain drive. So the fault law of elongation of chain pitch state is that when the chain pitch is elongated the energy of each component signal increases to varying degrees.

4.2 Wavelet Packet Analysis of Signal in Roller Wearing State

The lateral vibration acceleration signals of driven shaft under normal state and roller wearing state are processed based on wavelet packet decomposition and reconstruction. And the energy of each component is calculated and counted. The statistical chart is shown in Fig. 8.

Comparing this two histograms (a) and (b), It can be seen that after the roller wearing, the time-domain energy of each component signal changes and increases to varying degrees, among which the energy of 0 component increases the most, indicating that the signal amplitude at the low-frequency band changes the most

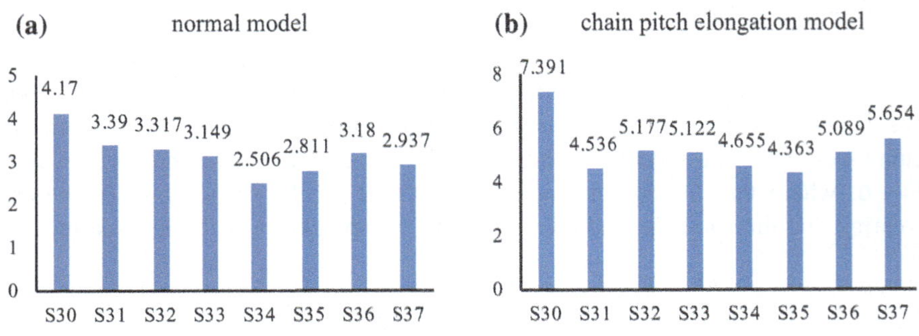

Fig. 7 Time-domain energy distribution histogram

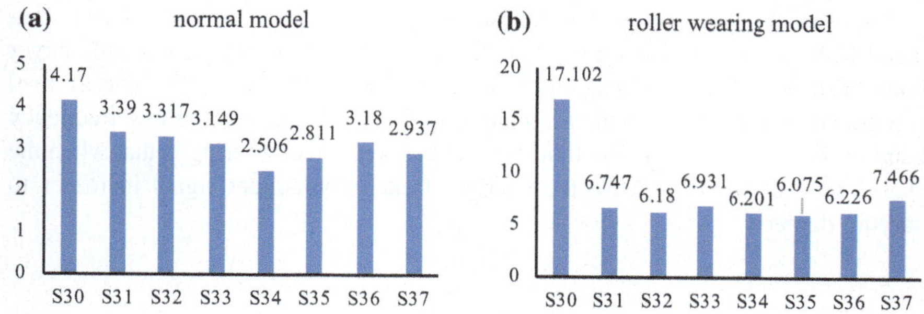

Fig. 8 Time-domain energy distribution histogram

after the roller wearing. In both models, normal model and roller wearing model, the energy of 0 component is also the largest, which is matched with the feature that the meshing frequency of chain drive is in low-frequency band. So the fault law of roller wearing state is that when the chain roller is worn the energy of each component signal increases to varying degrees.

4.3 *Wavelet Packet Analysis of Signal in Sprocket Wheel Wearing State*

The lateral vibration acceleration signals of driven shaft under normal state and sprocket wheel wearing state are processed based on wavelet packet analysis. The energy of each component is calculated, and the energy distribution histogram is shown in Fig. 9, in which (a) is wavelet component time-domain energy distribution histogram of the normal model signal and (b) is the wavelet component time-domain energy distribution histogram of the sprocket wheel wearing model signal.

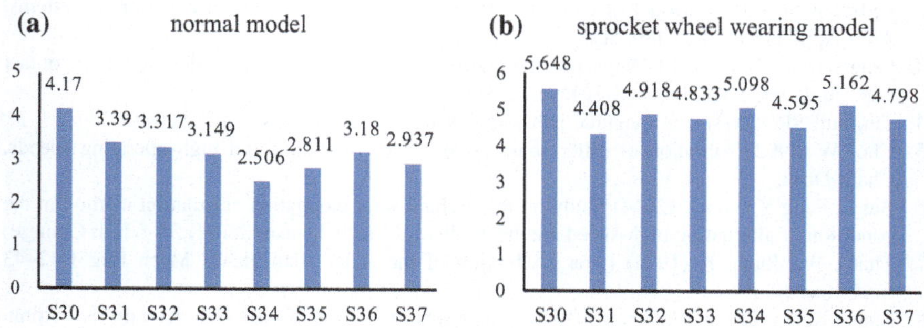

Fig. 9 Time-domain energy distribution histogram

The time-domain energy distribution histograms of the components are compared with each other. It observed that the components in (b) are significantly larger than those in (a), and the magnitude of the increase is different. The energy of 0 component is the largest, which is matched with the feature of the low-frequency band of the chain drive. So the fault law of sprocket wheel wearing is that when the chain sprocket wheel is worn the energy of each component signal increases to varying degrees.

5 Conclusion

In this paper, the fault forms of escalator step chain are analyzed, respectively, and the step chain models are established based on Simpack to simulate the running of the chain under normal state, chain pitch elongation state, roller wearing state and sprocket wheel wearing state, respectively, and they are integrated offline to output the lateral vibration acceleration signals of the driven shaft. The signals under normal state and three-step chain fault states are processed based on wavelet packet analysis. The time-domain energy of each component under normal state and three-step chain fault states are compared, in which the energy of 0 component is largest, which is matched with the characteristic that the meshing frequency of step chain drive is in the low-frequency band. And after the failure of step chain, the energy of each component increases in varying degrees, which is of great significance to the study of step chain fault law.

Acknowledgements This work is partly supported by National Key R&D Program of China (Contract No. 2017YFB1201203).

References

1. Chen Peng (2011) Detection and fault diagnosis technology of the scraper conveyor chain. New Technol New Prod China 16:159 (in Chinese)
2. Pedersen SL (2005) Model of contact between rollers and sprockets in chain-drive systems. Arch Appl Mech 74(7):489–508
3. Pedersen SL, Hansen JM, Ambrósio JAC (2004) A roller chain drive model including contact with guide-bars. Multibody Syst Dyn 12(3):285–301
4. Turnbull SR (2003) The dynamic behaviour of roller chain drives. Univ Newctle
5. Choi W (1992) Vibration of roller chain drives at low, medium and high operating speeds. Chain Drive
6. Bai L, Feng Y, Xue X (2014) Study on roller chain wear elongation calculation method in the condition of alternating high-speed and heavy-load. J Mech Transm 38(02):58–62 (in Chinese)
7. Zhang W, Zhang K (1994) Lateral vibration of the roller chain drive. Mech Eng 1:42–43 (in Chinese)
8. Zhang Keren, Zhang Wenxiang (1995) Longitudinal vibration of the roller chain drive. J Anhui Univ Sci Technol (Nat Sci) 1:51–56 (in Chinese)

Development Demand and Trend Analysis of Train Intelligence

Guoqin Zhang, Changqing Liu, Miao Sha, Zhenchen Chang, Yan Zhu, Hongwei Li, Hai Deng and Haibo Zhao

Abstract This paper describes the demand and trend analysis of train intelligence development. Firstly, the background of the industry is introduced, and the recent developments of intelligent in railway vehicles in abroad are reviewed. Secondly, the innovation opportunities brought by the development of intelligent technology in the new era, the development of comprehensive transportation system and the demands of the users for trains are discussed. Future development of intelligent trains is addressed in three aspects: Train operation, maintenance and service are also discussed in this paper with emphasis on the characteristics of perception, control, expression, response and interaction.

Keywords Train · Intelligence · Data integration · Intelligent driving · Intelligent service · Intelligent maintenance

1 Introduction

By the end of 2018, the total mileage of China's operational railway reached 131,000 km, including 29,000 km high-speed railway, ranking the world's first. The number of passengers transported is 3.37 billion, nearly half of the world's total. It is urgent to improve the capacity of railway transportation and transport service level in China.

China is building a modernized comprehensive transport system, adhering to the innovation-driven development strategy, promoting the integration development of "Internet Plus" and transportation, using intelligence to drive traffic modernization.

This paper aims at the demand for innovation and development of trains, lists typical intelligent technologies and applications used in the new generation trains of developed countries, analysis the opportunities brought by the development of intelligent technologies and the requirements of comprehensive transport system for

G. Zhang (✉) · C. Liu · M. Sha · Z. Chang · Y. Zhu · H. Li · H. Deng · H. Zhao
CRRC Changchun Railway Vehicles Co., Ltd., Changchun, China
e-mail: zhangguoqin@cccar.com.cn

Y. Qin et al. (eds.), *Proceedings of the 4th International Conference on Electrical and Information Technologies for Rail Transportation (EITRT) 2019*, Lecture Notes in Electrical Engineering 639, https://doi.org/10.1007/978-981-15-2866-8_42

train's development giving the suggestions for improvement and development trends for transport service, train's operation and maintenance, etc.

2 Innovative Development of Intelligent Train in the Developed Countries

2.1 Japan

The intelligent train control technology is one of the key research directions of Japan Railway Integrated Technology Research Institute in the next 15 years [1, 2]:

Japan is promoting high-speed railway applications and testings of 5G. The dynamic 5G test of trains was completed by the end of 2017, and 5G service is planned to launch in 2020.

Focus on the research and development of safety protection measures in transit, including anti-derailment systems, vehicle shake dampers, integrated disaster systems, earthquake early warning systems, etc. High-speed trains N700S and Alfa-X are all equipped with them.

Customized service information will be provided to passengers, such as passenger flow, train equipment information, other transportation and weather information. There are 250 stations which can push the information to parents' mobile phones after their children checking, and 70 stations launched automatic translation systems for railway staff in 2018.

Innovation of maintenance. Gradually promote the application of the "state repair" system and accelerate the research of autopilot and driving assistant technologies.

Effective utilization of energy. Pay attention to whole-process energy management and study comprehensive renewable and energy-saving technologies.

2.2 European Union

Joint route map of digital railway was published by EU in April 2016 for realizing interconnection and sharing, promoting the development of digital technology represented by big data and cloud computing and solving the problems of digital information interconnection and standardization. At present, the EU has included the intelligent train into the future 《EU railway technology development strategy》 [3].

Germany. Carrying out "railway digitization strategy"—railway 4.0 action in German, for aiming at improving passenger satisfaction, carrying out the R&D and application of intelligent equipment for autopilot, electronic ticketing and maintenance automation.

France. Focuses on the construction of SNCF industrial Internet, striving to build a network ecosystem connecting the three regions of train, road network and station and realizes the digital reconstruction of railway business process and the real-time information exchange. It enables to complete the inspection of more than 200 running trains at the same time and identifies the faults or potential problems in the door, air conditioning, gearbox oil temperature, water level of wiper and so on in real time. Big data mining and machine learning are used to form maintenance demand forecasting and analysis system.

Britain. Uses intelligent train technology to improve the utilization level of the railway network and uses the Internet of Things to make the trains perceive each other and reduces the operation number of train and delay.

2.3 The North American [4]

American railway mainly focuses on freight transportation and conducts digital technology exploration in predictive maintenance, customer service and other aspects.

3 The Intelligent Development of Trains in the New Era is Faced with New Trends

3.1 Innovation Opportunities Brought by the Development of Intelligent Technology

In May 2016, the National Development and Reform Commission and other departments jointly formulated the “Internet plus” three-year action plan for AI [2, 5]. We will implement the develop concept of innovation, coordination, green, open and sharing, and accelerate the R&D and industrialization of intelligent speech processing, biometric recognition, complex environment recognition, new human–computer interaction, natural language understanding, machine translation, intelligent decision control, network security and other technologies based on AI. Technological innovation will change our trip method intelligently.

3.2 New Requirements of Comprehensive Transport System for Train's Development

In order to meet the demand for urban development, realize the "zero distance" transfer and seamless transmits between various types of vehicle, realize interconnection between the flow of population, goods and information in the time-spatial level and put forward some new requirements for rail transit equipment:

Business Integration. With the gradual breaking of travel business barriers, further development of integrated business and the operational management level. The traffic equipment is developing toward the digitization, networking and intellectualization.

Safety and Reliability. In order to ensure the normal operation of comprehensive transport system, improve the active safety level of trains and the safety of train–ground information and data interaction and the overall level of counter-terrorism.

Green and Economy. Adopts energy-saving devices and optimizes control strategy to realize the energy conservation of trains. Under this premise, by enhancing information exchange, establishing the great transport concept, using operational planning and matching technology, to realize the change of management level from train to great transport system, the energy-saving and emission–reduction and resource recycling, the integrative effectiveness improving and sustainable development of low carbon traffic system.

3.3 New Requirements for Train Development from Users

Train is a terminal facing "users" (passengers, cargo owners, operations, drivers and conductors). The relationship between train and users is a long-term coordinated and common development. The intelligent experience from this terminal directly determines the functional requirements of intelligent train design, research, use and maintenance in the future:

Comfortable and humanity. For passengers, to meet the growing desire for a better travel including the improvement of the quality of sensory comfort, interference-free service, information accuracy push, appropriate instructions, barrier-free experience for disabled, mobile charging service, etc. For train conductors, customer service efficiency and management quality can be improved. Trains can provide customization and networking services for them in respect of the management for seating, onboard facility and information more convenient. For cargo owners, he can get the goods information timely by the application of the Internet of things.

Efficient Operation. Facing the needs of operations, according to the principle of "advanced, mature, reliable, economic and necessary", precise push is carried out. With multidimensional identification, comprehensive analysis, navigation and flexible display technology, users can fully understand the train operation status, the failure rate and processing time will be reduced.

Precision Maintenance. Under the premise of fully sensing the train status, based on the collection, mining and analysis of big data, accurate fault location and early warning are achieved to improve the accuracy of life-cycle management. At the same time, intelligent maintenance suggestion strategies, spare parts inventory recommendations, and replacement reminders are provided.

4 Fuxing Intelligent Train

Based on the above demand researches of multi-user for train, intelligence is upgraded in "Jingzhang intelligent high-speed train", including three aspects of driving, service and maintenance on the basis of CR400BF "Fuxing" EMU.

4.1 Intelligent Driving

Autopilot. It is the first time to realize autopilot at 350 km/h. On the basis of retaining the existing driving mode and train control system, C3+ATO technology is adopted and integrated with train technology to realize automatic departure of station, automatic operation of section, automatic parking of station, automatic opening of door and linkage control of vehicle door/platform door.

Onboard Safety. Emergency traction systems is added to ensure the train can be driven to the nearest station of "Beijing-Zhangjiakou high-speed railway" in the event of a overhead line failure or high-voltage system is unavailable. Lithium battery technology, environmental perception and matching technology are used to ensure train travel requirements. The integrated safety monitoring system is applied, the comprehensive monitoring and analysis from single system and single component status monitoring to multiple systems and multiple related factors are achieved.

4.2 Intelligent Services

Environmental Perception Adjustment. Through intelligent adjustments in temperature, pressure waves, lighting, and active noise reduction, passenger's sensory comfort is improved.

Rich Information Tips. Through the LCD display of the information beside the door, the instructions of seat numbers can guide passengers to find seats quickly, and guide the flow to realize live broadcast of Olympic Games.

WIFI Business Expands. Expanding the business scope of WIFI onboard, realize the interaction and sharing the data of operation, ticketing and passenger management and further improve the efficiency of integration of train–ground business.

4.3 Intelligent Operation and Maintenance

It is composed of onboard equipment, vehicle–ground data transmission and ground system. Interaction and sharing of data such as R & D, test, operation, maintenance, and history is realized among train manufacturing enterprises, application departments, and component suppliers. With fault prediction and health management as the core, intelligent evaluation of key component performance status, fault diagnosis and early warning, accurate fault location, dynamic prediction of spare parts, and recommendations for operation and maintenance decisions are realized.

5 Technological Innovation and Prospect

Intelligent train is a high-tech product which combines modern technology such as AI with traditional train. It realizes self-perception, self-diagnosis, self-decision, self-adaptation, self-repair, automatic and cooperative operation of train by means of omnidirectional environmental perception, automatic driving, operation control, fault prediction and health management. It has the characteristics of perceptible, controllable, expressible, responsible and interactive. Through the intellectualization of operation, service and maintenance, it can lead technology, meet people's growing demand for better travel and provide strong support for transportation [6–9].

5.1 Characteristics

Perceptible. In order to ensure the safety of train operation and have intelligent-related functions, advanced sensing technology and detection equipment are used to realize the state detection and perception of different dimensions of train and environment.

Controllable. On the basis of omnidirectional perception, the business processes of autopilot, fault-oriented safety, sudden disaster response and operation scheduling can be managed and controlled by self-diagnosis and self-decision.

Expressable. Using large capacity, high bandwidth onboard, vehicle–ground network, image recognition, speech recognition, information display and other technologies, the train operation process can be observed, expressed and understood.

Responsible. Intelligent technology is used to implement functions such as automatic driving, fault prediction and health management (PHM), and intelligent travel services, meeting the needs of precise train control, life-cycle management, and improved service level.

Interactive. Under the premise of ensuring information security, information collaborative sharing and integration between vehicle–ground, vehicle–vehicle, vehicle–person, vehicle–comprehensive transport system is realized.

5.2 *Intelligent Train Demand Framework*

Based on the intelligent features of "perceivable, controllable, expressive, responsive and interactive", the framework of functional requirements for intelligent trains is decomposed (see Fig. 1). The top-down sequence is perception layer, data layer, fusion layer, transmission layer, application layer and target layer [10, 11].

Perception Layer. By means of intelligent sensing technology, Internet of Things, antenna radar, AI identification, two-dimensional code and other modern electronic monitoring means, vehicle status data, operation data, environmental data and fault data are identified, which provides data basis for precise control.

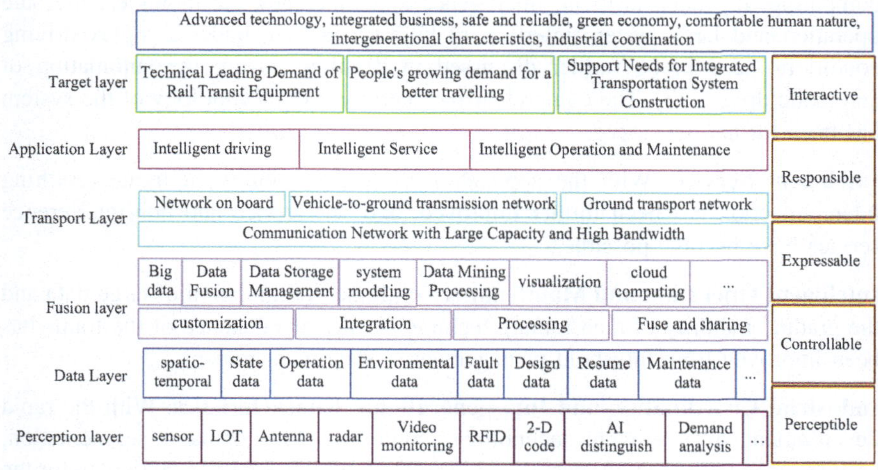

Fig. 1 Intelligent train demand framework

Data Layer. It contains three data sources, provides massive data resources for data fusion: self-perception data of train; early input or historical accumulation of data, including design data, maintenance data, biodata, external data through the transmission layer to the train, including operation data, ticketing data.

Fusion Layer. Through the integration, storage management, mining and processing of large data, customization, process and integration of intelligent technology are used to support the management and decision-making of application layer.

Transport Layer. Vehicle–vehicle and vehicle–ground input interactions are realized through large bandwidth and large capacity network and provides data resources for fusion layer.

Application Layer. Establish a core platform for operation, service and maintenance based on intelligent technology.

Target Layer. It represents the needs and objectives of intelligent train design. Intelligent trains, as the strong support of "Traffic Power, Railway First", need to meet the needs of rail transit equipment to achieve technology-led, to achieve the people's growing demand for better travel and to achieve the development needs of integrated transportation.

5.3 *Intelligent Train Prospect*

Intelligent Driving. With the further development and application of the next-generation communication technology, Beidou satellite navigation, cloud computing, big data, artificial intelligence and other new technologies, the safe operation and large system interface of high-speed train under complex driving conditions need to be further discussed in detail to ensure the automation of automatic driving [7]. The comprehensive energy-saving technology of the system has been further developed.

Intelligent Service. With the application of 5G, the ability of train-everything interconnection has been further improved, and self-service and non-interference service have become possible.

Intelligent Operation and Maintenance. With the accumulation of large data and the gradual maturity of data mining technology, the maintenance of the trains has been improved from threshold management to state management.

Industrial Coordination and Intergenerational Characteristics. With the rapid development of intelligent technology, combined with demand segmentation, upgrade speed of train will increase rapidly. Therefore, it is necessary to plan for the intelligent train platform construction and intergenerational upgrade of intelligent related functions in the early design stage. The standardization of intelligent technology application is imminent.

References

1. Aoki J, Yangchun L (2017) Japan's research and development for sustainable railway development. Foreign Railw Veh 54(04):1–5 (in Chinese)
2. Shuxin G, Huifeng J (2017) Digital railway development in Europe. Integr Transp 39(10):31–34, 75 (in Chinese)
3. Zhuang D (2018) Lu dongfu: pushing forward the key scientific and technological problems of intelligent high-speed railway. China's Strat Emerg Ind 13:70 (in Chinese)
4. Ping L, Sai S, Rui X, Xiaodong Z (2019) Research on the development trend of foreign railway digitization and intelligence. China Railw 02:25–31 (in Chinese)
5. Tongjun W (2019) Research on China's intelligent high-speed railway development strategy. China Railw 01:9–14 (in Chinese)
6. Limin J, Yong Q, Ping L (2015) Overall framework and key technologies of the new generation rail intelligent transportation system. China Railw (04):14–19, 60 (in Chinese)
7. Tongjun W (2018) General architecture and development prospect of intelligent railway. Railw Comput Appl 27(07):1–8 (in Chinese)
8. Tianyun S, Peng S (2017) Application status and development of railway internet of things. China Railw 12:1–6 (in Chinese)
9. Hongxia L (2019) Advances in the application of intelligent high-speed automatic driving technology in China. Railw Stand Des 63(06):151–155 (in Chinese)
10. Junling S (2016) Technical characteristics and development trend of foreign high-speed trains. China Railw 01:95–98 (in Chinese)
11. Development and Reform Commission issued "Internet+" three year action plan of artificial intelligence. Inf Technol Informatiz 2016(06):8–9 (in Chinese)

H_∞ Mixed Sensitivity Robust Control of Lateral Train Vibration

Xiusheng Xue, Yuan Lin, Jiaxin Ji and Baogui Wu

Abstract In order to suppress the lateral vibration and improve the ride comfort of the train, a method of solving the lateral vibration problem with H_∞ robust controller was proposed. Through reasonable simplification, a lateral dynamic model of a certain subway train is established, and a reasonable weighting function is selected for the sensitive frequency segment of human body and the resonance frequency point of the car body. H_∞ robust controllers based on mixed sensitivity PS/T were designed for yaw and lateral. The actuator selection in full active control is carried out and the optimal actuator transfer function for a metro train is selected through simulation. Simulink and SIMPACK co-simulation model were established to simulate the actual working conditions. The results showed that H_∞ robust controllers could effectively suppress the lateral vibration of the vehicle body, improve the running quality and ride comfort of the train, and the controller took both stability and robustness into account.

Keywords Rail trains · Lateral vibration · H_∞ robust control · Mixed sensitivity · Weighted function

1 Introduction

Improving the running speed is an effective way to enhance the competitiveness of railway vehicles. However, the higher speed will lead to the increase of vibration acceleration of the vehicle body, which will affect the safety of vehicle operation and passenger comfort [1]. The key to the performance of active suspension is the

---

X. Xue · Y. Lin · J. Ji (✉) · B. Wu (✉)
China University of Petroleum, Qingdao 266580, China
e-mail: Jijiaxin0226@163.com

B. Wu
e-mail: baoguiwu@163.com

Y. Qin et al. (eds.), *Proceedings of the 4th International Conference on Electrical and Information Technologies for Rail Transportation (EITRT) 2019*, Lecture Notes in Electrical Engineering 639, https://doi.org/10.1007/978-981-15-2866-8_43

selection of control algorithm. Different control strategies lead to different suspension characteristics and vibration reduction effects. Based on this, the active suspension control algorithm was studied.

Binda et al. [2] applied the optimal control theory to the semi-active suspension system to improve the performance of the suspension system, but their control algorithm had poor robustness because it depended on the accuracy of the vehicle model, so it was rarely applied to the active suspension system. Hong and Yildiz et al. [3–5] applied adaptive control to semi-active suspension system and achieved certain control effect. However, because the control algorithm needs to identify a large number of parameters online, its real-time performance is poor, and the system is greatly affected by the impact, its robustness is poor, so it is difficult to apply to the train suspension system. Yin et al. [6–8] studied fuzzy control of semi-active suspension, and the designed controller has good comfort and control efficiency. Zapateio et al. [9–11] developed semi-active suspension system by using neural network control algorithm and predicted the optimal control force through the algorithm. The research shows that the designed controller can improve the vehicle running stability to some extent. However, some control algorithms in intelligent control have high computational complexity due to their complex models, and some algorithms need a lot of theoretical research and experience summary, so it is difficult to achieve ideal control effects.

Based on this, based on the existing control algorithm and model, this paper develops a two-system full active lateral shock absorber for a metro train by using H_∞ robust hybrid sensitivity control algorithm. The H_∞ robust control system is designed for the yaw and rolling conditions by establishing the whole train model and the joint simulation model, selecting reasonable weighting function. So as to improve the train running quality and ride comfort.

2 H_∞ Robust Control of Mixed Sensitivity

The robust control converts the conventional control design problem into the standard H_∞ control problem to solve by selecting the appropriate weighting matrix, and the method to achieve the controller performance is called the mixed sensitivity H_∞ method. For vehicle systems with multiple inputs and outputs, it is one of the effective methods to design multivariable robust controller by mixed sensitivity H_∞ optimization method [12]. In order to satisfy the requirement that the controller is designed to meet the requirement that the closed-loop system has both the ability to suppress interference and the robustness stability, it is necessary to weighted the T_{WZ} and transform the standard H_∞ problem into the mixed sensitivity H_∞ problem as shown in Fig. 1.

Figure 1 shows the construction of a generalized control object model with mixed sensitivity PS/T [13], where w is the interference input signal, z is the controlled vector, u is the control signal, and y is the measurement signal. For the subway vehicle system, the measurement signal is generally the signal measured by

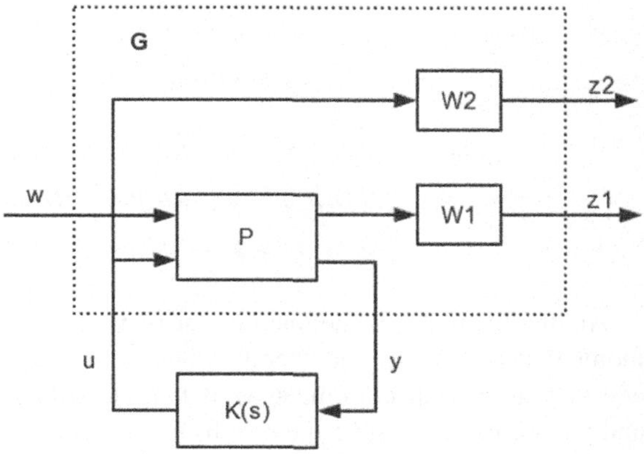

Fig. 1 Mixed sensitivity PS/T generalized object model

the sensor. *K(s)* represents the controller, and *P* is the controlled object (in this paper, the lateral dynamic model of subway train). The closed-loop transfer function $\|T_{WZ}\|_\infty$ from w to z can be expressed as:

$$\|T_{WZ}\|_\infty = \left\| \begin{matrix} z_1 \\ z_2 \end{matrix} \right\| = \left\| \begin{matrix} W_1 P(I - K(s)P)^{-1} \\ W_2 KP(I - K(s)P)^{-1} \end{matrix} \right\|_\infty \leq \gamma \tag{1}$$

where W_1 is the weight function of disturbance attenuation performance used to suppress external disturbances. W_2 is a weight function, which is selected for all possible high order unmodeled uncertainties. γ is a given constant greater than zero. In fact, for H∞ mixed sensitivity problem, G includes the controlled object P and the weighted matrices W_1 and W_2 that meet the design indexes.

3 Design of H∞ Controller with Mixed Sensitivity

3.1 Establishment of Lateral Dynamic Model of Railway Vehicles

In view of the lateral motion of the vehicle body, the lateral dynamic model of railway vehicles with three degrees of freedom including yaw, lateral motion, and lateral roll is established. The motion process is described and three motion differential equations related to the lateral motion of vehicles are obtained:

$$\begin{aligned} I_r\ddot{\theta}_r = {} & c_L l\left(l\dot{\theta}_r - \dot{y}_{10}\right) - c_L l\left(l\dot{\theta}_r + \dot{y}_{20}\right) - 2k_L l(l\theta_r - y_{10}) \\ & - 2c_L l(l\theta_r - y_{20}) + u_1 l - u_2 l \end{aligned} \tag{1}$$

$$M_B\ddot{y}_L = -c_L(\dot{y}_L - \dot{y}_{10} + h_2\dot{\theta}_R) - c_L(\dot{y}_L - \dot{y}_{20} + h_2\dot{\theta}_R) - 2k_L(y_L - y_{10} + h_1\theta_R) - 2k_L(y_L - y_{20} + h_1\theta_R) + u_1 + u_2 \quad (2)$$

$$I_R\ddot{\theta}_R = -c_Lh_2(\dot{y}_L - \dot{y}_{10} + h_2\dot{\theta}_R) - c_Lh_2(\dot{y}_L - \dot{y}_{20} + h_2\dot{\theta}_R) - 2k_Lh_1(y_L - y_{10} + h_1\theta_R) - 2k_Lh_1(y_L - y_{20} + h_1\theta_R) - 4c_Zb^2\dot{\theta}_R - 4k_Zb^2\theta_R + u_1h_3 + u_2h_3 \quad (3)$$

Among them: M_B is the mass of car body. I_r and I_R, are respectively for the car moment inertia of yaw and lateral; θ_r and θ_R, for car body roll angular displacement and yaw angular displacement. y_L is the lateral displacement of the car body; y_{10} and y_{20} are lateral displacement of front and rear bogies. h_1 is the vertical distance between the two lateral springs and the center of gravity of the car body; h_2 is the vertical distance from the center of the two lateral damper to the car body; h_3 is the vertical distance from the center of the lateral actuator to the car body. l is for the vehicle of distance in half; b is half of the secondary vertical suspension system distance of the left and right of the bogie.k_L and c_L are secondary lateral suspension system of lateral stiffness and damping, k_Z and c_Z are the secondary vertical suspension system of vertical stiffness and damping.u_1 and u_2 are the control force exerted on the vehicle body by the lateral actuators on the front and rear bogies.

Suppose the lateral displacement between the front and rear ends of the car body and the actuator is y_1 and y_2 respectively, then the following relations can be obtained:

$$\begin{cases} y_L = (y_1 + y_2)/2 \\ \theta_r = (y_1 - y_2)/2l \end{cases} \quad (4)$$

where y_L is the lateral displacement of the car body, and θ_r is the angular displacement of yaw.

In order to simplify the design of the controller and reduce the coupling between the motion, the controller is designed for the yaw motion and lateral motion of the vehicle body, respectively. In order to ensure signal acquisition and processing, the following calculation is made:

$$\begin{cases} y_Y = (y_1 - y_2)/2 \\ y_{YO} = (y_{10} - y_{20})/2 \\ \Delta y_Y = y_Y - y_{YO} \\ u_Y = (u_1 - u_2)/2 \end{cases} \quad \begin{cases} y_L = (y_1 + y_2)/2 \\ y_{LO} = (y_{10} + y_{20})/2 \\ \Delta y_L = y_L - y_{LO} \\ u_L = (u_1 + u_2)/2 \end{cases} \quad (5)$$

For yaw motion, the selected system variables are expressed as follows: state variables $X_Y = [\dot{y}_Y \quad \Delta y_Y]^T$, control variables $Z_Y = [\ddot{y}_Y \quad \Delta y_Y \quad u_Y]^T$, measuring the output $y_Y = [\ddot{y}_Y \quad \Delta y_Y]^T$, interfere with the vector $W_Y = [\dot{y}_{YO}]^T$. For lateral motion, take the state variable as $x_L = \begin{bmatrix} \dot{y}_L & \Delta y_L & \dot{\theta}_R & \theta_R \end{bmatrix}^T$, the measured signal $y_{yL} = \begin{bmatrix} \ddot{y}_L & \Delta y_L \end{bmatrix}^T$, the controlled signal $z_L = \begin{bmatrix} \ddot{y}_L & \Delta y_L & u_L \end{bmatrix}^T$, and the interference vector $w_L = \dot{y}_{L0}$.

By establishing the state-space equation from the above variables, $G_Y(s)$ and $G_L(s)$ of generalized controlled objects of yaw motion and rolling motion are obtained.

3.2 Establishment of Actuator Model

The full active ceiling control is to replace the semi-active shock absorber with the actuator. The advantage is that the actuator can provide the driving force in any direction without being affected by the motion direction of the mounting end.

The amplitude and frequency characteristic curve of the actuator is obtained through a lot of experimental research on the actuator. Therefore, a first-order inertia link can be used to replace the amplitude and frequency characteristic of the actuator. The first-order transfer function used in this paper is as follows:

$$H(s) = \frac{1}{(1/60)s + 1} \tag{6}$$

3.3 Weight Function Selection

In this paper, the evaluation index of control system performance mainly observes the attenuation performance of acceleration signal, so the selection of acceleration weight function is the key to improve the control system performance.

Assuming that the resonance frequencies of the vehicle yaw and rolling pendulum controllers are f_1 and f_2 respectively, the weighted function can select the bandpass transfer functions W_{ya_c} and W_{ya_c1} in the following Eqs. (7) and (8) to achieve the purpose of suppressing the vibration in this frequency band.

$$W_{ya\_c} = 0.05 \times \frac{s^2 + 32 \times 2\pi f_1 \times s + (2\pi f_1)^2}{s^2 + \frac{1}{3} \times 2\pi f_1 \times s + (2\pi f_1)^2} \tag{7}$$

$$W_{ya\_c1} = 0.03 \times \frac{s^2 + 8 \times 2\pi f_2 \times s + (2\pi f_2)^2}{s^2 + \frac{1}{3} \times 2\pi f_2 \times s + (2\pi f_2)^2} \tag{8}$$

On the other hand, the dynamic travel of the suspension must not exceed its physical limit. In addition, W_{ya_m} and W_{act} are selected as weighting functions respectively for the vehicle body relative displacement and control input:

$$W_{ya\_m} = 0.001; W_{\mathrm{act}} = 0.002 \quad (9)$$

According to the weight function and actuator selected above, the mixed sensitivity models of yaw and lateral H_∞ are established respectively. DGKF method is adopted to solve the problem through the robust control toolbox in MATLAB, and H_∞ robust controller of the system is obtained.

4 Control System Simulation

H_∞ controller of yaw and rolling is designed by MATLAB, and five-order yaw controller and eight-order lateral controller are obtained, respectively. Simulink, the simulation tool in MATLAB, was used to establish the mathematical model of the vehicle system. The model consists of two parts, one is the active vehicle system mathematical model with H_∞ controller, the other is the passive vehicle system mathematical model without control. Sinusoidal signal $u_i = 0.01 \sin 7.18t$ is selected as the interference input, thus obtaining the time response of the vehicle system under sinusoidal excitation input.

Figure 2 shows the time-domain simulation results of yaw and lateral controllers. Simulation results show that the control effect of the two controllers is obvious, and the lateral vibration acceleration of the car body is greatly suppressed, and the system has good performance robustness.

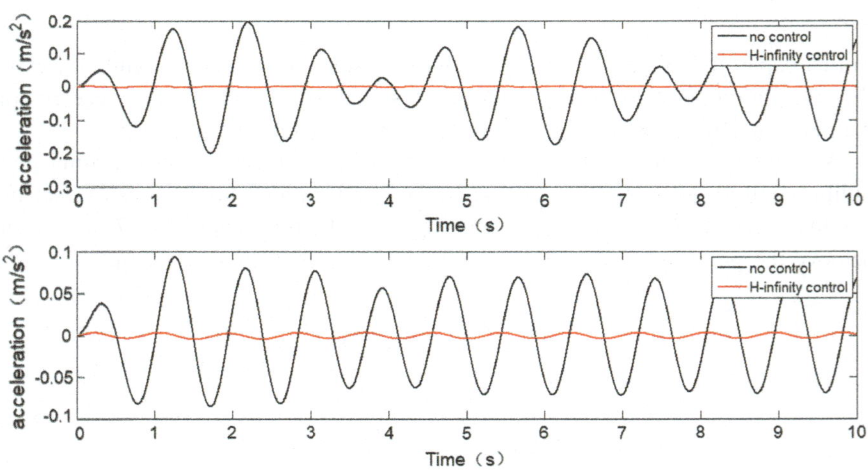

Fig. 2 System simulation under sinusoidal excitation

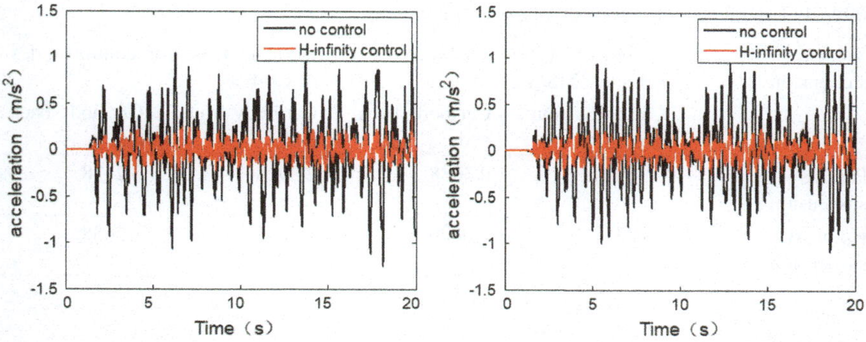

(a) Front end vibration acceleration **(b) Rear end vibration acceleration**

Fig. 3 Acceleration of vibration

In order to further verify the control effect, grade 5 us orbital spectrum was used as external orbital excitation. The speed of subway train is set at 100 km/h, and Simulink and SIMPACK are applied for co-simulation. Among them, no control in Fig. 3 is the traditional passive suspension without control. The acceleration control effect of the passive suspension and the active suspension with H_∞ control is compared and analyzed.

Figure 3 demonstrates the lateral vibration acceleration comparison curves of the front and rear ends of the car body. The vibration acceleration of the front end of the car body decreases from about 1.26 m s^{-2} passive to about 0.37 m s^{-2} active, which decreases by 70.63%. The vibration acceleration of the rear end of the car body decreased from the passive value of about 1.09 m s^{-2} to the active value of about 0.29 m s^{-2}, which decreased by 73.39%. Figure 4, taking the vibration acceleration of the front end of the car body as an example, analyzes its frequency domain, and obtains the power spectral density curve of the vibration acceleration

Fig. 4 The PSD of the front vibration acceleration of the car body

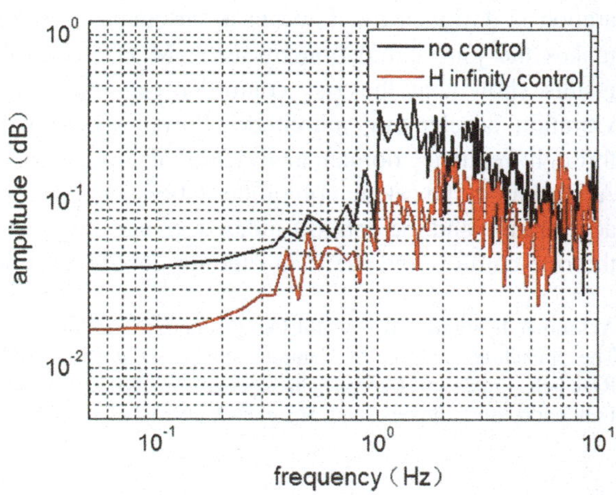

Table 1 Lateral stability index of vehicle system

Vehicle suspension	Lateral acceleration of the vehicle RMS/(m s^{-2})		Horizontal sperling comfort index of car body	
	Car body front end	Car body back end	Car body front end	Car body back end
Passive suspension	0.3694	0.3688	2.3928	2.2438
H_∞ active suspension	0.0835	0.0890	1.3251	1.3638

of the front end of the car body. It can be seen from Fig. 4 that under H_∞ robust control, the acceleration energy under 10 Hz decreases obviously, which has a good inhibitory effect on vehicle body vibration.

By solving the root-mean-square value and Sperling value of lateral vibration acceleration of front and rear end of vehicle body in Fig. 3, the index of lateral smoothness of vehicle system is shown in Table 1. It can be seen from Table 1 that both root-mean-square value and Sperling value of lateral vibration acceleration of vehicle body are reduced to some extent by active control.

By comparing the simulation results above, it suggests that using the robust H_∞ control design of lateral controller and the yaw controller has a good effect for the lower body lateral vibration and improvement of ride comfort. The system has good robustness stability. It can realize the control of any frequency band by changing the controller of the weighted function. So it has the broader scope of application.

5 Conclusion

Based on the existing research H_∞ robust control, this paper set up the subway train vehicle dynamics model. The H_∞ controller with shake and transverse mode separation is designed by choosing a reasonable weighting function. This paper also makes the joint simulation by using MATLAB and SIMPACK. The simulation results show that the H_∞ controller can effectively reduce the vehicle lateral vibration acceleration amplitude. Power spectrum energy effectively restrained through frequency domain analysis in the human body sensitive range of 0.2–2 Hz. According to the calculation of the stationarity index, the root-mean-square value of acceleration and the comfort index decrease obviously, which further indicates that the ride comfort and the train running quality are improved.

Acknowledgements We would like to thank the following supporters of this research project: The Shandong Provincial Natural Science Foundation (ZR2017LEE016), the Fundamental Research Funds of Central Universities (18CX02018A and 19CX02018A), the State Key Laboratory of Tribology, and Tsinghua University (SKLTKF16B14).

References

1. Xiukun W, Ming Z, Limin J (2016) A semi-active control suspension system for railway vehicles with magnetorheological fluid dampers. Veh Syst Dynamics 54(7):1–22
2. Binda MB, Marimuthu R (2014) Investigations on vehicle rollover prevention using LQG regulator. J Theor Appl Inf Technology 69(2):275–281
3. Le HN, Keum-Shik H, Seonghun P (2010) Road-frequency adaptive control for semi-active suspension systems. Int J Control Autom Syst 8(5):1029–1038
4. Yıldız AS, Sivrioğlu S, Zergeroğlu E (2015) Nonlinear adaptive control of semi-active MR damper suspension with uncertainties in model parameters. Nonlinear Dyn 79(4):2753–2766
5. Jing N, Yingbo H, Xing W (2017) Active adaptive estimation and control for vehicle suspensions with prescribed performance. IEEE Trans Control Syst Technol 99:1–15
6. Yin S, Shi P, Yang H (2016) Adaptive fuzzy control of strict-feedback nonlinear time-delay systems with unmodeled dynamics. IEEE Trans Cybern 46(8):1926–1938
7. Wei H, Yuhao C, Zhao Y (2016) Adaptive neural network control of an uncertain robot with full-state constraints. IEEE Trans Cybernetics 46(3):620–629
8. Yanjun L, Shaocheng T (2017) Barrier Lyapunov functions for Nussbaum gain adaptive control of full state constrained nonlinear systems. Automatica 76:143–152
9. Zapateiro M, Luo N, Karimi HR (2019) Vibration control of a class of semiactive suspension system using neural network and backstepping techniques. Mech Syst Signal Process 23 (6):1946–1953
10. Yingbo H, Jing N, Xing W (2018) Approximation-free control for vehicle active suspensions with hydraulic actuator. IEEE Trans Ind Electron 65:7258–7267
11. Yanjun L, Qiang Z, Shaocheng T (2019) Adaptive neural network control for active suspension systems with time-varying vertical displacement and speed constraints. IEEE Trans Ind Electron pp(99):1–1
12. Nardon M, Pianca P (2018) European option pricing under cumulative prospect theory with constant relative sensitivity probability weighting functions. Comput Manag Sci
13. Shujun F (2010) Simulation research of H_∞ mixed sensitivity design approache. Harbin Inst Technol (in Chinese)

Research on IGBT Voltage Drop Detection Method Based on Urban Rail PHM System

Zihan Sun, Ruichang Qiu and Fei Gao

Abstract With the development of intelligent industrial equipment, online fault detection technology has become the direction of industrial development in the new era. In this context, the power devices for the subway also hope to be able to predict the failure without stopping the machine. The paper proposes a method for measuring the voltage drop of an IGBT tube based on a switching signal and a voltage signal of a plurality of measuring points. The method is based on a city rail traction transmission system, capable of real-time inference and detection and does not require the installation of redundant sensors. The obtained detection value can be used for the system input of the urban rail PHM system, can find its short circuit fault in time, is used as a feature parameter for judging the aging degree of each tube of the IGBT and can thereby perform the health condition detection of the whole device.

Keywords PHM · Urban rail transit · Traction drive system · IGBT

1 Introduction

Transportation is the lifeblood of urban development. With the advancement of urbanization, the development of efficient and convenient transportation system is the primary goal of urban development. It is imperative to develop diversified cities. In this context, the new type of transportation mode of the subway has gradually

Z. Sun (✉) · R. Qiu
School of Electrical Engineering, Beijing Engineering Research Center of Electric Rail Transportation, Beijing Jiaotong University, 100044 Beijing, China
e-mail: 17121494@bjtu.edu.cn

R. Qiu
e-mail: rchqiu@bjtu.edu.cn

F. Gao
Beijing Dahua Radio Instrument CO., LTD, Beijing, China

Y. Qin et al. (eds.), *Proceedings of the 4th International Conference on Electrical and Information Technologies for Rail Transportation (EITRT) 2019*, Lecture Notes in Electrical Engineering 639, https://doi.org/10.1007/978-981-15-2866-8_44

become the focus of urban traffic construction, and more and more urban traffic has begun to build subway projects [1]. The construction of urban subway traffic can not only solve the problem of urban traffic congestion but also occupy the above-ground space and become the best choice for solving urban traffic problems [2].

The urban rail transit traction drive system is an important part of the entire urban rail transit system and plays a central role in energy conversion. The urban rail traction system is complex and contains many equipment. It belongs to a complex large system with multiple input, multiple output and nonlinear time-varying. If the traction drive system fails, it is often difficult to locate and analyze the fault location [3]. At the same time, the current urban rail traction system is still undergoing planned repairs, and daily inspections and planned maintenance waste a lot of manpower and material resources. Therefore, for a industrial system that can implement expert systems, fault diagnosis, life prediction and health status, it is important to be able to predict life and critical health of critical components.

The diagnosis of key components is the basis of the PHM system. PHM is the abbreviation of prognostic and health management, which is fault prediction and health management. PHM is widely used in various fields. For the entire metro PHM system, from the hub, fan, platform door to train, pantograph, the fault diagnosis and health prediction of the equipment are based on small component. From the point of view of mechanical, the wear of the gears and the aging of the track are included; from the electrical characteristics, the change of the capacitance value, the change of the resistance value and the change of the pressure drop of the switch tube are included.

Insulated gate bipolar transistor (IGBT) is the core device for energy conversion, transmission IGBT is an indispensable component in urban rail power supply system, and in traction converter system, IGBT is the most wearable device [4, 5], as shown in Fig. 1. Therefore, from partial to overall, life prediction of IGBT plays a vital role in the entire urban rail PHM system.

The failure of IGBT is a complex process related to its dynamic characteristics, involving many factors such as heat, electricity and machinery. It is the result of the internal fatigue gradually accumulating and interacting with various external factors such as the external operating environment. Binding leads, bonding points and solder layer package structure are the most vulnerable parts of IGBT, so the failure

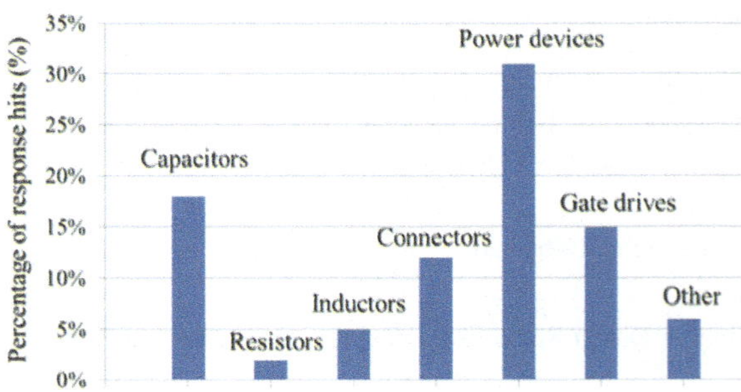

Fig. 1 Distribution of the most fragile components of power converters

Fig. 2 Examples of IGBT solder layer failure

is mainly determined by the ability and strength of bonding leads, bonding points and solder layers to withstand thermal stress and deformation [6]. Figure 2 is shown. The tube voltage drop of the IGBT will gradually increase with the age of use until it is completely damaged [7]. The increase of the collector voltage V_{ce} is an important indicator to judge the failure of the IGBT, and the V_{ce} is increased by 5% as the criterion for determining the failure of the IGBT lead-off [8]. Therefore, if the tube voltage drop of the IGBT can be detected, the life of the IGBT can be predicted by analyzing the comparison, and then, the overall health prediction of the device can be performed.

2 Algorithm Principle

The SVPWM modulation method is a pulse width modulated wave generated by a specific switching mode composed of six power switching elements and is applied to a three-phase full-bridge inverter. As shown in Fig. 3, a fully-controlled switching device can be used to enable an output current waveform for the ideal sinusoidal waveform. Compared with SPWM, the SVPWM technology has a small harmonic component of the winding current waveform, which reduces the motor torque ripple, the rotating magnetic field is closer to a circular shape, and the utilization of the DC bus voltage is greatly improved.

SVPWM is a widely used PWM control method and has been widely used in urban rail systems. The online detection method for IGBT is based on the SVPWM modulation method, based on the switching strategy of the switching tube, and is obtained by combining different conduction and combination of IGBT.

The key issue with online detection is the ability to obtain the value of the tube pressure drop through different IGBT switching sequences by known conditions by existing sensors.

The IGBT switching status and current flow direction of the three-phase full-bridge inverter circuit at a certain moment are as shown in Fig. 4. Generally

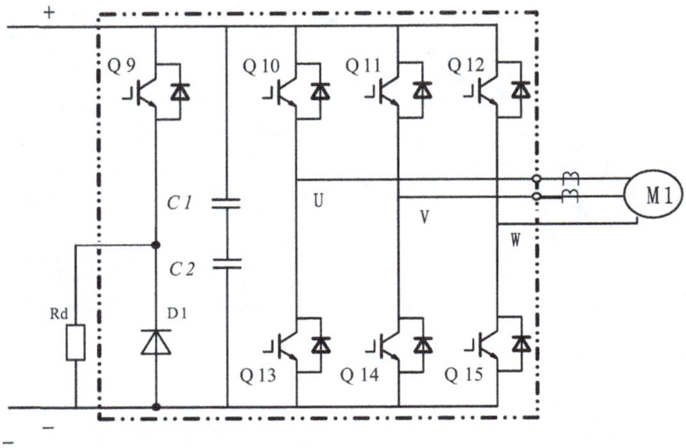

Fig. 3 The subway traction drive system's main circuit diagram

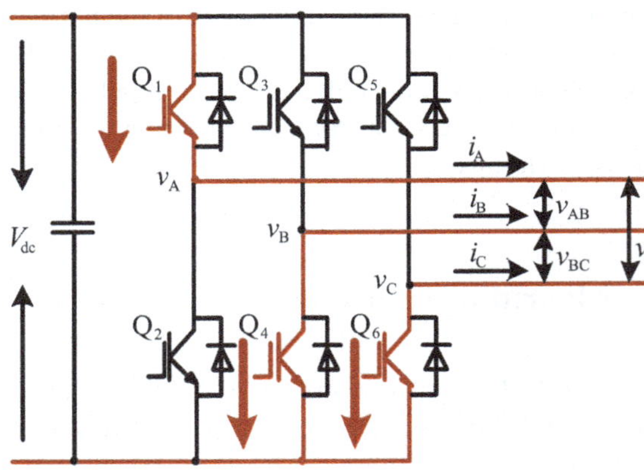

Fig. 4 IGBT switching of a three-phase full-bridge inverter circuit at a certain time

speaking, the three-phase full-bridge circuit has a total of eight switching states regardless of the special external conditions. The upper arm and the lower arm cannot be simultaneously turned on, otherwise a short circuit phenomenon may occur, which in turn causes a protection action of the device trip [9, 10]. In practical applications, the corresponding dead time is generally set to prevent this phenomenon.

The basic principle of realizing the real-time detection method of IGBT tube voltage drop is that according to the SVPWM modulation switch state, six groups of equations can be combined to solve the conduction voltage drop of each tube in real time. The following formula lists the formulas that can be obtained by juxtaposing the switching modes of Q1Q3Q6, Q2Q3Q6, Q2Q3Q5, Q2Q4Q5, Q1Q4Q5 and Q1Q4Q6:

$$\begin{cases} V_{dc} - v_{AC} = V_{Q1} + V_{Q6} \\ V_{dc} - v_{BC} = V_{Q3} + V_{Q6} \end{cases} \tag{1}$$

$$\begin{cases} V_{dc} - v_{BA} = V_{Q2} + V_{Q3} \\ V_{dc} - v_{BC} = V_{Q3} + V_{Q6} \end{cases} \tag{2}$$

$$\begin{cases} V_{dc} - v_{BA} = V_{Q2} + V_{Q3} \\ V_{dc} - v_{CA} = V_{Q2} + V_{Q5} \end{cases} \tag{3}$$

$$\begin{cases} V_{dc} - v_{CA} = V_{Q2} + V_{Q5} \\ V_{dc} - v_{CB} = V_{Q4} + V_{Q5} \end{cases} \tag{4}$$

$$\begin{cases} V_{dc} - v_{AB} = V_{Q1} + V_{Q4} \\ V_{dc} - v_{CB} = V_{Q4} + V_{Q5} \end{cases} \tag{5}$$

$$\begin{cases} V_{dc} - v_{AB} = V_{Q1} + V_{Q4} \\ V_{dc} - v_{AC} = V_{Q1} + V_{Q6} \end{cases} \tag{6}$$

Therefore, if the line voltages (V_{ab}, V_{bc}, V_{ac}) are known in the circuit, or the line voltage can be obtained indirectly from the phase voltage, and the switching state of the IGBT can be obtained from the modulation mode, the above equation can be solved in real time. The conduction voltage drop of the six switching tubes of the full-bridge circuit is obtained so that it can be used as an aging criterion for the IGBT.

3 Simulation Experiment

A three-phase full-bridge inverter circuit simulation model was built on the MATLAB/Simulink simulation experiment platform, and the SVPWM modulation method was used for modulation. Under the actual simulation conditions, the DC side voltage, phase voltage and switching state are taken as known parameters to calculate the online detection method of the above IGBT conduction voltage drop. The simulation results select the Q1 bridge arm IGBT for simulation experiment where the ordinate is the voltage value, as shown in Fig. 5.

The calculated IGBT turn-on voltage waveform is as shown in the above figure. It can be seen that the calculated waveform curve will show a fluctuating state after being unprocessed and is generally a waveform from violent fluctuation to steady development. Smoothing the calculated waveform through a Gaussian first-order filter yields a stable waveform, as shown in Fig. 6.

As can be seen from the above figure, the waveform after the first-order Gaussian filtering is smoothed, the variance is significantly reduced, and the calculated value gradually stabilizes with the passage of time. After the IGBT

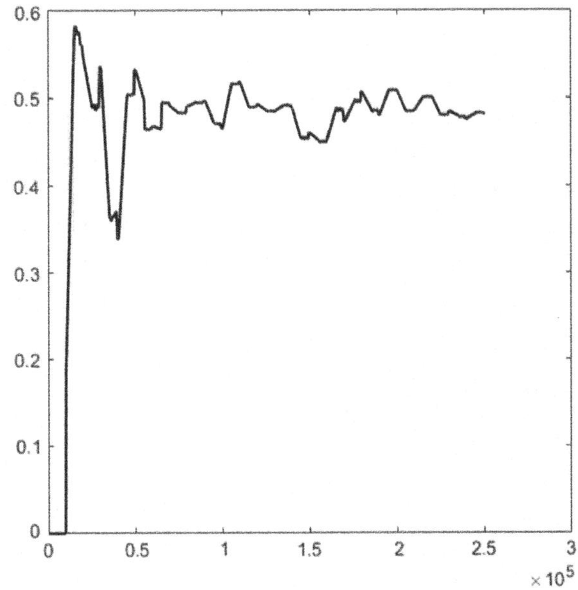

Fig. 5 On-voltage calculation waveform of a bridge arm IGBT

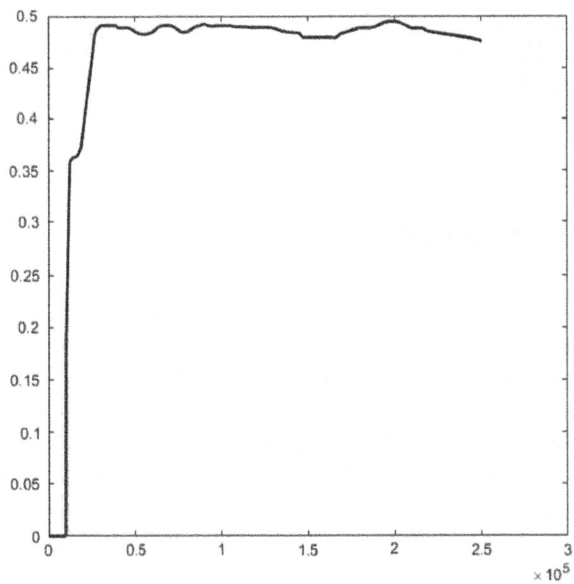

Fig. 6 IGBT turn-on voltage calculation waveform after first-order Gaussian filtering

conduction voltage drop algorithm and the first-order Gaussian filtering, the conduction voltage drop value after stabilization can be effectively obtained, which plays an important role in building a complete urban rail PHM system.

4 Conclusion

This paper first briefly introduces the importance of PHM system and IGBT in PHM system and then proposes that the conduction voltage drop can be used to measure the aging degree of IGBT, and the IGBT conduction voltage drop detection scheme is proposed. The feasibility of the proposed method is verified by MATLAB/Simulink simulation. Finally, the experimental waveform is smoothed to obtain the ideal smooth waveform.

Acknowledgements This work is supported by the National Key Research and Development Program (2017YFB1200802-01).

References

1. Jie R (2019) Analysis on the status quo and development strategy of urban subway construction in China. Enterp Technol Dev (01):278–279 (in Chinese)
2. Wei W (2017) Construction and development of urban subways in Beijing-Tianjin-Hebei region. In: The 4th national conference on smart cities and rail transit and rail transit group annual conference proceedings, 4. China Urban Science Research Association Digital City Professional Committee Rail Transit Group, Tianjin Rail Transit Group Co., Ltd., Midtown Branch (Beijing) Smart City Planning and Design Research Center (in Chinese)
3. Kai L, De qiang H, Hong sheng X, Rui G (2019) Research on fault scene analysis and diagnostic system structure of fully automatic driving train. Control Inf Technol (02): 59–63 + 81
4. Ya ling Z (2012) Research on maintenance mode selection of urban rail transit fixed equipment. Beijing Jiaotong University
5. Ling hui M (2017) Research on fault prediction and health management (PHM) and reliability assessment technology of traction converter. Beijing Jiaotong University
6. Sheng you X (2013) Research on IGBT condition monitoring and reliability evaluation method of power module. Chongqing University
7. Lu Y, Christou A (2019) Prognostics of IGBT modules based on the approach of particle filtering. Microelectron Reliab 92
8. Xiong Y, Cheng X, Shen J et al (2008) Prognostic and warning system for power-electronics modules in electric, hybrid electric, and fuel-cell vehicles. IEEE Trans Industr Electron 55 (6):2268–2276
9. Mao gang W, Rong xiang Z, Xin zhou T (2006) Dead space effect analysis and compensation method of space vector PWM inverter. J Zhejiang Univ (Engineering Edition) (03):469–473
10. Diao L, Zhao L, Jin Z, Wang L, Sharkh SM (2017) Taking traction control to task: high-adhesion-point tracking based on a disturbance observer in railway vehicles. IEEE Ind Electronics Mag 11(1):51–62

Vehicle-mounted Dynamic Detection and Prejudgment of Railway Snow Cover Conditions in Remote Cold Regions

Ji'ao Wei, Bing Wang, Wenzhe Wang, Jianping Hu, Xuezhao Fang and Enjie Meng

Abstract In view of the safety threat of sudden snowfall in the remote cold regions to railway transportation, China Railway Corporation has formulated strict speed limit regulations for trains in snow and ice weather. Real-time detection of snow cover on railway during train operation has become a key technology. In this regard, through the analysis of the detection methods and methods of railway snow cover status under heavy snow conditions in the remote cold regions, a dynamic detection technology of vehicle-mounted railway snow cover status is proposed. Based on real-time detection of local railway snow cover status, the dynamic prediction of the snow cover status in front of the railway is made, so as to provide effective information for railway transportation dispatching and safety management departments.

Keywords Remote cold region · Railway · Snow cover status · Real-time detection · Dynamic · Prejudgment

1 Introduction

Railway lines in remote cold regions are long, and sudden snowfall can easily lead to various types of train traffic accidents. And emergency rescue is very difficult, which poses a security threat to railway transportation [1]. In view of the snow and ice weather, China Railway Corporation has compiled "Regulations for Railway Technical Management" [2]. Referring to the Regulations, all kinds of trains in remote cold regions should reduce their speed or even temporarily stop operation in case of sudden snowfall. Railway transportation is a system engineering, if the trains slow down or stop by the sudden fall of heavy snow, which will lead to randomly adjust the transportation system and trigger the emergency rescue response. In order to provide the railway transportation dispatching and safety

J. Wei · B. Wang (✉) · W. Wang · J. Hu · X. Fang · E. Meng
School of Mechanical Engineering, Xinjiang University, Urumqi 830047, China
e-mail: 451619037@qq.com

Y. Qin et al. (eds.), *Proceedings of the 4th International Conference on Electrical and Information Technologies for Rail Transportation (EITRT) 2019*, Lecture Notes in Electrical Engineering 639, https://doi.org/10.1007/978-981-15-2866-8_45

management departments with the information of snow cover status quickly and effectively, under the condition of railway transportation characteristics in remote cold regions, it is particularly crucial to detect the local railway snow cover status in real time and predict the state of snow cover in front of the railway dynamically. In this regard, through the analysis of the detection methods and methods of railway snow cover state under the condition of heavy snow in cold regions, a dynamic detection technology of vehicle-mounted railway snow cover state is proposed. Based on real-time detection of snow-covering conditions of the railway, the dynamic Predict the snow cover of the railway ahead.

2 Analysis of Detection Modes and Methods of Railway Snow Cover Under Characteristic Conditions

In remote cold regions, such as northeast and northwest China, the railway is long and spans widely, and the weather environment is bad in winter. It is normal for heavy snowfall to occur abruptly. When a train is running on a snowy railway, the friction between wheel and rail decreases due to the snow cover on the railway, and the heavy snow weather blocks the train attendant's sight, which easily leads to various train traffic accidents. In this regard, in order to ensure safe driving in remote cold regions, scholars all over the world have done some researches on snow cover condition detection along the railway. Li and Ren [3] cooperated with meteorological monitoring station to realize a set of systematic methods of snow depth detection, image shooting, gale prediction and wireless transmission of information along the railway. Du [4] proposed a snow depth detection method based on the intensity attenuation difference caused by infrared light through snow, and Yanwei and Haihe [5] adopt the scheme of ultrasonic sensor ranging. It is concluded that the influence of temperature and wind speed on the ultrasonic ranging is the main factor leading to the low measurement accuracy. So, the temperature compensation method is proposed to improve the accuracy. Kan and Gao [6] proposed a method of snow depth detection based on laser multi-point ranging to improve the accuracy of snow depth detection.

To sum up, first, the above-mentioned literature reflects that in the selection of snow condition detection methods, scholars generally adopt the way of configurating monitoring stations along railways to monitor the snow cover state. However, due to the characteristics of remote cold areas, monitoring stations are difficult to configure, expensive and difficult to maintain, resulting in a series of obstacles, leading to scattered monitoring stations, distant from each other. The real-time performance and accuracy of snow cover condition detection need to be further verified and improved. For the selection of snow cover depth detection methods, the methods used in the above literature include ultrasonic, infrared and laser ranging methods, but these three methods are vulnerable to smoke, dust, raindrops and snow throwing

Table 1 Advantages and disadvantages of various ranging methods

Ranging methods	Strengths	Weaknesses
Ultrasonic	Short-distance measurement has good effect, mature technology, low cost and is insensitive to light and electromagnetic field	Slow speed, large emission angle and susceptibility to environmental interference
Infrared	Distance measurement is farther than ultrasonic, and emission angle is smaller than ultrasonic	Easy to be disturbed by sunlight or other similar wavelength light sources, as well as by smoke and dust
Laser	High accuracy, long distance, the farthest can reach tens of kilometers, the smallest launch angle	High cost, easy to be disturbed by smoke, dust and raindrops
Microwave	High accuracy, insensitive to dust and mud, negligible temperature effects, rarely affected by meteorological environment, good directionality and transmission characteristics	Poor anti-electromagnetic interference ability (this paper does not involve magnetic field interference)

during driving process. It leads to the decrease of measurement accuracy. The advantages and disadvantages of various sensing methods are shown in Table 1.

According to the characteristics of remote cold regions, this paper proposes a dynamic detection method of vehicle-mounted sensors to deal with the shortcomings of monitoring stations. According to Table 1, microwave has the characteristics of penetrating rain and fog, small propagation attenuation and can work all-weather with high measurement accuracy [7]. Therefore, according to the characteristics of microwave sensors, this paper uses vehicle-mounted microwave sensors to detect the snow-covered status, real-time detection of railway snow-covered status information in the process of train travel, overcomes the shortcomings of other sensor detection methods.

3 Vehicle-Mounted Real-Time Detection of Snow Covering State on Railway

Microwave sensor achieves the purpose of ranging by indirectly measuring the phase difference between transmitted and received signals [8]. When the signal transmitted by the instrument is as follows:

$$s = U_{\mathrm{m}} \cos(\omega t + \varphi_0) \tag{1}$$

Under ideal conditions, the signal received after returning from the measured object is as follows:

$$s = U_{\mathrm{m}} \cos(\omega t + \varphi_0 - \omega t') \tag{2}$$

where U_{m} is the signal amplitude, ω is the signal angular frequency, φ_0 is the signal initial phase, $\omega t'$ is the phase change produced after the round trip over the measured distance.

Through the following formula:

$$\begin{cases} \varphi = \omega t \\ \omega = 2\pi f \\ L = \frac{1}{2} vt \end{cases} \tag{3}$$

Thus, the distance of the measured object can be measured as

$$L = \frac{C}{4\pi f} \cdot \varphi \tag{4}$$

where L is the distance to be measured, C is the microwave propagation rate (near the speed of light in the atmosphere), f is the frequency of electromagnetic wave signal, and φ is the phase difference.

Meanwhile, φ is can be measured as

$$\varphi = N \cdot 2\pi + \Delta\varphi \tag{5}$$

According to (4) and (5), it can be concluded as follows:

$$L = \frac{C}{4\pi f} \cdot (2N\pi + \Delta\varphi) \tag{6}$$

Finally, L can be converted to be as follows:

$$L = \frac{\lambda}{2}(N + \Delta N) \tag{7}$$

where N is a positive integer and ΔN is a true fraction less than one. The measured target distance can be obtained only by measuring the phase difference between the transmitted signal and the received signal. Therefore, the microwave sensor has less influence on the detection of snow cover state during train traveling, which is less affected by rain, snow, fog and dust and can satisfy the detection of railway snow cover state information by trains traveling at high speed in bad weather.

A real-time detection scheme for railway snow-covered state during traveling is to install a microwave sensor 1 at L_1 from the center of the first wheel of the train head and vertically emit electromagnetic waves to the railway surface. Because

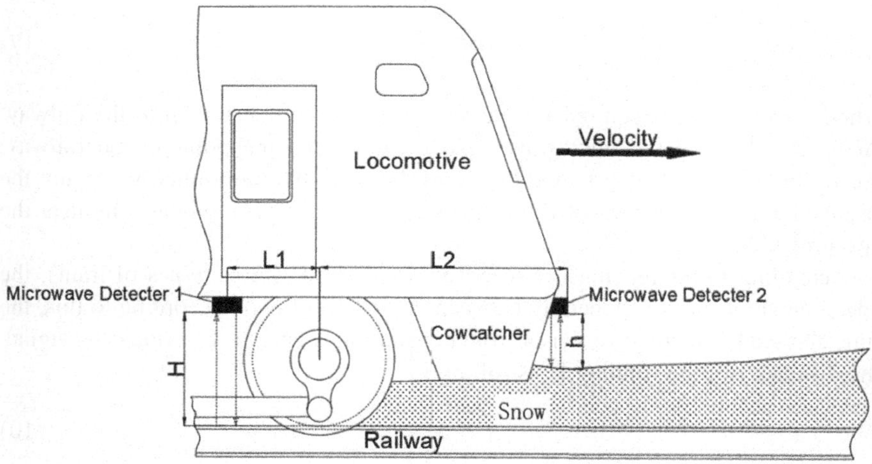

Fig. 1 Installation diagram of microwave sensor

there is a certain thickness of the cowcatcher under the head of the locomotive, so another microwave sensor 2 is installed at the front of the cowcatcher L_2 from the center of the wheel of the head. And emit electromagnetic waves vertically to the snow surface. The installation diagram is shown in Fig. 1.

When the train passes through the snow-covered railway at high speed, the snow on the railway is squeezed out of the railway at the same time, so the snow on the railway behind the wheel is covered with residual snow, but it does not affect the measurement of the sensor. The microwave sensor 2 is used to detect the distance from the snow surface to the emitter point of the sensor, and the microwave sensor 1 is used to measure the vertical distance from the sensor to the clean railway. The emitter points of the two sensors are located at the same horizontal position.

According to the position relation of the sensor in Fig. 1, the snow thickness can be obtained as follows:

$$h_0 = H - h \tag{8}$$

where h_0 is the thickness of the snow, H is the distance from the microwave sensor 1 to the railway, L_0 is the distance from the microwave sensor 2 to the snow surface.

In the process of measuring snow thickness during train traveling, it is necessary to consider the error relationship between the train travel speed and the real-time performance of microwave detection. Since the microwave belongs to the electromagnetic wave, its propagation speed approaches the speed of light in the atmosphere, and the sensor is mounted on the train in the manner shown in Fig. 1 above. By querying the information, the height of the train chassis from the railway is mostly not higher than 500 mm. Therefore, we can get formulated as follows:

$$\Delta t = \frac{\Delta h}{C} \tag{9}$$

where Δt is the time required for the microwave to be transmitted to the railway, Δh is the distance the microwave sensor emits from the point to the railway. According to the calculation, the time error between the transmitted wave and the received wave can be neglected for the electromagnetic wave propagating near the speed of light.

According to referencing the speed information of various types of trains, the speed range of trains is generally between 20 and 100 (m/s). According to this, the time elapsed by microwave sensor from transmitting signal to receiving echo signal, the distance of train running is as follows:

$$s = \Delta t \cdot V_r \tag{10}$$

where s is the distance travelled by the train from microwave launch to reception, V_{r} is the train speed. According to the actual calculation of the vehicle speed, the train only travels less than 1 μm after receiving the echo information. Therefore, the longitudinal acquisition error of the microwave sensor and the time error can be neglected.

4 Dynamic Prejudgment of Front Railway Snow Covering

According to the "Technical Regulations for Railway Technology," there are different regulations on the mode of operation of trains under different snow cover conditions. The stipulations for snowfall weather are as follows:

In snowfall weather, when snow falls heavily or snow cover the sleepers or ballast surface in operation section, the speed limit of ballastless track section is 250 km/h or less, and ballast track section is 200 km/h or less; when snowfall or snowstorm occurs in operation section, the speed limit of ballastless track section is 200 km/h or less, and ballast track section is 160 km/h or less.

When the thickness of sleeper snow in ballastless track is over 100 mm, the speed limit is 200 km/h or less; when the thickness of snow in ballast track is over 50 mm, the speed limit is 160 km/h or less.

Therefore, when the information of snow cover condition is detected, the measured snow thickness information is compared with the speed limit threshold set by the railway department in real time so that the train attendants can take manual speed reduction or stopping trains in advance according to the early warning information to ensure the safety of running.

Through the installation method of the microwave sensor mentioned above, it is obviously that the snow cover thickness of the railway at a certain time can be easily obtained by using the principle of microwave ranging in the course of a train traveling at a certain time.

If the speed of the train on the snow-free railway is V_t, according to the "Technical Regulations for Railway Technology," when the snow thickness exceeds h', the train will stop running. It can be considered that the speed of the train is 0 when the train stops, so a mathematical relationship can be established as follows:

$$h = k \cdot V + b \tag{11}$$

When $H = h'$, $V = 0$; $H = 0$, $V = V_t$, the mathematical relationship between train speed and snow cover thickness is obtained as follows:

$$V = 1 - \frac{h}{h'} \cdot V_t \tag{12}$$

where V_t is the train speed on the snow-free railway, h is the snow cover thickness, h' is the snow cover thickness when the train stops operation.

If the snow falls at a uniform rate, the snow thickness accumulates over time. At t_0, the snow cover thickness on the railway is h_0, and at t_1, the snow cover thickness on the railway is h_1. Therefore, the speed reduction rate can be obtained as follows:

$$V_s = \frac{h_1 - h_0}{t_1 - t_0} \tag{13}$$

From above, the mathematical relationship between snow thickness and time can be obtained as follows:

$$h(t) = h_0 + \frac{h_1 - h_0}{t_1 - t_0} \cdot t \tag{14}$$

where V_s is the snowfall rate, h is the snowfall thickness at time t, and h_0 is the initial snowfall thickness. Thus, $h(t)$ can be substituted into Eq. (12):

$$V = 1 - \frac{h_0 + \frac{h_1 - h_0}{t_1 - t_0} \cdot t}{h'} \cdot V_t \tag{15}$$

According to the relationship between train speed and snow thickness deduced above, in ideal condition, the snow cover state of the railway ahead can be pre-judgment according to the relationship between snow thickness and time, and the forecast information is transmitted to the railway transportation dispatching and railway safety management department to change the reorganization of the train and take necessary measures such as deceleration and parking for the running train.

5 Epilogues

This paper analyzes the advantages and disadvantages of various detection methods, and puts forward a dynamic detection technology of vehicle-mounted type railway snow cover state, which is based on the real-time detection of local railway snow cover conditions. According to the mathematical relationship, the snow-covered state of front railway is dynamically prejudgment, and the snow-covered state of railway is provided to railway transportation dispatching and safety management departments. The state information is used to ensure the rapid response and adjustment of railway transportation system under the condition of heavy snow in remote cold regions. Due to the lack of relevant research and literature on vehicle-mounted snow cover detection in remote cold region, it is expected that this paper can provide reference for the research of railway snow cover detection and automatic control technology of train dynamic speed limit.

Acknowledgements This paper is supported by the Natural Science Foundation of Xinjiang in 2018 (No. 2018D01C054).

References

1. Wang B, Shi X, Wang W (2018) Analysis of traffic impedance effects of highway rescue for passenger train accidents. Lecture Notes in Electrical Engineering, v 483, p 149–156. In: Proceedings of the 3rd international conference on electrical and information technologies for rail transportation, EITRT 2017-Transportation
2. China Railway Corporation (2014) Regulations for Railway Technical Management. China Railway Publishing House, Beijing (in Chinese)
3. Hong L, Li R (2014) Design and implementation of railway snow depth monitoring system. China Railw 03:67–69 (in Chinese)
4. Du C, Wang Q, Liu X, Zhao Y, Deng X, Cui L (2017) Research and application of ice thickness and snow depth automatic monitoring system. IEEE Trans Instrum Meas 66(2):325–331
5. Yanwei L, Haihe L, Berlin W (2012) Comparative experimental analysis of snow depth measurement and artificial observation with ultrasonic sensors. Meteorol Sci Technol 40 (02):198–202 (in Chinese)
6. Kan C, Gao X, Xing Z, Zhang Y, Shi F (2017) Research on multi-point measurement of snow depth of high-speed railway based on laser sensor. Mach Manuf Autom 46(05):214–218 (in Chinese)
7. Zheng D, Chen W, Zhang P, Li C (2014) Microwave displacement measurement method based on phase measurement. Sensors Microsyst 33 (06): 55–57+68 (2014) (in Chinese)
8. Dayong W, Chaofeng Z, Jianyi P, Ming L (2010) Analysis of the factors affecting the ranging accuracy of microwave rangefinders. Electronic Measur Technol 33(04):42–44 (in Chinese)

An Online Monitoring Scheme for Circuit Breakers Contact System

Jiao Guo, Jie Chen, Ting Li, Heping Fu and Ruichang Qiu

Abstract In recent years, the prognostic and health management (PHM) technology has been effectively developed as a key technology for the intelligent development of power grids. The online monitoring system for circuit breaker (CB) lifetime has significant research value. Based on the requirement for intelligent monitoring of CBs' electrical life, this paper provides an online monitoring scheme for CB contact system. This scheme adopts the improved arc current weighted algorithm, which calculates the contact electrical wear per interrupted online and estimate residual life of CB with the theoretical allowable wear deduced by the least squares method. The wavelet algorithm is used to find the accurate time point where the arc starts. The TMS320F28335 digital signal processor (DSP) + AD7656 + ESP8266 chip architecture is used for AD sampling, online computing, and wireless communication.

Keywords Circuit breaker (CB) · Electrical life · Online monitoring · PHM

J. Guo (✉) · J. Chen · T. Li · H. Fu · R. Qiu
School of Electrical Engineering, Beijing Engineering Research Center
of Electric Rail Transportation, Beijing Jiaotong University, 100044 Beijing, China
e-mail: 18126089@bjtu.edu.cn

J. Chen
e-mail: jiechen@bjtu.edu.cn

T. Li
e-mail: 18117010@bjtu.edu.cn

H. Fu
e-mail: 17121428@bjtu.edu.cn

R. Qiu
e-mail: rchqiu@bjtu.edu.cn

Y. Qin et al. (eds.), *Proceedings of the 4th International Conference on Electrical and Information Technologies for Rail Transportation (EITRT) 2019*, Lecture Notes in Electrical Engineering 639, https://doi.org/10.1007/978-981-15-2866-8_46

1 Introduction

Circuit breakers (CBs) are widely used in small and medium-sized power supply and distribution terminals, which are useful for protecting circuits and equipment.

The reliability of CBs directly affects the stability and safety of the power grid. In some application fields, it is more valuable to know exactly the total wear and residual life of the CB rather than simply improving the performance and prolonging the lifetime of products [1]. It will help power grid to mitigate possible safety concerns and accelerate the process of networking and intelligence by real-time monitoring for the operational performance and health status of the CB.

Some test results show that the operating mechanism, the thermal-magnetic tripping system, and the contact system in the CB have higher failure rates than other components [2]. During the interrupted process, the CB is eroded by the arc generated due to electrical load, and as a result, the contact material may be worn or directional transferred. As the number of interruptions increases, CB loses efficacy gradually with the contact surface roughness, gap and pressure changing. On these grounds, many scholars have focused on the research of the failure mechanism of CB contact system. Another fact is that the mechanical life of the CB is much longer than its electrical life. Therefore, the electrical life of the contacts is selected as the main technical index to estimate the residual life of the CB in this paper.

The residual life of the contact system can be predicted by monitoring various factors such as the coil current waveform, contact travel curve, dynamic resistance, arc current, and operating time. References [3, 4] propose that the coil current waveform and contact travel curve contain abundant information which can be used to diagnose some mechanical faults of CB. However, online analysis of current waveform and travel curve is difficult, and special expensive sensors need to be added. Reference [5] indicates that the dynamic resistance measurement is effective to assess the condition of contacts in the CB, but not adaptable to online monitoring schemes of circuit breakers. The arc current weighted accumulation method, which ignores the effect of arc time dispersion on contact electrical wear, is widely used in practice [6, 7]. In order to improve the accuracy of life model, an improved arc current weighted algorithm is adopted in this paper.

2 Improved Arc Current Weighted Algorithm

2.1 The Electric Wear Per Interrupted

The contact material is eroded by the arc generated due to the electrical load; each time, the CB is interrupted; and the lost material is proportional to the integration of current during the arcing period. Therefore, the electric wear per interruption of the CB can be defined by

$$q_n = K\int_0^{t_n} |i_n|^{\alpha}\mathrm{d}t \tag{1.1}$$

where K is the proportional constant, n = 1, 2, 3... is the sequence of interrupted cycle, i_n and t_n are the arc current and arc time during the nth interruption, respectively. α is the weighted index of the arc current, which is related to the contact material and usually within 1–2. Figure 1 shows the common arc current waveform of AC circuit breakers.

It can be seen from Fig. 1 that the arc current varies in a wide range, and the calculation accuracy and magnetic saturation of current sensor coil need to be considered when measuring the current. In practical applications, multi-level current sensors can be installed according to system requirements and various restructure algorithms of current are used to calculate actual current values.

2.2 *Determination of Formula Parameters and Allowable Wear*

According to the analysis in the previous section, the proportional constant K and the arc current weighted index α need to be determined. At the same time, in order to estimate the residual life of the CB, it is necessary to determine the theoretical allowable wear. The least squares method is used to determine the arc current weighted index α and the total allowable wear Q, and the proportional constant K is calculated according to the electrical life curve and the sampling rate [8, 9].

The manufacturer will provide the electrical life curve according to type of CB. Figure 2a shows an electrical life curve of a certain CB model provided by Schneider. The curve and formula in Fig. 2b can be deduced by using the least squares method.

Therefore, the relationship between the total wear Q and the weight α is as follows:

$$Q = \frac{1162.5}{I^{1.588}} I^{\alpha} \tag{1.2}$$

Fig. 1 Arc current waveform AC system

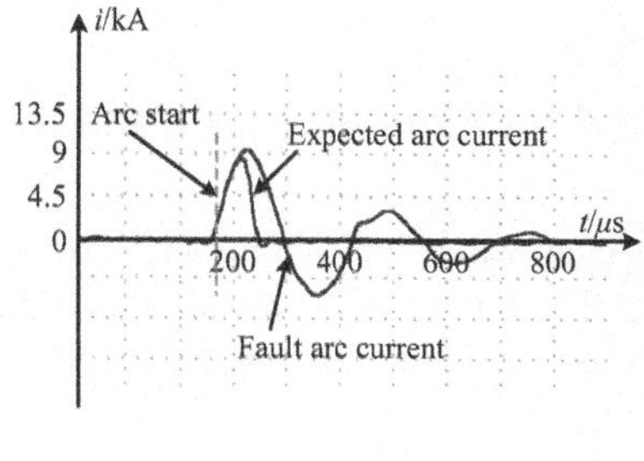

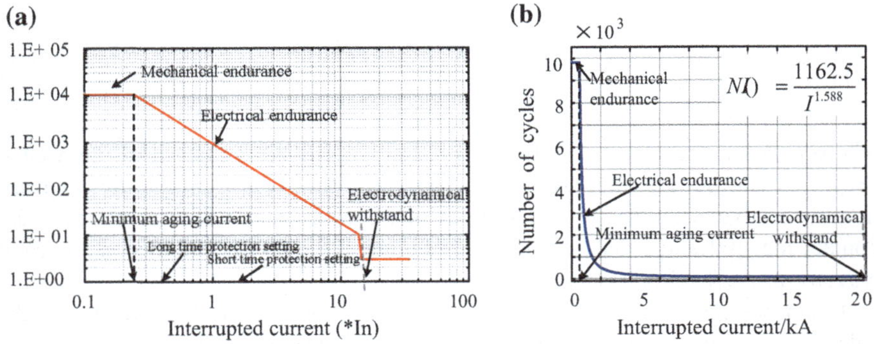

Fig. 2 Endurance curve for a circuit breaker

When the material, structure, and arc-extinguishing mode of the contact are designed, the weighted index α and total allowable wear Q should be determined. In this example, $Q = 1162.5$, $\alpha = 1.588$. Combining Eqs. (1.1) and (1.2), we can get

$$Q = K\sum_{n=1}^{N}\int_{0}^{t_n} |i_n|^{\alpha}\mathrm{d}t \tag{1.3}$$

where N is the allowable number of cycle under the arc current i_n.

Under the simplified condition, the three-phase arc time calculation function can be obtained with the effect of arc time dispersion on contact electrical wear be ignored, and then the proportional constant K under ideal conditions can be calculated by taking in (1.3). The electrical life prediction model calculated based on this assumption tends to be conservative, which will cause some waste. In order to obtain a more accurate proportional constant K, it can be determined by experiment or simulation.

3 Calculation Method of Arc Time

As mentioned above, the improved calculation strategy needs to know the arc starting and extinguishing time accurately. In this section, the wavelet algorithm is used to process arc current sampling data online, and the feasibility is verified by simulation and experiment.

3.1 Calculation of the Arc Time Using Wavelet Algorithm

As an extension of Fourier algorithm, wavelet algorithm is widely used in various fields [10]. For a signal with limited energy $f(t) \in L^2(R)$, its discrete wavelet transform is defined as

$$W_{\psi}f(m,n) = a_0^{-\frac{m}{2}} \int_R f(t)\bar{\varphi}(a_0^{-m}t - nb_0)\mathrm{d}t \tag{1.4}$$

where $\bar{\varphi}(t)$ is the complex conjugate function of sub-wavelet function, $a = a_0^m$ is the approximation coefficients and $b = nb_0a_0^m$ is the detail coefficients.

In this article, we choose Daubechies wavelet function as a sub-wavelet function to analyze the arc current waveform, as shown in Fig. 3, where waveform appears obvious distortions. As a singular point in the signal, the distortion point can accurately indicate the starting time of arc. Since the zero-crossing point of current is the extinguishing time of arc, the arc time can be easily calculated.

3.2 Simulation and Experiment Verification

In order to verify the feasibility and accuracy of the algorithm through simulation, the *CB* (*with arc*) *module* in MATLAB is used to build the model to simulate the arc phenomenon of the circuit breaker. The simulation model is shown in Fig. 4. The arc time constant of the CB is set to 15 μs, the initial arc conductance is set to 10 kS.

At the same time, in order to verify the accuracy of wavelet algorithm in practical application, a small power prototype and AD sampling circuit are used to obtain a set of experimental data, and the data is processed by MATLAB wavelet toolbox. Current waveforms and reconstructed detail coefficients curves in simulation and experiment are shown in Fig. 5.

It can be seen from the picture, for the *simulating waveform*, there is no jamming signal in the current, so the arc start point can be deduced accurately through wavelet analysis. However, the *experiment waveform* contains a large number of high frequency interference signals, which makes it difficult to find out the arc start point from the detail coefficients waveform directly. We used Butterworth low-pass filter to process the current signal, and the results from wavelet analysis become clearly readable. Both simulation and experimental results show that the wavelet algorithm can meet the accuracy requirements of calculating arc time.

Fig. 3 Arc current curve

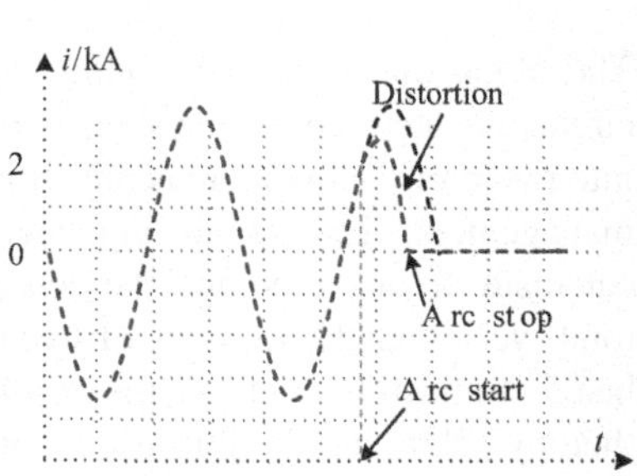

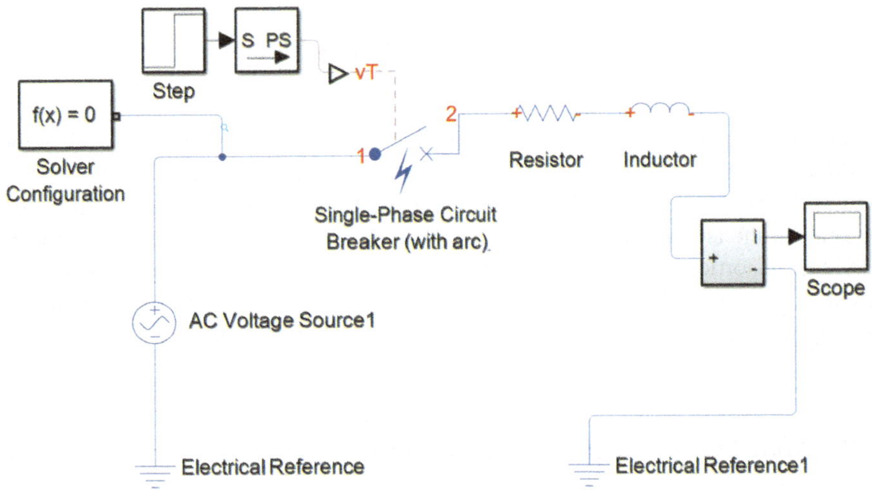

Fig. 4 Circuit breaker simulation model

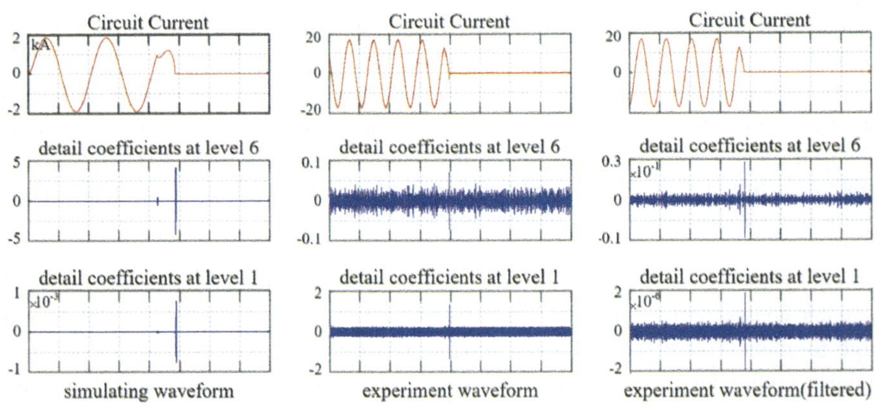

Fig. 5 Simulation and experimental results

4 Overall Framework of Monitoring System

The block diagram of the online monitoring system for the electrical life of the circuit breakers contact system is shown in Fig. 6. The system is mainly divided into two parts: semaphore acquisition processing and upper computer's analysis and management. The monitoring device collects the three-phase contact current to calculate the arc time and the contact wear, and stores historical data such as the total wear and the number of cycle. The high-speed WIFI module transmits the health indicators such as the residual electrical life of the CB to the terminal, thereby achieving the purpose of online monitoring.

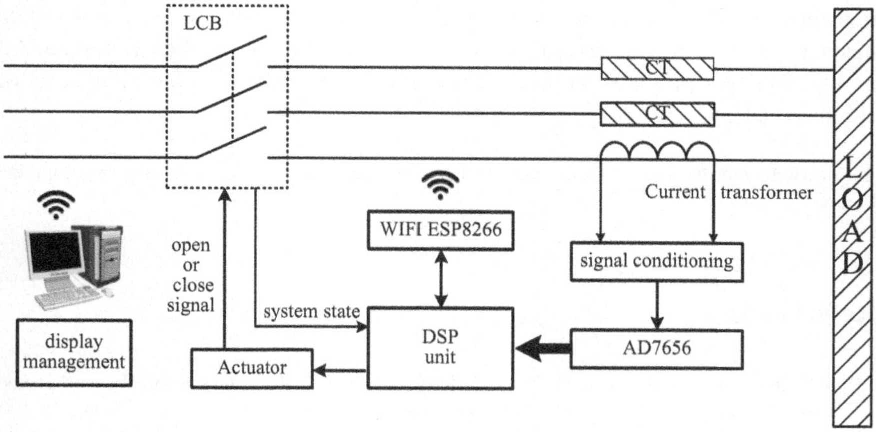

Fig. 6 The CB online monitoring system structure diagram

The contact electrical life model CB used in this paper only need to measure the three-phase line current. The multi-level current transformers and signal conditioning circuit convert the signal into a voltage range acceptable to the AD7656. The AD7656 has six independent ADC inputs and communicate with the DSP through high-speed data bus and can achieve a throughput rate of 250 kSps. The selected TMS320F28335 DSP chip has powerful digital signal processing capabilities, and can meet the requirements of data storage and online analysis by equipped with peripheral memory chips. The DSP communicates with the ESP8266 WIFI module through the serial peripheral interface (SPI) and uses the TCP protocol to achieve fast and accurate data transmission. After receiving the wireless transmission signal from the monitoring device, the host computer displays and evaluates the electrical and mechanical status of the CB and implements intelligent management of the CB system.

5 Conclusion

This paper provides a system scheme for online monitoring of the electrical life of circuit breakers. The scheme adopts an improved electric wear cumulative weighting model and gives a method for determining relevant parameters. In addition, the wavelet algorithm is used to calculate the arc time, which avoids the installation of device to detect the starting time of arc. And the feasibility of the algorithm is analyzed through simulation and experiment. Finally, based on the theoretical analysis described above, the overall framework of the online monitoring device is given. In this paper, real-time online processing of high sampling frequency and large data volume can be realized by adopting advanced AD sampling and digital processing system. At the same time, the high-speed wireless transmission scheme, the host computer's interface display function, and online management function are combined to realize the intelligence of the CB system. All in all, this paper provides a reference for online monitoring and fault diagnosis of circuit breakers.

Acknowledgements This work is supported by the National Key Research and Development Program (2017YFB1200802-01).

References

1. Razi-Kazemi AA, Vakilian M, Niayesh K, Lehtonen M (2014) Circuit-breaker automated failure tracking based on coil current signature. IEEE Trans Power Delivery 29(1):283–290
2. Kim BC, Kim ST, Ahn KY, Lee JH (2012) Development of a vacuum circuit breaker with breaking and switching capability. In: International symposium on discharges & electrical insulation in vacuum. IEEE
3. Wang C, Liu C (2015) Experimental study of the online monitoring of the mechanical characteristics of a hydraulic high-voltage circuit breaker. In: 2015 5th international conference on electric utility deregulation and restructuring and power technologies (DRPT). IEEE
4. Cheng CY, Chen LJ, Kao WS (2008) Diagnosing medium voltage GIS circuit breaker by observing trip coil current with simulation and actual tests. Electric Mach Power Syst 36(2):14
5. Zhang X, Zhang J, Gockenbach E, Borsi H (2009) Life management of sf6 circuit breakers based on monitoring and diagnosis. IEEE Electr Insul Mag 25(3):21–29
6. Borkowski P, Hasegawa M (2007) A computer program for the calculation of electrode mass loss under electric arc conditions. IEICE Trans Electron C 90(7):1369–1376
7. Chen G, Lan L, Pan Z, Wen X, Wu Y, Wang Y (2017) Electrical erosion test and condition assessment of sf6 circuit breaker contact sets. IET Gener Transm Distrib 11(8):1901–1909
8. Hu Q, He J, Zhang X, Fan X (2017) The on-line monitoring application study of vacuum circuit breakers contact system. J Electr Eng 13(1):43–48 (in Chinese)
9. Guan Y, Huang Y, Qian J (2001) On-line monitoring of the electrical wear of high voltage circuit breaker using an improvement method. High Voltage Appar 37(4):1–3 + 6 (in Chinese)
10. Zheng H, Qing Q (2000) The signal singularity index calculation method and its application based on wavelet theory. Electric Power Autom Equip 20(3):11–12 (in Chinese)

Research on the Segmentation and Extraction of Scenes Along Railway Lines Based on Remote Sensing Images of UAVs

Lei Tong, Limin Jia, Zhipeng Wang, Yunpeng Wu and Ning Wang

Abstract At present, the manual inspection along railway lines is still a major method to ensure railway operation safely, but the cost is high and work efficiency is low. Therefore, unmanned aerial vehicles (UAVs) patrol inspection is required. This paper presents the effective segmentation of scenes along railway lines (SRL) from remote sensing perspective of UAVs based on the full convolutional networks (FCN). Firstly, the datasets needed in this research are collected and produced from Langfang section of the Beijing–Shanghai high-speed railway. The datasets are expanded by using data augmentation to constrain the overfitting in the training process. Secondly, the segmentation model FCN-8s for SRL is developed and trained. The related setting and hardware environment in the training process are described in this paper. The experimental results show that a single image prediction needs 151.2 ms, to achieve 6.6 fps when input size is 384 × 384. Good accuracy is obtained on the test dataset, i.e., 55.8% MIoU and 70.2% MPA, which meets the expectations of FCN. At the same time, it is also found that the segmentation of railway area achieves the best result thus the railway area is extracted accordingly.

Keywords Railway scenes segmentation · Semantic segmentation · UAVs

L. Tong · L. Jia · Z. Wang (✉) · Y. Wu · N. Wang
State Key Laboratory of Rail Traffic Control and Safety, Beijing Jiaotong University, Beijing, China
e-mail: zpwang@bjtu.edu.cn

L. Tong · L. Jia · Z. Wang
National Engineering Laboratory for System Safety and Operation Assurance of Urban Rail Transit, Beijing Jiaotong University, Guangdong, China

L. Tong · L. Jia · Z. Wang
Beijing Research Center of Urban Traffic Information Sensing and Service Technologies, Beijing Jiaotong University, Beijing, China

Y. Qin et al. (eds.), *Proceedings of the 4th International Conference on Electrical and Information Technologies for Rail Transportation (EITRT) 2019*, Lecture Notes in Electrical Engineering 639, https://doi.org/10.1007/978-981-15-2866-8_47

1 Introduction

In recent years, the UAVs remote sensing technology has been applied to the transmission line inspection [1], and it also provides a fast and effective means for the inspection of railway lines. While in the patrol inspection work, images collected by UAVs are broad, rich in content and high in resolution, and both railway infrastructure and surroundings have a vital impact on the safe operation of the railway. Therefore, the segmentation and extraction of SRL based on UAVs remote sensing play an important role in the safety of monitoring railway lines.

In the future, computer vision will become a key tool for the analysis of images collected by patrol inspection of UAVs along railway lines. As a branch of computer vision, semantic segmentation technology will provide strong support for the segmentation of scenes along railways. Long et al. proposed FCN [2] for image semantic segmentation by adapting the structure of VGG [3] model and add some upsampling layers. On this basis, many models are developed to improve accuracy with different techniques [4–11]. In the work of segmentation of railway scenes using semantic segmentation techniques, Wang et al. proposed an architecture for segmentation of railway regions and optimized contours of orbital regions using polygonal fitting method [12]. He et al. also proposed a semantic segmentation network for segmenting railway scenes and verified the superiority of the network over UNet and FCN networks [13].

However, all these works do not put too much attention on ensuring the safety of railway operation. Furthermore, railway scenes defined in Refs. [12, 13] are very different from this research. We concentrate on SRL from the perspective of UAVs while they define railway scenes from the perspective of train cabs. Because FCN is the base model of many of the above models, this study aims to achieve effective segmentation and further extraction of SRL by using FCN model.

2 Dataset

2.1 Build of Dataset

Three predefined locations, i.e., areas A, B and C of the Langfang section of the Beijing–Shanghai high-speed railway, are selected for image collection and the weather conditions are very well on the day of collecting images. According to the requirements of visual range and resolution of the captured images, the flying height of UAVs is set between 80 and 200 m. After images collection, images with resolution 3648 $\times$ 5472 ($h \times w$) of SRL are selected, removing inapplicable ones that are repeated, blurred and angularly offset. 220 images from 430 selected images are used to create semantic segmentation datasets for segmentation of SRL, in which 200 images collected from A and B are used to build the training and validation dataset, and 20 images left collected from C are used to build test dataset.

In the built dataset, scenes are classified into five categories: background, buildings, vegetation, railways and roads, i.e., every pixel in images from the dataset is labeled as one of the above categories. In order to improve the efficiency of labeling process, all original images are scaled to 512 × 768 and all annotations are implemented using the image annotation tool LabelMe.

In the annotated images, color black represents background, and colors red, green, yellow and blue represent buildings, vegetation, railways and roads, respectively. These annotated images need to be converted to grayscale images when training the model.

2.2 Data Augmentation

At present, there are still many difficulties to collect images of SRL with UAVs and the amount of 220 images in the built dataset is far from enough for the training of deep learning model. In order to improve the training effect, constrain the over-fitting in the training process and improve generalization ability of the model, the annotated dataset is expanded by using data augmentation.

The built dataset can be expanded to 17 times of original size with rotating and horizontal mirroring operation (8 types), brightness transformation operation (4 types) and cropping operation (5 types) of the original images and annotations, respectively. Among them, the rotation and horizontal mirroring operation include 8 types in all, i.e., horizontal mirrors (yes, no) × rotation angle (0, 90, 180, 270) where the initial image is represented when the rotation angle is 0 and horizontal mirroring is not used; brightness transformation operation includes 5 types in all, i.e., brightness (0.5, 0.8, 1.2, 1.5); cropping operation includes 5 types in all, i.e., cropping position (bottom left, bottom right, center, top left and top right). Part of the data augmentation process is shown in Fig. 1. For ease of display, the images are scaled to a square shape. Through the above operations, 3400 images are obtained for training and validation and 340 images for test.

3 Segmentation Model for SRL

3.1 Network Architecture of the Model

The FCN model can use a variety of network architecture as its encoder [11], the key part of which is the convolutionalization[1] of fully connected layers and upsampling layers. In Ref. [2], the authors use different network architectures to compare achieved accuracy, among which the FCN network based on VGG [3]

[1]For further understanding about the word "convolutionalization" in Fig. 2 of Ref. [2].

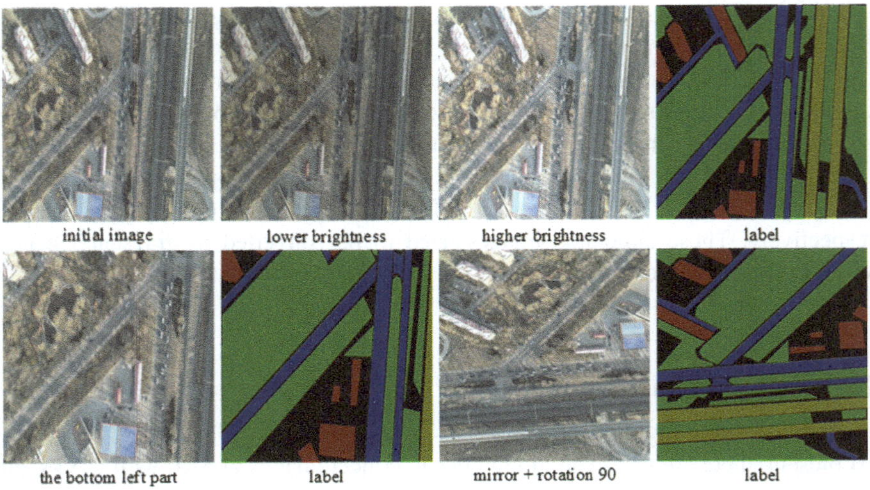

Fig. 1 Examples of data augmentation

model performs best. Besides, the authors adopt different skip structures in the process of upsampling to the resolution of the input image, i.e., FCN-32s, FCN-16s and FCN-8s, and point out FCN-8s achieves best accuracy. Therefore, FCN-8s model based on VGG-19 is implemented in this paper, which means there is no difference in the first five convolutional and pooling modules between VGG-19 and FCN-8s here, i.e., modules A to E are shown in Fig. 2. Also, it is modules from A to E to fine-tune in the transfer learning process which will be mentioned further in the following Sect. 3.2. Subsequent three modules F1 to F3 are all adapted from fully connected layers in VGG-19.

As illustrated in Fig. 2, three upsampling layers are built in the model to increase the resolution of feature maps. The first two upsampling layers restore the resolution of feature maps to 1/16 and 1/8 times of resolution of the original input images, respectively, while the third one restores the resolution of input image completely. Obviously, we know that there is the same resolution between the input image and the output image. The results of upsampling layers are fused with the corresponding pooling layers in the encoder [2, 11], combining fine local features in the lower layers and coarse global features in the higher layers to achieve precise feature extraction. The two fuse layers are obtained by adding the upsampling layer and corresponding pooling layer. Transposed convolution is an important way to achieve upsampling. In this paper, upsampling is implemented with transposed convolution and deep learning framework TensorFlow is used to implement FCN-8s model.

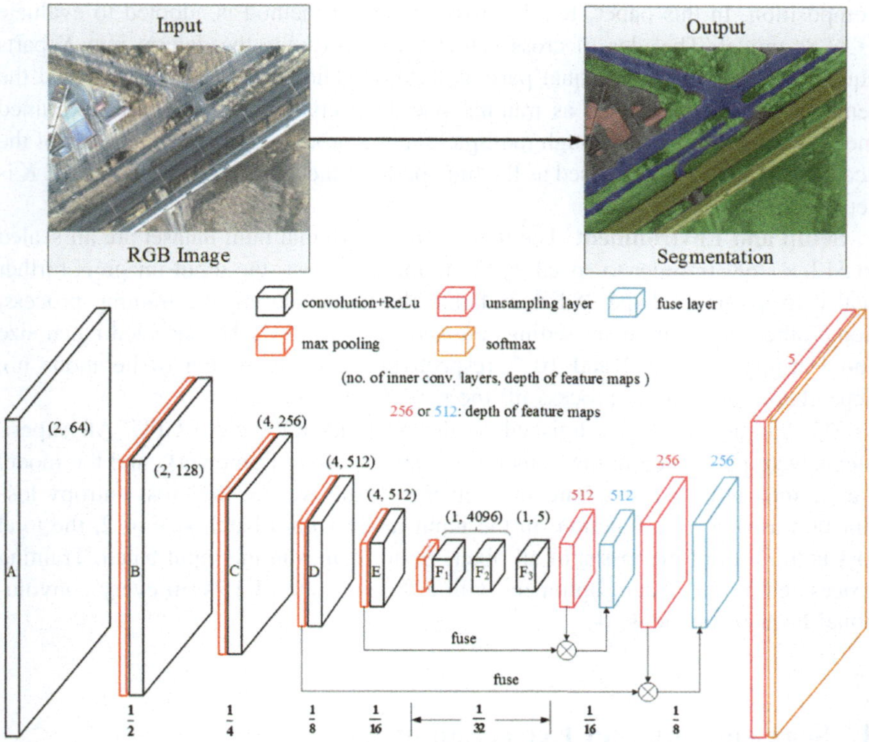

Fig. 2 Network architecture of segmentation model for SRL

3.2 *Training of the Model*

Transfer Learning and Cross-Validation. Constantly fine-tuning the weights from a pre-trained model is one of the main ways of transfer learning [14]. In addition, it is higher layers to be fine-tuned not lower layers since the lower one tends to contain more generic features that most models share. It is also important to choose an appropriate learning rate, generally a smaller one, when using transfer learning.

In this paper, the pre-trained model VGG-19 is used to accelerate the training process. The initialization of the parameters of the FCN-8s network model is finished by transforming fully connected layers of VGG-19 to convolutional ones and randomly initializing other newly built higher layers, e.g., unsampling layers, etc. In the subsequent training process, training of the model is completed by continuously fine-tuning the parameters in the network with smaller learning rate, especially the high-level parameters that are randomly initialized.

Cross-validation is a method to evaluate the performance of a model. It is mainly achieved by dividing dataset into training part and validation part with different composition. In this paper, K-fold cross-validation method is adopted to evaluate FCN-8s model. The idea of cross-validation is to divide the dataset into K parts equally, then each of the equal parts is used as validation dataset in turn and the remaining equal parts used as training one, respectively. In such case, K trained models can be obtained through multiple training processes. And the average of the accuracies of K models is used as the final performance indicator. In this paper, K is set to 10.

Setup and Environment. The images in the original built dataset are all scaled to 512 × 768. In order to speed up the training process, the input image is further scaled to a square shape 384 × 384 into the model during the training process. Some other hyperparameter settings are shown in Table 1. The selected batch size and learning rate are 2 and 10^{-4}, respectively, and the setting of iterations no. depends on the training process till the model converges.

The training process is finished on the NVIDIA RTX 2080 GPU. All experiments were performed in the TensorFlow deep learning framework, and the model was trained until the loss function converges. And we use the cross-entropy loss function as objective function. In the input mode with a batch size of 2, the total loss is the sum of the losses of all pixels of all images in the input batch. Training process employed Adam optimizer with a dropout rate of 15% in every convolutional layer of the network.

4 Segmentation and Extraction of SRL

4.1 Experimental Results

Runtime of the Model. After calculation, the running speed of the model built in this paper is summarized in Table 2. A single image prediction needs 151.2 ms, achieving 6.6 fps when input size is 384 × 384.

Accuracy on Validation Dataset. As shown in Tables 3 and 4, mean values of the ten trained models' evaluation metrics are obtained through cross-validation and segmentation performance of SRL is perfect. Since MIoU is a standard metric for

Table 1 Hyperparameters setting in the training process

Hyperparameters	Batch size	Learning rate	Iterations no.
Value	2	10^{-4}	Convergence

Table 2 Runtime and environment of the model

GPU	Framework	Input size	Run time	fps
NVIDIA RTX 2080	TensorFlow	384 × 384	151.2 ms	6.6

Table 3 Cross-validation results on validation dataset

Metrics	PA	MPA	MIoU	FWIoU
Mean value	91.3	90.5	81.9	84.5

Table 4 Cross-validation results of all classes on validation dataset

Class	Background	Building	Plant	Railway	Road
Mean of IoU	75.7	72.6	88.0	92.2	81.8

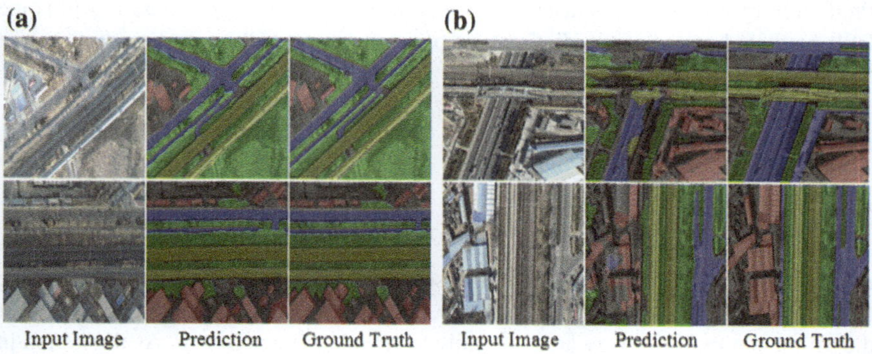

Fig. 3 Prediction results on validation and test dataset. **a** Validation part. **b** Test part

semantic segmentation tasks, cross-validation results for different classes of IoU are listed here, in particular, as shown in Table 4. From Table 4, it is illustrated that IoU of the railway area is the highest and the segmentation performance is the best, which has a close relationship with more regular geometry shape and specific color characteristics of the railway area. Prediction results on validation dataset are shown in Fig. 3a and perfect performance on segmentation work are realized as we can see.

Accuracy on Test Dataset. Since the images in test dataset are collected in area C, which is different from the images from training set and validation dataset (A and B), the evaluation on the test dataset seems to be more objective and accurate, as shown in Table 5. Compared with the accuracy metrics on the validation dataset, the accuracy metrics on the test dataset has a certain decline, but still meet the expectation on the accuracy FCN architecture can achieve itself, which is 56% [3]. As illustrated in Table 6, the IoU value of the railway area on the test dataset is still much better than the other classes, which maintains a high level compared to the results on the validation dataset. This objectively and realistically shows that the

Table 5 Cross-validation results on test dataset

Metrics	PA	MPA	MIoU	FWIoU
Mean value	70.9	70.2	55.8	56.9

Table 6 Cross-validation results of all classes on test dataset

Class	Background	Building	Plant	Railway	Road
Mean of IoU	53.3	39.8	57.2	78.6	52.3

built model is better for segmentation of the railway area than other classes. Prediction results on test dataset are shown in Fig. 3b. However, there is still some fault predicted part, e.g., predicting part of road area as railway and predicting part of building area as road.

4.2 Extraction of Railway Area

According to the segmented image that segmentation model for SRL predicts, the different SRL can be extracted separately. Since the model has a high accuracy for the segmentation of railway area, the railway area is extracted here. As illustrated in Fig. 4, the left image is an annotation of some image in the dataset; the right image shows that three areas filled in white are different regions of the same class: green rectangle represents a rectangle that can include a target subarea along horizontal and vertical axis directions of the image, respectively; red rectangle represents a rectangle that can include the a target subarea and has also the smallest area. Some concepts covered in this section are defined as follows:

Rectangular Subgraph: A rectangular part obtained by cropping images in the horizontal and vertical axis directions, e.g., rectangles in green in Fig. 4b.
Target Subarea: A rectangular part which includes one of the regions representing the same class and has the smallest area, e.g., rectangles in red in Fig. 4b.
Mask: A grayscale image with the same resolution with the initial image, in which only specific region's values of pixels are set to 1 and the remaining set to 0.
Submask: *Mask* corresponding to a specific *rectangular subgraph.*
Sublabel: The label image or predicted image of a specific *target subarea*, that is, the corresponding part that *target subarea* reflects on the original label image or the predicted image.

It is easy to find that the *target subarea* is contained in its corresponding *rectangular subgraph.* Suppose original image as *IMG* and corresponding label image or predicted image as *LBL.* For a specific *target subarea*, suppose the *mask* corresponding to the inner area of its green rectangle as MG_i, the *mask* on corresponding to the inner area of its red rectangle as MR_i and the *mask* on *LBL*

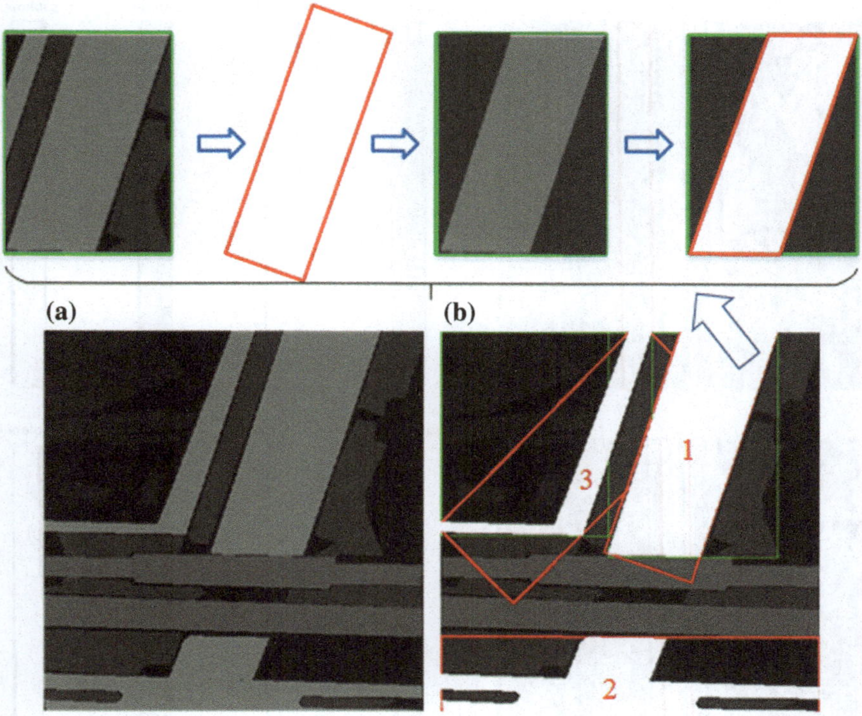

Fig. 4 Concepts for *rectangular subgraph*, *submask*, *target subarea* and *sublabel*. **a** A grayscale label image from the built dataset. **b** Key rectangular areas with different meanings. The extraction process of *target subarea* no. 1 is displayed on the top area

corresponding to its all pixels belonging to the current class as ML_c. As shown in Fig. 5, the extraction process of all the railway regions can be expressed as follows:

Step 1: Extract the *rectangular subgraph* from the original image, i.e., *subgraph*.

$$subgraph(i) = Crop_{MG_i}(IMG) \tag{1}$$

Step 2: Extract the part that MR_i falls on the *subgraph*, i.e., *submask*:

$$submask(i) = Crop_{MG_i}(MG_i \cap MR_i) \tag{2}$$

Step 3: Extract the *target subarea*, i.e., *subarea*:

$$subarea(i) = submask(i) \cap subgraph(i) \tag{3}$$

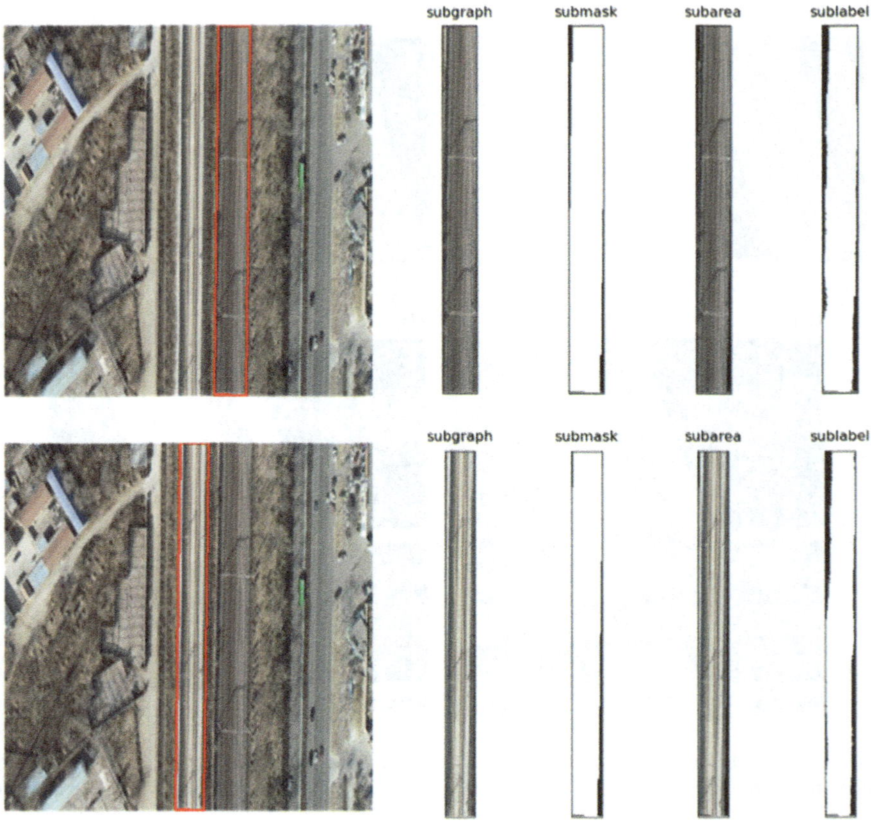

Fig. 5 Extraction of railway area. The *subgraph*, *submask*, *subarea* and *sublabel* here corresponding to the concepts in step 1–4

Step 4: Extract the *sublabel* corresponding to the *target subarea*, i.e., *sublabel*:

$$sublabel(i) = submask(i) \cap Crop_{MG_i}(LBL \cap ML_c) \tag{4}$$

In all above formulas, $Crop_{M(i)}(\bullet)$ represents cropping the image in parentheses according to the rectangle indicated by $M(i)$ and the rectangle is required to be in the horizontal and vertical axis directions, while i represents the ith aggregation area of the pixels belonging to class C. The process of extracting the railway area is illustrated in Fig. 5. Two corresponding sets of *rectangular subgraph*, *submask*, *target subarea* and *sublabel* are, respectively, extracted.

5 Discussion and Conclusions

Efficient monitoring of SRL can ensure effective and safe operation of railway. This research combines patrol inspection of UAVs and semantic segmentation techniques with safety monitoring along the railway for the first time. Segmentation and extraction of SRL based on UAVs remote sensing images are realized in this paper and the experimental results have demonstrated the effectiveness of the built model. In further work, the accuracy of the model needs to be improved. At the same time, the runtime of the model needs to be further reduced to meet the practical needs.

Acknowledgements This research is supported by the National Key R&D Program of China (No. 2016YFB1200203).

References

1. Zhao Z, Li S, Qi Y et al (2018) A semantic segmentation method for aerial image of transmission line with improved FCN model. J China Sciencepaper 13(14):1614–1620 (in Chinese)
2. Long J, Shelhamer E, Darrell T (2015) Fully convolutional networks for semantic segmentation. In: Proceedings of 2015 IEEE conference on computer vision and pattern recognition (CVPR), pp 3431–3440. IEEE, Piscataway
3. Simonyan K, Zisserman A (2015) Very deep convolutional networks for large-scale image recognition. In: Proceedings of the 3rd international conference on learning representations (ICLR 2015), arXiv preprint arXiv:1409.1556v6
4. Chen LC, Papandreou G, Kokkinos I et al (2015) Semantic image segmentation with deep convolutional nets and fully connected CRFs. In. Proceedings of the 3rd international conference on learning representations (ICLR 2015). arXiv preprint arXiv:
5. Chen LC, Papandreou G, Kokkinos I et al (2016) DeepLab: semantic image segmentation with deep convolutional nets, atrous convolution, and fully connected CRFs. J IEEE Trans Pattern Anal Mach Intell 40(4):834–848
6. Yu F, Koltun V (2016) Multi-scale context aggregation by dilated convolutions. arXiv preprint arXiv:1511.07122v3
7. Paszke A, Chaurasia A, Kim S et al (2016) Enet: a deep neural network architecture for realtime semantic segmentation. arXiv preprint arXiv:1606.02147
8. Roy A, Todorovic S (2016) A multi-scale CNN for affordance segmentation in RGB images. In: European conference on computer vision, pp 186–201. Springer, Heidelberg
9. Eigen D, Fergus R (2015) Predicting depth, surface normals and semantic labels with a common multi-scale convolutional architecture. In: Proceedings of the IEEE international conference on computer vision, pp 2650–2658. IEEE, Piscataway
10. Bian X, Lim SN, Zhou N (2016) Multiscale fully convolutional network with application to industrial inspection. In: 2016 IEEE winter conference on applications of computer vision (WACV), pp 1–8. IEEE, Piscataway
11. Badrinarayanan V, Kendall A, Cipolla R (2017) SegNet: a deep convolutional encoder-decoder architecture for image segmentation. J IEEE Trans Pattern Anal Mach Intell 39(12):2481–2495
12. Wang Z, Wu X, Yu G et al (2018) Efficient rail area detection using convolutional neural net-work. J IEEE Access 6:77656–77664

13. He Z, Tang P, Jin W et al (2018) Deep semantic segmentation neural networks of railway scene. In: 2018 37th Chinese control conference (CCC), pp 9095–9100. Elsevier, Amsterdam
14. Garcia AG, Escolano SO, Oprea SO et al (2017) A review on deep learning techniques applied to semantic segmentation. arXiv preprint arXiv:1704.06857

Rail Surface Defect Recognition and Classification Method Based on Deep Forest

Ying Zhou, Limin Jia, Zhipeng Wang and Yunpeng Wu

Abstract The main goal of this study is to propose a suitable method in order to reduce the detection cost of orbital defects, improve the efficiency of defect detection, and effectively achieve rail health assessment by combining UAV image acquisition and digital image processing technologies. The characteristics of the images collected by the UAV using the proposed appropriate method to deal with the defects, which Hough transform and horizontal projection method are used to extract the rail area. In the process of image enhancement for rail images, an improved local normalized image enhancement method is proposed, and a defect segmentation method based on maximum entropy threshold value is used for defect segmentation. According to the calculation and extraction of defect characteristics, a classification model is proposed based on the deep forest method to classify the two types of defects, namely spalling and crack. The results show that the proposed method can accurately and effectively classify the rail surface defects in a small sample and has a certain practical reference value.

Keywords Rail surface defect · UAV image · Defect detection · Deep forest

1 Introduction

The existence of track surface defects leads to unstable train operation, which greatly damages the wheel, wheelset and bogie components; it not only shortens the service life of train parts, but may even threaten the safe driving of trains, such as train derailment accidents. Therefore, the effective detection of track surface defects is an indispensable measure to ensure the safe and reliable operation of the railway system [1]. Due to the high frequency and high strength of rail, some wear and tear problems will still occur [2]. According to the summary of experience, various

---

Y. Zhou · L. Jia · Z. Wang (✉) · Y. Wu
State Key Lab of Traffic Control and Safety, Beijing Jiaotong University, No.3 Shangyuancun, Beijing, Haidian District, China
e-mail: zpwang@bjtu.edu.cn

Y. Qin et al. (eds.), *Proceedings of the 4th International Conference on Electrical and Information Technologies for Rail Transportation (EITRT) 2019*, Lecture Notes in Electrical Engineering 639, https://doi.org/10.1007/978-981-15-2866-8_48

defects on rail surface mainly refer to the various damage states on rail surface [3], such as dropping, peeling and cracking, which affect the normal service performance of trains.

Visual inspection has the advantages of high efficiency and low cost. Due to its intuitive and convenient characteristics, visual inspection has been widely used in the field of engineering detection [5, 6]. In order to reduce the artificial detection cycle and improve the detection accuracy, many universities and companies have carried out the subject of using machine vision technology [7]. A laboratory of the University of Florida in the USA has realized the detection of missing fasteners and crack peeling in recent studies. Ensco company in the USA also recently developed a rail integrated detection system, which can realize the detection of connecting parts and other aspects. However, these universities and companies have not carried out in-depth analysis and research on the surface defects of the rail, and have not completely and systematically solved various problems in the detection [8]. In recent years, deep learning has been gradually applied in various fields. However, the requirement of deep learning for sample size and hyperparameter greatly limits its application in image processing. Due to the small number of surface defects in normal operation railways and the limited range of data acquisition, it is not possible to obtain sufficient samples. Therefore, it is necessary to find a method suitable for small sample classification.

Therefore, we propose to use UAV for image acquisition and combine the deep forest method with the image processing method to identify the surface defects of the rail. This method provides a new idea for rail health assessment and ensures the safe and effective operation of the railway system. The advantages are as follows: (1) It can solve the blind zone problem of the traditional detection method in the field of rail surface defect detection. (2) The use of the UAV for rail surface defect detection can greatly reduce the cost of inspection and improve the detection efficiency. (3) It can improve the status quo of manual detection and provide a new idea and idea for automated detection. (4) Using the identification method based on small sample data, the problem of insufficient sample data existing at present can be solved on the basis of ensuring the correct rate.

2 Methodology

In this paper, we mainly studied two types of defects, spalling and cracking. We divide the classification process of rail surface defect identification into three steps: (1) Image acquisition based on the UAV. (2) Defect extraction. Defect extraction is divided into three parts: rail extraction, image enhancement and defect segmentation. According to the characteristics of the UAV-based images, a suitable method is proposed to process the image. (3) Defect classification based on the deep forest.

Fig. 1 Data collection site

2.1 Image Acquisition

The Matrice 600 UAV platform and the Zenmuse X5 PTZ camera were used in image acquisition. In order not to affect normal rail operations, we selected a freight line in Bao Ding, Hebei Province (Fig. 1).

2.2 Defect Extraction

2.2.1 Rail Track Extraction

In order to avoid extraction of non-rail areas [9], we use the combination of Hough transform and horizontal projection method to add the orbit direction correction step when extracting the rail region.

Hough transform to extract the longest line: (1) performing Hough transform on the image to obtain the Hough matrix; (2) finding the peak point in the Hough matrix, detecting the line segment; (3) calculating the edge length of the line segment, extract the longest line by comparison and mark it.

According to line, the angle between the line and the vertical direction can be confirmed, and rotate the image to θ. Let (x_1, y_1) and (x_2, y_2) be the position coordinates of the two endpoints of the line, respectively.

$$\theta = 90^\circ - \tan^{-1}\frac{y_2 - y_1}{x_2 - x_1} \tag{1}$$

After obtaining the appropriate image, a coordinate system is established to count the sum of the gray values of the pixels in the horizontal direction of the image [10]. Let the image be an $M * N$ matrix and give an N-dimensional matrix G [11]. Where $D_{i(N-1)}$ is the gray value of the coordinates (x, y).

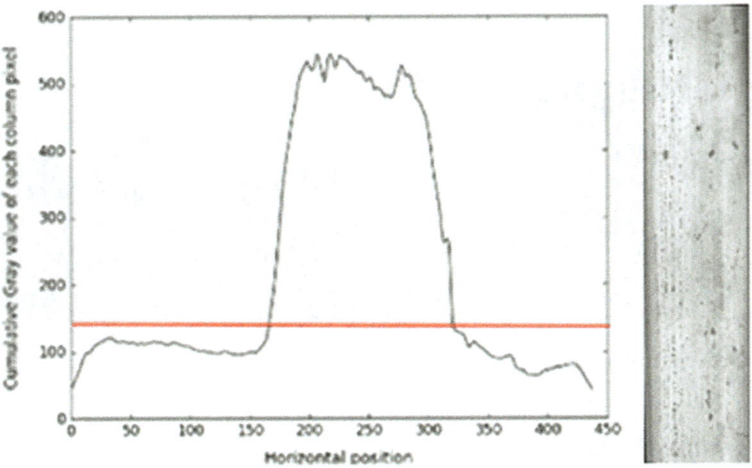

Fig. 2 Horizontal gray scale distribution image and extraction result

$$G = \left[\sum_{i=0}^{M-1} D_{i0}, \sum_{i=0}^{M-1} D_{i1}, \ldots \sum_{i=0}^{M-1} D_{i(N-1)} \right] \quad (2)$$

According to the statistical results, gray distribution images in horizontal direction can be obtained [12] (Fig. 2).

2.2.2 Image Enhancement

Li and Ren [13] proposed a local normalization enhancement method (LN) according to the background of the rail surface, the characteristics of the defect and the surface reflection characteristics. This paper proposes to use the improved local normalization method (LNI):

$$\mathrm{LNI}_{(x,y)} = \frac{F_{(x,y)} - E(F_{(x',y')})}{F_{(x,y)} + E(F_{(x',y')})}, (x', y') \in W \quad (3)$$

where $E(\bullet)$ is the average pixel value of all pixels in window W; $F_{(x,y)}$ represents the pixel value of the image.

Normally, in a partial window A, the illuminance is considered to be equal, the light which is reflected by defect is less than the other parts, and the gray value of the defect is lower than the gray value of the background. Thus, if the pixel value of one pixel is smaller than the average of the window, this point can be regarded as a defect point; otherwise, it is treated as a background point. Based on the above reasons and Formula (4), we can convert all non-defective points into a consistent background point through a dynamic threshold $E(F_{(x',y')})$.

$$\mathrm{LNI}_{(x,y)} = \begin{cases} \mathrm{LNI}_{(x,y)} = \dfrac{F_{(x,y)} - E(F_{(x',y')})}{F_{(x,y)} - E(F_{(x',y')})}, & \text{if } I_{(x,y)} < E(F_{(x',y')}) \\ 0, & \text{otherwise.} \end{cases} \tag{4}$$

2.2.3 Defect Segmentation

The maximum entropy (ME) algorithm is proposed in [16]. The (ME) algorithm can confirm a threshold which maximizes the amount of information provided by the pre-attribute distribution ϕ_o and the background point distribution ϕ_b.

$$\phi_o = \sum_{n=0}^{T-1} p_n, \quad \phi_b = 1 - \phi_o, \tag{5}$$

where p_n is the probability of the gray level n of an image. Assume that the grayscale range of an image is [0,255], the foreground entropy $H_o(T)$ and background entropy $H_b(T)$ are defined as the (6–9):

$$H_o(T) = -\sum_{n=0}^{T-1} \left(\frac{P_n}{\phi_o(T)} \ln \frac{P_n}{\phi_o(T)} \right) \tag{6}$$

$$H_b(T) = -\sum_{n=T}^{255} \left(\frac{P_n}{\phi_b(T)} \ln \frac{P_n}{\phi_b(T)} \right) \tag{7}$$

$$P_n = \frac{f_n}{M}, \quad n \in [0, 255] \tag{8}$$

$$P_{\mathrm{G}} = \{p_n, \ n \in [0, 255]\} \tag{9}$$

where M is the total number of pixels of image F; f_n is the frequency of gray value n in the image; P_{G} is the global distribution of image F.

Because of the affection of the vibration, illumination and noise, P_{G} is single-peak, which greatly affects $H_o(T)$ and $H_b(T)$; besides, in a rail, defects often only account for a small proportion of the rail surface. When using ME to segment rail surface defects, the ideal segmentation threshold should be able to make ϕ_o smaller while ϕ_b larger. Based on the above reasons, [14] proposed a proportional-enhanced maximum entropy threshold algorithm, as shown in Eq. (10). In this paper, the method is used as a method for segmentation of rail surface defects.

$$T^* = \arg\max(H_o(T) \bullet (\phi_b)^a), \quad T \in [0, 255] \tag{10}$$

2.3 *Defect Classification*

In this section, the characteristics of cracks and spalling are discussed. A deep forest classifier is used to classify the two types of defects.

2.3.1 Defect Feature Extraction

The features of the defect generally include geometric features, shape features and grayscale features. Since it is impossible to distinguish between crack and spalling according to the gray feature of the defect, we select several features with obvious differences [17] (Table 1).

2.3.2 Defect Classification

Deep forest (DF) method is a decision tree aggregation method. The structure of DF consists of two parts: Multi-grained scanning and cascaded forest. Multi-grained scanning generates feature vectors, and cascade forest is the result of classification or prediction through multiple forests and multi-level cascades. The specific steps of the two phases are as follows:

Phase 1: (1) Set the size of the sliding window, use the sliding window to divide the feature into multi-instance feature vectors and output the category probability vector through the forest transformation. (2) Combine the generated class probability vectors and generate new features.
Phase 2: (1) Enter the resulting category probability vector and then connect it to the original input as the output of the next layer of forest. (2) The data of the last category probability vector can be obtained by transforming the data through multiple cascaded forest processes. (3) Find the average probability vector of each

Table 1 Defect feature

Feature	Definition	Meaning
Area	The size of defect area	Reflects the severity of the defect
Squareness	The ratio of the defect area to the area of its minimum circumscribed g rectangle	Describes how much a defect fills its circumscribed rectangle
Aspect ratio	The ratio of the length to the width of the smallest circumscribed rectangle of the defect	Reflects the shape of the defect
Circularity	The ratio of defect circumference to area	Reflects the complexity of the boundary of the defect area

category in the category probability vector of multiple forest outputs, where the largest category probability is the result of prediction or classification.

3 Experimental Analysis

3.1 Defect Extraction Results

A comparative experiment using histogram equalization and LNI method in Matlab2015 was carried out. The result showed that the LNI method was more suitable for our study (Fig. 3).

The detection result is shown in Figs. 4 and 5. The image enhancement results with LNI are shown in Fig. (b), the defect segmentation results with threshold

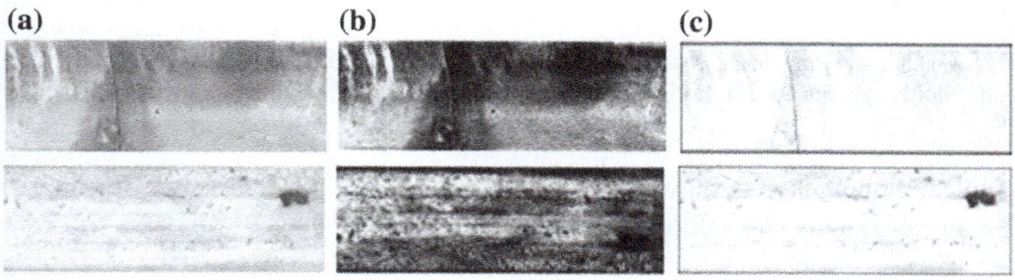

Fig. 3 Image enhancement results of different methods: **a** original image; **b** histogram normalization method; **c** LNI method

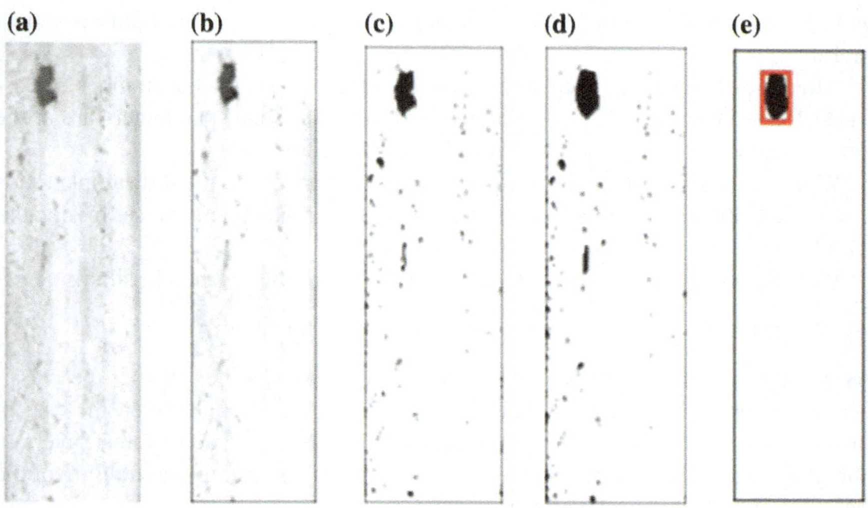

Fig. 4 Detection and location of peeling defects: **a** original image, **b** image enhancement, **c** threshold segmentation, **d** on operation, **e** front spot removal

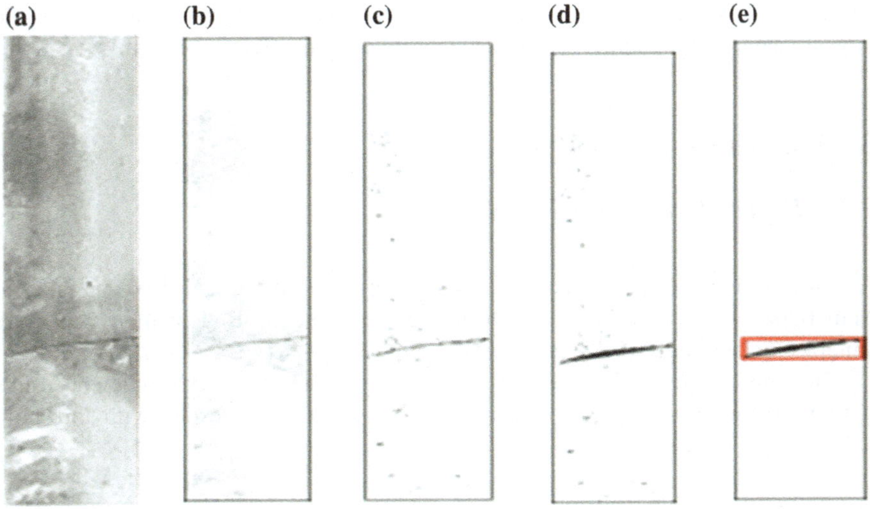

Fig. 5 Detection and location of crack defects: **a** original image, **b** image enhancement, **c** threshold segmentation, **d** on operation, **e** front spot removal

segmentation method is shown in Fig. (c). To regularize the defects, we operated on the image. Finally, we removed the small foreground points.

3.2 Defect Classification Results

We collected a total of 112 sets of data. The data set containing the four features and defect categories of defect area, rectangle, aspect ratio and circularity is shown in Table 2.

After establishing the deep forest classification model, set the number of cascaded forest layers $N = 2$, 3, 4, 5 and 6, and the other super-parameters are at the default settings. The classification results are shown in Table 3.

When the number of cascaded forest layers is set to $N = 2$, the test sample ratios $a = 10$, 20, 30, 40 and 50%, and the remaining hyperparameters are set by default, the classification results are shown in Table 4.

We can perform a simple analysis based on the results of the classifier.

(1) When the number of cascaded forest layers *N* changes.

As *N* increases, the accuracy of classification decreases. When $N = 2$, the classification accuracy is the highest. It is indicated that for the DF classifier, the current sample data volume does not need to adopt a complex cascade of forest layers to obtain a better classification effect. Too many layers may lead to overtraining of the classifier and affect the classification effect.

Table 2 Defect feature

Area	Rectangularity	Aspect ratio	Circularity	Defect type
255	0.4983	1.12	19.5	Peeling
315	0.5123	1.94	31.1	Peeling
591	0.5344	22.57	174.4	Crack
274	0.6166	7.21	50.8	Crack
433	0.5543	10.24	89.2	Crack
345	0.6545	11.41	78.4	Crack
400	0.4344	1.42	12.3	Peeling
...	...	...	...	...

Table 3 Classification accuracy in different *N*

N	2	3	4	5	6
Accuracy (%)	100	92.7	90.9	89.2	81.7

Table 4 Classification accuracy in different a

a (%)	10	20	30	40	50
Accuracy (%)	84.6	84.6	94.7	100	84.6

(2) When the ratio a of the test sample changes.

As a increases, the accuracy of the classification increases first and then decreases. When $a = 40\%$, the classification accuracy is the highest. When the test sample is 40% of the sample data and the training sample is 60% of the sample data, there is a better classification effect. Because the total sample data is small, the proportion of test samples is too small to reduce the accuracy; while the proportion of test samples is too large, the training sample data is not enough, and the classifier is not fully trained, which will reduce the accuracy of classification.

Therefore, we choose $N = 2$, $a = 40\%$ for classification experiments. We also compared the accuracy and running time of the traditional BP neural network classifier and SVM classifier using the same data. The results are shown in Table 5.

It can be seen from the results that the classification of DF is better than the other two methods when classifying the same small sample data. This fully embodies the superiority of the DF in small sample classification.

Table 5 Classification accuracy and model construction time

Method	Model construction time (s)	Classification accuracy (%)
DF	0.45	100
BP	3.6	87.4
SVM	2.3	96

4 Conclusion

This paper mainly discusses the research on the classification method of rail surface defect identification. According to the characteristics of the images acquired by the UAV, we selected the appropriate method to extract the rail defects, and also improved the image enhancement algorithm. Finally, we established a deep forest classifier to classify the defects and compared the results with the classification results of traditional classifiers. The experimental results show that the deep forest method has great advantages in small sample classification. This method provides a research direction for the identification and classification of small sample data and lays a foundation for the automation of rail surface defect detection.

Acknowledgements This work was supported by the National Key R&D Program of China (No. 2016YFB1200402).

References

1. Shen L (2012) Analysis of the status quo and countermeasures of railway transportation safety. Heilongjiang Sci Technol Inf 22:124 (in Chinese)
2. Pan G (1910) Discussion on the impact of railway track irregularity on driving safety and stability. Road Bridge Construct (18) (in Chinese)
3. Liu F (2015) Research on rail irregularity detection and disease identification based on operating vehicles. Beijing Jiaotong University (in Chinese)
4. Zhang B, Fu X (2001) Types and formation mechanism of tread stripping of railway wheels and wheels. China Railw Sci 02:76–81 (in Chinese)
5. Yang Jinfeng, Tao Wei (2011) An efficient direction field-based method for the detection of fasteners on high-speed railways. Sensors 11:7364–7381
6. Utrata D, Clark R (2003) Groundwork for rail flaw detection using ultrasonic phased array inspection. Am Inst Phys, 799–805
7. Yu H (2013) Research on the detection technology of rail surface defects based on machine vision. Hunan University, Hunan (in Chinese)
8. Li Q, Ren S (2012) Avisual detection system for rail surface defects. IEEE Trans Syst Man Cybern Part C 42(6):1531–1542
9. Zhai Y, Yin C, Dang J et al (2016) Rapid identification method of rail region based on image hue value mutation characteristics. J Traffic Transp Eng 16(1) (in Chinese)
10. Jesussek M, Ellermann K (2015) Fault detection and isolation for a railway vehicle by evaluating estimation residuals. Procedia Iutam 13(1)
11. Tang X, Wang Y (2013) Vision detection and recognition algorithm for rail surface defects. Comput Eng 39(3) (in Chinese)
12. Li Q, Ren S (2012) A real time visual inspection system for discrete surface defects of rail heads. IEEE Trans Instrum Meas 61(8)
13. Li Q, Ren S (2012) A real-time visual inspection system for discrete surface defects of rail heads. IEEE Trans Instrum Meas 61(8):2189–2199
14. Pun T (1980) A new method for gray-level picture thresholding using the entropy of the histo-gram. Signal Process 2(2):223–237
15. Pun T (1981) Entropic thresholding: a new approach. Comput Vis Graph Image Process 16 (2):210–239

16. Pun T, Pun T (1980) A new method for gray-level picture threshold using the entropy of the histogram. Sig Process 29(3):273–285
17. Lu G, Yang K (1996) Morphological characteristics and failure mechanism of rail damage. J China Railw Soc 18(3) (in Chinese)
18. Liu G, Qiu X (2018) Finger vein recognition based on multi-mode LBP and deep forest. Comput Technol Dev 07:1–6 (in Chinese)

Fault Diagnosis of Railway Turnout Based on Random Forests

Huiyue Zhang, Zhipeng Wang, Ning Wang, Jing Long and Tao Tao

Abstract The turnout is a key component of the railway infrastructure systems and is considered as a critical issue about the train operation safety. Therefore, the fault diagnosis research of the turnout is important. However, the existing methods of the fault diagnosis for the railway turnout have the problems such as low efficiency, inability to meet timeliness, and insufficient accuracy. To solve these problems, this paper presents a fault diagnosis method based on random forests. The random forests algorithm builds many CART decision tree classifiers, and introduces two random procedures: i.e., random samples and random features, to enhance the diversity of each decision tree classifier. The final classification result is obtained by majority voting method, which improves the execution speed and classification accuracy. In this paper, a case study is also presented by using the electric power data of the S700K switch machine, and the random forests classification model is constructed. The result shows that the random forests algorithm can accurately and quickly give the diagnosis results for the status of the railway turnout.

Keywords Turnout · Random forests · Fault diagnosis · Ensemble learning

H. Zhang · Z. Wang (✉) · N. Wang
State Key Lab of Rail Traffic Control and Safety, Beijing Jiaotong University, Beijing, China
e-mail: zpwang@bjtu.edu.cn

H. Zhang · Z. Wang · N. Wang · J. Long · T. Tao
National Engineering Laboratory for System Safety and Operation Assurance of Urban Rail Transit, Beijing Jiaotong University, Guangdong, China

H. Zhang · Z. Wang · N. Wang
Beijing Research Center of Urban Traffic Information Sensing and Service Technologies, Beijing Jiaotong University, Beijing, China

Y. Qin et al. (eds.), *Proceedings of the 4th International Conference on Electrical and Information Technologies for Rail Transportation (EITRT) 2019*, Lecture Notes in Electrical Engineering 639, https://doi.org/10.1007/978-981-15-2866-8_49

1 Introduction

The railway turnout can change the direction of the train and is critical to railway safety. With the continuous expansion of China's railway network, the demand for turnout applications is increasing, and the requirements for the status of turnout are also higher, so safety is especially important. However, due to the complicated structure, low mechanical strength and being susceptible to external environment, turnouts are prone to aging and failure. The frequent failures of the turnout and the long time taken to deal with the failure have caused a great impact on the train operation. Thus, it is necessary to perform accurate and timely fault detection on the turnout.

Therefore, many experts have conducted a lot of researches on the turnout fault diagnosis. In [1], the normal and two failure modes of the turnout are given by the expert system to realize the simulation analysis of the turnout fault diagnosis. Reference [2] introduces the gray correlation analysis method in the turnout fault diagnosis. Reference [3, 4] applies SVM classifiers and improved SVM classifiers to identify different types of turnout fault current curves. Reference [5–7] uses neural networks to classify and identify turnout faults. However, the diagnostic method based on expert system requires a huge knowledge base and has domain limitations and poor adaptability. SVM is often insufficient for multi-classification problems and takes up too much storage space when the amount of data is too large. Neural network is easy to fall into local optimum, resulting in poor performance, and requires a large number of samples for training, which has high requirements for data acquisition, and the calculation amount is large, which takes a long time and is difficult to meet the timeliness. Therefore, it is necessary to optimize the turnout fault diagnosis method to make the fault identification more accurate and fast and have strong generalization ability.

Obviously, most of the current fault diagnosis methods for the turnout focus on the single classifier model, and the experimental verifications are mostly based on simulation data. The accuracy of the model often depends on the completeness and accuracy of the simulation data. However, the actual data is inevitably related to equipment maintenance and environmental factors and has problems, such as noise interference and incomplete sample data, for which, the idealized simulation data cannot verify the applicability and accuracy of the algorithm model for real data. Moreover, in practical applications, the timeliness and accuracy of a single classifier model are difficult to improve, and it is often difficult to meet the diagnostic accuracy or speed requirements. Therefore, a more optimized classification model needs to be established. Compared with the traditional classification model based on single classifier, ensemble learning combines multiple classifiers and generally achieves better results than single classifier in accuracy and generalization ability [8–11]. The random forests algorithm [12] is one of the ensemble learning methods, and it has been favored by many scholars for its simple algorithm and powerful performance. The random forests algorithm used for classification can get high accuracy [13] and good stability [14], and there is no over-fitting [15], which means good generalization. It runs fast with low operation complexity, and this advantage is more obvious on larger datasets [16]. And it is insensitive to data noise, and can also get good results on datasets with some missing values [15]. These advantages of random forests make it widely used. Therefore, random forests algorithm used for turnout fault diagnose can effectively improve the accuracy, stability, and generalization of classification.

Therefore, this paper presents a turnout fault diagnosis method based on random forests algorithm. Firstly, a number of CART decision tree classifiers are constructed, and the classification results of each tree are obtained through training. In the training process, the training sample perturbation and attribute perturbation are used to enhance the diversity of the decision tree classifier, and then the training results are output by majority voting method [12]. Finally, the proposed algorithm is validated with real data. The real data is more in line with the actual turnout status, and can better verify the applicability and accuracy of the algorithm.

The rest of the paper is organized as follows: Sect. 2 introduces the random forests algorithm; Sect. 3 describes the application of the proposed method on turnout fault diagnosis with real data; Sect. 4 presents the conclusions of this paper.

2 Methodology

2.1 Classification and Regression Tree (CART)

Decision tree [17] is a common machine learning algorithm. It is a tree structure with root node, internal nodes, and leaf nodes. The decision tree is built by splitting nodes from the root node to the leaf nodes according to a splitting rule. When splitting downwards, we hope that the child nodes obtained by each split will contain as much pure data as possible. Classification and regression tree (CART) [17, 18] is one of the decision tree algorithms. To find the best split, it uses the Gini index as the splitting rule, and the Gini value is used to measure the purity of the dataset contained in a node. The larger the Gini value, the lower the data purity. The Gini value is defined as follows:

$$\mathrm{Gini}(D) = \sum\nolimits_{k=1}^{|y|} \sum\nolimits_{k' \neq k} p_k p_{k'} = 1 - \sum\nolimits_{k=1}^{|y|} p_k^2 \tag{1}$$

D is the dataset and $p_k(k = 1, 2, \ldots, |y|)$ is the proportion of the kth observations in D. On this basis, the Gini index is defined as:

$$\mathrm{Gini\_index}(D, a) = \sum_{v=1}^{V} \frac{|D^v|}{|D|} \mathrm{Gini}(D^v) \tag{2}$$

D is the dataset, a is the attribute that divides D, and a has V possible values, then it can get V child nodes after division, D^v is the observations contained in the vth child

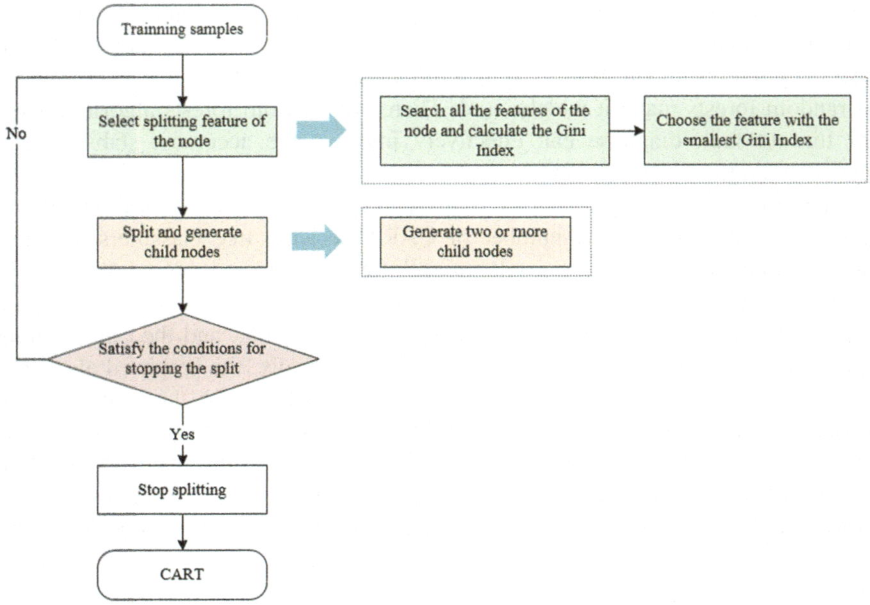

Fig. 1 CART decision tree algorithm flow chart

node, and $\mathrm{Gini}(D^v)$ is the Gini value of the dataset D^v. The smaller the Gini index, the better the splitting effect.

The CART decision tree algorithm is shown in Fig. 1. More details about the CART algorithm can be seen in [17, 18]. The generation progress is as follows:

Input: training set D and preset stop conditions.
Step 1: Calculate the Gini index of each attribute of the node;
Step 2: Find the feature with the smallest Gini index for splitting;
Step 3: Repeat steps 1 and 2 for the obtained child nodes until the preset stop conditions are satisfied.
Output: CART decision tree.

2.2 *Diversity Enhancement*

Random Input Training Samples. Random forests algorithm uses the bootstrap method to generate multiple different sample subsets to train different CART decision tree classifiers. The bootstrap method is as follows: for a dataset containing m samples, each time randomly select a sample and place it in the sample subset, and then put it back to the original dataset. Each time each sample is selected with a probability of $1/m$, and the probability that a sample is never extracted by sampling

m times is $(1-1/m)^m$. When $m \to \infty$, $(1-1/m)^m \to 1/e \approx 0.368$, which means that when the original dataset is large enough, about 36.8% of the samples never appear in a sample subset containing m samples, while other samples appear multiple times. In this way, the sample subsets used for training each decision tree classifier have great differences and the whole ensemble model involves all the samples at the same time, so that there are differences between each trained classifier, thus, the accuracy and generalization of the ensemble model are improved.

Random Input Features Selection. The splitting feature selection of random forests is different from the ordinary decision tree algorithm. In the training process, for the ordinary decision tree, it needs to search all the features of the current node to find the best feature for splitting; while for the random forests algorithm, it just searches a feature subset, which includes only several features selected randomly from the feature set contained in the current node, to find the best splitting feature.

The bootstrap method makes the training set used to train each tree different, and the random feature selection makes the splitting feature of each tree node also different. Therefore, after training, there is a certain difference between individual decision tree classifiers, which greatly reduces the over-fitting phenomenon that may occur in the decision tree and effectively improves the generalization ability.

2.3 *Majority Voting*

Majority voting is one of the ensemble strategies. Random forests algorithm uses this method to get the final ensemble results. That is, if a class label has the most votes, the label is the final classification result. If more than one class label gets the highest number of votes, select one randomly.

2.4 *Random Forests*

The random forests algorithm is shown in Fig. 2. The specific process is as follows:

Suppose the number of samples in the original training dataset is N, the number of sample features is M, the number of random input features is m, and the number of CART decision tree classifiers is T.

Step 1: Generating a subset from original training dataset: use the bootstrap method to randomly select n samples from the original dataset to form a subset;
Step 2: Construction of individual decision tree classifier: the decision tree classifier is trained by CART algorithm using the sample subset. During the tree classifier being trained, select m features from all the features of the splitting node to form a subset, and then find the best splitting feature from the subset using Gini index as the splitting rule. The tree is grown fully without pruning, and finally generates a CART decision tree classifier;

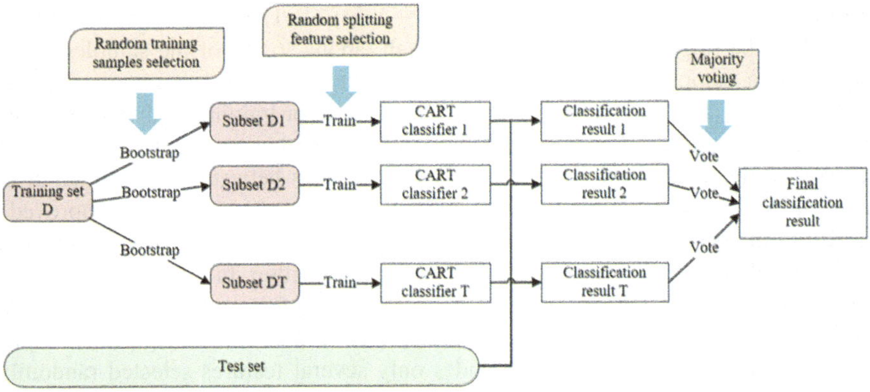

Fig. 2 Random forests algorithm flow chart

Step 3: Perform steps (1) and (2) T times, and train T CART decision tree classifiers, and each classifier is trained to obtain a classification result;
Step 4: Majority voting: vote for the training result of each classifier, and the class label with the most votes is the final classification result.

3 Experimental Analysis

3.1 Experimental Setup

In order to verify the proposed method, this study used the electric power data of the S700K turnout switch machine collected from the turnout testbed of Guangzhou Metro. The electric power data could directly reflect the working status of the turnout. A total of 13 abnormal data were collected including a single fault (abnormal friction, tight over-tightness, inclusion of foreign matter) and multiple faults.

To consider the overall movement of the turnout and better identify the turnout status, the data used in the experiment was the complete power curve data of turnout conversion process from normal to reverse and then to normal. The experimental data is shown in Table 1.

3.2 Fault Diagnosis Based on Random Forests

Random forests model was established to diagnose the turnout faults. Simulations were performed using MATLAB and PRTools toolbox. PRTools is a MATLAB toolbox for pattern recognition, which has a large number of basic classifiers for

Table 1 Experimental data

No.	Status	Number	Description
1	Normal	9	Normal
2	Nfault5	50	0.5 mm over-tightness
3	Nfault15	43	1.5 mm over-tightness
4	Ffault0	223	Abnormal friction
5	Ffault3	127	Abnormal friction and 0.3 mm over-tightness
6	Ffault10	127	Abnormal friction and 1.0 mm over-tightness
7	Ffault13	104	Abnormal friction and 1.3 mm over-tightness
8	Ffault20	109	Abnormal friction and 2.0 mm over-tightness
9	Ffault25	61	Abnormal friction and 2.5 mm over-tightness
10	Ffault30	61	Abnormal friction and 3.0 mm over-tightness
11	Fforeign2	12	Abnormal friction and 2.0 mm foreign matter
12	Fforeign4	10	Abnormal friction and 4.0 mm foreign matter
13	locked25	22	Abnormal friction and 2.5 mm over-tightness (unsmoothed unlocking)
14	locked30	4	Abnormal friction and 3.0 mm over-tightness (unable to unlock)

pattern recognition. Fault diagnosis is essentially a pattern recognition matter, identifying different fault modes, and classifying them.

Firstly, the dataset was randomly divided into two parts: the training set and the test set, with 680 sets of data in the training set and 282 sets of data in the test set. Then, the random forests algorithm presented above was applied. The number of CART decision trees in the random forests was set to 1, 11, 21, …, 901, and during splitting, the number of features in the candidate feature subset was the first integer smaller than $\log_2 M + 1$ (M is the number of sample features, here is 296), which was introduced by Breiman [12]. The training set was used as input to train the random forests model, and the test set was classified by the trained model. Then, we calculated the accuracy of fault classification. The result was shown in Fig. 3.

In Fig. 3, the horizontal axis is the number of CART decision tree classifiers, and the vertical axis is the classification accuracy of random forests. When the number of decision trees in the forest is small, the classification accuracy of random forests is not good due to the random selection of training samples and splitting features. However, as the number of decision trees increases, the performance of the random forests is significantly improved and gradually stabilized. It can be seen from Fig. 3 that when the number of decision trees reaches 21, a high fault diagnosis rate (96.82%) has been obtained, and when the number of decision trees exceeds 161, the accuracy of fault classification approaches stability. When the number of decision trees reaches 901, it still has stable classification accuracy, and no over-fitting occurs.

To illustrate the diagnostic performance of the proposed algorithm exactly, we calculated the average classification accuracy, maximum accuracy, and variance

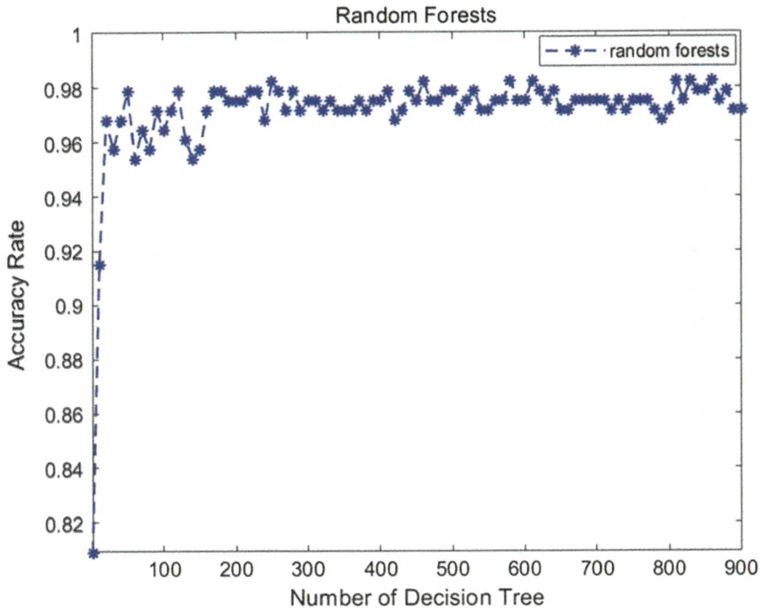

Fig. 3 Classification accuracy of random forests

Table 2 Classification results of turnout fault diagnosis using random forests algorithm

Method	Average accuracy (%)	Maximum accuracy (%)	Variance	Model construction time (901 decision trees) (s)
Random forests	97.38	98.23	0.000036	41

after 21 decision tress, and construction time of each classification model in the case of 901 decision trees, as shown in Table 2. Obviously, the random forests can get high accuracy in turnout fault recognition and also has good stability and fast operation speed.

In order to prove the performance of the proposed method in turnout fault diagnosis, the random forests were further compared with the classification model based on CART and bagging in model construction time and generalization ability. Here, the number of decision trees was set to 1, 21, 41, …, 501. The results were shown in Figs. 4 and 5. In the figures, the horizontal axis is the number of CART decision tree classifiers and the vertical axis is the classification accuracy. The red, green, and blue lines represent the model construction time (Fig. 4) and classification accuracy rate (Fig. 5) of single CART decision tree, bagging and random forests, respectively.

From the perspective of the model construction time, it is obviously that operation based on random forests is much faster than bagging method, especially with the increase of the number of CART decision trees. When the number of decision

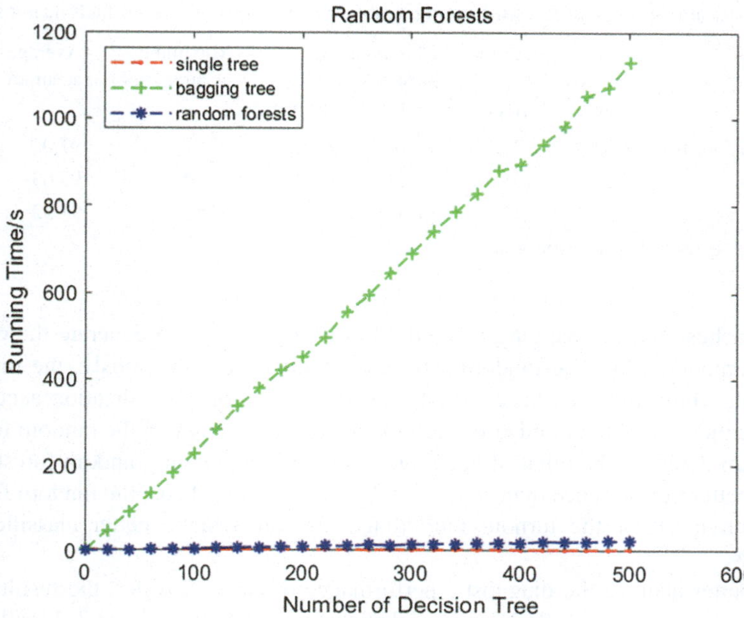

Fig. 4 Comparison of model construction speed based on random forests, bagging, and CART

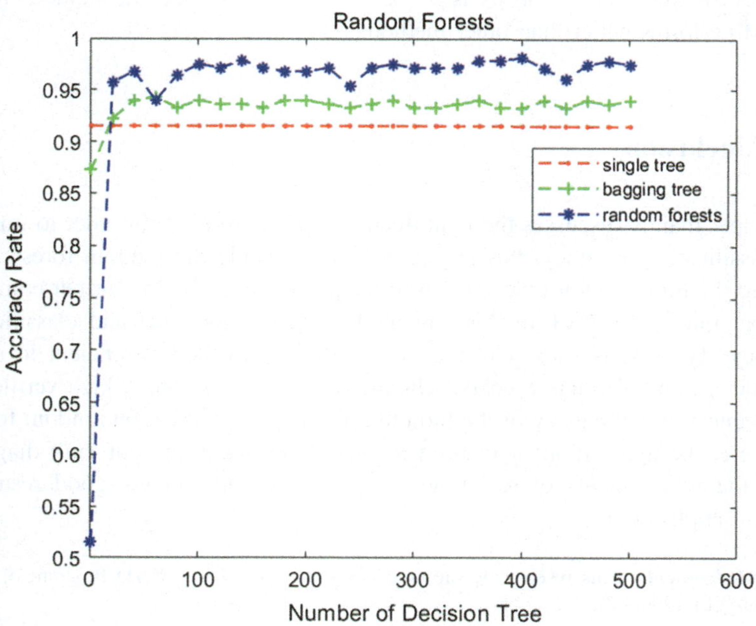

Fig. 5 Comparison of test accuracy based on random forests, bagging, and CART

Table 3 Results of random forests, bagging, and CART methods of turnout fault diagnosis

Method	Model construction time (s)		Classification accuracy (%)		Maximum accuracy (%)	Average accuracy (%)
	21 DTs	501 DTs	21 DTs	501 DTs		
Random forests	1.69	23.75	96.76	97.53	98.23	97.02
Bagging	47.17	1131.50	92.23	93.99	94.35	93.63
CART	2.54	2.54	91.52	91.52	91.52	91.52

Note DTs refers to the decision trees

trees reaches 501, the bagging method takes nearly 19 min to generate the classification model, while the random forests costs only 24 s. Obviously, the random forest algorithm operates faster. From the perspective of generalization accuracy, with the increase of the number of decision trees, the accuracy of the random forests algorithm is higher than that of bagging, and both bagging and random forests can obtain better performance than single CART decision tree. Thus, the random forests algorithm used for the turnout fault diagnosis can achieve better classification accuracy.

To better analyze the diagnostic performance of each classifier, the results (the number of decision trees ≥ 21) were calculated and shown in Table 3. Finally, the fault classification accuracy of single CART decision tree is 91.52%, the average classification accuracy of the bagging is 93.63%, and the average classification accuracy of the random forests is 97.02%. It is evident that the random forests method performs better than other methods.

4 Conclusion

This paper mainly discusses the fault diagnosis of the turnout. In order to improve the classification accuracy, this paper proposes to apply the random forests algorithm to the turnout fault diagnosis. And the proposed method is further compared with bagging and CART methods in model construction time and classification accuracy. The results show that the random forests method is efficient in model construction and also has excellent classification accuracy rate, which verifies the effectiveness and efficiency of the turnout fault diagnosis based on random forests. Therefore, the application of random forests algorithm in turnout fault diagnosis meets the requirements of real time and accuracy and can get good results in practical applications.

Acknowledgements This research is supported by the National Key R&D Program of China (No. 2016YFB1200402).

References

1. Atamuradov V, Camci F, Baskan S et al (2009) Failure diagnostics for railway point machines using expert systems. In: 2009 IEEE international symposium on diagnostics for electric machines, power electronics and drives (SDEMPED 2009). IEEE Computer Society, Cargese
2. Lin HZ, Qiao L (2014) Method of turnout fault diagnosis based on grey correlation analysis. J China Railw Soc 36(02):69–74 (in Chinese)
3. Mo XM, Fang Y, Yang YG (2013) Method on the fault detection and diagnosis for the railway turnout based on the current curve of switch machine. In: 2nd international conference on mechanical engineering, industrial electronics and informatization (MEIEI 2013), vol 427–429, pp 1022–1027. Trans Tech Publications Ltd
4. Tao H, Zhao Y (2015) Intelligent fault prediction of railway switch based on improved least squares support vector machine. J Metall Min Ind 7(10):69–75
5. Zhang K (2014) The railway turnout fault diagnosis algorithm based on BP neural network. In: IEEE international conference on control science and systems engineering (CCSSE 2014), pp 135–138. Institute of Electrical and Electronics Engineers Inc
6. Cheng Y, Zhao H (2015) Fault detection and diagnosis for railway switching points using fuzzy neural network. In: 10th IEEE conference on industrial electronics and applications (ICIEA 2015), pp 860–865. Institute of Electrical and Electronics Engineers Inc
7. Zhang K, Du K, Ju Y (2015) Algorithm of railway turnout fault detection based on PNN neural network. In: 7th international symposium on computational intelligence and design (ISCID), vol 1, pp 544–547. Institute of Electrical and Electronics Engineers Inc
8. Hu Q, He Z, Zhang Z et al (2007) Fault diagnosis of rotating machinery based on improved wavelet package transform and SVMs ensemble. J Mech Syst Signal Process 21(2):688–705
9. Liang T, Wu S, Duan W et al (2018) Bearing fault diagnosis based on improved ensemble learning and deep belief network. In: International conference on mechanical, electric and industrial engineering (MEIE), vol 1074. Institute of Physics Publishing
10. Tian J, Azarian MH, Pecht M et al (2017) An ensemble learning-based fault diagnosis method for rotating machinery. In: 8th IEEE prognostics and system health management conference. Institute of Electrical and Electronics Engineers Inc
11. Sikder N, Bhakta K, Al Nahid A et al (2019) Fault diagnosis of motor bearing using ensemble learning algorithm with FFT-based preprocessing. In: 1st international conference on robotics, electrical and signal processing techniques (ICREST), pp 564–569. Institute of Electrical and Electronics Engineers Inc
12. Breiman L (2001) Random forests. J Mach Learn 45(1):5–32
13. Yao Q, Wang J, Yang L et al (2016) A fault diagnosis method of engine rotor based on random forests. In: 2016 IEEE international conference on prognostics and health management (ICPHM 2016). Institute of Electrical and Electronics Engineers Inc
14. Chen X, Cui H, Luo L (2011) Fault diagnosis of transformer based on random forest. In: 4th international conference on intelligent computation technology and automation (ICICTA 2011), vol 1, pp 132–134. IEEE Computer Society
15. Yang B-S, Di X, Han T (2008) Random forests classifier for machine fault diagnosis. J Mech Sci Technol 22(9):1716–1725
16. Yan R-W, Cai J-D (2009) Fault diagnosis of power electronic circuit based on random forests algorithm and AR model. In: 2009 2nd international conference on information and computing science (ICIC 2009), vol 1, pp 285–288. IEEE Computer Society
17. Li G, Hu Y, Chen H et al (2017) Identification and isolation of outdoor fouling faults using only built-in sensors in variable refrigerant flow system: a data mining approach. J Energy Build 146:257–270
18. Miao L (2017) Application of CART decision tree combined with PCA algorithm in intrusion detection. In: Wenzheng L, Babu MSP, Xiaohui L (eds) 2017 8th IEEE international conference on software engineering and service science, pp 38–41

Turnout Health Assessment Based on Dynamic Time Warping

Ning Wang, Hong guang Wang, Limin Jia, Zhipeng Wang and Huiyue Zhang

Abstract Turnouts are vital components in the rail system. Assessing the healthy state of turnout and ensuring normal operation is essential. However, the faults of rolling element bearings under running occur suddenly. Thus, it is essential to develop a practicable and real-time method to put forward the assess the turnouts' running state, which could insure the normal operation. This paper presents a novel method based on the dynamic time warping (DTW) to assess the running status. DTW have two superiorities: (1) DTW could calculate the distance between the different state by using calculating the total cost (minimum distance) of two time series, it makes DTW algorithm fast and simply, so the state is assessed timely. (2) The total cost not only be used to diagnose the fault of turnout, but also could be used to transfer to CV (confident values) to assess the state of the turnout. Hence, this method uses DTW to calculate the distance between the different turnout's states and CV to assess the healthy state by using the distance. Finally, this method is verified by using the power data collected from Guangzhou Metro turnout testbed and Line 4. The result shows that it is accurate for turnout health assessment and suitable for practical scenarios.

Keywords Turnout · Dynamic time warping · Health assessment · Confident values

N. Wang · L. Jia · Z. Wang (✉) · H. Zhang
State Key Lab of Rail Traffic Control and Safety, Beijing Jiaotong University, Beijing, China
e-mail: zpwang@bjtu.edu.cn

N. Wang · L. Jia · Z. Wang · H. Zhang
National Engineering Laboratory for System Safety and Operation Assurance of Urban Rail Transit, Beijing Jiaotong University, Guangdong, China

N. Wang · L. Jia · Z. Wang · H. Zhang
Beijing Research Center of Urban Traffic Information Sensing and Service Technologies, Beijing Jiaotong University, Beijing, China

H. g. Wang
Operation Headquarters, Guangzhou Metro Group, Guangdong, China

Y. Qin et al. (eds.), *Proceedings of the 4th International Conference on Electrical and Information Technologies for Rail Transportation (EITRT) 2019*, Lecture Notes in Electrical Engineering 639, https://doi.org/10.1007/978-981-15-2866-8_50

1 Introduction

The turnout is an important part of the rail system, which is composed of the switch machine, the switch rail, the guardrail, the wing rail, and the direction of the train is changed by these components. Compared with other equipment in the rail system, the working environment of the turnout is complex and variable. Turnout is contacted directly with the train wheelset, so, its capability of the safety assurance is higher than other equipment, and also it has a greater impact on the train's safe operation. However, the mechanical strength of the turnout is generally not high, it is easy to have fatigue damage, performance degradation, and eventually lead to failure in daily operation [1]. Therefore, it is necessary to monitor the running status of the turnout [2, 3], identify and evaluate the health status of the turnout. The machine health assessment is used to indicate the evolutionary process form the health status to the fault status; it describes the healthy declining trend of the machine. The health assessment algorithm, on the basis of fault diagnosis algorithm, which is extending form distinguishing the normal and fault states of the machine to using state values to describe the state of machine. The original turnout fault diagnosis and health assessment algorithm is based on the dynamic model [4, 5]. This method requires professional skill and abundant mathematical and physical knowledge, so it is difficult to use. At present, many researchers have conducted extensive research on data-driven method. The turnouts' state is determined by analyzing the power data in the data-driven method. SVM is successfully applied to turnout fault diagnosis by using the power data [6], current carve [7] and audio data [8]. PNN neural network also has talent performance to diagnose the fault of turnout [9]. Auto-associative residual also has high fault diagnosis accuracy in the fault diagnosis [10]. Self-organizing Map (SOM) is useful to evaluate the healthy state [11] by using the power data. Information fusion has a good application prospect in the turnout fault diagnosis [12]. BP neural network [13], time delay neural networks [14] also are effective in diagnosing the fault.

Although the above algorithms have high accuracy, those also need a large amount of training data. Considering the real-time and randomness of the turnout fault, the turnout fault should be detected timely, thus the fault diagnosis method should be simple, and also a small amount of training data is needed. At the same time, the real operating conditions are complicated, and multiple faults may occur concurrently. It is impossible for the test sample to contain all fault types. In addition, some algorithms also need to pre-process the data, extract features, and it is possible to eliminate the effective information. Dynamic time warping [15] is a real-time algorithm; it could measure the difference between the different samples by calculating the total cost between two time series. The differences are calculated only by a small amount of data, the process of calculation is simple, convenient and real-time. At the same time, due to individual differences, the conversion time will be randomly changed during the conversion process, and DTW can effectively solve this problem. Therefore, DTW is a useful method to diagnose the turnout fault [16–18].

Thus, this paper uses the DTW to evaluate the turnout state by analyzing the power data. A few normal samples are used as the benchmark data, and the distance is respectively obtained by calculating the total cost between the different power curves and the benchmark curve, then the health value is obtained by using the evaluation function based on the CV [19].

The structure of this paper is as follows, the second part introduces the algorithm, the third part is experimental verification, and the fourth part is the conclusion.

2 Methodology

2.1 Process of Method

The process of method is shown in Fig. 1. The specific process is as follows:

1. The total costs [20] among the different states are calculated by DTW;
2. The parameter is calculated by the formula, the healthy states are evaluated by calculating the CV of states.

2.2 Dynamic Time Warping

Dynamic time warping (DTW) is a nonlinear warping technique that combines time warping and interval measurements to find the optimal alignment between two sequences. One of the sequence curves is adjusted by lengthening or shortening to match the length of the other curve, and then the similarity of the two sequences is calculated. The essence of DTW is that under the guidance of the dynamic programming idea, the minimum distances between the corresponding points of the two sequences are selected, and the shortest path is accumulated according to the local optimal method. On this shortest path, the vector distance between the two sequences is the shortest. The distance is higher than others, the similarity is smaller than others and the minimum distance is called total cost.

The DTW could be divided into two steps. The first step, the cost matrix is obtained by calculating the distance between the corresponding points of the time series. The second step, the shortest path is obtained in the distance matrix. The specific process is as follows:

Two time series are set as $X = \{x(1), x(2), \ldots, x(m)\}$ and $Y = \{y(1), y(2), \ldots, y(n)\}$, the number of time series point is m and n, $x(m)$ and $y(n)$ represents data point of time series. The distance of different data is calculated by Euclidean distance, the distance is defined as:

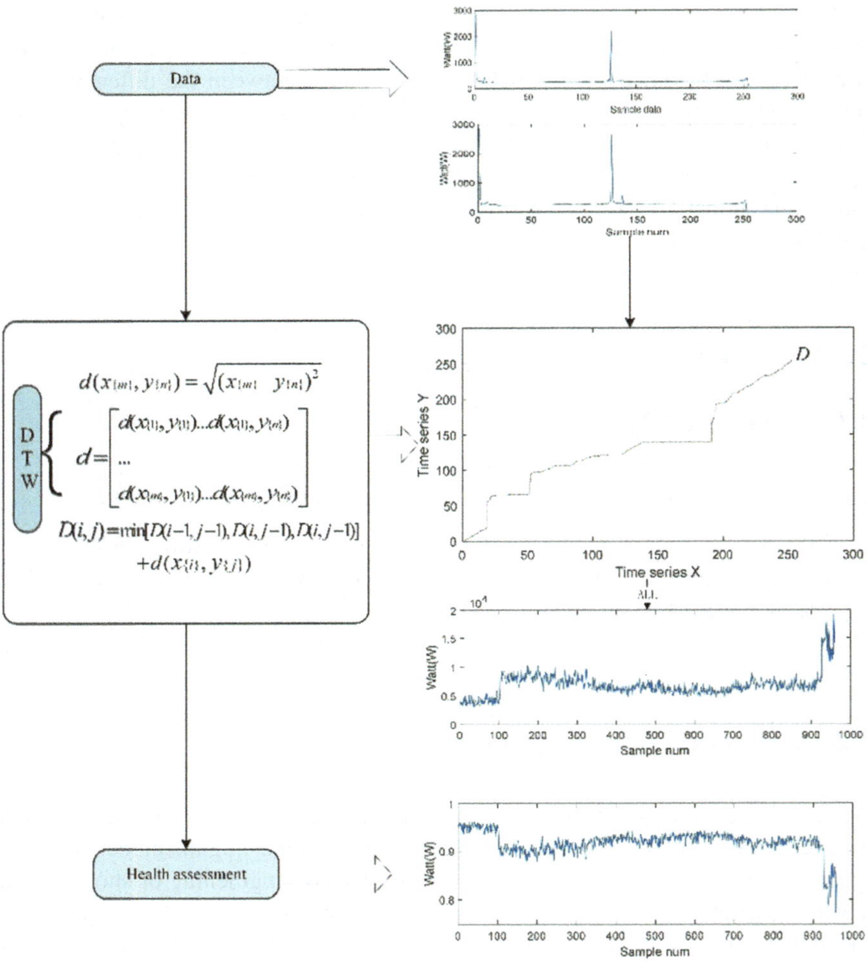

Fig. 1 Process of method

$$d(x_{\{m\}}, y_{\{n\}}) = \sqrt{(x_{\{m\}} - y_{\{n\}})^2} \tag{1}$$

Then the cost matrix is constituted by all the data, and the matrix is presented as follows:

$$d = \begin{bmatrix} d(x_{\{1\}}, y_{\{1\}})\ldots d(x_{\{1\}}, y_{\{n\}}) \\ \ldots \\ d(x_{\{m\}}, y_{\{1\}})\ldots d(x_{\{m\}}, y_{\{n\}}) \end{bmatrix} \tag{2}$$

Then the shortest path of the two time series is selected in the cost matrix, and the constraints of shortest path as follow:

1. The adjacent points $(x_{\{a\}}, y_{\{b\}})$ and $(x_{\{a'\}}, y_{\{b'\}})$ on the shortest path are satisfied as follow:

$$\begin{cases} a' - a \geq 0 \\ b' - b \geq 0 \\ a' - a \leq 1 \\ b' - b \leq 1 \end{cases} \tag{3}$$

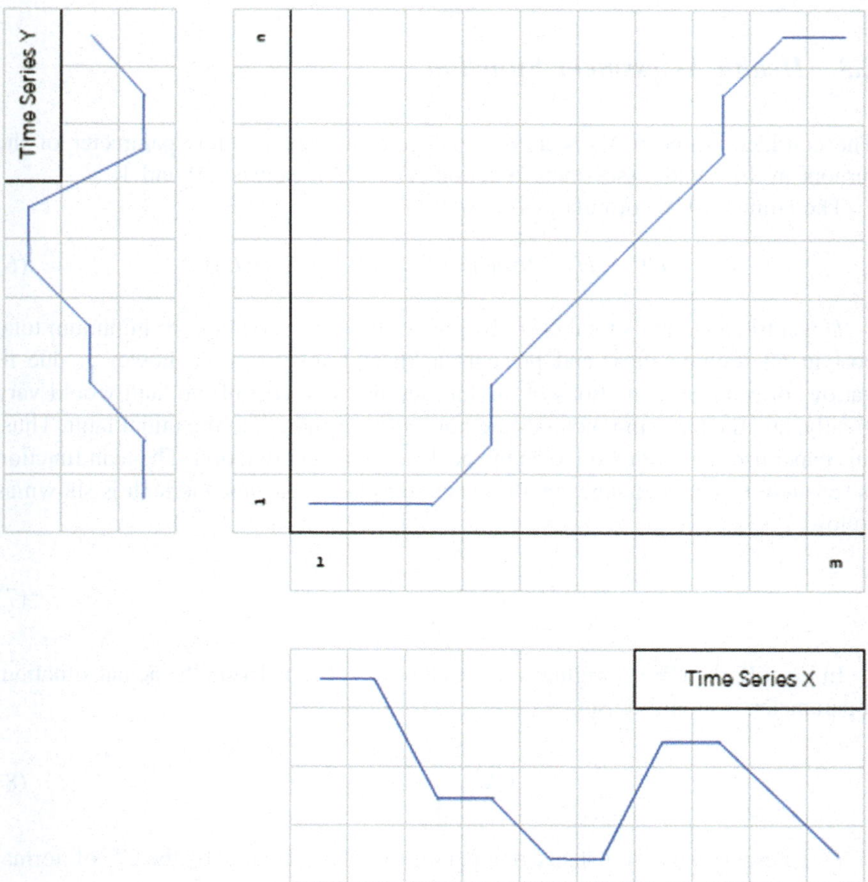

Fig. 2 DTW-shortest path

2. $(x_{\{1\}}, y_{\{1\}})$ and $(x_{\{m\}}, y_{\{n\}})$ are the begin and end point

Then different total costs are obtained, the cost distance of the next point is obtained by accumulating the distance from the previous point. The iterative process is represented by the following formula:

$$D(i,j) = \begin{cases} D(i-1,j-1) + d(x_{\{i\}}, y_{\{j\}}) \\ D(i,j-1) + d(x_{\{i\}}, y_{\{j\}}) \\ D(i-1,j) + d(x_{\{i\}}, y_{\{j\}}) \end{cases} \tag{4}$$

The path can be selected by the global constraint, and the shortest path is the path in which the sum of the cost distance is the smallest, the shortest path is calculated by the following formula, and shortest path is shown as Fig. 2:

$$D(i,j) = \min[D(i-1,j-1), D(i,j-1), D(i,j-1)] + d(x_{\{i\}}, y_{\{j\}}) \tag{5}$$

2.3 *Health Assessment Function*

The confident value (CV) is used to indicate the health status parameter of the turnout in the health assessment function. The CV is between 0 and 1.

The common CV formula as follow:

$$CV = (D - \mathrm{Min}(D))/(\mathrm{Max}(D) - \mathrm{Min}(D)) \tag{6}$$

D is a training data's total cost, the max-min CV is maximum or minimum total cost in all training data, and presents a linear transformation, however, due to various operational conditions of the turnout, the total cost of the fault would vary greatly, and the max-min methods are not sensitive enough to the initial fault. Thus, this paper uses tanh function to be the health assessment function. The tanh function is sensitive to initial changes and insensitive to later changes, the tanh is shown as follow:

$$\mathrm{CV} = \frac{2}{e^{2x} + 1} \tag{7}$$

In Eq. (7), the CV of original state is close to 1. Considering the actual situation, a parameter α should be set:

$$\mathrm{CV} = \frac{2}{e^{2\alpha D} + 1} \tag{8}$$

D represents the cost distance, the parameter α is calculated by the CV of normal state and $\mathrm{CV}_{\mathrm{pre}}$, the $\mathrm{CV}_{\mathrm{pre}}$ usually ranges from 0.99 to 0.90.

$$CV_{pre} = \frac{2}{e^{2\alpha D\mathrm{Normal}} + 1} \tag{9}$$

D_{Normal} represent the mean of total cost among the normal state, α is calculated as follow:

$$\alpha = \ln\left(\frac{2}{CV_{pre}} - 1\right)/2 \times D_{\mathrm{Normal}} \tag{10}$$

3 Results

In this paper, the experimental data was from Guangzhou Metro Line 4 and turnout testbed. The electric power data was collected from S700K-C turnout when it's switching. This experiment involved six different faults: locking of grease of sliding table, over-tightness, faulted unlocking (could not unlock), both locking of grease and over-tightness, both foreign matter and locking of grease, both unsmoothed unlocking (could unlock finally) and locking of grease. This paper defined the cycle data was that the data from the Positioning to the Inversion and from the Inversion to the Positioning. The experimental data is shown in Table 1.

The cycle data was shown in Fig. 3, the process of cycle data was the Position to the Inversion: T1-T2 (start), T2-T3 (unlocking), T3-T4 (conversion), T4-T5 (locking); the Inversion to the Position: T5-T6 (start), T6-T7 (unlocking), T7-T8 (conversion), T8-T9 (locking).

Table 1 Experimental data

No.	Label	Number	Explanation
1	Normal	59	Normal
2	Nfault15	43	1.5 mm over-tightness
3	Ffault0	223	without grease
4	Ffault3	127	Locking of grease and 0.3 mm over-tightness
5	Ffault10	127	Locking of grease and 1.0 mm over-tightness
6	Ffault13	104	Locking of grease and 1.3 mm over-tightness
7	Ffault20	109	Locking of grease and 2.0 mm over-tightness
8	Ffault25	61	Locking of grease and 2.5 mm over-tightness
9	Ffault30	62	Locking of grease and 3.0 mm over-tightness
10	Fforeign2	12	Locking of grease and 2.0 mm foreign matter
11	Fforeign4	10	Locking of grease and 4.0 mm foreign matter
12	Ulocked	22	Locking of grease and unsmoothed unlocking
13	Locked	4	Faulted unlocking

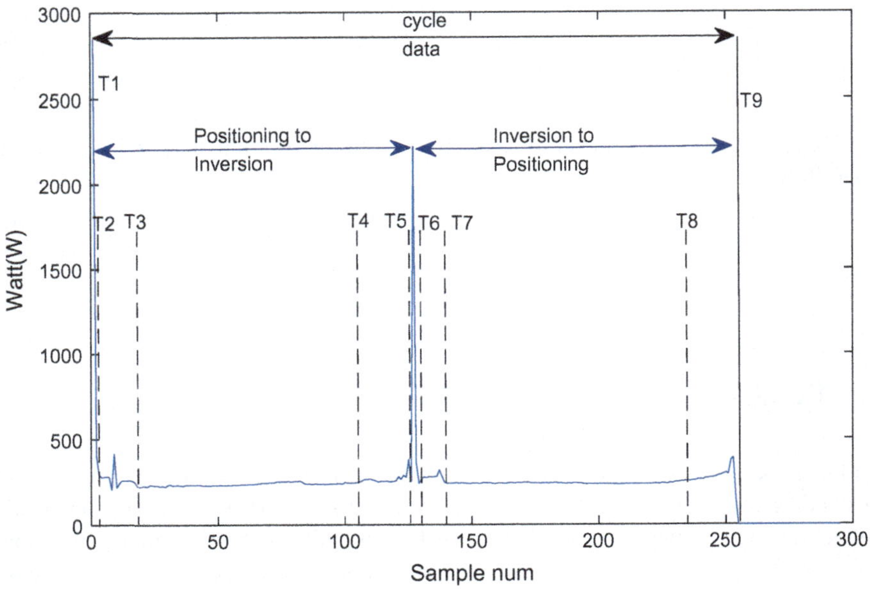

Fig. 3 Cycle data

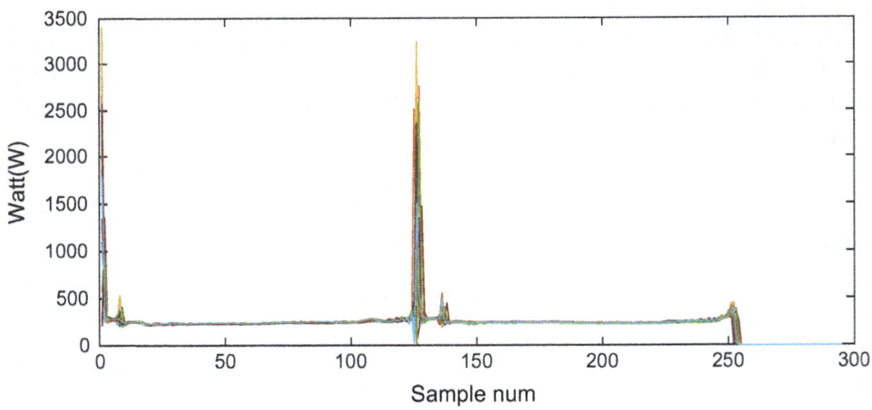

Fig. 4 Normal cycle data

The normal data was shown in Fig. 4.

The Sample num was the number of sampling point, and the Watt was the unit of electrical power.

The process of the method was shown:

1. Calculated the mean distance of normal state.

The mean total cost of normal state was calculating by DTW, the mean distance is $D = 3962.489$.

2. Determined parameter

After the distance was calculating, the CV_{pre} was set a value of 0.95, the parameter α was calculating: $\alpha = 1.31 \times 10^{-5}$, the parameter was set as $\alpha = 1.3 \times 10^{-5}$, so the health assessment function was as follow:

$$CV = \frac{2}{e^{2.6*10^{-5} \times D} + 1} \tag{11}$$

3. Calculated the distances between normal state and other states

The total costs between normal state and other states were calculated by DTW, and the values of total costs were shown in Fig. 5, the Sample num was corresponded to the Number in Table 1.

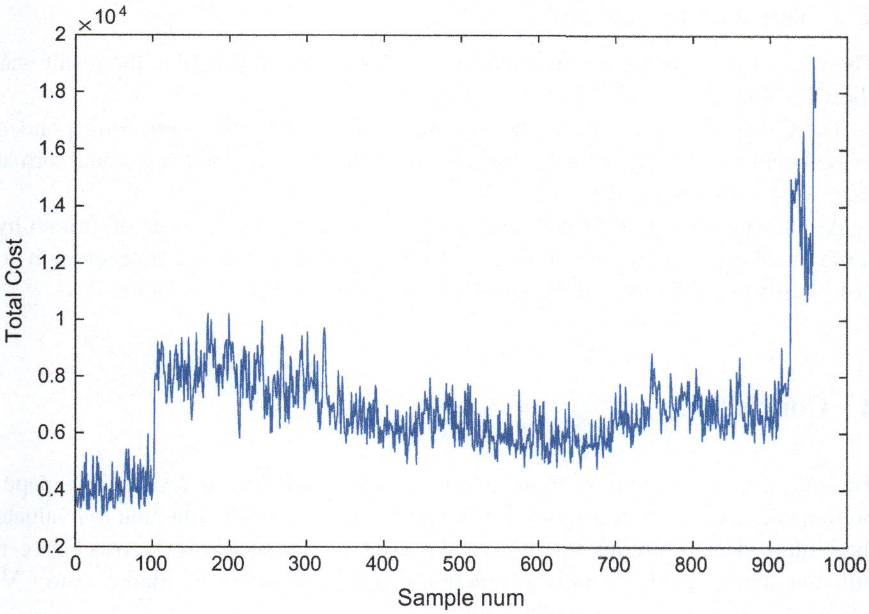

Fig. 5 Total costs of all states

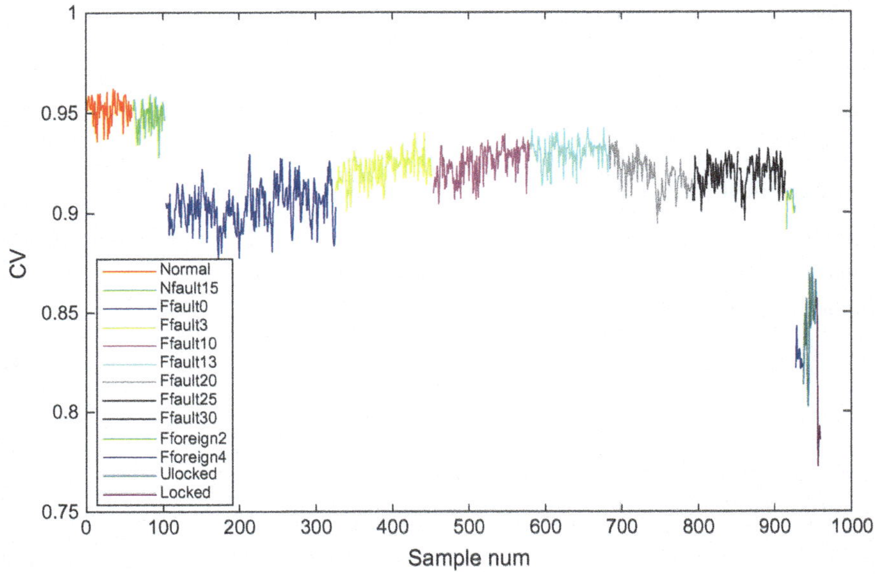

Fig. 6 CVs of all states

4. Evaluated the healthy state

The CVs of all state were calculated by health assessment function, the result was shown in Fig. 6.

The CVs of all state were shown in Fig. 6, and the all states were corresponded to the Label in Table 1. In Fig. 6, the deviations between the fault states and normal state were shown form the CV.

As shown in Fig. 6, this method accurately evaluated healthy state of turnout by using the health assessment function. However, it was not accurate enough to distinguish the difference of the similar state under the actual situation.

4 Conclusion

Turnouts are one of the most frequently used components in rail system. This paper presents the method based on the DTW and health assessment function to evaluate the healthy state of turnout. Firstly, DWT is used to calculate the total costs between different states, health assessment function is used to transfer the total cost to CV, CV is used to evaluate the healthy state.

However, in the actual situation, there exist some outliers or random data that is generated by the unstable current or improper maintenance, etc. Therefore, additional experiments should be done to improve the method.

Acknowledgement This research is supported by the National Key R&D Program of China (No. 2016YFB1200402).

References

1. Oyebande BO et al (2002) Condition monitoring of railway electric point machines. IEE Proc Electric Power Appl 149(6):465–473
2. Atamuradov V et al (2018) Degradation-level assessment and online prognostics for sliding chair failure on point machines. IFAC PapersOnLine 51(24):208–213
3. Silmon JA et al (2010) Improving railway switch system reliability with innovative condition monitoring algorithms. Proc Inst Mech Eng Part F-J Rail Rapid Transit 224(F4):293–302
4. Kaewunruen S et al (2014) Monitoring structural deterioration of railway turnout systems via dynamic wheel/rail interaction. 1:19–24
5. Marquez FPG et al (2007) Unobserved component models applied to the assessment of wear in railway points: a case study. Eur J Oper Res 176(3):1703–1712
6. Asada T et al (2013) An algorithm for improved performance of railway condition monitoring equipment: alternating-current point machine case study. Transp Res Pt C-Emerg Technol 30:81–92
7. Zhou FF, Xia L et al (2016) Fault diagnosis of high-speed railway turnout based on support vector machine, pp 1539–1544. IEEE, New York
8. Lee J et al (2016) Fault detection and diagnosis of railway point machines by sound analysis. Sensors 16(4):12
9. Zhang K et al (2014) Algorithm of railway turnout fault detection based on PNN neural network. In: 2014 seventh international symposium on computational intelligence and design, pp 544–547, IEEE, New York
10. Shi Z et al (2018) An auto-associative residual based approach for railway point system fault detection and diagnosis. Measurement 119:246–258
11. Ardakani HD et al (eds) (2012) PHM for railway system—a case study on the health assessment of the point machines. In: 2012 IEEE conference on prognostics and health management, IEEE, Denver
12. Liu MM, Yan X et al (2016) Fault diagnosis method for railway turnout control circuit based on information fusion, pp 315–320. IEEE, New York
13. Zhang K (2014) The railway turnout fault diagnosis algorithm based on BP neural network, pp 135–138. IEEE, New York
14. Yilboga H, Eker ÖF et al (2010) Failure prediction on railway turnouts using time delay neural networks. In: 2010 IEEE international conference on computational intelligence for measurement systems and applications, IEEE
15. Keogh E et al (2005) Exact indexing of dynamic time warping. Knowl Inf Syst 7(3):358–386
16. Zhen D et al (2013) Fault diagnosis of motor drives using stator current signal analysis based on dynamic time warping. Mech Syst Signal Proc 34(1–2):191–202
17. Kim H et al (2016) Fault diagnosis of railway point machines using dynamic time warping. Electron Lett 52(10):818–819
18. Huang S et al (2017) Turnout fault diagnosis through dynamic time warping and signal normalization. J Adv Transport
19. Qiu H et al (2003) Robust performance degradation assessment methods for enhanced rolling element bearing prognostics. Adv Eng Inform 17(3–4):127–140
20. Müller M (2007) Dynamic time warping. In: Information retrieval for music and motion, pp 69–84. Springer, Berlin

Research on Risk Consequence Severity Evaluation Based on Fuzzy Analytic Hierarchy Process and Evidence Theory

Lijuan Shi, Liquan Chen, Huize Sun, Zhiyu Wang, Jian Wang and Yiwei Wu

Abstract In view of the uncertainty of the severity of the consequences of the Chinese train control system level 2 (CTCS-2) risk events, the paper puts forward a combined approach to evaluate the consequence severity of a risk event by using the fuzzy analytic hierarchy process and the discount evidence theory on the basis of expert scoring. Firstly, the evaluation index system of consequence severity is established, and the weights of the evaluation indices are calculated by using the fuzzy analytic hierarchy process. Secondly, the membership degree of each evaluation index's severity grade is calculated by using discount evidence theory. Finally, the weights are multiplied with the membership degree, and the severity grade of risk consequence is determined according to the maximum membership principle. In this paper, taking the CTCS-2 train control system 'train overspeed unprotected' risk event as an example, based on expert scoring, the consequence severity of the risk event is evaluated by using fuzzy analytic hierarchy process (FAHP) and discounted evidence theory, and the consequence severity of the risk event is assessed as 'very serious,' which is consistent with our common sense.

Keywords Chinese train control system · Risk consequence severity · Fuzzy analytic hierarchy process (FAHP) · Evidence theory

L. Shi (✉) · H. Sun · Z. Wang · J. Wang · Y. Wu
Shanghai Key Laboratory of Rail Infrastructure Durability and System Safety, Tongji University, 201804 Shanghai, China
e-mail: shilijuan150@tongji.edu.cn

L. Chen
China Railway Guangzhou Group Co., Ltd, 510088 Guangzhou, China

Y. Qin et al. (eds.), *Proceedings of the 4th International Conference on Electrical and Information Technologies for Rail Transportation (EITRT) 2019*, Lecture Notes in Electrical Engineering 639, https://doi.org/10.1007/978-981-15-2866-8_51

1 Introduction

Risk is the uncertainty of loss or gain. Losses or gains refer to the consequences of the risks. This paper studies the safety of train operation, so it refers to the losses caused by risks [1]. Due to the difficulty in obtaining historical fault data, many scholars use semiquantitative analysis such as expert scoring method and fuzzy mathematics to obtain the required data. Xuanzi Hu used the analytic hierarchy process (AHP) to establish an index system and proposed an evaluation method of railway station based on the combination of gray clustering theory and AHP [2]. Zhenning Guo proposed a bid evaluation method based on the gray system theory to overcome the shortcomings of traditional bid evaluation methods, which transformed qualitative evaluation into quantitative evaluation and reduced the influence of subjective factors [3]. Wenjie Peng adopted a nonlinear combination weighting method based on AHP and entropy weight method to complete the risk grade assessment of station risk factors [4]. Other scholars contributed to the solutions of the combination of the cloud model, FAHP, network AHP, and evidence theory to achieve the risk assessment of the objects [5–9].

It can be seen that the approaches of AHP, entropy weight method, cloud model, and evidence theory were mostly used to establish a mathematical model to complete a risk assessment. However, at present, there are few studies on the calculation method of consequence severity. The calculation results of the AHP are difficult to verify, and its accuracy is low. Evidence theory has been widely used; but in practice, there may be major conflicts between multiple pieces of evidence. Therefore, this paper attempts to improve the above methods to complete the calculation of the severity of the risk consequences.

This paper establishes a mathematical model based on FAHP and discount evidence theory to calculate the severity of risk consequences. The specific chapters are as follows: (1) Summarize the previous research methods; (2) based on expert scoring method, establish a mathematical model by using fuzzy analytic hierarchy process and discount evidence theory to calculate the consequence severity of risk; (3) taking the 'train overspeed unprotected' in CTCS-2 train control system as an example, apply the mathematical model proposed in this paper to evaluate the severity of the risk consequence.

2 Computation Model Based on FAHP and Evidence Theory

2.1 Fuzzy Set Theory [10]

Definition 1 Existence $x \in X$ makes $\mu_A(x)$ satisfy the following function

$$\mu_A(x,a,b,c) = \begin{cases} 0, & x<a \text{ Or } x>c \\ \frac{x-a}{b-a}, & a\leq x\leq b \\ \frac{c-x}{c-b}, & b<x\leq c \end{cases} \tag{1}$$

Among them, $\mu_A(x)$ is called the degree of membership of A, and $A = (a, b, c)$ is a triangular fuzzy number on universe X.

Definition 2 The mean of triangular fuzzy numbers $A = (a_1, b_1, c_1)$ and $B = (a_2, b_2, c_2)$ can be expressed as follows:

$$1/2(A \oplus B) = 1/2((a_1, b_1, c_1) \oplus (a_2, b_2, c_2)) = 1/2(a_1 + a_2, b_1 + b_2, c_1 + c_2) \tag{2}$$

According to the expansion principle of fuzzy numbers, the average of $A \oplus B$ is still triangular fuzzy numbers.

2.2 *Definition of Evidence Theory*

Evidence theory is based on basic probability allocation and evidence fusion [11]. Uncertainty is described by recognition framework, basic probability assignment (BPA) and other related concepts.

Recognition framework.

Definition 3 Assume that there is a problem that needs to be distinguished. The set of all possible discriminant results is represented by Θ. If all elements in Θ are mutually exclusive, then Θ is called the recognition framework and expressed as

$$\Theta = \{\theta_1, \theta_2, \ldots, \theta_j, \ldots, \theta_N\} \tag{3}$$

In this formula, θ_j is called an element of the recognition framework; N represents the number of elements; $j = 1,2, \ldots, N$.

Basic probability assignment function.

On the basis of establishing the recognition framework, the evidence handlers analyze each evidence, determine the support degree of each evidence to each element in the recognition framework, and express it with the basic probability assignment function m.

$m(A)$ is called the basic probability allocation function of event A ($A \subseteq \Theta$, representing any subset of Θ), also known as mass function, which represents the initial allocation of evidence's trust in A.

Evidence fusion.

When there are multiple evidences to judge the same problem, all the evidences are fused by the established method. Among them, Dempster's evidence fusion criteria are as follows:

Definition 4 Assume that there are multiple evidences E_1, E_2, …, E_n, the corresponding mass functions are m_1,m_2, …, m_n, and the rule of synthesis is as follows:

$$m(A) = \begin{cases} \frac{\sum\limits_{\cap A_j = A} \prod\limits_{1 \le i \le n} m_i(A_i)}{1-K}, & A \neq \varnothing \\ 0, & A = \varnothing \end{cases} \tag{4}$$

Among them, $K = \sum\limits_{\cap A_j = \varnothing} \prod\limits_{1 \le i \le n} m_i(A_i)$ indicates the degree of conflict between the various evidences.

2.3 *Classification of Severity of Risk Consequences*

In the national standard of GB/T 21562, the severity is divided into four levels [12]. In order to calculate the severity of risk more accurately, this paper divides into six levels, which are catastrophic, very serious, severe, critical, light, and irrelevant.

2.4 *Model Construction*

Establishment of the evaluation index system. Once a risk event occurs, it will have multiple effects. From these aspects, the evaluation index system is established. A typical secondary indicator system is shown in matrix (5). The primary index set is $V = (v_1 \ldots v_j \ldots v_n)$. Among them, v_j is called an element of the indicators; n represents the number of elements; j = 1, 2, …, n. A secondary evaluation index system is constructed from the primary index set to assess the severity of the consequences of a risk event. Among them, v_{ji} (j = 1, …, n, i = 1, …, x) represents the jth element of the indicators at the ith severity level. Different elements of the indicators can have various severity levels, and x can take different values, such as m, k, or h.

$$[\text{Consequence severity index } V] = \begin{bmatrix} v_1 \\ \vdots \\ v_j \\ \vdots \\ v_n \end{bmatrix} = \begin{bmatrix} [v_{11} \cdots & v_{1i} \cdots & v_{1m}] \\ & \vdots & \\ [v_{j1} \cdots & v_{ji} \cdots & v_{jk}] \\ & \vdots & \\ [v_{n1} \ldots & v_{ni} \cdots & v_{nh}] \end{bmatrix} \tag{5}$$

Weight Calculation Based on FAHP.

Establishment of evaluation criteria.

FAHP calculates the weights of indices by constructing a comparison matrix between indices. The comparison is based on expert's qualitative language description, and each language description corresponds to a set of triangular fuzzy number, as shown in Table 1 [7].

Constructing fuzzy comparison matrix.

The relative importance of each index in the index set V is determined by comparing two indices. The result is expressed by triangular fuzzy function, and the $n \times n$ order fuzzy comparison matrix M is formed.

$$M = [m_{i,j}] = \begin{bmatrix} m_{1,1} & m_{1,2} & \cdots & m_{1,n} \\ m_{2,1} & m_{2,2} & \cdots & m_{2,n} \\ \vdots & \vdots & \ddots & \vdots \\ m_{n,1} & m_{n,2} & \cdots & m_{n,n} \end{bmatrix}, \quad i,j = 1,\ldots,n \tag{6}$$

Among them, $m_{i,j} = (a_{i,j}, b_{i,j}, c_{i,j})$, which means $m_{j,i} = \frac{1}{m_{i,j}} = \left(\frac{1}{c_{i,j}}, \frac{1}{b_{i,j}}, \frac{1}{a_{i,j}}\right)$.

$$\begin{aligned} & a_{i,k} = a_{i,j} \times a_{j,k}, b_{i,k} = b_{i,j} \times b_{j,k}, c_{i,k} = c_{i,j} \times c_{j,k}, \quad i<j<k; \\ & m_{i,j} \odot m_{j,k} = (a_{i,j} \times a_{j,k}, b_{i,j} \times b_{j,k}, c_{i,j} \times c_{j,k}) \\ & \quad = (a_{i,k}, b_{i,k}, c_{i,k}) = m_{i,k} \end{aligned}$$

Calculating the fuzzy weight of evaluation index.

The arithmetic mean matrix of the ith index is obtained by calculating geometric mean.

Table 1 Criteria for the importance of evaluation indicators

Linguistic values	Description	Triangular fuzzy numbers
Equally important	Two indices are equally important	(1,1,2)
Slightly important	The former is slightly more important than the latter	(2,3,4)
Obviously important	The former is obviously more important than the latter	(4,5,6)
Very important	The former is much more important than the latter	(6,7,8)
Absolutely important	The former is absolutely more important than the latter	(8,9,9)

$$\overline{M}_i = (\bar{a}_i, \bar{b}_i, \bar{c}_i) = \left(\sqrt[n]{\sum_{j=1}^{n} a_{i,j}}, \sqrt[n]{\sum_{j=1}^{n} b_{i,j}}, \sqrt[n]{\sum_{j=1}^{n} c_{i,j}} \right) \tag{7}$$

Through normalization, the fuzzy weight of the ith index is obtained as follows:

$$\mathrm{FW}_i = (a_i, b_i, c_i) = \left(\frac{\bar{a}_i}{\sum_{j=1}^{n} \bar{c}_j}, \frac{\bar{a}_i}{\sum_{j=1}^{n} \bar{b}_j}, \frac{\bar{c}_i}{\sum_{j=1}^{n} \bar{a}_j} \right) \tag{8}$$

Calculating the weight of the evaluation index.

The fuzzy weight mentioned above is defuzzified [13], and the weight of the ith index is obtained by normalization.

$$w_i' = \frac{\int \mu_{FW_i}(x)x\mathrm{d}x}{\int \mu_{FW_i}(x)\mathrm{d}x}, \quad w_i = \frac{w_i'}{\sum_{i=1}^{n} w_i'} \tag{9}$$

There are k indicators in total, and then the weight set is: $W = [w_1, w_2, \ldots, w_k]$.

Computation of membership degrees of the risk rank based on evidence theory.

Constructing a refognition Framework——comment set.

The expert fuzzy comment set is defined as CS = $(cs_1, cs_2, cs_3, cs_4, cs_5, cs_6)$ = (catastrophic, very serious, severe, critical, light, irrelevant).

Constructing the initial matrix of mass function of evaluation index.

Chose an index and five experts $E = (e_1, e_2, e_3, e_4, e_5)$ were invited to provide five evaluations to determine the probability of the severity of the index and to form the probability distribution matrix T as the initial evidence matrix of evidence fusion.

$$T = \begin{bmatrix} t_1 \\ t_2 \\ \vdots \\ t_5 \end{bmatrix} = \begin{matrix} \\ e_1 \\ e_2 \\ \vdots \\ e_5 \end{matrix} \begin{matrix} cs_1 & cs_2 & cs_3 & cs_4 & cs_5 & cs_6 \\ \left[\begin{matrix} t_{1,1} \\ t_{2,1} \\ \vdots \\ t_{5,1} \end{matrix}\right. & \begin{matrix} t_{1,2} \\ t_{2,2} \\ \vdots \\ t_{5,2} \end{matrix} & \begin{matrix} t_{1,3} \\ t_{2,3} \\ \vdots \\ t_{5,3} \end{matrix} & \begin{matrix} t_{1,4} \\ t_{2,4} \\ \vdots \\ t_{5,4} \end{matrix} & \begin{matrix} t_{1,5} \\ t_{2,5} \\ \vdots \\ t_{5,5} \end{matrix} & \left.\begin{matrix} t_{1,6} \\ t_{2,6} \\ \vdots \\ t_{5,6} \end{matrix}\right] \end{matrix}, \quad (0 \le t \le 1) \tag{10}$$

Among them, $t_{i,j}(i = 1, \ldots, 5, j = 1, \ldots, 6)$ represents the ith expert's probabilistic evaluation of the index at the jth severity level.

Calculating discount coefficient of the evidence.

The evidences provided by experts for each indicator are different; that is, there is a conflict. In order to solve this problem, this paper uses the distance Formula (11) to measure the distance between these evidences [14], uses Formula (12) to calculate the similarity, forms the similarity matrix (13), and finally calculates the discount coefficient of each evidence according to Formula (14). The distance between expert e_i and expert e_j is:

$$d(e_i, e_j) = \sqrt{\frac{\|t_i\|^2 + \|t_j\|^2 - 2\langle t_i, t_j\rangle}{2}}, \quad (i,j = 1, 2, \ldots, 5) \tag{11}$$

where $\|t_i\|^2 = \langle t_i, t_i\rangle$; $\|t_j\|^2 = \langle t_j, t_j\rangle$; $\langle t_i, t_j\rangle$ is the inner product of two vectors. According to Formula (12), the similarity between evidences is calculated and the similarity matrix is formed.

$$S(e_i, e_j) = 1 - d(e_i, e_j), \quad S = \begin{bmatrix} 1 & S_{e_1,e_2} & \cdots & S_{e_1,e_5} \\ S_{e_2,e_1} & 1 & \cdots & S_{e_2,e_5} \\ \vdots & \vdots & \vdots & \vdots \\ S_{e_5,e_1} & S_{e_5,e_2} & \cdots & 1 \end{bmatrix} \tag{12}$$

The degree to which the evidence is supported by other evidences by summing up each row of the similarity matrix:

$$\sup = [\sup(e_1), \sup(e_2), \ldots, \sup(e_5)] = \left[\sum_{\substack{j=1 \\ j \neq 1}}^{5} S_{e_1,e_j}, \sum_{\substack{j=1 \\ j \neq 2}}^{5} S_{e_2,e_j}, \ldots, \sum_{\substack{j=1 \\ j \neq 5}}^{5} S_{e_5,e_j} \right] \tag{13}$$

The discount coefficient matrix is formed by calculating the discount coefficient according to Formula (14):

$$\mathrm{Crd} = [\mathrm{Crd}_{e_1}, \ldots, \mathrm{Crd}_{e_5}] = \left[\frac{\sup(e_1)}{\sum_{j=1}^{5} \sup(e_j)}, \ldots, \frac{\sup(e_5)}{\sum_{j=1}^{5} \sup(e_j)} \right] \tag{14}$$

Update the membership matrix:

$$T' = \mathrm{Crd}^T \times T = \begin{bmatrix} t'_{1,1} & t'_{1,2} & t'_{1,3} & t'_{1,4} & t'_{1,5} & t'_{1,6} \\ t'_{2,1} & t'_{2,2} & t'_{2,3} & t'_{2,4} & t'_{2,5} & t'_{2,6} \\ \vdots & \vdots & \vdots & \vdots & \vdots & \vdots \\ t'_{5,1} & t'_{5,2} & t'_{5,3} & t'_{5,4} & t'_{5,5} & t'_{5,6} \end{bmatrix} \tag{15}$$

Evidence fusion.

According to Formula (12), the new membership matrix is fused with evidence, and the degree of membership of the severity of the consequences of the index is obtained.

$$\mu = [t(cs_1), t(cs_2), \ldots, t(cs_s)] \tag{16}$$

There are k indicators, and the membership degree set of all indicators is:

$$M' = [\mu_1, \mu_2, \ldots, \mu_k]^{\mathrm{T}} \tag{17}$$

Grading of serious consequence.

By using the weighted method, the membership degree of risk events subordinated to the severity level is calculated $\Psi = W \cdot \mu'$, and the severity level is determined according to the principle of maximum membership.

3 Example Computation—Calculation of Risk Consequence Severity of Train Control System

3.1 Train Control System Risk Event Analysis

Train control system consists of ground subsystem and onboard subsystem, which is used to control train speed and ensure safety and efficient operation of train. In CTCS-2 level train control system, the principle is that the onboard safety computer generates the permitted speed curve according to the forward driving conditions, temporary speed limit information, and vehicle information. When the train runs over speed, the onboard safety computer outputs the braking command [15]. If the train runs at excessive speed and the onboard equipment fails to take protective measures, the result will be unimaginable. Therefore, this paper regards 'train overspeed unprotected' as a risk event to be studied and calculates the severity of the consequences by taking the risk event as an example.

3.2 Calculation of the Severity of the Consequences of Risk Events

Establishment of the evaluation index system.

According to the determination of accident grade in the ***Regulations on Emergency Rescue and Investigation of Railway Traffic Accidents***, combined with the

Table 2 Evaluation index system of the severity of consequences

	Primary indices	Secondary indices
Indicators of the severity of consequence V	Security impact v_1	Death toll v_{11}
		Serious injury v_{12}
		Slight injury v_{13}
	Service impact v_2	Line service interruption v_{21}
		Interruption of station service v_{22}
		Interruption of train operation v_{23}
	Economic loss v_3	Direct loss v_{31}
		Consequential loss v_{32}

characteristics of train operation in China, a secondary evaluation index system is constructed from security impact, economic loss, and service impact, as shown in Table 2, and the consequence severity of risk events is calculated and analyzed.

Calculate the weight of evaluation indicators.

According to the index set of Table 2 and the evaluation criteria of Table 1, five experts compared the indices of the primary index set $V = (v_1, v_2, v_3)$, aggregated the fuzzy numbers according to the definition 3, and obtained the evaluation matrix of Table 3.

According to Formulas (1), the fuzzy comparison matrix is constructed.

$$M = \begin{bmatrix} (1,1,1) & (3.6,4.6,5.6) & (11.52,19.32,29.12) \\ (0.18,0.22,0.28) & (1,1,1) & (3.2,4.2,5.2) \\ (0.03,0.05,0.09) & (0.19,0.24,0.31) & (1,1,1) \end{bmatrix} \quad (18)$$

According to Formulas (7–9), the fuzzy weight of three indexes and the final weight of a single index are calculated.

$$\mathrm{FW}_0 = \begin{bmatrix} (0.40,0.51,0.63) \\ (0.26,0.30,0.36) \\ (0.17,0.19,0.21) \end{bmatrix}, \quad W_0 = [w_{v_1}, w_{v_2}, w_{v_3}] = [0.5071, 0.3040, 0.1889] \quad (19)$$

Similarly, according to the evaluation of 5 experts for other secondary indicators, the evaluation matrix is built.

According to Formula (6–9), the relative weight of each index is calculated as follows:

$$W_1 = [w_{v_{11}}, w_{v_{12}}, w_{v_{13}}] = [0.5027, 0.2928, 0.2045]$$
$$W_2 = [w_{v_{21}}, w_{v_{22}}, w_{v_{23}}] = [0.4895, 0.3186, 0.1919]$$
$$W_3 = [w_{v_{31}}, w_{v_{32}}] = [0.5846, 0.4154]$$

Table 3 Comparative judgment of indexes v_1, v_2, and v_3

Indices comparison	Linguistic judgment	Triangular fuzzy number	Aggregated triangular fuzzy numbers
v_1 versus v_2 ($m_{1,2}$)	Between slightly and obviously important	(3,4,5)	(3.6,4,6,5,6)
	Obviously important	(4,5,6)	
	Obviously important	(4,5,6)	
	Obviously important	(4,5,6)	
	Between slightly and obviously important	(3,4,5)	
v_2 versus v_3 ($m_{2,3}$)	Slightly important	(2,3,4)	(3.2,4.2,5.2)
	Between slightly and obviously important	(3,4,5)	
	Obviously important	(4,5,6)	
	Between slightly and obviously important	(3,4,5)	
	Obviously important	(4,5,6)	

The final weight of the secondary index is obtained by multiplying the weight of the primary index and the secondary index.

$$\begin{aligned} W &= \left[w'_{v_{11}}, w'_{v_{12}}, w'_{v_{13}}, w'_{v_{21}}, w'_{v_{22}}, w'_{v_{23}}, w'_{v_{31}}, w'_{v_{32}}\right] \\ &= [0.2549, 0.1485, 0.1037, 0.1488, 0.0968, 0.0584, 0.1104, 0.0785] \end{aligned}$$

Calculate the degree of severity of evaluation indices.

Five experts were invited to evaluate the possibility of the severity grade of the secondary evaluation index. Taking 'death toll v_{11}' as an example, the initial membership matrix is as follows: T_{v11}.

$$T_{v_{11}} = \begin{matrix} \\ e_1 \\ e_2 \\ e_3 \\ e_4 \\ e_5 \end{matrix} \begin{matrix} \begin{matrix} cs_1 & cs_2 & cs_3 & cs_4 & cs_5 & cs_6 \end{matrix} \\ \begin{bmatrix} 0.3 & 0.4 & 0.3 & 0 & 0 & 0 \\ 0.2 & 0.5 & 0.2 & 0.1 & 0 & 0 \\ 0.2 & 0.4 & 0.2 & 0.2 & 0 & 0 \\ 0.3 & 0.5 & 0.2 & 0 & 0 & 0 \\ 0.3 & 0.5 & 0.1 & 0.2 & 0 & 0 \end{bmatrix} \end{matrix}$$

The similarity matrix and reliability calculated by Formula (12–14) are as follows:

$$S_{v_{11}} = \begin{bmatrix} 1 & 0.86 & 0.83 & 0.90 & 0.79 \\ 0.86 & 1 & 0.90 & 0.90 & 0.88 \\ 0.83 & 0.90 & 1 & 0.83 & 0.88 \\ 0.90 & 0.90 & 0.83 & 1 & 0.84 \\ 0.79 & 0.88 & 0.88 & 0.84 & 1 \end{bmatrix}, \quad \mathrm{Crd}_{v_{11}} = \begin{bmatrix} 0.1962 \\ 0.2057 \\ 0.1996 \\ 0.2017 \\ 0.1969 \end{bmatrix}$$

The membership degree of consequence severity grade of v_{11} is calculated by Formula (14–16):

$$\mu_{v_{11}} = [0.0507, 0.9381, 0.0113, 0, 0, 0]$$

Similarly, the degree of severity of other secondary indicators is calculated in the degree of membership.

$$\begin{aligned} \mu_{v_{12}} &= [0.8142, 0.1809, 0.0048, 0, 0, 0], \\ \mu_{v_{13}} &= [0.3525, 0.6266, 0.0209, 0, 0, 0], \\ \mu_{v_{21}} &= [0.3255, 0.6510, 0.0217, 0.0018, 0, 0], \\ \mu_{v_{22}} &= [0.0623, 0.9339, 0.0029, 0010, 0, 0], \\ \mu_{v_{23}} &= [0.2857, 0.5357, 0.1429, 0.0357, 0, 0], \\ \mu_{v_{31}} &= [0.3038, 0.5696, 0.1013, 0.0253, 0, 0], \\ \mu_{v_{32}} &= [0.1233, 0.4932, 0.3288, 0.0548, 0, 0]. \end{aligned}$$

Assessment of severity of consequences.

The results of 3.2 are weighted to obtain the membership of the ultimate consequence severity grade.

$$\begin{aligned} \Psi = W \cdot \mu' &= \left[w'_{v_{11}}, w'_{v_{12}}, \ldots, w'_{v_{32}}\right] \left[\mu_{v_{11}}, \mu_{v_{12}}, \ldots, \mu_{v_{32}}\right]^T \\ &= [0.2847, 0.6512, 0.0546, 0.0095, 0, 0] \end{aligned}$$

According to the maximum membership principle, the consequence of 'train overspeed unprotected' is 'cs_2: very serious.'

4 Conclusion

In this paper, expert scoring, FAHP, and discount evidence theory are used to establish an evaluation model of consequence severity of the risk events. A proposed model is proposed to evaluate the severity of the consequence of the risk event of 'train overspeed unprotected' of CTCS-2 train control system. The severity of the consequence is judged as 'very serious,' which is consistent with our common sense. The main contributions of this paper are as follows:

(1) Based on the expert scoring method, a mathematical model is established by using the FAHP. The weights of the evaluation indices are obtained through the defuzzification calculation.
(2) The mathematical model of consequence severity is put forward by using evidence theory. By calculating the discount coefficient, evidences provided by all experts are discounted to reduce the conflict between different evidences and therefore making the calculation more realistic.
(3) The model established in this paper is universal and suitable for the index of the risk consequence severity evaluation based on the expert scoring method.

For the risk event of 'train overspeed unprotected' is 'very serious,' measures should be taken from two aspects. One is decreasing the frequency of key equipments failures, such as track circuit and balise transmission module (BTM); the other is reducing the severity of consequences which can be started with both quick detection of the risk event and establishment of more prompt response to the emergency plan.

Acknowledgements Authors would like to acknowledge the support of the research program of Comprehensive Support Technology for Railway Network Operation (2018YFB1201403), which is a subproject of Advanced Railway Transportation Special Project belonging to the 13th Five-Year National Key Research and Development Plan funded by Ministry of Science and Technology of China.

References

1. Song S (2013) Core theory and key technologies of railway safety risk management (in Chinese)
2. Hu X, Xie C (2006) Application of grey clustering theory combined with AHP in railway station evaluation. In: 2006 IEEE international conference on information acquisition, Weihai, pp 1289–1294
3. Zhen-ning G (2011) Evaluation bid method of railway construction project based on gray situation decision. In: 2011 international conference on consumer electronics, communications and networks (CECNet), XianNing, pp 2191–2194
4. Peng W, Lin Z (2018) research on risk assessment of railway freight station based on nonlinear combination of Ahp-entropy. In: 2018 15th international conference on service systems and service management (ICSSSM), Hangzhou, pp 1–6
5. Zhang Y, Li Y (2016) Risk assessment of railway signal system based on cloud model and evidence theory. J Railw (1) (in Chinese)
6. An M, Chen Y, Baker CJ (2011) A fuzzy reasoning and fuzzy-analytical hierarchy process based approach to the process of railway risk information: a railway risk management system. Inf Sci 181(18):3946–3966
7. An M, Qin Y, Jia LM et al (2016) Aggregation of group fuzzy risk information in the railway risk decision making process. Saf Sci 82:18–28

8. Majumder D, Debnath J, Biswas A (2013) Risk analysis in construction sites using fuzzy reasoning and fuzzy analytic hierarchy process. Procedia Technol 10:604–614
9. Yingying X, Chao L, Tao T (2016) Research on risk assessment of CTCS based on fuzzy reasoning and analytic hierarchy process. In: International conference on intelligent information processing. ACM
10. Zhang X, Pei D, Daijianhua (2013) Fuzzy mathematics and rough set theory. Tsinghua University Press, 1–99 (in Chinese)
11. Yang F, Wang X (2010) Conflict evidence synthesis method of D-S evidence theory. National Defense Industry Press (in Chinese)
12. GB/T 21562–2008 (2008) Specification for reliability, usability, maintainability and safety of rail transit (in Chinese)
13. Wang T, Li M, Xu Z (2003) Fuzzy risk assessment of safety-critical software. Comput Appl 23:182–184 (in Chinese)
14. Deng Y, Shi WK, Zhu ZF, Liu Q (2004) Combining belief functions based on distance of evidence. Decis Support Syst
15. Tang T (2014) Train operation control system. China Railway Press, 1–7 (in Chinese)

Research on Dynamic Analysis of Equipment Fault for High-Speed Railway Integrated with Train Operation Diagram

Dongxiu Ou, Yusen Wu and Bowen Gao

Abstract Equipment fault can make a major impact on the normal operation in high-speed railway. Such events cause delay to passengers and lead to a deviation from the scheduled train diagram. Delay impact can be described as an absolute value of delay time reflecting the overall operation status. Obtaining a dynamic impact prediction can potentially help to monitor the status and manage the adjustment measures. To investigate the impact of the resulting passengers' delay due to the equipment fault, this research proposes a dynamic analysis method to compute the total delay impact efficiently. Integrated with the train operation diagram, the duration of the impact is divided into three phases, and the delayed trains are tracked continuously. This approach allows updating the delay at regular intervals. Moreover, a gray prediction model is used to make a time series prediction on delay impact of the fault event. The results show that the prediction has a great accuracy. The presented method is significant for real-time monitoring and short-term prediction in delay analysis.

Keywords Equipment fault · Operation diagram · Delay analysis · Prediction

D. Ou (✉) · Y. Wu · B. Gao
The Key Laboratory of Road and Traffic Engineering, Ministry of Education, Shanghai, China
e-mail: ou.dongxiu@tongji.edu.cn

Shanghai Key Laboratory of Rail Infrastructure Durability and System Safety, Tongji University, Shanghai, China

Y. Wu
e-mail: 1831361@tongji.edu.cn

B. Gao
e-mail: 1810960@tongji.edu.cn

Y. Qin et al. (eds.), *Proceedings of the 4th International Conference on Electrical and Information Technologies for Rail Transportation (EITRT) 2019*, Lecture Notes in Electrical Engineering 639, https://doi.org/10.1007/978-981-15-2866-8_52

1 Introduction

In recent years, under the conditions of high-density and high-speed railway operation, the safety and efficiency requirements of Chinese high-speed railway network are increasing. Equipment fault can make a major impact on normal operation, causing different degrees of delay.

Internal mechanisms and external regulation are the two main points of relevant research on the impact of railway incidents. Many studies start with historical data and explore the rules to analyze the relationship between the delay and other factors [1, 2]. When delays occur, passengers are directly affected groups, and some studies begin to consider the role of 'passenger flow' as a risk carrier in delay propagation [3, 4]. Since the ultimate goal of analyzing the internal mechanism of delay is to reduce the impact loss, some approaches to dynamic rescheduling have emerged [5, 6]. It requires a quantitative analysis of the whole process of the event. Therefore, this paper focuses on the process of train delay, proposes a dynamic analysis method and predicts the impact results in the short term.

2 High-Speed Railway Equipment Fault Analysis

2.1 High-Speed Railway Equipment Fault Event

High-speed railway equipment fault event refers to an incident that affects the normal operation due to equipment reasons, causing delay or endangering driving safety. From a large number of the realistic fault event data, we can find that it mainly involves signal equipment, communication equipment, power supply equipment, monitoring equipment, etc. The spatial location of equipment fault can be divided into three types: single train, station and section, as shown in Table 1.

Figure 1 shows the statistic result on the historical equipment fault events for a line of Chinese high-speed railway from 2016 to 2018.

For a specific equipment fault event, the description should include:

- The number of the event (ID).
- The location or range of the event (Loc).
- The reason of the event (R).

Table 1 Types of equipment fault in different spatial location

Location	Description
Single train	Affect only single train operation initially
Station	Affect all trains passing through the station during the fault
Section	Affect all trains passing through the section during the fault

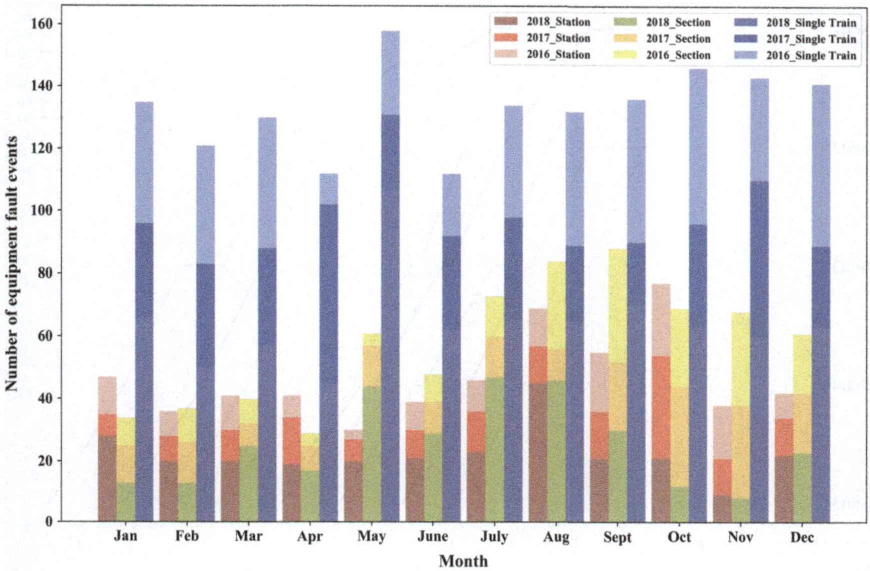

Fig. 1 Statistic result on the historical equipment fault events

- The duration from the beginning to the end of the fault impact (D).
- The impact of the event (I).

2.2 Impact of High-Speed Railway Equipment Fault Event

There is a constraint relationship between the trains because of the time and space occupation of the track. Therefore, the impact of high-speed railway equipment fault events usually involves spatial and temporal dimension. Due to the high density and short buffer time, the delay propagation may occur when the trains are delayed, causing a propagation chain in time and space. As the event continues to affect, delays will spread throughout the entire line.

The impact of equipment fault will directly affect the train operation diagram. The primary delay is the deviation of the schedule caused by the interference during the operation, and the secondary delay is the deviation caused by the conflict of access resource request between trains or waiting for the delayed train due to the equipment fault.

As shown in Fig. 2, the dotted line is the scheduled operation line and the solid line is the actual operation line. We can see that G1 is the first train affected, as the fault continues, subsequent trains are affected by the propagation chain. The fulfillment rate of the entire operation diagram is very low, which has a significant impact on the normal train operation.

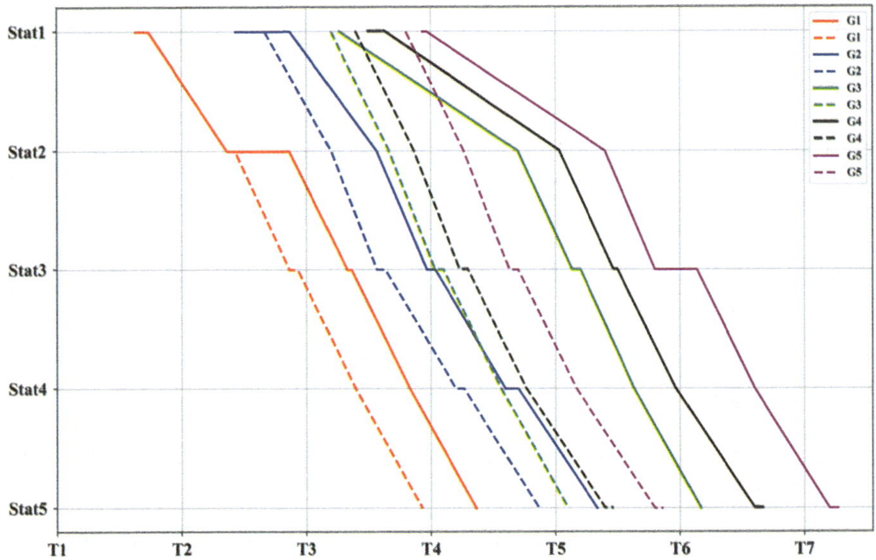

Fig. 2 Deviation between actual operation diagram and scheduled operation diagram due to the equipment fault

3 Dynamic Analysis Model for Equipment Fault Event

3.1 Model Hypothesis

In order to analyze the impact of high-speed railway equipment fault events quantitatively, based on the temporal and spatial dimension of the operation diagram, we propose the delay loss dimension using the total delay impact and establish a dynamic analysis model for Chinese high-speed railway equipment fault. It is assumed that only the single event factor plays a role in the impact duration of the event.

3.2 Variable Definition

To analyze the dynamic impact of equipment fault quantitatively, we define the variables in Table 2.

Table 2 Variable definition

Variable	Description
ID	Train number
Stat	Station name
Arr	Actual arrival time
Dept	Actual departure time
Arr*	Scheduled arrival time
Dept*	Scheduled departure time
M	The set of all trains m
N	The set of all stations n
K	The set of all time intervals k
V	The set of all train operation records
$P_{mn} \subseteq V$	The train operation record for train m in station n
$\mathrm{TP}_m \subseteq V$	The set of train operation records for train m
$\mathrm{SP}_n \subseteq V$	The set of train operation records for station n
$\mathrm{GP}_k \subseteq V$	The set of train operation records for time interval k
NP_{mn}	The affected passengers for record P_{mn}
TD	The total delay impact

3.3 *Train Operation Diagram*

The train operation diagram reflects the position information of the train in time and space, and it should be mathematically described by different dimensions. Usually, we use the split method as Ref. [7].

As shown in Fig. 3, P_{mn} is a specific train operation record for train m in station n. P_{mn} consists the information including ID, Stat, Arr, Dept, Arr* and Dept*. Through the extraction of all train operation records by the dimension of train, station and time interval, respectively, the set of records TP_m, SP_n, GP_k are obtained.

3.4 *Delay Impact Model*

We divide the impact duration D of the event into three phases, as shown in Fig. 4.

1. First phase (P1): Fault time. During this phase, the equipment fault occurred and was not fixed.
2. Second phase (P2): Impact growth time. The equipment fault was repaired, and the operation diagram began to adjust and recover. Due to the large density of the adjusted train, the impact of this phase was rapidly increasing.
3. Third phase (P3): Impact remission time. As the train operation diagram gradually recovers, the impact eases gradually.

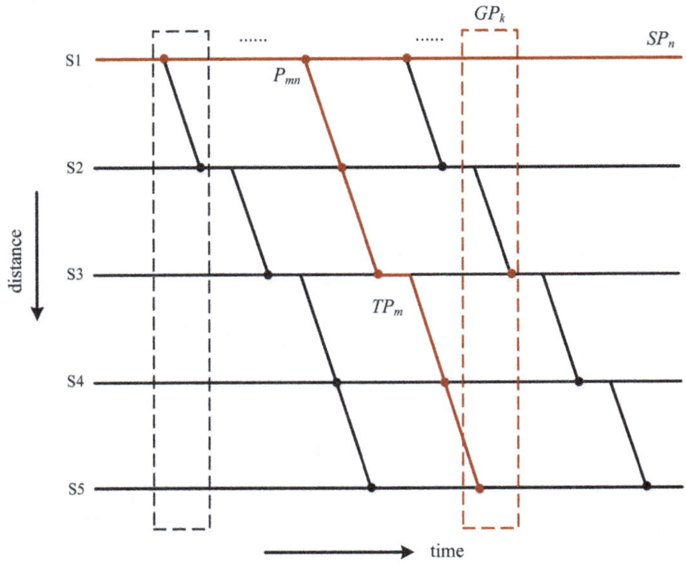

Fig. 3 Split of the train operation diagram

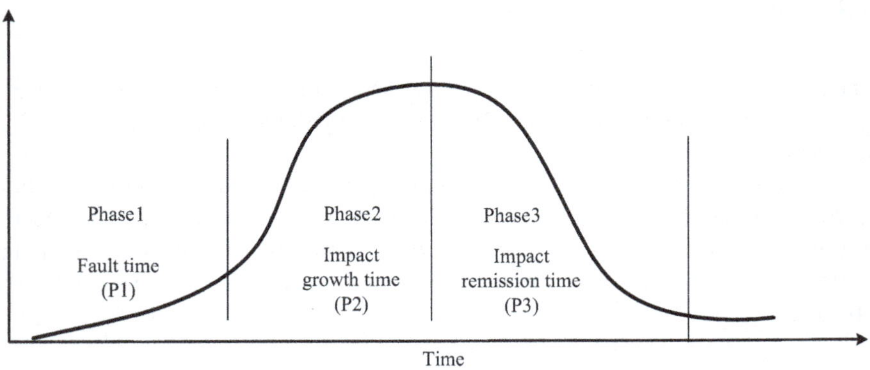

Fig. 4 Three phases of the delay impact

During the impact duration, the delayed trains need to be tracked for the impact computation. Due to the dynamic fluctuation of the punctuality rate of the train, the train that delayed by more than 2 min is considered to be affected by the event. Assuming that there are w trains affected, we can get a set of delayed train records Train_D.

From the passenger's point of view, the delay impact depends on the degree of train delay when they get off the train. Therefore, the number of passengers affected NP_{mn} is defined as the number of passengers leaving the train for record P_{mn}.

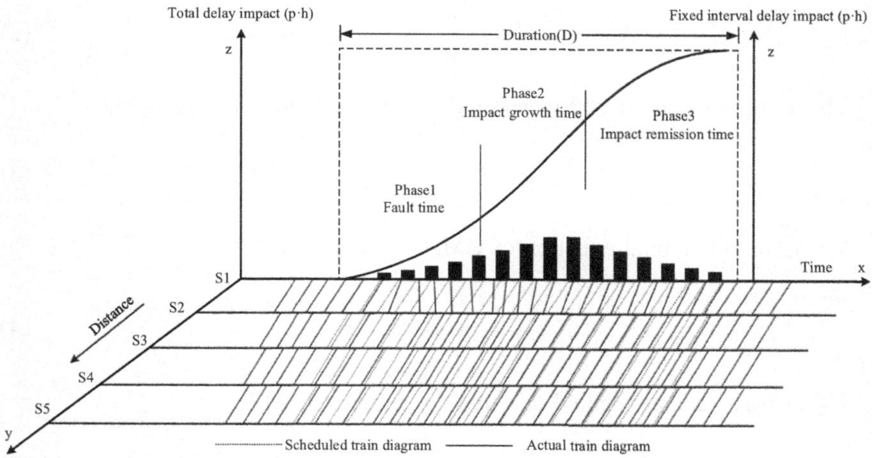

Fig. 5 Delay impact model. The *xoy* plane is the train operation diagram, the *x*-axis represents the time, the *y*-axis is the station, the *z*-axis (*left*) is the total delay impact and the *z*-axis (*right*) is the fixed interval delay impact

The total delay impact TD is the result of delay time of the train at each station, and the number of people affected by the delay. During the impact duration, it has the dynamic growth characteristics in the time series, as shown in Fig. 5.

$$\mathrm{TD} = \sum_{m,n} D_{mn} \times \mathrm{NP}_{mn}, \quad m \in [1, w], \ n \in [1, N_m] \tag{1}$$

where D_{mn} is the delay time for record P_{mn}, and N_m is the number of stops for train m.

3.5 Prediction Model

As the whole system hierarchy is fuzzy, the gray prediction model GM(1, 1) can be used to predict the continuous development of delay impact. Referring to Ref. [8], the method of the model is to process the original data, generate the data sequence with strong regularity and establish the corresponding differential equation, so as to predict the impact of delay in the next period.

Since the total delay impact is a dynamic change curve obtained by tracking delayed trains, it is not suitable to directly input GM(1, 1) model due to the characteristics of sequence correlation.

According to time granularity, we use the fixed interval delay impact as the original input data of GM(1, 1) model:

$$X^{(0)} = \left\{x^{(0)}(1), x^{(0)}(2), \ldots, x^{(0)}(n)\right\} \tag{2}$$

The 1-AGO (accumulated generating operation) sequence is as below:

$$X^{(1)} = \left\{x^{(1)}(1), x^{(1)}(2), \ldots, x^{(1)}(n)\right\} \tag{3}$$

Then, the differential equation of GM(1, 1) model is:

$$\frac{\mathrm{d}X^{(1)}}{\mathrm{d}t} + aX^{(1)} = b \tag{4}$$

Using least squares method to solve the parameters a, b:

$$[a, b]^T = (B^T B)^{-1} B^T Y \tag{5}$$

where B, Y can be calculated as below:

$$B = \begin{bmatrix} -\frac{1}{2}\left[x^{(1)}(1) + x^{(1)}(2)\right] & 1 \\ -\frac{1}{2}\left[x^{(1)}(2) + x^{(1)}(3)\right] & 1 \\ \vdots & \vdots \\ -\frac{1}{2}\left[x^{(1)}(n-1) + x^{(1)}(n)\right] & 1 \end{bmatrix} \tag{6}$$

$$Y = \begin{bmatrix} x^{(0)}(2) \\ x^{(0)}(3) \\ \vdots \\ x^{(0)}(n) \end{bmatrix} \tag{7}$$

GM(1, 1) prediction model can be obtained by solving the differential equation:

$$x^{(0)}(k+1) = \left[x^{(0)}(1) - \frac{b}{a}\right] e^{-ak} + \frac{b}{a} \tag{8}$$

4 Case Study

4.1 Case Description

Take a certain line L of Chinese high-speed railway as an example. The line L has 10 stations, which are marked as {Stat1, Stat2, …, Stat10} in descending direction. We conduct dynamic analysis on signal equipment fault event E in station Stat2, and the specific description of event E is shown in Table 3.

Table 3 Specific description of event *E*

Field	Description
ID	*E*
Loc	Stat2
R	Signal equipment fault, the white light band left behind by the passing train cannot be unlocked automatically
P1	[8:33, 10:42]

4.2 Delay Impact Result

Due to the difficulty in statistics of OD passenger flow in specific delay event, according to historical experience, we use the average number of passengers at each station to approximate the number of affected passengers. Calibrate that line *L* adopts CRH380AL EMU trains, with a total of 16-segment grouping, a capacity of 1061 and a full load rate of 1.

By tracking the delayed trains, the delay of multiple time segments is computed. The heat map is shown in Fig. 6. It can reflect the distribution of delays in time and space intuitively. We can find that the impact caused by the event *E* is transmitted to other stations through the movement of the train.

According to the delay impact model, the total delay impact is plotted as a function of time, as shown in Fig. 7. The first phase is the fault time. This time is fixed for a specific event. In this case, the first phase (P1) ranges from 8:33 to 10:42. It means that the equipment fault is repaired at 10:42. In the second phase (P2), the delay impact grows rapidly because of the high density and the short tracking train

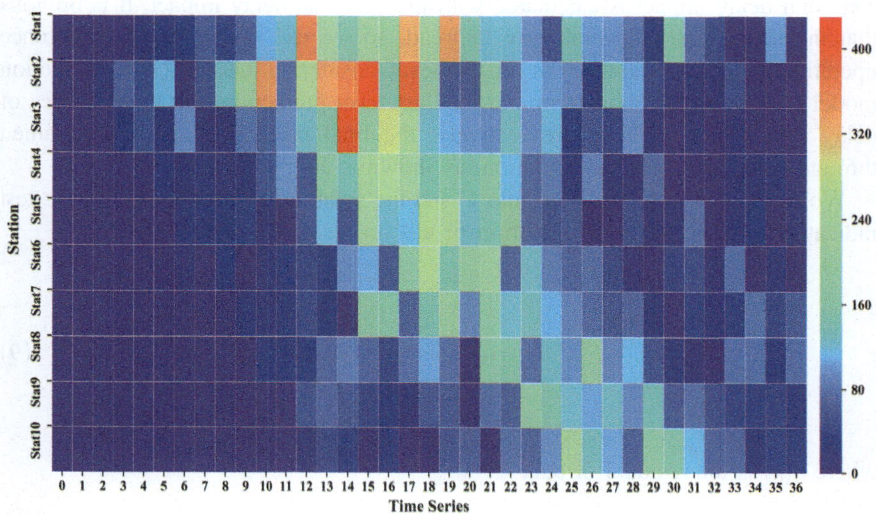

Fig. 6 Delay impact heat map

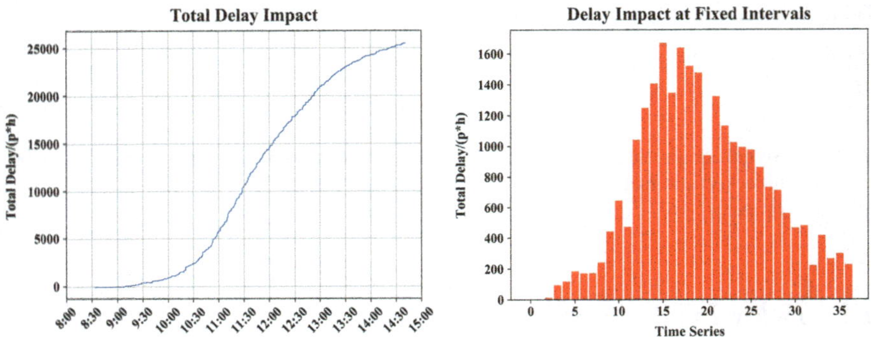

Fig. 7 Delay impact. (*left*) Total delay impact. (*right*) Interval delay impact

interval. Referring to the delay impact at fixed intervals, the P2 comes to the end when the impact growth rate reaches the maximum (the end of the P2 is 11:30 in this case). In the third phase (P3), the growth rate starts to slow down. Generally, the duration of third phase is relatively long due to the interference in the actual operation. For the convenience of calculation, referring to Ref. [4], we approximate that P3 comes to the end when the impact duration D is three times of the P1. After the third phase, it can be thought that the impact of the event has eased and tends to be stable.

4.3 Prediction Model Performance

The total delay impact is calculated with the previous delay impact. It is obvious that the series has an overall growth trend, so we use the first-order difference operation method to stabilize the time series. Considering that GM(1, 1) prediction model is suitable for short-term and medium-term prediction, the three phases of delay impact are predicted, respectively, the final prediction result is obtained through differential inverse operation, as shown in Fig. 8.

We select the posterior error ratio C and the small error frequency P as the test indicators to perform the posteriori error test on the prediction results.

$$C = \sqrt{\frac{\frac{1}{n}\sum_{k=1}^{n}\left(x^{(0)}(k) - \overline{x}\right)^2}{\frac{1}{n}\sum_{k=1}^{n}\left(q(k) - \overline{q}\right)^2}} \tag{9}$$

where $q(k)$ is the residual of data k.

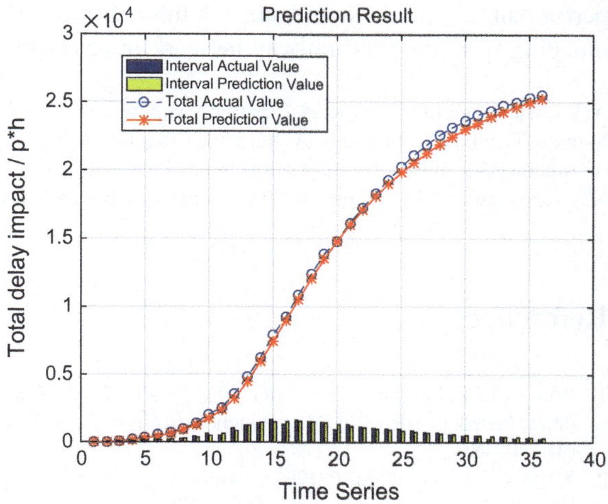

Fig. 8 Delay impact prediction result

Table 4 Accuracy standard of posteriori error

Level	*C*	*P*
1 (great)	$C < 0.35$	$P > 0.95$
2 (good)	$C < 0.45$	$P > 0.80$
3 (normal)	$C < 0.50$	$P > 0.70$
4 (bad)	$C \geq 0.65$	$P \leq 0.70$

$$P = P\left(|q(k) - \overline{q}| < 0.6745 \times \sqrt{\frac{1}{n}\sum_{k=1}^{n} (x^{(0)}(k) - \overline{x})^2} \right) \tag{10}$$

It should be noted that the test sequence is the input and output sequence of the predictive model. Finally, we obtain the posterior error ratio C is 0.0496, small error frequency P is 1. Referring to the accuracy standard of posteriori error in Ref. [9], as shown in Table 4, the prediction result has a great accuracy.

5 Conclusion

In this paper, the impact dimension of total delay is added to the real-time drawing process of train operation diagram. We propose a calculation method of total delay impact, which can dynamically reflect the operation state of the line by tracking the delayed trains and calculating the compound impact of delay time and affected passengers. The presented approach has a good versatility. It can be extended and applied to other lines in high-speed railway.

Moreover, the gray prediction model is also used to make short-term prediction on the impact of delay. The result shows the prediction model has a good performance. This is significant for the real-time monitoring and adjustment formulating of high-speed railway network in case of equipment fault operation risk.

Acknowledgements This work is partially supported by the research program of Comprehensive Support Technology for Railway Network Operation (2018YFB1201403), which is a sub-project of Advanced Railway Transportation Special Project belonging to the 13th Five-Year National Key Research and Development Plan funded by Ministry of Science and Technology of China.

References

1. Wen C, Li Z, Lessan J, Fu L, Huang P, Jiang C (2017) Statistical investigation on train primary delay based on real records: evidence from Wuhan-Guangzhou HSR. Int J Rail Transp 5 (3):170–189
2. Xu P, Corman F, Peng Q (2016) Analyzing railway disruptions and their impact on delayed traffic in Chinese high-speed railway. IFAC-PapersOnLine 49(3):84–89
3. Ghaemi N, Zilko A, Yan F, Cats O, Kurowicka D, Goverde R (2018) Impact of railway disruption predictions and rescheduling on passenger delays. J Rail Transp Plan Manag 8 (2):103–122
4. Chen WY, Yang JY, He SW (2018) Research on risk propagation laws of subway operation sudden accident. J Saf Sci Technol 14(7):86–91 (in Chinese)
5. Kecman P, Corman F, D'Ariano A, Goverde R (2013) Rescheduling models for railway traffic management in large-scale networks. Public Transp 5(1–2):95–123
6. Zhan S, Kroon L, Veelenturf L, Wagenaar J (2015) Real-time high-speed train rescheduling in case of a complete blockage. Transp Res Part B: Methodol 78:182–201
7. Qin Y (2016) Analysis and dynamic emergency management of rail transit emergencies. Science Press, Beijing (in Chinese)
8. Liu SF, Deng JL (2000) The range suitable for GM(1, 1). Syst Eng Theory Pract 5:121–124 (in Chinese)
9. Ning XX, Liu SF (2009) Management forecasting and decision-making methods. Science Press, Beijing (in Chinese)

Health Management of Subway Bogie System Based on Multiplex Network and Dempster–Shafer Evidential Theory

Xiang Li, Yong Qin, Zhipeng Wang and Ning Wang

Abstract Subway bogie is an important part of subway. It plays a vital role in the safe and comfortable operation of the subway. Therefore, the reliability analysis of subway bogies has become the focus of attention. In order to solve the problem that different connections between components cannot be well reflected when using the network-based method to analyze the importance of the system, we propose an approach based on multiplex network to analyze the importance of the subway bogie system. After building a multiplex network, we consider the importance of the components by solving the values of degree, closeness centrality and eigenvector centrality. Then, for the problem of too many importance indicators, we use D-S evidential theory to fuse these indicators into a comprehensive importance indicator. Finally, we sort the components according to the comprehensive importance index. The result of the sort will become the basis for our daily maintenance.

Keywords Bogie · Multiplex network · Dempster–Shafer evidential theory · Reliability analysis

1 Introduction

In recent years, with the rapid development of Chinese cities, people's dependence on the subway is also increasing. Subway bogie system is one of the important components of rail vehicles, it can not only ensure the safe operation of vehicle, but also improve the stability of vehicle operation.

Nowadays, researchers have done a lot of research in this field. S. Kabir provides an overview of fault tree analysis [1]. Pickard et al. introduced a useful method to analyze multiple failures for complex systems simultaneously [2]. Jin Zhou et al.

X. Li · Y. Qin · Z. Wang (✉) · N. Wang
State Key Laboratory of Rail Traffic Control and Safety,
Beijing Jiaotong University, 100044 Beijing, China
e-mail: zpwang@bjtu.edu.cn

Y. Qin et al. (eds.), *Proceedings of the 4th International Conference on Electrical and Information Technologies for Rail Transportation (EITRT) 2019*, Lecture Notes in Electrical Engineering 639, https://doi.org/10.1007/978-981-15-2866-8_53

propose a railway faults spreading model which improves the SIR model and made it suitable for analyzing the dynamic process of faults spreading [3]. Yanhui Wang et al. propose a network model of urban rail transit operation safety impact factors based on complex networks [4].

2 Multiplex Networks

The steps of the importance analysis method based on the traditional network model are as follows:

Step 1: Building a single-layer network. Take the different components in the system as nodes of the networks, and the connect edge between nodes is determined by the connection between the components.
Step 2: Importance analysis based on the constructed network.

However, the reliability analysis method based on the traditional network model does not consider the difference between the components. If multiple connections between nodes are considered to have the same status and role, it is likely that the results of the research will be wrong. In view of the problems existing in the traditional network, we propose a new method based on multiplex networks for importance analysis. Multiplex networks can better describe the complex characteristics of systems than single-layer networks [5].

2.1 Introduction of the Multiplex Networks

We first build many single-layer networks. Different single-layer networks together form a multiplex network. The nodes in each layer of the multiple networks are the same, but the connections of each layer are different. Figure 1 is the schematic diagram of multiplex networks.

In the figure, G represents each layer of network in multiple networks. Proj(M) is a combination of all layers. The solid line is the inner connect edge, which represents the connection between different parts, and it both has weights and direction. The dotted line is the connect edge between the layers.

The connect relationship of each layer of the multiplex networks can be represented by an adjacency matrix. Assuming that the amount of nodes is a, then the adjacency matrix can be expressed as a matrix $A^{a \times a} = a_{ij}$. Where i represents the starting point of connection, j represents the end of the connection and a_{ij} is the weights of the connect edge between i and j.

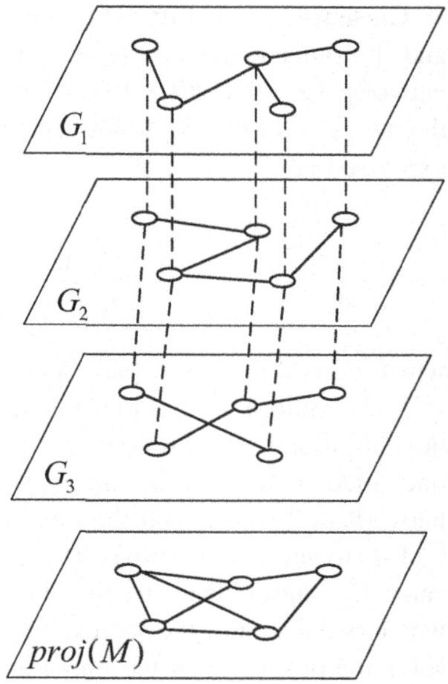

Fig. 1 Schematic diagram of multiplex networks

2.2 *Importance Indicators*

After drawing the multiplex networks, we need to calculate the importance of each point and rank them. The importance indicators of multiplex networks are degree, clustering coefficient and eigenvector centrality.

Degree. The degree of a node means the number of edges connected from the considered node to other nodes. The degree value can reflect the influence of other nodes when the considered node fails. The formula for calculating degrees in the single-layer network is:

$$k_i = \sum_{j=1}^{N} a_{ij} \tag{1}$$

where N is the number of nodes in the network, and k_i is the degree of node i.

To the multiplex network, the degree can be defined as the sum of degrees of each layer, the formula is:

$$k_i = k_i^{[1]} + k_i^{[2]} + \cdots + k_i^{[m]} \tag{2}$$

The degree can reflect the importance of a node intuitively and simply. However, the degree thinks that the importance of the nodes connected with the considered node is the same which does not match the fact. Therefore, the degree cannot accurately express the importance of nodes [6].

Closeness Centrality. The sum of the shortest path between the considered node and the other nodes can reflect the location of the considered node in the entire network. The smaller the sum of the shortest paths is, the greater the importance of the nodes is. In the single-layer network, the formula of closeness centrality can be expressed as [7]:

$$M = (N - 1)\bigg/\sum_{j}^{N} d_{ij} \tag{3}$$

where d_{ij} is the shortest path between node i and node j.

In the multiplex networks, the definition of the closeness centrality is the same as the definition in the single-layer network. When considering the shortest path from one node to another in multiplex networks, we should consider the multiplex network as a whole, and the inter-layer connections are paths.

Eigenvector Centrality. The eigenvector centrality is another important indicator. Its central idea is to give the considered node in the network a relative score, which is the sum of the scores of the neighbor nodes of the considered node. It can solve the problem that the importance of nodes connected to the considered nodes cannot be taken into account. In single-layer network, the formula of the eigenvector centrality is

$$E = \frac{1}{\lambda}\sum_{j=1}^{N} a_{ij}x_j \tag{4}$$

where λ is the maximum eigenvalue of the adjacency matrix, and x is the eigenvector corresponding to the maximum eigenvalue of the adjacency matrix.

Recently, there have been many methods to extend the calculation method of eigenvector centrality in single-layer complex networks to multiplex networks. Among them, the simplest method is to consider the eigenvector centrality of each layer separately. After calculating the eigenvector centrality of each layer, we can get the eigenvector centrality of the multiplex network:

$$E_i^T = \left(E_i^{[1]^T} \middle| E_i^{[2]^T} \middle| \cdots \middle| E_i^{[m]^T}\right) \tag{5}$$

After obtaining the eigenvector centrality of each node in the multiplex network, we can compare and sort by summing them.

3 D-S Evidential Theory

There are many indicators for evaluating the importance of nodes in complex networks. They portray the importance of nodes from different angles. They have certain advantages, but at the same time, there are some defects. In order to evaluate the importance of the nodes comprehensively, we use D-S evidential theory to combine the multiple indicators. Next, we will introduce several important concepts of evidence theory [8–10].

3.1 Identification Framework

Identification framework, θ is composed of a series of complete and mutually exclusive discriminant assumption which represents a collection of all possible propositions for a problem. Its form of expression is as follows:

$$\theta = \{\theta_1, \theta_2, \theta_3, \ldots, \theta_n\} \tag{6}$$

The set which is composed of all subset of the identification framework θ is the power set of θ.

$$2^{\theta} = \{\emptyset, \{\theta_1\}, \{\theta_2\}, \ldots, \{\theta_1, \theta_2\}, \ldots, \theta\} \tag{7}$$

3.2 Basic Probability Assignment Function

Assuming that θ is an identification framework, the elements in the 2^{θ} are endued with value between 0 and 1, which is $m(X) : 2^{\theta} \to [0, 1]$.

$m(X)$ needs to meet the following conditions: (1) $m(\emptyset) = 0$ means that the impossible events are endued with 0; (2) $\sum_{X \in 2^{\theta}} m(X) = 1$ means that the sum of the assumptions' basic probability allocation is 1, which ensures the normality of all hypothetical trust assignments.

3.3 Combination Rule

Assuming that the E_1 and E_1 are two evidence under the identification framework. The basic probability assignment functions of them are m_1 and m_2.

$$k = \sum_{A_i \cap B_j = \emptyset} m_1(A_i) m_2(B_i) < 1 \tag{8}$$

$$M(A) = \frac{\sum_{A_i \cap B_j = A} m_1(A_i) m_2(B_j)}{1 - k} \tag{9}$$

The above formula reflects the rule of combining two evidences, $1/1-k$ is called regularization parameter. If m is used to represent the probability assignment function, the orthogonal sum of two probability assignment functions m_1 and m_2 is recorded as $m_1 \oplus m_2$. When $k = 1$, it is considered that m_1 and m_2 are contradictory. Under these circumstances, we cannot synthesis basic probability assignments by combination rule; when $k \neq 1$, the combination rule is effective. Combination rules satisfy exchange laws and conjunction laws, etc. We can use the combination law to combine more than two pieces of evidence [11].

The steps of the multi-index comprehensive evaluation algorithm based on D-S evidential theory are as follows:

Step 1: Standardizing each indicator;
Step 2: Calculating the fault distribution function m corresponding to each indicator;
Step 3: We can get the comprehensive evaluation index by combining multiple pieces of evidence.

We can use this method to get the comprehensive importance indicator of the nodes in the multiplex network. The larger the indicator is, the greater the importance is.

4 Instance

Table 1 shows the components of the bogie and their numbers [12, 13].

To the bogie system, in addition to mechanical connections there are electrical and informational connections between components, so we build a three-tier complex network as shown in Fig. 2.

After building the complex network of bogie system, we can calculate the degree, closeness centrality, eigenvector centrality and comprehensive importance indicator of each node. In this case, the weight of the connected edge between each node is one. Table 2 shows the calculated indicators of each component.

Table 1 Bogie components

Number	Component	Number	Component
1	Gearbox	2	Architecture
3	Coupling	4	Wheel
5	Axle box	6	ATC support
7	Air spring (second spring)	8	Rim lubrication device
9	Traction rod	10	First spring (herringbone spring)
11	Anti-rolling torsion bar	12	Traction motor
13	Junction box	14	Temperature sensor

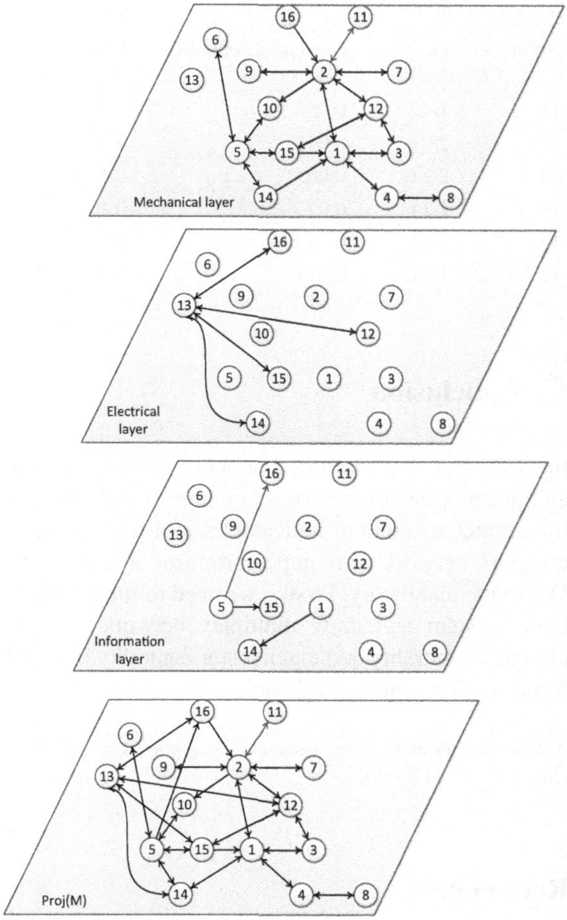

Fig. 2 Complex network of bogie system

Table 2 Probability distribution function *m* of indicators and comprehensive importance of nodes

Number	Degree	Closeness centrality	Eigenvector centrality	Comprehensive importance indicator
1	3/42	0.083	0.061	0.062
2	1/7	0.083	0.093	0.190
3	1/21	0.057	0.050	0.023
4	1/21	0.059	0.026	0.013
5	1/7	0.069	0.045	0.076
6	1/42	0.048	0.000	0.000
7	1/42	0.054	0.033	0.007
8	1/42	0.041	0.009	0.002
9	1/42	0.054	0.033	0.007
10	1/42	0.048	0.006	0.001

(continued)

Table 2 (continued)

Number	Degree	Closeness centrality	Eigenvector centrality	Comprehensive importance indicator
11	1/42	0.054	0.033	0.007
12	2/21	0.071	0.143	0.167
13	2/21	0.065	0.130	0.139
14	1/14	0.07	0.104	0.090
15	2/21	0.073	0.132	0.158
16	1/21	0.071	0.100	0.058

5 Conclusion

In this paper, the multiplex network is used to analyze the subway bogie system, and its components are sorted by importance. Aiming at the problem that the single importance evaluation indicator cannot fully reflect the importance of the nodes in complex network, this paper proposes a multi-indicator fusion method based on D-S evidential theory. Firstly, we need to fully understand the structure of the subway bogie system and draw multiplex networks. Secondly, we calculate the degree, closeness centrality and eigenvector centrality of the nodes. Finally, the indicators are fused by D-S evidential theory.

Acknowledgements This research is supported by the National Key R&D Program of China (No. 2016YFB1200402).

References

1. Kabir S (2017) An overview of fault tree analysis and its application in model based dependability analysis. Expert Syst Appl 77:114–135
2. Pickard K, Müller P, Bertsche B (2005) Multiple failure mode and effects analysis: an approach to risk assessment of multiple failures with FMEA. In: Reliability and maintainability symposium, pp 457–462
3. Zhou J, Xu W, Guo X, Ma X (2015) Railway faults spreading model based on dynamics of complex network. Int J Mod Phys 29:150038
4. Wang Y, Li M, Shen R (2013) Research on evaluation method and application of safety degree of urban rail operation based on emerging and entropy. J China Railw Soc 35(4):1–8
5. Li W (2016) Operational safety assessment method of high-speed train and the case study. Beijing Jiaotong University (in Chinese)
6. Ren XL, Lü LY (2014) Review of ranking nodes in complex networks. Chin Sci Bull 59 (13):1175–1197 (in Chinese)
7. Agostino GD, Scala A (2014) Networks of networks: the last frontier of complexity. Springer, Berlin
8. Wu G, Fang L-g, Li Z (2016) Node importance estimation in complex networks based on multi-index comprehension. Comput Eng Des 37(12):3146–3150 (in Chinese)

9. Yager RR (2009) Comparing approximate reasoning and probabilistic reasoning using the Dempster-Shafer framework. Int J Approx Reason 50(5):812–821
10. Basir O, Yuan X (2005) Engine fault diagnosis based on multi-sensor information fusion using Dempster-Shafer evidential theory. Inf Fusion 8(4):379–386
11. Wang Y (2014) Research of motor fault diagnosis based on D-S evidential theory and Bayesian network. Shenyang University of Technology (in Chinese)
12. Zeng Q (2015) Construction and solution of subway bogie configuration model. Southwest Jiaotong University (in Chinese)
13. Zhu S (2000) Guangzhou metro bogie. Rolling Stock 67–68 (in Chinese)

Research on Environment Perception and Obstacle Detection for Unmanned Vehicle Based on Machine Vision

Rui Wang, Yong Chen and Hugao Wang

Abstract The Technology of Environment Perception and Obstacle Detection for Unmanned Vehicle Based on Machine Vision is a technology that senses the rail, obstacle, and surrounding environment based on the multi-sensor system. In this paper, a new machine vision technology for automatic vehicles is proposed. Through the perception of surrounding environment by machine vision, the obstacles in front of the vehicle can be detected, so that the vehicle can locate the obstacle and compare it with the gauge. Thus, the vehicle can classify obstacle by its type and size in order to achieve autonomous driving.

Keywords Machine vision · Obstacle detection · Rail transportation · Vehicle gauge · Autonomous driving · ATO

1 Introduction

Rail transit system refers to the transportation system that operates vehicles on a specific track [1]. At present, the mainstream technology for rail transit system that senses the surrounding environment in China is still in an embryonic stage. The application of image recognition technology and multi-sensor fusion technology for urban rail transit system is relatively rare. Thus, the application of image recognition for urban rail system is still in the research and development stage. Currently, the main obstacle detection methods include non-contact detection method and contact detection method installed on the steering mechanism [2]. However, both of these two systems have some drawbacks.

With the development of the technology of machine learning and artificial intelligence, how to senses the surrounding environment through machine vision has become an urgent problem for intelligent transportation. Therefore, this paper

R. Wang (✉) · Y. Chen · H. Wang
The State Key Laboratory of Heavy Duty AC Drive Electric Locomotive System Integration, 412001 Zhuzhou, Hunan, China
e-mail: wangrui199111@126.com

Y. Qin et al. (eds.), *Proceedings of the 4th International Conference on Electrical and Information Technologies for Rail Transportation (EITRT) 2019*, Lecture Notes in Electrical Engineering 639, https://doi.org/10.1007/978-981-15-2866-8_54

demonstrates how binocular stereo vision or multi-camera stereo vision combined with multi-sensor fusion technology can be used to detect obstacles, to detect the distance and surrounding environment of obstacles and to provide the data for implementing traction, braking, and other functions of vehicles. Finally, this technology will be the bedrock for realizing the fully self-perception, self-recognition, and self-driving.

2 Automatic Driving and Rail Obstacle Detection

2.1 Unattended Train Operation System

Unattended train operation technology is a vehicle technology that senses the surrounding environment based on environmental awareness to control the vehicle through the motion planner and decision maker layer and finally reaches the destination. It is the highest level of automatic driving technology for planning the train at present [3]. As early as the 1970s, automatic train control (ATC) was introduced into the metro design. ATC system is divided into fixed blocks and mobile mode; fixed blocks are divided into a speed code system and distance-based system. Fixed ATC system is an automatic block system based on traditional line condition. This block system mainly determined by the line conditions where the corresponding traction plan can be formulated. Once the line condition is determined, it would generally remain unchanged [4]. The block area is taken as the minimum dividing unit, and the ATC system realizes the automatic control of vehicle command and operation-based block area. Moving block transmits signals to the vehicle controller through wireless communication, induction ring, and microwave signals. Thus, the train operation can synchronize the vehicle status according to the information of line conditions, vehicle status, and vehicle parking position.

With the opening of Beijing Yanfang Line and the designing of Shanghai Metro Line 18, it marks the mature of automatic train operation of urban transit system from with staff to without staff. It can be predicted that more and more unmanned vehicles will be put into operation shortly.

2.2 Obstacle Detection Method

At present, there are mainly two kinds of obstacle detection methods for the urban transit system; one is the contact obstacle detection method, and the other is a non-contact obstacle detection method [5].

Contact obstacle detection method. Contact obstacle detection method is a method that mainly through the detection device installed in front of the vehicle to detect the obstacle. With the pressure sensors and other sensors, the contact obstacle detection can identify obstacles in front of the vehicle, remove small obstacles, and detect large obstacles. Thus, the contact obstacle detection system can provide basic data for braking and support the vehicle to prevent train derailment and to reduce losses. At present, the main contact obstacle system has been widely used in the vehicle under the automatic train operation and unattended train operation. In recent years, with the requirement for unmanned rail transit vehicle, major railway cooperation use bogie detection devices installed in the front of the vehicle to remove obstacles and send signals when detecting large obstacles and starting the automatic braking system of trains. However, the detection system installed on the bogie detects the obstacle by contacting the obstacle, which has a short detection distance and does not have the function of early warning. Therefore, it is difficult to make a prediction and reduce losses, and it is also difficult to meet the requirement of unmanned driving in the era of unmanned and intelligent.

Non-contact obstacle detection method. Non-contact obstacle detection method mainly uses a multi-sensor system to measure the distance of the obstacle by receiving the image signal, the microwave signal, and the laser signal for providing basic data for braking and traction. In 2013, Beijing Institute of Technology began to develop an autonomous non-contact detection system, which can implement anti-collision radar system [6]. In recent years, some major metro companies in China have begun to use non-contact obstacle detection. For example, the vehicle in Shenzhen Metro Line 10 combines vision technology and secondary microwave to identify obstacles in front of the vehicle. The vehicle in Shanghai Metro Line 7 implements radar and vision technology to identify obstacles ahead comprehensively. However, non-contact obstacle detection technology is a new obstacle detection technology whose robustness and accuracy are not high. However, compared with the complex urban environment, the surrounding environment of the rail transit system is simpler. The surrounding environment is shown in Figs. 1 and 2.

Fig. 1 Surrounding environment of the tunnel line

Fig. 2 Surrounding environment of the non-tunnel line

As shown in Figs. 1 and 2, the surrounding environment of the mass transit system is relatively simple. It can be predicted that with the further development of technology and the improvement of intelligent level, non-contact obstacle detection method will be more widely used in the field of the mass transit system in the future.

3 Surrounding Environment Perception

3.1 Pattern Recognition, Matching, and Distance Measuring

The image taken by the camera can be preprocessed in two ways, one is to convert the image into a binary image which is an image which has only two possible values, the other is to process the pixel information of the image through RGB model. The RGB model is a model for image segmentation based on additive colors (red, green, blue). RGB is among the most important model because it is what our computer monitors use to display images. Thus, it has been widely used in computer vision.

Further, edges need to be detected for edge detection. At present, the mature method is to detect the edge by describing the data brightness for considering the first derivative and second derivative. The first derivative is used to calculate the gradient of data brightness, and the second derivative is used to calculate the rate of change of brightness gradient. At present, the operators used for detection include: first-order: Kirsch operator, Prewitt operator, second-order: Canny operator, Laplacian operator, and so on. The most widely used operator methods include the Kirsch operator, the Prewitt operator, and the Canny operator, and other methods include median filtering and so on [7].

Kirsch operator. R. Kirsch proposed the Kirsch operator to detect edges. As presented in Fig. 3, the gradient magnitude and the gradient direction are determined by using 3 × 3 templates or 5 × 5 templates in this direction. The calculated form is shown as below:

$$K_n(x,y) = \sum_{i,j} M_n(i,j) \times V(x+i, y+j) \tag{1}$$

where $K_n(x, y)$ is magnitude value, $M_n(i, j)$ is weight value, and $V(x + i, y + j)$ is pixel value.

Where compass direction kernels M could be determined by eight compass directions. By weighting the formulas, eight or twenty-four K values are found to determine $K_{\max}$, $K_{\max} = \max(K_1(x,y), K_2(x,y), K_3(x,y), \ldots)$.

By setting the threshold T_v, we could determine whether $K_{\max} > T_v$. If the condition is satisfied, regard it as the edge point where edge value is 1.

The advantage of the Kirsch algorithm is that it can keep the value of 360° in each direction. Thus, it has a good effect in detail preservation and anti-noise, but the continuity of edges is poor, and clutter is easy to occur.

Canny operator. Canny edge detection operator is a multi-level edge detection algorithm, which is mainly used for edge matching and detection. The goal of the Canny operator is to find an optimal edge detection algorithm. Detection steps are shown as follows:

Grayscale. The algorithm of the grayscale image is determined as the following:

$$\text{Gray} = P_R \times R + P_G \times G + P_B \times B \tag{2}$$

V_1	V_2	V_3
V_8		V_4
V_7	V_6	V_5

V_1	V_2	V_3	V_4	V_5
V_{16}	V_{17}	V_{18}	V_{19}	V_6
V_{15}	V_{24}		V_{20}	V_7
V_{14}	V_{23}	V_{22}	V_{21}	V_8
V_{13}	V_{12}	V_{11}	V_{10}	V_9

Fig. 3 Templates of Kirsch operator (left side is a 3 × 3 template, and the right side is 5 × 5)

where P_G is the red coefficient of the Input image, R is the proportion of red in the input image, P_G is green coefficient of the input image, G is the proportion of green in the input image, P_B is blue coefficient of the input image, and B is the proportion of blue in the input image.

Gauss filtering. The algorithm of one-dimensional Gauss filtering:

$$K = \frac{1}{\sqrt{2\pi\sigma}} e^{-\frac{x^2}{2\sigma^2}} \tag{3}$$

where K is pixel value, x is the value of x coordinate, and σ is standard deviation.

The algorithm of two-dimensional Gauss filtering:

$$K = \frac{1}{2\pi\sigma} e^{-\frac{x^2+y^2}{2\sigma^2}} \tag{4}$$

where K is the pixel value, x is the value of x coordinate, y is the value of y coordinate, and σ is the standard deviation.

Non-maximum suppression. In order to find the local maximum of the pixel, the gradient values can be compared along the gradient direction. Linear interpolation and nonlinear interpolation can be implemented.

Double Threshold Analysis. The color threshold is set to filter clutter signal or noise. Take the canny operator as an example. The Canny operator sets double thresholds, which are a high threshold and a low threshold to distinguish edge pixels. When the gradient value of edge pixels is greater than the high threshold, edge pixels can be considered as the strong edge points. When the gradient value of edge pixels is smaller than a high threshold and greater than the low threshold, edge pixels can be considered to be weak edges. When the gradient value of edge pixels is smaller than the low threshold, edge pixels can be considered as a cluster.

Prewitt operator. Prewitt operator is a first-order detection operator which mainly uses the gray difference between the upper and lower, left and right adjacent points to determine the edge. The threshold T is set at the edge, and the false edge is removed to eliminate the noise.

4 Obstacle Detection and Automatic Train Operation Based on Machine Vision

4.1 *Automatic Train Operation Based on Machine Vision*

When a train moves on the track, it is necessary to detect the track and obstacles automatically and brake according to the location and size of the obstacles. Usually, for rail transit, we think that objects intruding into the gauge act as obstacles, so we need to compare the relationship between obstacles and gauge.

Image Capturing. The threshold segmentation method is used to analyze the rail transit system under different conditions. Firstly, the image is captured [8]. Vehicles can identify the signal light and traffic sign by a camera installed in front of the vehicle. Vehicles are moving forward, as shown in Fig. 4.

The image taken in front of the vehicle is shown in Fig. 5.

When the vehicle captures the image in front of it, the figure needs to be recognized. Take the signal light in front of it as an example; the image is gray processing, as shown in Fig. 6.

After implementing operator to complete gray processing, the signal light can be extracted. Considering that the surrounding environment of the subway is relatively simple, the clutter mainly comes from the surrounding fence and the antenna, the fire hydrant, and pillar near the track. Choosing the region of interest or the filter

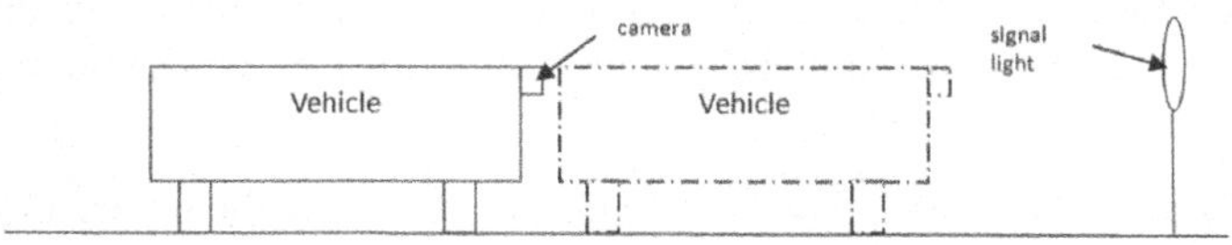

Fig. 4 Diagram of vehicle moving (the solid line is the original position of the vehicle, and the dotted line is the final position of the vehicle)

Fig. 5 Image taken by the vehicle in a different location

Fig. 6 Image processed by gray processing

Fig. 7 Figure for signal light in a different location

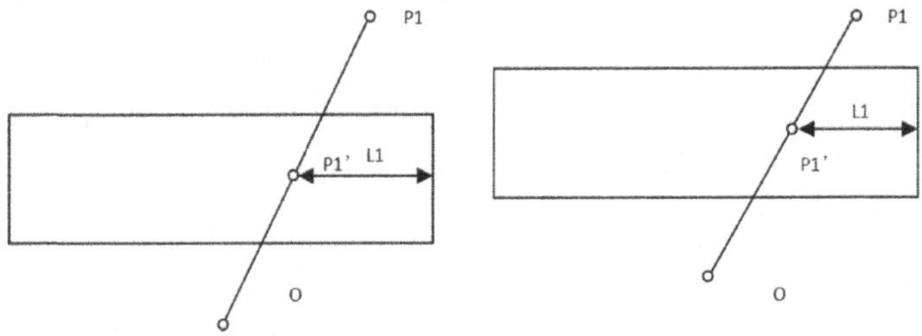

Fig. 8 Diagram of imaging theory and corresponding location

method can eliminate the interference signal directly. For example, for the filtering method, the clutter can be eliminated directly by setting the threshold as shown in Fig. 7.

Location Corresponding Theory. The camera can capture the information in front of the vehicle synchronously and display it on the image. The principle of the camera is shown in Fig. 8.

As shown in the figure, P_1 is the object point, and the point P_1' on the image is the image point. When the vehicle is moving forward, the image point P_1' moves forward to the centerline, and the length of L_1 decreases. Because there are several signal lights or traffic signs on each section of the road, when the vehicle passes through a signal light or traffic signs, it can complete recording. Thus, the position of P_1 in the picture can be determined by P_1' under the condition of the whole road condition.

Information Comparing. Information comparing refers to the comparison between the image in memory, captured image by finding a specific traffic sign or signal light and then the corresponding position by analyzing the image. That is to say, by comparing the vehicle information L_{2O} from P_1 to image edge and the position L_{2E} from P_2 to image edge during driving, i is the area where the vehicle is located [9].

$$D_{2O} = f(L_{2O}) \tag{5}$$

$$D_{2O} - D_{2E} = N(L_{2O} - L_{2E})\forall N \in N^{+} \tag{6}$$

$$L_{2E} = L_{2O} \Rightarrow D_{2E} = D_{2O} \tag{7}$$

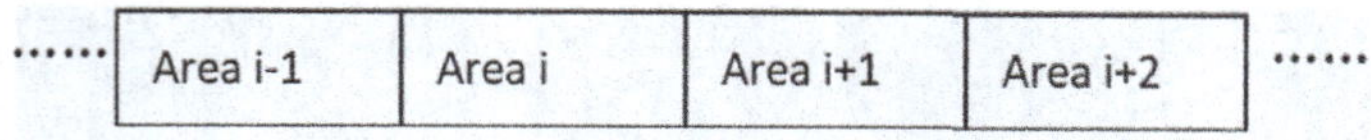

Fig. 9 Schematic illustration of the driving area partition

Fig. 10 Procedure of image capturing and processing

where L_{2O} is the length from P_2 to the edge in memory, L_{2E} is the length from P_2 to the edge when the vehicle enters the station, D_{2O} is the distance from the vehicle to traffic light in memory, D_{2E} is the distance from the vehicle to traffic light when the vehicle enters the station, and N is positive integer.

When there are no traffic signs in the vehicle identification area for some time, the vehicle is considered to have left the area, and $i = i + 1$ is recorded. Thus, the whole railroad is divided into N regions. By synthesizing the relationship between i and D, the position of the vehicle can be determined uniquely, and the driving behavior of the vehicle can be determined in the controller. The schematic illustration of the driving area partition is shown in Fig. 9.

4.2 *Rail Detection and Obstacle Detection*

When an obstacle appears in front of the vehicle during the vehicle drives based on the image processing in the railroad automatically, the obstacle can be found by machine vision. According to the size and distance of the detected obstacle, the traction and braking can be implemented by traction and braking system.

When a train moves on the rail, it is necessary to detect the track and obstacles automatically and implementing braking according to the location and size of the obstacles. As shown in the figure, the vehicle extracts the front obstacle by capturing the image, processing the image, and taking the sphere recognition on the track as an example. The procedure of image capturing and processing is shown as Fig. 10.

Usually, for rail transit system, we think that objects intruding into the gauge are regarded as obstacles, so we need to compare the relationship between obstacles and gauge. As shown in Fig. 11, after capturing the image, the vehicle detects the

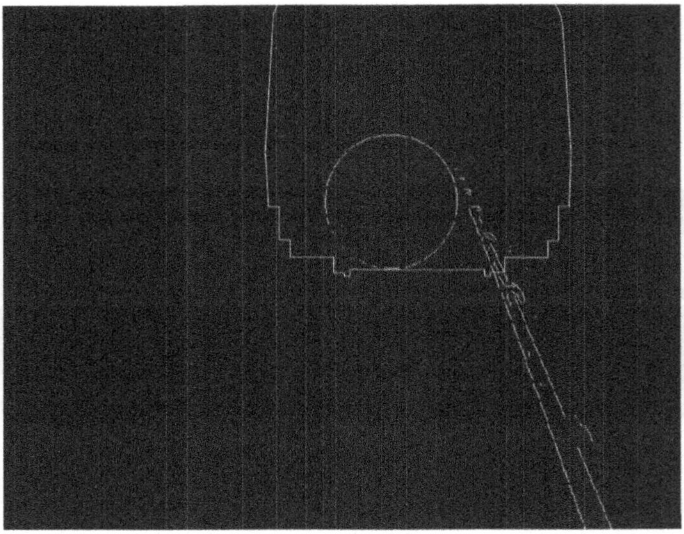

Fig. 11 Comparison between obstacle and gauge

track based on Canny or Prewitt operator and Hough line test. The gauge can be determined by rail and obstacle to locate the gauge section. According to the data stored in the vehicle, the distance between the obstacle and the vehicle can be determined, and the size and distance of the obstacle can be found. As the figure is shown, the farther the rail is, the closer the rail is near to the centerline of the image. Therefore, the equation can be constructed as $D = f_d(L)$ and $A_o = f_o(A_g)$ where D is the distance between the vehicle and obstacle, L is the distance between the vehicle and rail in the diagram, A_o is the size of the obstacle, G is the size of obstacle in gauge.

Take Fig. 11 as an example, the area of this sphere is 2.5 m^2, and the distance between the vehicle and the obstacle is 10.6 m.

Therefore, the braking instructions are implemented according to the distance between the vehicle and the obstacle and size of the obstacle intruding into the gauge. According to the rail transit criteria, when the actual size of the obstacle exceeds 30 cm × 30 cm × 30 cm, the speed of rail vehicle should decrease; when the actual size of the obstacle exceeds 60 cm × 60 cm × 60 cm, the rail vehicle will implement emergency braking. According to the condition of straight line and curve, the straight line can analyze directly, and the curve can be transformed into a straight line (which can be processed in segments) so that the braking distance can be determined.

5 Conclusion

This chapter introduces an automatic driving system base on machine learning for mass transit vehicle to detect the obstacle and to drive automatically. Through detecting the signal light and traffic sign in each section, the vehicle can determine the location of the vehicle and detect the obstacle based on the machine vision automatically. To determine the size of the obstacle and the distance between the vehicle and obstacle, this article put forward nonlinear mathematics model to compare with the vehicle gauge. According to the size and the distance of the obstacle, the traction system and braking system of vehicle implement the instruction of braking and traction, allowing the vehicle running stable. Currently, this method only put forward the distance algorithm and the detection operator while further optimization and filter method need to be implemented to improve the accuracy and robustness of this method. Furthermore, the experiment only considers the condition of the night; scene analysis must be considered in the future work to allow it commercialized. All in all, high-level unmanned mass transit vehicle is one of the parts of national 2025 planning, and the mass transit vehicle with machine vision and machine learning will be practical importance for commercial implementation of high-level unmanned mass transit vehicle in the future.

References

1. Jiang M (2019) Introduction to development and trend of rail traffic signal system. Railw Signal Commun Eng 101(5):93–98 (in Chinese)
2. Wei L(2013) A rail vehicle obstacle contact detection device. CN201310243984.1, Jun 2013 (in Chinese)
3. Zhang W, Dewang C, Zhenyu Y (2011) Performance evaluation of urban rail transit ATO system. Urban Rapid Rail Transit 24(4):26–29 (in Chinese)
4. Ke H (2008) Implementation of data query subsystem for automatic train operation. Beijing Jiaotong University, Beijing (in Chinese)
5. Haonan F, Kai F, Hongwei D, Hongwei D (2018) Research on interconnection framework of urban transit signal system. Urban Rapid Rail Transit 11:63–68 (in Chinese)
6. Haibo L, Mengmeng S, Xiaoqian Y (2013) A study of MMW collision avoidance radar system for trains. J Radars 02:234–238 (in Chinese)
7. Yanpeng H, Xiaolan J, Lin Y, Hao L, Ruiming H (2017) Evaluation of natural icing condition for in-service insulators based on image segmentation. High Volt Eng 43(1):285–292 (in Chinese)
8. Quan Y, Chong Z, Feng M (2017) Application of binocular vision range measuring in humanoid robots. J Wuhan Inst Technol 39(2):193–198 (in Chinese)
9. Rui W (2019) Research on benchmarking parking system of urban rail transit based on machine vision. Electr Drive Locomot 02:107–110 (in Chinese)

Study on Fault Diagnosis for Bearing Based on Hierarchical Extreme Learning Machine

Yakun Zuo, Limin Jia, Zhipeng Wang, Ning Wang and Xinan Chen

Abstract Rolling bearings are widely used in mechanical systems but have a high damage rate. Its running state is related to the production safety and stable operation of various industries. Nowadays, scholars have applied so many signal processing methods such as differential entropy, energy entropy, and empirical mode decomposition methods in conjunction with various algorithms which likes particle swarms and neural networks to implement pattern classification in the process of the vibration signals of rolling bearings (Qin et al. in Mech Des Manuf 08:11–14, 2018 [1]). On this basis of it, this paper presents the variational mode decomposition–singular value decomposition (VMD-SVD) method based on the previous studies by other scholars with good verification effect that is developed and used to extract the characteristics of different IMF components under different operating conditions in order to establish the characteristic matrix. The latest and better effect of hierarchical extreme learning machine (H-ELM) is applied for training and verification. Besides, by comparing with the traditional ELM method, it verifies its superiority in rolling bearing fault diagnosis.

Keywords Rolling bearing · SVD-VM · H-ELM

1 Introduction

In industrial production, there are often rolling bearings in the field of power, and as a type of important forced component, the damaged rate always be high. Under these circumstances, the research on the fault diagnosis of rolling bearings has never stopped.

The literature [2] proposed a composite diagnosis method for rolling bearings based on MCKD and Teager energy operators. The literature [3] introduced the

---

Y. Zuo · L. Jia · Z. Wang (✉) · N. Wang · X. Chen
State Key Laboratory of Rail Traffic Control and Safety, Beijing Jiaotong University, Beijing, China
e-mail: zpwang@bjtu.edu.cn

Y. Qin et al. (eds.), *Proceedings of the 4th International Conference on Electrical and Information Technologies for Rail Transportation (EITRT) 2019*, Lecture Notes in Electrical Engineering 639, https://doi.org/10.1007/978-981-15-2866-8_55

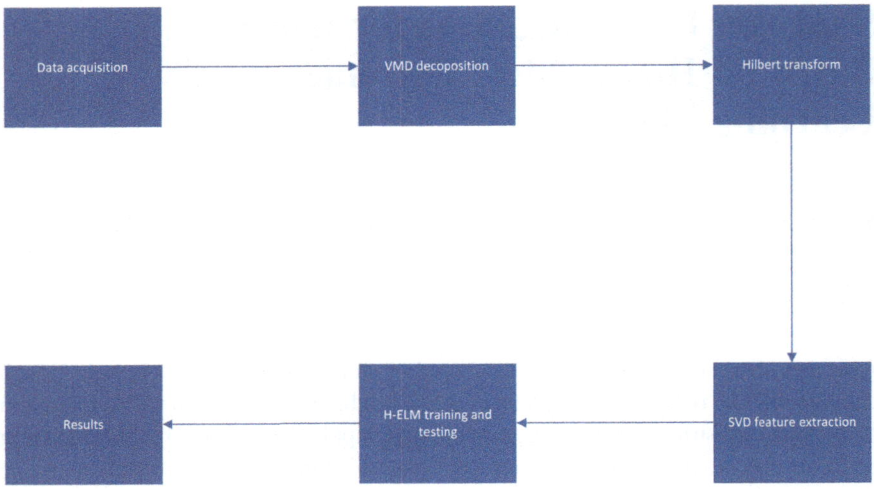

Fig. 1 Specific implementation process

widely used particle swarm optimization algorithm into the fault diagnosis of rolling bearings in recent years. The literature [4] proposed a fault diagnosis method for rolling bearings based on CEEMD-SVD and ELM. After comparing, the VMD method [5] derived from EMD [6] avoids the problem of modal aliasing, which shows better signal division accuracy in practical applications. Therefore, this paper selected VMD cooperation. Then, SVD [4] method was used to extract the features of the vibration signal. In the final fault identification, different fault identification methods affect the precision and speed, while fault diagnosis method based on the extreme learning machine had been verified the quickness and the effectiveness of it on the pattern recognition; this paper therefore put forward by the selection of the latest hierarchy extreme learning machine to make the final pattern recognition, and in the specific implementation of below, it demonstrates its superiority with the traditional extreme learning machine method in the identification of rolling bearing vibration signals. The specific implementation process is shown in Fig. 1.

2 Method

2.1 Hierarchical Extreme Learning Machine

The original hierarchical extreme learning machine [7] is a multilevel nesting mode based on the extreme learning machine. In the extreme learning machine algorithm, after specifying the input data and the output data and specifying the classification type of the data (0 regression or complex classification), it can adjust the weight and threshold of the neural network by itself, so as to achieve unsupervised learning; at

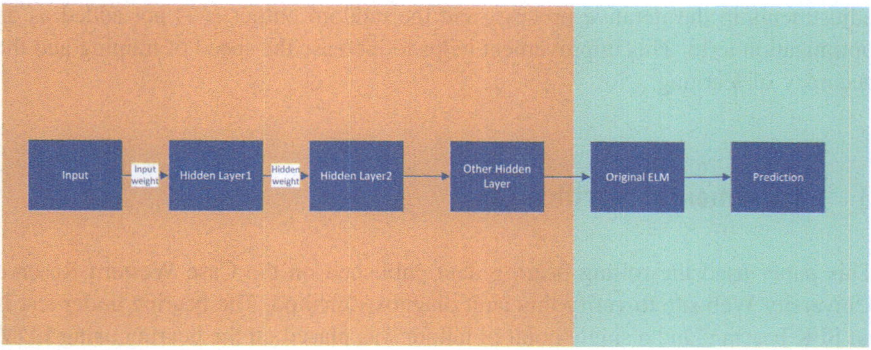

Fig. 2 Learning or testing structure

the same time, it satisfies the orthogonal condition when adjusting the weight and the threshold. The hierarchical extreme learning machine used in this paper improves the structure and ELM automatic coding method [8], and its structure is shown in Fig. 2.

It divides the entire network into two separate subsystems, considering parallelism in a more comprehensive way. Its features:

(1) Using the random projection of the feature extraction result as the input of the feature classification subsystem.
(2) Using the penalty factor instead of the 2-norm method in the simple cascaded limit learning, a more compact and sparse hidden information is obtained.
(3) Since the orthogonal constraint is unreasonable when the number of input nodes is different from the number of output nodes, orthogonal initialization of the ML-ELM is avoided [9].

It converts the input raw data into a random feature space before feature learning, which can effectively mine hidden information between training samples. Then, perform N-layer learning on the nodes to obtain advanced sparse features. The output of each hidden layer can be expressed as:

$$H_i = g(H_{i-1} \cdot \beta) \tag{1}$$

where H_i represents the output of each layer, g is the activation function, and β is the output weight. The optimized model of the improved sparse autoencoder is summarized as follows:

$$O_\beta = \arg\min_{\beta}\left\{ \|H\beta - X\|^2 + \|\beta\|_{\ell 1} \right\} \tag{2}$$

X represents input data, H represents a random map output, β represents hidden layer weights. Compared with the original DL algorithm, X is not necessary to make adjustments in the iterative process, and the random output H is not added as an optimization term. This improvement helps to increase the speed of training and the accuracy of learning.

3 Experimental Verification

This paper used the rolling bearing data published on the Case Western Reserve University Web site to verify this fault diagnosis method. The bearing under test is an SFK bearing, and a single point of failure was placed on the bearing using EDM technology. The fault diameter was chosen to be 0.007 mils, and the speed was 1772 rpm [10]. And the data of four states of rolling bearing including normal state, inner ring fault, outer raceway fault, and rolling unit fault were analyzed under the condition of the sampling frequency of 12 kHz. Each state selected 120,000 points, half of which was used as training data, the other half was used as test data.

Firstly, the sample data were read and the spectrogram was drawn. After observation, the main frequency is 4, so the modal decomposition number should be bigger than or equal to 5. After testing, it was found that the modal number set to 6 showed the best decomposition effect. The parameters of the VMD were then set, and the signal was input to obtain the decomposed IMF component. Considering the spectrum map, center frequency, and equipment speed, the first four IMF components containing the main fault information were selected, as shown in Figs. 3, 4, and 5.

Then, the Hilbert matrix was established for each IMF component to reduce the influence of the disturbance, and the feature extraction effect was enhanced, and then the matrix was decomposed into the SVD to extract the feature values.

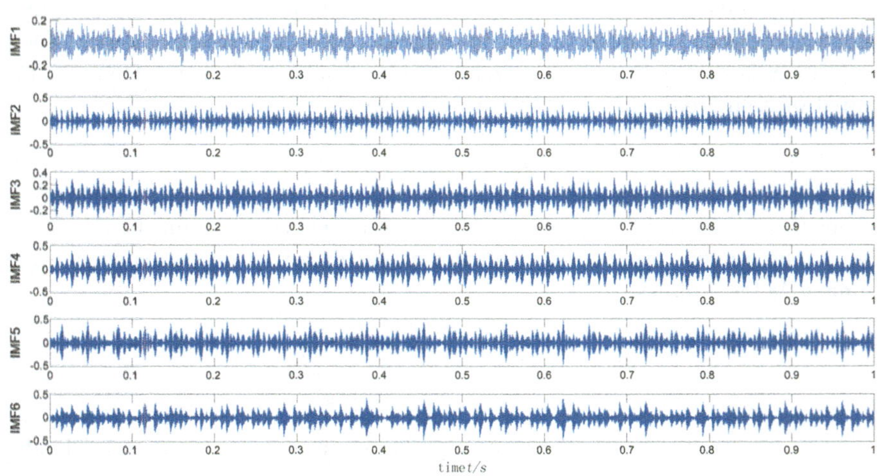

Fig. 3 Inner ring failure

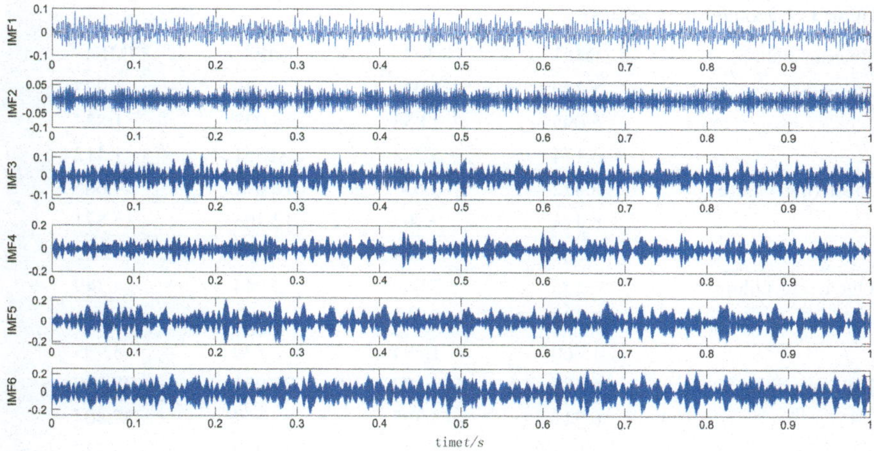

Fig. 4 Outer ring failure

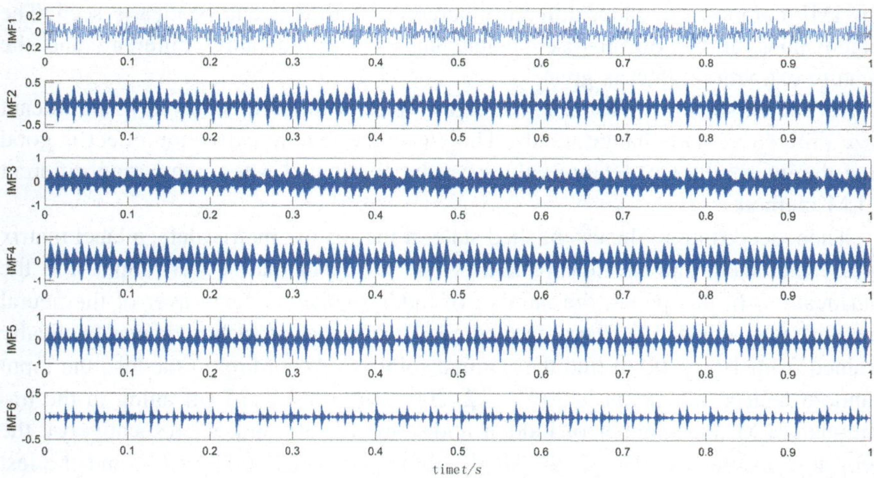

Fig. 5 Rolling unit failure

Since the feature matrix was established according to the number of values in many documents, it was found that the ELM and H-ELM have been able to achieve nearly 100% test accuracy through parameter adjustment, and in the test, the accuracy in H-ELM is similar to in the traditional ELM when the test points below 50. And both of them can maintain accuracy above 95% with a difference of less than 5%. Therefore, each group of 50 eigenvalues was selected, and a total of 200 points were used to establish a feature matrix. The following three sets of eigenvalues of vibration signals under different conditions were listed below. From this,

Table 1 Decomposition results

Bearing status	IMF1	IMF2	IMF3	IMF4
Normal	5.459	5.029	4.036	1.927
	5.095	4.858	4.027	1.834
	3.174	3.278	0.351	1.412
Inner ring failure	8.472	10.038	12.612	14.665
	8.414	9.962	12.108	13.871
	3.631	8.674	7.552	11.632
Outer ring failure	2.555	2.069	2.944	4.749
	2.495	2.021	2.711	4.283
	1.760	1.094	2.529	3.995
Rolling unit failure	6.385	18.180	38.951	38.009
	6.311	14.365	37.657	37.404
	4.834	13.743	15.979	33.537

the IMF components under different operating conditions can be clearly seen. The eigenvalues of the decomposed eigenvalues are significantly different, and the feature extraction effect is great.

As can be seen from Table 1, the eigenvalues in different states have significant size differences and change trends. Therefore, the feature extraction effect is good and the interference of feature values can be eliminated when comparing different ELM methods.

Then, according to the established state of the feature matrix data, a label matrix was established, and the feature matrix and the label matrix were input into the verification. In this paper, the number of hidden nodes in each layer of the neural network was adjusted by stepping method. The number of each node was cyclically trained from 10 to 100 to find the optimal solution. According to the test, the input random matrix dimension is 200×12, 200×30, 200×35 (meaning in the traditional ELM, the number of hidden nodes are 12, 30, and 35, respectively), the loop test accuracy is 97.5%, the single training time is 0.0023087 s, and the test time is 0.00078823 s. And then, when the activation function is tribas, and the number of hidden nodes is 300, the ELM test accuracy is 91.5%, and the single training total time is 0.0938 s.

The results show in Fig. 6 and Fig. 7 as follows.

The specific comparison results are shown in Table 2.

It can be seen from Table 2 that the improved H-ELM fault identification method has obvious advantages in terms of recognition accuracy and running time, compared with the traditional ELM, which fully demonstrates the superiority of the algorithm.

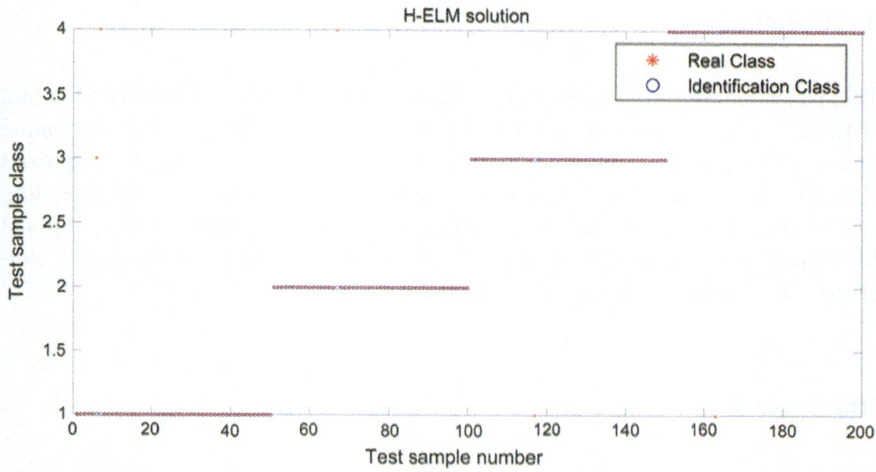

Fig. 6 H-ELM solution

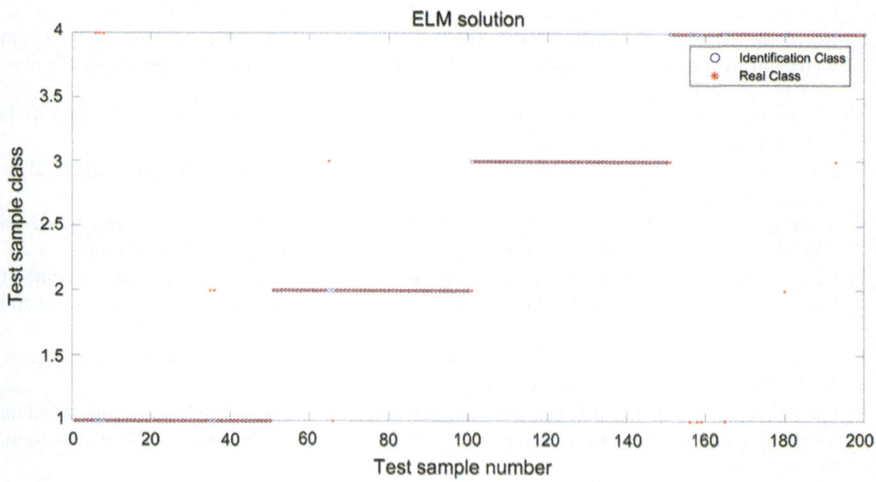

Fig. 7 ELM solution

Table 2 Comparative results

Type of network	The number of training samples	The number of testing samples	Running time	Classification accuracy (%)
H-ELM	200	200	0.002316523	97.5
ELM	200	200	0.0938 s	91.5

4 Conclusion

This paper introduced a bearing fault diagnosis method based on VMD-SVD and H-ELM. For the combined use of VMD and SVD, it is possible to obtain the feature matrix with an obvious difference in different states. And then, it used the improved H-ELM to train the characteristic data of the bearing under different operating conditions and achieved a higher classification accuracy. Compared with traditional ELM fault identification, H-ELM used in bearing fault diagnosis has higher precision, faster running speed, and fewer hidden nodes.

References

1. Qin B, Sun G, Wang J (2018) Application of H-K-ELM in fault diagnosis of rolling bearings. Mech Des Manuf 08:11–14 (in Chinese)
2. Qi Y, Liu F, Gao X, Li Y, Liu L (2019) Combined fault diagnosis of rolling bearing based on MCKD and teager energy operator. J Dalian Univ Technol (01):35–44. https://kns-cnki-net.webvpn.bjtu.e-du.cn/kcms/detail/21.1117.N.20190129.1715.012.html (in Chinese)
3. Lv M, Su X, Chen C, Liu S (2019) Application of improved particle swarm optimization algorithm for support vector machines in fault diagnosis of rolling bearings. Mach Electron (01):42–48 (in Chinese)
4. Wu M, Huang G, Zhou W (2018) Fault diagnosis of rolling bearing based on CEEMD-SVD and ELM. Chem Autom Instrum 45(10):784–788 (in Chinese)
5. He Y, Lu Y, Liu J (2019) Fault diagnosis of rolling bearing based on VMD and particle filter. Renew Energy 37(01):126–131 (in Chinese)
6. Yang B, Zhang J, Wang J, Zhang C (2019) Early diagnosis of rolling bearings based on CEEMD and adaptive MCKD. J Beijing Polytech Univ 45(02):111–118 (in Chinese)
7. Huang GB, Zhu QY, Siew CK (2004) Extreme learning machine: a new learning scheme of feedforward neural networks. In: 2004 IEEE international joint conference on neural networks, 2004. Proceedings, vol 2. IEEE, pp 985–990
8. Tang J, Deng C, Huang G-B (2015) Extreme learning machine for multilayer perceptron. IEEE Trans Neural Netw Learn Syst 27:809–821
9. Qin B, Sun G, Wang J (2017) Research on fault diagnosis method of rolling bearing based on VMD and layered extreme learning machine. Comb Mach Tool Autom Process Technol (4) (in Chinese)
10. Case Western Reserve University Bearing Data Center. Bearing Data Center fault test data. http://csegroups.case.edu/bearingdatacenter/pages/12k-drive-end-bearing-fault-data. 10 Feb 2019

Design and Implementation of Simulation Software of DC Traction Power Supply System for Urban Rail Transportation

Yangyang Song, Bo Wu, Jie Chen, Ruichang Qiu and Zhigang Liu

Abstract For the power flow calculation of the DC side traction power supply system for urban rail transportation, the DC traction calculation is carried out first of all, then the mathematical model of each part of the DC side is established to give the equivalent model of the DC traction power supply system, and then the DC power flow calculation method based on Newton–Raphson method is derived. As for software development, using the C# language in the Visual Studio 2013 environment to get the overall design of the software, the software can visualize the modeling and simulation results and finally combine the actual lines to perform the operation of the interface, and display the results.

Keywords Urban rail transportation · DC traction power supply system · Simulation software

Y. Song (✉) · J. Chen · R. Qiu · Z. Liu
School of Electrical Engineering, Beijing Engineering Research Center of Electric Rail Transportation, Beijing Jiaotong University, 100044 Beijing, China
e-mail: 18121490@bjtu.edu.cn

J. Chen
e-mail: jiechen@bjtu.edu.cn

R. Qiu
e-mail: rchqiu@bjtu.edu.cn

Z. Liu
e-mail: zhgliu@bjtu.edu.cn

B. Wu
Taiyuan Railway Transit Development CO., LTD, Taiyuan, China
e-mail: 94586124@qq.com

Z. Liu
Beijing Jiaotong University, Beijing, China

Y. Qin et al. (eds.), *Proceedings of the 4th International Conference on Electrical and Information Technologies for Rail Transportation (EITRT) 2019*, Lecture Notes in Electrical Engineering 639, https://doi.org/10.1007/978-981-15-2866-8_56

1 Introduction

At present, China's urban rail transit construction is in full swing, and the DC side traction power supply system is an important part of urban rail transit. Simulation software such as MATLAB is used in the simulation of urban rail transportation traction power supply system, but during the application process, the graphics of the software were found to be too weak. In response to this situation, our team is dedicated to the simulation software of urban rail traction power supply system [1]. In this team, the author is mainly responsible for the design of DC side traction power supply system simulation software.

When establishing the simulation model, the output of the traction calculation will be used as the raw data for the analysis of DC traction power supply system for urban rail transportation. The traction calculation is required to output the running distance of the train as a function of time and the relevant data such as electric power received from the contact network [2].

For the simulation software design of the DC traction power supply system for the urban rail transportation, firstly, the key part of the DC side is mathematically modeled, and then, the equivalent model of the DC side traction power supply system is obtained. Then use the Newton–Raphson method to write the DC power flow calculation algorithm and finally display the power voltage of each node.

In the software design, the function of the train simulation software is divided by modular method to provide a good user interface, rich configuration parameters, and basic data recording and analysis functions. I will introduce the preparation process of the traction power supply system simulation software of the DC side below.

2 Modeling of DC Traction Power Supply System

The DC traction power supply system includes the traction network, the traction substation, the rectifier unit, the energy feedback device, and the train. They coordinate and unify with each other [3] as shown in Fig. 1.

The traction substation converts the AC high voltage to DC 750 V or DC 1500 V through the rectifier unit and the energy feedback device and then supplies power to the train through the traction network or contact rail. The feeder wire sends the direct current of the traction substation to the catenary. The catenary is a special power supply line erected along the train travel track, which may be the over-head contact network or the ground contact rail. The train obtains electricity through the catenary. The rail is also an important part of the traction network.

Regarding the modeling of the traction network, the catenary and the rail are replaced by different resistance [4]. The modeling of the rectifier unit is using piecewise linearization, and the equivalent model is an ideal voltage source connected in series with internal resistance and diode. Regarding the modeling of the

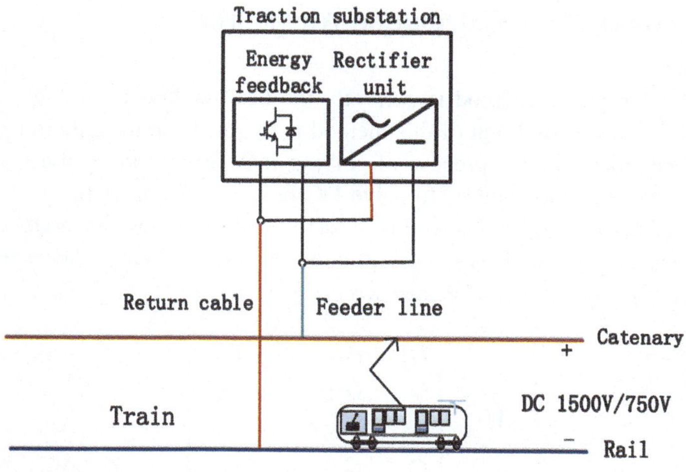

Fig. 1 DC traction power supply system schematic

train, the equivalent power of the train is independent of the voltage fluctuation of the catenary, but its equivalent current varies with the pressure of the catenary [5]. Therefore, the train is equivalent to the constant power model. The core of the energy feedback device is the three-phase voltage-type inverter, which adopts double closed loop feedforward decoupling control. The voltage outer loop function is to control the DC voltage of the three-phase converter, and the function of the current inner loop is to control the current according to the current instruction output by the voltage outer loop. The current command controls the current, which is equivalent to the no-load voltage and equivalent DC impedance of the traction capable device [6].

The equivalent model of the DC traction power supply system is shown in Fig. 2. U_d is the ideal no-load DC voltage of the diode rectifier unit, r_d is the equivalent resistance of the diode, and U_o and r are the no-load voltage and the equivalent DC impedance of the energy feedback device. U_i is the node i voltage, and R_{ij} is the line resistance between nodes i and j.

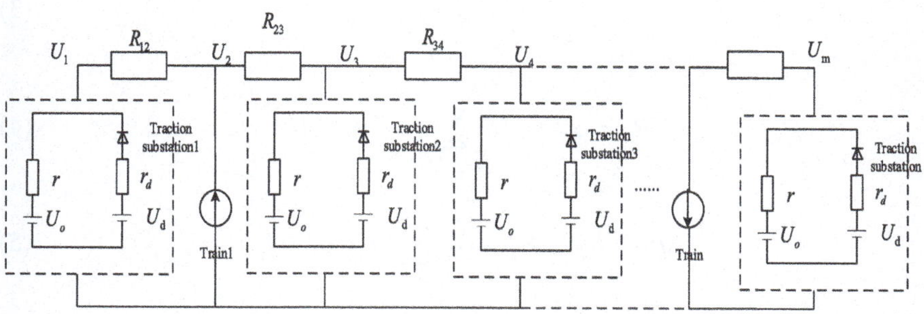

Fig. 2 DC traction power supply system equivalent model

3 DC Power Flow Calculation Algorithm

The Newton–Raphson method is currently the best method for solving nonlinear equations [7]. The main point of this method is to transform the solution process of nonlinear equations into a process of solving the corresponding linear equations repeatedly. We use this method to solve DC power flow calculation.

For the DC traction power supply system with n nodes, its node voltage is $U = [U_1, U_2, \ldots, U_n]$, and its node current is $I = [I_1, I_2, \ldots, I_n]$. According to the admittance formula $G = \frac{I}{U}$, and then we can get

$$G = \begin{bmatrix} G_{11} & G_{12} & \cdots & G_{1n} \\ G_{21} & G_{22} & \cdots & G_{2n} \\ \vdots & \vdots & \ddots & \vdots \\ G_{n1} & G_{n2} & \cdots & G_{nn} \end{bmatrix} \tag{1}$$

We know that the unified power flow model of the power grid is:

$$P_i + jQ_i = V_i \sum_{j=1}^{n} G_{ij} U_j \quad (i = 1, 2, \ldots, n) \tag{2}$$

Among them, P_i is the active power injected by the node i, and Q_i is the reactive power injected by the node i. G_{ij} is the admittance between node i and node j, and U_j is the voltage of node j. When the Newton–Raphson method is used to calculate the DC power flow calculation, the left side of the equation is the injected power for the given node, and the right is the node injection power obtained by the voltage of each node. The difference between the two is the unbalanced power of the node power [8].

The flow distribution of the DC network is determined by the voltage of the traction node and the line impedance. With the direction of injection power being positive, the formula [9] for the injection current and active power and power deviation of node i are:

$$I_i = \sum_{j=1, j \neq i}^{n} G_{ij}(U_j - U_i) \tag{3}$$

$$P_i = U_i I_i = U_i \sum_{j=1, j \neq i}^{n} G_{ij}(U_j - U_i) \tag{4}$$

$$\Delta P_i = P_{di} - P_{ti} - U_i \sum_{j=1, j \neq i}^{n} G_{ij}(U_j - U_i) \tag{5}$$

U_i is the voltage of node i; I_i is the current of node i; G_{ij} is the admittance between node i and node j; P_{di} is the active power injected into the node i; P_{ti} is the load power of node i; and ΔP_i is the load power variation of node i.

Therefore, the nonlinear equations of the n-node system are simultaneous. After the Taylor series is developed and the high-order term is omitted, the modified equation expressed in matrix form is obtained as follows: According to the Taylor series expansion, after omitting the higher-order terms, the modified equation in matrix form is as follows:

$$\begin{bmatrix} \Delta P_1 \\ \Delta P_2 \\ \vdots \\ \Delta P_n \end{bmatrix} = \begin{bmatrix} U_1 H_{11} & U_2 H_{12} & \cdots & U_n H_{1n} \\ U_1 H_{21} & U_2 H_{22} & \cdots & U_n H_{2n} \\ \vdots & \vdots & \ddots & \vdots \\ U_1 H_{n1} & U_2 H_{n2} & \cdots & U_n H_{nn} \end{bmatrix} \begin{bmatrix} \Delta f_1 \\ \Delta f_2 \\ \vdots \\ \Delta f_n \end{bmatrix} \tag{6}$$

Among them, $H_{ij} = \frac{\partial \Delta P_i}{\partial U_j}$, $\Delta f_n = \frac{\Delta U_n}{U_n}$, ΔP_n represents the voltage-active deviation of node n; and ΔU_n represents the voltage correction of node n. This modified equation is the basic equation for repeated iteration in Newton–Raphson power flow calculation. The modified equation is solved by Newton–Raphson method, and the corresponding Jacobian matrix should satisfy:

$$\mathbf{\Delta P} = \mathbf{J}\frac{\Delta \boldsymbol{U}}{\boldsymbol{U}} = \left(\mathbf{U}\frac{\partial \Delta \boldsymbol{P}}{\partial \boldsymbol{U}}\right)\frac{\Delta \boldsymbol{U}}{\boldsymbol{U}} \tag{7}$$

$$\begin{cases} U_j \frac{\partial \Delta P_i}{\partial U_j} = -U_i G_{ij} U_j \\ U_i \frac{\partial \Delta P_i}{\partial U_i} = -P_i - U_i^2 G_{ii} \end{cases} \tag{8}$$

In the formula, $\mathbf{\Delta P} = [\Delta \boldsymbol{P}_1, \Delta \boldsymbol{P}_2, \ldots, \Delta \boldsymbol{P}_n]^T$ is the active deviation vector of each node; $\boldsymbol{U} = [\boldsymbol{U}_1, \boldsymbol{U}_2, \ldots, \boldsymbol{U}_n]^T$ is the DC voltage vector of each node; and $\Delta \boldsymbol{U} = [\boldsymbol{U}_1/\Delta \boldsymbol{U}_1, \boldsymbol{U}_2/\Delta \boldsymbol{U}_2, \ldots, \boldsymbol{U}_n/\Delta \boldsymbol{U}_n]^T$ is the DC voltage correction amount of each node.

Solve the modified equation and get $\Delta \boldsymbol{U}$. After the node voltage is obtained according to the above formula, iterative is performed, and the iteration formula is as follows:

$$U_i^{(k+1)} = U_i^k + U_i^k \left(\frac{\Delta U_i^k}{U_i^k}\right), \quad i = 1, 2, \ldots, n \tag{9}$$

In the formula, ΔU_i^k is the value of the kth iteration.

Then the flow chart of DC power flow calculation can be obtained as shown in Fig. 3.

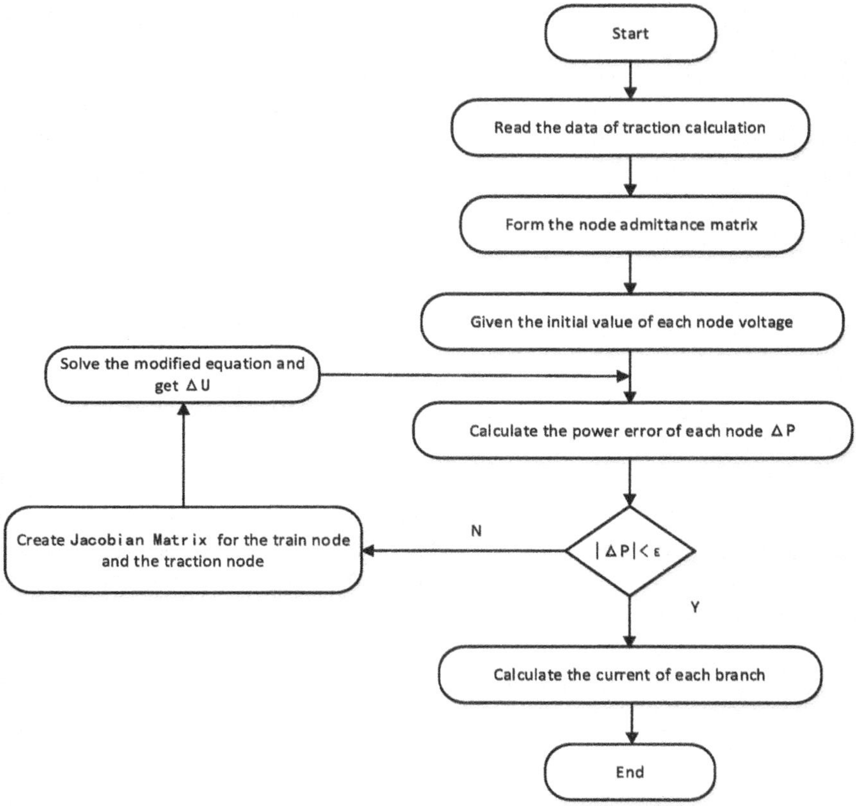

Fig. 3 DC power flow calculation flow chart

4 Simulation Software Design of DC Traction Power Supply System

The software development is written in the C# programming language in the Visual Studio 2013 environment. It adopts the Windows Form style, and the user can perform related operations by simply identifying the functions of the interface. Take a subway line as an example below, and its parameters are given in Table 1.

Table 1 System simulation parameters

Name	Number
No-load voltage of traction station	1650 (V)
Catenary resistance	0.014 (Ω/km)
Maximum operating speed	100 (km/h)
Maximum traction power	5000 (kW)
Maximum braking power	−5000 (kW)

According to the parameters of the subway line, the vehicle information and line parameters of the subway line are input to the traction calculation interface, and the data of the traction calculation is obtained. The calculation is performed after

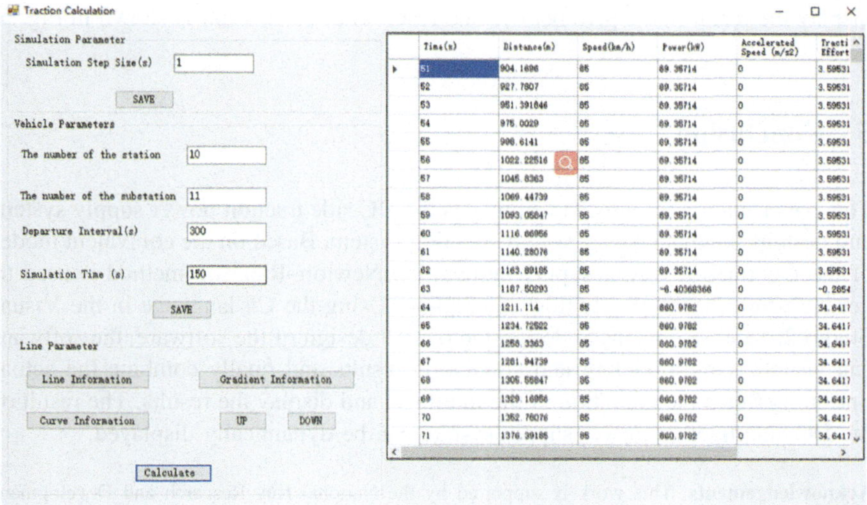

Fig. 4 Traction calculation interface

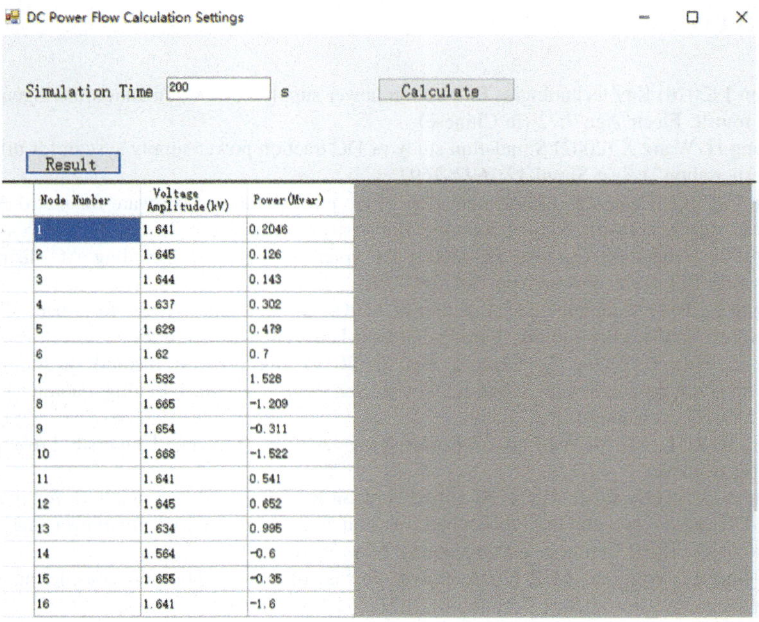

Fig. 5 DC power flow calculation result

selecting the up or down state of the train, and the calculation result will be displayed in the right grid. The traction calculation interface operation is completed as shown in Fig. 4.

Then, go to the DC power flow calculation interface, import the traction calculation data, input the simulation time, perform DC power flow calculation, and display the result in the grid below. The result after calculation is shown in Fig. 5.

5 Conclusion

This paper mainly designs and simulates the DC side traction power supply system and designs the main part of the DC traction system. Based on the equivalent model of the DC traction power supply system, the Newton–Raphson method is used to get the result of DC power flow calculation. Using the C# language in the Visual Studio 2013 environment to design the overall design of the software, the software can visualize the modeling and simulation results and finally combine the actual lines to perform the operation of the interface, and display the results. The result of the DC side traction power supply system can be dynamically displayed.

Acknowledgements This work is supported by the National Key Research and Development Program (2017YFB1200802-17).

References

1. Chen J (2018) Key technologies for traction power supply of new generation intelligent urban rail transit. Electr Age 7:72 (in Chinese)
2. Zhang H, Wang X (2002) Simulation study of DC traction power supply system for urban rail transportation. J Syst Simul 12:1692–1697
3. Yan F (2017) Design and implementation of the joint simulation software for metro AC–DC power supply network. Beijing Jiaotong University (in Chinese)
4. Pires C, Nabeta SI, Cardoso JR (2009) DC traction load flow including AC distribution network. IET Electr Power Appl 3(4):287–298
5. Zhang J (2017) Modeling and simulation of AC/DC power supply system for urban rail transit based on MATLAB/Simulink. Beijing Jiaotong University (in Chinese)
6. Liu Z, Hao F, Zhang G, Chen J, Liu Z (2019) Optimization method of train-ground coordination parameters for urban traction power supply system. J Beijing Jiaotong Univ 43 (01):79–87 (in Chinese)
7. Lang B, Xu L, Xia M, Wu J (2014) Power system analysis. Tsinghua University Press, pp 76–84 (in Chinese)
8. Panosyan A, Oswald R (2004) Modified Newton-Raphson load-flow analysis for integrated AC/DC power system. In: Proceedings of 39th international university power engineering conference. IEEE Conference Publications, New York, pp 1222–1229
9. He Jinghan, Wang X, Li Z (2016) Power flow algorithm for DC grid considering various control modes. Power Syst Technol 40(3):712–719 (in Chinese)

A Study on Fault Diagnosis Technology Based on Failure Mode Analysis of EMU System

Huabo Shen, Jianbo Zhao, Xiaodong Sun and Yuan Gao

Abstract With the rapid development of railway, the production and operation quality of high-speed EMU is an important factor to ensure the safety of passengers. This paper takes EMU as an overall system, conducts FTA failure mode analysis and FMEA failure mode analysis, obtains a comprehensive system failure mode model, generates a system function failure chain, integrates the failure chain into the fault expert knowledge base, and optimizes the expert knowledge base. According to the system function of EMU, the functional logic diagram of EMU is analyzed and obtained. The field fault information and vehicle status information are analyzed together with the system failure mode model and fault expert knowledge base. The system failure component nodes, failure mode chain, and fault suggestion processing measures are, respectively, output to complete on-site fault processing guidance. The whole process information will be fed back to optimize the accumulation of fault expert knowledge base. The system can rapidly diagnose on-site faults, guide on-site fault handling, effectively improve fault handling efficiency, and play an auxiliary decision-making role.

Keywords EMU · Fault diagnosis · Fault tree · FTA · FMEA

1 Introduction

With the intensification of market competition at home and abroad, equipment manufacturing enterprises have further explored how to effectively guarantee quality while improving production efficiency and reducing production costs. In recent years, high-speed EMU has developed into an intelligent large-scale complex system with complex structure, high technology content, and high integration. Its

H. Shen (✉) · J. Zhao · X. Sun · Y. Gao
CRRC Qingdao Sifang Co., Ltd., 266111 Qingdao, China
e-mail: shenhuabo.sf@crrcgc.cc

J. Zhao
e-mail: zhaojianbo.sf@crrcgc.cc

Y. Qin et al. (eds.), *Proceedings of the 4th International Conference on Electrical and Information Technologies for Rail Transportation (EITRT) 2019*, Lecture Notes in Electrical Engineering 639, https://doi.org/10.1007/978-981-15-2866-8_57

reliability is directly related to the safety and benefits of high-speed and heavy-haul railway transportation [1]. For high-speed EMUs, the debugging link is the key link to ensure the quality. On the other hand, due to the high professionalism and complexity of the debugging work itself, it is necessary to study the digital debugging technology to realize efficient and reliable functional inspection of EMU system. However, the operation safety of EMU is related to the national economy and the people's livelihood. Professional and strict debugging and inspection should be carried out on the system functions of EMU before newly built and advanced repairs leave the factory, so as to eliminate hidden troubles and ensure driving safety, driving order, and passenger comfort [2].

The system includes all the elements in the system and the relationship between each element, and all the elements in the system are associated in a certain way to form a whole system. In this paper, the EMU is taken as a whole system for fault analysis. The EMU system structure refers to the sum of the correlation relationships between the components of the EMU. When the differences between various elements in the EMU system cannot be ignored, the elements cannot be integrated with a single mode. It needs to be divided into different parts and integrated according to different modes to form subsystems.

This paper takes EMU as a system, analyzes the overall frame of EMU, and forms the EMU failure mode element library. Based on this, combined with fault tree analysis (FTA) theory, failure mode and effects analysis (FMEA)theory, functional logic model, fault expert knowledge base analyzes the EMU failure information and on-board network data, outputs the EMU failure causes and proposes treatment measures, guides the vehicle failure treatment, and plays an auxiliary decision-making role.

2 Failure Mode Analysis of Traditional EMU

2.1 Failure Mode Analysis of EMU Based on FTA Theory

FTA is a technology of system analysis. It identifies all possible causes from a single potential failure mode and analyzes system errors. FTA considers interrelated and independent reasons. In addition to the fault tree structure and all logical associations, FTA usually includes the identification of failure possibility, so that the system reliability can be calculated from the reliability of components [3]. The steps for establishing the fault tree are as follows:

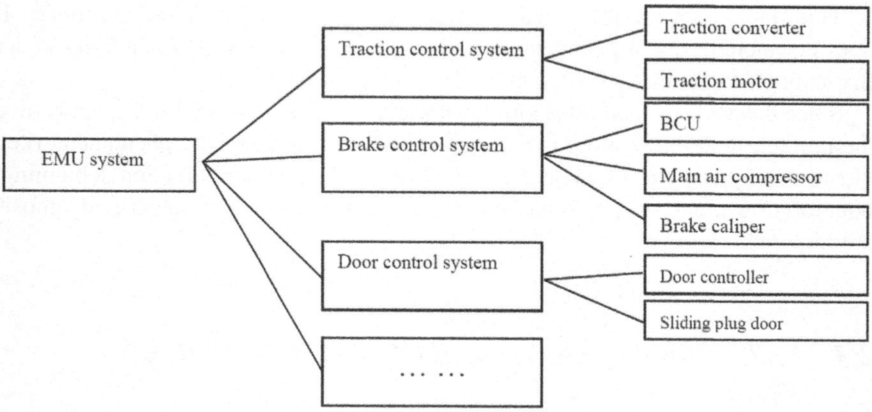

Fig. 1 FTA failure mode structure diagram of EMU system

(1) Select and determine the top event (which can be failure mode, system failure code) [3];
(2) Top events are analyzed to find out the direct, necessary, and sufficient causes for top events. Top events are taken as output events, and all direct causes are taken as input events and are connected with appropriate logic gates according to the logical relationship of these events [3];
(3) Decompose step by step down until all input events can no longer be decomposed or need not be decomposed [3].

Combined with FTA theory, the failure mode of EMU is analyzed, and it is determined that the failure top event of EMU is that the EMU system cannot operate, and the direct cause of the failure of EMU function is the failure of its subsystems [4], including traction control system, brake control system, and door control system. The direct cause of subsystem function failure is the failure of a component in the subsystem. After analysis, the FTA failure mode structure diagram of EMU system is finally formed, as shown in Fig. 1.

2.2 *Characteristics and Limitations of FTA Method*

The traditional FTA fault tree analysis method [5] requires that the bottom event and the top event of the system are deterministic events [6], i.e., failure or normality, so as to determine whether the top event is in a normal state. It cannot solve the problem of uncertainty of the occurrence probability of the top event and the bottom event and need to rely on the system function tree to analyze the reliability of the system.

Traditional FTA is constructed according to human experience (failure mode). It does not include signal parameters of system equipment, and its granularity is not fine enough to be applied to system and equipment fault diagnosis.

Since the analysis method is top-down analysis [7], traditional FTA cannot solve the problem that one bottom event corresponds to multiple fault phenomena (i.e., the intersection between fault trees). For system and equipment function debugging, both top-down fault diagnosis analysis and bottom-up functional integrity diagnosis are required.

2.3 Failure Mode Analysis of EMU Based on FMEA Theory

FMEA is a design method to improve the reliability of systems and equipment in the system [8]. It is an activity to analyze the systems, subsystems, and components that make up a product in the product design phase and the process design phase one by one, to find out all potential failure modes (which can also be called failure modes in the equipment application phase), and to analyze their possible consequences so as to take necessary measures in advance to improve the quality and reliability of the product. Detailed statistics of FMEA provide favorable data support for reliability analysis. FMEA usually includes the following contents: (1) What are the functions, features or requirements (which objects); (2) what are the problems (failure modes of specific objects); (3) what are the consequences (effects of failure modes); (4) how bad is it (the severity of the impact *S*); (5) what is the cause (failure cause/mechanism); (6) current control methods (current prevention and detection methods); and (7) what can be done (suggested measures).

Based on the FTA failure mode structure of EMU system, FMEA failure mode analysis is carried out on EMU system in combination with FMEA theory [9], and the fault tree is qualitatively analyzed by minimum cut set method [10, 11]. The failure phenomena of EMU subsystem and its components are analyzed qualitatively. The failure mode library of the failure subsystem and its components is finally generated according to the causes of failure caused by different on-site analysis and its treatment methods. The FMEA failure mode analysis structure diagram is formed by taking door control system and gate controller as examples in this paper, as shown in Fig. 2.

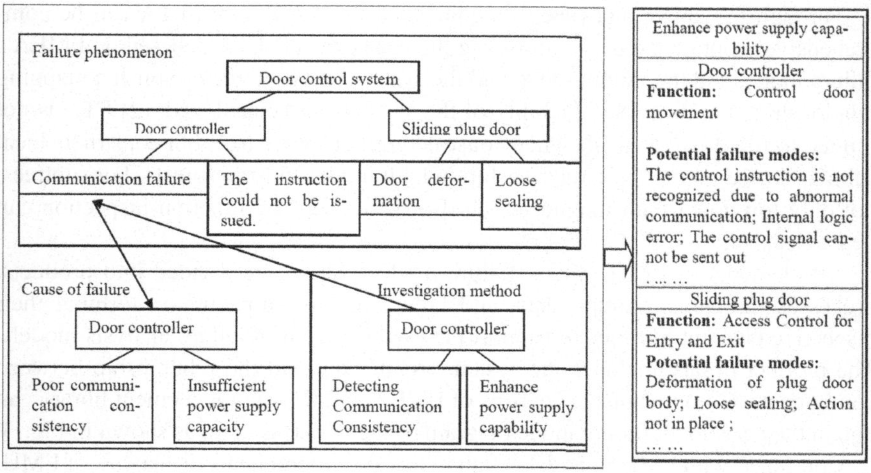

Fig. 2 FMEA failure mode analysis structure diagram of EMU system

2.4 *Characteristics and Limitations of Traditional FMEA*

Traditional FMEA considers the corresponding function points for failure modes (failure modes) and analyzes the causes and impacts of failure modes for individual function points in more detail, but does not consider the composition relationship of complex system functions [12].

Traditional FMEA can evaluate the risk level (risk sequence number RPN = $S * O * D$) according to the risk level (severity S), occurrence probability (frequency O) and hidden danger detection means (difficult detection degree D) and provide an analysis model for failure modes of a single function point. However, the analysis of failure causes is relatively scattered, and the same failure mode is often encountered in practical application, which is not only the result but also the cause of other failure modes.

3 Failure Mode Analysis and Optimization of EMU Based on FTA and FMEA Principles

3.1 *Comprehensive Analysis of EMU Failure Mode Based on FTA and FMEA Principles*

Traditional FMEA and FTA both have inherent limitations and need to be expanded. However, the expanded model is a more complex and systematic deduction method. The analysis process is complicated and requires a large amount of calculation. It needs to be completed by computer to model the data.

Through the above analysis, it is concluded that FMEA and FTA can be comprehensively optimized in the following three aspects: (1) FTA is based on FMEA's failure mode, and the failure mode and the system function tree establish a mapping relationship, and thus, the reliability of the system can be analyzed. (2) FTA is not limited to top-down analysis when constructing fault logic tree, but constructs local subtrees based on the logic relationship of fault modes to form forests. For subtrees, not only top-down fault diagnosis but also bottom-up full-function inspection can be carried out.

All elements of EMU system failure mode analysis are divided into products, processes, functions, components, and failure modes, respectively, forming their respective element libraries to support the establishment of failure analysis models. The product library includes the system architecture and the relationship between components. Comprehensive analysis of FMEA and FTA in the element library can obtain the equipment failure mode diagram of the whole system, as shown in Fig. 3.

Combined with FMEA and FTA failure modes' comprehensive analysis of EMU, not only the structural relationship of failure modes among various levels of components and subsystems can be obtained from the system structure, but also the corresponding relationship among various failure modes at various levels can be obtained, and finally the system failure chain is generated. This paper takes the door control system as an example to generate the door control system failure chain as shown in Fig. 4. From the system structure aspect, the failure of the component layer gate controller leads to the failure of the subsystem layer door control system, which ultimately leads to the failure of the system layer EMU. According to the failure modes of various components, the failure mode of "control signal cannot be sent out" of the door controller and the failure mode of "door does not act" in the door control system form an internal correlation relationship. After the accumulation of various failure chains, a failure chain library is formed to guide fault diagnosis.

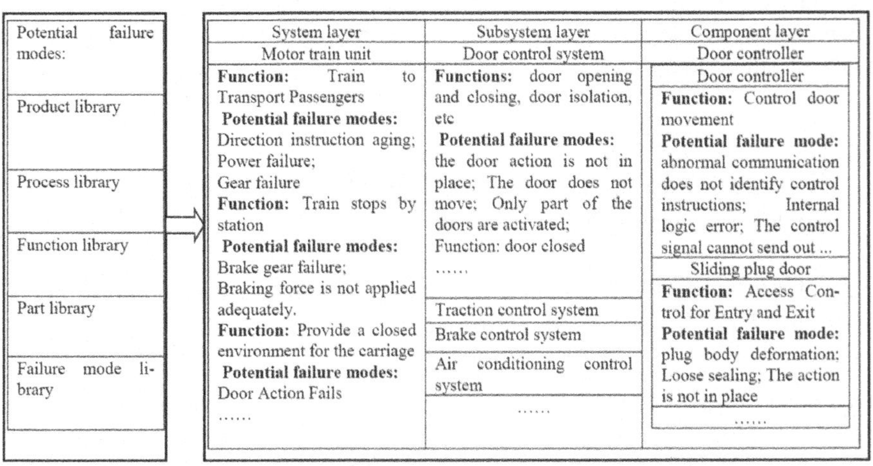

Fig. 3 Comprehensive analysis of FMEA and FTA failure modes of emus

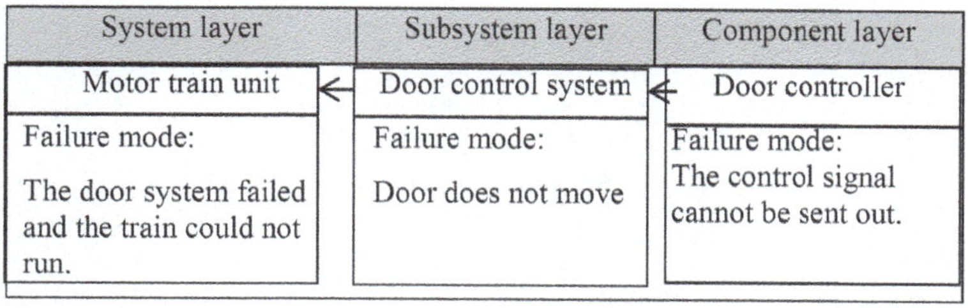

Fig. 4 Door control system failure chain

3.2 *Fault Knowledge Base of EMU Based on FTA and FMEA Principles*

Based on FTA and FMEA principles, the EMU failure mode comprehensive analysis extended model framework includes system function objects and failure modes, identifies the types of FTA nodes (failure modes, failure codes, signal parameters), and manages the external mapping relationship of FTA nodes. The data object relationship is shown in Fig. 5.

The analysis of the new fault diagnosis model focuses on the system, subsystem, sub-subsystem, or component. Through the top-down layer-by-layer analysis of all signal nodes under the focus, all direct and indirect causes leading to the occurrence of the focus event are found, and thus, the logical relationship between signal nodes is established and expressed in tree form.

The key table structure of the data model is shown in Fig. 6 after expanding the contents and organizational structure of the comprehensive failure mode analysis model for emus based on FTA and FMEA principles.

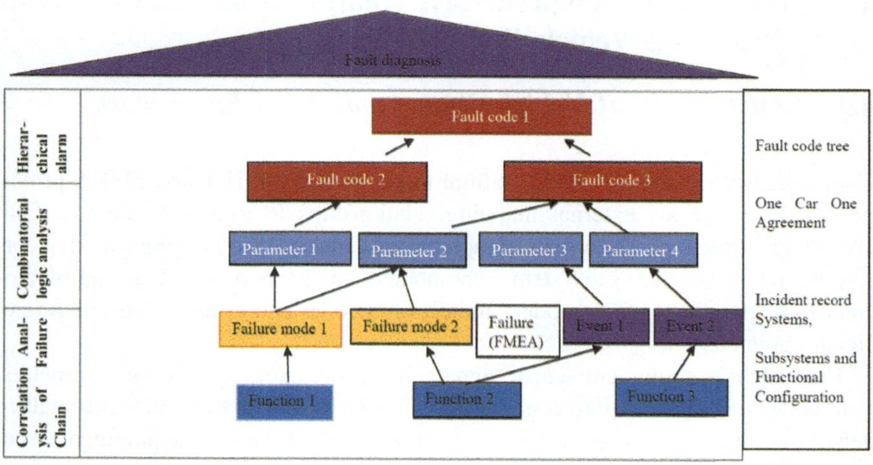

Fig. 5 Extended framework for failure mode analysis

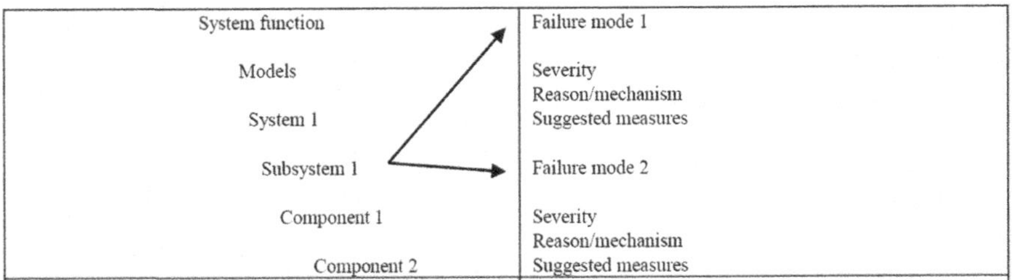

Fault model tree		Fault model signal logic relation table	
ID	INTEGER	ID	INTEGER
Model code	VARCHAR2(255	Fault tree model ID	INTEGER
Model ID	INTEGER	Signal ID	INTEGER
Fault model name	VARCHAR2(255	Signal type	VARCHAR2(255
Fault model coding	VARCHAR2(255	Abbreviations of Signal Names	VARCHAR2(255
Abnormal results show	VARCHAR2(255	Normal value of signal results	VARCHAR2(255
Severity of failure	VARCHAR2(255	Abnormal results show	VARCHAR2(255
System function	VARCHAR2(255	Car number	VARCHAR2(255
Parent ID	VARCHAR2(255	Logical relation between signals	VARCHAR2(255
Logical Relationship of Fault Model	VARCHAR2(255	(and/or)	VARCHAR2(255
(and/or)	VARCHAR2(255	Version number	VARCHAR2(255
Remarks	VARCHAR2(255	Parent ID	VARCHAR2(255
Founder	DATE	founder	DATE
Creation time	INTEGER	Creation time	INTEGER
Delete flag		Delete flag	

Fig. 6 Key table structure of fault data model

(1) The "Fault Model Tree" table is used to manage the basic structure of FTA and map the functional structure of the system.
(2) "Fault Model Signal Logic Relation Table" is used to manage signal parameters of train equipment and map to fault modes in "Fault Model Tree."

4 Application of Comprehensive Analysis Model Based on Real-Time Vehicle Data

4.1 Construction of Vehicle Data Application Framework

Comprehensive analysis of EMU failure modes based on FTA and FMEA principles extends various external mapping relationships to establish effective fault knowledge structures to support comprehensive diagnosis: (1) failure modes corresponding to systems, subsystems, components, and functions; (2) signal parameters corresponding to fault modes; and (3) fault codes and events corresponding to signal parameters.

Data such as equipment sensor signals, logical relations of control parameters, fault codes and signal parameters are added to the failure analysis model to realize debugging function inspection (from bottom to top) and fault cause judgment (from top to bottom), as shown in Fig. 7.

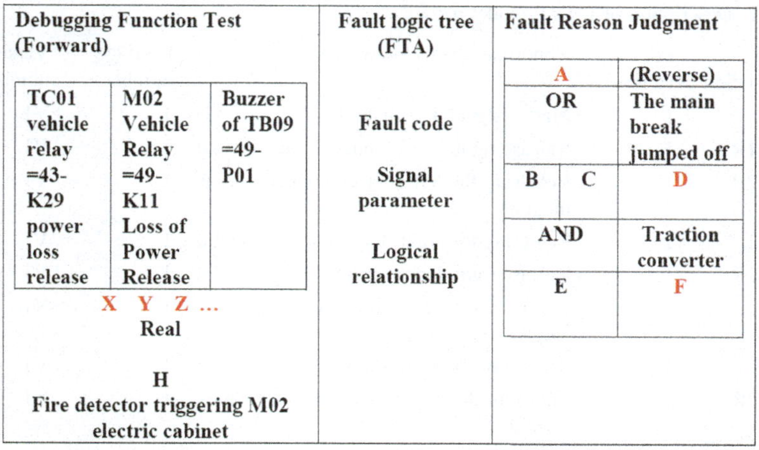

Fig. 7 Functional inspection and fault judgment after integrating fault codes and signal parameters

4.2 *Functional Logic Model Based on Vehicle Data*

By establishing the above model, the train system functions can be automatically debugged digitally with high efficiency. Taking the door system of a certain EMU as an example, the fault mode is mapped to the logic tree of signal parameters of each relay. The node signal logic model can better show the logical causal relationship between the door fault and each signal node of the door and can effectively improve the accuracy and efficiency of door fault diagnosis, as shown in Fig. 8. Table 1 illustrates the definition of signal nodes in Fig. 8.

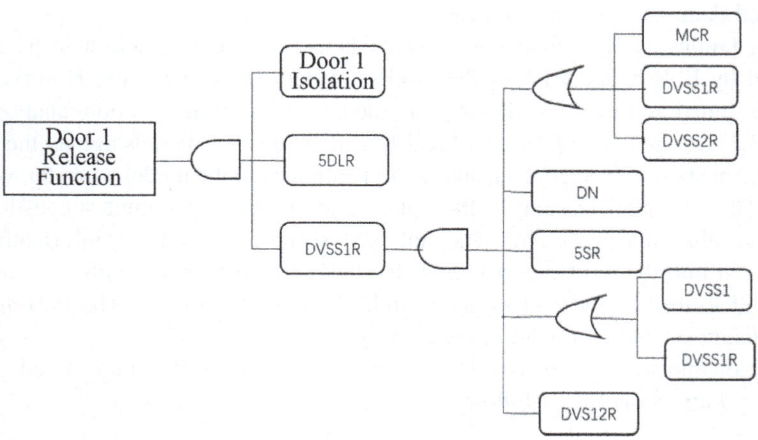

Fig. 8 Functional logic model

Table 1 Description of logical model names

Signal abbreviation	Logical value	Abnormal results show	Carriage number	Version number
MCR	1	Master signal 1 is invalid	1	V1
DVSS1R	1	Release right door command relay invalid	1	V1
DVSS2R	1	Releasing the left door command relay is invalid	1	V1
DVSS1	1	Releasing the right door button is invalid	1	V1
DVSS1R	1	Release right door command relay invalid	1	V1
DN	1	It is invalid to close the door when it is empty	1	V1
5SR	1	Speed greater than 5 km	1	V1
DVS12R	1	The right door closing command relay is invalid	1	V1
5DLR	0	Speed greater than 5 km	1	V1
DVSS1R	1	The right door release command relay of vehicle 1 is invalid	1	V1
Door 1 isolation	0	1 door 1 isolation invalid	1	V1

5 Functional Analyses and Verification

5.1 Functional Model Diagnosis Process

The following system platform is built to verify the EMU failure analysis model and logic model in combination with EMU abnormal information. The model completes the analysis of on-board data through the on-board data matching protocol. The information input of the model includes the status of EMU equipment, detected data, and fault phenomena.

The failure analysis model is a comprehensive system failure analysis model formed by FMEA and FTA of the product system element library. However, the logical function model is the logical model of division function analyzed in Sect. 3.2. On the basis of on-board real-time data, the two models upload the EMU equipment status, detected data, and fault phenomena to the model when the vehicle fails. The two models analyze the data, respectively, and output suggestions to guide the abnormal processing. The fault analysis process and result information are optimized into the fault expert knowledge base, which together with the guidance obtained from the model processing guide the fault processing. The flowchart of fault diagnosis and treatment is shown in Fig. 9.

The diagnostic logic of the door control system function failure based on the function logic model is as follows:

(1) In the process of automatic debugging and testing, if there is a fault, the fault mode and other information shall be defined according to the interface specification of a certain type of EMU and stored in the database.

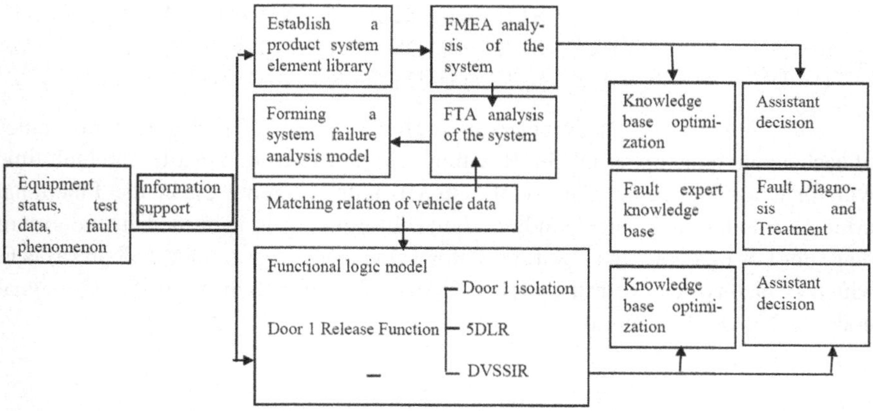

Fig. 9 Fault diagnosis flowchart

(2) The fault data is automatically matched with the fault model in the fault model tree according to the fault mode, and a signal node logical relation model configured under the fault model is found in association, for example, a signal node logical relation model of a plug gate releasing a right gate.
(3) Find the top-level event of the logic relation model of the signal node of the plug door releasing the right door—"door 1 releasing."
(4) Find the lower signal nodes of the top-level event "Door 1 Release": 5DLR, DVSS1R, and door 1 isolation; The normal state values of signal nodes are 5DLR = 0, DVSS1R = 1, and door 1 isolation = 0, respectively.
(5) The real-time state values of signal nodes 5DLR, DVSS1R, and door 1 isolation in n minutes before and after the time of the fault occurrence are obtained.
(6) The real-time state value of the signal node obtained in step (5) is compared with the normal value of each signal node in step (4) one by one. If it is detected that a certain signal state value in step (5) is inconsistent with the signal state value in step (4), for example, the real-time state value of DVSS1R in step (5) is inconsistent with the normal state value of DVSS1R in step (4), it indicates that there is a problem with the signal of DVSS1R node and the step (7) of continuously searching for the lower-level signal node of DVSS1R signal node. If both data are detected to be consistent, it means that the signal node is not a fault point, continues to analyze other signals at the same level in step (4), and repeats the operation of step (5) and subsequent steps.
(7) Find and analyze six kinds of logical relationships of signal nodes under DVSS1R, and the logical relationships are as follows:

(a) DN = 1, 5SR = 1, DVS12R = 0, MCR = 1, DVSS1 = 1
(b) DN = 1, 5SR = 1, DVS12R = 0, DVSS1R = 1, DVSS1 = 1
(c) DN = 1, 5SR = 1, DVS12R = 0, DVSS2R = 1, DVSS1 = 1

(d) DN = 1, 5SR = 1, DVS12R = 0, MCR = 1, DVSS1R = 1
(e) DN = 1, 5SR = 1, DVS12R = 0, DVSS1R = 1, DVSS1R = 1
(f) DN = 1, 5SR = 1, DVS12R = 0, DVSS2R = 1, DVSS1R = 1.

Nodes with inconsistent comparison result data compared real-time status values of each signal node under DVSS1R within N minutes before and after the fault time with that normal status value of the signal node are reflected in the functional principle model, and corresponding signal information is identified in the circuit diagram. For example, the real-time status value of the signal node 5SR in (a) is 0, which is inconsistent with the normal status value of 1, indicating that the signal node 5SR is the fault point.

5.2 Verification of Functional Model Test Results

In this paper, the test data of the actual automatic debugging process of China's standard EMU are selected as the test data, and the fault model of the released right door is called for diagnosis and analysis. In order to ensure the accuracy and effectiveness of the analysis process, the actual recorded data of CR400AF EMU door debugging in a EMU main engine factory are selected on the spot, and the analysis is carried out by combining the functional model and the system failure mode comprehensive analysis model established in this paper.

After the driver's cab is put into the main control, a door release test is carried out. After the door release button is pressed, a set of real data of the door test is obtained, and the release right door fault diagnosis model is called. The comparison shows that the 5DLR signal node is inconsistent with the normal value, and the 5DLR signal node has no child nodes. Therefore, it can be concluded that the possible cause of the door release operation failure is the damage of the relay 5DLR signal node, which verifies that the functional logic model is feasible and effective.

Under the condition of abnormal 5DLR, the 5DLR signal acquisition circuit is analyzed. The failure modes include 5DLR relay sticking, 5DLR relay coil continuity, and 5DLR contact virtual connection. The system failure chain is automatically generated and the new failure chain is optimized into the failure library.

6 Conclusion

In this paper, the EMU configuration diagram is obtained through FTA failure mode analysis of the EMU overall system, and FMEA failure mode analysis of the equipment elements in the configuration diagram is continued, finally forming the EMU system failure mode library and failure expert knowledge base. Combined with the occurrence situation of EMU faults, the vehicle information when EMU faults occur is collected, the functional logical framework of EMU is built, the

EMU fault information is analyzed together with the fault mode library, the system failure chain is generated, the fault expert knowledge base is optimized, the fault handling opinions are output, and the field fault fast processing is guided. The conclusion is as follows:

Through FTA failure mode analysis and FMEA failure mode analysis of EMU system, the comprehensive failure mode framework of EMU is obtained, and a variety of system failure chains are generated, which can guide the construction and optimization of fault knowledge base to a great extent.

In the aspect of fault diagnosis, analysis of fault information combined with system functional logic, fault expert knowledge base, and system failure mode framework can quickly and effectively diagnose the cause of the fault and play an auxiliary decision-making role on site.

References

1. Jiao FCh, Wang BJ (2013) The management and maintenance of locomotive rolling stock. Beijing Jiaotong University Press, Beijing (in Chinese)
2. Xu J (2016) Editorial Department of *China Journal of Highway and Transport*. A summary of academic research on China's traffic engineering. 2016. China J Highway Transp 29(6) (in Chinese)
3. Yang WD, Zhao G (2018) Failure mode management of rail transit equipment. Informatiz China (07):86–87 (in Chinese)
4. Wang FH, Guo JL, Suo Q, Zhang Q (2016) Research on the application of complex networks in quality management. Complex Syst Complex Sci 13(2) (in Chinese)
5. Heydt GT, Graf TJ (2010) Distribution system reliability evaluation using enhanced samples in a Monte Carlo approach. 1EEE Trans Power Syst 25(4):2006–2008
6. Song LL, Wang TY, Song XW, Xu L, Song DG (2014) Fault diagnosis of EMU pantograph based on petri net modeling and FTA. Chin J Sci Instrum 35(9):1990–1997 (in Chinese)
7. Lv L (2018) Overall design of fault diagnosis expert system for high speed EMUs. Res Urban Rail Transit 21(02) (in Chinese)
8. Hu HT, Gao ZH, He ZY, Yuan L (2012) Reliability evaluation of metro traction power supply system based on FTA and FMEA. J Railw 10 (in Chinese)
9. Zhang J, Huang W (2000) The integrated reliability analysis method using FMEA and FTA. Mech Des Manuf (1):1–3 (in Chinese)
10. Fussell JB (1976) Fault tree analysis—concepts and techniques. In: Henley EJ, Lynn JW (eds) Generic techniques in reliability assessment. Nordoff, Leyden, Holland
11. Xiang Xp, Shao Yk. Reliability evaluation of distribution systems based on (in Chinese)
12. Zhang H, Zhou Z, Sun Y, Zeng Z (2016) Failure analysis of intelligent brake based on FMEA and FTA. J China Constr Mach 14(02):109–113 (in Chinese)

Huabo Shen male, born in 1983, studies the debugging technology of EMU.

Jianbo Zhao male, born in 1986, studies EMU commissioning technology.

Risk Analysis of Train Control System Based on Fuzzy Mathematics and Bayesian Network

Lijuan Shi, Anping Hang, Zuoning Jia, Mingshu Tian, Jian Wang and Yiwei Wu

Abstract This paper takes the 'train overspeed unprotected' event of Chinese train control system level 2 (CTCS-2) as the studied risk event and analyzes the weak links leading to the risk event. Fault tree, expert scoring, fuzzy mathematics, and Bayesian network (BN) are applied in combination to conduct this study. Firstly, by analyzing the structure and function of CTCS-2 system, all the possible risk sources are identified, and a fault tree is established by taking the 'train overspeed unprotected' as the top event. Secondly, based on expert scoring method, a calculation model is established by using fuzzy mathematics to calculate the frequencies of risk sources. Thirdly, the fault tree is transformed into a BN, and the frequency of risk sources is used as the input of BN leaf nodes. The frequency of the risk event and the posterior probability of each risk source are calculated to identify the main weak links that lead to the risk event. The three main risk sources are track circuit, balise transmission module (BTM), and train control center (TCC), which provides the basis for taking further decision support for risk reduction control measures.

Keywords Train control system (TCS) · Risk analysis · Fault tree · Expert scoring · Fuzzy mathematics · Bayesian network

L. Shi (✉) · Z. Jia · M. Tian · J. Wang · Y. Wu
Shanghai Key Laboratory of Rail Infrastructure Durability and System Safety, Tongji University, 201804 Shanghai, China
e-mail: shilijuan150@tongji.edu.cn

A. Hang
China Academy of Railway Sciences Corporation Limited, 100081 Beijing, China
e-mail: h_anping@sina.com

Y. Qin et al. (eds.), *Proceedings of the 4th International Conference on Electrical and Information Technologies for Rail Transportation (EITRT) 2019*, Lecture Notes in Electrical Engineering 639, https://doi.org/10.1007/978-981-15-2866-8_58

1 Introduction

The main technical feature of the CTCS-2 system is of implementing the transmission of ground control information to the vehicle based on track circuits and balises. The data from trackside is passed to an on-board decoding and safety processor on vehicle. The permitted speed is checked against the actual speed, and if the permitted speed is exceeded, a brake application is initiated. Once the 'train overspeed unprotected' risk event occurs, the consequences of running will be extremely dangerous. This paper takes the 'train overspeed unprotected' of CTCS-2 system as the risk event, establishes fault tree, seeks appropriate mathematical modeling method to calculate the frequency of risk base events, and analyzes the safety related weakness of CTCS-2 system.

Considering of the historical fault data of CTCS-2 system is usually difficult to be obtained, the analysis methods combined with expert knowledge and fuzzy mathematics are widely used for risk analysis. Prof. An M of Birmingham University has conducted an in-depth study on the application of fuzzy mathematics. He calculated the weight of risk assessment index by using expert scoring method and Fuzzy Analytic Hierarchy Process (FAHP), and then evaluated the risk of railway transportation [1]. Yingying used FAHP to calculate the weight of each risk in the TCS by constructing the trigonometric membership function [2]. Kabir built a dynamic fault tree with the water pump on the ship as the research object, established the trigonometric membership function, mapped the expert evaluation to each interval of the trigonometric membership function, and obtained the failure frequency of each component of the pump by defuzzification [3].

The purpose of risk analysis in this paper is to analyze the weakness in the system that lead to the occurrence of the risk event, and to rank the risk severity of all the basic events to provide decision support for risk management. There are main three methods to find the weakness. Firstly, calculate the weights of each index based on expert evaluation according a proposed method [4]. Secondly, build a fault tree and calculate the probability importance of the basic events of a fault tree to sort the evaluation indexes [5]. Thirdly, build BN and calculate the posterior probability of leaf nodes and inter-mediate nodes, to sort the indexes [6–8]. For the above three methods, the first method is greatly influenced by expert knowledge and the comparison criteria formulated. The calculation amount of the second method will be larger when the fault tree is more complicated. The third method needs to input the prior probability leaf nodes; calculation amount is smaller, and more accurate analysis results are compared with the other two methods.

Based on the above analysis, this paper establishes a fault tree with the risk top event of 'train overspeed unprotected' and establishes the mathematical model with fuzzy mathematics on the basis of expert scoring method for the occurrence frequency of risk basic events. Finally, BN is applied to analyze the risk weakness of CTCS-2 system leading to the 'train overspeed unprotected.'

2 Analysis of Risk Events

2.1 CTCS-2 Structure and Function

CTCS-2 system includes ground subsystem and on-board subsystem. The communication between the two subsystems is carried out by balise and track circuit. The structure is as shown in Fig. 1.

The on-board subsystem takes the safety computer as the center, generates dynamic speed control curve according to the principle of speed–distance control mode and monitors the safe operation of vehicle according to the information of track occupation, rail line parameters, temporary speed limit, and vehicle parameters provided by ground subsystem. Track circuit reader (TCR) receives information from track circuit through antenna, and BTM receives information from ground balise by connecting with balise antenna. Train interface unit (TIU) provides the interface between automatic train protection (ATP) related safety computer and

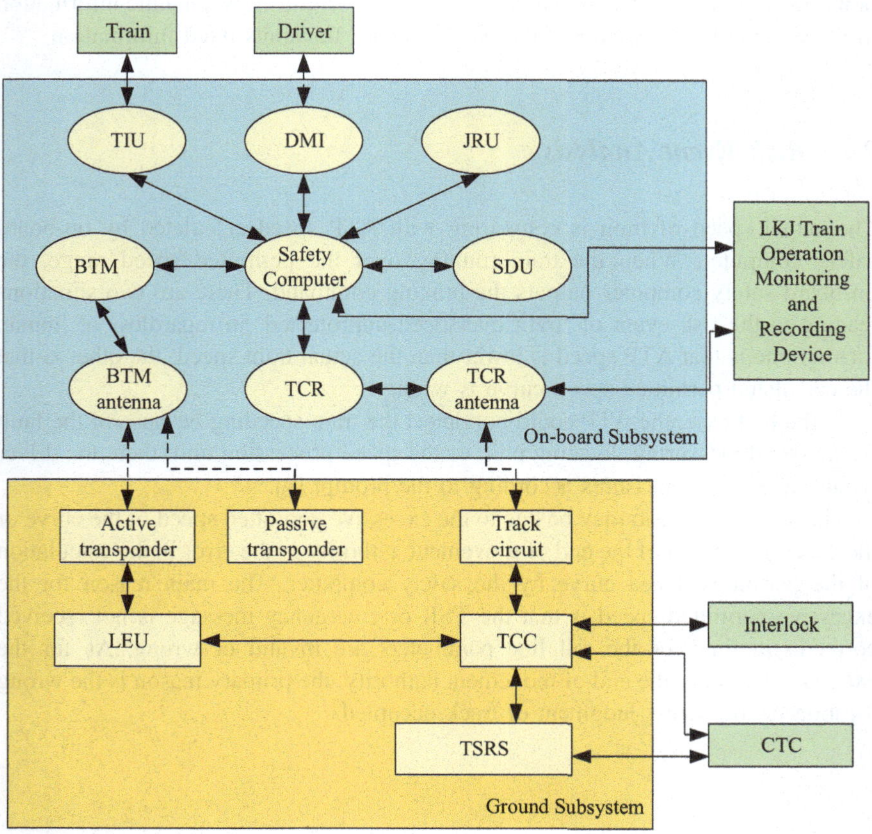

Fig. 1 Physical structure of CTCS-2 system

train itself system. Speed distance unit (SDU) converts the pulse signals of speed sensor and Doppler radar into digital data and transmits them to speed distance process (SDP) to calculate train operating speed and distance. Driver–machine interface (DMI) provides information interface and access between driver and on-board equipment. Considering that the failure of Juridical recording unit (JRU) does not affect the running safety of the train, failure analysis of JRU would not be made in the follow-up.

The ground subsystem takes TCC as the center and transmits information between components. Temporary speed restriction system (TSRS) centrally manages temporary speed restriction (TSR) and state, and transfers TSR through TCC and balise. The main functions of TCC include encoding the track circuit, receiving the TSRS temporary speed limit command and feedback the execution status, and transmitting TSR and route status to the on-board subsystem through the electronic unit beside the track and the active balise. The track circuit checks the occupancy of trains and the automatic block of sections and provides the idle information of the front sections for the on-board subsystem. Line-side electronic unit (LEU) generates the message needed by balise from the information provided by TCC. Balise is a point transmission device, in which the active balise transmits dynamic information to the on-board subsystem and the passive balise transmits fixed information.

2.2 *Risk Event Analysis*

The actual speed of train is comparing with ATP speed calculated by on-board safety computer. When the train running over the permitted speed range, the on-board safety computer outputs the braking command. There are two situations leading to the risk event of 'train overspeed unprotected' in regardless of human factors: one is that ATP speed is lower than the actual train speed; the other is that the calculated permitted speed curve is wrong.

In the first case, the ATP could not detect the train speeding because of the fault of the speed measuring, locating unit, or the speed processing unit; then, the driver would take safety measures according to the prompt [9].

The second scenario may be due to the excessive permitted speed of the curve or the excessive length of the end of movement authority or the error in the calculation of the permitted speed curve by the safety computer. The main reason for the excessive permitted speed is that the TSR or emergency message is not received correctly in time, or the rail line parameters are invalid or wrong. As for the excessive length of the end of movement authority, the primary reason is the wrong location or the wrong judgment of track occupied.

2.3 Risk Event Fault Tree

According to the above preliminary analysis, a fault tree is constructed based on the 'train overspeed unprotected' as the top event, as shown in Fig. 2.

The unit of velocity and distance measurement includes radar and speed sensor as well as SDU. Low-speed measurement of radar and speed sensor or wrong calculation of SDU may cause ATP to get lower speed than actual operation speed.

The primary reasons for the high permitted speed curve are the faults of BTM or BTM antenna, which are used to transmit the temporary speed limit and emergency information to the on-board safety computer.

The main causes for the excessive permitted speed curve caused by ground equipment failure are the faults of TSRS, TCC, LEU, or balise. Temporary speed limit and emergency messages are transmitted to the on-board system through the four modules in turn, and any failure of the dangerous side of any one of the modules may lead to excessive permitted speed curve.

The data for train positioning are from the fixed position parameters of passive balise and train travel distance. The parameters of passive balise are transmitted to on-board safety computer through BTM transmission module, and the data for calculating travel distance mainly come from SDP and SDU.

Section cleared information is mainly used to determine the end of movement authority, which is transmitted from the track circuit and then by TCR module to the on-board safety computer ultimately.

3 Risk Analysis Model Based on Fuzzy Mathematics and Bayesian Network

3.1 Calculation Model on Risk Source Frequency Based on Fuzzy Mathematics

This paper establishes a mathematical model by using fuzzy mathematics to solve the problem that it is difficult to obtain the field data of the occurrence frequency of risk base events of CTCS-2 system. Firstly, five experts are invited to make a judgment on the frequency of risk base events occurrence qualitatively, and then the expert's comments are converted into numerical value by using the calculation method proposed in this paper.

Determining membership functions. The risk matrix listed in national standard (GB/T 21562) divides the occurrence frequency into six levels: frequent, probable, occasional, rare, impossible, and incredible [10]. Considering the requirement for the integrity of the safety-stringent system, according to the characteristics of low failure rate of each module in TCS, the occurrence frequency is further refined into nine levels, as shown in Table 1. Determine the membership

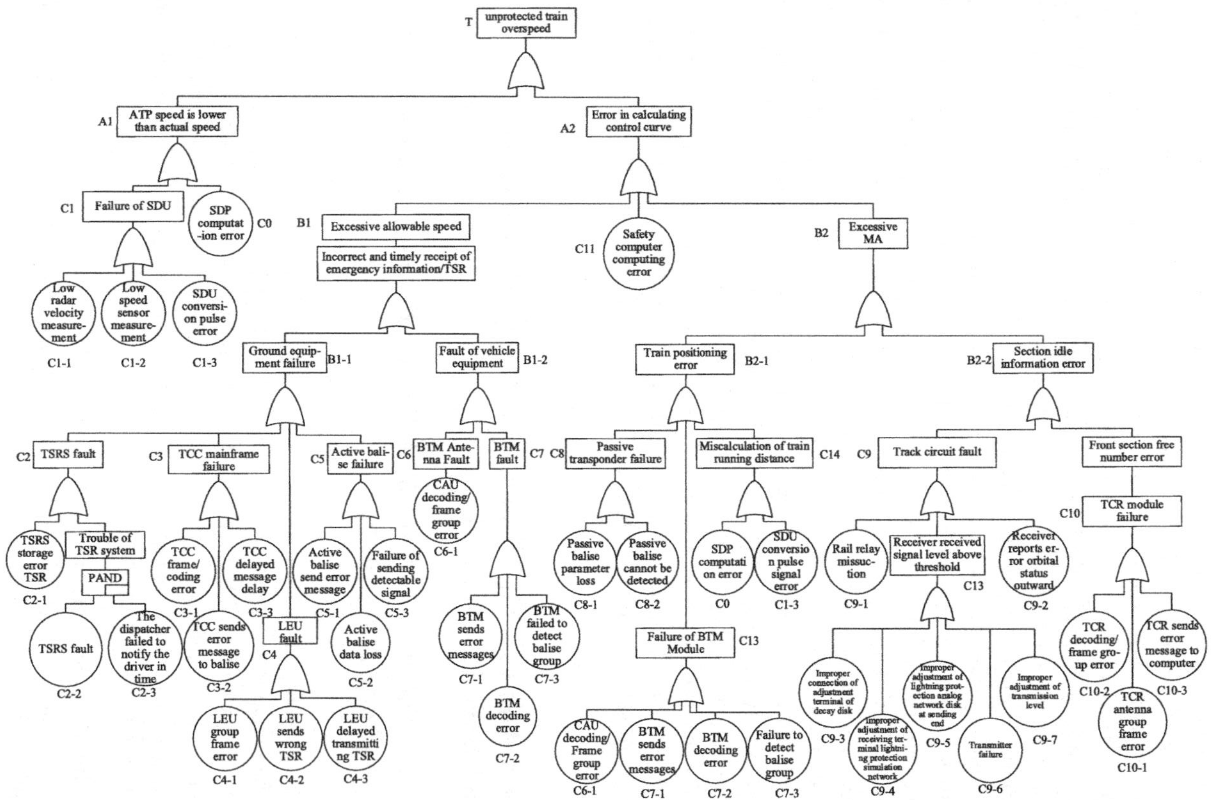

Fig. 2 Fault tree of risk event

Table 1 Ladder fuzzy functions

Grade	Linguistic value	Failure frequency (times/h)	Fuzzy number *B*
F9	Incredible	[0, 1.14 × 10^{-10})	(0, 0, 8.83 × 10^{-11}, 3.705 × 10^{-10})
F8	Impossible	[1.14 × 10^{-10}, 1.14 × 10^{-9})	(8.83 × 10^{-11}, 3.705 × 10^{-10}, 8.83 × 10^{-10}, 3.705 × 10^{-9})
F7	Rare	[1.14 × 10^{-9}, 1.14 × 10^{-8})	(8.83 × 10^{-10}.3.705 × 10^{-9}, 8.83 × 10^{-9}, 3.705 × 10^{-8})
F6	Very few	[1.14 × 10^{-8}, 1.14 × 10^{-9})	(8.83 × 10^{-9}, 3.705 × 10^{-8}, 8.83 × 10^{-8}, 3.705 × 10^{-7})
F5	Very seldom	[1.14 × 10^{-7}, 1.14 × 10^{-6})	(8.83 × 10^{-8}, 3.705 × 10^{-7}, 8.83 × 10^{-7}, 3.705 × 10^{-6})
F4	Occasional	[1.14 × 10^{-6}, 1.14 × 10^{-5})	(8.83 × 10^{-7}, 3.705 × 10^{-6}, 8.83 × 10^{-6}, 3.705 × 10^{-5})
F3	Often	[1.14 × 10^{-5}, 1.14 × 10^{-4})	(8.83 × 10^{-6}, 3.705 × 10^{-5}, 8.83 × 10^{-5}, 3.705 × 10^{-4})
F2	Frequent	[1.14 × 10^{-4}, 1.14 × 10^{-3})	(8.83 × 10^{-5}, 3.705 × 10^{-4}, 8.83 × 10^{-4}, 3.705 × 10^{-3})
F1	Extremely frequent	[1.14 × 10^{-3}, ∞)	(8.83 × 10^{-4}, 3.705 × 10^{-3}, 1, 1)

function and its fuzzy number. According to the definition of trapezoidal membership function [11] and the fuzzy number in Table 1, the expression of trapezoidal membership function is as follows, taking F_1 as an example.

$$\mu_{F_1}(x) = \begin{cases} 0, & x < 8.835 \times 10^{-4} \\ \frac{x - 8.835 \times 10^{-4}}{2.8215 \times 10^{-3}}, & 8.835 \times 10^{-4} < x \le 3.705 \times 10^{-3} \\ 1, & 3.705 \times 10^{-3} < x \le 1 \\ 0, & x > 1 \end{cases}$$

Collection of expert comments. Five experts were invited to judge the frequency level of risk sources by referring to the frequency of occurrence in the risk matrix. The five expert comments are corresponded to fuzzy numbers one by one. Formula (1) is used to aggregate expert comments.

$$A_i = 1/5 \times (B_1 \oplus B_2 \oplus \cdots \oplus B_5) \tag{1}$$

where A_i represents the fuzzy number corresponding to the *i*th ($i = 1, 2, \ldots, n$) risk source, and B_j represents the fuzzy number corresponding to the comment of the *j*th ($j = 1, 2, \ldots, 5$) expert on the risk source.

Defuzzification. Defuzzification with gravity center method is to get exact value [12], such as (2).

$$\lambda_i = \frac{\int \mu_{A_i}(x)x\mathrm{d}x}{\int \mu_{A_i}(x)\mathrm{d}x} \quad (2)$$

In the formula, λ_i denotes the exact frequency of the occurrence of the *i*th (i = 1, 2, …, n) risk source, and $\mu_{Ai}(x)$ denotes the membership function of the *i*th risk source.

3.2 Quantitative Risk Source Analysis Model Based on Bayesian Network

BN is a directed acyclic graph composed of node I and line E. Nodes represent events, and wires connect causal events [13]. Let $X = (X_i)_{i \in I}$ be the random variable represented by node i in the network; then, the joint probability distribution of node X can be expressed as:

$$P(x) = \prod_{i=1}^{n} P(X_i = x_i | X_{i+1} = x_{i+1} | X_n = x_n) \quad (3)$$

1. Cause–result reasoning, e.g., (3) the conditional probability between nodes is input, and the prior probability of nodes is used to calculate the probability of top-level nodes.
2. Result–cause reasoning, e.g., (3) the conditional probability between nodes is input, the prior probability of nodes is assumed to be 1, and the posterior probability of each node is calculated.

4 Experimental Results

4.1 Calculation of Frequency of Risk Sources

Assemblage of Expert Comments. Based on their own work experience and historical data, experts make a linguistic evaluation of the frequency of risk sources in Fig. 1. This section takes the node ‘C4-2: LEU Sending Error TSR’ in Fig. 2 as an example to calculate.

Five expert comments are “impossible”, “impossible”, “rare”, “rare” and “impossible” respectively. Five comments are corresponded to fuzzy numbers one by one. Formula (1) is used to aggregate five expert comments. Then, the ambiguity number of ‘C4-2: LEU sending error TSR’ is:

$$\begin{aligned} A_{4-2} &= 1/5 \times (B_1 \oplus B_2 \oplus B_3 \oplus B_4 \oplus B_5) \\ &= 1/5 \times ((3 \times (8.835 \times 10^{-11}, 3.705 \times 10^{-10}, 8.835 \times 10^{-10}, 3.705 \times 10^{-9})) \\ &\quad \oplus (2 \times (8.835 \times 10^{-10}, 3.705 \times 10^{-9}, 8.835 \times 10^{-9}, 3.705 \times 10^{-8}))) \\ &= (4.064 \times 10^{-10}, 1.704 \times 10^{-9}, 4.064 \times 10^{-9}, 1.704 \times 10^{-8}) \end{aligned}$$

The corresponding membership function is:

$$\mu_{A_{4-2}}(x) = \begin{cases} 0, & x < 4.064 \times 10^{-10} or x > 1.704 \times 10^{-8} \\ \frac{x - 4.064 \times 10^{-10}}{1.298 \times 10^{-9}}, & 4.064 \times 10^{-10} \le x < 1.704 \times 10^{-9} \\ 1, & 1.704 \times 10^{-9} \le x < 4.064 \times 10^{-9} \\ \frac{1.704 \times 10^{-8} - x}{1.298 \times 10^{-8}}, & 4.064 \times 10^{-9} \le x \le 1.704 \times 10^{-8} \end{cases}$$

Calculating the frequency of risk sources. Defuzzification with gravity center method is used to get exact value. The formula is as (2).

$$\begin{aligned} \lambda_{4-2} &= \frac{\int_{4.064 \times 10^{-10}}^{1.704 \times 10^{-9}} \frac{x^2 - 4.064 \times 10^{-10} x}{1.298 \times 10^{-9}} dx + \int_{1.704 \times 10^{-9}}^{4.064 \times 10^{-9}} x \, dx + \int_{4.064 \times 10^{-9}}^{1.704 \times 10^{-8}} \frac{1.704 \times 10^{-8} x - x^2}{1.298 \times 10^{-8}} dx}{\int_{4.064 \times 10^{-10}}^{1.704 \times 10^{-9}} \frac{x - 4.064 \times 10^{-10}}{1.298 \times 10^{-9}} dx + \int_{1.704 \times 10^{-9}}^{4.064 \times 10^{-9}} 1 \, dx + \int_{4.064 \times 10^{-9}}^{1.704 \times 10^{-8}} \frac{1.704 \times 10^{-8} - x}{1.298 \times 10^{-8}} dx} \\ &= 6.536 \times 10^{-9} \end{aligned}$$

Therefore, the exact value of C4-2 is $\lambda_{4-2} = 6.536 \times 10^{-9}$. Similarly, the frequency of other risk sources can be calculated.

4.2 *Quantitative Analysis Based on Bayesian Network*

Criteria for Transforming Fault Tree to Bayesian Network. Assuming that C is the top event and B1 and B2 are the bottom events, logic gates are used to connect top event and base events. The transformation criteria are shown in Fig. 3, where state = 0 indicates that the bottom event did not occur, and state = 1 indicates that it did.

Construct Bayesian Network. Based on these above transformation criteria, the fault tree of Sect. 3 is transformed into the BN as shown in Fig. 4 in GeNIe, and the probability of risk sources is used as the input of the leaf nodes of the BN.

Quantitative Analysis. *Cause–effect reasoning*. According to the calculation of Bayesian network, the frequency of some important nodes is shown in Table 2.

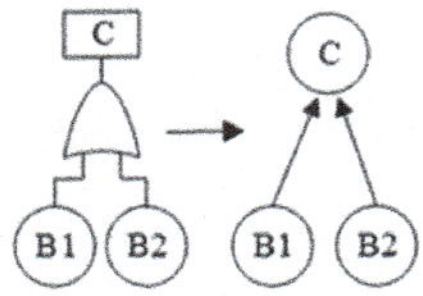

B1	state=0		state=1	
B2	state=0	state=1	state=0	state=1
state=0	1	0	0	0
state=1	0	1	1	1

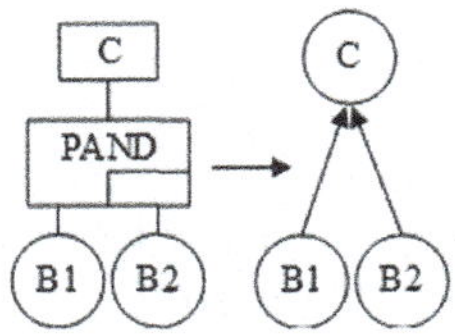

B1	state=0		state=1	
B2	state=0	state=1	state=0	state=1
state=0	1	1	1	0
state=1	0	0	0	1

Fig. 3 Criteria for transforming

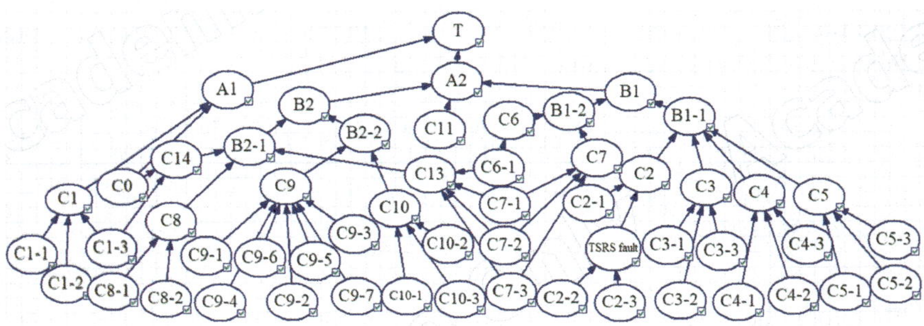

Fig. 4 Bayesian network

Table 2 Occurrence frequency of main crucial nodes

Node number	Node name	Character of nodes	Occurrence frequency (times/h)
T	Unprotected train overspeed	Root node	1.377×10^{-9}
A1	ATP speed is lower than actual speed	Second-level node	8.204×10^{-9}
A2	Control curve errors	Second-level node	4.541×10^{-7}
B1	Allowable excessive speed	Third-level node	2.903×10^{-7}
B2	Excessive driving permit	Third-level node	2.670×10^{-7}

Result–Cause Reasoning. Assuming that the state of the upper node is state = 1, the posterior probabilities of the lower nodes are calculated based on the reasoning algorithm of BN, and then the weaknesses of TCS can be analyzed.

1. Posterior probability of secondary nodes

As can be seen from Table 3, the direct factor leading to the occurrence of risk events is 'error in calculation of permitted speed curve.' The causes of A2 are found in ground and on-board subsystems. Therefore, the weak modules in ground and on-board subsystems are analyzed.

2. Posterior probability of weak modules

The modules of on-board and ground subsystems include C0–C11. Assuming that the error of permitted speed curve calculation occurs at the upper node, i.e., state = 1 of A2, the posterior probability of each module is shown in Fig. 5.

As can be seen from Fig. 5, the track circuit, BTM, and TCC are the important factors leading to the error of control curve calculation.

3. Posterior probability of blade nodes

From Fig. 5, it can be concluded that the posterior probability of dangerous side faults occurring in track circuit is the greatest, which is the most important factor leading to 'permitted speed curve calculation error,' followed by the occurrence of

Table 3 Posterior probabilities of A1 and A2

Node number	Node name	Character of nodes	Posterior probability
T	Unprotected train overspeed	Root node	1
A1	ATP speed is lower than actual speed	Second-level node	0.018
A2	Permitted speed curve errors	Second-level node	0.982

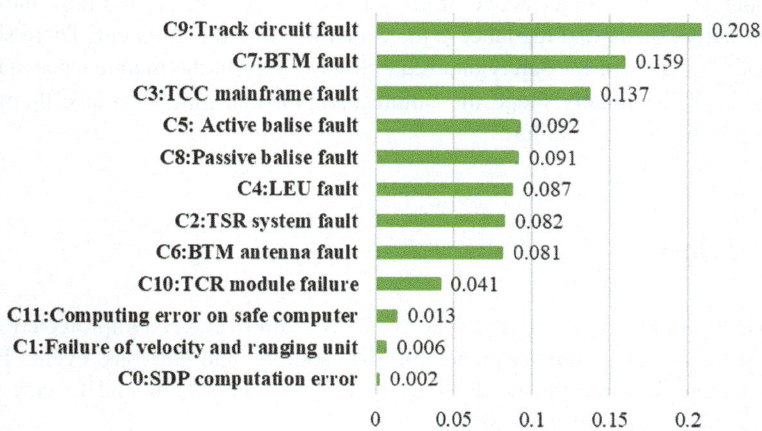

Fig. 5 Posterior probability diagram of system module

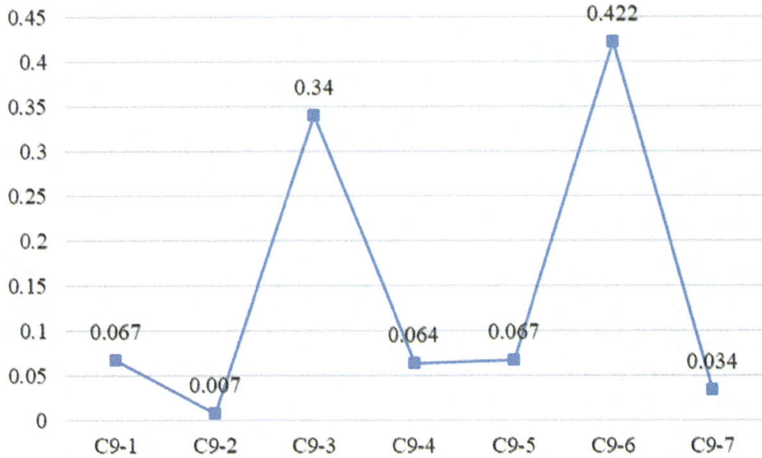

Fig. 6 Posterior probability of C9 nodes

BTM and TCC dangerous side faults. This part takes track circuit as an example to further analyze the causes of C9 (which leads to dangerous side fault of track circuit). According to Fig. 4, the next level nodes of C9 include C9-1 to C9-7. Assuming C9's state = 1, the posteriori probability of the next level node is calculated and plotted as shown in Fig. 6. It is illustrated that the two most important factors causing C9 to occur are the dangerous side faults of transmitters and receivers in track circuits, which account for the largest proportion.

From above analysis, it seems that the frequency of risk events is 'rare,' but if only from the point of view of equipment, the frequency of dangerous side faults in A0 (equipment) is 4.589×10^{-7} times/h, which is 'very rare.' According to the above analysis, the relevant measures should be taken from the track circuit first, and the transmitter and receiver should be paid more attention to. The second is BTM and TCC; the former is one of the entrances to receive ground information in on-board equipment, and the latter is the center of ground equipment. There should be higher demand for the safety and reliability of both. In the maintenance stage, it is required to accurately judge the failure components, timely replace them, and reduce the incidence of failure.

5 Conclusion

This paper takes the risk event of CTCS-2 TCS 'train overspeed unprotected' as an example to calculate the frequency of risk sources and the risk event. BN is established to identify the weaknesses in CTCS-2 TCS that lead to risk event occurring. The main research results are as follows:

1. For the problem that it is difficult to obtain the historic fault data, a mathematical model of the frequency of risk sources is established by making full use of expert knowledge and adopting the theory of fuzzy mathematics.
2. The frequency of 'train overspeed unprotected' risk event is 4.589×10^{-7} times/h, which belongs to the 'rare' level.
3. By constructing BN, the weaknesses of CTCS-2 TCS are analyzed. The results show that the posterior probability of dangerous side faults in track circuit is the largest, which is the most important factor leading to 'permitted speed curve calculation error,' followed by BTM and TCC. Among them, the transmitter and receiver in the track circuit are the main causes of the dangerous side failure.

Acknowledgements Authors would like to acknowledge the support of the research program of Comprehensive Support Technology for Railway Network Operation (2018YFB1201403), which is a subproject of Advanced Railway Transportation Special Project belonging to the 13th Five-Year National Key Research and Development Plan funded by Ministry of Science and Technology of China.

References

1. An M, Qin Y, Jia LM et al (2016) Aggregation of group fuzzy risk information in the railway risk decision making process. Saf Sci 82:18–28
2. Yingying X, Chao L, Tao T (2016) Research on risk assessment of CTCS based on fuzzy reasoning and analytic hierarchy process. In: International conference on intelligent information processing. ACM
3. Kabir S, Walker M, Securius P et al (2016) Fuzzy temporal fault tree analysis of dynamic systems. Int J Approx Reason 77(C):20–37
4. Benarieh D, Chen Z (2006) Linguistic-labels aggregation and consensus measure for autocratic decision making using group recommendations. IEEE Trans Syst Man Cybern Part A: Syst Hum 36(3):558–568
5. Ma J, Ruan D, Xu Y et al (2007) A fuzzy-set approach to treat determinacy and consistency of linguistic terms in multi-criteria decision making. Int J Approx Reason 44(2):165–181
6. Bobbio A, Portinale L, Minichino M et al (2001) Improving the analysis of dependable systems by mapping fault trees into Bayesian networks. Reliab Eng Syst Saf 71(3):249–260
7. Zhang C, Ma C, Hu Y et al (2005) On quantitative analysis of fault tree based on Bayesian network. J Missile Guidance 25(2) (in Chinese)
8. Su H, Che Y, Zhang Y (2014) Dependability assessment of CTCS-3 on-board subsystem based on Bayesian network. China Railw Sci 35(5):96–104 (in Chinese)
9. Yue C, Guo S (2013) Typical failure cases of high-speed railway signal equipment. China Railway Press 1:134–135 (in Chinese)
10. GB/T 21562-2008 specification for reliability, usability, maintainability and safety of rail transit (2008) (in Chinese)
11. Zhang X, Pei D, Dai j (2013) Fuzzy mathematics and rough set theory. Tsinghua University Press, pp 1–99 (in Chinese)
12. Wang T, Li M, Xu Z (2003) Fuzzy risk assessment of safety-critical software. Comput Appl 23:182–184
13. Wang S (2010) Bayesian network learning, reasoning and application. Lixin Accounting Publishing House, Shanghai (in Chinese)

Analysis of the Influence of Track Irregularity on Comfort Performance of Maglev Train Based on Site Monitoring

Jingyu Huang and Xiong Zhou

Abstract Track irregularity is an important factor that causes the vibration of maglev train and has a significant impact on the comfort of the train. In order to accurately reflect the characteristics of the track irregularity of maglev traffic and provide data support for the follow-up research, the track irregularity of Beijing S1 maglev line was tested. And a multi-dof medium-low speed maglev train-track dynamics simulation model was established to obtain the dynamic response of the train. Through the analysis of comfort index, the influence of track irregularity on ride comfort was studied.

Keywords Maglev traffic · Rail irregularity · Comfort

1 Introduction

Track irregularity is an important factor that causes the vibration of maglev train and has a significant impact on the comfortable performance of the vehicle during operation. Due to the lack of engineering application of maglev transportation at present, the actual measured irregularity data of maglev track are insufficient. Most relevant studies refer to the data of wheel-rail irregularity, which cannot accurately reflect the characteristics of the irregularity of maglev track [1–3].

Huang et al. obtained the data of track irregularity of maglev by on-site detection of maglev traffic lines, and carried out subsequent correlation analysis, proving that the track irregularity of maglev cannot be simply referred to the wheel-rail traffic system [4]. Wu et al. proposed an onboard dynamic detection system based on the

J. Huang (✉)
National Maglev Transportation Eegineering R&DCenter, 4800 Caoan Road, Shanghai, China
e-mail: huangjingyu@tongji.edu.cn

J. Huang · X. Zhou
Department of Civil Engineering, Tongji University, Shanghai, China
e-mail: xiaoxiong@tongji.edu.cn

Y. Qin et al. (eds.), *Proceedings of the 4th International Conference on Electrical and Information Technologies for Rail Transportation (EITRT) 2019*, Lecture Notes in Electrical Engineering 639, https://doi.org/10.1007/978-981-15-2866-8_59

principle of inertial reference method for the vertical irregularity detection of magnetic levitation track [5]. Shi et al. proposed the treatment method of high-speed magnetic levitation track irregularity, analyzed the irregularity characteristics, fitted the track spectrum, and provided a reference for line space maintenance [6]. Based on site monitoring, zhang et al. studied the dynamic response of high-speed maglev vehicle/guideway system [7].

In this work, the site monitoring on Beijing maglev line was performed and a multi-dof medium-low-speed maglev train-track dynamics simulation model with measuring irregularity data were established. The dynamic response of the train under the excitation of irregularity was obtained by simulation calculation. Through the analysis of comfort index, the influence of track irregularity on ride comfort was studied, that can provide reference for the research of track maintenance and train vibration control.

2 Site Monitoring

In order to study the characteristics of uneven maglev track and provide data support for train ride comfort, the track irregularity of Beijing S1 maglev line was tested on site. The test characteristics include the repeatability of each parameter under the same speed and direction and the repeatability of each parameter under the same direction and different speed. The measured orbit parameters include left and right height, right and left orbit direction, level, gauge and curvature. Some of the results are shown in Fig. 1.

3 Simulation Model

Based on the above measured track irregularity data, a multi-dof maglev train-track dynamics simulation model was established (Fig. 2). This simulation model was modeled by general dynamics software, and was composed of car body, travelling mechanism and the track. The car body and travelling mechanism were built with reference to the 3d model of maglev train, and the corresponding constraint connections were imposed on each part according to the actual structure. The relevant parameters of the first and second suspension systems were set according to the actual situation. The track was constructed according to the above measured data, and the vehicle load is the weight of the vehicle body. Through the simulation of the whole model, the ride comfort performance of the vehicle under the excitation of track irregularity was obtained.

According to China GB5599-1985 [8] stationary evaluation index, the calculation formula of passenger vehicle stationary index W_i is as follows:

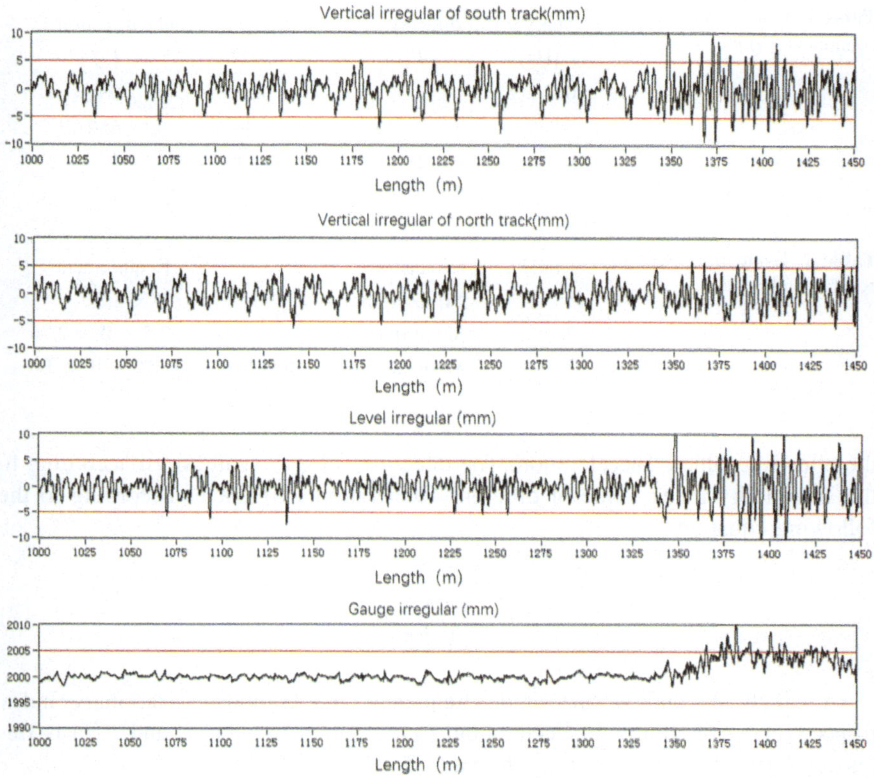

Fig. 1 Track irregularity test results

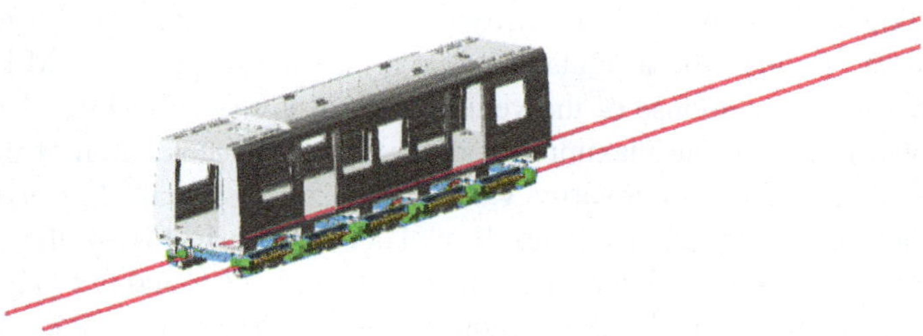

Fig. 2 Vehicle-track dynamics simulation model

$$W_i = 7.08\sqrt[10]{A^3F(f)/f} \tag{1}$$

The above equation is the stationarity index of a single frequency. However, the actual vibration of a vehicle contains many frequencies, so the acceleration needs to be grouped according to the frequency of vibration acceleration, so as to calculate

Table 1 Correction coefficient $F(f)$

The vertical direction		The horizontal direction	
f (Hz)	$F(f)$	f (Hz)	$F(f)$
0.5–5.9	$0.325f^2$	0.5–5.4	$0.8f^2$
5.9–20	$400/f^2$	5.4–26	$650/f^2$
>20	1	>26	1

Table 2 Evaluation grade of comfort index

Grade	Evaluation	Requirement
1	No discomfort	$W < 2.5$
2	Slight discomfort	$2.5 < W < 2.75$
3	Moderately uncomfortable	$2.75 < W < 3.0$

the effective value of acceleration. The effective value is synthesized according to the stationarity index value of different acceleration frequencies according to the following equation:

$$W = \sqrt[10]{\sum_{i=1}^{n} W_i^{10}} \tag{2}$$

In which W_i and w are single frequency and composite stationary index respectively, A is vibration acceleration, f is vibration frequency, and $F(f)$ is frequency correction coefficient, as shown in Table 1.

The evaluation standard of the comfort index grade (GB5599-1985 standard) was shown in Table 2. The riding comfort index of maglev train should not be lower than grade 1 of this standard.

According to the above multi-body dynamics simulation and analysis model, the vehicle passed through the irregular track at a maximum speed of 120 km/h. The calculated vibration response of the vehicle body was shown in Figs. 3 and 4.

As shown in Fig. 3, the maximum vertical vibration acceleration of the vehicle was 0.62 m/s^2, and the stationarity value was 1.53, less than 2.5, had met the requirements of the stationarity index level 1. As shown in Fig. 4, the maximum acceleration of lateral vibration of the vehicle was 0.15 m/s^2, and the stability value was 1.42, less than 2.5, had met the requirements of the stability index level 1.

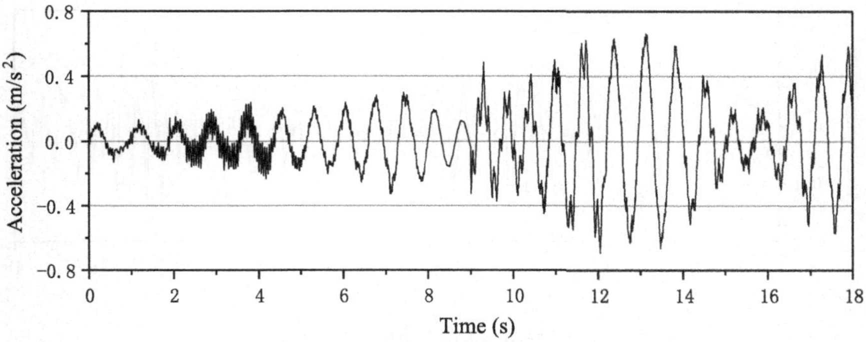

(a) at the time domain

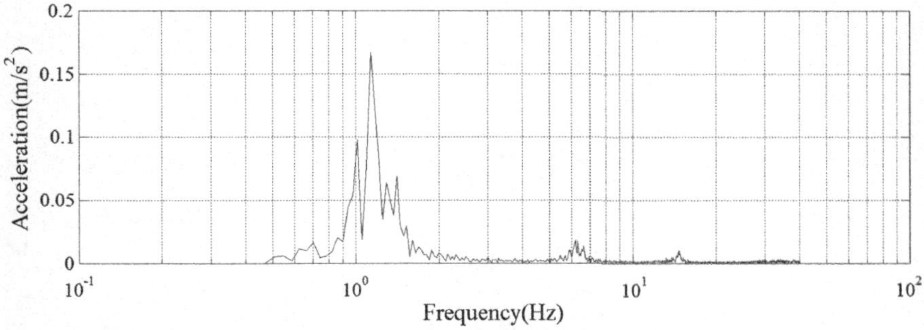

(b) at the frequency domain

Fig. 3 Vertical acceleration of vehicle body vibration

4 Conclusion

In this paper, the field measured track irregularity data of Beijing S1 maglev line were analyzed and processed, and a multi-body dynamics simulation model was established to analyze the influence of track irregularity on the operation comfort of maglev train, and the following conclusions were drawn:

Based on the analysis and research of the measured track irregularity data, this paper concluded that the track of maglev transportation system has good operation quality because of its special structural form, strict requirements for the manufacturing and installation accuracy of track beam, and the overall small amplitude of track irregularity of the track itself.

Based on the modeling of the existing maglev vehicle body and the analysis of its dynamic response under the excitation of track irregularity, this paper concluded that the vibration response of the vehicle body could meet the related standard requirements of train ride stability under such conditions of track irregularity, and the ride comfort of the vehicle body could meet the first level requirements of GB5599-1985 standard.

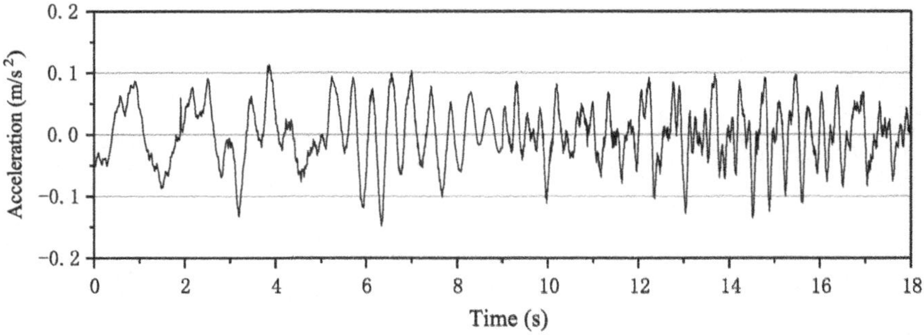

(a) at the time domain

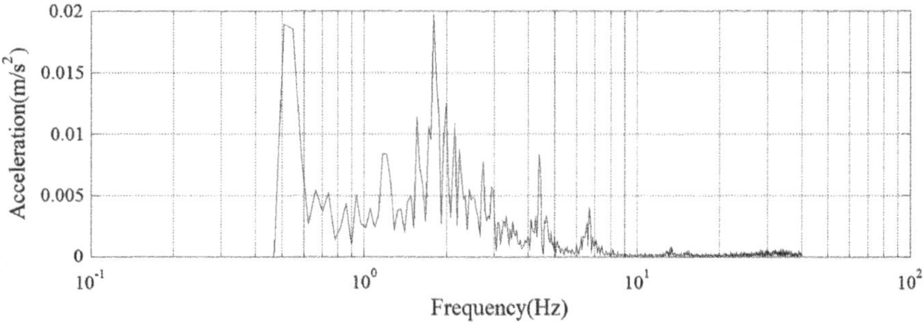

(b) at the frequency domain

Fig. 4 Lateral acceleration of vehicle body vibration

The maglev rail transit system is quite different from the traditional wheel-rail transit system because of its unique track form. In the aspect of track irregularity, the characteristic curve of the wheel-rail system cannot accurately reflect the characteristics of track irregularity of the maglev traffic system, and the site monitoring for the track of the maglev traffic system is of great significance.

Acknowledgements The research described in this paper was financially supported by the National 13th Five-Year Science and technology support program (Project No.: 2016YFB1200602).

References

1. Ju SH, Leong CC, Ho YS (2015) Control of maglev trains moving on bridges during foundation settlements. In: International conference on computer information systems and industrial applications, 28–29 Jun 2015
2. Li P, Hu W (2016) Influence analysis of magnetic track beam irregularity on the suspended state. Electr Drive Locomot 2016:65–68 (in Chinese)

3. Fan Q (2012) Discussion on the demand of high-speed EMS vehicle on guideway irregularity. China Railw Sci 2002(23):73–76 (in Chinese)
4. Huang JY, Wu ZW, Shi J, Gao Y, Wang D-Z (2018) Influence of track irregularities in high-speed maglev transportation systems. Smart Struct Syst 21(5):571–582
5. Wu j, Tang j, Hong X (2018) Dynamic measurement system design for vertical irregularity of high speed maglev track. J Tongji Univ (Nat Sci) 46(5)
6. Shi J, Fang W-s, Wang Y-j, Zhao Y (2014) Measurements and analysis of track irregularities on high speed maglev lines. J Zhejiang Univ Sci A: Appl Phys Eng 15(06):385–394
7. Zhang L, Huang J (2019) Dynamic interaction analysis of the high-speed maglev vehicle/guideway system based on a field measurement and model updating method. Eng Struct 180:1–17
8. Railway vehicles specification for evaluation the dynamic performance and accreditation test (GB5599-1985)

Information Service Frequency of Urban Rail Transit Based on Passenger Satisfaction

Bei Li, Huijuan Zhou and Yu Liu

Abstract With the rapid development of urban rail transit, the phenomenon of outburst passenger flows flocking to stations is occurring much more frequently. This paper studies the optimal frequency of information service from the perspective of passengers. Firstly, we obtain the research data by questionnaire and put forward the influence factors of passenger satisfaction under outburst mass passenger flow. Then we use survey data to screen them by SPSS software and further determined the effective influence factors. To determine the specific relationship between information service frequency and passenger satisfaction, questionnaire data is used to curve each factor and comprehensively consider these factors to establish a combined optimization satisfaction model. Finally, the optimal information service frequency per hour is determined by the model solution.

Keywords Urban rail transit · Information service frequency · Outburst mass passenger flow · Passenger satisfaction

1 Introduction

Congestion problems in China's large cities have created an urgent need for the construction and expansion of the metro system. More and more cities will open a new situation in urban rail transit network operation. The high complexity and

B. Li · H. Zhou (✉)
Beijing Key Lab of Urban Intelligent Traffic Control Technology,
North China University of Technology, No. 5 Jinyuanzhuang Road,
Shijingshan District, 100144 Beijing, China
e-mail: zhouhuijuan@ncut.edu.cn

B. Li
e-mail: li0807bei@163.com

Y. Liu
CCCC Railway Consultants Group Co., Ltd., Zhongjiao Railway Building,
Building No. 19, No. 43, North Third Ring Road, Haidian District, 100088 Beijing, China
e-mail: liuyu0921smile@163.com

Y. Qin et al. (eds.), *Proceedings of the 4th International Conference on Electrical and Information Technologies for Rail Transportation (EITRT) 2019*, Lecture Notes in Electrical Engineering 639, https://doi.org/10.1007/978-981-15-2866-8_60

strong correlation of the network enhance the blindness of passengers in the travel process. The guidance of official information is more important to passengers. High-quality information services provide passengers with more effective information to meet the needs of passengers and thus effectively improve the level of traffic operation management.

Passengers in different environments have different requirements for information services. When rail transit is in an exceptional circumstance, there will be higher requirements for information services. Outburst mass passenger flow refers to the situation that the passenger flow in the subway station suddenly reaches the maximum capacity of 70% or more under the influence of the external environment, which emphasizes the sudden increase of passenger flow in a short period of time. This situation brings about an unbalanced between transportation capacity and passenger demand, which increases the risk of safety.

In order to improve the safety and operation efficiency of rail transit, domestic and foreign researchers have studied it from different aspects. Zhang and Chen [1] studied the key factors affecting emergency management by constructing the Markov chain model under a Petri net. Zhao et al. [2] proposed a subway emergency information dissemination model. Yin [3] studied from the scope and content of the information service and proposed a passenger flow induction model; Oña et al. [4] analyzed the behavior trend of transfer passengers through the orderly regression model; Bababeik et al. [5] solved the problem of passenger congestion in emergencies from the perspective of dispatching.

We find that there are fewer studies from the perspective of information service frequency. This paper takes passenger satisfaction as an indicator to optimize information service frequency, which gives passengers a clearer guide under outburst mass passenger flow.

2 Influence Factors of Passenger Satisfaction

We obtain the factors affecting passenger satisfaction at different frequencies by questionnaire to provide data support for determining the optimal frequency. Since passenger satisfaction shows the subjective feelings of passengers, the questionnaire method is a common method for researchers to ask for opinions from the respondents [6].

There are 14 questions in the questionnaire. We investigate passenger satisfaction from the aspects of gender, age, education level, outburst mass passenger flow experience, travel time urgency, and travel purpose. A total of 2693 questionnaires were issued, of which 2607 were valid questionnaires and 83 were invalid questionnaires.

According to the proposed influence factors of passenger satisfaction, we use the SPSS software to perform a Levene variability equality test on each factor, which determines whether equal variance is used. Next, we perform independent sample

T test on different variables in order to decide whether the factor has a significant moderating effect on information service satisfaction.

Finally, the test results demonstrate that gender and education levels have no significant adjustment effect on the relationship between information service frequency and passenger satisfaction, while age, purpose of travel, travel time urgency, and experience of outburst mass passenger flow can moderate the relationship.

3 Satisfaction Model

3.1 Model Establishment

According to the test results, the passengers' age, travel purpose, time urgency, and the experience of outburst mass passenger flow all have an impact on the passenger satisfaction under outburst mass passenger flow. In order to comprehensively consider the impact of four factors on passenger information service satisfaction, a relationship model between each factor and satisfaction is proposed:

$$S_{a(f)} = \partial\left[F_{1(f)}, F_{2(f)}, F_{3(f)}, F_{4(f)}\right] \tag{1}$$

Among them, $F_{1(f)}$, $F_{2(f)}$, $F_{3(f)}$, $F_{4(f)}$ represent the relational functions between passenger's age, travel purpose, travel time urgency, and the experience of outburst mass passenger flow and satisfaction, respectively.

According to the literature [7], when considering user satisfaction, the relationship between factors and satisfaction is often similar to the inverted "U"-type distribution. So we consider the relationship as a quadratic function in this study, then the functional models of $F_{1(f)}$, $F_{2(f)}$, $F_{3(f)}$, $F_{4(f)}$ are, respectively,

$$F_{1(f)} = \varphi_1 f^2 + \omega_1 f + \theta_1 \tag{2}$$

$$F_{2(f)} = \varphi_2 f^2 + \omega_2 f + \theta_2 \tag{3}$$

$$F_{3(f)} = \varphi_3 f^2 + \omega_3 f + \theta_3 \tag{4}$$

$$F_{4(f)} = \varphi_4 f^2 + \omega_4 f + \theta_4 \tag{5}$$

The satisfaction degree model of combinatorial optimization is as follows:

$$S_{a(f)} = \alpha F_{1(f)} + \beta F_{2(f)} + \gamma F_{3(f)} + \delta F_{4(f)} \tag{6}$$

Among them, α, β, γ, δ are the weight coefficients of four kinds of factors.

3.2 Model Solution

In order to obtain the optimal satisfaction value, this study takes different influence factors functions as different objective functions and uses an improved genetic algorithm [8] to solve the objective function under the influence of multiple factors. The solution steps are as follows:

Firstly, the population size is M, and N is the number of optimal solutions.

Step 1: Population initialization. Firstly, the initial population with M size is randomly generated.
Step 2: Update the temporary solution set. The initial population of M scale in Step 1 is brought into the objective function to obtain p objective function values, and p function values are stored as temporary P solution sets.
Step 3: Choose. $M–N$ paternal generations were selected: According to the random weights of four influence factors, fitness values were calculated and paternal individuals were selected.
Step 4: Hybridization. The offsprings were produced by crossing the selected individuals of each parent generation.
Step 5: Variation. Variation was performed on each offspring produced by Step 4.
Step 6: Optimal choice. N individuals were randomly selected from the temporary P solution set, and the selected N individuals were directly added to $M–N$ individuals to form a new population.
Step 7: Stop the operation. When the constraints are satisfied, stop the algorithm and output the optimal results.

The flow chart is shown in Fig. 1.

Under the constraint of the optimal information service frequency interval, the improved genetic algorithm is used to solve the objective function.

4 Determine the Optimal Frequency of Information Services

4.1 Satisfaction Relation Under Different Influence Factors

In order to determine the specific relationship between the frequency of information services and passenger satisfaction, we carry out a more specific classification of different influence factors. According to the results of the questionnaire, the frequency range of information service that passengers can accept under outburst mass passenger flow is 11–30 times per hour. Then, a satisfaction degree corresponding to different frequencies is averaged and performed regression fitting and parameter estimation on them. Finally, the relationship function between passenger satisfaction degree and frequency of information service is obtained.

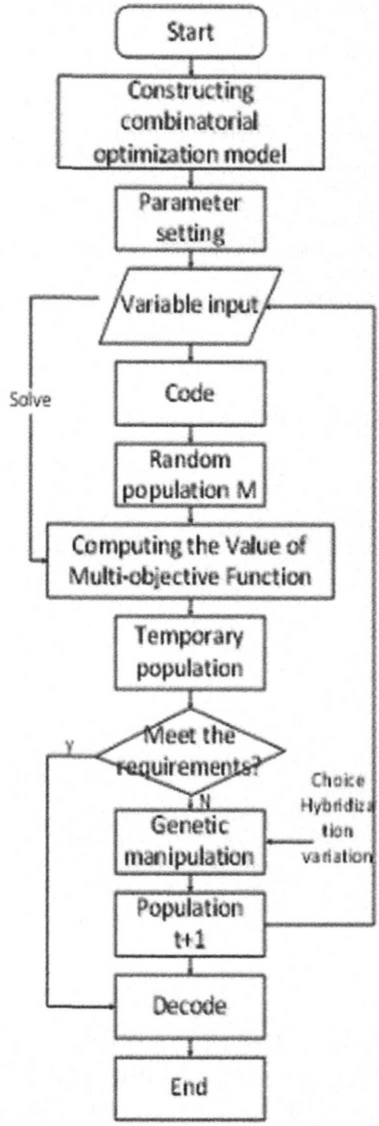

Fig. 1 Flow chart of genetic operation

When the variable is age, 40-year old as an important watershed of age, people's emotions, perceptions, and attitudes to deal with problems will change to a certain extent. Therefore, the survey respondents are divided into two categories with 40 as the age limit. Their parameter estimation results and satisfaction fitting curves are shown below.

Table 1 shows that their fitting effects are 84.5% and 74.6%, respectively, which reach a significant level ($P < 0.01$). The functional relationship under the age of 40

Table 1 Model statistics and parameter evaluation

Equation	Influence factor	Model summary				Parameter estimation			
	Age	R^2	F	df1	df2	Significance	Content	b1	b2
Linear	Under 40 years old	0.071	1.769	1	23	0.196	3.700	0.105	
Quadratic curve		0.845	60.086	2	22	0.000	−2.561	1.496	−0.054
Linear	Over 40 years old	0.430	17.321	1	23	0.000	1.725	0.257	
Quadratic curve		0.746	32.299	2	22	0.000	−2.278	1.146	−0.034

Dependent variable: satisfaction; independent variable: frequency

Fig. 2 Under the age of 40

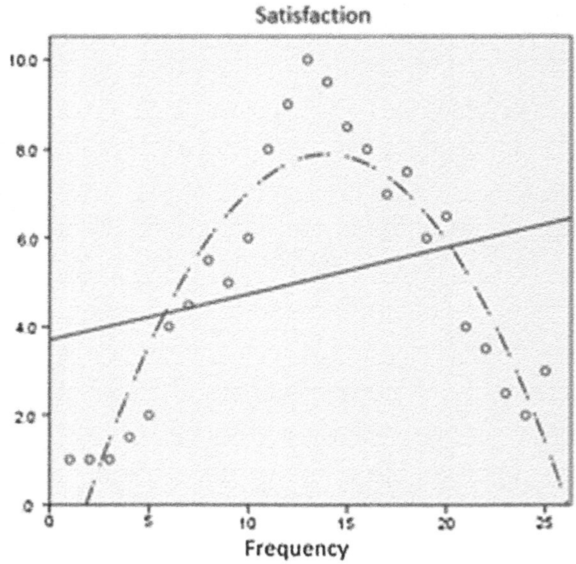

is $Y = -0.054X^2 + 1.496X - 2.561$, and the relationship over the age of 40 is $Y = -0.034X^2 + 1.146X - 2.278$, the fitting curve is shown in Figs. 2 and 3.

When the variable is travel purpose, we found that survey respondents can be roughly divided into two categories: commuting travel and non-commuting travel passengers. We conduct model statistics and parameter evaluation for the data of them, respectively. The functional relationship of non-commuting passengers is $Y = -0.016X^2 + 0.744X - 1.12$, and the functional relationship of commuting passengers is $Y = -0.053X^2 + 1.364X - 0.879$. The fitting effect reached a significant level with a fitness of 79.2% and 77%, respectively.

When the variable is travel time urgency, we found that passengers' sensitivity to information will be enhanced under outburst mass passenger flow. Especially when passengers have different urgency, passengers' satisfaction with different

Fig. 3 Over the age of 40

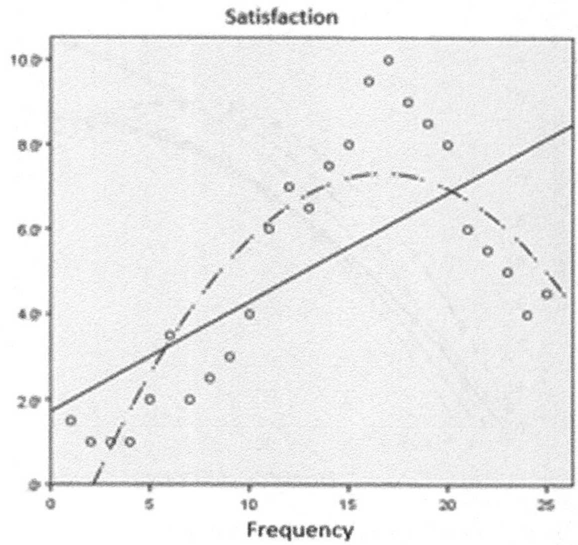

frequencies will be significantly different. Therefore, we divide them into two categories: urgent travel time and loose time passengers. The functional relationship of urgent time passengers is $Y = -0.038X^2 + 1.29X - 3.372$, and the functional relationship of loose time passengers is $Y = -0.033X^2 + 1.122X - 2.204$. The fitting effect reached a significant level with a fitness of 76.8% and 74%, respectively.

Whether there is an outburst mass passenger flow experience will lead to different psychological states of passengers facing the unexpected situation. The passenger data will be classified according to whether there is an outburst mass passenger flow experience. Model statistics and parameter evaluation are carried out for data with outburst mass passenger flow experience and without experience. The former functional relationship is $Y = -0.012X^2 + 0.669X - 0.91$, and the latter functional relationship is $Y = -0.051X^2 + 1.301X - 0.637$. The fitting effect reached a significant level with a fitness of 83.2% and 75.1%, respectively.

4.2 *Analysis of Satisfaction Result Under Different Influence Factors*

According to the fitting analysis results in Sect. 4.1, there is an inverted "U" relationship between information service frequency and passenger satisfaction under different influence conditions. So the corresponding quadratic function relationship between them is established. The results under different classification conditions are shown in Fig. 4.

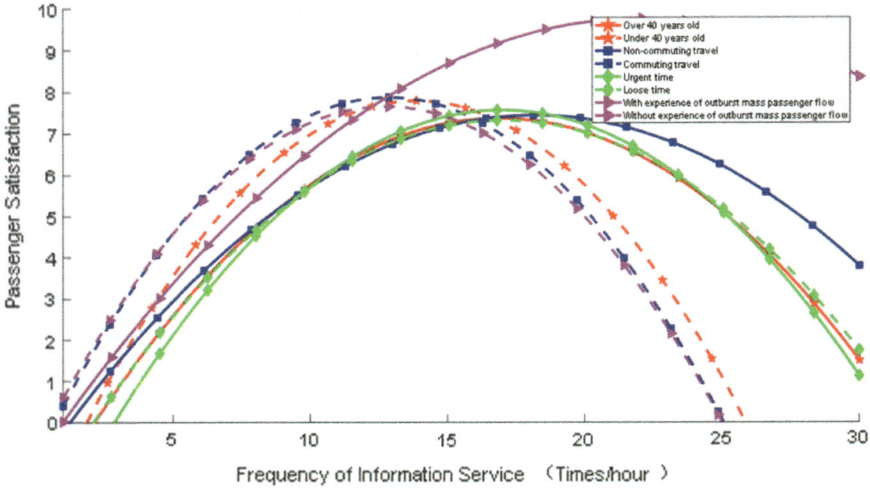

Fig. 4 Satisfaction curve of different influence factors

As shown in the Fig. 4, the information frequency of passenger satisfaction varies with different classification conditions, but it is basically stable at 11–25 times per hour. When the passenger is over and under 40 years old, the satisfactory information service frequency is 18 and 14 times per cycle, and the highest satisfaction is 7.38 and 7.8, respectively, as shown by the red line in the above figure; when passengers for commuting and non-commuting, the expected frequency of information service is 13 and 17 times per cycle, and the satisfaction is 7.88 and 7.53, respectively; when passengers travel time is tight or slack, their satisfactory information service frequency is 19 and 17 times per cycle, and the highest satisfaction rate reaches 7.58 and 7.33, respectively; passengers with experience of outburst mass passenger flow show a stronger need for official information. The optimal frequency of satisfaction with information service is about 20 times per cycle, and the satisfaction rate is as high as 8.4; passengers without the experience of outburst mass passenger flow only need information service 15 times per cycle, and their satisfaction is 7.66.

4.3 Determining the Optimum Frequency of Information Service

Based on the detailed analysis in Sect. 4.2, we select the one with lower satisfaction in each indicator and establish a combination optimization satisfaction model. So we choose the satisfaction functions of passengers over 40 years old, non-commuting passengers, passengers with tight time and passengers without sudden large passenger flow experience. The functions corresponding to each factor are as follows:

$$F_{1(f)} = -0.034f^2 + 1.146f - 2.278 \quad (7)$$

$$F_{2(f)} = -0.016f^2 + 0.744f - 1.12 \quad (8)$$

$$F_{3(f)} = -0.038f^2 + 1.29f - 3.372 \quad (9)$$

$$F_{4(f)} = -0.051f^2 + 1.301f - 0.637 \quad (10)$$

Formulas 7–10 are introduced into Formula 6 and solved by genetic algorithm proposed in Sect. 3. In this study, the weight coefficient was 0.25, respectively. After many experiments, the population size M is set at 20, the crossover probability is 0.8, and the mutation probability is 0.5%.

After solving, when the final outcome of information service frequency is $f = 16.24$, passenger satisfaction reaches the maximum. Considering the actual situation, the frequency of information service should be taken as an integer. So the satisfaction value of information service frequency should be compared with that of 16 and 17 times per hour. When $f = 17$, S_a is 7.09, and when $f = 16$, S_a is 7.14; so the final value is $f = 16$. The solution results under different information service frequencies are shown in Fig. 5.

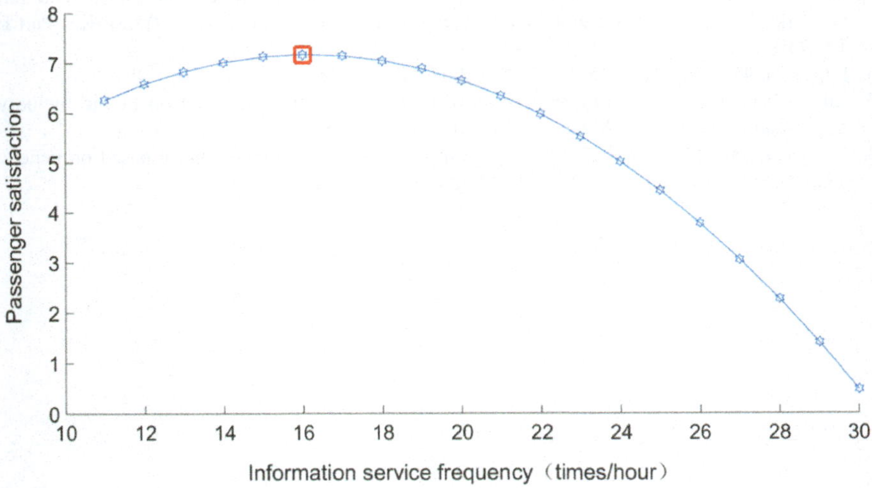

Fig. 5 Relationship between the frequency of information services and satisfaction

5 Conclusion

As the ultimate service object of urban rail transit, passenger satisfaction plays a vital role in optimizing the information service system. Taking passenger satisfaction as an index, considering the age, the travel purpose, the time urgency, and the experience of outburst mass passenger flow, this paper establishes a passenger satisfaction model of combination optimization and uses genetic algorithm to solve the model to obtain the optimal information service frequency per hour. This study makes up for the research gap of information service frequency in information service system and has important significance to improve the quality of rail transit information service under outburst mass passenger flow.

Acknowledgements The authors would like to acknowledge the support of the National Key R&D Program of China (2016YFB1200402).

References

1. Zhang YF, Chen WZ (2015) Markov chain analysis of emergency response mode of sub way operation emergency. Chin J Saf Sci 25(2) (in Chinese)
2. Zhao HF, Sun YQ, Tse E (2017) Information dissemination of subway emergency events based on bounded trust model. Syst Eng Theory Pract 37(12):3244–3252 (in Chinese)
3. Yin HD (2017) Research on passenger travel behaviour modelling and passenger flow induced optimization of urban rail transit under operational interruption (in Chinese)
4. Oña J, Oña R, Laura E et al (2018) An ordered regression model to predict transit passengers' behavioural intentions. Case Stud Transp Policy 6:449–455. S2213624X17301372
5. Bababeik M, Khademi N, Chen A (2018) Increasing the resilience level of a vulnerable rail network: the strategy of location and allocation of emergency relief rains. Transp Res Part E 119:110–128
6. https://wenku.baidu.com/view/cc376db8ccbff121dc36832b.html
7. Liu Z, Zhang Y (2011) Emergency information release strategy based on public memory manipulation theory. Res Manag 32(9):100–107 (in Chinese)
8. Li RM, Lu HP (2009) Multi-objective optimization of traffic signal control based on genetic algorithm. J Chang'an Univ: Nat Sci Ed 3:85–88 (in Chinese)

Traceable Correction Method for the Online Measurement System of Train Wheelset Profile Based on Neural Network

En-quan Fang, Jie Qin, Li-li Geng, Xiang Li and Zhipeng Wang

Abstract Online measurement system of train wheelset profile can automatically measure the wheelset size of all trains passing through the detection zone and timely pre-alarm the oversized wheelset to ensure the safe and smooth operation of the train. In this paper, the main technologies of wheelset size online detection at home and abroad are retrieved, and the problems are analyzed, such as data accuracy is difficult to maintain for a long time, the detection index is imperfect, and the dynamic and static calibration of the system has different in practical applications. Also through analyzing the main sources of detection error from the detection principle and application conditions, expounding the influence of radial runout, and axial displacement on measurement results emphatically, a parameter tracing method based on neural network is proposed and key calibration steps and engineering application examples are also given.

Keywords Wheelset profile · Online measurement · Neural network · Parameter correction

1 Introduction

Train wheelset is an important part of rail vehicle, wheelset profile is closely related to wheel–rail dynamics, the size of wheel and rail is related to the safety and smooth operation of vehicles, because of which the change of real-time monitoring of wheelset profile has attracted much attention. At present, the main method for the

E. Fang (✉)
Guangzhou Metro Group Co., Ltd., 510330 Gguangzhou, China
e-mail: fensar@163.com

J. Qin · L. Geng
Guangzhou Auto-Transport Measure & Control Technology Ltd., 510020 Guangzhou, China

X. Li · Z. Wang
State Key Laboratory of Rail Traffic Control and Safety, Beijing Jiaotong University, Beijing, China

Y. Qin et al. (eds.), *Proceedings of the 4th International Conference on Electrical and Information Technologies for Rail Transportation (EITRT) 2019*, Lecture Notes in Electrical Engineering 639, https://doi.org/10.1007/978-981-15-2866-8_61

online measurement system of train wheelset profile includes optical intercept image technology, laser ranging technology, automatic vision measurement technology; many domestic scholars have conducted related research [1–3]. Zhang et al. [4] analyze the causes of image distortion in the method for the dynamical measurement system of train wheelset profile based on optical intercept image measurement technology, and the calibration algorithm is given in combination with the measurement system. A new method for online measurement of train wheel size based on 2D laser displacement sensor has been proposed by Li et al. [5]. The system structure and online measurement principle of the method are clarified. Chen and Ren [6] introduce a measurement method for measuring wheelset profile, tread and near-surface defects, which uses charge-coupled device (CCD) camera, laser displacement sensor, and electromagnetic ultrasonic testing technology. However, in terms of the method for the online measurement system of train wheelset profile, there are still the following problems to be solved:

(1) The long-term stability of the line wheel pair size measurement still has problems. Due to the influence of the installation conditions of the rail side sensors and the factors such as train running oscillation, vibration, and ambient temperature, there are many problems such as complex calibration, low accuracy, and poor long-term stability.
(2) The dynamic measurement of wheelset size was not raised to the height of train moving part safety monitoring. A large number of security warning data show that the main state indexes of the train moving part safety warning are size, vibration, and temperature [7, 8], and there is a certain relationship between the changes of the three-state indicators when the running department fails. At present, most people believe that the main reason for inaccurate size measurement is speed rather than measurement method. The position of the measuring device is designed in the lower velocity area of the access section which lost the basic conditions of big data application, because there is no load and low speed; also, there is no acceleration or deceleration force at that time, and it is difficult to find the early abnormal of the running part.
(3) The method for the online measurement system of train wheelset profile is carried out when the train is running; however, the system can only be statically calibrated, which makes differences between conditions of use and calibration; this difference will have a certain impact on online system measurements. Although artificial intelligence system parameter correction method based on neural network has been studied and attempted, it is difficult to achieve the system parameter traceability correction due to the small amount of data, the large number of input variables and the existence of a variety of interference. It is also impossible to track changes in dynamic and static calibration parameters due to environmental parameters (such as temperature, dust, and foundation settlement) during train operation [9].

Based on the above problems that combine the principle of online detection technology of wheelset profile, this paper analyzes the error compensation method in the online wheelset measuring system, considering the influence of the radial runout, and axial displacement on the measurement results; also, the method and key steps of parameter tracking and correction based on neural network are proposed.

2 Traceable Correction Method for the Online Measurement System of Train Wheelset Profile

2.1 Main Theoretical Basis and Key Characteristic Quantity

The online measurement system of train wheelset profile mainly measures the wheel diameter, rim height, rim thickness, rim comprehensive value (QR value), and wheelset inner distance; Fig. 1 is the schematic of the measurement method. Left picture mainly shows the scanning process of the rim tread curve, complete measurement of rim thickness, rim height, and QR value, and right picture shows the measurement method of the medial distance.

The system uses laser displacement sensor to measure wheel diameter early. However, mirror reflection will occur due to the surface machining or wear, which will affect the signal of the laser displacement sensor and thus affect the stability of wheel diameter measurement. Figure 2 is the schematic diagram of the innovative method.

Chord length measurements are obtained by projecting light from two parallel lasers; the design makes the strong correlation between the chord length measurement and velocity become weak correlation. The simple geometric formula for the radius of a vertex circle is as follows:

$$R_M^2 = h^2 + D_1^2 \tag{1}$$

$$R_M^2 = (h + \Delta H)^2 + D_2^2 \tag{2}$$

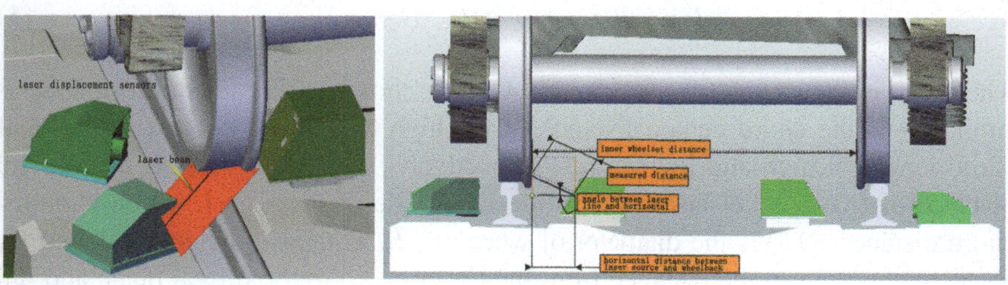

Fig. 1 Schematic of measurement method

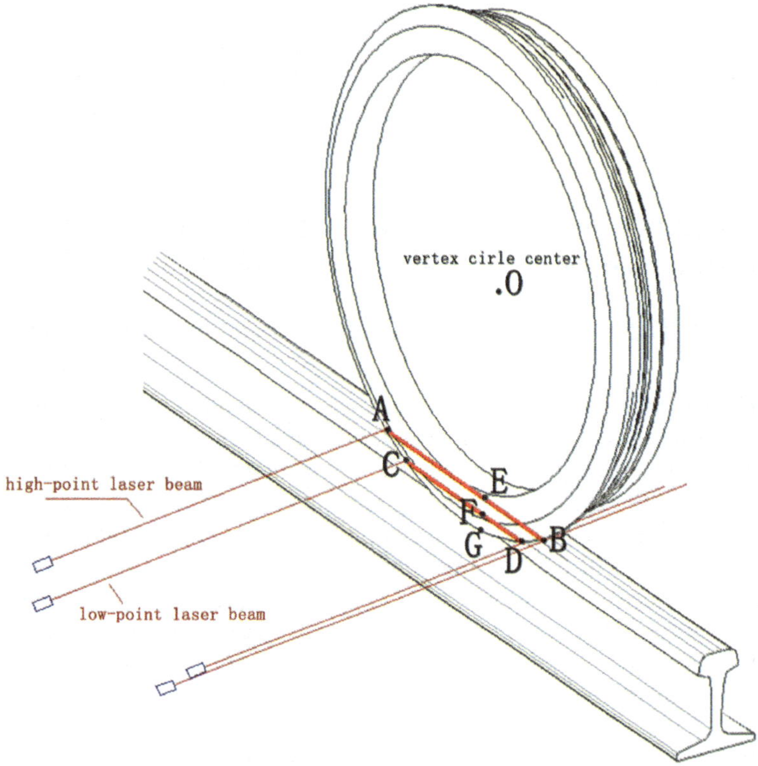

Fig. 2 Schematic diagram of vertex circle measurement method

Using (1) and (2), we can get:

$$h = \frac{(D_1^2 - D_2^2 - \Delta H^2)}{2\Delta H} \quad (3)$$

$$D70 = 2(R_M + H_y) \quad (4)$$

$$H_y = R_M - h - \Delta H - D_2h - \mathrm{IF} \quad (5)$$

where R_M is the radius of vertex circle OA, as shown in Fig. 2, the following is similar; h is the chord center distance of the high chord OE; D_1 and D_2 are half chord length of high and low chords, respectively, AE and CF; ΔH is the vertical distance of two parallel laser rays to the inner side of the wheel, which is called chord distance; $D70$ is the diameter of wheelset; H_y is the rim height; D_2h and IF, respectively, refer to the distance from the lower chord to the orbital plane and rail settlement during passing.

Sensor static calibration parameters, such as D_2h and ΔH, cannot accurately describe the train before produce deformation and relative running wheel sensors on the basis of deviation; the error is expressed as dD_2h_i and $d\Delta H$. The former is mainly affected by axial displacement, and the latter is mainly affected by foundation deformation and sensor dust shielding. $d\Delta H$ can be obtained through big data by neural network algorithm, and the system will detect D_1, D_2, D_2h, IF, and ΔH directly. The heart distance of the high chord h is derived from (3), wheel diameter D_{70} is derived from (4), and the rim height H_y is derived from (5).

Dynamic calibration should be introduced because the field environment is disturbed by temperature and dust. It is possible to produce irregular wear on the wheelset, which will increase the radial runout of the rolling circle, and the radius of the vertex circle is determined by the maintenance process and has the invariant property (the position does not wear out) of the maintenance cycle (such as 12 months), which makes dynamic tracking of sensor parameters possible. Dynamic calibration is obtained by the characteristics of normal distribution based on big data sample. Firstly due to the processing in different time (about 12 months) which makes the statistical average of the left and right radii of the vertex circle in this epicycle is zero, because of this, the consistency error of laser position sensors on both sides of the orbit was calibrated. Secondly, when the standard deviation of the big data measured by the sensor is in line with the empirical threshold compared with the measured values by manual or rotary table (such as 1 mm before rotation and 0.5 mm after rotation), the absolute error was also corrected, and the rotary error and long-term machining stability were also evaluated.

The axial displacement (*Based*) is measured by the laser displacement sensor inside the track, and the distance between the inner side of the finger wheel and the inner edge of the rail can be used to track and calibrate the displacement sensor for a long time, due to the inside distance of single wheelset is constant when its working fine (e.g., 10-year service life): First, find the average value of the axial displacement of big data ($Based_0$), add dx to left axial displacement, reduce dx to right axial displacement; then find the wheelset axial displacement $Based_i$, the dx with the smallest standard deviation of the big data samples. In other words, the consistency error of displacement sensors on both sides is solved and compensated, which is because the medial distance is measured by both the left and right side displacement sensors, the consistency error of left and right sensors can be obtained by iterative algorithm; that is, the dx of the left and right measured values with the minimum standard deviation after compensation is solved. Similarly, the compensation weight factor K_y for edge height measurement was obtained by iterative method, and also the error compensation value of wheel diameter measurement caused by axial displacement can be obtained by $dD_2h_i = (Based_0 - Based_i)K_y$.

2.2 *Critical Path and Characteristics of Correction Methods*

Figure 3 is the system parameter tracking and correction neural network graph; in the figure, laser displacement sensor in the input characteristic scans the wheelset

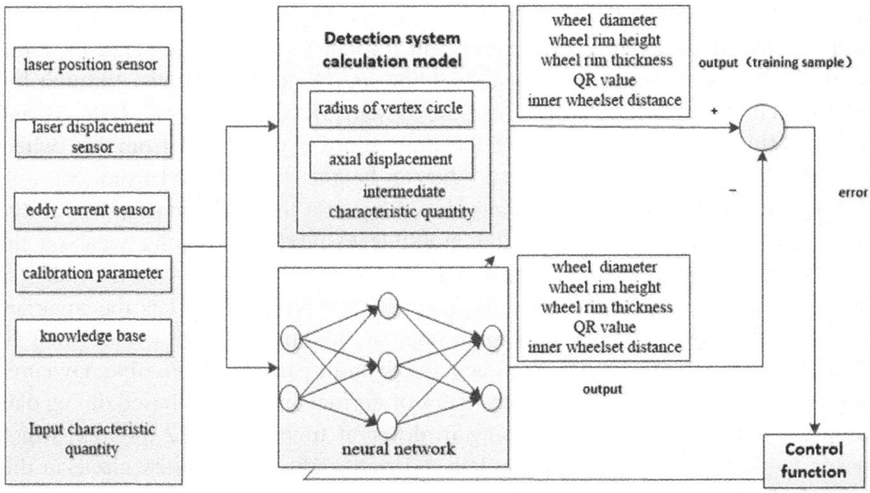

Fig. 3 System parameter tracking and correction neural network graph

tread and wheel back curve. Laser position sensor scans the condition of the wheelset motion and vertex circle, and eddy current sensor scans the rail dynamic settlement.

The key path of the tracking and correction of the wheelset online measurement system is as follows: Take large samples of raw measured data static calibration parameters and empirical parameters, measurement output, maintenance processing data as large closed-loop feedback system; two important intermediate characteristic quantities—radius of vertex circle and displacement of wheel of axial—were extracted critically; through automatic recognition and the classification of output results and acquisition of optimal experience threshold, adaptive filtering and system parameter calibration are realized; by accurately measuring all kinds of interferences, the modified values of dynamic parameters are obtained to ensure the accuracy and long-term stability of the output results of the measurement system.

The main characteristics of the tracking and correction of the wheelset online measurement system are as follows: Describing multi-input and multi-output nonlinear relationship of the system based on the analysis of big data multilayer neural network; automatic recognition of data preprocessing classification; the middle layer has priority weight characteristic variable and experience threshold; integration of database and knowledge base.

2.3 Steps of System Parameter Correction

Aiming at Fig. 3, the system parameter correction steps are specified as follows:

(1) Wheelset measuring dimension system is made up of eight laser displacement sensors, eight laser position sensors, two eddy current sensors, and a wheel sensor. The input initial value of the neural network is obtained through the spot parameter calibration and spot measurement of the original large sample data.
(2) The measured rim height, rim thickness, QR value, wheel diameter, and wheel-to-wheel internal ranging of the sample were taken as the output values of the neural network. At the same time, the output value is classified; normal value and abnormal value are got from experience threshold.
(3) Adjust the parameter calibration values of the input values one by one, and also the intermediate characteristic quantity between the radius of vertex circle and the weight of each input change on the intermediate characteristic quantity and the output change.
(4) According to engineering design theory and empirical model, such as the known two-chord length circle, new car standard parameters, repair processing standard parameters. We analyze the systematic error and standard deviation of statistical samples.
(5) The radius of vertex circle is established as the first intermediate feature quantity. Static calibration cannot simulate the basic deformation of the tunnel track relative sensor when the train passes; this is the main factor of the error. Adjusting the chord distance input to minimize the systematic error and standard deviation of the vertex circle, it is especially important to identify the consistency error between the left and right wheels diameter of the new car and the newly repaired car in the large sample. At this time, it is assumed that the driving direction's distance calibration error or error variation of the position sensor is zero (the base of sensor in this direction is not affected by the motion force, and the change is small). It is worth paying special attention that the feature quantity of vertex circle is invariant to the whole repaired cycle. This provides a reliable basis for the parameter tracking and correction of the position sensor and the measurement of wheel out of roundness.
(6) The calibration of high and low chords does not affect the vertex circle; together with eddy current sensor, it is the key parameter that affects wheel diameter measurement. The input cannot exceed a reasonable range, if beyond the well-adjusted calibration driving distance of the driving direction, making the input–output within a reasonable range to achieve optimal parameter matching.
(7) Establishing the intermediate characteristic quantity of the second layer as axial displacement, this is an important running part safety index [10]. From the perspective of measurement, this is the third factor affects wheel diameter measurement; in severe cases, it is equivalent to the influence of item 6.
The characteristic quantity of axial displacement is mainly used to measure the inner distance of wheelset, and the inner pitch has the invariant characteristic of wheelset life cycle; the problem of long-term tracking and calibration is solved.

The calibration steps of the inner displacement sensor are as follows:

(a) The large sample axial displacement plus and minus a small dx, adjusting the dx to minimize the standard deviation of the distance from the left and right measurements to the sample values; at that time, the dx value is the error relative correction of the left and right inner displacement sensors.
(b) Absolute correction of error was obtained by comparing the instrumental measurement and online measurement.
(c) Plugging in the weighting factor and compensating to the chord height variable, it makes the standard deviation of left and right wheels diameter smallest. This step is adjusted repeatedly with the previous several to the best output. Finally, the modified weight factor was determined.

(8) Using similar methods to calibrate the lateral displacement sensor, due to the accurate measurement of the associated wheel diameter and the calibration of the inner sensor, the originally difficult to measured edge thickness and QR value can be successfully completed.

2.4 Technique of Correction Method

(1) Classification and pretreatment of original data: automatically selecting and adjusting the normal data in line with the empirical threshold into the optimal stratification calculation; recording abnormal data and distinguish the reason of sensor and wheel pair under test.
(2) The invariant feature of vertex circle radius and inner distance of wheelset is used; vertex circle radius is used for laser position sensor to track and calibrate; the inner wheelset distance is used for laser displacement sensor to track and calibrate. The use of this invariant feature increases the effectiveness of the output from 80% to more than 95%; the comparison of the data is shown in Fig. 4 in a certain subway line, which has been handled with this invariance or not.

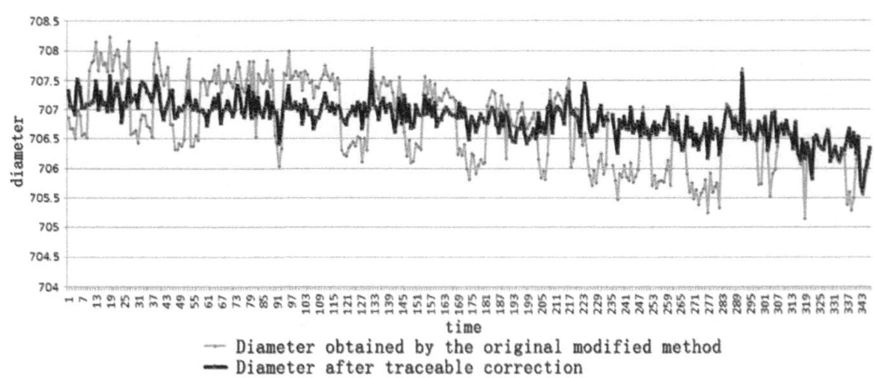

Fig. 4 Comparison of the validity of former and after data in a certain railway line

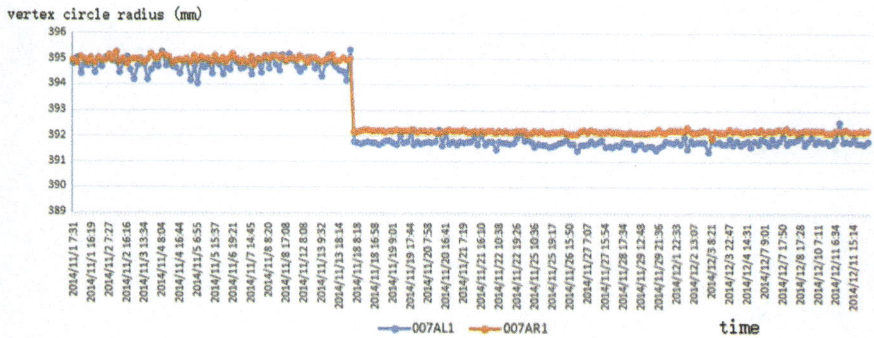

Fig. 5 Schematic diagram of wheelset repaired features

(3) Using historical database for measuring raw data and dynamic tracked and calibrated process records, 'breakpoints' can automatically identify the repair or change wheelsets, namely step change record of wheelset size. Figure 5 is schematic diagram of wheelset repaired features of automatic identification.

3 Engineering Example of Correction Method Application

3.1 Example in A Certain Line

Figure 6 shows the data of the correction method applied to the online measurement of the wheelset size in A certain line. The figure clearly shows the size of the machined sample—the thickness of the rim is 32 mm, the height of the rim is 27 mm, QR value is 10 mm, which is consistent with the height measured by the online system. Firstly, the absolute value is consistent; secondly, the difference between the left and right measurements tends to zero.

3.2 Example of B Certain Line

This method is applied to the online measurement system of wheelset size on a certain line in B subway. Due to the improvement of the detection accuracy, the measuring error of the diameter in the repaired wheel was found. Table 1 shows the statistics of the wheelset lathe error and online measurement data in a certain subway, 3264 data of 34 trains which run in forward and inverted directions were selected on Feburary 15, 2016.

(1) The adjustment of lathe refers to the calibration of sensor for lathe measuring wheel diameter.

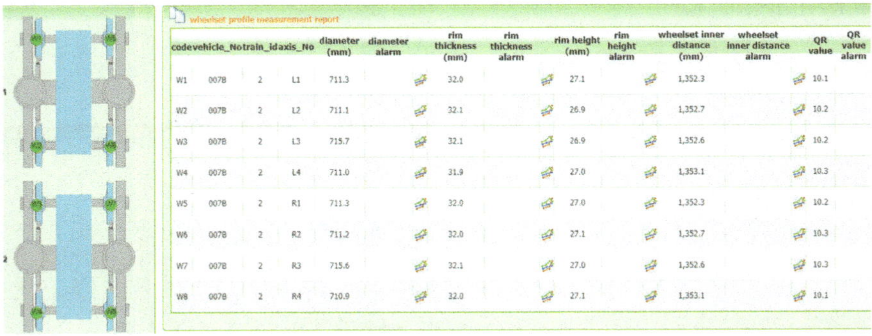

code	vehicle_No	train_id	axis_No	diameter (mm)	diameter alarm	rim thickness (mm)	rim thickness alarm	rim height (mm)	rim height alarm	wheelset inner distance (mm)	wheelset inner distance alarm	QR value	QR value alarm
W1	007B	2	L1	711.3		32.0		27.1		1,352.3		10.1	
W2	007B	2	L2	711.1		32.1		26.9		1,352.7		10.2	
W3	007B	2	L3	715.7		32.1		26.9		1,352.6		10.2	
W4	007B	2	L4	711.0		31.9		27.0		1,353.1		10.3	
W5	007B	2	R1	711.3		32.0		27.0		1,352.3		10.2	
W6	007B	2	R2	711.2		32.0		27.1		1,352.7		10.3	
W7	007B	2	R3	715.6		32.1		27.0		1,352.6		10.3	
W8	007B	2	R4	710.9		32.0		27.1		1,353.1		10.1	

Fig. 6 Data of online measurement of wheelset dimensions using the correction method in some certain line

(2) Forward and reverse running means that the line was designed with U-turn for the train, and there are two programming sequences when passing through the online detection area.
(3) Forward repair means that the trains with odd numbers enter the repair position first, and the reverse is the trains with even number enter first.
(4) The online detection device uses the same train wheels to pass through the detection sensors in a short period of time for consistent absolute calibration.
(5) Self-calibration accuracy of big data online detection device reaches 0.1 mm, and the error of lathe wheel diameter measurement is verified too.

4 Conclusion

The method for the online measurement system of train wheelset profile calculates the wheelset size after fusing and refactoring the data get from many sensors, and the detection signal and the result are strongly nonlinear. The neural network method is applied to the measurement and correction of wheelset size, which can solve the nonlinear response problems of the output and input for the online measurement system of train wheelset profile. Analysis method of traceability and correction provides some useful safety warning indexes, such as axial displacement, radial runout, and axial movement of the running part of the train under load and high speed for the users, and a new method for evaluating the data of lathe was developed. Traceable correction method for the measurement system of train wheelset profile based on a neural network can solve the difference between static parameters and dynamic parameters, enhance the adaptability of the system running environment and the stability of measurement, and meet the requirements of system online detection.

Table 1 Statistics of the wheelset lathe error and online measurement data (unit: mm)

Left and right wheels diameter difference after the lathe is adjusted			Left and right wheels diameter difference before the lathe is adjusted					
24 trains	Forward running	Reverse running	4 trains	Forward running	Reverse running	6 trains	Forward running	Reverse running
Adjustment of lathe			Forward repair			Reverse repair		
Average of wheel diameter error	−0.15	−0.09	Average of wheel diameter error	−0.85	1.06	Average of wheel diameter error	1.03	−0.92
Standard deviation of wheel diameter error	0.46	0.44	Standard deviation of wheel diameter error	0.90	0.90	Standard deviation of wheel diameter error	0.89	0.79

Acknowledgements This research is supported by the Science and Technology Planning Project of Guangdong Province of China (No. 2013B010134007), the National Key R&D Program of China (No. 2016YFB1200402).

References

1. Li Z, Chen X, Xing Z (2015) On-line dynamic measuring system for wheel pairs of rail vehicles. Mod Urban Transit 12(3):5–8 (in Chinese)
2. Wang J, Mao L, Wen S (2006) Study on wheelset profile real time and dynamic detection system of railway wagons. Coal Mine Mach 27(6):1043–1045 (in Chinese)
3. Ying H, Meng J, Jiang X (2008) Algorithm research of measurement method of wheelset size. Autom Instrum 1:74–76 (in Chinese)
4. Zhang Y, Peng J, Yang K et al (2011) Camera calibration algorithm for wheel profile inspection. Opto-Electron Eng 39(2):189–193 (in Chinese)
5. Li H, Cheng X, Su Z et al (2016) Online measurement system of urban rail transit wheel set size based on laser displacement sensor. Urban Mass Transit 19(9):10–14 (in Chinese)
6. Chen G, Ren G (2012) On-line detecting system of wheel-set state. Urban Mass Transit **15**(10):79–81, 124 (in Chinese)
7. Han G, Song C, Li G et al (2017) Tracing test of the relationship between high-speed train wheel OOR and vibration noise. Urban Mass Transit **20**(2):1–9 (in Chinese)
8. Wang Y, Fei C, Sun H (2018) Fault diagnosis method for bogie bearings of the metro vehicle based on improved weighted D-S evidence theory. J Beijing Jiaotong Univ **42**(6):75–82, 90 (in Chinese)
9. Zou Z, Xiang Y, Qin J et al (2016) An error tracing and correction method for online monitoring system of train wheelset. Chinese patent (in Chinese)
10. Zhang L, Liu J, Zhang J et al (2017) Influence of wheels lateral displacements on wheel/rail wear on metro straight line. J Dalian Jiaotong Univ 38(5):11–14 (in Chinese)

Analysis of the Factors Influencing on Bird Nesting and Its Impact on Railway Operation

Lijuan Shi, Zuoning Jia, Huize Sun, Tian Mingshu and Liquan Chen

Abstract This paper studies the affecting factors on bird nesting on electronic railway catenary lines, and the impact of bird nesting events on railway operation. Firstly, with one year's bird nest events in form of unstructured natural language collected from Shanghai Railway Bureau, the records were structured with the help of Python software tool. Secondly, the method of root cause analysis (RCA) was used to identify all the possible influencing factors which are inclined to affect the probability of bird nesting. Thirdly, the possible factors then were fallen into two categories to meet subsequent analysis separately; category one was Outside Factors (i.e., geographical conditions-related factors), and the other was Inside Factors (i.e., railway-related factors). As to Outside Factors, such as temperature, vegetation coverage, population, longitude and latitude, nearby water, etc. by using some geographical APIs offered by Amap and the tools of ArcGIS, geographical data were processed in relation to nest events on the visual map and their correlationship was studied. It was observed that factors of city population, geographic position affect nesting observably. As to railway-related Inside Factors, statistic analysis method was used to disclose the significant influencing factors to bird nesting. Statistical result indicated that factors of season, rail location (station or section) and the specific equipment part of railway catenary lines had significant effects on nest events. Finally, it was demonstrated that both location and nesting on equipment part have no correlation with delay, while railway type had a significant but low correlation with delay. Two lognormal stochastic models were put forward which are the best-fit models for delay caused by bird nesting events of both normal-speed and high-speed railway, which discloses the principle of impacts of nest events on railway operation.

L. Shi · Z. Jia (✉) · H. Sun · T. Mingshu
Shanghai Key Laboratory of Rail Infrastructure Durability and System Safety, Tongji University, Shanghai 201804, China
e-mail: 674767670@qq.com

L. Shi
e-mail: shilijuan150@tongji.edu.cn

L. Chen
China Railway Guangzhou Group Co., Ltd., Guangzhou 510088, China

Y. Qin et al. (eds.), *Proceedings of the 4th International Conference on Electrical and Information Technologies for Rail Transportation (EITRT) 2019*, Lecture Notes in Electrical Engineering 639, https://doi.org/10.1007/978-981-15-2866-8_62

Keywords Railway operation · Risk management · Bird nest · RAC analysis · Statistic analysis · Geographical information · Lognormal stochastic model

1 Introduction

With the development of high-speed railway, it plays an increasingly important role in national transportation system. The safety and efficiency of high-speed train operation are getting seriously concerned. Short-circuit trip of railway catenary lines caused by bird nesting, which is a very common kind of risk event, generally exerts a significant negative impact on train operation and ultimately decreases the efficiency and safety of railway transport.

The risk event caused by bird nesting is next to lightning strike and external force destruction, ranking the third in the total number of power supply line faults. According to the investigation and statistical analysis of various types of catenary bird damage faults, the most prominent bird damage fault is nesting, which is called nest event in this paper. The specific parts of the catenary lines equipment of electrified railway are very suitable for birds nesting, such as supporting column, hard horizontal beam and other positions. Due to the selection of a variety of bird nesting materials, including steel wire, wooden rods, etc., nesting in these locations is easy to cause a short circuit, trip-off and other faults in the railway power supply line, affecting the normal railway transport order, and even damage the equipment. Since 2010, 1671 nests have been treated in one power supply section, and in 2011 alone, 516 bird's nest locations were found and treated in a section of the Wenzhou-Fuzhou and Fuzhou-Xiamen high-speed railway, causing five trip-offs and affecting power supply for up to one hour and 12 min at a time [1].

There are few studies on the risk events caused by bird nesting and the impacts of nest events on train operation quantitatively [2, 3]. Generally, empirical judgment was applied in practice.

Duan Wangwang proposed the HOG feature-based detection of the nests. The method was adopted that first roughly observing, then fining the detection process of nesting, and finally determining the bird's nest area through probability density estimation [4]. Li Tengfei took analyzed the shortage of the traditional bird-driving method in monitoring and management and studied the relationship between structure and nest phenomenon of catenary device, and his work put forward further of the catenary bird-driving system developed on the detection, communication, data processing and other function, providing a reference for further research and development of similar system [5].

Wang Mengyang, Lin Jiansheng, Wang Rui and many other researchers drew conclusions on the habit of birds and influenced factors of the nesting behaviors through analysis in terms of both time and space dimensions, including topography and geomorphology, time characteristics, location characteristics, repeated features and material characteristics, which to some extents inspires the catalog of characteristics in this paper [2, 3, 6, 7].

In the following sections of this paper, firstly, with the obtained one year's risk events in the form of natural language from Shanghai region railways, the nest event-related records were selected and possible affecting factors were extracted. Secondly, all the possible factors which are inclined to bird nesting were identified with the method of RCA. Then, with the help of SPSS, ArcGIS and other statistic methods, analyze the correlation of Outside Factors (geographical characteristics factors data from Internet) and Inside Factors (railway-related factors data from Shanghai Railway Bureau) with bird nesting events, respectively. Finally, modeling the effects of nest events on railway operation.

2 Data Collection

The data collected in the study were from two sources, one is from Shanghai Railway Bureau, and the other is geographic data collected from the Internet.

There are more than 2500 risk event records formed in natural language obtained from Shanghai region railways during almost one year, the nest events were selected and converted into structured data partly with the help of Python software modules such as XLRD and XLWT. There are 275 structured nest event records with all possible valid attribute information for further analysis, such as nesting location (station or section), nesting time, nesting on equipment part of catenary lines, train type, and the delay of the nest events on train operation.

It is observed that geographical attributes, such as temperature, vegetation coverage, population, longitude and latitude (geographical position), nearby water, , may influence bird nesting. Geographical APIs of POI searching provided by Amap were used in this study to search longitude and latitude according to the nesting locations (stations or sections along the railway lines). High-resolution raster data such as China tree distribution map, population density map of the sixth population census and China water system map are collected from major research Web sites for further analysis.

3 Analysis of the Possible Influencing Factors on Nesting with RCA Method

RCA is a structured problem-solving method, which is used to gradually find out the causes of the problem and then analyze and solve it. The approach of RCA method consists of seven steps: selecting team members, actions to prevent the problem from spreading, identifying root causes, further solutions, implementing solutions, and evaluating performance. The root cause identification step is the key of RCA method. During the process of nesting factors analysis, we focused more on the steps of identifying root causes and further solutions. In the root cause analysis stage, RCA method provides four root cause analysis tools, including causality diagram, brainstorming method, fishbone diagram and why-why diagram [8].

The fishbone diagram was chosen to help the analysis of the possible influencing factors on nesting. The application process of fishbone diagram was as follows: (1) state the problem of bird nesting; (2) identify major cause categories. According to the difference in data sources and analysis methods, two categories are decided, i.e., railway-related factors and geographical characters; (3) use the brainstorm method to fill the reasons under each category, including nesting time, nesting on the equipment part, railway type, the delay of the nest events on train operation, nesting location (station or section), geographical position and environment; (4) consider the factors that may contribute to each cause and put them on a line from the cause. For instance, environment cause can be divided into temperature, population and vegetation coverage factors. So lines were drawn from environment to the tree factors; (5) discuss how the factors affect bird nesting; (6) discuss the way to prove the causes and factors using data analysis. According to the collected data, the factors probably affecting bird nesting were depicted in the fishbone diagram as shown in Fig. 1.

For convenience of expression, the factors extracted from the nest event records are given the general term 'Inside Factors', including nesting location (station, section), event time (discovery time, responding time, end time), nesting on equipment parts of catenary lines (supporting column, hard horizontal beam, compensation device, pedestal, protecting mesh, disconnector switch, return line

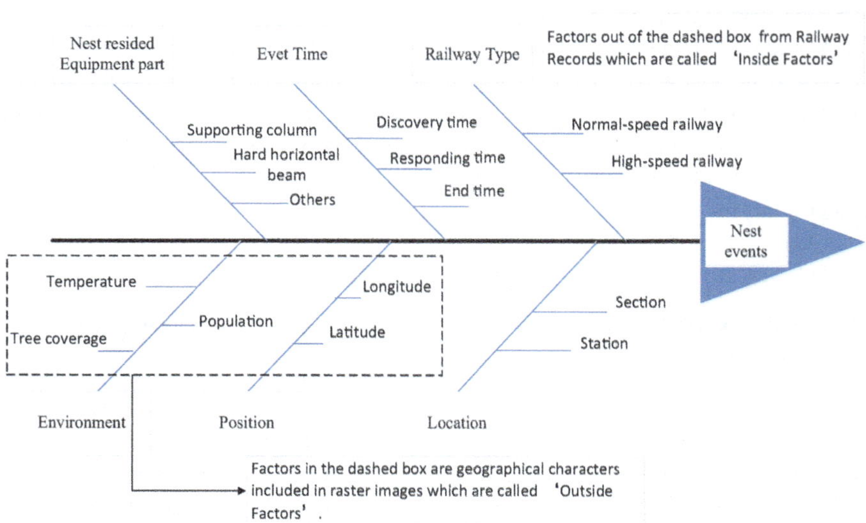

Fig. 1 Fishbone diagram

shoulder, insulator, etc.), railway type (normal-speed railway, high-speed railway), the delay of the nest events on train operation, etc. the factors obtained from the geographical data are named 'Outside Factors', such as temperature, vegetation coverage, population, longitude and latitude (geographical position), nearby water.

4 Analysis of Outside Factors

In the stage of RCA analysis, the 'Outside Factors' contain the following parameters: population density, vegetation coverage and water distribution. All the nest events data include time and position of each event, which can be expressed in the form of points on the map. To expend data such as population density, vegetation coverage and water distribution to the points, raster data shall be collected into point data. The buffer is established with ArcGIS software. The event points are taken as the center of the circle of the buffers, and the buffers with 10, 20, and 50 km radius are created separately. With ArcGIS statistics tools, vegetation coverage, population density, and water distribution within the buffers are calculated for analysis. The visual results are shown in Fig. 2a–d. For example, in Fig. 2a, the base map is China's population distribution map, the yellow circles are 10 km buffers, green

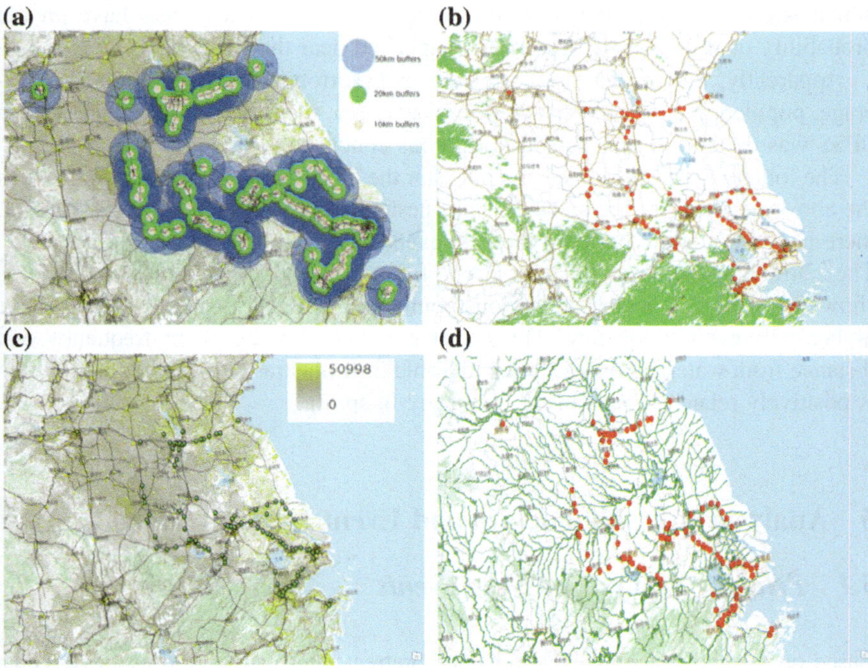

Fig. 2 **a** Buffers with 10/20/50 km radius. **b** Vegetation coverage. **c** Population density. **d** Water distribution

Table 1 Correlation between the nest event frequency and vegetation coverage

Correlation coefficient	10 km radius	20 km radius	50 km radius
Spearman	0.051	0.022	0.121
Significance level	0.01	0.01	0.01

Table 2 Correlation between the nest event frequency and population density

Correlation coefficient	10 km radius	20 km radius	50 km radius
Spearman	0.272	0.272	0.251
Significance level	0.01	0.01	0.05

Table 3 Correlation between frequency and water distribution

	Correlation coefficient	*P*-value	Significance level
Spearman	0.211	0.036	0.05

circles are 20 km buffers and blue circles are 50 km buffers. The summation of population is calculated in each circle.

In Fig. 2a–d, some characteristics of these possible affecting 'Outside Factors' could be learned intuitively. Nest events occur in areas with sparse tree distribution, which is different from the previous studies. Some populated areas have greater probability of nest events. Nest events happened near the water areas.

Apparently, nest events are correlated with surroundings like vegetation coverage, population density, distance from water area, then statistical data analysis in SPSS was done to get the scientific rigorous conclusions.

The following conclusions come out after the analysis of the correlation between the above data and the frequency of bird nest events. As demonstrated in Table 1, there is no correlation between the nest event frequency and vegetation coverage. The correlation between the nest event frequency and population density is as shown in Table 2. It makes an obvious point that the higher population density, the higher nest event frequency. The correlation between the event frequency and distance from water area is as shown in Table 3. It illustrates that water distribution is positively related to nest event frequency in spring.

5 Analysis of Inside Factors and Event Lasting Time

5.1 Primary Analysis of Nest Events

Based on the above analysis, the obtained pretreated 275 structured nest event data records include multidimensional parameters, such as time, location, railway type and nest resided equipment part. Each parameter has a number of different values.

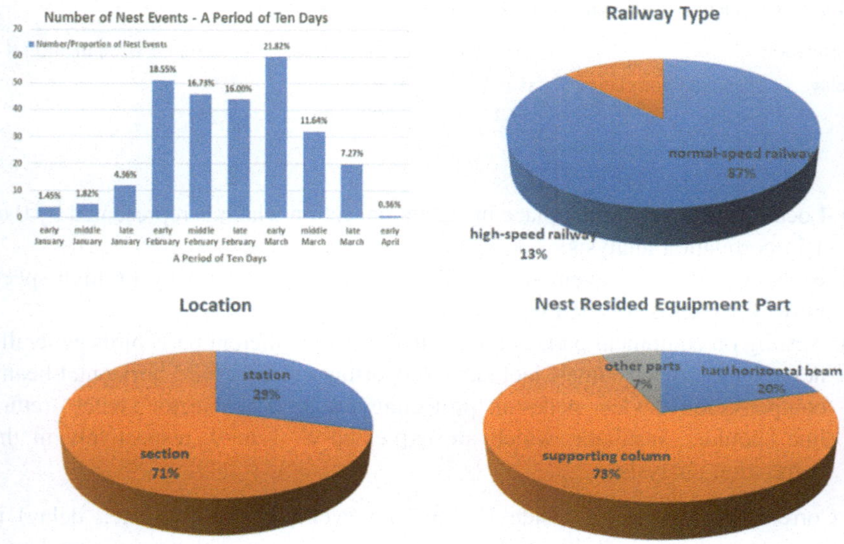

Fig. 3 Nest events spatiotemporal distribution characteristics

The statistical result of the 275 records indicates that the nest events spatiotemporal distribution characteristics in proportion as follows, as shown in Fig. 3: (1) The nest events generally happen in spring, covering the period from January to early April, and ninety-two percent of nest events occurred in February and March. Bird nesting event accounts for a third of all railway risk events in spring. (2) 71% of the nest events took place in the sections, while 29% in stations. (3) 13% of the nest events took place in high-speed railways, while 87% took place in normal-speed railways. (4) As to birds' favorite nesting equipment part, 73% of bird nests resided on supporting column of catenary, while 20% on hard horizontal beam, and sum of other cases accounts for 7%.

5.2 *Correlation Analysis of Inside Factors on Event Lasting Time*

Bird nesting event lasting time (hereinafter also called delay) is given by event end time minus discovery time, which is important to determine the severity of impact of nesting event on train operation. All the possible Inside Factors are enumerated to conduct the correlation analysis with event lasting time to determine the significant influence factors. The explanation for the possible Inside Factors is as follows:

Table 4 Correlation coefficient for Inside Factors and delay

Correlation		Location	Railway type	Nesting on equipment part
Delay	Spearman correlation	0.294	0.294	−0.067
	p-value	0.645	0.000	0.271

(a) Location. The event took place in station or section which is represented by 0 or 1 in correlation analysis.
(b) Railway type. The event happened on normal-speed railway or high-speed railway which is denoted by 0 or 1, respectively.
(c) Nesting on equipment part. There are mainly nine different parts birds generally nest on the catenary lines, including supporting column, hard horizontal beam, compensation device, pedestal, protecting mesh, disconnector switch, return line shoulder, insulator, which are expressed as 1 to 9, respectively in the correlation analysis.

Correlation analysis of Inside Factors and event lasting time (i.e., delay) is conducted, as shown in Table 4. There are two values for each pair of correlation analysis. The first value in upper cell is the Spearman correlation coefficient. The second value in lower cell is the p-value, $\alpha = 5\%$. Italic Figures are statistically insignificant. It is observed that both location and nesting on equipment part have no correlation with delay, while railway type has a significant but low correlation with delay. The Spearman correlation coefficient was 0.294.

5.3 *Delay Stochastic Model of Different Railway Type*

Delay Frequency Distribution

Data of all records were preprocessed by mathematical statistics method. The statistical delays caused by bird nesting events with more than 3 times of variance added to the mean were removed as abnormal data. The remaining 233 normal-speed railway event records and 34 high-speed railway event records are statistically analyzed in the following.

Summary statistics for delay of different railway types are shown in Table 5, and delay frequency distributions of normal-speed railway and high-speed railway are shown in Fig. 4, respectively.

Delay Stochastic Model Fit

From the visual inspection of Fig. 5, two stochastic models seem plausible: log-normal and Weibull. Consequently, two models are proposed in this section to fit the delay frequency distributions of two-type railways, respectively.

Table 5 Summary statistics for delay of different railway types

Railway type	Normal-speed railway	High-speed railway
Number of observations	233	34
Mean delay (min)	28.2	47.3
Standard deviation of mean delay (min)	20.9	26.2
Standard error of mean delay (%)	1.4	4.5
Skewness	3.9	0.6

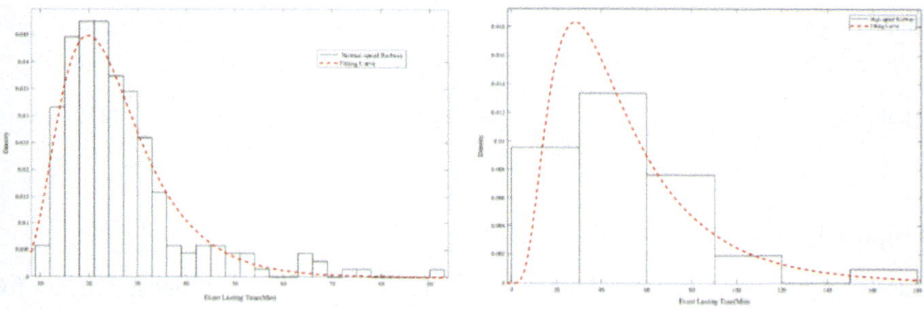

Fig. 4 Frequency distributions for delay of different railway types

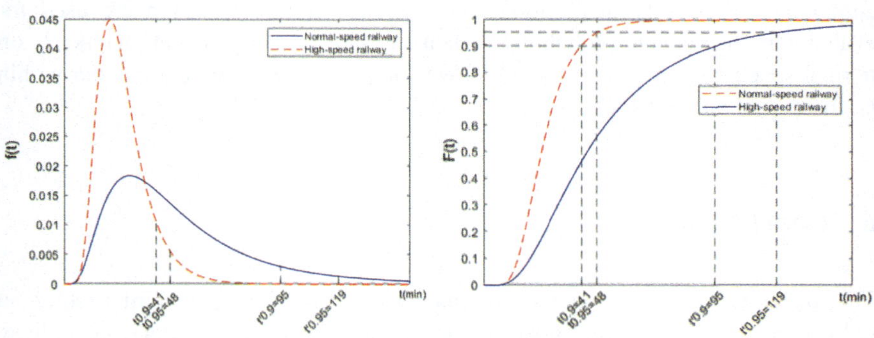

Fig. 5 Probability density and cumulative probability distributions

The results of delay frequency distributions fit shown in Table 6 are measured by the squares sum of error (SSE) and R-square. The fit is more realistic when the SSE is smaller as well as the R-square is correspondingly closer to 1.

Thus, we could tell from Table 6 that the lognormal stochastic model is the best-fit model for delays caused by bird nesting events of both normal-speed railway and high-speed railway. K–S test is then applied for hypothesis testing the lognormal models of the two railway types. The results imply that delays under each railway type obey the lognormal distribution at the 5% significance level.

Table 6 Delay frequency distributions fit under different railway types

Distribution fitting	Stochastic model	Sum of squared errors (SSE)	R-square
Normal-speed railway	Lognormal	0.00026294	0.9637
	Weibull	0.0015	0.7916
High-speed railway	Lognormal	0.00002848	0.8420
	Weibull	0.00005367	0.7022

Table 7 Fitted delay lognormal probability density functions of two railway types

Railway type	Lognormal probability density functions	$t_{0.9}$ (Min)	$t_{0.95}$ (Min)
Normal-speed railway	$f(t) = \frac{1}{0.4249 \times t\sqrt{2\pi}} e^{\frac{1}{2}\left(\frac{\ln t - 3.163}{0.4249}\right)^2}$	41	48
High-speed railway	$f(t) = \frac{1}{0.6319 \times t\sqrt{2\pi}} e^{\frac{1}{2}\left(\frac{\ln t - 3.739}{0.6319}\right)^2}$	95	119

Proposed Model

Based on the above analysis, lognormal stochastic model is the proposed model for the delay frequency distribution under two railway types. The delay lognormal probability density functions of normal-speed railway and high-speed railway respectively are shown in Table 7. And the corresponding probability density and cumulative probability distributions are depicted in Fig. 5, which are used for evaluating the delay probability when a bird nesting event happens on normal-speed railway or high-speed railway. $t_{0.9}$ means 90% of delays occur within $t_{0.9}$ time, and so does for $t_{0.95}$.

6 Conclusion

This paper is to analyze all the possible factors influencing on bird nesting on railway and to disclose the model of nest events impacting on delay of railway operation. There are several findings as follows:

(1) A fishbone diagram is constructed to disclose all possible 'Inside Factors' and 'Outside Factors' affecting bird nesting on railway.
(2) Both population density and water distribution in spring from 'Outside Factors' have positively significant influence on nest events.
(3) The nest events generally happen in spring. Most of the nest events took place in the sections of railway. Supporting column of catenary is the birds' favorite nesting equipment.
(4) Lognormal stochastic models are the best-fit models for delays caused by bird nesting events of both normal-speed and high-speed railway, which are used for evaluating the delay probability when a bird nesting event happens on normal-speed railway or high-speed railway.

Acknowledgements Authors would like to acknowledge the support of the research program of Comprehensive Support Technology for Railway Network Operation (2018YFB1201403), which is a subproject of Advanced Railway Transportation Special Project belonging to the 13th Five-Year National Key Research and Development Plan funded by Ministry of Science and Technology of China.

References

1. Wang P (2013) Discussion on using sliding ladder under the condition of catenary fault. China Venture Capital 250 (in Chinese)
2. Lin J (2016) Research on the prevention and control measures of bird pests in high-speed rail catenary. Straits Sci 08:24–27 (in Chinese)
3. Feng B, Li P, Yang L (2017) Analysis of bird's nest disease of high-speed railway contact network based on spatio-temporal big data. China Railw 2017(01):29–34 (in Chinese)
4. Duan W (2016) Detection and abnormal recognition of catenary equipment for monitoring video. Southwest Jiaotong University (in Chinese)
5. Li T (2017) Research and development of distributed bird driving system for railway catenary based on wireless communication. Hefei University of Technology (in Chinese)
6. Wang M (2018) Discussion on anti-bird measures of high speed railway catenary. Telecom Power Technol 07:237–238 (in Chinese)
7. Wang R, Yu S, Gao L (2008) Fault analysis and prevention of bird damage on 220 kV line. In: Progress in power grids and hydropower (in Chinese)
8. Xu D (2006) RCA/CA: root cause analysis and corrective action. Sci Educ Collects 2006 (04):185–186 (in Chinese)

A Study on the Combustion Behaviors of Interior Materials of Metro Trains

Zhiqiang Li, Taolin Zhang, Zhiyuan Jiao, Wei Cui, Weilian Ma and Qilong Shi

Abstract In this work, the combustion behaviors of three kinds of polymer and composite materials used as interior materials of metro trains, including PVC floorcloth, PC lampshade and painted aluminum plate, were invested by a cone calorimeter. The tests were performed with the cone calorimeter in State Key Laboratory of Fire Science, University of Science and Technology of China. This cone calorimeter is based on the oxygen consumption method and meets ISO 5660 Standard. We focused on the combustion behaviors of three kinds of polymer and composite materials, which are used as metro train interior materials. This study can provide basic fire information for performance-based fire protection of metro trains. Heat release rate, CO_2 and CO concentrations were obtained. The results indicated that PVC floorcloth is the most flammable, and PC lampshade presents the highest both thermal and toxic hazard.

Keywords Combustion behaviors · Heat release rate · Interior material · Metro train

1 Introduction

Urban rail transit (URT) plays a more and more important role in citizens' transportation because of the advantages of convenience, on time, vase capacity, etc.

In a practical field, polymer and composite materials have been shown to express some advantages, for example, lightweight, high-strength, anti-rust and easily processable. So, more and more polymer and composite materials have been used

Z. Li · Z. Jiao · W. Cui · W. Ma · Q. Shi
CRCC Qingdao Sifang CO., LTD., No. 88 Jinhongdong Road, Chengyang District, Qingdao, Shangdong, China

T. Zhang (✉)
School of Automotive and Traffic Engineering, Jiangsu University of Technology, No. 1801 Zhongwu Road, Changzhou, Jiangsu, China
e-mail: ustcfire@126.com

Y. Qin et al. (eds.), *Proceedings of the 4th International Conference on Electrical and Information Technologies for Rail Transportation (EITRT) 2019*, Lecture Notes in Electrical Engineering 639, https://doi.org/10.1007/978-981-15-2866-8_63

as interior materials of metro trains [1]. But, polymer and composite materials are mostly flammable. And material combustion emits hazard gas, such as CO_2 and CO [2, 3]. There are few researchers who illustrate fire test performance of PVC [4–7] and PC [8], and these PVC and PC materials are not used as metro train interior materials.

In this work, we focused on the combustion behaviors of three kinds of polymer and composite materials, which are used as metro train interior materials. This study can provide basic fire information for performance-based fire protection of metro trains.

2 Experimental

The tests were performed with the cone calorimeter in State Key Laboratory of Fire Science, University of Science and Technology of China. This cone calorimeter is based on the oxygen consumption method and meets ISO 5660 Standard.

Three kinds of interior material, including PVC floorcloth, PC lampshade and painted aluminum plate, were prepared according to cone test requirement and tested in a horizontal orientation under three levels of the incident heat flux 25, 35 and 50 kW/m^2.

3 Results and Discussions

3.1 Ignition Properties and Combustion Phenomena

The ignition and combustion phenomena of PVC floorcloth, PC lampshade and painted aluminum plate are shown in Figs. 1, 2 and 3, respectively.

As shown in Fig. 1, the samples of PVC floorcloth can be ignited under all the heat fluxes. As shown in Figs. 2 and 3, the samples of PC lampshade and painted aluminum plate cannot be ignited under the heat fluxes of 25 kW/m^2. The results indicate that the PVC floorcloth is more flammable, and the fire risk is greater.

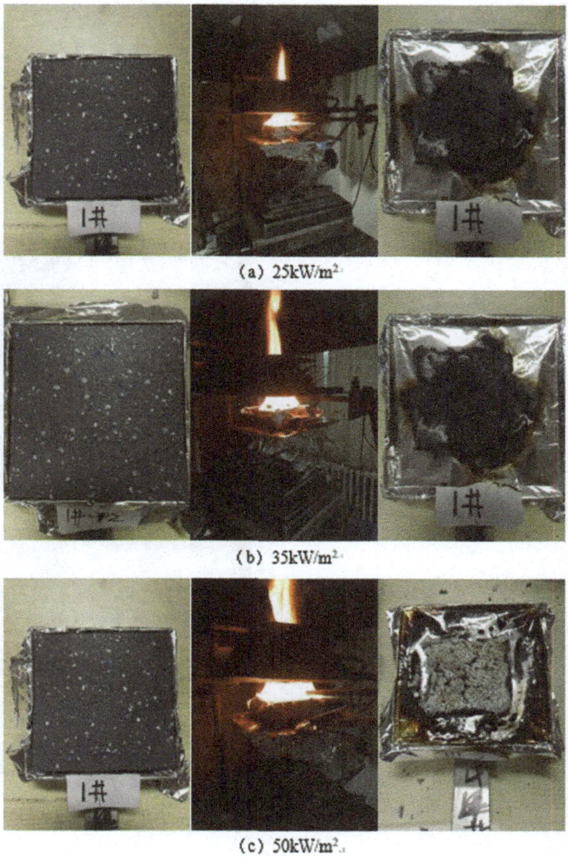

Fig. 1 Test phenomena of PVC floorcloth

3.2 *Heat Release Rate (HRR)*

The heat release rate of materials refers to the heat released under certain heat fluxes. At the same time, the peak heat release rate (pHRR) is one of the important parameters of fire characteristics. It is an important parameter to evaluate the thermal hazard of materials used in practice. The HRR vs. time curves of PVC floorcloth, PC lampshade and painted aluminum plate are presented in Figs. 4, 5 and 6, respectively.

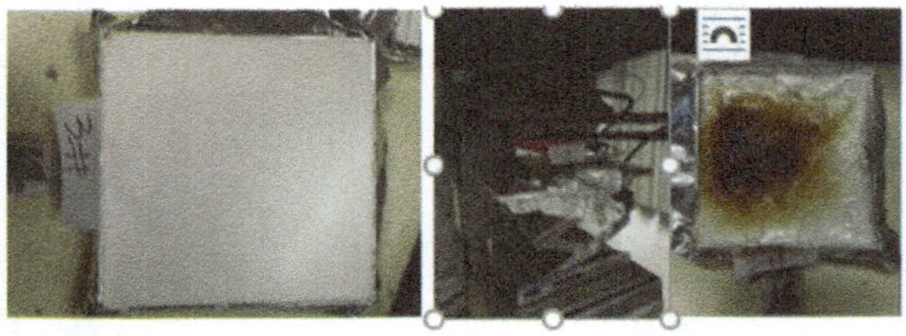

(a) $25kW/m^2$

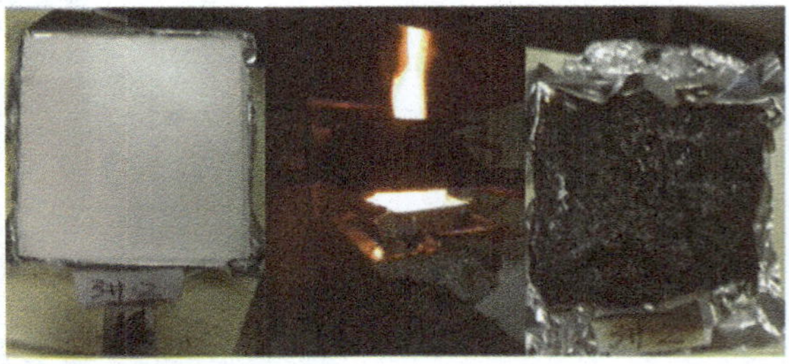

(b) $35kW/m^2$

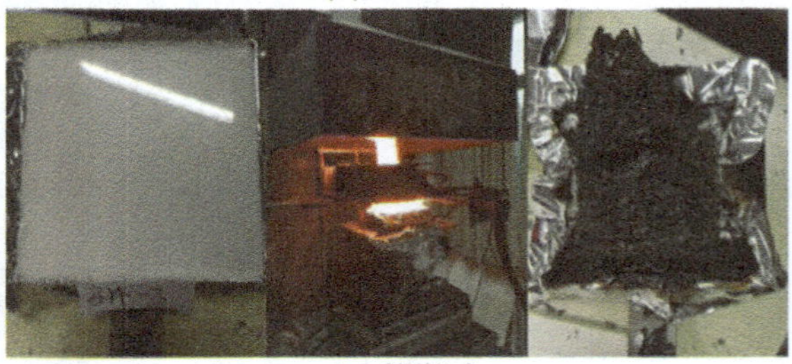

(c) $50kW/m^2$

Fig. 2 Test phenomena of PC lampshade

It can be seen from Fig. 4 that the PVC floorcloth exhibits a nearly same heat release rate curve under different heat fluxes. The heat release rate under external heat flux shows a rapid increase and then decreases sharply. The pHRR also increases when the heat flux increases. As shown in Figs. 5 and 6, the HRRs of PC lampshade and painted aluminum plate have the same varying tendency with PVC floorcloth. The pHRRs of PVC floorcloth, PC lampshade and painted aluminum plate under different heat fluxes are listed in Table 1.

3.3 Smoke Toxicity Analysis

For polymer and composite materials, carbon dioxide and carbon monoxide are the main toxic gas in the gas produced in the process of combustion, which is considered as the major cause of victims during fire. The variations of the CO_2 and CO

Fig. 3 Test phenomena of painted aluminum plate

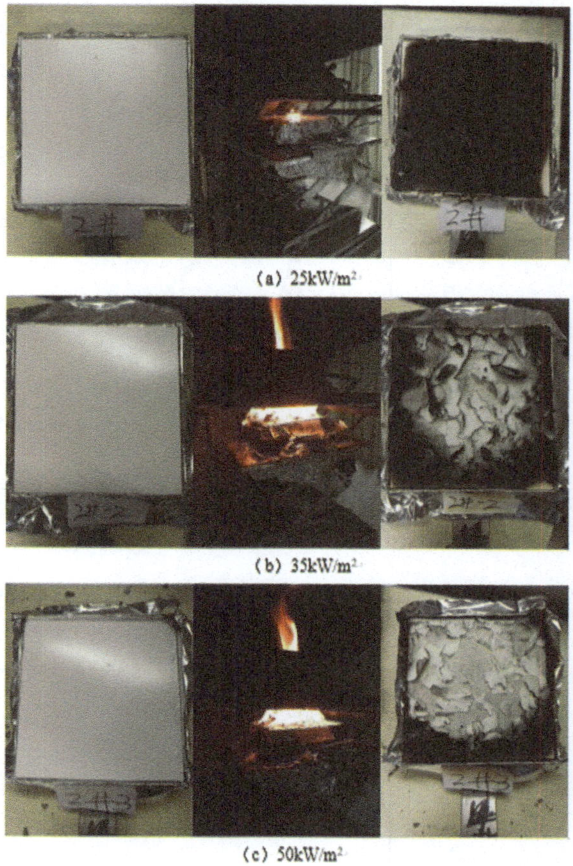

concentrations with time and the of PVC floorcloth, PC lampshade, painted aluminum plate under 25, 35 and 50 kW/m^2 are shown in Figs. 7, 8 and 9, respectively. And the total outputs CO_2 and CO can be derived either. Under the same external heat flux, the total CO_2 and CO releases of the PC lampshade are the greatest, which means the smoke produced by PC lampshade is more toxic than that of PVC floorcloth and painted aluminum plate.

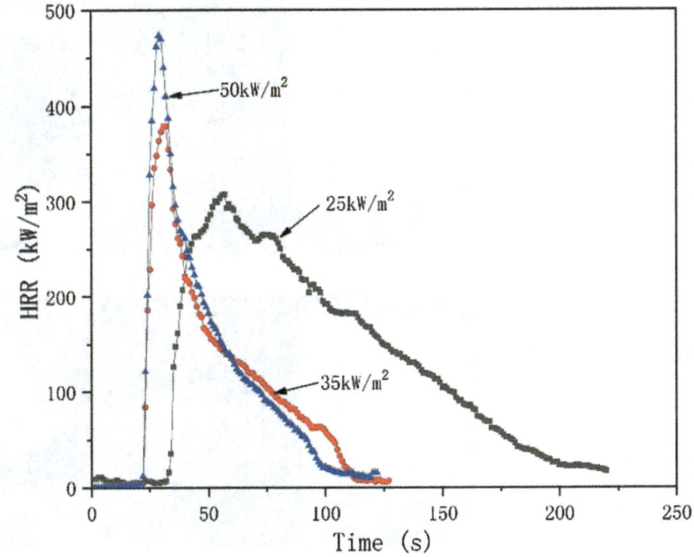

Fig. 4 HRR versus time curves of PVC floorcloth at different fluxes

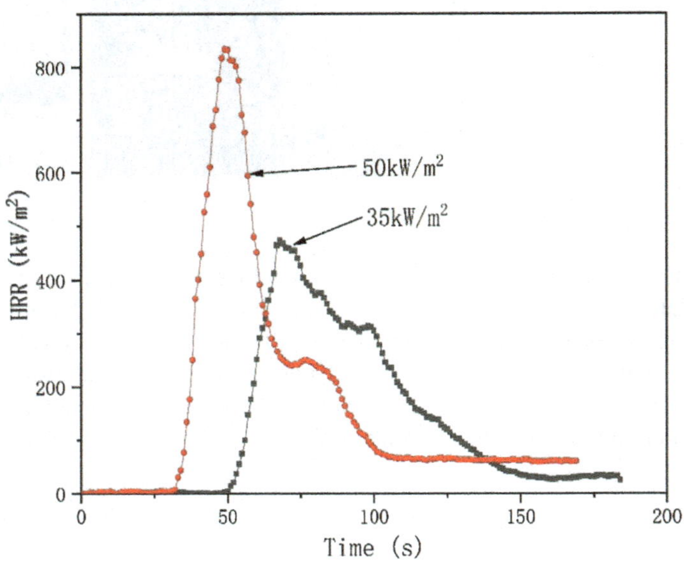

Fig. 5 HRR versus time curves of PC lampshade at different fluxes

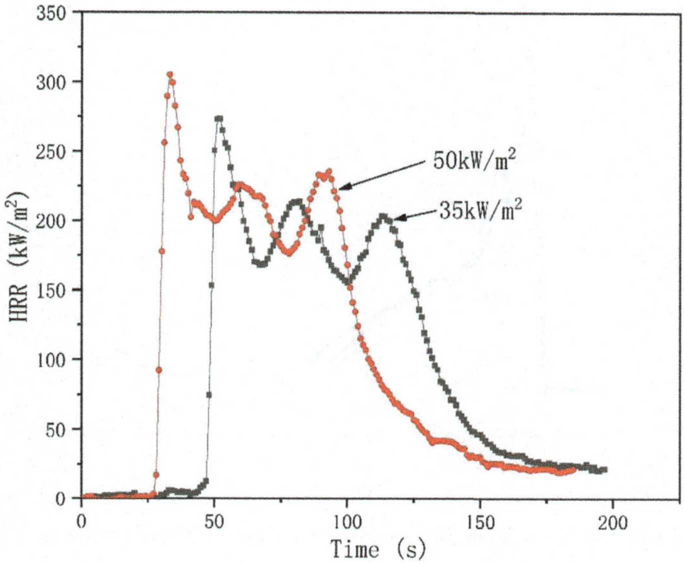

Fig. 6 HRR versus time curves of painted aluminum plate at different fluxes

Table 1 pHRRs under different heat fluxes

Sample	Heat flux (kW/m^2)	pHRR (kW/m^2)
PVC floorcloth	25	307.0
	35	379.3
	50	473.5
PC lampshade	35	474.0
	50	834.2
Painted aluminum plate	35	273.1
	50	305.4

4 Conclusions

In this work, the combustion behaviors of three kinds of polymer and composite materials used as interior materials of metro trains are invested by a cone calorimeter, including combustion phenomena, heat release rate, CO_2 and CO concentrations. The results indicated that PVC floorcloth is the most flammable, and PC lampshade presents the highest both thermal and toxic hazard.

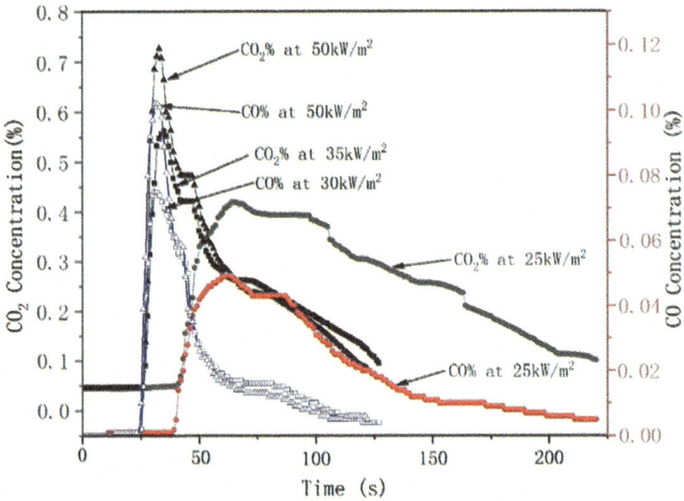

Fig. 7 CO_2 and CO concentration curves of PVC floorcloth at different fluxes

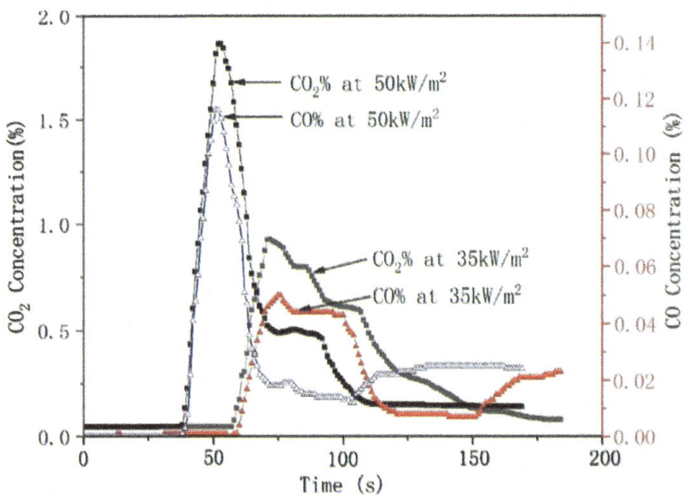

Fig. 8 CO_2 and CO concentration curves of PC lampshade at different fluxes

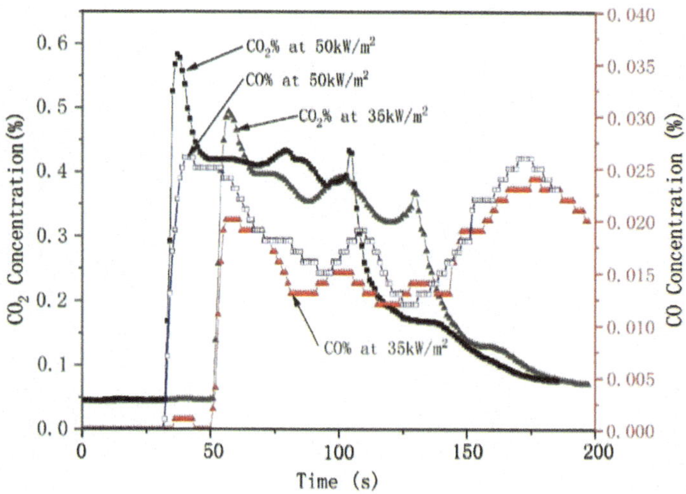

Fig. 9 CO_2 and CO concentration curves of painted aluminum plate at different fluxes

References

1. Ito M, Nagai K (2008) Degradation issues of polymer materials used in railway field. Polym Degrad Stabil 93:1723–1735
2. Ito M, Enari T, Sakihata Y et al (2002) Recycling of polymer materials used in railway vehicles. RTRI Report 16(10):33–38
3. McNeil IC, Memetea L, Cole WJ (1995) A study of the products of PVC thermal degradation. Polym Degrad Stabil 49:181–191
4. Liu H, Wang ZX, Li HH et al (2015) Experimental study on heat release rate of PVC desktop decoration materials. Sci Technol Rev 33(1):81–85 (in Chinese)
5. Wang W, Zhang HP, Wan YT (2008) A study on the burning behavior of PVC cables in the conic calorimeter test. J Safety Environ 8(2):117–120 (in Chinese)
6. William HS, Robert DP, Jenine RC et al (2003) Cone calorimetric study of copper-promoted smoke suppression and fire retardance of poly (vinyl chloride). Polym Degrad Stabil 82:15–24
7. Yang H, Fu Q, Cheng XD et al (2013) Investigation of the flammability of different cables using pyrolysis combustion flow calorimeter. Procedia Eng 62:778–785
8. Xie F, Su ZL, Wen YF (2014) The thermal decomposition characteristics of polycarbonate and its relationship with heat release. Chin Plast Ind 1:55–58 (in Chinese)

Research on Fault Diagnosis of Platform Protection Equipment Based on FTA

Fan Li, Chao Zhou, Chengxin Du, Zhifei Wang and Junhua Zhao

Abstract The platform door is an important protection device for passenger safety in the process of getting on and off the urban rail transit. This paper applies the fault tree analysis (FTA) method to diagnose the fault of the platform door control system. Based on the analysis of the control principle of the platform door, the safety function of the platform door is obtained by the HAZOP (Hazard and Operability Analysis) method. The platform door cannot be closed as the top event, the downlink method is used to obtain the minimum cut set of the fault tree, and the fault tree that the platform door cannot automatically close the door is constructed. The faults include the optocoupler fault, the inverter fault, the motor sampling circuit fault, etc., and then conduct a comprehensive analysis of qualitative and quantitative. It provides assistant decision-making for the fault diagnosis of the platform door control system and has certain guiding significance for the R&D design of the platform door system.

Keywords FTA · Platform door · HAZOP · Assistant decision-making

1 Introduction

The platform door is a multi-professional and comprehensive product, involving mechanical, electronic, computer, control, materials, and other professional fields [1]. The working cycle of the urban rail station door is fast, and the automatic sliding door (ASD) keeps on switching, the key electrical components door control unit (DCU), PLC, drive motor, relay, and indicator light are used frequently. Especially during holiday peak hours such as vacations and commuting, the platform door is often

F. Li · C. Zhou (✉) · C. Du · Z. Wang
Institute of Computing Technology, China Academy of Railway Sciences, Beijing 100081, China
e-mail: zhou9468123@163.com

J. Zhao
Beijing Jingwei Information Technology Company, Beijing 100081, China

Y. Qin et al. (eds.), *Proceedings of the 4th International Conference on Electrical and Information Technologies for Rail Transportation (EITRT) 2019*, Lecture Notes in Electrical Engineering 639, https://doi.org/10.1007/978-981-15-2866-8_64

influenced by passengers, extrusion, irregular support, and external environmental temperature, humidity, strong wind, and other factors, resulting in frequent failures of the platform door system, various types, difficult to handle, inconvenience to passengers, affecting the operation and management of station staff, even major accidents involving passenger casualties [2] have had a serious impact on the safe operation of urban rail transit. Therefore, it is very important to conduct fault diagnosis research on the platform door system to maximize the stability of the equipment and shorten the maintenance time.

Domestic and foreign scholars have studied the FTA method. In 1960, the American Bell Laboratory first proposed the FTA analysis method when studying the automatic control system. In 1970, Boeing designed the computer software based on FTA, and the University of Tokyo analyzed the fault tree. The system of common failure rate and repair rate is obtained by probability model and solving equation [3]. Northeast Petroleum University proposed a new T-S fuzzy fault tree analysis method [4], which introduces fuzzy logic and T-S fuzzy model into fault tree analysis, so that the fault tree has the ability to process fuzzy information. Beijing Institute of technology research the fault tree and simulation of train gearbox, the operation characteristics, and failure mode of high-speed rail gearbox are analyzed in depth, and the fault tree model of high-speed train gearbox system failure is established.

In this paper, the FTA method is applied to the fault diagnosis research of the platform door system. The fault gate of the platform door cannot be automatically closed as the top event. Through qualitative and quantitative analysis, the causes why platform door system cannot automatically open and close the door are found out, which provides auxiliary decision-making for system fault diagnosis.

2 Control Principle of the Platform Door

The platform door system consists of two parts, mechanical and electrical. The mechanical part is divided into a door structure and a door machine system. The electrical part is divided into a monitoring system and a power system. The door structure is composed of a sliding door, an end door, an emergency door, a fixed door or a fixed railing, a top box, and a supporting structure. The door machine system is composed of a driving motor, a speed reducer, a transmission device and a locking device, and the central control system is composed of PSC (Platform Doors Central Control Panel), PSL (Platform Doors Local Control Panel), PEC (Platform Doors Emergency Control Panel), LCB (Local Control Box). The power supply system is composed of a driving power source, a control power source, and a battery.

The control system has three levels of control: system level, station level (including PSL control and emergency mode PEC control), and manual operation (including LCB control). In the system-level and platform-level control mode, the platform door system is issued by the core component logic control unit (including PEDC control) of the PSC to the door control unit (including DCU control). The DCU receives the start signal and stop signal and drives the motor to move, and the motor drives the door. The drive device drives the sliding door to complete the specified switch door action.

In the PSC, the interface with the external device uses a safety relay group. When the PSL, PEC, or signal system is operated, the coil of the interface relay group is driven, and the output signal enters the logic processing relay group on the one hand and enters the logic of the PLC on the other hand. The relay group and the PLC can issue the signal command and the signal command is transmitted to each door DCU through the hard-wire transmission, and the DCU accepts the command and controls the door to open or close. The control principle of the platform door is shown in Fig. 1.

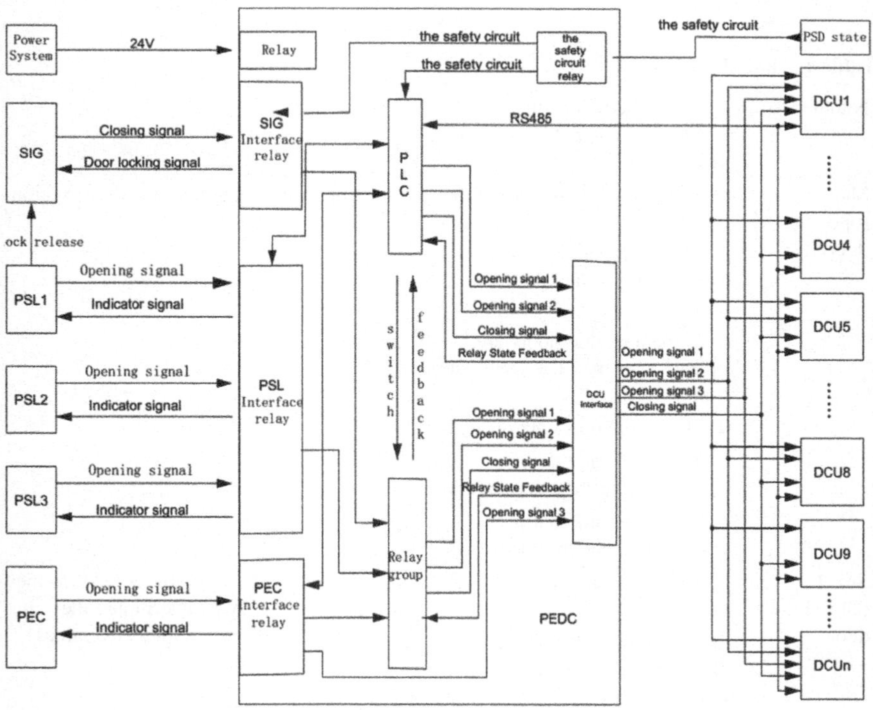

Fig. 1 PSD control schematic

3 Security Features

The safety function is defined in IEC61508-4-2010 for functional safety: "part of the overall safety relating to the EUC and the EUC control system that depends on the correct functioning of the E/E/PE safety-related systems and other risk reduction measures." The logic control unit (PEDC) and the door control unit (DCU) in the platform door control system are the most important components of the safety door control system. The safety function of the corresponding product is related to the opening and closing functions of the system. In the system-level and station-level control mode, through the HAZOP method, combined with a large number of guiding words, Table 1 analyzes for the function of the platform door opening and closing. It is concluded that the FTA analysis is performed when the train arrives at the station and the platform door cannot be closed.

Table 1 Hazard analysis table

Numbering	Hazard description (in system-level and station-level control mode)	Consequence
Sha-1 (open the door)	1. The entire side doors cannot be opened	Affect passengers getting on and off, station service interruption
	2. The entire side doors are opened earlier than the specified time	
	3. The entire side doors are opened later than the specified time	
	4. The partial doors are opened	
	5. The entire side doors are opened before the train arrives	
	6. The entire side doors are opened after the train leaves	
Sha-2 (closed the door)	1. The entire side doors cannot be closed	Affecting passengers getting on and off the bus, the station service interrupted the passengers to fall on the track, casualties
	2. The entire side doors are closed earlier than the specified time	
	3. The entire side doors are closed later than the specified time	
	4. The partial doors are closed	

4 Fault Tree Analysis

Fault tree analysis is from top to bottom deductive formula failure analysis. The method uses Boolean logic to combine low-order events to analyze undesired states in the system [5]. Fault tree analysis is mainly used in safety engineering and the field of reliability engineering to understand the system invalid and find the best way to reduce risk or confirm the incidence of a security incident or a specific system failure. The basic form of the fault tree and the graphical representation of the logic gate are as follows (Fig. 2).

4.1 Confirm Top Event

According to the analysis of the second section of this paper, analyze the system security function, consider the security function requirements of the system, and select the top event of the fault tree analysis:

(1) The platform door is abnormally opened;
(2) When the train arrives at the station, the platform door cannot be opened;

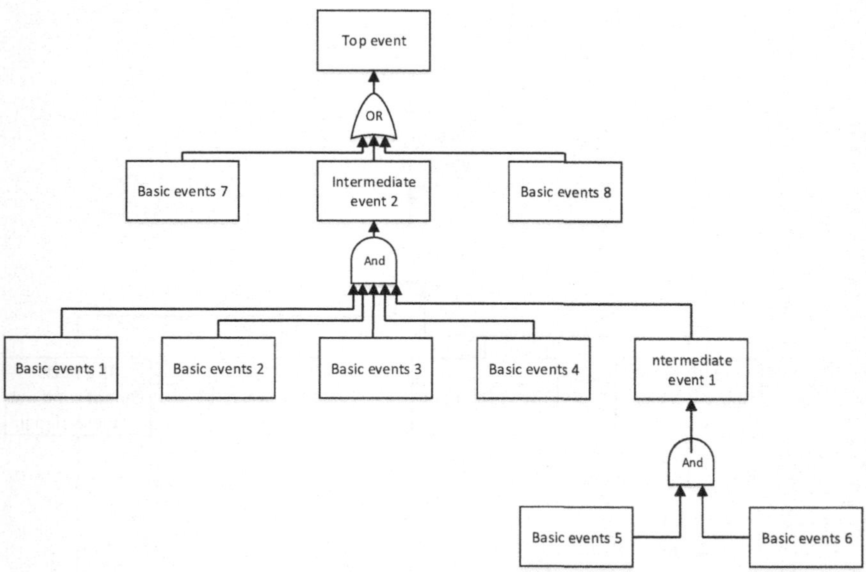

Fig. 2 Fault tree typical analysis structure

(3) When the train leaves the station, the platform door cannot be closed;
(4) The anti-pinch function is invalid, and the passenger is pinched;
(5) The door is closed, and the safety circuit cannot be conducted.

5 Building a Fault Tree

When the train exits from the station, but the platform door cannot be closed as the top event to construct the fault tree. By analyzing the control principle of platform door, the cause of failure is obtained, including DCU fault, PEDC fault, and power system failure. Analyzing the importance of each bottom event is to determine the impact of each minimum cut or bottom event probability, change the top event probability, and determine the minimum cut set [6]. Using the downlink method, all the minimum cut sets of the above fault trees are obtained as shown in Fig. 3, and all bottom events are single points of failure. For single point of failure, it should be avoided as much as possible, some certain compensation measures should be taken, such as adding redundant components, quality assurance measures, and adequate testing to reduce the probability of occurrence of single point of failure [7] (Figs. 4 and 5).

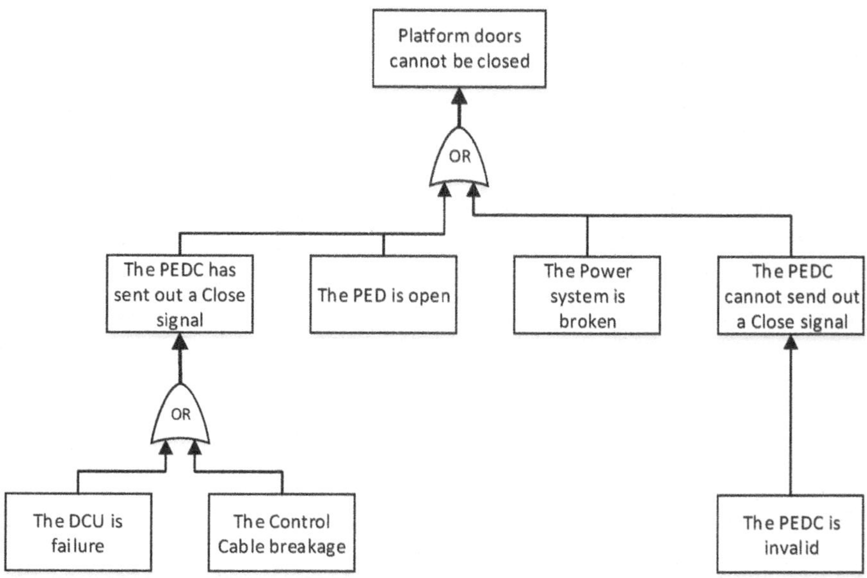

Fig. 3 Fault tree of PSD

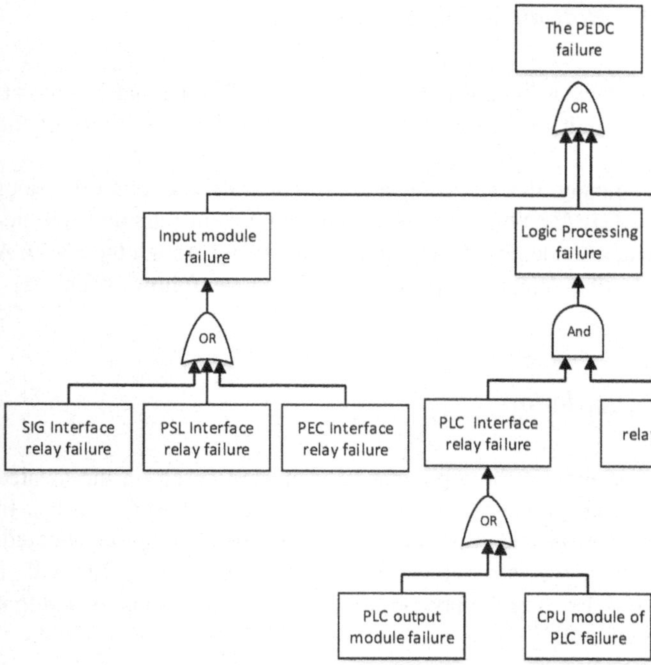

Fig. 4 Fault tree of PEDC

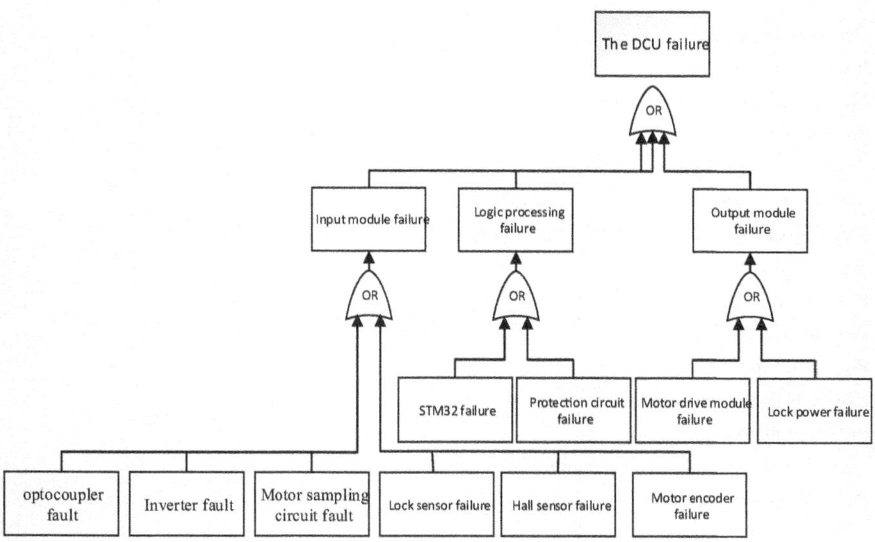

Fig. 5 Fault tree of DCU

5.1 Determine the Bottom Event Failure Rate

The probability that a product has not failed within a certain time and the probability of failure within a unit time after that time is called the failure rate of the product [8].

According to a company design schematic diagram of the platform door equipment, refer to the GJB/Z+299A-1991 electronic equipment reliability prediction manual and the data provided by the equipment supplier. Through the FMECA analysis, the probability of calculating the bottom events is as follows (Table 2).

5.2 Fault Tree Calculation

The probability of occurrence of the top event is calculated based on the created fault tree and the calculated probability of occurrence of all bottom events [9]. In this analysis, all the devices provided by the external interface are not considered. In the calculation process, we don't think it will happen, which event with a probability less than 10^{-9}, because the probability of occurrence is too low and that is an exponential gap with 10^{-6}.

Table 2 Minimum cut set bottom event table

Bottom event	Probability of occurrence (10^{-6})
Optocoupler failure	0.0657
Inverter failure	0.0655
Motor sampling circuit failure	0.067
Stm32 failure	0.1323
Protection circuit failure	0.0707
Motor drive module failure	0.0534
Electromagnetic lock power module failure	0.0826
Sig interface relay failure	0.1
PSL interface relay failure	0.1
PEC interface relay failure	0.1
PLC input module failure	2.76
CPU OF PLC module failure	2.76
PLC output module failure	2.76
Logic processing relay group failure	0.3
DCU interface relay failure	0.1

The failure rate is expressed in mathematical notation

$$\lambda(t) = \frac{\mathrm{d}r(t)}{N\_s(t)\mathrm{d}t} \tag{1}$$

The conditional probability that a product that has not expired at any time will fail within a unit time of $t \sim t + \Delta t$.

$$\lambda(t) = \lim_{\Delta t \to 0} \frac{1}{\Delta t} P(t < T \le t + \Delta t | T > t) \tag{2}$$

The above equation reflects the rate of failure (fault) at time t, called the instantaneous failure (fault) rate.

$$P(t < T \le t + \Delta t | T \rangle t) = \frac{P(t < T \le t + \Delta t)}{P(T > t)} \tag{3}$$

Derivation of failure rate formula

$$\begin{aligned} \lambda(t) &= \lim_{\Delta t \to 0} \frac{P(t < T \le t + \Delta t)}{P(T > t)\Delta t} \\ &= \lim \frac{F(t + \Delta t) - F(t)}{R(t) \cdot \Delta t} \\ &= \frac{\mathrm{d}F(t)}{\mathrm{d}t} \cdot \frac{1}{R(t)} = \frac{f(t)}{R(t)} = -\frac{\mathrm{d}R(t)}{\mathrm{d}t \cdot R(t)} = -\frac{R'(t)}{R(t)} \end{aligned} \tag{4}$$

The $\lambda(t)$ represents the failure rate, and the $\mathrm{d}r(t)$ represents that the number of products that failed during the time. The $N\_s(t)$ represents the number of remaining products, the R(t) represents the reliability function, and the $F(t)$ represents the unreliable function.

Table 3 Calculation formula

Logic gate	Calculation formula
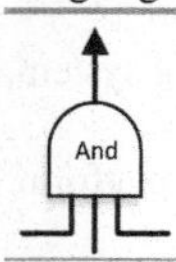	$F_S(t) = E\left[\sum_{i=1}^{n} x_i(t)\right] = F_1(t) \cdot F_2(t) \cdots F_n(t)$
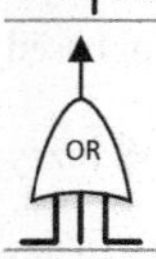	$F_s(t) = E\left[1 - \sum_{i=1}^{n} (1 - x_i)\right] = 1 - [1 - F_1(t)][1 - F_2(t)] \cdots [1 - F_n(t)]$

In the calculation of the failure rate of the top event of the platform door, in order to strengthen the safety assessment, the calculation is performed with the maximum failure rate occurrence mode. Combined with the following formula, the failure rate is $0.64 \cdot 10^{-6}$ and meets the design requirements (Table 3).

6 Conclusion

In this paper, the important functions related to the safety of the platform door are carried out by the FTA, and we calculate the failure rate and have a more in-depth discussion about the safe function. Firstly, the electrical principle of opening or closing of the platform is analyzed, and then, the different guiding words are used according to the HAZOP method to obtain the safety function of the platform door. FTA analysis is carried out for the failure of the platform door to close, and the fault tree that the platform door cannot automatically close the door is constructed. In the electrical system, it mainly includes optocoupler fault, inverter fault, motor sampling circuit fault, STM32 fault, protection circuit fault, motor drive module fault, electromagnetic lock power module fault, and sig interface relay fault, PSL interface relay fault, PEC interface relay fault, PLC input module fault, CPU module fault, PLC output module fault, logic processing relay group faults, and DCU interface relay faults, so R&D designers need to focus on these items and consider safety design, and finally calculate the failure rate of the top event. This result is basically consistent with laboratory test data and engineer experience and can provide a certain reference value for R&D design and fault diagnosis of PSD electrical system.

Acknowledgements This work was financially supported by the Science and technology research program of China Railway (J2019X005), the Science and technology research program of China Railway Network business Co. Ltd (DFYF19-13), and Foundation of China Academy of Railway Sciences (1851DZ0602).

References

1. Yang ZH (2013) Research and design of platform screen doors monitoring and control system. Lanzhou Jiaotong University, Lanzhou, 10–12 (in Chinese)
2. Li B (2017) On the function and common faults of the screen door system of subway platform. Technol Market 24(4): 187–188 (in Chinese)
3. Chen CY, Zhang DS, Ren PH (2003) Study on automobile fault diagnosis expert system based on fault tree analysis. Trans Chin Soc Agric Mach 34(5):130–133 (in Chinese)
4. Li WF, You QH, Liao Q (2018) Research on remote fault diagnosis method based on T-S fuzzy FTA. Control Eng 2018(9):1703–1708 (in Chinese)
5. Vaurio JK (2017) Ideas and developments in importance measures and fault-tree techniques for reliability and risk analysis. Reliability Eng Syst Safety 95(2):99–107

6. Luo YI, Wei O, (2017) Computing minimal cut sets of fault tree using SAT solver. Comput Eng Sci 39(4):725–733 (in Chinese)
7. Pereira-Santos L, Nazar GL, Carro L (2017) Exploring redundancy granularities to repair real-time FPGA-based Systems. Microprocess Microsyst 51.
8. Lan TY (2017) Research on vehicle failure rate prediction based on bp neural network. Scientist 3:63–64 (in Chinese)
9. Volk M, Junges S, Katoen JP (2017) Fast dynamic fault tree analysis by model checking techniques. IEEE Trans Industr Inf (99):1–1

Research on Temperature Prediction for Axles of Rail Vehicle Based on LSTM

Xiaoming Yang, Honghui Dong, Jie Man, Fuzhao Chen, Lejing Zhen, Liming Jia and Yong Qin

Abstract With the continuous improvement of railway speed and the continuous increase of operating mileage, automatic fault detection technology of railway vehicle equipment is becoming more and more meaningful. As a key component of the rail vehicle, the running state of bearing directly affects the safe operation of trains. Since bearing faults are closely related to the increase of axles' temperature during the running process, bearing temperature detection has become an important method for bearing fault diagnosis. This paper presents a data-driven bearing temperature prediction framework based on the deep neural network LSTM for rail vehicle to predict the bearing temperature during operation. The data used in this study are derived from the sensor data generated by the rail vehicle during the actual operation. The accuracy of the experimental results indicates that the framework provided in this paper is a feasible prediction method of axle temperature.

Keywords Rail vehicle · Axles' temperature prediction · LSTM

1 Introduction

Bearings are the key components which ensure the safe operation of trains. They are distributed in many key positions uch as axle box, gearbox and motor on the bogie. They play an important role in supporting the rotating body, reducing friction loss and ensuring rotation accuracy. The bearings on the running gear keep working in high-speed and large-load environment, and they are subjected to alternating stress for a long time. The operating environment is terrible, which

---

X. Yang · H. Dong (✉) · J. Man · F. Chen · L. Zhen · L. Jia · Y. Qin
State Key Laboratory of Rail Traffic Control and Safety, Beijing Jiaotong University, No.3 Shangyuancun, Beijing, Haidian District 100044, People's Republic of China
e-mail: hhdong@bjtu.edu.cn

Y. Qin et al. (eds.), *Proceedings of the 4th International Conference on Electrical and Information Technologies for Rail Transportation (EITRT) 2019*, Lecture Notes in Electrical Engineering 639, https://doi.org/10.1007/978-981-15-2866-8_65

easily leads to wear, peeling, cracks and other faults, so it becomes a huge threat to the safety of train operation. Abnormal changes of axle temperature are an important indicator of train bearing failure, and the severity of bearing failure is closely related to the axial temperature. If the axle temperature change can be predicted in advance, the subsequent changes can be tracked and predicted, so more observation time, processing time and processing means can be obtained to better ensure the safety of train operation.

Experts and scholars in various fields have studied temperature prediction methods. Deng et al. established a mathematical model of the thermal characteristics of the spindle bearing system using the heat source method [1]. Bing Chen-yang et al. designed a digital temperature-based CRH EMU axle temperature alarm system [2]. W. Ma et al. used statistical methods based on historical data to study the precursors of the hot axis [3]. In recent years, the large-scale development of machine learning algorithms and the significant improvement in the performance of computing platforms have also brought new possibilities for the prediction of axial temperature. As a settlement to solve gradient explosion or diminish during backpropagated error in learning process of RNN, LSTM was proposed as a transmutation of recurrent neural network [4, 5] models [6]. Duan Y et al. proposed a LSTM-based structure to predict the travel time of each link and got a small error on the test dataset [7]. AbdElRahman ElSaid et al. used LSTM to predict excess vibration events in aircraft engines and got a promising means for the warning system [8]. C. Luo et al. propose an LSTM-based approach to forecast the temperature trend of an axle, and the forecasting precision promises a referable result [9].

The structure of this paper is roughly arranged as follows: Sect. 2 introduces the existing rail vehicle axle temperature monitoring system and time series prediction model LSTM. Section 3 introduces our axle temperature prediction model architecture. We carried out the experiments and evaluated the results with the actual vehicle data in Sect. 4. Section 5 concludes the work of this paper.

2 Axle Temperature Monitoring System and LSTM Model

2.1 The Rail Vehicle Axle Temperature Monitoring System

The rail vehicle axle temperature monitoring system collects the temperature data of the train axle box, gearbox and traction motor in real time through the dual-channel PT100 temperature sensor to realize the uninterrupted monitoring of the bearing temperature [10]. There are nine different temperature sensors on each axle box of the power carriages, including the axle head temperature, the large gearbox wheel side temperature and so on, as shown in Fig. 1. In the actual driving process, the power carriages provide operational power and deceleration braking force, which causes a large load on the bearing. This paper predicts the bearing temperature

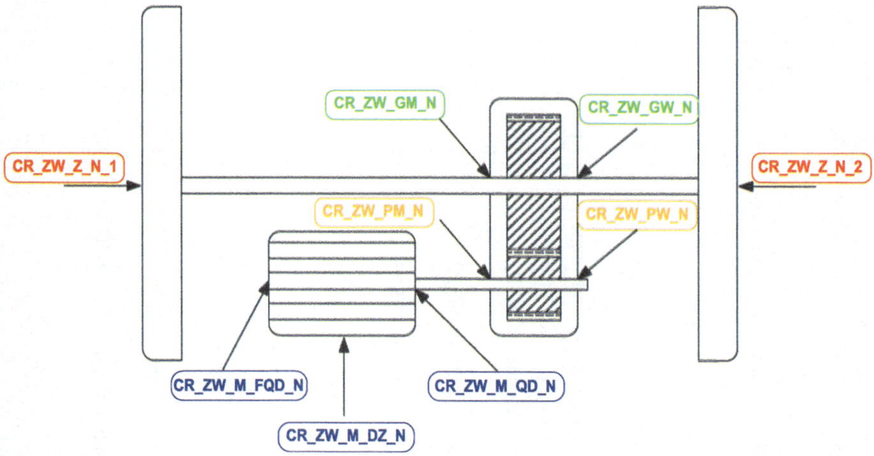

Fig. 1 Bearing temperature sensor mounting position

installed at the bottom of the power carriages to realize the fault warning. The correspondence between different sensor codes and their meanings is shown in Table 1.

With the power transmission on the "motor-gearbox-axle box" drive chain, the measured maximum value of the axle temperature data is decreasing, the intensity of the fluctuation is weakening, the temperature fluctuation of the axle box is the slowest, and the motor stator temperature fluctuations are the most severe [11]. There is a phenomenon of exponential growth and attenuation in the process of the axle temperature rising and falling. Figure 2 shows the temperature variation of different measuring points recorded by one of the on-axis temperature detection systems installed on the bottom of a power carriage which belongs to the rail vehicle.

2.2 LSTM Model

LSTM is a special RNN network structure. The hidden layer of the original RNN has only one state *h*, which is sensitive to short-term input, and LSTM essentially implements an improvement to the RNN hidden layer through the memory unit. Based on the hidden layer of the RNN, in the LSTM network model, the hidden layer adds a state *C*, called the cell state, to preserve the long-term state [12].

The forget gate first reads the previous hidden layer output information h_{t-1} and the current input x_t, then outputs a specific value to participate in the part of the cell state C_t.

Table 1 Features table in the prediction

Feature code	Description	Unit	Code	Feature code	Description	Unit	Code
CR_SPEED	Train's running speed	km/h	A_1	CR_ZW_M_QD_N	Temperature at point 7 on axle *N*	°C	A_{14}
CR_KT_WD_OUT	Environment temperature	°C	A_2	CR_ZW_PM_N	Temperature at point 8 on axle *N*	°C	A_{15}
CR_ZD_LCG	The train pip pressure	kPa	A_3	CR_ZW_PW_N	Temperature at point 9 on axle *N*	°C	A_{16}
CR_ZD_TFG	Park cylinder pressure	Kpa	A_4	CR_ZD_KH_1	Air spring pressure of bogie 1	Kpa	A_{17}
CR_QY_LJ	Wheel diameter	mm	A_5	CR_ZD_KH_2	Air spring pressure of bogie 2	Kpa	A_{18}
CR_ZD_ZFG	Pressure of main air duct	kPa	A_6	CR_ZD_WSP1	Axial velocity of WSP1	Km/h	A_{19}
CR_ZD_ZDG	Brake cylinder pressure	kPa	A_7	CR_ZD_WSP2	Axial velocity of WSP2	Km/h	A_{20}
CR_ZW_Z_N_1	Temperature at point 1 on axle *N*	°C	A_8	CR_ZD_WSP3	Axial velocity of WSP3	km/h	A_{21}
CR_ZW_Z_N_2	Temperature at point 2 on axle *N*	°C	A_9	CR_ZD_WSP4	Axial velocity of WSP4	Km/h	A_{22}
CR_ZW_GM_N	Temperature at point 3 on axle *N*	°C	A_{10}	CR_QY_TCU1	Rotation frequency of TCU1	HZ	A_{23}
CR_ZW_GW_N	Temperature at point 4 on axle *N*	°C	A_{11}	CR_QY_TCU2	Rotation frequency of TCU 2	HZ	A_{24}
CR_ZW_M_FQD_N	Temperature at point 5 on axle *N*	°C	A_{12}	CR_QY_TCU3	Rotation frequency of TCU 3	HZ	A_{25}
CR_ZW_M_DZ_N	Temperature at point 6 on axle *N*	°C	A_{13}	CR_QY_TCU4	Rotation frequency of TCU 4	HZ	A_{26}

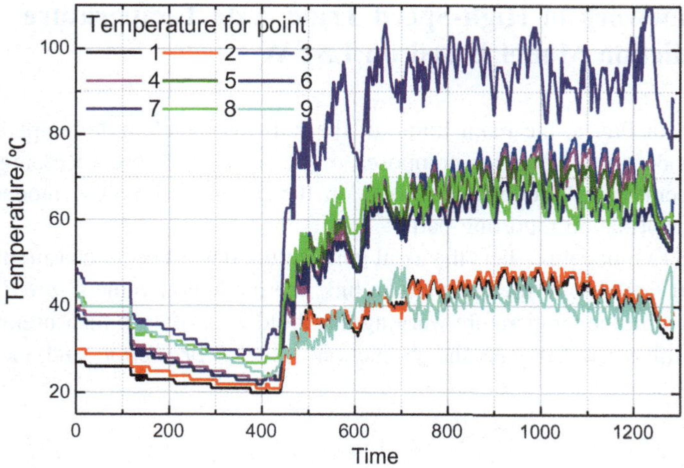

Fig. 2 Sample of nine measuring points on a certain axis N

$$f_t = \sigma\left(W_f \cdot [h_{t-1}, x_t] + b_f\right) \tag{1}$$

The input gate processes the information input of the current sample, determines which information will be applied to the cell structure and uses the sigmoid activation function to output i_t:

$$i_t = \sigma(W_i \cdot [h_{t-1}, x_t] + b_i) \tag{2}$$

The control gate determines the update of the cell state from C_{t-1}–C_t:

$$C_t' = \tan h(W_c \cdot [h_{t-1}, x_t] + b_c) \tag{3}$$

$$C_t = f_t * C_{t-1} + i_t * C_t' \tag{4}$$

The output gate is responsible for generating the output of the cell unit O_t and updating the hidden layer state vector h_{t-1}:

$$O_t = \sigma(W_o \cdot [h_{t-1}, x_t] + b_o) \tag{5}$$

$$h_t = O_t * \tan h(C_t) \tag{6}$$

where W_f,W_i, W_c and W_o are weight matrices corresponding to each gate structure, b_f, b_i, b_c and b_o are bias vectors corresponding to each gate structure.

3 Framework of High-Speed Train Axle Temperature Prediction Model Based on LSTM

As shown in Fig. 3, the main frame of the railway vehicle axle temperature prediction model in this paper is composed of the sensor data preprocessing part, the axle temperature prediction part based on the pre-trained LSTM model and the warning information reporting part.

The LSTM model predicts the axial temperature data of each measurement point in the next k cycles. In the comparison link, the prediction value *T_prediction* was compared with the temperature warning threshold *T_m* of each measurement point, and the difference between the prediction value *T_prediction* and the external

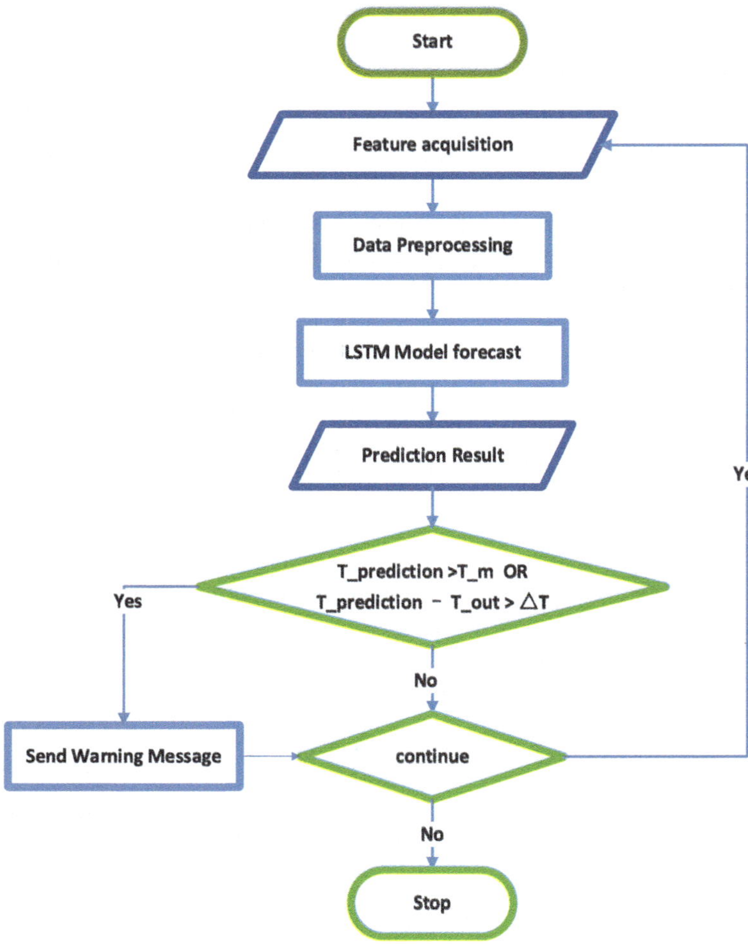

Fig. 3 Sample of nine measuring points on a certain axis N

environment temperature *T_prediction* was compared with the temperature warning threshold ΔT of each measurement point. When any one of the prediction values exceeded the threshold, the warning signal would be sent. This paper mainly introduces the data preprocessing part and the LSTM model part within the framework.

3.1 Data Preprocessing

The data presented in this paper are all from the recorded data recorded during the actual operation of the high-speed train. There are many factors affecting the axle temperature of each measuring point of the axle, such as the train's running speed, external ambient temperature, braking gear position, acceleration magnitude, air spring pressure, WSP speed, TCU frequency and so on [10]. We selected 26 typical features (A_1–A_{26}) related to the axle temperature of each measuring point as framework's inputs. Table 1 lists the specific meanings and codes of these features.

Since the sampling frequency of different sensors can be different, and the collected raw data usually contain a lot of noise, a series of data preprocessing operations are needed to improve the quality of the samples. In the experiment, the obvious outliers and invalid data are first removed, and then a modified sample period (30 s in the experiment) is determined. Next, the redundant data of the same feature in a single cycle are eliminated. Finally, linear interpolation is used to complete the interpolation of some sensor data which is missed for a short term (less than three sample periods). In this way, we get the time series data suitable for training LSTM models.

For the axis *N*, after a series of data preprocessing, the sample structure shown in Table 2 is obtained for training by the LSTM model, and each sample contains 19-dimensional features. Where χ_1–χ_{16} are, respectively, corresponding to the measured value A_1–A_{16} in Table 1.

Assume that the position of the axis *N* is on the first bogie of a certain carriages, then χ_{17} will be assigned by the measured value A_{17} in the sample period, otherwise the measured value A_{18}. Besides, let the position of the axis *N* be the *i*th axle (numbered 1–4) of the carriages, and then χ_{18} and χ_{19} will assign with A_{18+i} and A_{22+i}, respectively.

Table 2 Proposed data samples of axle *N*

Time	χ_1	…	χ_{16}	χ_{17}	χ_{18}	χ_{19}
2018/04/11 06:01:00	35	…	43	500	95.75	0.21
2018/04/11 06:01:30	35	…	43	502	95.70	0.21
…	…	…	…	…	…	…
2018/04/11 16:00:00	36	…	37	486	83.97	0.26

3.2 LSTM Prediction Model Structure

The network structure designed in this paper is shown in Fig. 4. The input is N sets of time series data $\boldsymbol{\chi}_n = \{\boldsymbol{\chi}^{(1)}, \boldsymbol{\chi}^{(2)}, \ldots, \boldsymbol{\chi}^{(T)}\}_n$, where N is the number of axles, and T is the length of the specified input sequence window. This paper assumes that there is no difference in temperature prediction between different axles of the same train, that is, the N axis temperature prediction requirements of the same train share the same model. We only need to input the sampling data of all axles into the model in proper order, and then we would obtain the corresponding prediction results sequence of all axles. Any sample point $\boldsymbol{\chi}^{(t)} \in R^m$ of each axis is an m-dimensional vector $\{\chi_1^{(t)}, \ldots, \chi_4^{(t)}, \chi_6^{(t)}, \ldots, \chi_{19}^{(t)}\}$, with m equals to 19 such that each element $\boldsymbol{\chi}^{(t)}$ corresponds to each sample data in Table 2. χ_5 represents the diameter of the wheel diameter, which is a factor affecting the change of the axle temperature, but its value does not change during the running of the train. The model's output is a sequence of N sets $\mathbf{Y}_n = \{\mathbf{Y}^{(T+k)}\}_n$, where $\mathbf{Y}^{(T+k)} \in R^l$ is a l dimension vector $\{y_1^{(T+k)}, \ldots, y_9^{(T+k)}\}$ representing the prediction results of the nine points of the axis n after the next k cycles.

For the Nth axis of a train, its input contains T sample points in one cycle, each sample point itself is a 19-dimensional vector, and its corresponding output is a nine-dimensional vector. The input layer and the output layer contain two hidden layers composed of LSTM64 units, and then divided into two branches. The first branch passes through a fully connected layer and directly outputs the predicted data $\mathbf{Y}^{(T+k)}$ after k cycles. The other branch then, the second layer LSTM output, is merged with the feature χ_5, and then the prediction data $\hat{\mathbf{Y}}^{(T+k)}$ after k cycles are

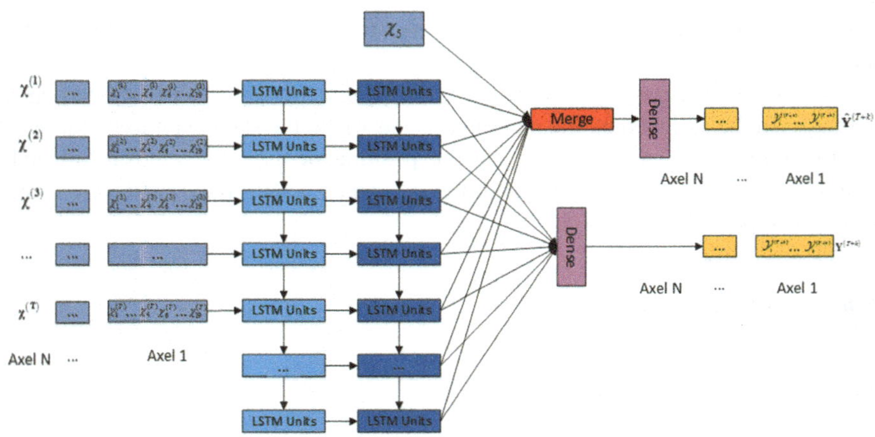

Fig. 4 LSTM network architecture diagram

output through a fully connected layer. This design ensures that the first branch can converge normally even if the second branch has a gradient dispersion during training.

In this paper, $k = 1$, 5, 10 and 15 were chosen, so as to predict the axle temperature at different measuring points for the next 30 s, 2.5, 5 and 7.5 min. In addition, we set the parameter T as 36 in this paper based on the experimental analysis.

4 Experiments and Evaluation

4.1 Experimental Scheme

In the experiment, a high-standard computation device fitted with four Intel CPUs, two 32G DDR4 flash memories and four Nvidia Tesla M40 graphics cards was used to design our axle temperature prediction model. The specific hardware and software parameters are shown in Table 3.

We collected 26 characteristics A_1–A_{26} related to axial temperature of five different rail vehicles in 10 days, among which A_5 only related to the model type and did not change by the time. Then we integrate sensor data into sample format as shown in Table 2 according to data preprocessing requirements. On consideration of smoothness, mean squared error (MSE) was selected in the experiment as the loss function in the training process, and the weight of two branch losses was set as 1.0 and 0.1, respectively. Adam was selected as the optimization function, while the batch_size and the learning_rate were set as 100 and 0.0001, respectively.

4.2 Results and Evaluation

The test data which have passed the data preprocessing operation are put into the axial temperature prediction model to obtain the predicted values of axial temperature $\hat{\mathbf{Y}}^{(T+k)}$, $k = 1, 5, 10, 15$ at nine measuring points in the future 30 s, 2.5, 5 and 7.5 min of a certain axle.

Table 3 Software and device details

Item	Configuration	Item	Configuration
CPU	(Intel E5-2643v4 @3.4 GHZ/8 Core) * 4	CUDA version	v9.0
RAM	(32G) * 2	TensorFlow version	v1.9.0
GPU	(Nvidia Tesla M40 24G) * 4	CuDNN version	v7.0.5
OS	Ubuntu 16.04	Keras version	v1.4.4

The evaluation criteria indicate the overall prediction ability of the models by comparing the test datasets $\tilde{\mathbf{Y}}^{(T+k)}, k = 1, 5, 10, 15$ and the predicted value $\hat{\mathbf{Y}}^{(T+k)}, k = 1, 5, 10, 15$. In order to evaluate the performance of the deep prediction model of axial temperature, we introduce three indicators, which are

Root Mean Square Error (RMSE):

$$\mathrm{RMSE}_i^k = \sqrt{\frac{\sum\limits_{\mathrm{Num}} (\hat{y}_i^{T+k} - \tilde{y}_i^{T+k})^2}{\mathrm{Num}}} \tag{7}$$

Correlation Coefficient (CC):

$$\mathrm{CC} = \frac{\sum (\hat{Y}^{(T+k)} - \hat{Y}_{\mathrm{mean}}^{(T+k)})(\tilde{Y}^{(T+k)} - \tilde{Y}_{\mathrm{mean}}^{(T+k)})}{\sqrt{\sum (\hat{Y}^{(T+k)} - \hat{Y}_{\mathrm{mean}}^{(T+k)})^2} \sqrt{\sum (\tilde{Y}^{(T+k)} - \tilde{Y}_{\mathrm{mean}}^{(T+k)})^2}} \tag{8}$$

where $\hat{Y}_{\mathrm{mean}}^{(T+k)}$ and $\tilde{Y}_{\mathrm{mean}}^{(T+k)}$, respectively, represent the mean of $\hat{\mathbf{Y}}^{(T+k)}$ and $\tilde{\mathbf{Y}}^{(T+k)}$.

Fit:

$$\mathrm{Fit} = \left(1 - \frac{\sqrt{\frac{1}{\mathrm{num}} \sum (\tilde{Y}^{(T+k)} - \hat{Y}_{\mathrm{mean}}^{(T+k)})^2}}{\sqrt{\frac{1}{\mathrm{num}} \sum (\tilde{Y}^{(T+k)} - \tilde{Y}_{\mathrm{mean}}^{(T+k)})^2}}\right) \times 100\% \tag{9}$$

According to the experimental scheme, the test dataset is put into the model, and the corresponding prediction result is obtained. Table 4 presents the prediction performance with the evaluation criteria.

5 Conclusion

As an important part of rail vehicle health management system, the research direction of bearing temperature prediction is of great significance. In this paper, a bearing temperature prediction framework of rail vehicle based on the LSTM model is proposed and tested with real train data. The test results show that the prediction error of the framework is within the acceptable range. This study provides a feasible solution for bearing failure prediction of the rail vehicle and a research idea for bearing temperature warning in wind power and other related fields.

Table 4 Prediction performance of LSTM method

RMSE				
Axle position	*K* =1	*K* =5	*K* =10	*K* =15
Position 1	0.8235	1.0173	1.1364	1.3933
Position 2	0.9073	1.0208	1.1246	1.3813
Position 3	1.4028	1.9113	2.3453	2.7847
Position 4	1.6308	1.8130	2.3840	2.8867
Position 5	1.3402	1.4567	1.8642	2.0080
Position 6	1.0204	1.6680	1.8946	2.1924
Position 7	2.9804	2.8177	3.2435	3.9005
Position 8	2.0697	2.3430	2.6254	3.0142
Position 9	1.3779	1.3490	1.6391	1.9323
CC				
Axle position	*K* =1	*K* =5	*K* =10	*K* =15
Position 1	0.9963	0.9953	0.9942	0.9909
Position 2	0.9973	0.9956	0.9934	0.9922
Position 3	0.9979	0.9961	0.9946	0.9921
Position 4	0.9972	0.9962	0.9945	0.9905
Position 5	0.9982	0.9975	0.9951	0.9946
Position 6	0.9985	0.9970	0.9948	0.9937
Position 7	0.9955	0.9959	0.9943	0.9926
Position 8	0.9928	0.9907	0.9883	0.9853
Position 9	0.9819	0.9807	0.9731	0.9582
Fit				
Axle position	*K* =1	*K* =5	*K* =10	*K* =15
Position 1	98.8844	98.6095	98.4357	98.0936
Position 2	98.7455	98.5848	98.3624	98.0917
Position 3	98.7722	98.3206	97.9864	97.5492
Position 4	98.5469	98.3787	97.9582	97.4036
Position 5	98.7594	98.6457	98.4268	98.1256
Position 6	99.0275	98.3915	98.0435	97.8965
Position 7	98.1233	98.2251	97.9651	97.5639
Position 8	98.1491	97.9006	97.7086	97.3193
Position 9	98.2213	98.2515	97.8962	97.4802

Acknowledgements This work was funded by the National Key Research and Development Plan of China (Grant No. 2016YFB1200103).

References

1. Deng X, Fu J, Zhang Y (2015) A predictive model for temperature rise of spindle-bearing integrated system. J Manuf Sci Eng Trans ASME
2. Bing C et al (2016) Design of CRH axle temperature alarm based on digital potentiometer. In: Control conference IEEE
3. Ma W et al (2017) A prediction method based on stepwise regression analysis for train axle temperature. In: International conference on computational intelligence & security, IEEE
4. Hochreiter S, Schmidhuber J (1997) Long short-term memory. Neural Comput 9(8):1735–1780
5. Gers FA, Schraudolph NN, Schmidhuber J (2003) Learning precise timing with LSTM recurrent networks. J Mach Learn Res 3.1(2003):115–143
6. Kolen JF, Kremer SC (2001) Gradient flow in recurrent nets: the difficulty of learning longterm dependencies
7. Duan Y, Lv Y, Wang FY (2016) Travel time prediction with LSTM neural network. In: 2016 IEEE 19th international conference on intelligent transportation systems (ITSC)
8. ElSaid A, Wild B, Higgins J, Desell T (2016) Using LSTM recurrent neural networks to predict excess vibration events in aircraft engines. In: 2016 IEEE 12th international conference on e-Science (e-Science), IEEE, pp. 260–269
9. Luo C et al (2017) LSTM-based temperature prediction for hot-axles of locomotives. In: ITM web of conferences, vol 12
10. Chen D (2003) Study of the rule of train axle temperature and infrared detection mode of axle temperature. Harbin Engineering University (In Chinese)
11. Zhang W et al (2012) Axle temperature detecting system of new-generation high-speed EMUs CRH380C. Electr Drive Locomotives J 6(2012) (In Chinese)
12. Olah C (2015) Understanding LSTM networks. [Online]. Available: http://colah.github.io/posts/2015-08-Understanding-LSTMs/

Bearing Fault Diagnosis Method Based on Graph Fourier Transform and C4.5 Decision Tree

Yuze Wang, Yong Qin, Xuejun Zhao, Shunjie Zhang and Xiaoqing Cheng

Abstract Bearing is a key component of rail vehicles. Its operational status greatly affects the safety of passengers and cargo on the train. Therefore, it is especially important to find a fault diagnosis method suitable for train bearings. In order to adapt to the railway application background, the graph Fourier transform (GFT) is introduced into its fault diagnosis. As the foundation of graph signal processing (GSP), GFT is the expansion of graph signal in terms of the eigenfunctions of graph Laplacian matrix. The vibration signal is converted into a path graph signal. Using GFT to extract the graph spectrum domain feature as fault feature set combines with the C4.5 classification algorithm to identify the fault of the rolling bearing. Taking into account the practicality of the application, train axle rolling bearings vibration signal with real faults have been collected from on-site trains to verify its validity. By comparing with the time domain feature and the frequency domain feature classification result, it reflects the practicability of the method under the complex operation conditions of the train.

Keywords Train bearing fault diagnosis · Graph Fourier transform · Graph spectrum feature · C4.5 decision tree

1 Introduction

For the nonlinear and nonstationary characteristics of vibration signal of rolling bearings, traditional signal processing methods are often difficult to extract effective fault features, which is the key to diagnosis of mechanical faults successfully. Therefore, it is necessary to further study the method of extracting fault characteristics of rolling bearings.

Y. Wang (✉) · Y. Qin · X. Zhao · S. Zhang · X. Cheng
State Key Laboratory of Rail Traffic Control and Safety, Beijing Jiaotong University, No. 3 Shangyuan Village, Haidian District, Beijing, China
e-mail: 17120894@bjtu.edu.cn

Y. Qin et al. (eds.), *Proceedings of the 4th International Conference on Electrical and Information Technologies for Rail Transportation (EITRT) 2019*, Lecture Notes in Electrical Engineering 639, https://doi.org/10.1007/978-981-15-2866-8_66

In recent years, various graph signal processing (GSP) methods based on graph signals have been vigorously developed. The main research content is to extend the traditional signal processing method to the analysis and processing of graph signals. The graph structure [1] is a mathematical abstraction of the network, which can naturally represent the sensor network, the transportation network, and the social network. It is one of the important tools for processing network-related data. The graph signal processing technology maps the data set to the graph structure, in which the data elements are abstracted into node structure, and the association relationship between the data elements is represented by the edge connecting two nodes, so that the data obtained from the network is mapped to a signal on the graph structure. The most important theoretical basis for graph signal processing techniques is the graph spectrum theory [2, 3]. Graph Fourier transformation (GFT) [4] is a data transformation method similar to Fourier transformation (FT), which focuses on the interaction between the internal structure (Laplace matrix) of the graph and the corresponding graph feature (graph signal), providing an analysis perspective of GSP eigenvalue spectrum.

Shuman et al. [5] gave a general overview of the main challenges and research contents in the field of GSP; Zhu and Michael [6] conducted a detailed theoretical analysis of why the Laplacian eigenbasis of the graph is used as the Fourier transform base of the graph and discussed whether the Laplacian feature vector base makes sense for all graphs. From the above research, it is found that there is a corresponding relationship between the GFT and the FT of the bearing signal relative to the bearing vibration signal, and the graph Laplacian eigenbasis of the GFT is similar to the Fourier characteristic base of the FT. Therefore, the GFT can also be used to extract feature of the rolling bearing vibration signal.

In this paper, the application of GFT to fault feature extraction of rolling bearings is proposed. A fault diagnosis method for rolling bearings based on GFT feature extraction and C4.5 decision tree classification and recognition is proposed. In order to further verify the superiority of the method for bearing fault feature extraction in the context of rail transit, the frequency domain features and time domain features are combined with the C4.5 algorithm to compare the results.

2 Bearing Vibration Signal and GFT

In the graph spectrum theory (directed graph is not considered in this article), the undirected, connected, weighted graph $G = (V, E, W)$ is defined, where V is a finite set of all vertices, E is the set of all vertices connected into the edge e_{ij}, and W is the weighted connection matrix; the weight of the edge e_{ij} is w_{ij}. In practical applications, the weights connecting the two vertex edges are generally defined as

$$\begin{cases} \text{if two vertices connected} & w_{ij} = 1 \\ \text{otherwise} & w_{ij} = 0 \end{cases} \quad (1)$$

Considering the representation of the vibration signal in the time domain, it is found to correspond to the path graph. The rolling bearing vibration signal has a structural correspondence with the path graph signal as a time series signal: The sequence of the time series signal sequentially corresponds to the graph structure of the path graph.

The Laplacian matrix of the graph is defined as

$$L = D - W \quad (2)$$

where D is the degree diagonal matrix and the degree of the vertex n is $d_n = \sum_{m \neq n} w_{mn}$ [7].

The Laplacian matrix can be orthogonally decomposed into

$$L * \lambda_l = \lambda_l * x_l \quad (3)$$

where λ_l and x_l are eigenvalues and eigenvectors.

The time series signal can correspond to the path graph vertex field as a path graph signal. Conceptually, the GFT [8] of the vertex function f of G is an extension of the eigenfunction of the Laplacian, similar to the definition of FT as Eq. (4). And the GFT of path graph signal $f(n) \in R_n$ is its expansion of according to the eigenfunction (characteristic vector) of the Laplacian matrix. The GFT of the graph signal $f(n)$ is represented by $\hat{f}(\lambda_l)$ [9], and the expression is Eq. (5)

$$\hat{f}(\omega) = f(t), \exp(i\omega t) = \int_{-\infty}^{+\infty} f(t) \exp(-i\omega t) \mathrm{d}t \quad (4)$$

$$\hat{f}(\lambda_l) = f(n), x_l = \sum_{n=1}^{N} x_l^* f(n) \quad (5)$$

where N is the number of vertices of the path graph.

Therefore, the path graph signal in the vertex domain can be converted to the graph spectrum domain by GFT.

3 Fault Diagnosis Algorithm

3.1 Feature Extraction

For the algorithm of mechanical fault identification, the time domain features and frequency domain features of the vibration signal are generally extracted as the feature set of fault recognition. In the literature [10], several frequency domain fault features are introduced. The representative six characteristics of the frequency domain fault features include the mean amplitude frequency (F_1), the center of gravity frequency (F_2), the standard deviation frequency (F_3), the square root frequency (F_4), the skewness frequency (F_5), and the kurtosis frequency (F_6). Find the corresponding feature indicators in the time domain features, including the mean (T_1), the standard deviation (T_2), the root mean square (T_3), the skewness (T_4), and the kurtosis (T_5). Similar to classical signal processing, the GFT of the graph signal makes it possible to represent in the "graph spectrum". The corresponding feature indicators in the graph spectrum domain include the graph spectrum mean amplitude feature (G_1), the graph spectrum center of gravity feature (G_2), the graph spectrum standard deviation feature (G_3), the graph spectrum root mean square feature (G_4), the graph spectrum skewness feature (G_5), and the graph spectrum kurtosis feature (G_6).

In Table 1, $f(i)$ is the time domain signal, n is the time domain signal point number, $y(k)$ is the frequency spectrum of $f(i)$, K is the value of spectral, f_k is the frequency value of the kth spectral line, $\hat{f}(\lambda_l)$ is the eigenvalue spectrum of the graph signal, N is the number of fixed points of the path graph, and λ_l is the characteristic value.

As shown in Table 1, the physical quantities based on statistical features are used as fault feature sets, which can reflect the amplitude, energy of the signal feature range, and the variation of fluctuations, and can effectively characterize the fault data. Considering the correspondence, this article applies the comparison in Sect. 4 and selects T_1, T_3, T_4, T_5 as the time domain fault feature set, F_1, F_4, F_5, F_6 as the frequency domain fault feature set, and G_1, G_4, G_5, G_6 as the graph spectrum domain fault feature set for fault detection classification.

Table 1 Characteristic indicator expressions

Time domain feature	Frequency domain feature	Graph spectrum domain feature
$T_1 = \frac{\sum_{i=1}^{n} f(i)}{n}$	$F_1 = \frac{\sum_{k=1}^{K} y(k)}{K}$	$G_1 = \frac{\sum_{l=0}^{N-1} \hat{f}(\lambda_l)}{N}$
\	$F_2 = \frac{\sum_{k=1}^{K} f_k y(k)}{\sum_{k=1}^{K} y(k)}$	$G_2 = \frac{\sum_{l=0}^{N-1} \lambda_l \hat{f}(\lambda_l)}{\sum_{l=0}^{N-1} \hat{f}(\lambda_l)}$
$T_2 = \sqrt{\frac{\sum_{i=1}^{n} (f(i)-T_1)^2}{n}}$	$F_3 = \frac{\sqrt{\sum_{k=1}^{K} (f_k - F_2)^2 y(k)}}{\sum_{k=1}^{K} y(k)}$	$G_3 = \frac{\sqrt{\sum_{l=0}^{N-1} (\lambda_l - G_2)^2 \hat{f}(\lambda_l)}}{\sum_{l=0}^{N-1} \hat{f}(\lambda_l)}$
$T_3 = \sqrt{\frac{1}{n} \sum_{i=1}^{n} \left(\frac{f(i)-T_1}{T_1}\right)^2}$	$F_4 = \sqrt{\frac{\sum_{k=1}^{K} f_k^2 y(k)}{\sum_{k=1}^{K} y(k)}}$	$G_4 = \sqrt{\frac{\sum_{l=0}^{N-1} \lambda_l^2 \hat{f}(\lambda_l)}{\sum_{l=0}^{N-1} \hat{f}(\lambda_l)}}$
$T_4 = \sqrt{\frac{n \sum_{i=1}^{n} (f(i)-T_1)^3}{(n-1)(n-2)T_2}}$	$F_5 = \frac{\sum_{k=1}^{K} (f_k - F_2)^3 y(k)}{K F_3^4}$	$G_5 = \frac{\sum_{l=0}^{N} (\lambda_l - G_2)^3 \hat{f}(\lambda_l)}{N G_3^4}$
$T_5 = \frac{1}{n} \sum_{i=1}^{n} \left(\frac{f(i)-T_1}{T_1}\right)^4$	$F_6 = \frac{\sum_{k=1}^{K} (f_k - F_2)^4 y(k)}{K F_3^4}$	$G_6 = \frac{\sum_{l=0}^{N} (\lambda_l - G_2)^4 \hat{f}(\lambda_l)}{N G_3^4}$

3.2 C4.5 Decision Tree

Decision trees are a widely used classification model in machine learning. There are many decision tree algorithms. They include ID3, classification and regression tree (CART) [11], C4.5 [12], and so on. In most decision tree algorithms, the decision tree is constructed in a top-down and recursive manner, and in each iteration, the training examples are divided according to the values of the selected attributes present in the training example set.

In 2006, the IEEE International Data Mining Conference (ICDM) nominated up to 10 of the most famous data mining algorithms. Due to its outstanding performance, the C4.5 decision tree was first listed as a classification algorithm [13]. Therefore, this paper chooses C4.5 algorithm for fault classification.

Therefore, combined with the feature extraction in Sect. 3.1, a fault classification algorithm based on the C4.5 classification algorithm is established as Fig. 1.

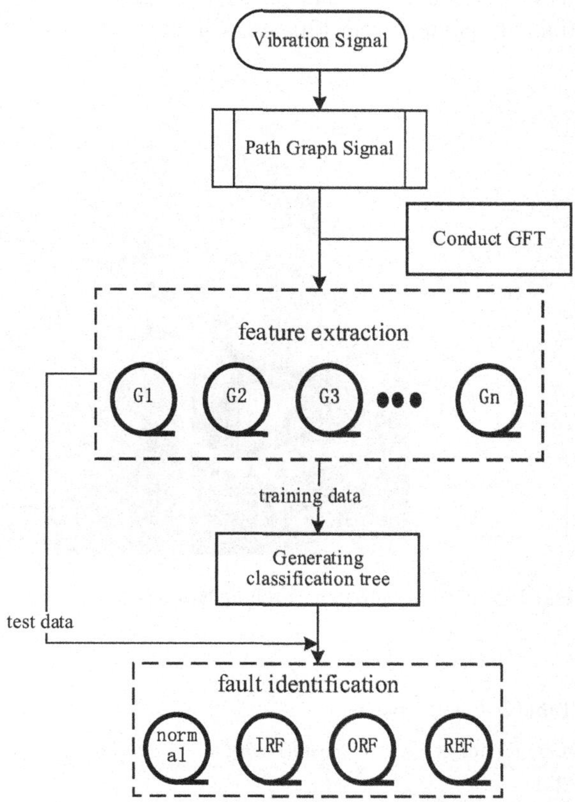

Fig. 1 Proposed diagnostic process

4 Applications

In order to verify the superiority of GFT and the adaptability of railway application background. The axle rolling axle with real faults collected from the on-site work train is simulated on the train test bench, as shown in Fig. 2, collecting the vibration signal on different working conditions. The vibration data of the normal, inner race fault (IRF, stripping fault), rolling element fault (REF, crack fault), and outer race fault (ORF, stripping fault) of the SKF197726 rolling bearing are used. The specific working conditions are grouped as shown in Table 2.

According to the time domain, frequency domain, and graph spectrum domain feature indicators introduced in Sect. 3 and the proposed algorithm, the detection accuracy of faulty bearings under each working condition is analyzed based on time domain features T_1, T_3, T_4, T_5, frequency domain features F_1, F_4, F_5, F_6, and graph spectrum domain features G_1, G_4, G_5, G_6. The signal sampling frequency is 12.8 kHz, and the sample data length is 1024. For each failure mode, 40 consecutive data samples were randomly selected, among which 10 data samples were used as training sets and 30 data samples were used as test sets. A total of 40 training samples and 120 test samples were selected. In order to eliminate the

Fig. 2 Test rig of railway axle rolling bearings

Table 2 Case2 grouping

Group number	Speed/(km/h)	Vertical load/(kN)	Lateral load/(kN)
2–1	60	56	0
2–2	90	56	0
2–3	120	56	0

randomness of the samples, the classification results in this paper are the average results of 100 random sampling and fault classification.

Under different working conditions, the detection results based on different characteristics are shown in Fig. 3. Each set of histograms represents the fault detection results of the three different types of characteristic indicators in the same working condition in Table 2. The detection accuracy of each fault type is shown in Fig. 4. Each histogram shows the detection accuracy of the same fault type under different working conditions and different characteristic indicators.

Figure 3 shows that on the basis of real train fault bearing vibration data, the GFT feature is significantly higher than the FT feature and the time domain feature fault detection rate by 5–20%, and the time domain feature has the worst detection effect. It shows that under the conditions of large bearing, complex conditions, and low signal-to-noise ratio, the detection effect of GFT features is the best, still staying at about 90%, and the detection effect of FT features is decreased about 5%. Time domain feature detection is the worst, because the higher noise affects its classification effect.

The detection accuracy of different working conditions and fault types is shown in Fig. 4. Obviously, the GFT feature has a high detection accuracy for almost all fault types and working conditions, showing the stability of the method. As the speed increases, the detection accuracy of each fault type increases; that is, at the same load, higher speed means higher detection accuracy. The rolling element fault detection effect is the most unstable. This phenomenon can be explained by the different degrees of random sliding of the rolling elements and the inner and outer race of the bearing under different working conditions.

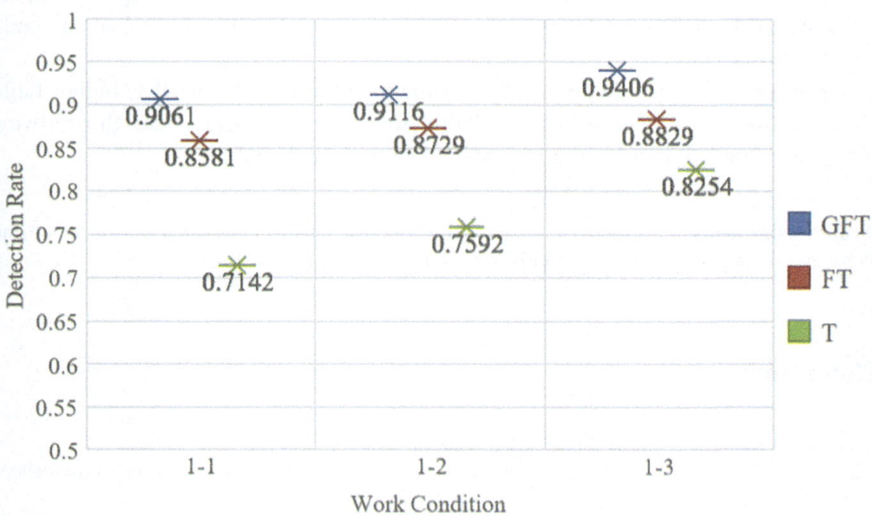

Fig. 3 Classification results under different characteristic indicators (GFT, FT, and T represents the classification results of the feature of the graph spectrum domain, the frequency domain, and the time domain, respectively)

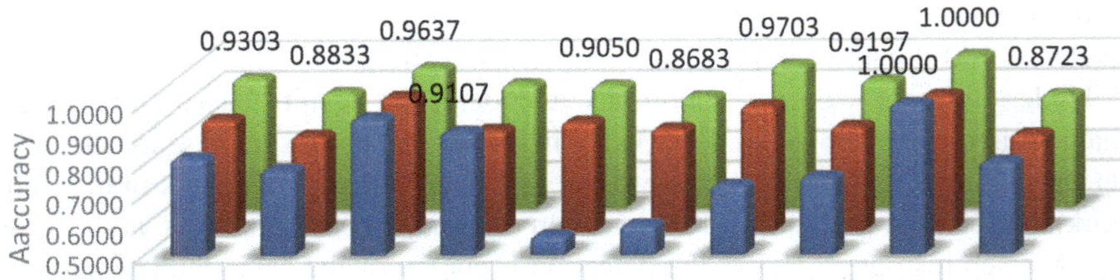

	REF	ORF	nor mal	IRF	REF	ORF	nor mal	IRF	REF	ORF
			1-2				1-3			
T	0.825	0.787	0.955	0.910	0.567	0.603	0.732	0.760	1.000	0.808
FT	0.874	0.820	0.948	0.837	0.864	0.842	0.915	0.847	0.949	0.819
GFT	0.930	0.883	0.963	0.909	0.905	0.868	0.970	0.919	1.000	0.872

T FT GFT

Fig. 4 Detection accuracy of different fault types in case2

5 Conclusion

The results show that the proposed method is an effective algorithm. The main conclusions are as follows:

(1) This paper adopts a relatively novel idea—using the C4.5 decision tree classification algorithm combined with the characteristics of the map domain to identify fault bearing types and enrich the research path of mechanical fault diagnosis.
(2) The results of the example analysis show that the GFT feature has higher fault detection effect and is more suitable for fault diagnosis under the railway background. It can handle low SNR data with high stability.

Acknowledgements This research is supported by the National Key Research and Development Program of China (Grant No. 2016YFB1200402).

References

1. Diestel RF (2000) Graph theory. Math Gazette 173(502):67–128
2. Chung FRK (1997) Spectral graph theory. The American Mathematical Society, Providence, Rhode Island
3. Cvetkovic DF, Rowlinson PS, Simic ST (2009) An introduction to the theory of graph spectra. Comp Gen Pharmacol 2(6):217–224
4. Spielman DF (2009) Spectral graph theory. Yale University Press, Yale University, New Haven

5. Shuman DI, Narang SK, Frossard P (2012) The emerging field of signal processing on graphs: extending high—dimensional data analysis to networks and other irregular domains. IEEE Signal Process Mag 29(5):633–634
6. Zhu X, Michael R (2012) Approximating signals supported on graphs. In: International conference on acoustics, speech, and signal processing (ICASSP), Kyoto, Japan, pp 3921–3924
7. Sandryhaila A, Moura JMF (2013) Discrete signal processing on graphs: graph fourier transform. In: International conference on acoustics, speech, and signal processing (ICASSP), Vancouver, BC, Canada
8. Hu W, Cheung G, Ortega A, Au OC (2015) Multiresolution graph fourier transform for compression of piece-wise smooth images. Trans Image Process 24:419–433
9. Ou L, Yu D, Yang H (2016) A new rolling bearing fault diagnosis method based on GFT impulse component extraction. Mech Syst Signal Process 81:162–182
10. Lei Y, He Z, Zi Y (2011) EEMD method and WNN for fault diagnosis of locomotive roller bearings. Expert Syst Appl 38(6):7334–7341
11. Quinlan JR (1993) C4, 5: programs for machine learning, vol 1 (1993)
12. Quinlan JR (1986) Induction of decision trees. Mach Learn 1:81–106
13. Wu X, Kumar V, Quinlan JR, Ghosh J, Yang Q, Motoda H, Zhou ZH (2008) Top 10 algorithms in data mining. Knowl Inf Syst 14(1):1–37

Overview of Optimization Models and Algorithms for Train Platforming Problem

Yinggui Zhang, Aliyu Mani Umar and Min An

Abstract In this paper, an overview of recent advances in the research on train platforming problem (TPP) is presented. The TPP is usually the last problem encountered in planning a railway system which occurs after a schedule of trains in a railway network (train timetable) has been determined. It aims to map a given train timetable to an existing station infrastructure. This process is critical as it determines the feasibility of an optimally generated train timetable along a railway line at station(s) to be visited by trains on the timetable. This optimization problem is in most stations solved manually, and it is a time consuming and error-prone process. Several computer programs are now being developed to aid infrastructure managers and train operators as decision support systems in solving this problem. This paper presents some of these solutions. However, due to variations in operating policies of railway industries in different countries, several variants of this problem exist in the literature. These variations could be seen in the solution approach through the importance attached to level of service, safety of operations, capacity utilization, etc. These variations and the various optimization techniques adopted by researchers are also discussed in this paper. Currently, most models and algorithms presented in literature are not ready for use as commercial systems. Integrating such systems into real-life planning and operations is crucial for efficient use of railway systems.

Keywords Train platforming problem · Optimization models · Optimization algorithms · Review

---

Y. Zhang · A. M. Umar
School of Traffic and Transportation Engineering, Central South University, Changsha 410004, China

A. M. Umar
Department of Civil Engineering, Ahmadu Bello University, Zaria 810107, Nigeria

M. An (✉)
School of Science, Engineering and Environment, University of Salford, Manchester M5 4WT, UK
e-mail: M.An@Salford.ac.uk

Y. Qin et al. (eds.), *Proceedings of the 4th International Conference on Electrical and Information Technologies for Rail Transportation (EITRT) 2019*, Lecture Notes in Electrical Engineering 639, https://doi.org/10.1007/978-981-15-2866-8_67

1 Introduction

In railway operations, trains operate in a railway network following a systematic predetermined schedule. One of such schedules is the train platforming plan. This essential component of railway operations planning provides information on the routing of trains at stations and platform each train will occupy for a definite period of time. Hence, the train platforming problem aims to solve, for a given train timetable and station topology, the allocation of platform and route to (and from) such platform for each train. This plan is crucial as it validates the feasibility of an optimal train timetable since most macroscopic-modelled timetables only contain an upper bound of the maximum number of trains that can simultaneously be present in a station.

The solution to such problem is usually an easy task when the station has fewer number of tracks and less traffic. However, the problem becomes complicated as the number of platforms and traffic increases; which is mostly the case, as many countries are promoting railway transport over other modes of transportation. As a result, using computer algorithms in solving the TPP becomes necessary as the conventional manual method is tedious and, in some cases, yielding infeasible solutions. Cardillo and Mione [1] highlighted how in a particular case, platforming 242 trains in a station with 16 platforms require 15 working days for an expert planner.

Capacity of stations to handle the TPP is usually determined by the number of platforms, station tracks, and the trains operations (coupling and uncoupling of trains, frequency, arrival and departure times, headway, dwell times, etc.). All these factors are known a priori and are considered in coming up with a station platforming plan. However, the occurrence of a disturbance in real-life railway operations is inevitable and when such happens, an existing platforming plan in most cases becomes infeasible. Hence, the TPP is a problem encountered at three levels of a railway system [2, 3]. The first level (strategic level) involves analysis of future infrastructural capacity requirements of station. The second level (tactical level) is during the timetabling stage; where the feasibility of a generated timetable at stations is determined. Lastly, during real-time operations (operational level) when a rescheduled timetable invalidates an existing train platforming plan.

This paper focuses only on recent optimization models and algorithms for solving train platforming problems at strategic, tactical, and operational levels. However, discussions on what the authors believe are fundamental older models and algorithms are included.

The paper is structured as follows; Sect. 2 gives a background on train timetabling and train platforming and how the two are related. Section 3 presents the different variants of TPP models and algorithms with their performances. Finally, Sect. 4 contains conclusions and suggestions on future research paths.

2 Background

2.1 Train Timetable and Station Infrastructure

In railway transportation, trains move along a network of rail lines (sections) and stations in a systematic way using a series of coordinated signals and communications systems. Due to capacity limitation and the need for a safe operation, these trains move along the network through predetermined schedule defined in a timetable. The timetable provides information such as arrival and departure times at stations to be visited by trains, direction of travel, etc. This information helps in providing a conflict-free operation across the railway network. The train timetable could be cyclic; in which it repeats itself every hour of a day or acyclic; in which trains are operated as per demand period and also repeats itself every day. Because transportation is a derived demand, demand for transportation could be low during some periods of the day and exceptionally high during other periods (peak demand). This makes the latter method of timetabling more accommodating to real-life situations. However, in many countries, the train timetable is periodic [4] technically because it makes the operation and management easier and is also easier for passengers to remember.

An optimum timetable (which is one of the two inputs in a train platforming plan) ensures that there are no conflicts along the sections in a railway network while making efficient use of available resources. This timetable is usually obtained after frequency and stopping patterns (line planning) have been defined [5]. Researchers developed several computer programs to generate these timetables while others provide a conflict-free timetable in the event of a perturbation in the system [6–8]. Because this paper is not focused on train timetabling, we will refer the reader to a review on mathematical models and algorithms involved in railway timetable scheduling [3] and railway timetable rescheduling [9].

The second input in a train platforming plan is the station infrastructure, usually presented in a form of station topology. The optimum train timetable ensures a conflict-free operation across the railway sections. The next task is to ensure a conflict-free operation at station. This is determined with the aid of the station topology. The station topology is a diagrammatical representation of physical elements in the station (platforms, turnouts, switches, track sections, etc.) with nodes and directional lines (Fig. 1).

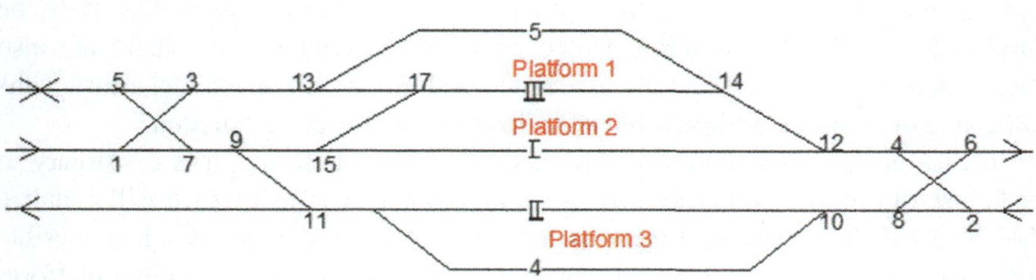

Fig. 1 Topology of a typical passenger railway station

2.2 General Train Platforming Problem and Mathematical Formulation

The assignment of trains to platforms in a station as per timetable schedule is what a platforming problem entails. This assignment has to also define the path that each train will follow to such platform while maintaining operational constraints. Usually, such problem requires the planner (either manual or automatic) to map the train traffic in a given train timetable unto station infrastructure. Over the years, computer programs have been developed to aid dispatchers solve such problem, some of which have been incorporated into real-train operations, example, RFI-Italy [10] and Ocapi-Belgium [11]. Like other railway operations, the platforming problem is mostly solved as a periodic event scheduling problem (PESP) since most train timetables are cyclic. In such problem, event times are confined within [0, C), where "C" represents the cycle length.

Different mathematical formulations were provided by researchers; which will be discussed briefly in section three. However, as a representative example, we will present a description of a general and encompassing mathematical formulation similar to that in Caprara et al. [10]. The station to be considered here (Fig. 1) has (one-way) double lines and a single (two-way) line, 6 platforms (or 3 shared platforms) with several arrival and departure paths. An arrival path, here, is a set of interconnected sections of track and switches a train will follow upon entering the station to its assigned platform. A departure path in this paper, defines the set of interconnected sections of track and switches a train will follow when leaving the station from its assigned platform. In the general version of the problem, we are given a set B of platforms and a set T of trains to be routed to a platform every day of a given time horizon. Moreover, for each train $t \in T$, we are given a collection P_t of possible *patterns*. Each pattern corresponds to a feasible route of train t within the station, including a stopping platform, an arrival path and an arrival time, a departure path and a departure time. Each train must be assigned a pattern that will be repeated every day of the time horizon.

Operational constraints forbid the assignment of patterns to trains if this implies occupying the same platform at the same time, or also using arrival/departure path that intersects at the same time or too close in time. In the general version, this is represented by defining a pattern-incompatibility graph with one node for each train-pattern pair (t, p) with $p \in P_t$, and an edge joining each pair (t_1, p_1), (t_2, p_2) of incompatible patterns. This graph models "hard" incompatibilities that must be forbidden in a feasible solution. However, in the general version, there are also "soft" incompatibilities, generally associated with the use of arrival/departure paths close in time that are admitted but penalized in the objective function.

In case not all trains could be assigned to regular platforms, it is customary to make use of dummy platforms; which are fictitious platforms that we will penalize their use (in the objective function) but may be necessary to obtain a feasible solution. For a strategic train platforming plan, the use of a dummy platform suggests enlarging the station, whereas for a tactical and operational train platforming plan, the use of a dummy platform suggests that not all trains can be platformed at the given instance. When such happens, the options are either to cancel trains, queue-up trains or relax some hard constraints imposed in the model.

The TPP requires the assignment of a pattern $p \in P_t$ to each train $t \in T$ so that no two incompatible patterns are assigned and the objective function defined by the following coefficients is minimized. There is a cost c_b for each platform $b \in B$ that is used in the solution, a cost $c_{t,p}$ associated with the assignment of pattern $p \in P_t$ to train $t \in T$, and a cost ct_1, p_1, t_2, p_2 associated with the assignment of pattern $p_1 \in Pt_1$ to train t_1 and the assignment of pattern $p_2 \in Pt_2$ to train t_2 for $(t_1, t_2) \in T^2$, in case these two patterns have a "soft" incompatibility. Here, $T_2 \subseteq \{(t_1, t_2): t_1, t_2 \in T, t_1 \neq t_2\}$ denotes the set of pairs of distinct trains whose patterns may have a "hard" or "soft" incompatibility.

3 Train Platforming Models and Algorithms

3.1 Strategic Level Optimization Models and Algorithms

The TPP at this level is typically a station's infrastructure capacity assessment, with a view of determining the adequacy or otherwise of station infrastructure. Zwaneveld et al. [2] approached the routing of trains through stations based on a node-packing approach following their proof of the problem as NP-complete. The algorithm developed, which is based on the formulation of the problem as a node-packing problem, and on the application of preprocessing techniques, heuristics and a branch-and-cut procedure was implemented into the planning system, STATIONS.

Zwaneveld et al. [12] improved on the model and algorithm presented in Zwaneveld et al. [2]. Specifically, the model was improved by incorporating shunting decisions and preferences to allocation of trains to certain desired platforms and routes. The algorithm was improved by extending the preprocessing techniques and also investigating their characteristics with respect to propagation. These improvements proved promising as all the problem instances studied were resolved to optimality within an average computing time of about 1 min.

3.2 Tactical Level Optimization Models and Algorithms

At the tactical stage, it is believed that the platformer has all the organizational details of the railway system to plan for. These details include the train timetable, layout of stations along the line and other enterprises' policies that exist.

While Zwaneveld et al. [12] considered the general routing of trains through stations (which assigns trains to a complete path through a station; platform allocation being part of the task) other researchers solved the problem while putting emphasis on the allocation of platforms to arriving trains. One of such works was carried out by Cardillo and Mione [1]. They modelled the TPP as a graph coloring problem (the k L-list τ coloring problem).

An algorithm was developed based on the formulation of the problem as a graph coloring problem and application of a backtracking and heuristic technique to solve the problem. In one of their reported case studies, a station with 13 tracks (platforms) and 177 trains on a 24-hour cycle took a Linux Pentium at 166 MHz, 115 s of CPU time to yield a solution.

Billionet [13] suggests integer programming as an alternative solution approach to the TPP as formulated by De Cardillo and Mione [1]. The two ILP formulations he described aim to find at each time, whether an integer solution exists or not. These solutions, however, do not provide an optimization result of the TPP. To obtain that, Billionet [13] introduced into the more effective ILP an objective function which maximizes the assignment of trains to a particular platform. A station with up to 200 trains and 14 platforms could be solved using standard and commercially available ILP solver software.

Carey and Carville [14] presented a greedy heuristic solution to the TPP which aims to simulate the practical process of train operations in countries where there are competing train operating companies (TOCs) operating on common lines and stations. To overcome the difficulty in adding up the costs or penalties imposed on deviations from preferred train arrival and departure times and cost of choosing less preferred platform, Carey and Carville [14] introduced "lexicographic" cost functions or decision rules. To resolve conflicts, Carey and Carville [14] delay the trains rather than advance them in an effort to imitate the practice of traditional manual planners (especially in Britain).

The algorithm (which they call A1) proves promising when tested on the busy and complex Leeds station (in the North of England) with 12 main platforms (or 34 sub platforms) and 491 trains daily. The work of Carey and Crawford [15] extends the problem to consider a network of busy complex stations. This is particularly essential because a change in the planned arrival and/or departure time (s), dwell time at a station for a train will propagate to subsequent stations the train will visit especially when adequate buffer time is not available.

Caprara et al. [10] considered minimizing the number of dummy platforms used in the objective function. The model contains a quadratic term which results from the "soft" incompatibility constraints. This complexity in the model is relieved by using a novel linearization method that requires smaller number of variables and leads to a stronger linear programming relaxation instead of the conventional approach of introducing additional variables to represent the product of the original binary variables.

To assess the performance of their branch-and-cut-and-price method, they compared it with the current heuristic method used by Rete Ferroviaria Italiana. In the four cases they studied, their algorithm proved superior at all possible values of dynamic threshold (π) tested.

The ability of a train platforming plan to absorb the inevitable disturbances in railway operations is crucial. This led Dewilde et al. [16] to introduce an approach to improve the robustness in a complex station zone. To do that, they focused on three aspects of the planning; the routing of the trains through the station zone, the timetable at the stations within this zone, and the platform assignments. The algorithm developed has three modules, each to tackle an aspect of the planning.

In the platforming module, platform assignment of all trains is assumed to be fixed, as is the usual approach in solving platforming problems. To save computation time, only relevant candidate platforms are evaluated for a train (when assumption of a fixed platform assignment could not be made). A dominance rule is used to limit the number of candidate new routes and a restriction is placed on the amount of conflicts associated with the new route in comparison with the old one. The process described will yield for each candidate platform change, a solution for all the train platforming at all the stations within the zone. The impact of each change is evaluated using the internal timetabling module and the best platform change is selected if it leads to an improved solution. Such cycle is repeated until the overall algorithm is not able to find an improved version of route, timetable, and platform assignment anymore.

Contrary to De Cardillo and Mione [1] and Billionet [13], Sels et al. [11] dropped the assumption that all routes in the station will require the same time to be traversed by trains. This assumption is impractical, considering the variation in speed limit at different switches, length of routes, train length and speed, etc. Optimality in the mixed-integer linear programming (MILP) model is attained by minimizing the total cost function, which comprises of penalty for assignment of a non-preferred (real) platform and an even higher penalty for assignment of a dummy platform. In the goal function, all hard constraints are forbidden. This is necessary so that more platforming options could be obtained when a preferred platform assignment could not be made.

The authors compared three solvers (CPLEX, Gurobi, and XPRESS) to determine which best solves the MILP model within a reasonable time. The computation times obtained are all satisfactory even without the use of variable reduction techniques. For the 10 station's one-day traffic tested, and results showed that about 30 s are required to platform all trains at the tactical level and below 9 min at the strategic level.

Petering et al. [17] modeled the train timetabling and platforming problems together by a mixed-integer linear programming (MILP) model and consider a single track, unidirectional rail line consisting of an origin, destination, intermediate stations laying between the origin and destination, and a set of parallel sidings (platforms) in each station that accommodate trains stopping in that station.

The MILP model has two parts objective function. The first aims at minimizing the cycle length, while the second minimizes the total journey time of all train-types using linear constraints and a linear objective function. The effectiveness of the model was demonstrated when it solved a large problem instance inspired by the Japanese Shinkansen train in less than an hour using IBM ILOG CPLEX 12.5 solver on a desktop computer with eight 3.4 GHz cores and 16 GB RAM. Due to the complexity of the model and the importance attached to computing times, preprocessing technique is used and this helps in reducing computing time.

3.3 Operational Level Optimization Models and Algorithms

The solution of TPP at the operational level is the most sought-after, since it is at this level that real-time management of operations is involved. To enhance the stability of a train platforming model, Miao et al. [18] present a model that omits the compatibility constraints of resource occupations. This restricts the assignment of only one resource to every operation. Stability according to Miao et al. [18] could be achieved by making the headway times among potential conflicting tasks as rationale as possible. In the two-component objective function, they propose, the first component (which is the primary objective) ensures the stability of the train platforming plan by maximizing the time interval between two adjacent occupations of track and the second component (secondary objective) ensures compatibility of the platform allocation plan in the station. The overall objective function aims to minimize the cost of changing arrival and departure times of trains to return a feasible solution. The stability enhancing train platforming model (SETPM) is solved using an ant colony optimization algorithm.

The assess of the effectiveness of the SETPM, they compared its performance with a model for minimizing resource allocation costs. This is comparison of cost minimization; therefore, the component of the SETPM objective function that measures stability is dropped and a penalty is introduced to ensure a feasible solution. The results of an experiment carried out on a high-speed train station in Changsha reveals that the SETPM is capable of increasing the stability of the train platforming plan by about 37%.

Chakroborty and Vikram [19] presented an optimum solution approach to TPP to take care of the uncertainties that occur during real-life operations. This according to them is necessary as most long-distance trains are often delayed by an hour or more (in their case study, India). This situation leads to some trains queuing up at the station entrance due to unavailability of platforms. They presented a model which takes into account the delay (that happens in real-life operations) and subsequent queuing up of trains as a result of such delays. The model is capable of resolving such problems provided the arrival of trains to stations is known at least an hour in advance. Because this is a solution at operational level, the authors do not want any adjustments to the arrival times of trains (since this will translate to even more delays or impractical advancement) and hence, arrival times of trains are direct inputs in the model (not variables). The key decision variables are the times trains (queue at the station entrance) will enter the station and the allocated platform for each train.

To obtain an optimum assignment, the costs on time a train spent waiting at station entrance, a non-preferred platform assignment and last-minute change to previously (announced) assigned platform are minimized. The MILP formulation using ILOG CPLEX 9.0.0 on a 400 MHz processor and 1 GB RAM is used to solve to optimality various problems related to a busy station in India; with 9 platforms and an average arrival rate of 55 trains per hour (specifically, 110 trains in 2-hour time horizon). In the 10 min computation time, all trains were platformed without any queue at the station entrance.

4 Conclusions and Further Research

In this paper, we discussed the train platforming problem, which is a problem of assigning platforms to arriving trains in a station while satisfying various constraints encountered in railway operations. We presented a general mathematical description of the problem and the various levels of railway system at which this problem is encountered. A great number of papers on TPP tackled the problem at the tactical level with an aim to provide an optimum (or at least a feasible) assignment of platforms to trains. Most models considered the optimum assignment of as many trains as possible to platforms and the unassigned trains will either be rescheduled or cancelled. Other models considered preferences in allocation of certain platforms to certain trains. In formulating such models, it is believed that some operations (in real time) will overlap and lead to infeasibility. Hence, buffer times are introduced to absorb such small discrepancies. However, perturbations in real-life railway operations are unpredictable and, in most cases, render an existing train platforming plan infeasible. This problem is addressed in TPP at operational level. This is perhaps the most demanding, since real-time management of train operations is involved and in the event of a disturbance which invalidates an existing train platforming plan, solutions will be required within short period of time. Unlike in the strategic and tactical levels, computing time for solving TPP at operational level is very important.

The use of combined approach in tackling the problem of perturbations at operational level is seen in most recent works on railway operations planning. This combined approach could involve incorporating the timetabling and platforming plans into one problem and solving the problem all together. Although, most TPP models and algorithms are developed as stand-alone solutions, others could be used as components for a more general system in scheduling a railway network. This approach makes the whole process much efficient and easier to manage.

In further research, more attention should be focused on improving the robustness of railway stations by considering an integrated approach of timetabling and platforming for even larger network of stations. This will ensure the stability of train timetabling and platforming plans to effects of disturbances and disruptions.

Also, the use of some hard constraints limits usable capacity in a station. This could be seen in models where a hard constraint is imposed on the occupation of a route in a station by two trains irrespective of the clearance between them. This is indeed not always true, as liberation points exist in real stations that allow two trains to occupy the same route at a time especially during peak periods or periods where the timetable is rescheduled. Subsequent research should explore the use of these flexible constraints that could improve the capacity of a station while maintaining safety of operations.

Acknowledgements The research is supported by the National Natural Science Foundation of China (Grant No. 71971220) and the Natural Science Foundation of Hunan Province, China (Grant No. 2019JJ50829).

References

1. Cardillo DDL, Mione N (1998) kL-list λ colouring of graphs. Eur J Oper Res 106(1):160–4
2. Zwaneveld PJ, Kroon L, Romeijn H, Salomon M, Dauzere-Peres S, Van Hoesel SPM et al (1996) Routing trains through railway stations: model formulation and algorithms. Transport Sci 1(30):181–194
3. Lusby RM, Larsen J, Ehrgott M, Ryan D (2011) Railway track allocation: models and methods. OR Spectrum 33:843–883
4. Sparing D, Goverde RMP (2017) A cycle time optimization model for generating stable periodic railway timetables. Transp Res Part B 98:198–223
5. Sels P, Cattrysse D, Vansteenwegen P (2016) Automated platforming & routing of trains in all Belgian railway stations. Expert Syst Appl 62:302–316
6. Schrijver A, Steenbeek A (1995) Timetable construction for Railned Amsterdam: CWI
7. Tormos P, Lova A, Barber F, Ingolotti L, Abril M, Salido MA (2008) A genetic algorithm for railway scheduling problems. In: Xhafa F, Abraham A (eds) Metaheuristics for scheduling in industrial and manufacturing applications, Springer, Berlin, Heidelberg, pp 255–76
8. Bettinelli A, Santini A, Vigo D (2017) A real-time solution algorithm for the train rescheduling problem. Transp Res Part B 106:237–265
9. Cacchiani V, Huisman D, Kidd M, Kroon P, Toth P, Veelenturf L et al (2014) An overview of recovery models and algorithms for real-time railway rescheduling. Transp Res Part B 63:15–37
10. Caprara A, Galli L, Toth P (2011) Solution of the train platforming problem. Transport Sci 45 (2):246–257
11. Sels P, Vansteenwegen P, Dewilde T, Cattrysse D, Waquet B, Joubert A (2014) The train platforming problem: infrastructure management perspective. Transp Res Part B 61:55–72
12. Zwaneveld PJ, Kroon LG, Romeijn HE, Salomon M, Dauzere-Peres S, Van Hoesel SP et al (1996) Routing trains through railway stations: model formulation and algorithms. Transport Sci 30(3):181–194
13. Billionet A (2003) Using integer programming to solve the train platforming problem. Transport Sci 2(37):213–222
14. Carey M, Carville S (2003) Scheduling and platforming trains at busy complex stations. Transp Res Part A 37:195–224
15. Carey M, Crawford I (2007) Scheduling trains on a network of busy complex stations. Transp Res Part B 41:159–178
16. Dewilde T, Sels P, Cattrysse D, Vansteenwegen P (2013) Robust railway station planning: An interaction between routing, timetabling and platforming. J Rail Transp Plann Manag 3:68–77
17. Petering MEH, Heydar M, Bergmann DR (2015) Mixed-integer programming for railway capacity analysis and cyclic, combined train timetabling and platforming. Transport Sci, Articles In Advance. 2015:1–18
18. Miao J, Yu Y, Meng L, Yang Z (2012) Stability-oriented optimization of train platforming problem on high speed railway. J Transpn Sys Eng IT 3(12):115–121
19. Chakroborty P, Vikram D (2008) Optimum assignment of trains to platforms under partial schedule compliance. Transp Res Part B 42:169–184

Empirical Wavelet Transform and Power Spectral Entropy for Rotating Machinery Fault Diagnosis

Shunjie Zhang, Yong Qin, Ge Xin, Yuze Wang, Xiaoqing Cheng and Xuan Sun

Abstract In order to extract the fault feature of rotating machines, a new method based on the empirical wavelet transform (EWT) and power spectral entropy (PSE) is proposed. EWT is introduced to first decompose the raw signal into several intrinsic mode function (IMF) signals. The power spectral entropy is used to quantify the complexity and uncertainty of each constructed component's spectra; the difference value (D-value) between the neighboring entropies is, therefore, calculated to indicate the most information of reconstructed signals. Finally, the real signal is tested by the proposed method, whose results show that it can effectively extract the most abundant fault characteristic information in machinery fault signals.

Keywords Rotating machines · Fault diagnosis · Empirical wavelet transform · Power spectral entropy

1 Introduction

Rolling bearings are widely used components in modern rotary machines, which are also the most malfunctioning parts for equipment maintenance. In particular, rolling bearings in railway locomotive bogie are one of the vulnerable parts, which directly affect the running safety of the train. Therefore, it is of great significance for railway operation to accurately detect and identify bearing faults, thus end users can take further actions to avoid or fix the faults. When a local defect occurs in a rolling bearing, periodic impulses are induced, whose information can be used for fault

S. Zhang · Y. Qin (✉) · G. Xin (✉) · Y. Wang · X. Cheng · X. Sun
State Key Laboratory of Rail Traffic Control and Safety, School of Trafficand Transportation, Beijing Jiaotong University, Beijing 100044, China
e-mail: 18120986@bjtu.edu.cn; yqin@bjtu.edu.cn

G. Xin
e-mail: gexin@bjtu.edu.cn

Y. Qin et al. (eds.), *Proceedings of the 4th International Conference on Electrical and Information Technologies for Rail Transportation (EITRT) 2019*, Lecture Notes in Electrical Engineering 639, https://doi.org/10.1007/978-981-15-2866-8_68

diagnosis [1, 2]. However, the crucial fault signature is often embedded in the strong background noise in the practice application.

Due to the vibration signals of rolling bearing fault are mostly nonlinear and non-stationary signals, time-frequency analysis methods, e.g., wavelet decomposition (WD) [3], instantaneous power spectrum (IPS) [4], empirical mode decomposition (EMD) [5, 6], local mean decomposition (LMD) [7], and local characteristic-scale decomposition (LCD) [8] had been applied to extract the fault signal from the measurement. They have been widely used in the fault diagnosis of rolling element bearing due to their ability to provide the local information of vibration signals in both the time and frequency domains.

EMD is an effective method for processing nonlinear signals [2, 5]. As an adaptive tool, EMD can decompose signals into multiple single-component signals, which does not require user-defined parameters and preset functions. However, EMD decomposition is prone to the result in modal mixing and excessive decomposition. In addition, there are a few reasonable mathematical explanation related to EMD [5, 6]. Based on EMD, Smith proposed an effective adaptive time–frequency analysis method, LMD [8]. Unlike EMD, LMD can decompose a multicomponent signal into a series of product functions (PFs), each PF being an amplitude modulation signal and an FM signal (AM–FM signal). As mentioned in Ref. [9], LMD is better than EMD in terms of modal mixing, end-point effects as well as the computational cost.

EWT is a new signal processing method proposed by Gilles [10], which divides the Fourier spectrum of the signal. As such, it establishes a set of wavelet filter banks to filter the divided spectrum, which obtains a group of single components. EWT has good mathematical support while avoiding end-point effects and modal aliasing, but it decomposes a lot of excess components.

Information entropy is a method used to quantitatively describe the probability and mathematical statistics of uncertainty and complexity of information. In order to measure information amount quantitatively, the method of power spectral entropy (PSE) is used to characterize the spectral distribution of vibration signals and measure the distribution of the signal energy in frequency domain. The distribution of the spectrum composition of signal of rolling bearing can be explained by this method [11, 12].

This study presents a new method for extracting the weaker fault feature of a rolling element bearing signal. The proposed of it is as follows. EWT is used to decompose the collected raw vibration signal at first, then the N intrinsic mode function (IMFs) can be obtained to reconstruct the new signal. For the reason that PSE can measure information quantitatively, the difference value (D-value) between the adjacent entropies can be used as a tool to select the number of IMFs, while restraining the interfering components.

2 Fault Diagnosis Algorithm

2.1 EWT Algorithm

The main procedure of EWT [12, 13] is based on the Fourier spectrum of the signal adaptive segmentation, constructing crops of filter group, obtaining the empirical wavelet function with orthogonal properties. By applying the empirical wavelet transform to the segmented interval in frequency domain, different AM–FM components of this signal with a compact support Fourier spectrum are extracted. In order to satisfy the rule of Shannon, the Fourier spectrum range of signal is normalized as $[0, \pi]$.

Assuming that the signal is composed of N single components of AM–FM, in order to extract all single components, the Fourier spectrum range $[0, \pi]$ is divided into N sub-intervals, where the required boundary number is $N + 1$. Let define ω_n as the boundary of the sub-intervals $\Lambda_n = [\omega_{n-1}, \omega_n]$ $(n = 1, 2, \ldots, N)$, which is obtained by taking the minimum value between adjacent maxima as the boundary, where $\omega_0 = 0$, $\omega_N = \pi$ and thus, $\cup_{n=1}^{N} \Lambda_n = [0, \pi]$.

The empirical wavelet is a band-pass filter defined on interval Λ_n. Based on the Meyer's and Littlewood-Paley wavelets, when $n > 0$, the empirical scale function $\phi_n(w)$ and the empirical wavelet function $\hat{\psi}_n(w)$ are, respectively, defined as follows,

$$\hat{\phi}_n(\omega) = \begin{cases} 1 & \text{if } |w| \le w_n - \tau_n \\ \cos\left[\frac{\pi}{2}\beta\left(\frac{1}{2\tau_n}(|w| - w_n + \tau_n)\right)\right] & \text{if } w_n - \tau_n \le |w| \le w_n + \tau_n \\ 0 & \text{otherwise} \end{cases} \quad (1)$$

and

$$\hat{\psi}_n(w) = \begin{cases} 1 & \text{if } w_n + \tau_n \le |w| \le w_{n+1} - \tau_{n+1} \\ \cos\left[\frac{\pi}{2}\beta\left(\frac{1}{2\tau_{n+1}}(|w| - w_{n+1} + \tau_{n+1})\right)\right] & \text{if } w_{n+1} - \tau_{n+1} \le |w| \le w_{n+1} + \tau_{n+1} \\ \sin\left[\frac{\pi}{2}\beta\left(\frac{1}{2\tau_n}(|w| - w_n + \tau_n)\right)\right] & \text{if } w_n - \tau_n \le |w| \le w_n + \tau_n \\ 0 & \text{otherwise} \end{cases} \quad (2)$$

When $\gamma < \min_n\left(\frac{w_{n+1} - w_n}{w_{n+1} + w_n}\right)$, a tightly supported frame $\{\varphi_1(t), \{\psi_n(t)Y_{n-1}^N\}$ is formed in $L^2(R)$ space. The arbitrary function $\beta(x)$ in Eqs. (1–2) are most often used as

$$\beta(x) = x^4(35 - 84x + 70x^2 - 20x^3) \quad (3)$$

According to the wavelet transform method, the detail coefficient $W_f^e(n,t)$ and approximate coefficient $W_f^e(0,t)$ can be calculated by inner product, and the specific formulas are as follows,

$$W_f^e(n,t) = \langle f, \psi_n \rangle = \int f(\tau)\overline{\psi_n(\tau - t)}\mathrm{d}\tau = (f(w)\overline{\hat{\psi}_n(w)})^\vee \tag{4}$$

$$W_f^e(0,t) = \langle f, \phi_1 \rangle = \int f(\tau)\overline{\phi_1(\tau - t)}\mathrm{d}\tau = (f(w)\overline{\phi_1(w)})^\vee \tag{5}$$

where the function $\hat{\psi}_n(w)$ and $\hat{\phi}_1(w)$ are the Fourier transform of $\psi_n(t)$ and $\phi_1(t)$, and $\overline{\psi_n(t)}$ and $\overline{\phi_1(t)}$ are the Complex conjugate function of $\psi_n(t)$ and $\phi_1(t)$, respectively.

Thus, the reconstruction expression of signal $f(t)$ is as follows,

$$\begin{aligned} f(t) &= W_f^\varepsilon(0,t) * \phi_1(t) + \sum_{k=1}^{N} W_f^\varepsilon(n,t) * \psi_n(t) \\ &= \left(\hat{W}_f^\varepsilon(0,w)\hat{\phi}_1(w) + \sum_{n=1}^{N} \hat{W}_f^\varepsilon(n,w)\hat{\psi}_n(w) \right)^\vee \end{aligned} \tag{6}$$

The single component $f_j(t)$ decomposed by EWT from $f(t)$ is $f_{j=0}(t) = W_f^\varepsilon(0,t) * \phi_1(t)$, $f_{j=k}(t) = W_f^\varepsilon(k,t) * \psi_k(t)$, where $j = 0, k$; $k = 1, 2, 3, \ldots$.

2.2 *Principle of Power Spectral Entropy*

PSE is a description of the uncertainty and complexity of a signal in the frequency domain [14, 15] and it is defined as follows.

For the signal $f(t) = \{x_i | i = 1, 2, \ldots, n\}$, its power spectrum estimation can be defined as,

$$\hat{S}(\omega) = -\frac{1}{n}|X(\omega)|^2 \tag{7}$$

where $X(\omega_i)$ is the FFT of $f(t)$, ω_i is one of spectra.

On the basis of the law of Parseval's theorem, the transformation of energy from time domain to the frequency domain is conserved, thus the transformation formula of this process can be written as follows,

$$\sum_{i=1}^{n} |x_n|^2 = \sum_{r=1}^{n} |S(r)|^2 \tag{8}$$

Therefore, $S_r(r = 1, 2, \ldots, n)$ can be regarded as energy apportionment of $f(t)$ in the frequency domain. Therefore, the definition of PSE is as follow:

$$\mathrm{PSE} = -\sum_{r=1}^{n} P_r \ln P_r \tag{9}$$

PSE can represent the spectrum distribution and energy distribution of a section of vibration signal in the frequency domain. What's more, it can quantitatively characterize the uncertainty and complexity of signal from the perspective of sub-band power distribution. It is noticed that when spectrum distributes uniformly, a bigger PSE value corresponds to more complexity and uncertainty of the signals.

2.3 *The Proportion Probability of Spectrum Energy*

In the work [16], it defines the proportion probability of spectrum energy as

$$p_r = S_r \bigg/ \sum_{r=1}^{n} S_r \tag{10}$$

where, p_r is the ratio of the energy of the rth spectrum to the total spectrum energy. This ratio is known as the proportion probability.

The larger p_r is, the higher the ratio of the energy at the characteristic frequency to the total energy in the fixed frequency segment is, indicating that more effective energy information is included.

2.4 *Method Flow Chart Based on EWT and PSE*

The fault signal of rotating machinery was sampled with a sampling length of n points. EWT is used to decompose the signal to get d IMFs, and then the $k\,(k = 1, 2, \ldots, d)$ IMFs are constructed used to construct new structural signals, respectively, to get d structural signals. Calculate the PSE value of each new build signal and subtract the PSE value of the previous build signal from the PSE value of the later build signal, defining the difference as a difference value (D-value). In this case, we define the turning point as the point where D-value is very close to 0, which means in terms of energy, the signals constructed by k IMFs before this point contain stronger feature information of the signal of interest. To the contrary, the residual parts indicate its weaker counterpart. The flow chart of the proposed method is shown in Fig. 1.

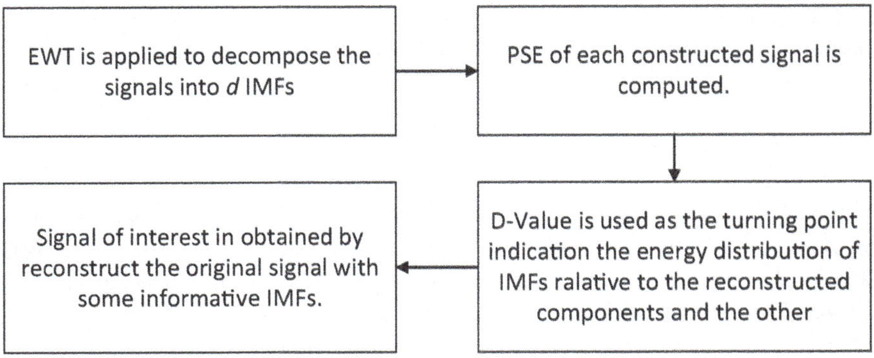

Fig. 1 Method flow chart of the proposed based on EWT-PSE

3 Application to Rotating Machine Signals

3.1 Experimental Scheme

In order to validate the effectiveness of the method, the experimental data was measured on a test-rig at the Vibrations and Acoustics Laboratory of the University of New South Wales (Sydney). The rotational speed of the shaft is 360 RPM. The signal sampling frequency is 48 kHz and the sample data length is 100,000. The fault feature frequency of the gears is 192 Hz. Since the fault feature frequency of bearing is 42.8 Hz.

3.2 Application to Real Signals

In this study, areal fault signal with inner race fault bearing and gear with 50,000 points (1.0 s) is decomposed by EWT. In order to clearly display the signal, the signal of 0.25 s is shown in Fig. 2.

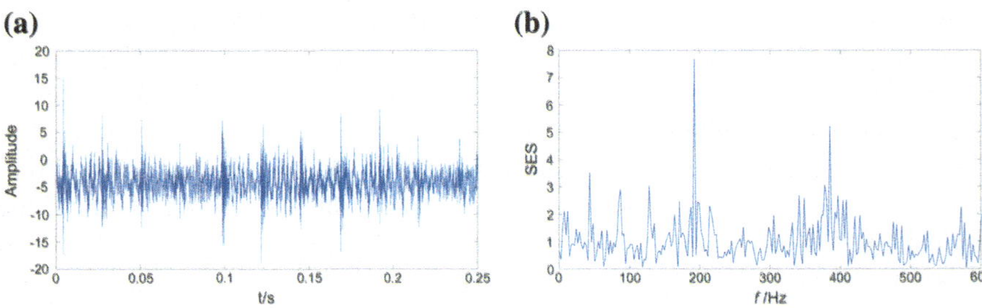

Fig. 2 Real fault signal. **a** Signal waveform in the time domain; **b** Hilbert square envelope spectrum of in the frequency

As shown in Fig. 2, we can only see the engagement frequency of gears at 192 Hz, but cannot observe the characteristic frequency of inner ring fault.

EWT is applied to the decomposed real rotating machine fault signal, obtaining IMFs of $d = 64$. The first k ($k = 1, 2, \ldots, 64$) IMFs successively accumulated to obtain 64 new structural signals. On the basis of the criterion of the value of PSE, select the fault diagnosis criterion with stronger characteristic information of interested signal, and put forward the fault diagnosis criterion based on PSE. Calculate the PSE for each newly built signal, and the result is shown in Fig. 3.

As we can see in Fig. 4, more fault feature information can be extracted with the increase of IMFs constructed. With the information becomes larger, spectra and the spectral value become bigger at these frequencies, and the distribution of the spectrum becomes more uniform, and complexity and uncertainty of the distribution are higher, thus the PSE value becomes bigger. However, more IMFs means that more noise information can be include.

The value of PSE constructed by the former minus the value of PSE constructed by the latter is the difference value (D-value), which is shown in Fig. 4.

As can be seen in Fig. 4, when k is in the range of [0, 23], the value of d varies greatly. However, within the range of [23, 64], D-value changes little and is very close to 0, thus choose 23 as the turning point. As can be seen from Fig. 3, PSE value of the newly constructed signal is not large when $k = 23$, which means that the first 23 IMFs constructed signals may have contained the most abundant fault feature information.

For the purpose of verifying the validity of the proposed IMFs criterion method, the energy ratio between combinations is used to calculate the energy of different combinations and the total energy of each component, and the proportion of the

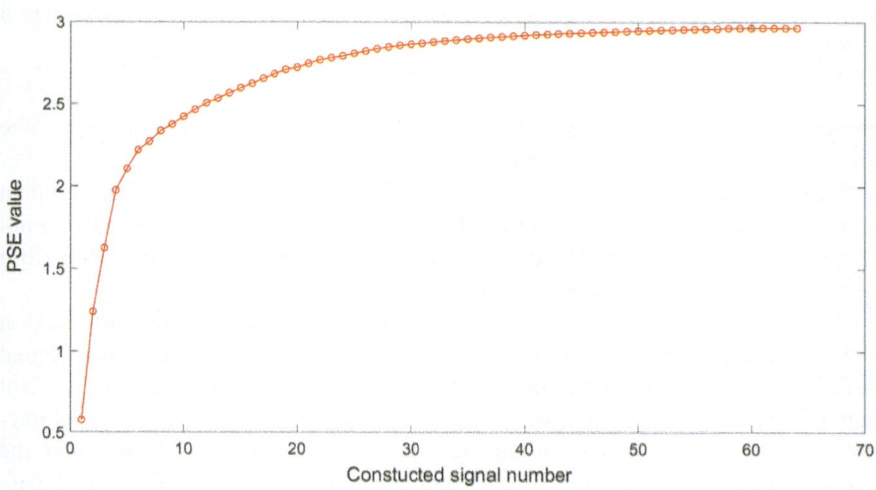

Fig. 3 Entropy of each reconstructed signal based on real fault signal

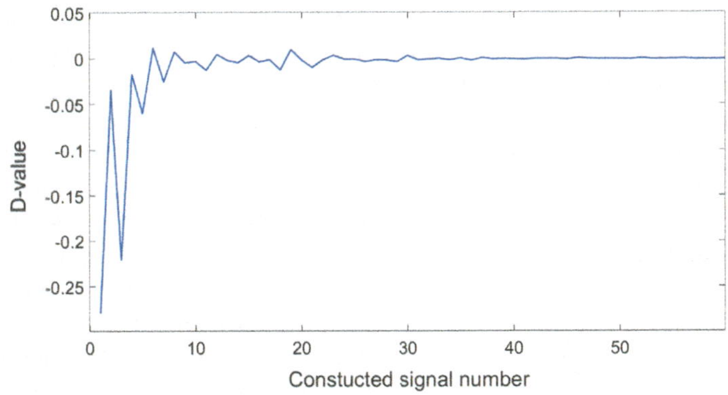

Fig. 4 D-value of each reconstructed signal based on real fault signal

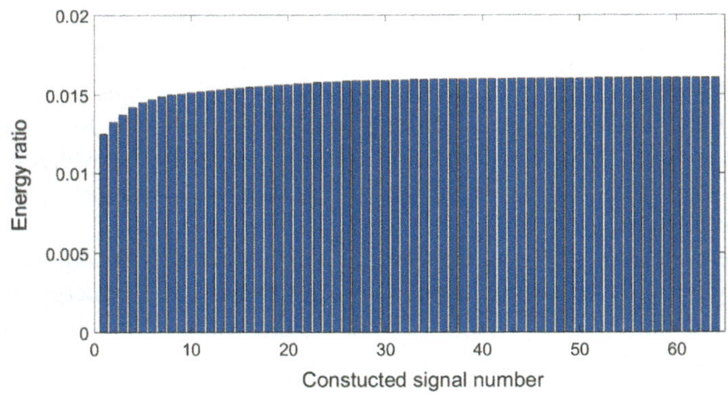

Fig. 5 Proportion probability of spectrum energy of each reconstructed signal based on real fault signal

energy of each component and the energy of all the components is obtained. The distribution of the 64 groups is shown in Fig. 5.

It can be concluded that the structural signals of the chosen first 23 IMFs contain almost all energy, so the constructed signals contain the most abundant fault characteristic information. The constructed signals' waveform and power spectral density of in the frequency domain are as shown in Fig. 6.

As can be seen in Fig. 6b, the frequency of gear fault feature information (192 Hz) and its harmonic can be extracted more obviously than the original signal before processing. To the contrary, in Fig. 7, it can be clearly seen that the fault feature frequency of bearing inner ring fault feature information after the subtraction of the original signal and the reconstructed signal is 43.25 Hz, and the extracted harmonic component is also more obvious than before. Compared with the real fault signal in Fig. 2b, more real fault feature information (gear and bearing inner ring fault) can be extracted from the constructed signal based on PSE.

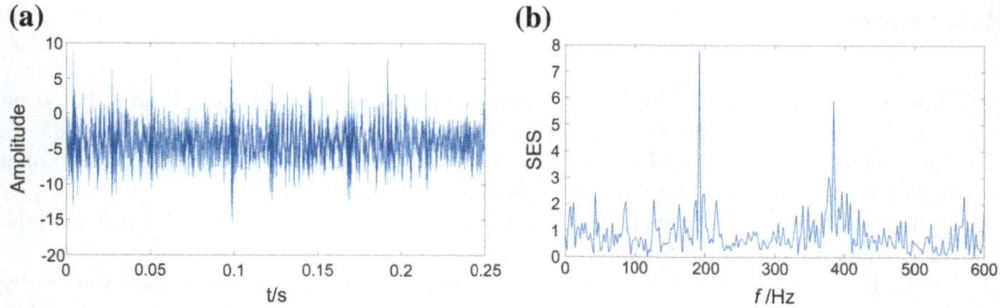

Fig. 6 Constructed signal based on PSE. **a** Waveform in the time domain; **b** Hilbert square envelope spectrum of in the frequency

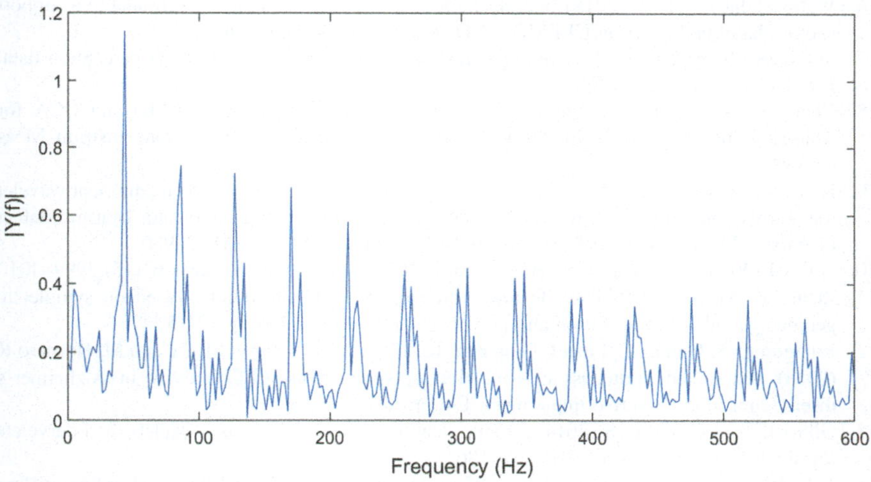

Fig. 7 Residual signal based on real fault signal

4 Conclusion

On the purpose of the fault feature extraction of rotating machinery, a new approach based on the EWT and PSE is proposed. It is beneficial to select the most informative IMFs by means of PSE, which can accurately evaluate the complexity and uncertainty of the raw signal. The performance of proposed method is validated by the real signals, and the weak bearing fault is further analyzed and reconstructed from the strong interfering signals. Although the performance of proposed method is based on some marked limitation, it intends to explore a new perspective to address the issue of feature extraction at an incipient stage of the fault signal.

Acknowledgements This research is supported by the National Key Research and Development Program of China (Grant No. 2016YFB1200402).

References

1. Ming Y, Chen J, Dong G (2011) Weak fault feature extraction of rolling bearing based on cyclic Wiener filter and envelope spectrum. Mech Syst Signal Process 25(5):1773–1785
2. Xin G, Hamzaoui N, Antoni J (2018) Semi-automated diagnosis of bearing faults based on a hidden Markov model of the vibration signals. Measurement 127:141–166
3. Daubechies I (1990) The wavelet transform, time-frequency localization and signal analysis. IEEE Trans Inf Theory 36(5):961–1005
4. Xin G, Hamzaoui N, Antoni J (2019) Extraction of second-order cyclostationary sources by matching instantaneous power spectrum with stochastic model—application to wind turbine gearbox. Renew Energy
5. Xiong Q, Xu YH, Peng YQ, Zhang WH, Li YJ, Tang L (2017) Low-speed rolling bearing fault diagnosis based on EMD denoising and parameter estimate with alpha stable distribution. J Mech Sci Technol 31:1587–1601
6. Wang Z, Jia L, Qin Y (2018) Bearing fault diagnosis using multiclass self-adaptive support vector classifiers based on CEEMD–SVD. Wireless Personal Commun
7. Smith JS (2005) The local mean decomposition and its application to EEG perception data. J R Soc Interface 2:443–454
8. Chen X, Xu XY, Liu A, Martin M (2018) The use of multivariate EMD and CCA for denoising muscle artifacts from few-channel EEG recordings. IEEE Trans Instrum Meas 67:359–370
9. Kedadouche M, Thomas M, Tahan A (2016) A comparative study between empirical wavelet transforms and empirical mode decomposition methods: application to bearing defect diagnosis. Mech Syst Signal Process S088832701600100X(1010):18120986
10. Gilles J (2013) Empirical wavelet Transform. IEEE Trans Signal Process 61(16):3999–4010
11. Zheng Z, Xin G (2019) Fault feature extraction of hydraulic pumps based on symplectic geometry mode decomposition and power spectral entropy. Entropy 21(5):476
12. Ruizgómez S, Gómez C, Poza J, Gutiérrez-Tobal GC, Tola-Arribas MA, Cano M, Hornero R (2018) Automated multiclass classification of spontaneous EEG activity in Alzheimer's disease and mild cognitive impairment. Entropy 20:35
13. Gilles J, Tran G, Osher S (2014) 2d empirical transforms. Wavelets, ridgelets, and curvelets revisited. SIAM J Imag Sci 7(1):157–186
14. Ji Y, Wang X, Liu Z, Yan Z, Jiao L, Wang D, Wang J (2017) EEMD-based online milling chatter detection by fractal dimension and power spectral entropy. Int J Adv Manuf Technol 92:1185–1200
15. Llanos F, Alexander JM, Stilp CE, Kluender KR (2017) Power spectral entropy as an information-theoretic correlate of manner of articulation in American english. J Acoust Soc Am 141:EL127
16. Jiang WL, Zheng Z, Zhu Y, Liu SY (2014) Hydraulic pump fault diagnosis based on optimal flat structure element length. Zhendong yu Chongji/J Vibration Shock 33(15):35–41 (in Chinese)

Switch Machine Fault Diagnosis Method Based on Kalman Filter and Support Vector Machines

Xiang Li, Yong Qin, Zhipeng Wang, Jiayu Kan and Xiaofeng Zhang

Abstract Switch machines are used for operating railway turnout; its error can cause delays, increase operating costs and may even lead to train accidents. Therefore, the fault diagnosis technology for the switch machine has received more and more attention. This paper proposes a fault diagnosis method based on the action current of switch machine. Firstly, the Kalman filter is used to preprocess the collected action current to reduce the influence of the unavoidable error of the measurement. In addition, we can further improve the accuracy of fault diagnosis by extracting the characteristics of the action current curve, like the maximum, minimum and average value, etc. Finally, we use DAG-SVMs to intelligently diagnose switch failures. Experiments show that the accuracy of classification after Kalman filter preprocessing is better than that of direct classification.

Keywords Switch machines · Feature extraction · Kalman filter · DAG-SVMs

1 Introduction

As one of the important basic components of the railway signal system, the switch machine is very frequently operated and is prone to accidents. If it is not repaired in time, it can easily lead to serious consequences. Therefore, timely and accurate detection of faults plays a very important role in the daily operation of the railway

X. Li · Y. Qin · Z. Wang (✉) · X. Zhang
State Key Laboratory of Rail Traffic Control and Safety, Beijing Jiaotong University, Beijing 100044, China
e-mail: zpwang@bjtu.edu.cn

X. Li
National Research Center of System Engineering of Railway Intelligent Transport, Beijing 100081, China

J. Kan
National Research Center of System Engineering of Railway Intelligent Transport, Signal & Communication Research Institute, Beijing 100044, China

Y. Qin et al. (eds.), *Proceedings of the 4th International Conference on Electrical and Information Technologies for Rail Transportation (EITRT) 2019*, Lecture Notes in Electrical Engineering 639, https://doi.org/10.1007/978-981-15-2866-8_69

signal system. Nowadays, if the switch fails during operation, it can send a simple alarm through the monitoring system, but it lacks the identification of the specific failure mode, and the early failure alarm lacks accuracy. With the development of machine learning and artificial intelligence, there is an urgent need for intelligent analysis and diagnosis of switch failures.

Diagnosis, the process of identifying the failure, is basically a classification problem. Scientists at home and abroad have proposed many methods for the problem of fault diagnosis of switch machines. Mo proposed an algorithm based on envelope and morpheme match to predict whether the curve would have fault or not, also SVM is used to diagnose the fault conditions intelligently [1]. Huang develops an intelligent diagnosis method for railway turnout through dynamic time warping (DTW) [2]. Yan explores a method to assess assets' performance and predict the remaining useful life [3]. Zhou studies an intelligent diagnosis method for high-speed railway turnout based on the support vector machine [4]. Based on particle swarm optimization, Chen proposes automatic parameters selection for least squares support vector machine (LS_SVM) [5]. A Spanish scholar, F P García Márquez makes pretreatment to the current curves by using Kalman filter [6–9].

2 Operating Current of Switching Machine

After a comprehensive analysis of the basic curves of various switching machine action currents, the action current curve can be divided into four steps for analysis [10]:

Step 1: Unlocking process. When the motor starts, it generates a large starting current and generates a large torque. At this time, the switch enters the unlocked state.
Step 2: Conversion process. In this process, the motor is decelerated by level 2, which drives the ballast to smoothly convert, and the action current curve is smooth. If the operating current is small, the conversion resistance is small; if the operating current is large, the conversion resistance is large; if the operating curve fluctuates greatly, it indicates that the switch has electrical or mechanical problems.
Step 3: Locking process. In this process, the switch rail is driven to the other side, the tip rail is closely attached to the basic rail, the motion gear locks curved sliders, and the automatic current switch cuts off the operating current. Its action current curve is a smooth and rapid return to the tail, and the tail is slightly upturned back to zero.
Step 4: The fourth period is the phase where the tail current of the curve is zero.

The basic curve of the operating current of the switch machine is shown in Fig. 1.

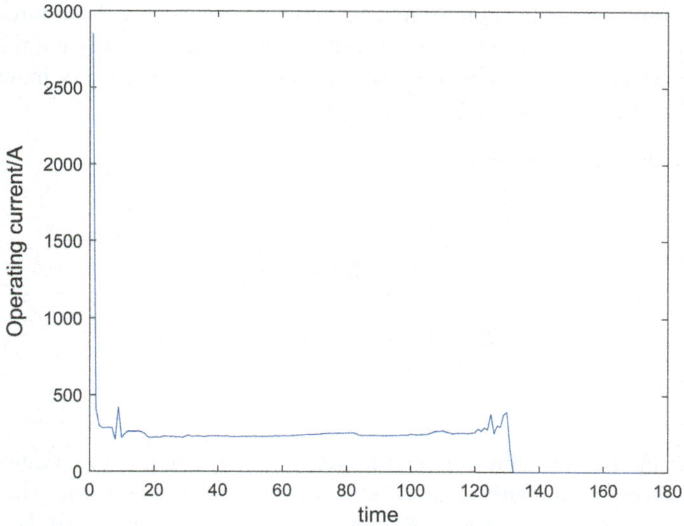

Fig. 1 Basic curve of the operating current of the switch machine

3 Kalman Filter

The filtering problem refers to how to filter the noise from the observed signal and reduce the influence of noise as much as possible. In some cases, the real signal is completely polluted by noise. The purpose of filtering is to filter the noise and extract the unknown real signal. Such problems are widely found in the fields of signal processing, communication, target tracking and control. Due to the randomness of common white noise and the colored noise of the ARMA model, the filtering problem essentially turns into a statistical estimation problem. The most common optimal estimation method is linear minimum variance estimation, that is, the true error variance of the required signal or state optimal estimation is minimized. This filtering is also called optimal filtering. In the Kalman filter algorithm, the Kalman filter uses a set of recursive equations to estimate the state vector x and update the estimate with each new observation. This process is carried out in two steps [11]:

Step 1: Forecasting process

$$x_t^- = Fx_{t-1} + Bu_{t-1} \tag{1}$$

$$P_t^- = FP_{t-1}F^{\mathrm{T}} + Q \tag{2}$$

where x_t^- is the a priori state estimation at time t. F is the state transition matrix, and it shows how we can estimate the state of the current moment from the state of the previous moment. B is the control matrix, and it indicates how the control quantity u acts on the current state. The second formula shows the transmission relationship of uncertainty at various times. P is the uncertainty of state at every moment. Q is the covariance matrix for predicting process noise.

Step 2: Status update process

$$x_t = x_t^- + K_t\left(z_t - Hx_t^-\right) \tag{3}$$

$$K_t = P_t^- H^{\mathrm{T}}\left(HP_t^- H^{\mathrm{T}} + R\right)^{-1} \tag{4}$$

$$P_t = (1 - K_t H)P_t^- \tag{5}$$

where x_t is the output value we need to get at time t. z_t is the observation at time t. H is the observation matrix. It associates x_t with the observed value z_t at time t. K_t is the Kalman coefficient at time t. R refers to the covariance matrix of the measured noise.

4 Support Vector Machines

The learning strategy of the SVM algorithm in the case of linear separability is to find the classified hyperplane with maximum edge distance. The SVM algorithm overcomes the shortcomings of overfitting and dimensionality disasters in traditional machine learning algorithms. SVM has strong learning and generalization ability. For a given training set, the essence of SVM algorithm training is to solve an optimization problem [12].

$$\begin{cases} \min = 1/2\|\omega\|^2 \\ \text{s.t.} \quad y_i(\omega^T x_i + b \geq 1) \quad (i = 1, \ldots, n) \end{cases} \tag{6}$$

where ω and b are hyperplane normal vector matrices and intercept vectors, respectively. This was originally a convex quadratic programming problem. We introduce the Lagrange multiplier and derive the parameters to find the dual problem corresponding to the original problem. The specific formula is as follows:

$$\begin{cases} \max = \sum\limits_{i=1}^{n} a_i - \frac{1}{2}\sum\limits_{i,j=1}^{n} a_i a_j y_i y_j x_i x_j \\ \text{s.t.} \quad a_i \geq 0\,(i = 1, \ldots, n, \ldots); \quad \sum\limits_{i=1}^{n} a_i y_i = 0 \end{cases} \tag{7}$$

where a is the hyperparametric matrix after transformation. This conversion has two purposes: (1) Dual problem is easier to solve than the original problem; (2) we can introduce kernel functions in the case of linear inseparability.

Finally, the dual problem is solved by SMO and other excellent methods to obtain the solution to the original problem. Since the parameters corresponding to the non-boundary samples are all 0, only the vector samples are supported to have an effect and contribution to the classification. We just need to calculate the inner product of the test sample and the support vector when testing.

$$f(x) = \omega^T x + b = \sum a_i y_i \langle x, x_i \rangle + b \tag{8}$$

where $\langle x, x_i \rangle$ is the inner product of x and x_i. For linear indivisible cases, a penalty function can be introduced. The penalty function can allow the presence of noise points and can effectively avoid overfitting. For nonlinear problems, the introduction of kernel functions can map samples from low-dimensional space to high-order space, which avoids the dimensionality problem after mapping. We use the kernel function to replace the inner product of the high-dimensional space after mapping and perform the inner product calculation directly in the original low-dimensional space while the classification effect is expressed in high-dimensional space. Commonly used kernel functions are Gaussian kernel, linear kernel and polynomial kernel. The general kernel function expression is

$$K(x, z) = \langle \sigma(x), \sigma(z) \rangle \tag{9}$$

where σ is the mapping of the original data to the high-dimensional space.

Traditional SVM can only divide data into two categories. In order to divide data into multiple categories, we have the following methods:

(1) One versus rest: The samples of a certain category are classified into one class, and the other remaining samples are classified into another class. In this way, k classes of samples are constructed to construct k SVMs. When classifying, the unknown sample is classified as the one with the largest classification function value.
(2) One versus one: The idea is to design an SVM between any two types of samples. Therefore, samples of k categories need to design $k(k-1)/2$ SVMs. When an unknown sample is classified, the category with the most votes is the category of the unknown sample.
(3) DAG-SVMs: DAG-SVMs are derived from the directed acyclic graph (DAG). The training process of this method is similar to the "one versus one" method. k class problem needs to solve $k(k-1)/2$ support vector machine classifier. These classifiers form a directed acyclic graph. The directed acyclic graph contains $k(k-1)/2$ internal nodes and k leaf nodes, and each node corresponds to a second-class classifier. Taking a problem which is to classify a sample into four categories as an example, the directed acyclic graph of it is shown in Fig. 2.

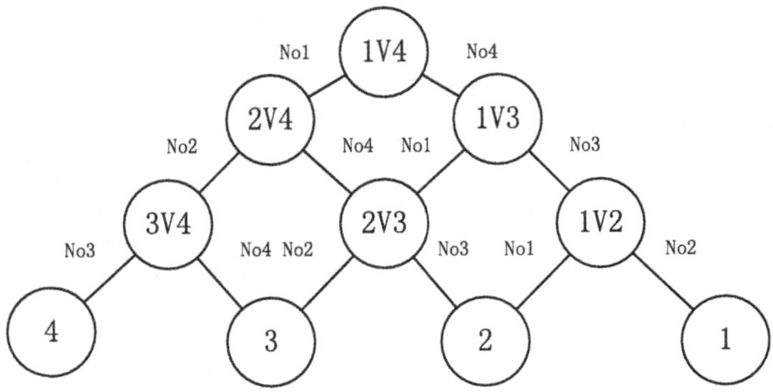

Fig. 2 Directed acyclic graph of the four classification problems

5 Instance

In this part, we first propose a model for troubleshooting and then classify the collected current data. Finally, we compare the accuracy of the classification directly with DAG-SVMs and with the model we proposed.

5.1 Model Establishment

First, we observe the action current curve to find out the characteristics of the curves and use these curve features together with the current data we collected as the basis for classification. Then, we average the data in each tag, use average number as our predicted data and perform Kalman filtering on each set of data. Finally, we use DAG-SVMs to classify. The model is shown in Fig. 3.

5.2 Data Verification

This article selected 962 sets of data. Each set of data is the operating current of a complete duty cycle of the switch from orientation to reverse and then from reverse to orientation. The description of experimental data is shown in Table 1.

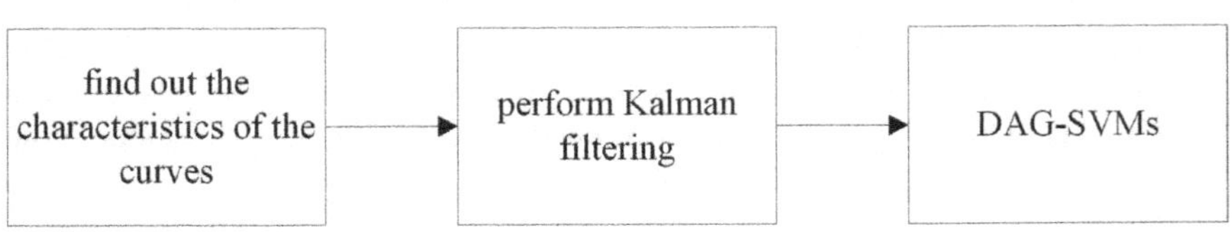

Fig. 3 Proposed model

Table 1 Description of experimental data

No.	Status	Number	Description
1	Normal	9	Normal
2	Nfault5	50	0.5 mm over-tightness
3	Nfault15	43	1.5 mm over-tightness
4	Nfault0	223	Abnormal fiction
5	Nfault3	127	Abnormal fiction and 0.3 mm over-tightness
6	Nfault10	127	Abnormal fiction and 1.0 mm over-tightness
7	Nfault13	104	Abnormal fiction and 1.3 mm over-tightness
8	Nfault20	109	Abnormal fiction and 2.0 mm over-tightness
9	Nfault25	61	Abnormal fiction and 2.5 mm over-tightness
10	Nfault30	61	Abnormal fiction and 3.0 mm over-tightness
11	Nforeign2	12	Abnormal fiction and 2.0 mm foreign matter
12	Nforeign4	10	Abnormal fiction and 4.0 mm foreign matter
13	Locked25	22	Abnormal fiction and 2.5 mm unlocked
14	Locked30	4	Abnormal fiction and 3.0 mm unlocked

Through the analysis of the current curve, we choose the maximum of the current and its corresponding time, current variance, average value of current at t = 10–100 and the maximum of current at t = 200–300 as the curve characteristics. Then, we preprocess the data by Kalman filtering.

After preprocessing the data, we all messed up the data and selected the first 600 of them as the training set and the rest as the test set. We selected Gaussian kernel function as the kernel function of the support vector machines. In order to verify the superiority of the algorithm we proposed, we performed two sets of experiments, respectively. One is using DAG-SVMs for classification, and the other is using feature extraction and Kalman filtering for data mathematical analysis before DAG-SVMs classification. Finally, the result is shown in Table 2.

From the results, we can see that the accuracy has been greatly improved after feature extraction and Kalman filtering.

Table 2 Fault detection result

	DAG-SVMs (%)	Preprocess + DAG-SVMs (%)
Training set	100	100
Test set	94.7658	98.895

6 Conclusion

Railway signal complex system is responsible for the transmission of the railway signal. With the rapid development of intelligence and information technology, the importance of railway signals is becoming more and more large. As one of the basic components of the complex system of railway signals, the failure of the switch can not only lead to train delays but also lead to more serious consequences. The more accurate intelligent fault diagnosis system is the direction of future research. This paper proposes an intelligent fault diagnosis method based on Kalman filter and DAG-SVMs. Through Kalman filtering, we can remove the noise, and the characteristics of the curves can be better reflected. In order to classify better, we can also extract the features of the curve and use it as the basis for classification. Finally, we use support vector machine for fault diagnosis. By comparing the method of directly using the DAG-SVMs and preprocessing data before using the DAG-SVMs, we can find that the accuracy has been improved greatly after data pretreatment.

Acknowledgements This research is supported by the open fund of National Research Center of System Engineering of Railway Intelligent Transport (No. RITS2018KF01).

References

1. Xuemei M, Yu F, Yunguo Y (2013) Method on the fault detection and diagnosis for the railway turnout based on the current curve of switch machine. Appl Mech Mater 1022–1027
2. Shize H, Fan Z, Rongjie Y (2017) Turnout fault diagnosis through dynamic time warping and signal normalization. J Adv Transport 1–8
3. Jihong Y (2004) A prognostic algorithm for machine performance assessment and its application. Prod Plann Control 769–801
4. Fenfang Z, Li X, Wei D, Xinya S, Xiang Y, Qianchuan Z (2016) Fault diagnosis of high-speed railway turnout based on support vector machine. In: Proceedings of the IEEE international conference on industrial technology, pp 1539–1544
5. Yongqi C, Zhanxin Z, Qijun C (2007) The research and application of LS_SVM based on particle swarm optimization. In: International conference on automation and logistics, pp 1115–1120
6. M′arquez PG, Schmid F (2007) A digital filter-based approach to the remote condition monitoring of railway turnouts. Reliability Eng Syst Safety 92:830–840
7. García FP, Schmid F, Collado JC (2003) A reliability centered approach to remote condition monitoring. A railway points case study. Reliability Eng Syst Safety 80:33–40
8. Márquez FPG, Tercero DJP, Schmid F (2007) Unobserved component models applied to the assessment of wear in railway points: a case study. Eur J Oper Res 176:1703–1712
9. Márquez FG, Roberts C, Tobias AM (2010) Railway point mechanisms: condition monitoring and fault detection. Proc Inst Mech Eng, Part F: J Rail Rapid Transit, 224:35–44

10. Hao Y (2018) Analysis of turnout action current monitoring curve. Urban Mass Transit 55–58 (in Chinese)
11. Jie X (2009) Accurate estimation of SOC of power battery pack based on Kalman filter. Hangzhou Dianzi University (in Chinese)
12. Tianyu Z, Yi L, Shiliang S (2017) Prediction of spontaneous combustion risk for multi-coal based on SVM. J Hunan Univ of Sci Technol (Natural Science Edition) 34(2):11–17 (in Chinese)

The Turnout Abnormality Diagnosis with Semi-supervised Learning Method

Zeng Shu Shi, Yi Man Du, Tao Du and Guo Chao Shan

Abstract The turnout abnormality is very easy to cause the traffic accident or influence the efficiency due to the operating environment of railway transportation in China. However, the existing monitoring means are relatively backward, and the more mature automatic diagnosis method is lacking. In this study, a method based on semi-supervised learning algorithm for abnormal state diagnosis of turnout action curve is proposed. The method is used to analyze and extract the electrical characteristics of the turnout by using the turnout action curve and the static and dynamic properties collected by the railway centralized monitoring system. The support vector machine model is used to construct the initial classifier with a small number of labeled samples, and the labeled samples are expanded from a large number of unlabeled samples. The diagnosis model is constructed by using unlabeled data with a small amount of labeled data, and the switch curve is analyzed and diagnosed. The experimental results show that the method can automatically diagnose turnout electrical characteristics with high accuracy. Compared with supervised learning, the cost is low, but it can achieve higher accuracy and improve the practicability of fault diagnosis of turnout.

Keywords Railway transportation · Turnout abnormality diagnosis · Semi-supervised learning method · SVM · Unmarked sampling

1 Introduction

Turnout is an important part of railway track and one of the equipments with the highest failure rate. Once the turnout breaks down, the required action cannot be completed, parking caused time delay and even casualties will be caused. Frequent

Z. S. Shi · G. C. Shan
The School of Information Science and Technology,
Southwest Jiaotong University, Chengdu, China

Y. M. Du (✉) · T. Du
Beijing SWJTU RichSun Tech Co., Ltd., Beijing, China
e-mail: duyiman@swjtu-richsun.com

Y. Qin et al. (eds.), *Proceedings of the 4th International Conference on Electrical and Information Technologies for Rail Transportation (EITRT) 2019*, Lecture Notes in Electrical Engineering 639, https://doi.org/10.1007/978-981-15-2866-8_70

railway transportation operation and changeable operation environment in China easily lead to turnout failure. However, the existing monitoring conditions and maintenance methods of turnouts are relatively backward, which can easily cause traffic accidents or affect the efficiency of driving. At present, in order to prevent such accidents, railway departments arrange professionals to check the electrical characteristic curve of turnouts every day and identify the abnormal or fault of turnouts by comparing with the monitoring data of normal turnouts. This manual monitoring method needs a lot of manpower and material resources and is prone to misjudgment and omission, which seriously affects the efficiency and cannot meet the needs of railway development.

In the existing technology, there are also some models and algorithms for predicting the working state of turnout based on the historical data such as the action curve of turnout and fault information collected. These include the prediction model based on experience and the prediction model based on machine learning algorithm [1–3]. Experience-based prediction models usually use simple formulas to define the correlation model. The modeling process is too direct and idealized, and many factors are not taken into account. Therefore, in practical applications, the prediction performance of these models is usually poor and their robustness is not good. There are many prediction models based on machine learning algorithm, for example, in reference [4], expert system is used for fault diagnosis of signal equipment, various analog and switching data collected by computer monitoring are input into expert fault diagnosis system, and in reference [5], Fisher-based kernel function method is used to analyze turnout action curve and turnout action electricity. In the paper [6–8], the support vector machine (SVM) is proposed as the core, and different algorithms are used to realize the fault diagnosis of turnout. Huang et al. [9] have studied the method of turnout fault diagnosis based on dynamic time warping (DTW). Five kinds of fault current curves are collected as fault data templates, and the state of turnout is judged according to the accumulated distance between the test data and the model. However, these prediction models based on machine learning algorithms are all supervised learning methods, which need to use expert knowledge to label a large number of historical data manually. The cost of labeling data is very high. At the same time, in order to achieve better generalization performance, a large number of labeled fault samples are usually needed. It consumes a lot of manpower and material resources. In reality, the inability to obtain large sample label data results in a great discount in the performance of the model in practical scenarios.

Based on the above situation, a fault diagnosis method of turnout curve based on semi-supervised learning algorithm is proposed in this paper. Using the turnout action curve collected by railway signal centralized monitoring system and the dynamic properties of turnout equipment, the characteristics of turnout action are analyzed and extracted, and the branches are selected based on the theory of semi-supervised learning. The diagnosis model is established based on SVM algorithm. A small number of labeled samples are used to construct the initial classifier, and a large number of unlabeled samples are used to train on the basis of a small number of labeled samples. This method can effectively improve the prediction performance.

2 Extraction and Preprocess of Characteristic Data of Turnout Action Curve

2.1 Dynamic Characteristic Analysis

The normal operation process of turnout can be divided into: unlock–switch–lock. The collection of turnout current curve starts with 1DQJ suction and stops falling. After the switch is switched over, the action current is cut off and the 1DQJ slow-down (slow-down time is not less than 0.4 s) falls. The acquisition period of switching action current is 0.04 s. The data points in the process of switching action are plotted as curves, which are the switching action curves. The normal switching current curve is shown in Fig. 1.

The switching action current curve of Fig. 1 can be divided into three periods for analysis. The first period is the unlocking process of the turnout, in which the current increases rapidly and falls back; the second period is the turnout conversion process, in which the action current curve is smooth; the third period is the construction presentation stage, the turnout control circuit will be cut off, and the turnout indication circuit will be connected, the current value will occur sharply. Decline.

If a part of the turnout fails, it will cause the movement current curve of the turnout to deform and become the abnormal movement current curve. For example, if the action current is small in the process of switching, it means that the resistance of switching is small; if the action current is large, it means that the resistance of switching is large; if the action curve fluctuates greatly, it indicates that there are electrical or mechanical problems in the turnout. Figure 2 shows the abnormal

Fig. 1 Normal current curve

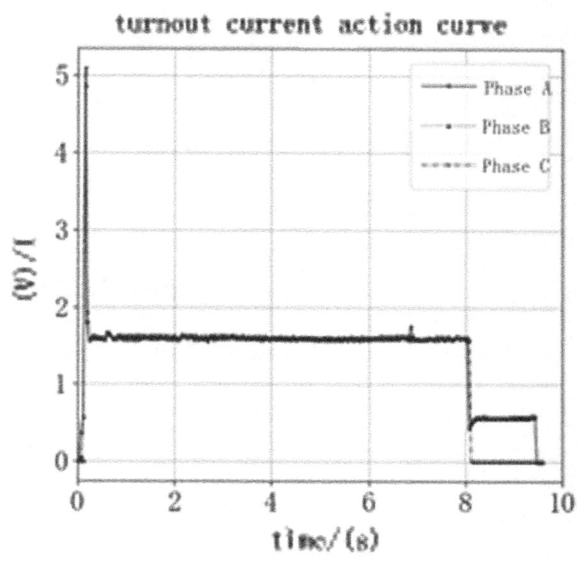

turnout action curve caused by the idling fault of the switch machine. From Fig. 2, it can be seen that the action current curve of the switch is normal in the unlocking stage and in the conversion stage, but there is no current step in the slow-release zone until the phase-breaking protector automatically cuts off the circuit, and the action time of the curve is proportional to that of the phase-breaking protector.

2.2 Dynamic Feature Extraction

To analyze and judge accurately and effectively, it is necessary to express the current curve of each fault mode effectively, that is, to change the character description and graphical display of the curve characteristics into the numerical expression that can be input into the diagnosis model. In this study, the current curve of turnout is analyzed by means of current range characteristics.

Firstly, the curve is divided into three sections according to the normal current curve of the turnout. With 1.9 and 0.6 A as the boundaries, the current greater than 1.9 A is classified as the unlocking stage current, the current less than 0.6 A is classified as the construction stage current, and the current between them is classified as the conversion stage current. Then eight statistical features are extracted for each segment of the same set of three-phase currents, which are maximum, minimum, average, variance, standard deviation, median, range and the number of current points. In this way, each phase curve has 24 features, and the three phases together have 72 features as a set of judgment curves.

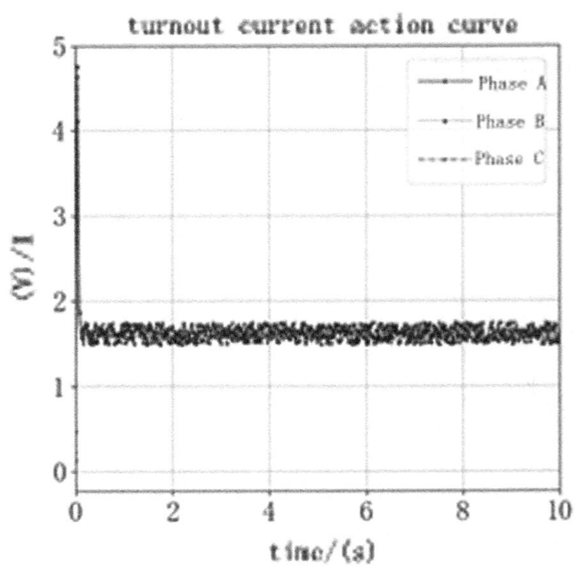

Fig. 2 Abnormal current curve

$$\text{Feature}_1 = \max(I) \tag{1}$$

$$\text{Feature}_2 = \min(I) \tag{2}$$

$$\text{Feature}_3 = \frac{1}{l}\sum_{k=1}^{l} I_k \tag{3}$$

$$\text{Feature}_4 = \begin{cases} I_{(l+1)/2} & \text{, if } l \text{ is odd number} \\ I_{((\frac{l}{2})+(\frac{l}{2}+1))/2} & \text{, if } l \text{ is even number} \end{cases} \tag{4}$$

$$\text{Feature}_5 = \frac{1}{l}\sum_{k=1}^{l} (I_k - \text{Feature}_3)^2 \tag{5}$$

$$\text{Feature}_6 = \sqrt{\text{Feature}_5} \tag{6}$$

$$\text{Feature}_7 = l \tag{7}$$

$$\text{Feature}_8 = \text{Feature}_1 - \text{Feature}_2 \tag{8}$$

where I—Current value; l—Sample point.

3 Feature Selection and Reduction

Too high dimensionality of feature data will lead to a large amount of computation and time-consuming of the algorithm model, and there may be redundant features in high-dimensional features. Only a small number of features can be used to represent the curve well. Feature selection and reduction are conducted to remove redundant information and form a set of more recognizable features.

3.1 Feature Selection

In this study, the filter method is used to conduct the feature selection. Each feature is scored according to divergence or correlation. After chi-square test, the p-values of each feature are calculated, and the p-values of each feature are expressed in the form of a histogram as follows (Fig. 3).

According to the p-value result of chi-square test, 0.001 is regarded as the limit. The feature whose p-value is less than 0.001, can be considered that its significance level is very high and the feature is related to the classification results. According to

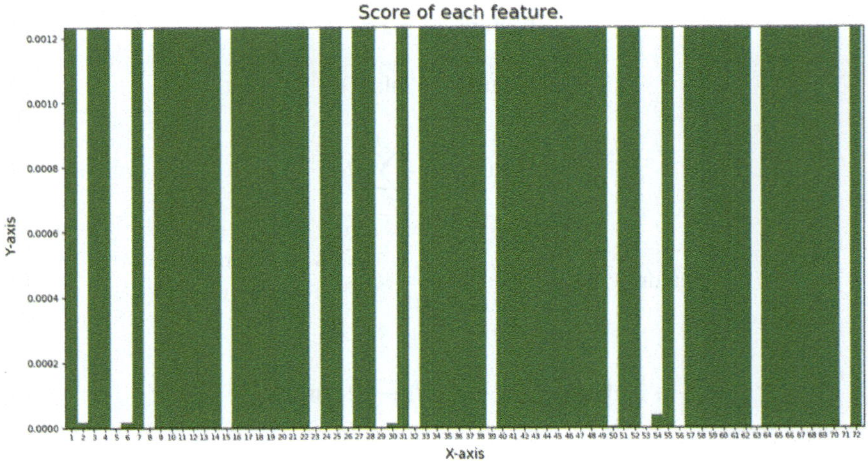

Fig. 3 Score of feature

the results of the graph, the feature numbers 15, 39, 63, 23, 71, 32, 8, 56, 29, 5, 53, 26, 50, 30, 2, 6, 54 are selected.

3.2 Feature Reduction

The semi-supervised discriminant analysis (SDA) is proposed to do the feature reduction in this study. The purpose of SDA algorithm is to find a projection matrix. The information contained in this projection matrix includes the discriminant structure in marked data points and the inherent geometric structure in marked data and unlabeled data. In the algorithm, a graph containing neighborhood information of data sets is constructed by using labeled data and unlabeled data points.

Given a labeled training sample set $\{(\boldsymbol{x}_i, \boldsymbol{y}_i)\}_{i=1}^{L}$, there are a total of C categories and an unmarked training sample set. Where $\boldsymbol{x}_i \in \boldsymbol{R}^m$, y_i is the mark of x_i. Assuming the number of K sample is l_k, there are $\sum_{k=1}^{c} \boldsymbol{l}_k = \mathbf{L}$. The following objective functions are established:

$$\max_{\alpha} \frac{\boldsymbol{\alpha}^{\mathrm{T}} \boldsymbol{S}_b \boldsymbol{\alpha}}{\boldsymbol{\alpha}^{\mathrm{T}} \boldsymbol{S}_w \boldsymbol{\alpha} + \boldsymbol{\beta} \sum_{i,j=1}^{n} \left(\boldsymbol{\alpha}^{\mathrm{T}} \boldsymbol{x}_i - \boldsymbol{\alpha}^{\mathrm{T}} \boldsymbol{x}_j\right)^2 \boldsymbol{S}_{ij}} \tag{9}$$

$$\boldsymbol{S}_b = \frac{1}{\boldsymbol{n}} \sum_{k=1}^{c} \boldsymbol{l}_k \left(\boldsymbol{\mu}^{(k)} - \boldsymbol{\mu}\right) \left(\boldsymbol{\mu}^{(k)} - \boldsymbol{\mu}\right)^{\mathrm{T}} \tag{10}$$

$$S_w = \frac{1}{n}\sum_{k=1}^{c}\left(\sum_{i=1}^{l_k}\left(x_i^{(k)} - \mu^{(k)}\right)\left(x_i^{(k)} - \mu^{(k)}\right)^{\mathrm{T}}\right) \tag{11}$$

where μ is the mean of total sample, $\mu^{(k)}$ is the mean of sample K, α is projection vector, S_b is Interclass Discreteness Matrix, S_w is Interclass Discreteness Matrix with Tag, β is coefficient of adjusting the proportion of the two parts, S_{ij} is elements in the weight matrix of adjacency graphs:

$$S_{ij} = \begin{cases} 1, & \text{if } x_i \in N_k(x_j) or\, x_j \in N_k(x_i) \\ 0, & \text{Otherwise} \end{cases} \tag{12}$$

4 Semi-supervised Diagnosis Model

Traditional methods based on fully supervised learning framework need to label a large number of unlabeled samples as training samples. Such methods often require a large number of training samples to obtain better prediction performance. However, acquiring a large number of training samples will inevitably take a lot of time, manpower and material resources, and even require a certain amount of expert knowledge to label the samples well. In the actual project development process, this is undoubtedly a huge cost.

In this study, semi-supervised support vector machine (SVM) is used to construct diagnosis model. This method uses a small number of labeled samples and some unlabeled samples to train the SVM classifier and then uses the trained classifier to classify the remaining samples.

The basic model of SVM algorithm is to find the best separating hyperplane in the feature space to maximize the positive and negative sample spacing on the training set. Combining SVM with semi-supervised method, the optimal hyperplane can be optimized using both labeled and unlabeled data. The algorithm process is as follows:

Input: $L = \{(x_k, y_k)\}$ training set with marked sample data, where $k = 1, 2, \ldots, 1$

$U = x_k$ training set with unlabeled sample data, where $k = 1 + 1, l + 2, \ldots, n$.

Process:

(a) The initial SVM classifier is trained by using the labeled data training set L.
(b) Create a data set U' and initialize it as an empty set, and add a part of the data to U' from the unmarked data set U. At the same time, create a data set L' and initialize it as an empty set.
(c) Mark the data in the data set U' with the trained SVM classifier.
(d) After labeling the data set U', all the data labeled as Y_j are searched for the same labeled data in the training set L. The Euclidean distance is calculated, and the pseudo-labeled data nearest to the same labeled data is screened out and added to the training set L'.

(e) A new SVM classifier is obtained by training the data set $L \cup L'$ with SVM classifier.
(f) Repeat this process until the data in U' is empty.

5 Example

According to the above algorithm, 300 samples of training set and 3000 samples of testing set are simulated. Among them, 30 data were labeled in training set, and the rest were unlabeled data.

According to the algorithm principle of SDA mentioned above, the characteristic data of 30 turnouts' operating current curves are marked, including three kinds of abnormal types: 1, 2 and 3. The dimension reduction process is carried out together with the unlabeled action current characteristic data. After SDA algorithm, the result dimensionality reduction of feature data to three-dimensional is shown in Fig. 4. It can be seen from the graph that four kinds of labeled data, 0, 1, 2, 3, form four small clusters in three-dimensional space. After dimension reduction, the interval between classes is obvious, and the clustering of same type of data is also relatively close, which is conducive to the following classification model to determine the classification boundary between classes, and more conducive to a high classification accuracy model. It can be explained that SDA algorithm can do feature reduction effectively.

The test results are shown in Fig. 5. The algorithm will end at 53 iterations, and all unlabeled samples will be transformed into labeled samples. According to the results of each iteration, the accuracy of the classifier for test set data can reach almost 99%. Thus, the designed SVM classifier has a good effect on the diagnosis

Fig. 4 Result of reduction of dimensions

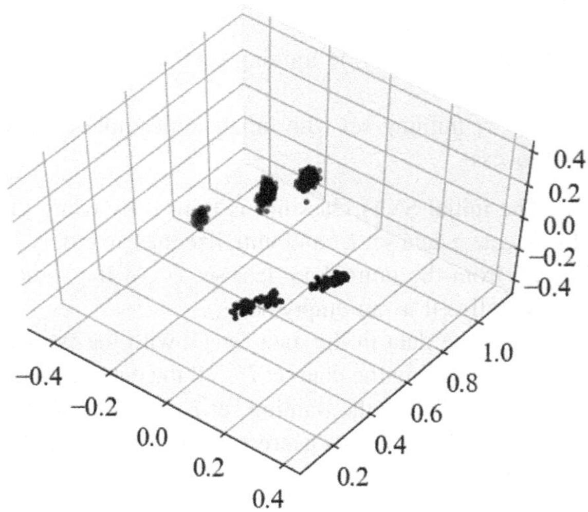

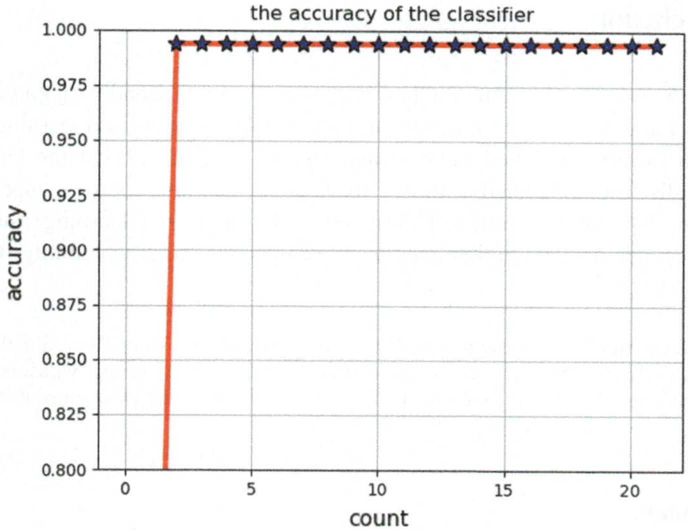

Fig. 5 Accuracy of diagnosis result

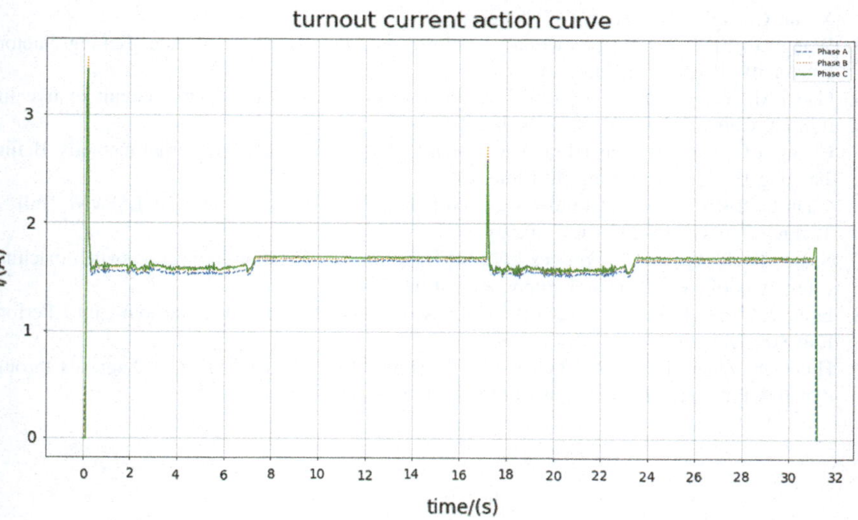

Fig. 6 Example of abnormality diagnosis

of four kinds of abnormal data and can be iterated. Convergence can be achieved within a limited number of times.

Figure 6 is one of the turnout current curves in testing sample. And, Fig. 6 is reported as abnormal after the diagnosis process.

6 Conclusion

In this study, machine learning method is adopted to automatically diagnose turnout electrical characteristics using historical data, and semi-supervised learning method is adopted to mix unlabeled samples and labeled samples so that the learner can automatically use unlabeled samples to improve learning performance without relying on external interaction. Compared with supervised learning, the cost is lower, but it can achieve higher accuracy and improve the practicability of turnout fault diagnosis.

Acknowledgements This research was funded by Research on Third-Party Simulation Test Technology of Train Control System of High-speed Rail Signal Engineering, which is supported by China Railway Corporation Science and Technology Research and Development Plan.

References

1. Zhang X (2016) The research and implementation of a method for recognizing switch fault current curve based on similarity. Lanzhou Jiaotong University, Lanzhou (in Chinese)
2. Zhang K (2016) Study of intelligent railway turnout fault diagnosis based on neural network. Xi'an: Chang'an University (in Chinese)
3. Wang G (2017) Failure prediction of railway turnouts using textual data. Beijing Jiaotong University, Beijing (in Chinese)
4. Gao LM, Yang J, Yang SZ (2014) Fault diagnosis method for control circuit of five line turnout. China Railw 2014(05):39–42 (in Chinese)
5. Cheng YJ (2016) High speed railway switch fault diagnosis based on kernel methods. Beijing Jiaotong University, Beijing (in Chinese)
6. Guan Q (2015) High-speed railway switch failure diagnosis based on FOA-LSSVM. Bull Sci Technol 31(04):230–232 (in Chinese)
7. Soares N, de Aguiar EP, Fonseca LG (2017) Failure classification in electric switch machines using symbolic concepts and computational intelligence
8. Eker Ö, Camci F, Kumar U (2012) SVM based diagnostics on railway turnouts. Int J Perform Eng 8(8):289–298
9. Huang S, Zhang F, Yu R, Chen W, Hu F, Dong D (2017) Turnout fault diagnosis through dynamic time warping and signal normalization. J Adv Transport 1–8

Fatigue Life Assessment of Rail Transit Electronics Based on Physics of Failure

Jing Long, Sheng yun Xu, Zhipeng Wang and Xiao qing Cheng

Abstract The failure prediction and life assessment method based on physics of failure (PoF) approach has been more widely used. This paper mainly describes the fatigue life assessment process design of board-level PCBA based on PoF approach. At first, the complete board-level PCBA fatigue life evaluation process is given, and the application of PoF model in various stages of the evaluation process is stated. Then a detailed description of their applicable conditions, as well as input and output parameters when they are applicated are given, thus providing a reference for the fatigue life evaluation and reliability design and evaluation of PCBA.

Keywords PoF model · Fatigue life assessment · Printed circuit board assembly

1 Introduction

In this paper, the physics of failure (PoF) method is used to evaluate the life of single plate; the basic characteristics of this method are as follows: the failure mechanism of products is different and subject to different physical processes. In different physical processes of product failure, environmental load is one of the main reasons for product life loss and ultimate failure. Different failure mechanisms and physical processes need to be evaluated by different failure physical models, and the final results are calculated as the result of various comprehensive actions.

J. Long · S. y. Xu
Guangzhou Metro Group Co. LTD, Guangdong, China

J. Long · S. y. Xu · Z. Wang (✉) · X. q. Cheng
National Research Center of System Engineering of Railway Intelligent Transport, Beijing, China
e-mail: zpwang@bjtu.edu.cn

Z. Wang · X. q. Cheng
State Key Lab of Rail Traffic Control and Safety, Beijing Jiaotong University, Beijing, China

Y. Qin et al. (eds.), *Proceedings of the 4th International Conference on Electrical and Information Technologies for Rail Transportation (EITRT) 2019*, Lecture Notes in Electrical Engineering 639, https://doi.org/10.1007/978-981-15-2866-8_71

2 Theoretical Foundation of PoF

The interconnect failure of welding joints between electronic components and circuit boards is a major factor affecting the reliability of electronic products. In the process of product design and production, the influence on welding joint reliability should be considered in the links of component packaging selection, PCB design, electrical installation auxiliary material inspection, process control and environmental test. In electronic packaging reliability research, most is welding joint. The research shows that the welding joint is usually the weakest part of the packaging structure in the failure of electronic packaging and assembly.

For most electronic products, the main forms of interconnection failure are thermal fatigue failure and vibration-induced high cycle fatigue failure.

2.1 Component Model Library

The component model library provides functions for component maintenance, component type maintenance, component package maintenance, component package model maintenance and component package alias maintenance. Users can accumulate component information and data, and the library can be used by other users with data sharing capabilities.

The component package library is mainly for the maintenance of the component package name.

The component package model library is mainly for the maintenance of complex modeling methods of components. For BGA package devices, it is mainly the modeling of different solder ball shapes (cylindrical solder balls and drum solder balls and square solder balls). For SOP and QFP packaged devices, it is mainly the modeling of different pin generation methods (stacking pins and stretching pins).

In the component package model maintenance, the local complex model is seen in the 3D view on the right side.

The component package alias library mainly maintains the package alias of the component, and it should be noted that the package alias library is an important reference library for three-dimensional information comparison and extraction, so information such as component package alias and component package type must be guaranteed.

Throughout the use process, the information in the library can be maintained, modified, updated, saved, etc. on the basis of the original component library, and then the subsequent calculations are performed.

2.2 Failure Mechanism Model Library

The failure mechanism model library contains many failure mechanisms. The management method of the failure mechanism model of pofPWA is mainly stored in the form of scripts in the data table of the server's failure mechanism model library. Different failure mechanisms are equipped with corresponding failures. A text file standardized by the mechanism model.

pofPWA and RelSIM faprlKB have a mechanism for sharing the failure mechanism model library. Therefore, the failure mechanism model can be maintained either on the pofPWA or the RelSIM faprlKB interface. The maintenance method mainly refers to the standardized text file of the established failure mechanism, and the script program is embedded. Go to the pofPWA or RelSIM program.

2.3 The Evaluation Process of Fatigue Life in pofPWA

See Fig. 1.

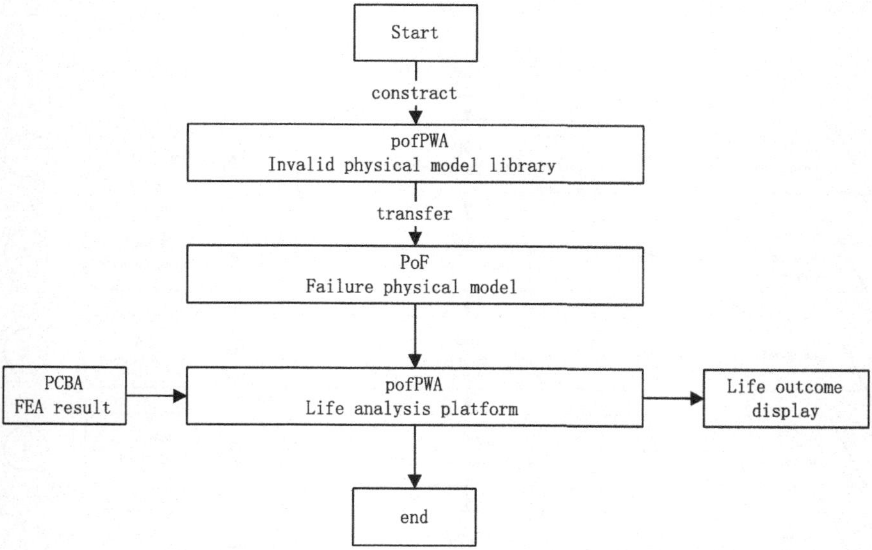

Fig. 1 The evaluation process of fatigue life in pofPWA

3 Automatic Fem Simulative Analysis of Single Board

3.1 Automated Simulation Analysis

Figure 2 depicts the main analysis process of the evaluation of fatigue life in PHM system for single-plate solder joints and the relationship between the analysis process and the applied failure physical models [1, 2] (Fig 3).

4 Life Analysis

4.1 Failure Physical Quantization Formula

By selecting the appropriate failure physical model, the solder joint failure principle can be better restored and the failure process can be quantified. Through inputting the stress and strain values calculated by the finite element software such as ANSYS into the physical model of the failure, the fatigue physical life of the solder joint can be accurately evaluated, which provides a reference for the life monitoring and reliability design of the single board.

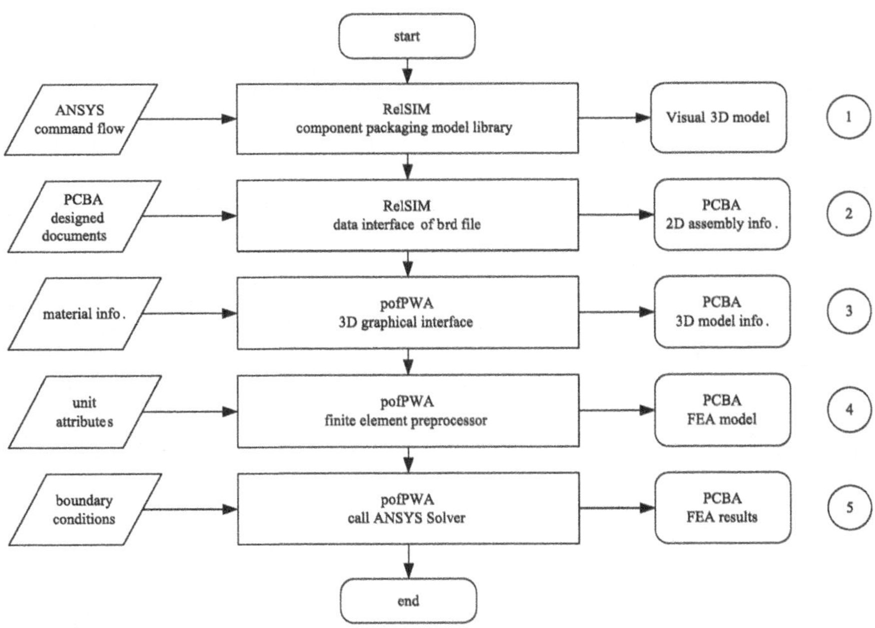

Fig. 2 Single-board automation simulation process of pofPWA

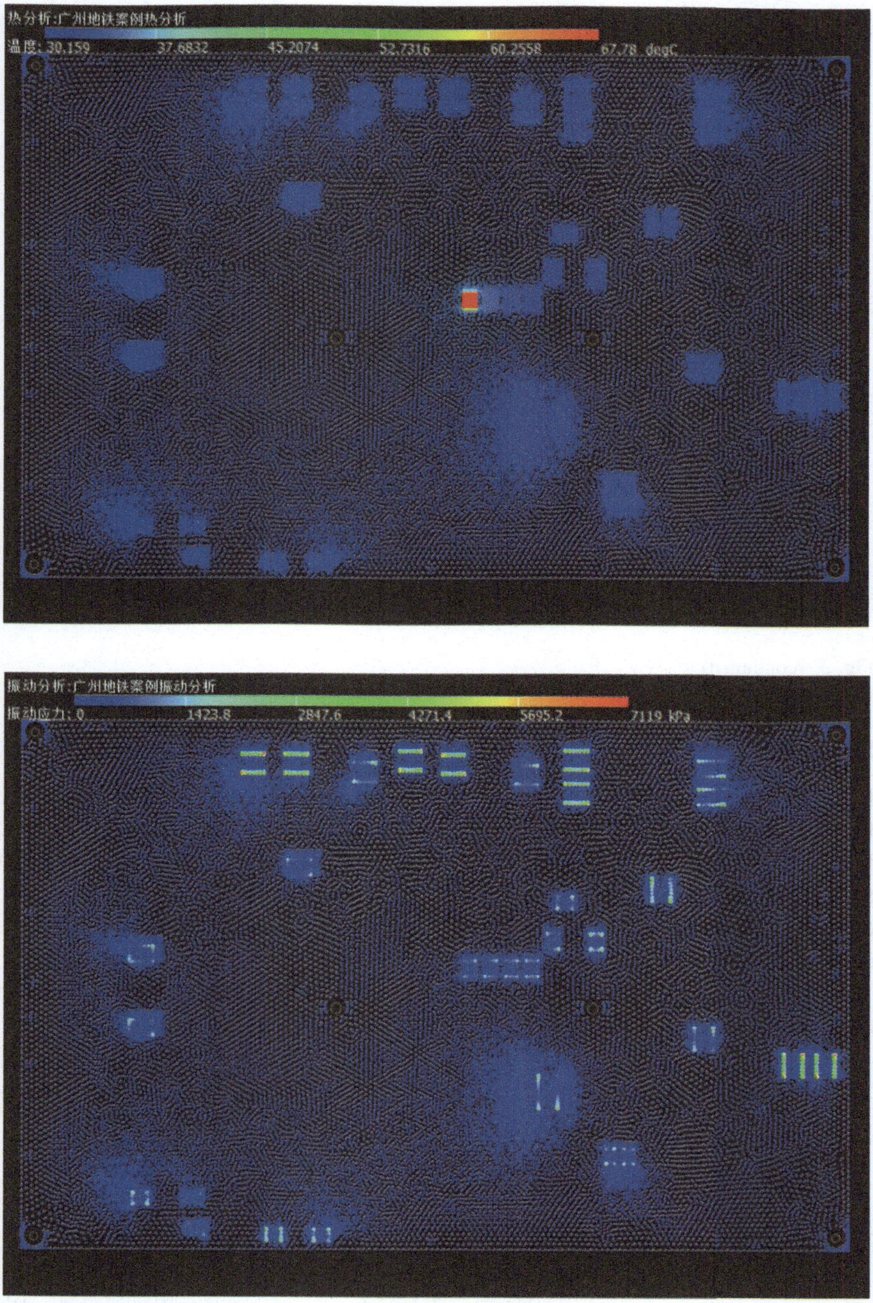

Fig. 3 The result of single board automation simulation

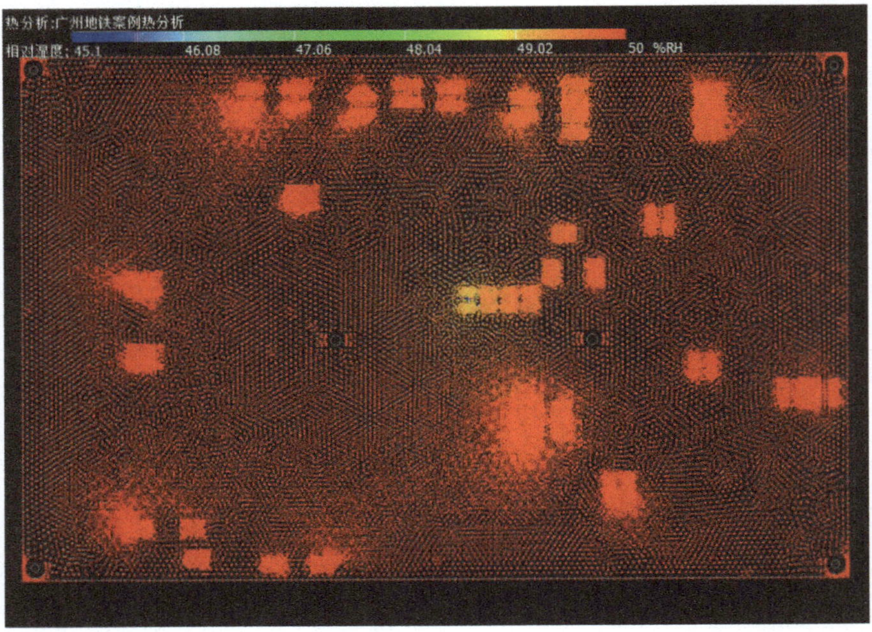

Fig. 3 (continued)

Steinberg Model

Most of the solder joint failure caused by random vibration is due to the cumulative fatigue damage caused by a series of variable cyclic loads. Under variable cyclic loads, fatigue damage can be accumulated linearly, and each stress is independent and irrelevant. When the cumulative damage reaches a certain value, fatigue damage will occur. The Steinberg model is typical for describing linear cumulative damage [3].

Steinberg model is mainly used to evaluate fatigue life under random vibration conditions. Its formula is as follows:

$$\mathrm{CDI} = \frac{0.683 f_0}{N_{1\sigma}} + \frac{0.271 f_0}{N_{2\sigma}} + \frac{0.043 f_0}{N_{3\sigma}} \tag{1}$$

$$t = \frac{1}{\mathrm{CDI}} \tag{2}$$

$$N_{1\sigma} = \frac{1}{2} \times \left(\frac{\sigma_0}{\sigma_f'} \right)^{1/b} \tag{3}$$

$$N_{2\sigma} = \frac{1}{2} \times \left(\frac{2\sigma_0}{\sigma'_f} \right)^{1/b} \tag{4}$$

$$N_{3\sigma} = \frac{1}{2} \times \left(\frac{3\sigma_0}{\sigma'_f} \right)^{1/b} \tag{5}$$

In the formula, CDI is called cumulative damage index; if it is single-stage loading, the damage occurs when the cycle ratio is equal to 1. If it is multi-stage loading, the total damage is equal to the sum of each cycle ratio (or damage ratio), and the damage occurs when the total cycle ratio is equal to 1 [4].

t is the predicted failure life of random vibration, assuming that the failure critical value of CDI is 1; $N_{i\sigma}(i = 1, 2, 3)$ is the number of fatigue failure cycles and corresponds to 1σ, 2σ and 3σ stresses, respectively, according to normal distribution; and the total duration of 1σ, 2σ and 3σ strains is 68.31%, 27.1% and 4.33%, respectively, which are taken into account as their coefficients in the formula.

f_0 is the first-order resonance frequency; σ'_f is the fatigue strength coefficient; σ_0 is the root mean square value of stress.

The 3σ criterion of normal distribution is involved here. In normal distribution, it can be considered that the probability values are almost all concentrated in the range of $(\mu - 3\sigma, \mu + 3\sigma)$. Therefore, when calculating cumulative damage, only the damage in 3σ is counted, and the error can be neglected. For the whole PHM analysis process, Steinberg model is mainly used to obtain the distribution information of cumulative vibration fatigue life of solder joints, and there are the following input and output relationships [5]:

Input: First-order resonance frequency of veneer, fatigue strength coefficient of material, fatigue stress and corresponding fatigue failure cycles of 1, 2 and 3σ under this stress level are analyzed. The first-order resonance frequency is determined by the properties of the veneer, and the fatigue strength coefficient is determined by the material properties, temperature and time, etc. The fatigue stress and the corresponding fatigue failure period can be obtained from the output of the Basquin model.

Output: cumulative damage index of solder joint. This characteristic quantity represents the random vibration fatigue life of solder joints.

Coffin-Manson Model

Thermal fatigue life is evaluated by using the modified Coffin-Manson formula, which is commonly used for low cycle fatigue. The formula has a wide range of applications and is well suited for both plastic and brittle materials. Its form is as follows [6]:

$$N_f = A_0 * (1/\Delta\varepsilon_p)^B \quad (6)$$

where N_f is the number of failure cycles and A_0 is the constant related to the material. $\Delta\varepsilon_p$ is the range of plastic strain and B is the empirical parameter.

The strain range $\Delta\varepsilon_p$ is proportional to the exponential form of temperature, i.e.,

$$\Delta\varepsilon_p \propto (\Delta T - \Delta T_0)^B \quad (7)$$

where ΔT is the range of variation of the overall temperature and ΔT_0 is the range of temperature variation of the elastic region of the material.

Assuming that the elastic temperature change of the material ΔT_0 is much less than ΔT, then ΔT_0 can be neglected. Therefore, the modified Coffin-Manson formula is as follows:

$$N_f = C_0 * (\Delta T)^{-q} \quad (8)$$

where C_0 is a constant related to the material. q is the Coffin-Manson index, which is mainly based on empirical values, and the values are different under different failure mechanisms.

For different integrated circuit materials, the empirical value q is shown in Table 1.

The input quantities are: material correlation constant, Coffin-Manson index and temperature difference at adjacent time. The material correlation constant is determined by the material properties. The Coffin-Manson index is mainly based on empirical values, and the values under different failure mechanisms are different. The temperature difference between adjacent moments is obtained by interpolation of the temperature file at the adjacent time [7].

The output is the number of fatigue failure cycles of solder joint. This feature represents the thermal fatigue cycle life of solder joint under the change of heat source temperature.

Table 1 The empirical value of q

Material	q
Plastic material (e.g., solder joints)	1–3
Hard metal alloy/intermetallic compound (e.g., Al–Au)	3–5
Brittle materials (such as silicon and SiO_2, Si_3N_4 and other electrolytes)	6–9

Arrhenius model

$$\frac{\partial x}{\partial t} = K = \Lambda e^{-\Delta E/kT} \tag{9}$$

For the above formula, if the characteristic value X is regarded as the end of life once it reaches the specified value X_{sp}, then we can have [8]:

$$X_{sp} = Kt + x_0 \tag{10}$$

If $x_0 = 0$, then:

$$L = \frac{X_{sp}}{K} \tag{11}$$

Substituting the Arrhenius equation into the above equation and taking the logarithm, we can get:

$$\text{In}L = A + \frac{\Delta E}{kT} \tag{12}$$

Or as follows:

$$L = Ae^{\frac{\Delta E}{kT}} \tag{13}$$

That is, the logarithm of life and $\frac{1}{T}$ holds the linear relationship.

X: characteristic value or amount of degradation;
K: the reaction rate is the rate of degradation;
Λ: frequency;
ΔE: activation energy (0.58–0.7 eV);
k: Boltzmann constant (8.62 e–5 eV/K);
T: absolute temperature;
A: constant;
L: life of chips.

Corrosion life model.

The failure time equation for integrated circuit failure caused by corrosion is:

$$\text{TF} = A_0(\text{RH})^{-n} \exp\left(\frac{Q}{K_g T}\right) \tag{14}$$

where:

A_0 is related to materials and processes;
n is the power law exponent, generally taken as 2.7;
RH is relative humidity;
Q is the activation energy of the material, taken as 0.7–0.8 eV when the chloride is corroded to the aluminum.

This model was built on the basis of chloride corrosion in plastic encapsulated chips. Chlorine etching is generally used in the manufacturing process of aluminum alloys. If the residual chloride adheres to the chips after etching, the moisture will increase and the chips will be corroded. The power law model has been widely used in the corrosion research of integrated circuit plastic packaging chips.

4.2 Life Estimation Under Single Load Conditions

Based on the life analysis of different failure mechanisms, some perform fatigue analysis on the solder joints, and some perform electromigration and corrosion analysis on the chips [9].

Vibration fatigue life

By selecting the appropriate failure physical model, the failure principle of solder joint can be better reduced and the failure process can be quantified. The fatigue life of welding joints can be accurately evaluated after the stress and strain values calculated by ANSYS and other finite element software are input into the failure physical model for calculation, which provides a reference for the monitoring of single plate life and reliability design (Fig. 4).

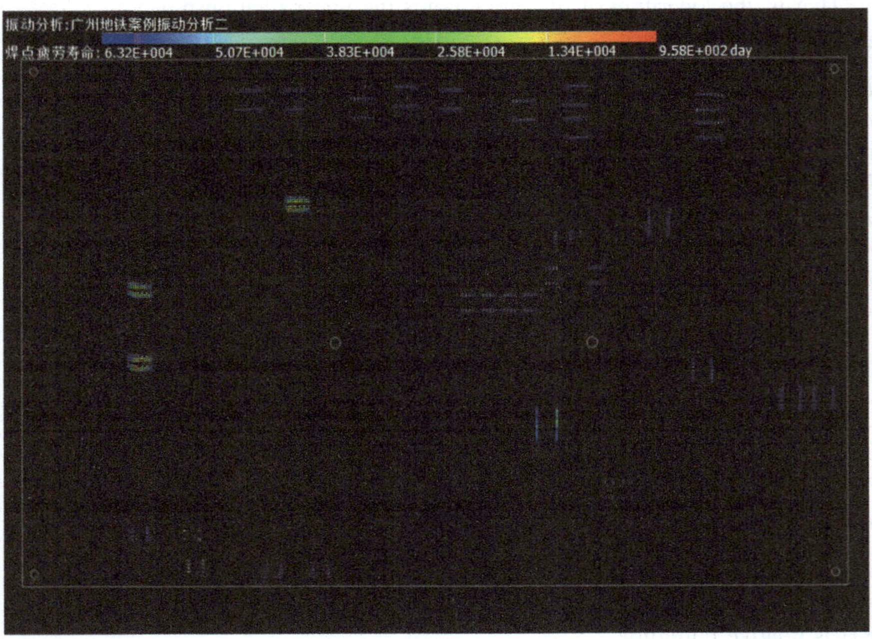

Fig. 4 Case study of vibration fatigue life

Thermal fatigue life

By selecting the appropriate failure physical model, the failure principle of solder joint can be better reduced and the failure process can be quantified. The fatigue life of welding joints can be accurately evaluated after the stress and strain values calculated by ANSYS and other finite element software are input into the failure physical model for calculation, which provides a reference for the monitoring of single plate life and reliability design (Fig. 5).

Electromigration life

By selecting the appropriate failure physical model, the failure principle of solder joint can be better reduced and the failure process can be quantified. The fatigue life of welding joints can be accurately evaluated after the stress and strain values calculated by ANSYS and other finite element software are input into the failure physical model for calculation, which provides a reference for the monitoring of single plate life and reliability design (Fig. 6).

Corrosion life

By selecting the appropriate failure physical model, the failure principle of solder joint can be better reduced and the failure process can be quantified. The fatigue life of welding joints can be accurately evaluated after the stress and strain values calculated by ANSYS and other finite element software are input into the failure physical model for calculation, which provides a reference for the monitoring of single plate life and reliability design (Fig. 7).

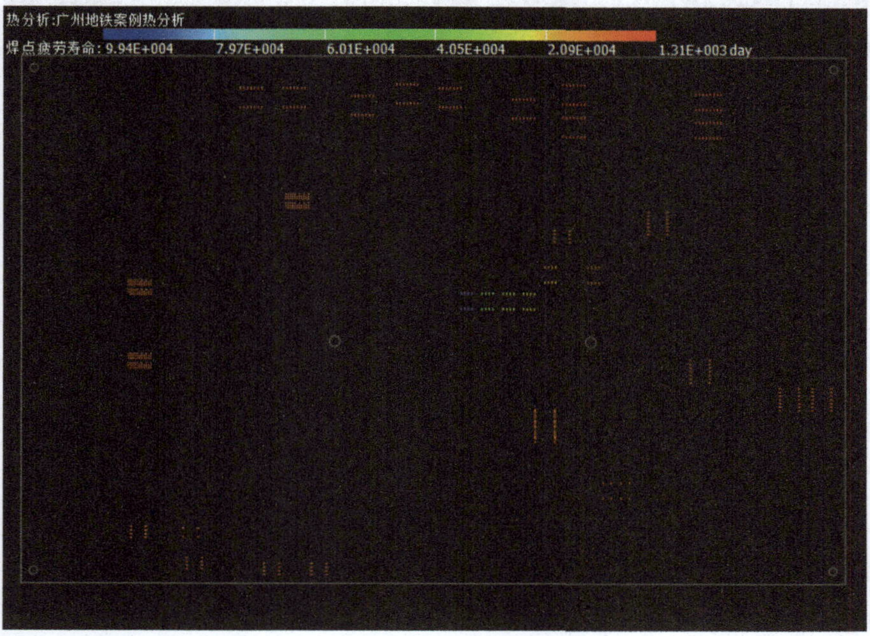

Fig. 5 Case study of thermal fatigue life

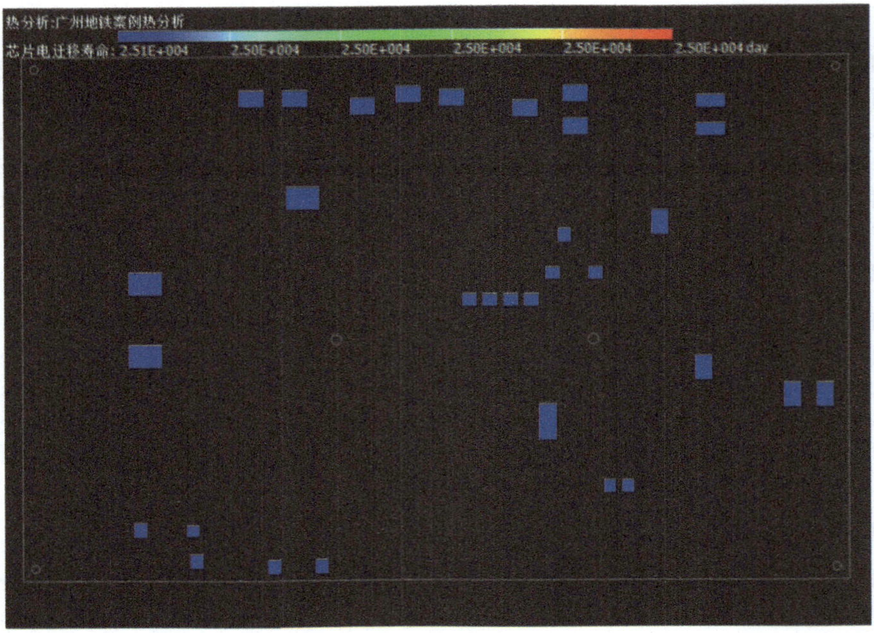

Fig. 6 Case study of electromigration life

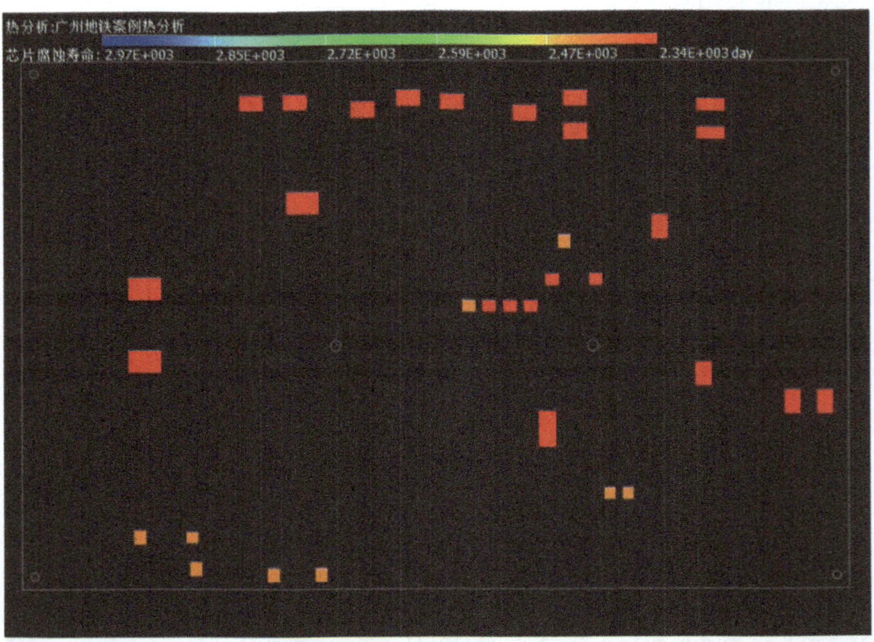

Fig. 7 Case study of corrosion life

4.3 Lifetime Estimation at the Device Level

By selecting the appropriate failure physical model, the failure principle of solder joint can be better reduced and the failure process can be quantified. The fatigue life of welding joints can be accurately evaluated after the stress and strain values calculated by ANSYS and other finite element software are input into the failure physical model for calculation, which provides a reference for the monitoring of single plate life and reliability design (Fig. 8).

Under the assumed environment and simplified model, the device life cloud chart of the engineering example obtained in this project, namely, the single-plate hazardous device, is estimated to have a life of about three years. According to this example, the main failure mechanism of these hazardous devices is vibration and thermal fatigue, and their detailed information is shown in Table 2.

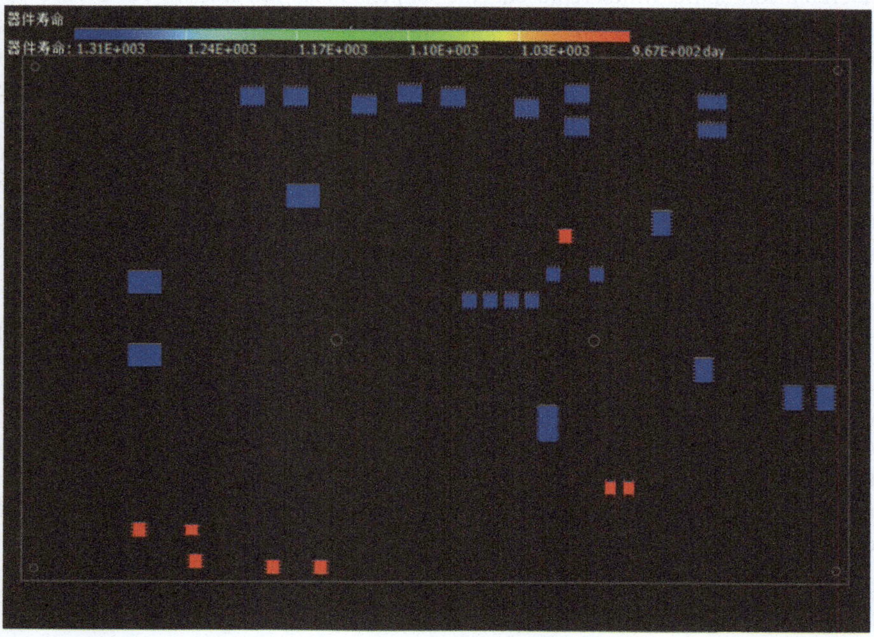

Fig. 8 A case study of device-level lifetime estimation

Table 2 Hazardous device details

The device type	Bit number	Type1	Type2	Failure mechanism
TLP126	52	SOP4	3 years	Vibration and thermal fatigue
	418	SOP4		
	450	SOP4		
HCPL-0201 IL710-3	532	SOP8		
	533	SOP8		
	534	SOP8		
	535	SOP8		
	536	SOP8		

4.4 Estimation of Cumulative Life Loss

Palmgren-Miner model is used to accumulate damage of different fatigue life in interconnection life, which represents the simplest linear integration. The model was established based on the Miner fatigue damage accumulation hypothesis and was based on the *s-n* characteristic curve of the material. According to the theory, when a part runs a force cycle n less than *N* under the stress which is higher than the fatigue limitation, each load cycle will cause partial damage, and during this cycle, each damage is the same, that is, the damage rate under the action of *A* is *n*/*N*. If the solder joint contains different stress levels during the operation, there will be a corresponding damage rate, and such damage can be cumulative. Failure can be predicted when the sum of these damage rates reaches 1. The equation has the following form:

$$\sum_{i=1}^{c}\left(\frac{N_i}{N_j}\right) = D \tag{15}$$

where c is the total number of cycles;
N_i , N_j correspond to the number of cycles and the number of failure cycles when the stress level is i;
D is the total damage of the material. According to the Palmgren-Miner linear cumulative fatigue damage theory, the fatigue life of the component is:

$$N = 1/D \cdot \Delta t \tag{16}$$

where N is the total number of fatigue failure cycles of the material, indicating the time of each cycle.

For the whole PHM analysis process, the Palmgren-Miner model, due to its simple linear cumulative property, is mainly used to obtain the distribution information of solder joint comprehensive fatigue life, and the following input and output relations exist:

The input value is: cycle number of welding joints under a certain stress level and fatigue failure cycle number of welding joints under this stress level, total number of cycles, duration of each cycle.

Output: total fatigue failure cycle number of solder joints. This characteristic quantity represents the comprehensive fatigue life of solder joint.

Assuming that the single plate works for 100 days under the above-mentioned vibration and thermal cycle conditions, in which the single plate works for the first 50 days under the vibration condition and the second 50 days under the vibration condition, the cumulative life loss of the most dangerous solder joint is:

$$S = S + \sum \frac{1}{L} = \frac{50}{958} + \frac{50}{200} + \frac{100}{1310} \approx 0.38 \tag{17}$$

5 Conclusions

By selecting the appropriate failure physical model, the failure principle of solder joint can be better reduced and the failure process can be quantified. The stress and strain values of the veneer are calculated by the pofPWA software. After the failure physical model is input and calculated, the fatigue life of the solder joint can be accurately evaluated, which provides a reference for the life monitoring and reliability design of the veneer.

Acknowledgements The research work of this work was funded by the National Key Research and Development Program (No. 2016YFB1200402), National Natural Science Foundation of China (Grant No. 61833002) as well as the open fund of National Research Center of System Engineering of Railway Intelligent Transport (NO. RITS2018KF01).

References

1. Xie J etc (2012) Reliability assessment of an electronic assembly with a large number of ball grid array packages and interconnects. In: ASME 2012 international mechanical engineering congress & exposition (IMECE 2012), Houston, Texas
2. Xie J etc (2012) Reliability assessment of a SMT assembly with prediction of its board-level interconnect life distribution. In: ASME 2012 international mechanical engineering congress & exposition (IMECE 2012), Houston, Texas
3. Wong TE, Reed BA, Cohen HM, Chu DW (1999) Development of BGA solder joint vibration fatigue life prediction model. In: 1999 IEEE electronic components and technology conference, pp 149–154
4. Zhou B, Qiu BJ (2011) Vibration durability modeling and dynamic response analysis of PBGA mixed solder joints. In: International conference on electronic packaging technology & high density packaging
5. Xue DJ, Wang CH, Xue JS (2013) Life analysis of solder joints of circuit boards under engineering environment conditions. Electr Mass 2013 (5):20–24 (In Chinese)

6. Rafanelli AJ (1992) Ramberg-Osgood parameters for 63–37 Sn-Pb solder. In: Transactions of the ASME, vol 114, pp 234–8
7. Pang JHL, Xiong BS, Che FX (2004) Modeling stress strain curves for lead-free 95.5 Sn-3.8 Ag-0.7 Cu solder. In: 5th international conference on thermal and mechanical simulation and experiments in microelectronics and microsystems, 2004. EuroSimE 2004. Proceedings of the, IEEE, pp 449–453
8. Engelmaier W (2009) Solder creep-fatigue model parameters for SAC & SnAg lead-free solder joint reliability estimation. In: Proceedings from IPC midwest conference and exhibition, pp 1–44
9. Barker D et al (1990) Combined vibrational and thermal solder joint fatigue—a generalized strain versus life approach. ASME J Electr Packag, vol 112, pp 129–134

The manufacturer's authorised representative in the EU is Springer Nature Customer Service Centre GmbH, Europaplatz 3, 69115 Heidelberg, Germany. If you have any concerns regarding our products, please contact ProductSafety@springernature.com

Printed and bound by CPI Group (UK) Ltd, Croydon, CR0 4YY

30/09/2026

02225292-0018